PARASITOLOGY

The Biology of
Animal Parasites

PARASITOLOGY

The Biology of Animal Parasites

ELMER R. NOBLE, Ph.D.

Professor of Zoology Emeritus
University of California, Santa Barbara
Santa Barbara, California

GLENN A. NOBLE, Ph.D.

Professor of Biology Emeritus
California Polytechnic State University
San Luis Obispo, California

GERHARD A. SCHAD, Ph.D.

Professor of Parasitology
School of Veterinary Medicine
University of Pennsylvania
Philadelphia, Pennsylvania

AUSTIN J. MacINNES, Ph.D.

Professor of Biology
Department of Biology
University of California, Los Angeles
Los Angeles, California

Sixth Edition

LEA & FEBIGER

PHILADELPHIA
LONDON

1989

Lea & Febiger
600 Washington Square
Philadelphia, Pa. 19106-4198
U.S.A.
(215) 922-1330

1st Edition, May, 1961
2nd Edition, February, 1964
    Reprinted, 1965
3rd Edition, April, 1971
    Reprinted 1973, 1974
4th Edition, October, 1976
    Reprinted, 1977
5th Edition, May, 1982
6th Edition, 1989

Library of Congress Cataloging-in-Publication Data

Parasitology: the biology of animal parasites.

   Rev. ed. of: Parasitology/Elmer R. Noble,
Glenn A. Noble. 5th ed. 1982.
   Includes bibliographical references and index.
   1. Parasitology. I. Noble, Elmer Ray, 1909-
II. Noble, Elmer Ray, 1909-   Parasitology.
QL757.P34   1989   591.52'49   88-9235
ISBN 0-8121-1155-9

Printed in the United States of America

Print Number   3   2   1

Preface for Students

This book is written primarily for the undergraduate college or university student who has completed one basic course in biology and one in general chemistry.

Our aim is to provide you with the opportunity to gain some understanding and appreciation of, and respect for, the vast and fascinating phenomenon of animal parasitism. Animal parasites, living on or in the bodies of other animals (the hosts), far outnumber the free-living animals. We can only initiate the study of parasitism and give merely a glimpse of what is taking place everywhere in the biologic world. Among the many hundreds of published papers that we have examined, we selected the conclusions of relatively few; the major criterion for selection was implicit in the question: "Is this of real significance for undergraduates taking their first (and probably only) course in parasitology? The frequently negative answer resulted in omitting much important detail. This detail, unfortunately, may include the area of your instructor's special interest.

You will find that there is too much information for anyone to digest and absorb during one short course of study. Your goals in reading this textbook should be to get the basic concepts (many of them in the last four chapters), to learn the chief distinguishing structures, life cycles, and functions of the *major* parasites that infect humans, and to familiarize yourself with typical examples of those parasites infecting other groups of animals. Your instructor will select the subjects, examples, and concepts that fit the particular class, geographic location, and department facilities where you are enrolled. In every class there are students who wish to spend extra time preparing for a special report or for anticipated further courses in parasitology, or who want to pursue an aspect of the subject simply because it appeals to them. For them the extra detail in this book will be of value. Moreover, we cite many references, if not specifically on a subject of special interest, at least on papers that themselves lead you to your objective.

Many of you are interested in medicine (human or veterinary), nursing, or medical technology. This book does not attempt to duplicate a text on clinical parasitology, but you will find considerable emphasis on such diseases as malaria, African sleeping sickness, amebic dysentery, schistosomiasis, leishmaniasis, filarial infections, including elephantiasis, and coccidiosis. There is little information, however, on general problems of public health, and no coverage of treatment.

This sixth edition represents a major revision. The authorship has been expanded from two to seven, a change reflected throughout the book. Some chapters have been rewritten and all have been expanded and updated. Biochemistry, physiology, immunology, and ecology have been given major attention. These important changes, coupled with an increase in the number and variety of illustrations, make this a modern resource for everyone beginning the study of animal parasites.

Santa Barbara, California   Elmer R. Noble
San Luis Obispo, California   Glenn A. Noble
Philadelphia, Pennsylvania   Gerhard A. Schad
Los Angeles, California   Austin J. MacInnes

Contributors

Ralph A. Barr, Ph.D.
Division of Infectious and Tropical Disease
School of Public Health
University of California, Los Angeles
Los Angeles, California

F.G. Hochberg, Ph.D.
Santa Barbara Museum of Natural History
Santa Barbara, California

Richard T. O'Grady, Ph.D.
Division of Worms
Department of Invertebrate Zoology
National Museum of Natural History
Smithsonian Institution
Washington, D.C.

Contents

1

Introduction to Parasitology

Because about three-quarters of all known animals are insects, an estimate that parasitic insects represent close to half the animals on earth does not seem unrealistic.[21] Add to this estimate the parasitic Protista, trematodes, tapeworms, annelids, parasitic crustacea, and other groups; the magnitude of animal parasites and the major role they play in the world's biomass is indeed astonishing and is still only partly understood and little appreciated by many biologists.

During the millions of years that animals and plants have competed among themselves for food and space, parasites have invaded practically every kind of living body. These bodies, called **hosts** (see Glossary for definitions of different kinds), generally provide food and shelter, and because hosts furnish different kinds of space in the form of external surfaces, organs, tissue, and fluids, they usually acquire more than one kind of parasite. Today, most animals have on or within their bodies several species of parasites, sometimes totaling hundreds or even millions of individuals. There are, therefore, more kinds and numbers of animal parasites than free-living animals. The major groups of animal parasites are found among the Protista, the helminths (flatworms and roundworms), and the arthropods. The host and its parasites constitute a community of organisms living in intimacy and exerting a profound effect upon one another.

To illustrate the kinds of parasites that an animal might support and some of the interrelationships and problems in parasitism, we select the fish as a typical vertebrate host. Almost any small vertebrate would be satisfactory, but fish may be secured easily by students for exami-

nation. Frogs are excellent alternatives (Smyth and Smyth).[24] Our book is concerned only with animal parasites, so we shall omit the multitudinous bacterial, fungal, and viral infections that plague all animal groups. A careful dissection of a single fish usually reveals three or four species of animal parasites, seldom more than five or six, and sometimes only one. There may be, however, thousands or even millions of members of a single species e.g., blood parasites, in one host.

An examination of any host for its parasites should start with the outside. The skin and scales of fish are commonly the home of copepods and other Crustacea, encysted larval stages of digenetic trematodes (flukes), adult monogenetic trematodes, leeches, and several kinds of protozoa. Copepods often have sharp claws that enable them to cling to skin, or anchoring devices that are deeply embedded under scales. Fish lice (Branchiura) are temporary parasites of the skin. Encysted larvae of digenetic trematodes are called metacercariae, and the fish must be eaten by another host (e.g., a bird or another fish) before larval fluke can develop into adult worms. Monogenetic trematodes may damage host skin by means of claw-like hooks on the posterior ends of their bodies. Leeches suck blood and may thereby transmit blood parasites from one fish to another. Protistan parasites on the skin occasionally cause so much damage that the fish dies.

The next place to look is inside the mouth and on the gills. Here may be found the same kinds of worms as on the skin; as well as additional kinds of parasites. Isopod crustaceans often cling to gills or mouth lining; sometimes

1

a single parasite is so large that it almost fills the mouth cavity. Hundreds of copepods may be partly embedded in the gills. Cysts of Myxozoa may appear as white spots or lumps.

If the fish is alive or freshly killed, blood smears should be made, and fresh blood should be scrutinized for such parasites as flagellated trypanosomes (related to those that cause African sleeping sickness in man) and another protozoan form, the haemogregarines, which live within red cells. Fish often have more internal parasites than external ones, and any organ may be infected. Many parasites have been described from the digestive tract, especially nematodes and flukes. Thorny-headed worms, called acanthocephalans, are common in the intestine. Several kinds of Protista may occur in the intestinal contents. Coiled, larval nematodes are easily seen in mesenteries and walls of the coelom as well as in muscles. Larval tapeworms of several kinds inhabit a variety of organs. Myxozoa are common in the gallbladder, urinary bladder, kidneys, muscles, and other organs. Microspora may infect the cells of most organs of the fish. Both Myxozoa and Microspora may cause fatal diseases although, generally, death and clinical disease seldom occur as the result of parasitic infections.

One of the first questions that a student may ask is, "How is it possible for any host to live in apparent good health with many parasites crowding its body?" The answer is a complicated one, involving a consideration of the results of gradual adaptations between hosts and parasites during their evolution together. After all, it is not to the advantage of the parasite to kill or even to injure its definitive host, because a healthy host means a healthy environment for the parasite.

Another question often asked by students is, "How do these parasites get into the host and what are their life cycles?" The answers are numerous and can be found throughout the pages of this book, in which the various kinds of parasite and host relationships are described in detail. Many parasites have a simple, direct life cycle whereby the infective stage (such as a cyst, spore, or motile larva) released by one host is directly taken up (often ingested) by another host, in which the parasite grows and develops. Other species of parasites may have a complicated, indirect life cycle requiring one or more additional hosts, often belonging to a different phylum, to complete their development. The timing of reproductive behavior and release of eggs or larvae, and the behavior of free-living stages of parasites, are closely related to the host life cycle, so that infective stages are ready to locate and to invade the host when the latter is available and when the environment inside and outside the host is best suited for dissemination of the parasite.[14]

Parasites do not simply lie in their hosts, absorbing food and laying eggs. They move about, constantly confronted with normal physiologic changes of host cells and tissues and changes due to host chemical and mechanical responses. The numbers, sizes, behavior, life cycles, and reactions to the host of any parasite species in any host species are regulated by the age, sex, and size of the host, by its community organization, by its inherent capability of resistance, by its previous experience with the parasite's nutritional and physiologic status, by the size of initial infection, by the presence of other species of parasites, and by the external climate, season of the year, and geographic location. A biologic description of the parasite community at one time and one place, therefore, will always change as the time and place change.

Specific questions that may be asked are: What are the important morphologic and physiologic features of parasites? How do parasites live within a host? How does a host respond to parasites? What are the parasite's nutritional requirements? Upon what factors in a host do parasites depend? Do parasites provide anything of value to a host? How does the life cycle and behavior of a host affect its parasites? How do the parasites of one species affect those of another species in the same host? What factors trigger each developmental change during the life cycle of a parasite? What genetic and developmental factors have particular significance in parasitism? What is the evidence for pathogenicity? We have enough information to answer some of these questions in part, but all of them and many others require much more attention, especially by experimental parasitologists.

Parasites of the same species but of different strains, often in widely separated geographic locations, have generally been differentiated according to minor morphologic differences, identity of susceptible hosts, serology and other characteristics that may vary with environmental factors. More reliable and accurate differentiations now commonly used are made with

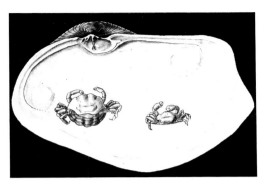

Fig. 1–1. Commensal male (small) and female pea crab, *Pinnixa faba,* in shell of gaper clam, *Schizothaerus nuttalii.*

the use of isoenzyme electrophoresis. Each different banding pattern is called a "zymodeme."

SYMBIOSIS

The term "symbiosis" was proposed in 1879 by de Bary[7] to mean the "living together" of two species of organisms. This term came to be used in a more restricted sense to connote mutual benefit, as exemplified by the termite and its gut protistans. Indeed, de Bary used a lichen as the clearest example of symbiosis. O. Hertwig defined symbiosis as " the common life, permanent in character, of organisms that are specifically distinct and have complementary needs." However, a cursory examination of **symbionts**—those organisms living together symbiotically—reveals a wide variation in permanency of the association, degree of intimacy, and degree of pathogenicity. Of special interest (but not covered in this book) are the endosymbionts in the cytoplasm and nuclei of many protozoa. Bacterial symbionts are called "xenosomes." See Lee and Corliss[16] for a discussion of this relationship and its impact on broader problems of cell biology and evolution.

Textbooks on parasitology frequently distinguish among the following three general kinds of symbiosis; commensalism, mutualism, and parasitism.

Commensalism. This occurs when one member of the associating pair, usually the smaller, receives all the benefit and the other member is neither benefited nor harmed. The basis for a commensalitic relationship between two organisms may be space, substrate, defense, shelter, transportation, or food (Fig. 1–1). If the association is merely a passive transportation of the commensal by the host, it is called **phoresy.**

Mutualism. This occurs when *each* member of the association benefits the other. For example, in the association between termites and their flagellates and between ungulates and their ciliates, the parasites digest the food (cellulose) of the host in return for board and lodging (see Chapter 23 for details). Similar to this kind of symbiosis is the ubiquitous association between animals and such parasites as bacteria, yeasts, and other fungi. These symbiotes provide essential vitamins for their hosts. They are a special category of parasites, as defined in the following paragraph, and rightly belong in a textbook on parasitology. According to MacInnis's definition,[19] mutualism occurs "when each of the interacting species functions as *both* host and parasite."

Parasitism. The original meaning of the word "parasite" (from the Greek *parasitos*) was "one who eats at another's table" or "one who lives at another's expense," and had no reference to pathogenicity. Some investigators include the killing of the host by the parasite in their definition of parasitism. Crofton[6] stated that the ability of the parasite to kill the host differentiates parasitism from commensalism. Anderson[1] believes that the inducement of host mortalities and/or a reduction in host reproductive potential is a necessary, but not sufficient, condition for the classification of an organism as parasitic. These investigators and others generally are considering *populations* of parasites and hosts, certainly not individual parasites. But where does one draw the line? How does one know that an invading species cannot kill the host? Are two dead hosts enough? How can one be sure that death was the fault of the parasites? How can we be certain that a symbiote does *not* affect its host in a way more subtle than causing obvious physical damage or change in behavior? Numerous parasites apparently act as commensals most of the time, but are pathogenic when their numbers become unusually high, or the host's resistance is depressed. *Entamoeba histolytica,* a well-known parasite of man, can cause dysentery, but most of the time it lives in the small intestine as a nonpathogen and becomes pathogenic only when certain physiologic changes take place in the host and probably also in the parasite. Textbooks on parasitology are not necessarily restricted to a study of pathogenic parasites, and a parasitologist is frequently not concerned with pathogenic parasites at all.

Trager[27] has described, in a delightful manner, the interplay between parasite and host as it occurs between protozoa living within host cells.

"In intracellular parasitism the host cell is a true and hospitable host. The parasite does not have to break in the door. It has subtle ways of inducing the host to open the door and welcome it in. One of the exciting fields in the future of parasitology is to find out what these ways are and why they are sometimes so highly specific that the cell that invites one parasite in will not open the door to another closely related species. Once inside, the parasite not only exploits nutrients already available in the cell and the cell's energy-yielding system, but it further induces the cell to assist actively in its nutrition. Like a bandit who has cajoled his way in, the parasite now forces his host to prepare a banquet for him. Finally it may destroy its host cell . . . or it may stimulate its host cell to abnormal increase in size or to have an altered metabolism with the formation of new products. Or it may even contribute some positive benefit to the host cell or to the multicellular organism of which the cell is a part, so that the two kinds of organisms then live together in a state of mutualism or symbiosis."

The concept of parasitism that clearly separates it from other categories of symbiosis is based on biochemical relationships between host and parasites. We have not been able to describe mutualistic and parasitic relationships with greater precision because we do not understand enough of the economics of these various associations. Such an understanding requires more precise knowledge of the biochemistry involved.

If a species of parasite has lived with its host species for millions of years, each partner must have had to adapt itself to the other in many ways. Among the morphologic and functional changes that a free-living organism must undergo to become a parasite are metabolic changes that require the presence of host tissues or fluids. Parasites, therefore, are metabolically dependent upon their hosts.

Cameron[3] stated that a parasite is "an organism which is dependent for some essential metabolic factor on another organism which is always larger than itself." Smyth[23] also described parasitism as an intimate association between two organisms in which the dependence of the parasite is metabolic. His definition, however, included an important addition when he said

that in parasitism "some metabolic by-products of the parasite are of value to the host."

A study by Anderson and May of a wide range of mathematical models showed that "coevolution between hosts and parasites (defined to include viruses, bacteria, protozoans, and helminths) may explain much of the polymorphism found in natural populations." The authors' "major conclusion is that a 'well-balanced' host-parasite association is *not necessarily* one in which the parasite does little harm to its host." Reproductive success is often correlated with parasite virulence or pathogenicity; "the parasite mode of life is the most commonly adopted of all modes of life in the animal kingdom." In 1971, Lincicome[17] developed the concept of "goodness" accompanying the parasite-host relationship that is "the structural foundation of this phenomenon."

We are well aware that for any definition there are exceptions and borderlines cases, but we are convinced that the criterion of metabolic dependence is the simplest and most applicable. MacInnis's definition is more precise than ours. He defined parasitism "as the case in which one partner, the *Parasite*, of a pair of interacting species is dependent upon a minimum of one gene or its product from the other interacting species, defined as the *Host*, for survival."[19]

We define parasitism as an *obligatory association between two distinct species in which the dependence of the parasite upon its host is a metabolic one involving mutual exchange of substances. This dependence is the result of a loss by the parasite of genetic information.* In addition to this basic genetic dependence, other characteristics are usually encountered in parasitism. Parasites are generally smaller than their hosts and depend upon hosts for nutrition, developmental stimuli, and habitat. Parasite species have a higher reproduction potential than do host species.

For a comparison of parasites with predators, parasitoids, and castrators, see Table 1–1.

ZOONOSES

Zoonoses are diseases or infections that are naturally transferable between vertebrate animals and man. In a broad sense, all animals are included in the defintion, but most studies of zoonoses involve only diseases of vertebrates.[26] The term **anthroponoses** means human diseases that are transmissible to animals.

Table 1–1. A Summarized Comparison of the Biological Features of Insect Parasitoids, Crustacean Parasitic Castrators, Typical Parasites, and Generalized Predators

	Biological Feature	Insect Parasitoids	Parasitic Castrators	Typical Parasites	Generalized Predators
1.	Outcome of a single act	results in death of the host	results in reproductive death of the host	no reduction in host viability	a single predator consumes many prey
2.	Summation effects: superinfection by single species	usually none	always none	frequently important, leading to a reduction in host viability	does not apply
3.	Summation effects: mixed infections	usually none	none indicated	frequent	does not apply
4.	Density-dependent effects on host populations	demonstrated in many cases	suspected herein	variable	variable, demonstrated in many cases
5.	Superinfection as it affects parasite (predator)	usually only one parasitoid emerges (except gregarious species)	usually rarer than expected; often only two parasitic isopods may mature per host	multiple infections common	does not apply
6.	Outcome of mixed-species infections	usually only one survives; first to arrive commonly destroys later animals	often less frequent than expected; in one case the second arrival causes the death of the first castrator	many mixed infections normally coexist	does not apply
7.	Density of parasite (predator) relative to host	at high densities parasitoid is implicated as a host population control agent; low and high densities common	commonly very low; occasionally high	variable, often high	almost invariably low
8.	Host (prey) specificity	most are moderately host-specific; some species with low or high specificity	typically highly specific; a few have low or moderate degrees of specificity	variable; often highly specific	variable; often very unselective in terms of prey species
9.	Size of mature parasite (predator) stage in relation to size of the mature host (prey) stage	parasitoid often less than one order of magnitude smaller than host	castrator often less than one order of magnitude smaller than host	parasite several orders of magnitude smaller than host	ranges from predator similar in size to prey to predator several orders of magnitude larger than prey
10.	Correlation of size of individual parasite with individual host	positive correlation demonstrated	positive correlation demonstrated	no correlation except in cases involving intermediate hosts	typically not correlated; sometimes a positive correlation

(From Kuris, A.M.: Quart. Rev. Biol. 49:129–148, 1974.[15])

The overall concept of zoonoses is complex. It involves man, another vertebrate, often an arthropod, the agent that causes the disease, and the environment—all forming a biologic whole. The interaction of these parts involves more than just a sum of the parts. A serious study of zoonoses should thus include the ecology of all organisms involved—parasite, animal, vector, and man.

Many zoonoses, such as balantidiasis (caused by an intestinal ciliate), fascioliasis hepatica (liver fluke disease), and tongue-worm infection, are found almost exclusively in animals and only rarely in man. Others, such as leishmaniasis (Oriental sore), flea infestation, trypanosomiasis (African sleeping sickness), and opisthorchiasis (Chinese fluke infection), are common in both animals and man. About 200 zoonoses may be grouped on the basis of the causative organisms: viruses, rickettsiae, bacteria, fungi, protozoa, nematodes, trematodes, cestodes, and arthropods.

Hydatidosis is an example of a parasitic zoonosis with worldwide distribution. Hydatid disease is caused by a larval stage of the minute tapeworm, *Echinococcus granulosus* (see Chapter 12 for a description of the life cycle).

The U.S. Department of Health and Human Services (and others, see references under: The Zoonoses) has prepared a list of zoonoses including some that are not true zoonoses but are environmental diseases common to man and animals, and some human diseases that are not found naturally in animals. The list contains almost 200 diseases. The domestic animal carrying the largest number of zoonotic diseases to man is the dog. For a thorough review of the zoonoses, see Fiennes,[8] Soulsby,[25] Hubbert et al.,[11] Jacobs,[12] and Mann.[20]

ECOLOGIC APPROACH TO THE STUDY OF PARASITISM

The whole assemblage of parasites associated with a host population, or a single host, may be called the **parasite-mix.** Such an assemblage is a small biocenose, and it includes all the viruses, bacteria, protozoa, molds, rickettsiae, worms, and arthropods that live on or in another organism. The small biocenose is a biologic entity that is constantly changing as it reacts with the environment. Parasitology is thus a study in ecology. Such an approach has been emphasized only in recent years.

When we label morphologic or physiologic features as specific adaptations to parasitism, we must bear in mind the universal need to adapt to the environment. Many characteristics described as hallmarks of the parasitic habit are also to be found among the free-living species. The hallmark is sometimes present in only one or two species, or it may even disappear during a phase in the life cycle of an individual parasite. For example, cyst formation, so characteristic of parasites, is common among free-living protozoa and metazoa. The complicated and significant alternation of sexual with asexual generations during life cycles of sporozoa, trematodes, and other parasites is duplicated in foraminifera, hydroids, and many other free-living species. The saprozoic form of nutrition can be found among soil-dwelling organisms as well as among parasites.

In order to understand more completely the ecology of parasitism, we must thoroughly review environmental variables. We must avoid the promulgation of too many broad generalizations inadequately supported by specific data. Although generalizations must be synthesized and elaborated, they must emerge from detailed, long-term studies, preferably with experimental work.

Following our introduction to the major groups and examples of specific parasites, Section VIII of this book provides more details on the ecology of parasites.

GENERAL PRINCIPLES

A principle is a fundamental doctrine, theory, or belief. Understanding the basic principles of ecology, evolution, genetics, morphogenesis, physiology, and immunology is tantamount to understanding the basic principles of parasitology. These principles, however, must be adapted to the needs of parasitologists because parasitism is much more than a combination of parasites and hosts. Associations of these organisms create a system that is unique. The components of the system can effectively be examined separately, but if principles of symbiosis are to be developed, the interrelations among all components of the system must be understood. The generalizations and hypotheses stated in the following paragraphs could constitute the beginnings of a statement of "principles" of parasitism. Other principles may be found throughout the book, especially at the ends of the last four chapters.

Parasites have lost the capacity for continuous free living and have become dependent for their survival upon one or more other living species. They have, in general, lost sense organs, locomotor abilities, and certain metabolic functions such as the elaboration of some digestive enzymes. These losses are compensated by various gains: a habitat that provides abundant food, shelter and some protection, a long individual life, specialized modes of reproduction and life cycles, and specialized organs of attachment. Many of the parasitic nematodes, such as *Ascaris lumbricoides*, have attained a size much greater than that found among the nonparasitic species.

The host has also lost some freedom. It must share its body with the parasites. The loss of food and the functions of resistance result in the diversion of energy. However, the host may benefit from the exchange of chemical substances with the parasite. In addition, the presence of one species of parasite may interfere with the establishment of another, perhaps injurious, species.

Parasites and their hosts must struggle to keep these gains. They must cooperate, so that the host remains in a healthy state and the parasite is not rejected. They must tolerate each other and resist each other, thereby becoming mutually adaptive and mutually beneficial. In this situation, the environment (the host) adjusts to the parasites. Because the host is the environment, the parasite must find a means of transport from environment to environment. A single host body provides limited space, and it eventually dies. To satisfy this need, parasites depend upon the food and habits of the host. Appropriate triggering mechanisms initiate the change from infective stages to parasitic stages. Once the parasite has begun its existence in a new host body, other triggering mechanisms initiate each change of the parasite during its development.

In a discussion of principles of parasitism, Read[22] emphasized the dependence of the parasite on the host for chemical compounds that are essential for initiating certain parasite functions or behavior (e.g., hatching of eggs, molting of larvae). From these considerations, he formulated the *Principle of Interrupted Coding*, which he defined thus: "The host must have genetic information and thus the capacity to furnish the necessary compounds and/or physical conditions to overcome a genetic block in the development of a given symbiote." He felt that Smyth's[23] concepts relating to different nutritional requirements at different stages of a parasite's development could be considered as related to Read's principle of interrupted coding. For example, the yolk sac of a chick egg is rich in nutrients and is often used as a cultivation medium for parasites. Certain tapeworm larvae, however, will not grow in the yolk sac. This failure may be attributed to a lack of chemical "signals." These considerations are speculative, but they deal with aspects of parasite-host relationships that should be included in any formulation of principles. MacInnis considers "Interrupted Coding" to be a subset of his definition of parasitism.[19]

FURTHER READING

Because many students studying animal parasitism are primarily interested in medicine (human and veterinary) or in medical technology, we list the following titles: Ash, L.R., and Orihel, T.C.: Atlas of Human Parasitology. Chicago, American Society of Clinical Pathologists, 2nd ed., 1984. Markell, E.K. and Voge, M.: Medical Parasitology. 6th ed., Philadelphia, Saunders, 1986; and from the references below, the following books: Brown & Voge (neuropathology); Georgi (for veterinarians); Katz, et al. (general diseases); U.S. Dept. Health & Human Services (the zoonoses). We also recommend Ash, L.R. and Orihel, T.D.: Parasites: A Guide to Laboratory Procedures and Identification. Chicago, Amer. Soc. Clinical Pathol., 1987.

REFERENCES

1. Anderson, R.M.: The regulation of host population growth by parasitic species. Parasitology, 76:119–157, 1978.
2. Brown, W.J., and Voge, M.: Neuropathology of Parasitic Infections. Oxford, Oxford University Press, 1982.
3. Cameron, T.W.M.: Parasites and Parasitism. New York, John Wiley & Sons, 1956.
4. Campbell, W.C.: The chemotherapy of parasitic infections. J. Parasitol., 72(1):45–61, 1986.
5. Cheng, T.C. (ed.): Parasitic and Related Diseases. Vol. 8, Basic Mechanisms, Manifestations, and Control. New York, Plenum Publ. Corp., 1985.
6. Crofton, H.D.: A quantitative approach to parasitism. Parasitology, 62:179–193, 1971.
7. de Bary, H.A.: Die Erscheinung der Symbiose. Strassburg. Karl J. Tübner, 1879.
8. Fiennes, R.: Zoonoses and the Origins and Ecol-

ogy of Human Diseases. London, Academic Press, 1979.

9. Lee, J.J., and Corliss, J.O.: Symposium on "Symbiosis in Protozoa." Introductory Remarks. Jour. Parasit. *32*(3):371–493, 1985.

10. Georgi, J.R.: Parasitology for Veterinarians. 3rd ed. Philadelphia, W.B. Saunders Co., 1981.

11. Hubbert, W.T., McCulloch, W.F., and Schnurrenberger, P.R. (eds.): Diseases Transmitted from Animals to Man. 6th Ed. Springfield, Ill., Charles C Thomas, 1975.

12. Jacobs, L.: Protozoan Zoonoses. CRC Handbook Series in Zoonoses, Sect. C: Parasitic Zoonoses, Vol. I. Boca Raton, Florida, CRC Press, 1982.

13. Katz, M., Despommier, D.D., and Gwadz, R.: Parasitic Diseases. New York, Springer-Verlag, 1982.

14. Kennedy, C.R.: Ecological Aspects of Parasitology. Amsterdam, North-Holland Publishing Company, 1976, pp. 143–160.

15. Kuris, A.M.: Trophic interactions: similarity of parasitic castrators to parasitoids. Quart. Rev. Biol., *49*:129–148, 1974.

16. Lee, J.J., and Corliss, J.O.: Symposium on "Symbiosis in Protozoa": Introductory Remarks. J. Parasit., *32*(3):371–372, 1985.

17. Lincicome, D.R.: The goodness of parasitism: A new hypothesis. *In* Aspects of the Biology of Symbiosis. Edited by T.C. Cheng. Baltimore, University Park Press, 1971, pp. 139–227.

18. Loke, Y.W.: Transmission of parasites across the placenta. Adv. Parasitol., *21*:155–216, 1982.

19. MacInnis, A.J.: How parasites find hosts: some thoughts on the inception of host-parasite integration. *In* Ecological Aspects of Parasitology. Edited by C.R. Kennedy. Amsterdam, North-Holland Publishing Company, 1976, pp. 3–20.

20. Mann, I.: Zoonoses. Insect Sci. Applic., *7*(3):337–348, 1986.

21. Price, P.W.: General concepts on the evolutionary biology of parasites. Evolution, *31*:405–420, 1977.

22. Read, C.P.: Parasitism and Symbiology. An Introductory Text. New York, Ronald Press, 1970.

23. Smyth, J.D.: Introduction to Animal Parasitology. 2nd Ed. New York, John Wiley & Sons, 1976.

24. Smyth, J.D., and Smyth, M.M.: Frogs as host-parasite systems. I. London, Macmillan, 1980.

25. Soulsby, E.J.L.: Parasitic Zoonoses: Clinical and Experimental Studies. New York, Academic Press, 1975.

26. The Zoonoses. Prepared by U.S. Dept. Health and Human Services, Centers for Disease Control and Infectious Diseases, and Office of Biosafety. Atlanta, Georgia, 1982.

27. Trager, W.: Some aspects of intracellular parasitism. Science, *183*:269–273, 1974.

2

Physiology, Biochemistry, Molecular and Immunoparasitology, Genetics: An Overview

Historically, the study of parasites has proceeded from observational to experimental approaches. One usually needs to know the morphology, the anatomy, and how to recognize and classify parasites before delving into the secrets of their life cycles, behavior, physiology, biochemistry, molecular biology, immunology, genetics, ecology, etc. The student beginning to study parasites may still profit from following this historic approach. Because so much information has accrued on parasites, and as model systems became available for experimentation in laboratories, the modern course in parasitology often includes experimentation concomitant with learning morphology and taxonomy. Likewise, it is now possible for research scientists and students who have never studied parasites to conduct experiments on them by using these model systems, without benefit of the classic approach. Thus, authors of introductory textbooks are often in a quandry as to how, when, and where to introduce the subjects touched on in this chapter. Accordingly, the instructor may assign this chapter at the beginning or end of the course. Specific details of some aspects of these topics are included in chapters on the major groups of parasites.

GOALS OF RESEARCH ON FUNCTIONAL ASPECTS OF PARASITES

In addition to the goal of acquiring basic knowledge about parasites, much of the research and information obtained on biologic functions of parasites has been directed and driven by the fact that many parasites produce disease. Hence accrual of information on the topics of this chapter, on both parasite and host, also have the goal of enabling us to control or eliminate the parasite, its vectors, and associated pathology. Because much of the information accrued is on major parasites of humans or model systems, generalities and concepts must be viewed cautiously.

Such basic research, it is hoped, will lead us to **rational approaches** to interruption of life cycles, to chemotherapy that kills the parasite, kills the vector, but does not permanently disable the host or the environment. Many ecologic studies also fall into this category. Such rational approaches have been used to break a link in the chain of a parasite's life cycle (albeit, often temporarily), for example, biologic control of vectors such as mosquitos or snails. The rational approach to chemotherapy has stimulated many of the studies on the biochemistry of parasites. The goal has been (and is) to identify enzymes and metabolic pathways, receptors, neurotransmitters, etc. in parasites that are different from those in the host, and therefore are susceptible to the rational design of drugs that affect such pathways in the parasite, but have no effect on the host. Others search for chemicals which may preferentially inhibit, for example, an enzyme in the parasite, and have negligible effect on the same enzyme in the host. A good, if not the best, example of rational chemotherapeutics was the design of anti-folate drugs in the treatment of malaria (discussed in Chap. 5).

Today, we are in a new era of studies on parasites. With the great advances made in molecular genetics, genetic engineering, and basic immunology, we are now entering a phase called **rational immunotherapy** (MacInnis, 1984, concept used at meeting of National Academy of Sciences on Future Research Approaches to Parasites, Los Angeles). Rational immunotherapy may be exemplified by the production of monoclonal antibodies to a genetically engineered antigen from a parasite; such a monoclonal antibody might temporarily protect the host, or the antigen may be used as a vaccine. The modern tools of genetic engineering are also being applied to diagnosis of parasites, which, in turn, opens new approaches to epidemiology.[2]

In spite of our optimism (perhaps also, a blow to our egos) and the usefulness of "rational approaches" to the study of parasites, we cannot overlook the success of **empirical approaches** such as random testing of the effects of chemicals on parasites, as illustrated by the discovery of the avermectins used to treat nematode infections.[44] Such discoveries also add to our understanding of the biology of parasites, as the mode of action of such drugs is elucidated.

Nor can we omit the role of **serendipity,** such as the chance discovery by Romanovsky, of the conversion of methylene blue by fungi to compounds subsequently used to produce the most useful Wright's and Giemsa stains still essential for the diagnosis of blood parasites.

The student should realize that, just as all life is a continuum, the approaches to the study of parasites (or any organism) are a continuum. For convenience, we give these approaches different names, e.g., biochemical, molecular, ecological research, etc., and often one such grouping of scientists refuses to communicate with others. Eventually all of this information will come together, and we will truly know and understand a parasite. That day is not yet here. But this emphasizes the opportunities that are available to the beginning student of parasites.

Meanwhile, the parasites are not resting, nor are their vectors. They continue to evolve, much to the chagrin of the chemotherapist, when drug resistance appears in a parasite or vector. We can safely predict that when some rational immunotherapy has been produced, the parasite will devise a way to escape.

Many persons will feel no remorse if a major parasitic disease is eradicated by modern technology or serendipity. Even though parasites are not yet endangered species, it is worthwhile noting the philosophy of a colleague, Brunko Kurlec, a parasitologist from Yugoslavia, who stated "Parasites are such exquisitely beautiful creatures, it would be a shame to eradicate even one of them before we understand it!"

As beautiful and exquisite as parasites appear in an artistically and colorfully stained slide, or alive, it is difficult to maintain such lofty ideals on the preservation of parasites if your mate or child is in the heaving fever, shuddering chills, or coma from an attack by a drug-resistant strain of falciparum malaria, or has died from the infection. When the time for eradication arrives, we may set aside some space in museums or a zoo for these killers.

REDUCTIONIST vs HOLISTIC STUDIES

The major emphasis of this book is ecologic. Indeed, as noted in Chapter 1, an ecologic approach may include all aspects of the environment and the biologic processes of parasite and host that interact to produce the union at any particular time. Together, host and parasite may create a biological environment which is different than the environment in the absence of a parasite. To study this union of host and parasite, it is often necessary (even for the ecologist) to attempt to reduce the system to parts or pieces to attempt to understand the whole. This **"reductionist approach"** is even more important to studies on the physiology, biochemistry, molecular biology, immunology, genetics, etc. of parasites. Parasites have been and are being rendered into their constituent parts, ions, elements, molecules, macromolecules, membranes, organelles, and cells. Then the pieces are slowly reconstructed in attempts to understand how the whole organism functions.

In the previous version of this chapter, Yoshino aptly emphasized the comments of Maegraith[46] on the union of host and parasite resulting in a physiologic condition that is different than that existing in the non-parasitized host. This illustrates the eventual need for a task more difficult than the reductionist approach: the **holistic study** of parasite and host together. We are just beginning to enter an era of development and use of techniques, such as nuclear magnetic resonance (NMR), that enables us to "see" and measure non-invasively, the biochemical and metabolic events in parasites living normally in their hosts.[47] Such studies undoubtedly will contribute to our already vast knowledge of host-parasite interaction, and the host-parasite interface.

The present state of our knowledge of parasites regarding their physiology and biochemistry rests on more than 50 years of study. Knowledge of molecular aspects has accumulated over at least 20 years of research. Indeed, there are now entire books devoted to nearly every major parasite of man and his domestic animals which include one or more chapters on these subjects. Thus we can only provide an introduction to these functional aspects of parasites, hoping to pique the curiosity, and stimulate the reader to further study, perhaps even research.

PHYSIOLOGY AND BIOCHEMISTRY

Physiologic and biochemical studies have been limited largely to parasites of special economic or medical importance and/or to parasites readily available and easy to maintain in the laboratory. Caution should be used, therefore, in applying resulting concepts to other parasites.

Biochemical studies of parasites and of parasite-host relationships have increased enormously during the past few years, but the field is still young and needs much more investigation. The biochemical approach to parasitology is solving some long-obscure problems, for example, species identity, and it is the key to understanding how parasites are able to live within other living organisms. Parasites that undergo various stages of development in their life cycles may have different biochemical requirements at each stage. When one stage is in a mammal and another is in an insect, for example, the difference in biochemical adjustments may be great.

Host-Parasite Interface

Basic physiologic and biochemical functions of parasites are similar to those of free-living organisms. There are, however, significant differences owing to the unique requirements of parasitism. Of prime importance is the fact that the outer surfaces of parasites are usually in contact with cellular or other membranes of their hosts.

"Because the two components of this system (host and parasite) are living organisms, and are themselves attempting to maintain equilibrium, there must exist continual interchange between the two in order to allow the basic relationships to continue. All physiological and biochemical attempts to visualize and understand this interchange must, sooner or later, involve some consideration of the host-parasite interface. The interface may be regarded as that surface through which exchange of material of physiological and immunological importance takes place. This involves the passage of substances, possibly antigenic, into the host in the form of excretions, secretions and egestions. The movement inward consists of the absorptions of nutrients, somatic and ionic interchange and the eventual entry of antibodies from the host."[21]

The importance of the outer surface of the body of the parasite, as indicated in the foregoing quotation, has prompted detailed investigations of the outer membranes of the numerous parasite species by means of such tools as electron microscopy, cytochemistry, or biochemical analysis. Parasite membranes are covered with the **glycocalyx,** which means "sweet husk." It is a sugar-rich component of the plasma membrane containing macromolecules and is probably present on all cells. Its diverse structure and composition are determined genetically and are associated with different functional properties of the cell surface, absorptive and immunologic.

The carbohydrates of glycocalyx are oligo- and polysaccharide moieties composed of neutral hexoses, acetylamino sugars, or possibly acidic derivatives. These sugars are associated with glycolipids and glycoproteins, which represent the major structural components of parasite plasma membranes. The functional significance of the glycocalyx is still poorly understood; however, modulation of host immune responses and protection of the parasite from host proteolytic degradation have been suggested as possible functions.

Under the glycocalyx of trematodes (flatworm flukes) and cestodes (tapeworms) is the rest of the tegument (described in Chap. 9, "General Considerations," and Chap. 11, "Introduction." In addition to its obvious protective function, the tegument absorbs compounds of low molecular weight that may serve as exogenous sources of parasite nutrition. The tegumental surface is covered with numerous microvilli, which increase the absorptive surface area.

Nutrients pass through the tegument by diffusion, facilitated diffusion, and active transport. With facilitated diffusion, the solute moves in relation to the prevailing concentration difference, as with diffusion, but it is not a linear function of the solute concentration. The rate of absorption follows saturation kinetics. Facilitated diffusion also is inhibited by chemicals of identical or similar structures. In contrast to active transport, no energy is expended by the parasite for diffusion or for facilitated diffusion. Flukes possess a gut; thus, absorption of nutrients may occur through the body surface and the intestinal surface. Tapeworms and Acanthocephala (Chaps. 9 and 13) have no gut, so all nutrient intake of low molecular weight must be absorbed through the cellular tegument. Salt and water balance in the flatworms also is maintained through the tegument and other specialized excretory organs called flame cells or flame bulbs (Chap. 9).

The outer covering of roundworms (nematodes, Chap. 14) is called the cuticle. It is noncellular and is impermeable to physiologically significant amounts of amino acids and sugars, except in entomophilic species. These worms, however, have a complete gut that is lined with a single layer of columnar cells bearing microvilli and is underlined with a basal lamina. Because there is no intestinal muscle layer, nutrients absorbed by the intestinal cells are easily transported to the pseudocoelomic fluid, which bathes the gut and other internal organs.

Biochemistry

Because many parasite habitats contain little oxygen, there has been much discussion of aerobiosis and anaerobiosis. Details are given later. One must remember that oxygen may be required for processes other than respiration (e.g., oxidation of amino acids). As one observer stated, parasites are metabolic opportunists. If a molecule is present it may be used. Thus, if oxygen is available (usually there is at least a trace of it) it may be used for respiration or for some other metabolic function. (See Bryan[9] for CO_2 fixation and involvement of carbon in intermediary metabolism.)

"More carbon dioxide is excreted by adults of the sheep nematode, *Haemonchus contortus, in vitro,* than any other catabolite (Ward & Huskisson, 1978) and two mechanisms appear to be involved in its formation. One is anaerobic, but it continues to function with little change in the presence of air. The other is the tricarboxylic acid cycle which the worm promptly utilizes whenever oxygen is available. *H. contortus* not only excretes large quantities of CO_2 but also readily fixes it and again the presence or absence of air has little effect on the process. Clearly CO_2 plays an important part in the energy metabolism of *H. contortus*."[75]

Parasitism results in, or is accompanied by, a loss of various structures (e.g., appendages in copepods). Biochemical functions may also be lost as an adaptation to living within another organism. Sometimes, however, the lack of a function may be due to the absence of the function in the ancestral free-living animal. For example, parasitic worms cannot synthesize cholesterol, so they must depend upon dietary intake. However, some free-living nematodes also do not synthesize cholesterol, so are also dependent upon an adequate exogenous source of sterols.[7]

The following quotation is from Davey.[17] It concerns nematodes, but may be adapted to other parasites.

"If there is a link between the environment and developmental and physiological events in nematodes, what is its nature? How are the environmental stimuli mediated? Given the relative lack of sophistication of the integrative aspects of the nematode central nervous system, and the apparent scarcity of peripheral connections, it is unlikely that there is much in the way of direct nervous control of developmental events. Another means of linking the environment to various developmental events is by endocrines, and since the study of endocrinology becomes increasingly the study of neurosecretion as one proceeds down the evolutionary scale, it is hardly surprising that nematodes have proved to contain nerve cells which exhibit the staining properties of neurosecretory cells . . . stimulation of the sense cells may lead directly to release of hormones without further integration, providing the possibility of a direct link between the environment and as yet undetermined target organs."

Pheromones

Parasitologists have made only a beginning in their understanding of the behavior of parasites. Considerable work has been done on site-finding conduct of parasitic insects and free-living stages of other kinds of parasites.

However, the whole realm of recognition and response of all parasites to host sites and host behavior and the precise methods of the parasite-host communication essential for establishment and survival of parasites, require much more study. In 1971, Ulmer[71] made the following statement, which is pertinent today: "The careful and critical analysis of adaptive behaviour for each life cycle stage, and the elucidation of trigger mechanisms including chemical, hormonal, sensory, and neurosensory stimuli, undoubtedly will provide challenging areas of inquiry for the intellectually curious helminthologist."

For many decades, parasitologists have recognized that "something" must attract a parasite to its host, or a parasite to another parasite (e.g., leading to mating behavior). Some kind of attractant within the host must initially stimulate a specific response. Such attractants are called **pheromones,** and they have been identified widely throughout the animal kingdom. Shorey stated that "pheromones are chemicals, either odors or taste substances, that are released by organisms into the environment, where they serve as messages to others of the same species."[64] The term has been expanded to include messages to other species, and various additional terms are used to indicate varieties of pheromones. For example, miraxone, released by snails *(Biomphalaria glabrata),* attracts schistosome miracidia (see Chap. 10). Probably, the stimulating factor for miracidia is the calcium-magnesium ratio and possibly certain amino acids. Specific examples of pheromone activity are given in other chapters of this book.

For further information on the physiology of parasites see: Anya,[1] Barrett,[3] Van den Bossche,[72] Bourns,[6] Bryant,[8] Chappell,[12] Fallis,[22] Gutteridge and Coombs,[27] Levandowski and Hutner,[37] Lumsden,[43] Müller et al.,[51] Slutzky.[67]

IMMUNITY

Immune Reactions in Vertebrates

The ability to ward off organisms invading body tissues has always been an essential component of internal defense mechanisms, collectively termed the "immune response." This response is equally important for the parasite because in order to become successfully established within the host, these defenses must be overcome. It is this dilemma that has served to bind firmly together the disciplines of parasitology and immunology, and it represents one of the most rapidly expanding fields of research in parasitology today.

The immune system of most vertebrates is traditionally divided into two major components: **humoral immunity,** including responses involving the production of **antibodies** (serum glycoproteins), and **cell-mediated immunity** (CMI), including responses mediated by specialized cells (T cells). Humoral immunity is initiated by a class of lymphocytes called B cells that, upon being activated by the recognition of a foreign body or substance, differentiate into antibody-secreting **plasma cells.** Antibodies produced in response to the foreign entity are then able to react against this material and to initiate a variety of elimination processes. Foreign substances that trigger the immune response mechanism by stimulating specific antibody production or specific CMI activity are called **antigens.** Actually, only a small portion of the antigen molecule, the **epitope,** may elicit production of antibodies specific for that particular determinant. Antibodies reacting with the antigen become attached to the antigenic determinant at specific combining sites. A large antigen may possess two or more distinctive determinants and therefore may be responsible for eliciting synthesis of as many different antibodies.

Many substances found in parasite tissues have the ability to stimulate antibody formation. The majority are proteins, occasionally with conjugated lipid, carbohydrate, or nucleic acid moieties associated with the molecule. Membrane-bound or secreted polysaccharides may also serve as antigens with broad specificities. In parasites, there may be a number of sources for antigenic complexes. They may be components on the surfaces of protozoa, eggs, or larvae, substances found in metabolic secretions (called "ES antigens"), or even shed cuticles of worms or pieces of tissue or cells from injured parasites (called "somatic antigens"). Somatic antigens may be identified experimentally in ground-up whole worms, in specific tissue extracts, for example, cuticle or muscles from whole worms, or in histologic sections of worm tissues subjected to labeled antibody techniques.

Compared with immune responses caused by bacteria and viruses, the antigen-antibody manifestations involved with metazoan parasites are exceedingly complex. This complexity stems from the multiplicity of antigen systems of each metazoan parasite. Because of the great variety of cells and tissues in the parasitic body, many kinds of antigens are produced. During the development of a helminth parasite, especially one that goes through two or more stages of development and requires one or more intermediate hosts, biochemical and physiologic changes constantly occur that add to the antigenic mosaic complexity. Heyneman[32] has stated,

"This is the essential difference between microbial and helminth immunity—the worm's size and its antigenic complexity . . . . An individual nematode larva, passing through various growth stages as it migrates through its host, presumably undergoing metabolic phases as well, sheds antigens not only as successive larval cuticles, but more importantly as a spewing out of metabolic waste products and a variety of other secreted and excreted antigenic substances."

Antibodies formed in respone to antigens from one species of parasite occasionally react with antigens of a different parasite species, producing immunologic **cross-reactions.** When albino mice, for example, are immunized against larvae of the nematode *Nippostrongylus brasiliensis,* and then followed by a challenge infection with cercarial stages of the trematode *Schistosoma mansoni,* the subsequent number of schistosomes is lower than in control mice. Immunization of these mice against nematode larvae conferred partial immunity to schistosome infection, probably through the induction of cross-reacting antibodies.

The central functional feature of the immune system, as indicated previously, is to distinguish between "self" and intruding foreign organisms or substances and to bring about the destruction, elimination, or isolation of these foreign substances in the body. The key to this ability for "self"-"foreign" discrimination lies with the antibody molecule that plays a central role in this process of internal defense. Antibodies belong to a group of structurally and functionally related serum proteins known as **immunoglobulins.** Each immunoglobulin molecule consists of four polypeptide chains, two short "light" chains, and two longer "heavy" chains, covalently bonded through disulfide linkages. In mammals, five classes of antibody molecules are distinguished on the basis of differences in heavy chain structure and are designated IgM, IgG, IgA, IgD, and IgE. These molecules are responsible for effecting an array of immunologic functions, which include the binding and inactivation of antigens (IgG, IgM, IgA, IgD, IgE); the enhancement of foreign particle ingestion and intracellular digestion by phagocytic white blood cells and macrophages (IgG, IgM, IgA); the lysis of foreign cell membranes through activation and binding of specialized serum proteins called complement (IgG, IgM); antibody-dependent cellular cytotoxicity reactions (IgG, IgE); and the initiation of immediate-type hypersensitivity or allergic reactions (IgE). Specific functions of these antibody classes as they relate to various parasitic organisms are discussed further in several chapters in this book.

CMI results from activities of lymphocytes, called **T cells,** which developed within the thymus gland. In contrast to **B cells,** which, upon antigenic stimulation, differentiate into antibody-secreting plasma cells, T cells, when contacting a specific antigen, differentiate into cells that are able to interact directly with foreign cells or tissues and destroy them; hence the designation cytotoxic or "killer" T cells. This "killer" function may be accomplished either through direct contact of effector T cells with surface membranes of target cells (e.g., tumor cells) or by elaboration of soluble, nonspecific, nonantibody mediators called **lymphokines,** which act in a variety of ways, for example, by poisoning foreign cells or tissues (lymphotoxins), by stimulating phagocytic activities in macrophages (macrophage activating factor), or by attracting inflammatory cells to a site of injury (chemotaxis). In addition, T cells can interact with B cells, occasionally in cooperation with macrophages, either to enhance (T helper cells) or to depress (T suppressor cells) antibody formation against various antigens. Some parasites, for instance, trypanosomes, apparently are able to escape immune destruction by interfering with this T cell regulatory function.

The establishment of a parasitic relationship with a host requires a delicate balance in immunologic function whereby both host and parasite are able to survive. **Concomitant immunity (= premunition)** represents such a compromise, in which the presence of a few parasites in the body continually stimulates an active immunity to reinfection by the same organism, thereby avoiding superinfection of the host. Factors that determine the degree of protective immunity or immune responsiveness in a particular parasite-host association are assumed to have a genetic basis involving both members, but have yet to be identified.

Dineen[19] has proposed that the immune response creates an environment for the selection of genetic variants during the evolution of the parasite-host relationship. He described the factors that might determine the mean threshold level of parasitic infections as: (1) the degree of antigenic disparity between host and parasite, and (2) the rate of flow of antigenic information. If an antigen does not stimulate a response influencing the survival (or "fitness") of the parasite, it is immunologically impotent. There is no immunologic selective pressure to modify the parasite that produced the antigen, and such a parasite may remain highly antigenic. This situation might explain the presence of antibody with little or no effect on infection. Dineen concluded that "the role of the immunological response in the 'adapted' host/parasite relationship is to control the parasite burden rather than to cause complete elimination of the infection."

Concurrent infections (see Chap. 23) with two or more species of parasites in one host body are common. Therefore, the similarities and dissimilarities of antigens of the different parasites must be considered, as well as the immune responses of the host. Schad[61] has proposed a hypothesis stating that "when co-occurring parasites are likely competitors, cross immunity may be a device evolved to limit the abundance of a competing species." In this theory, parasite species A produces an antigen that elicits an immunologic response against parasite species B, but not against A.

That parasitism is widespread in almost all species of animals would imply that parasites have developed the capacity to escape, or to render ineffective, host internal defense mechanisms. The subject of parasite immune evasion has lately become an area of intensive research for parasitologists and immunologists alike and has resulted in a number of hypotheses to explain this phenomenon. Damian,[16] who found that *Schistosoma mansoni* and its mouse host shared common antigens, proposed the concept of "eclipsed antigens" whereby the parasite antigens, in resembling those of the host, are not recognized as foreign and are therefore "eclipsed" or hidden from immune recognition. A similar method of protecting the parasite oc-

curs when a worm incorporates its host antigens into its body surface. These antigens might help to disguise the worm as part of the host and thus prevent its rejection as foreign tissue. Smithers and his co-workers[68,69] presented evidence that young adult schistosomes in the primary or initial host infection may be selectively contaminating their surface membranes with host molecules, which apparently protect these worms from a strong immune response that eventually develops. This mechanism of antigen masking offers at least a partial explanation for the development of concomitant immunity in schistosome infections.

Damian[15] also has proposed an interesting new concept of interaction between parasite and host, which he calls "**exploitation** of the host immune responses by parasites." Thus parasites may not only evade the immune response directed against them, they may also use these responses to their own benefit. As an example of such exploitation, Damian proposed that the CMI and granuloma formation around eggs of *Schistosoma mansoni* in the gut of infected mice may be used to "push" the eggs into the lumen of the gut, hence enabling eggs to leave the host and continue the life cycle of the parasite. It has long been a puzzle as to how an immotile egg manages to traverse the cellular layers of the gut. Perhaps Damian's concept is the answer. Other examples of possible exploitation of immune responses by parasites are also discussed in this important paper.

In trypanosomes, and perhaps in other protozoans, avoidance of immune destruction appears to be accomplished through a continual alteration of their surface membrane antigens. When antibodies are produced against the predominant surface antigen, the parasites synthesize and express a new variant antigen to which the host must again produce a new antibody. By continuing this process of producing new antigenic variations, the host immune system is rendered ineffectual.[4,5,33]

Several parasite species, for example, in the genera *Plasmodium, Toxoplasma, Trichinella*, and the trypanosomes, are able to suppress the ability of the host to respond immunologically. The exact mechanisms by which parasites are able to produce a state of host **immunosuppression** are not clearly understood. Parasite interference with macrophage functions, antigenic competition due to excessive amounts of parasite antigens, or the impairment of antibody regulatory activities have been suggested for some parasitic infections. Other mechanisms undoubtedly exist.

Immune reactions to parasites may occasionally produce undesirable effects, such as injury to host tisues or impairment of metabolic processes. Such reactions are termed **immunopathologic.** Circulating immune complexes (immunoglobulins + antigen + complement) may become localized in certain organs and may cause lesions. In a similar fashion, antibodies produced against a parasite antigen may cross-react with various host tissues, and again in the presence of cytolytic serum proteins, complement, tissue damage may result. One example is the formation of renal lesions during malaria. The production of **autoantibodies,** such as antiglobulins, antinuclear antibodies, or antitissue antibodies, may be responsible for pathologic changes such as anemia. Host responses against eggs of schistosomes (see Chap. 10) are an example of CMI-induced immunopathologic change. In this example, lesions take the form of large granulomas (cellular capsules surrounding isolated eggs) in the liver, as well as possible tissue necrosis resulting from the deposition of immune complexes in host kidney. Immunosuppression is a modified immune response that may be considered as a pathophysiologic mechanism. See Houba[34] for details of the pathophysiology of immune responses to parasites.

Interferon, a protein that can be extracted from cells (e.g., leukocytes), was originally found to inhibit the multiplication of viruses. The reproduction of some parasitic protozoa can also be inhibited by interferon and by interferon inducers. Interferon-containing serum from mice conferred protection against the malarial parasite, *Plasmodium berghei.* Mouse interferon, but not chick, gave protection against the widespread coccidian parasite, *Toxoplasma gondii.* When chick kidney cells infected with the coccidian *Eimeria* were treated with interferon in culture, the developmental stages of the parasite were decreased. The effects of interferon on other intracellular parasites seem to differ. For example, infections of mice with the protozoan *Leishmania donovani* were enhanced when the animals were injected with potent inducers of interferon. Probably, there is little, if any, effect on the flagellates belonging to the genus *Trypanosoma.*[30]

Infection-specific mediators, characterized as heat-resistant proteins of small molecular weight (4000 and 5000 daltons), have been identified in sensitized lymphocytes that protect macrophages, fibroblasts, and kidney cells in vitro from invasion and intracellular multiplication by the parasites *Toxoplasma gondii* and *Besnoitia jellisoni.*[13] These specific parasite growth-inhibiting factors are apparently distinct from interferon and lymphokines, and they further illustrate the diversity of protective mechanisms available to the vertebrate host.

For serodiagnosis of parasitic diseases, see Kagan,[35] and Walls and Schantz.[74] For a thought-provoking essay on the biologic significance of the immune response, see Davies et al.[18] Mitchell provides an outstanding analysis of immunity to parasites and the potential for vaccines.[50]

Immune Reactions in Invertebrates

Like the vertebrates already discussed, coelomate invertebrates (e.g., arthropods, molluscs, annelids) possess the ability to distinguish "self" from "nonself" materials and to react against foreign entities through a variety of mechanisms. Invertebrate internal defenses can be divided into two general categories: **cellular** and **humoral** immune responses. Cellular reactions involve small, motile amoeboid cells referred to as hemolymph cells, leukocytes, amoebocytes, or hemocytes, which occur in large numbers in the hemolymph or blood circulation. Humoral reactions involve soluble substances in the hemolymph that have bacteriolytic, agglutinating, or similar activity that limits the growth or viability of foreign organisms. Certain hydrolytic enzymes such as lysozyme or phenoloxidases that catalyze melanization reactions in insects and other invertebrates are specific examples of naturally occurring humoral substances. Although these kinds of reactions superficially parallel those in vertebrates, the absence of antibodies (immunoglobulins) and the immunocyte system responsible for antibody synthesis and immunologic memory in invertebrates has led to the general conclusion that the mechanisms responsible for the recognition of nonself in the latter group are *not* comparable to those operative in the vertebrates.

Because many parasites of medical, veterinary, or commercial importance use invertebrates as intermediate or definitive hosts, the study of internal defense mechanisms in these hosts is at present an area of considerable research interest to parasitologists and comparative immunologists. Cellular reactions are currently recognized as of primary importance in the isolation and/or elimination of foreign material in coelomate invertebrates. Small foreign particles, such as bacteria, viruses, or some protozoans, or foreign soluble substances such as parasitic secretions or excretions, are removed from the host by the process of endocytosis (including both phagocytosis and pinocytosis). If particles are too large for endocytosis, as is usually the case in larval helminth infections, blood cells respond by accumulating in layers around the parasite and forming an encapsulating nodule. Extracellular fibrils may also take part in capsule formation, as occurs in capsules formed by the American oyster, *Crassostrea virginica*, against larval cestodes. Encapsulation may involve different hemolymph cell types. For example, molluscan hyalinocytes (small, agranular hemocytes) are responsible for the encapsulation of renicolid (trematode) sporocysts in the marine prosobranch snail, *Cerithidea californica*,[77] whereas capsule formation in the pulmonate snail, *Biomphalaria glabrata*, around larval nematodes (e.g., *Angiostrongylus*)[28] or echinostome (trematode) sporocysts[38,39] involves primarily granulocytes. Encapsulation of larval helminths, nematodes in particular, in insect hosts is often accompanied by melanin pigment deposition in the granuloma (melanotic encapsulation).[57] The precise role of melanin in the internal defense system of insects, however, is still poorly understood.

Specific factors regulating immune reactions in the invertebrates are not well recognized. Regardless of the underlying mechanisms, however, it is obvious that most parasites and their hosts have evolved a benign immunologic relationship, one of peaceful coexistence. However, this state of "compatibility" hangs in a precarious balance. In their natural snail host, *Biomphalaria glabrata*, healthy larval stages of the blood fluke *Schistosoma mansoni* stimulate little or no cellular reaction. If a different geographic strain or a special, genetically selected refractory strain of the same snail species is used as the host, infecting larval schistosomes are met with rapid and lethal cellular response. Other factors, including the physiologic state of of the parasite or host, host age, infection "dosage," or multiple infections, also could affect the de-

gree of immune compatibility in a given association.

A case of acquired immunity to parasites in invertebrates has yet to be clearly demonstrated. This situation probably can be attributed to the absence in invertebrates of lymphoid cells involved in long-term immunologic memory, although several investigators have reported that a short-term memory component is indicated in graft rejection responses in earthworms, echinoderms, and sponges. A few studies have provided evidence that suggests an induced resistance in molluscs to infection by larval trematodes. For example, Lie et al.[38] have found that exposure of snails to irradiated echinostome miracidia stimulates in the host a resistant state to subsequent challenges of normal miracidia. The mechanisms underlying these reactions are not yet fully understood. Clearly, much more experimental work must be done, particularly with regard to reaction specificity, to the role of cellular and humoral components, and to the question of whether such reactions represent generalized mechanisms of responsiveness. For further reading on invertebrate immunity, see reviews edited by Maramorosch and Shope[48] and Gershwin and Cooper,[24] as well as a review by Lackie.[36]

GENETICS

The genetics of a parasitic relationship are inherently complex because not one but two or more genetic systems, those of the parasite and host(s), are operating in any particular association. Although progress has been made in the use of genetics in studying host resistance, in clarifying taxonomic problems, and in tracing evolutionary relationships, parasitologic genetics is still in its infancy. Problems dealing with gene identification, elucidation of meiotic division processes, and anaylses of extranuclear DNA or RNA functions have hardly been touched. The genetics of some parasite groups, e.g., Acanthocephala, remain unexplored. Students considering some aspect of parasitology for special study will find this area an open and rewarding field.

A population of bisexual parasites comprises the potentially interbreeding individuals in any locality. The genetic system of this population is the combined characteristics that determine hereditary behavior of the species, for example, mode of reproduction, chromosome cycle, breeding system, and population structure and dynamics. Independent variables are provided in the genetic cycle by the ecologic setting for a population. Parasites may normally live where few if any other members of the species exist, making bisexual reproduction difficult or impossible. In these areas, breeding is made possible by mating before dispersal of immature or new adults, or by development of hermaphroditism or parthenogenesis. These strategies are common among parasites.[58]

Chromosome numbers are being reported for numerous parasites. This information has been used in studies of parasite meiosis and mitosis, or for parasite species identification. For example, mature microgametocytes of the malarial organism *Plasmodium falciparum* contain at least 20 chromosomes. Microgametes contain about 7, the reduction being due to 3 mitotic divisions without replication. The microgametocyte is presumably haploid. In *Eimeria tenella*, a coccidian parasite of birds, and in *Toxoplasma* (Chap. 5), the zygote divides by reduction division (meiosis); thus, the postzygotic stages are haploid. Information regarding haploid chromosome numbers for different trematode species is being used to help determine the phylogenetic relatedness of the 3 major groups comprising the Trematoda: namely, the Monogenea, Digenea, and Aspidogastrea.[42] The cestodes (tapeworms) also have been subjected to chromosome analysis; for example, one species of caryophyllid has 3 sets of 8 chromosomes and thus is a triploid worm, the result of nondisjunction. In this tapeworm there are no functional sperm, owing to a breakdown of mitosis, so parthenogenesis occurs. The karyotype of a laboratory strain of the roundworm, *Trichinella nelsoni*, shows the female with a 2n (diploid chromosome number) of 6 and the male with a 2n = 5. The "odd" (univalent) fifth chromosome is probably the sex chromosome. Similar chromosome differences have been observed in other nematodes; for example, females of *Ascaris dissimilis* and *Ascaridia galli* have a 2n = 10 chromosome number, while males have a 2n = 9. Short has provided an elegant review of chromosomes of schistosomes, their sex-determining chromosomes, and interesting comments on the biology of "sex and the single schistosome."[65]

Identification of parasite or host species can be aided by genetic studies. Most species of the lung fluke *Paragonimus* have a haploid chromosome number of 11; however, one species, *P. westermani*, appears to be triploid (3n = 33), whereas other closely related species have a 2n = 22. Differences in chromosome morphology also have been reported among the various *Paragonimus* spp.[41] Identification of 25 species of the blackfly, *Simulium*, in Africa was accomplished through the use of chromosome analysis. The fly is an important vector of the filarial worm, *Onchocerca volvulus* (Chap. 18). The use of chromosome analysis also has been helpful in identifying snail vectors (*Bulinus* complex) of schistosomiasis haematobia in humans.[40]

The effectiveness of employing enzyme markers (e.g., the dehydrogenases or esterases) to analyze genetic variability within populations of parasites and free-living organisms alike is well established. Electrophoretic studies of enzymes (representing 13 gene loci) from 2 schistosome species, *Schistosoma japonicum* and the newly described *S. mekongi*, revealed an 82 to 91% divergence in enzyme electrophoretic mobility between these 2 species when compared with only a 17 to 36% divergence among 4 geographic strains of *S. japonicum*.[23] Strong evidence is thus provided for maintaining *S. mekongi* as a separate species. Studies of enzyme variation in malarial parasites infecting wild-caught rodents have provided important information on the genetic heterogeneity of natural parasite populations. These studies demonstrated that "natural malarial infections usually comprise genetically heterogenous populations of parasites. Nevertheless, the number of genetically distinct types of parasite of any one species present in a single infected host appears to be small. Generally not more than 2 or 3 clones of parasite of distinct genetic constitution are present in a single infected animal."[11] The degree of genetic polymorphism in intermediate host populations has also been investigated using enzyme analysis. This information is being used to identify geographic strains of host vectors and to determine their relationship with parasite transmission.

Host susceptibility to parasites may be regulated by genetic systems. Such a relationship has been shown between mosquitoes and the filarial worms and the malarial parasites. Infection by the filarid *Brugia malayi* is controlled by a single sex-linked recessive gene in its mosquito host *Aedes aegypti*. The gene exhibits incomplete dominance, and its expression prevents normal worm development.[45] Another *A. aegypti* gene, however, regulates mosquito susceptibility to the canine filarial worm *Dirofilaria immitis. Plasmodium gallinaceum* (avian malaria) susceptibility in *Aedes* is controlled by a single, autosomal allele. Stocks of the snail *Biomphalaria glabrata* have been developed that are genetically resistant to infection by larvae of the blood fluke *Schistosoma mansoni* (Chap. 10). In one snail stock, genes for parasite suceptibility in juvenile snails are regulated by a complex of four or more genetic factors, whereas in adult snails, a single gene apparently regulates schistosome susceptibility, with the trait of insusceptibility dominating.[59,60] It is obvious that these factors are intimately involved in determining parasite-vector host specificity, and it has been suggested that knowledge of the genetic basis for parasite susceptibility could lead to genetic control of arthropod and molluscan vectors of human disease.

In mammals, genetic differences in the host's ability to respond immunologically to parasitic helminths have been demonstrated. Mice, for example, which differ in their major histocompatibility types, have exhibited characteristically different levels of protective immunity to schistosomes when immunized with larval antigens.[52] Nonimmunologic factors also may be involved in parasite susceptibility. Racial differences in vivax malaria susceptibility in human populations appear to be related, at least in part, to the genetically determined presence or absence of specific blood group antigens. Hybridization between South African strains of *Schistosoma haematobium* and *S. matthei* occurs naturally in man, and the hybirds can be identified. They exhibit heterosis in increased infectivity to snails and hamsters, more rapid growth, and earlier maturation, and in increased daily egg production as compared with either of the parental species.[76] The book edited by Skamene et al.[66] provides an excellent overview of earlier studies on the genetic control of natural resistance to protozoan and helminthic parasites; of malignancies, which are similar to parasites in some aspects of their recognition or nonrecognition by the "host's" immune response; and of induction of autoimmunity.

Resistance of parasites to host defense mechanisms or to drugs used in chemotherapy also may be due to genetic factors. The development of chloroquine resistance by malaria-causing *Plasmodium* spp., most recently in the pathogenic *P. falciparum,* has been the result of a continuous process of genetic mutation and selection for chemical resistance in the parasite population. *Toxoplasma gondii* is another example of an important human-infecting parasite[56] that appears to have developed drug resistance in several mutant strains. The ability of parasites to escape host immunity is beautifully illustrated by the trypanosome hemoflagellates, which are able to alter their surface glycoprotein components by the sequential expression of genes for different surface antigens (called variant surface antigens). Infected hosts are unable to mount an effective immune response against a population of trypanosomes that are continually undergoing antigenic changes.[4,5,33]

MOLECULAR BIOLOGY

Investigations of the structure and function of parasite DNA or RNA are becoming more commonplace and represent a rapidly growing parasitologic subdiscipline referred to as molecular parasitology. *Trypanosoma,* a group of hemoflagellates of medical and veterinary importance, have received a large share of genetic study. In these flagellates and in other members of the family Trypanosomatidae, extranuclear DNA occurs in the kinetoplast and is called kDNA. The kinetoplastic DNA amounts to approximately 20% of the total cellular DNA. Replication of kDNA in general accompanies that of nuclear DNA. Its organization, however, is in the form of mini- and maxicircles. About 2×10^4 of the minicircles occur in a single kinetoplast and probably all replicate at each cell division. The function of kDNA is not fully clarified, but by analogy with mitochondrial DNA in mammalian cells, it may play a vital role in the control of morphogenesis by containing genetic information for the synthesis of some mitochondrial components. The ability of some trypanosomes (e.g., special strains of *T. equiperdum*) to function in the absence of a kinetoplast (dyskinetoplastic forms) remains one of the unsolved mysteries of kDNA research.

Using the tools of molecular biology (nucleotide sequencing, restriction enyzmes, and DNA hybridization techniques) parasitologists are now probing the basic mechanisms of parasite gene expression. For example, trypanosomes (*T. brucei*) are being employed in DNA,

RNA, and Western blots to investigate the expression of genes associated with the formation of variant surface antigens.[4,5,33] Of special interest is the fact that studies on the transcription and processing of mRNA for these variant antigens are breaking new ground in molecular biology. These studies, reviewed by Borst,[4,5] have led to new ideas on how splicing of mRNA may occur. This story is not yet complete, and many molecular parasitologists are actively pursuing this phenomenon. Such studies on variant antigens have also provided some of the first evidence for polycistronic messages in eucaryotes. These and other studies reveal that research on parasites also contributes basic knowledge fundamental to all of biology.

Another molecular approach that is currently being applied to most of the major parasites of humans and domestic animals combines genetic engineering with immunology. In this approach, DNA "libraries" are prepared from genomic DNA, or via cDNA prepared using reverse transcriptase and mRNA. Clones containing portions of a parasite's genome or genes are detected using expression vectors which synthesize the protein. Clones producing the antigen are detected with antibodies. The antibodies used may be monoclonals prepared, for example, against a purified variant surface antigen, or other antigen. Often, "polyvalent" antisera from naturally infected hosts are used as a first screen to detect any clone that reacts with the antisera. Such positive clones may be useful in that they may provide an ample source of antigen for use in serodiagnosis. Prior to the advent of genetic engineering, very small amounts of such antigens and their genes were available for study.

Even more exciting is the prospect that some of these clones will produce antigens which will afford some degree of protection for the host against the parasite. Indeed, we will be most fortunate if the product of a single gene, or portion of a gene such as a peptide, provides any protection. Even single-celled parasites may be composed of thousands of genes. But it is feasible in the next 5 to 10 years to clone and test many genes. Hopefully, combinations of three or four cloned genes from a parasite such as a schistosome, one *vs* cercariae, one *vs* schistosomules, and one each *vs* adult males and females, may provide worthwhile protection. Only future studies will reveal the requirements.[44]

Probably the most progress on a potential, "cloned vaccine" for a parasite has been made on malaria. These studies also provide an excellent example of "*Rational Immunotherapy*." Based on knowledge of the malarial life cycle, and the long-known fact that some protective immunity existed in malarious areas, it was decided that a vaccine against the sporozoite stage of malaria would be sought. It was reasoned that a vaccine that killed sporozoites, or prevented their successful development in hepatocytes, would prevent subsequent development of the erythrocytic stages, and hence the clinical and pathological aspects of the disease. The research of many scientists from many countries has been essential to achieve the present state of knowledge regarding the malarial vaccine.[25] We include here only a few references that will serve to introduce you to this literature.

The current status of attempts to produce a malaria vaccine are promising, but the goal of a protective vaccine has not yet been achieved. Cloning and sequencing of the circumsporozoite antigen revealed a protein that consisted, in part, of repeating units of 12 amino acids.[78] Peptides were engineered that contained different numbers of these repeated units, and they induced antibodies in model systems. Because much is yet to be learned about the effectiveness of such "peptide" vaccines, one such unit was selected for testing in primates and eventually in human volunteers.[31] This program to produce and test even one genetically engineered peptide requires vast resources and is being conducted by what is known as an "Industrial-Military-Academic Complex" or consortium. Even if the first peptide vaccine tested is not successful, much new information will be learned, and valuable progress will be made. Advances are also being made on similar antigens from other stages of malaria, the merozoites, and gametocytes, and, as with other eucaryotic parasites and viruses, a polyvalent vaccine may be required for effective protection.[29,62,63]

Progress toward a vaccine for schistosomiasis has been reviewed by Cox,[14] and for *Theileria* by Doherty and Mussenzweig.[20] A repeating, nonapeptide unit has recently been discovered in a surface antigen from *Trypanosoma cruzi*.[55] A summary of genes recently cloned from schistosomes and nematodes may be found in the report of the UCLA Symposium on Molecular Biology.[44] Pearson has reviewed antigens from parasites.[54]

Another new approach to rational immunotherapy of parasites was reported by Capron's group from France, where they attempted vaccination via anti-idiotype antibodies (AB_2). In this experiment antibodies (AB_2) were prepared against a monoclonal antibody (AB_1), which was known to react with an epitope (antigenic determinant) on the surface of a schistosome. AB_2, the anti-idiotype antibody, should "contain" an internal image of the epitope recognized by AB_1. When rats were immunized with AB_2, producing AB_3 (an internal image of the internal image, which should recognize the original epitope!), significant protection against challenge was achieved. Further discussion of this approach is available in references 10 and 26.

To further illustrate the levels at which antigens from parasites are now being investigated, attention is called to the recent study by x-ray crystallography of two variant surface antigens from *Trypanosoma brucei*.[49] Studies such as this, of course, were not possible until the genes for the antigens had been cloned. These and similar studies on other antigens may help derive a truly rational approach to immunotherapy and diagnosis of parasites.

It also has been suggested that DNA sequencing could be used as a basis for taxonomic studies. As Newton[53] wrote, "If we accept that the sequences in nucleotides in DNA determine all characteristics of a cell, it follows that a system of taxonomy might ultimately be based on similarities and differences in such sequences." In addition to species identification, information regarding the degree of DNA sequence similarity or dissimilarity could be used in determining evolutionary relationships among parasites of a particular group, or between parasites and closely related nonparasitic members. For further reading on parasite genetics, see Taylor and Muller[70] and Wakelin.[73] Pearson has summarized parasite antigens.[54]

Thus the vast array of parasites afford opportunities for investigations from the global level of populations, down to six namometers of resolution at the molecular level—truly exciting, fascinating, and with immense rewards and satisfaction to be gained by reducing the horrendous impact parasites have on mankind.

REFERENCES

1. Anya, A.O.: Physiological aspects of reproduction in nematodes. Adv. Parasitol., *14*:267–350, 1976.

2. Barker, R.H. Jr., et al.: Specific DNA probe for the diagnosis of *Plasmodium falciparum* malaria. Science, *231*:1434–1436, 1986.

3. Barrett, J.: Biochemistry of parasitic helminths. London, MacMillan Publishers LTD, 1981, p. 308.

4. Borst, P.: Discontinuous transcription and antigenic variation in trypanosomes. Ann. Rev. Biochem., *55*:701, 1986.

5. Borst, P., and Greaves, D.R.: Programmed gene rearrangements altering gene expression. Science, *235*:658–667, 1987.

6. Bourns, T.K.R.: Protozoa and parasitic helminths. *In* Biochemical and Immunological Taxonomy of Animals. Edited by C.A. Wright. London, Academic Press, 1974, pp. 387–395.

7. Brand, von T.: The biochemistry of helminths. Z. Parasitenkd., *45*:109–124, 1974.

8. Bryant, C.: The regulation of respiratory metabolism in parasitic helminths. Adv. Parasitol., *16*:311–331, 1978.

9. Bryant, C.: Carbon dioxide utilization and the regulation of respiratory pathways in parasitic helminths. Adv. Parasitol., *13*:35–69, 1975.

10. Capron, A., et al.: Anti-idiotypes and vaccines against schistosomiasis. *In* Molecular Paradigms for Eradicating Parasitic Helminths. Edited by A.J. MacInnis. New York, Alan Liss, Inc. UCLA Symposia Series *60*:45–54, 1987.

11. Carter, R.: Studies on enzyme variation in the murine malaria parasites *Plasmodium berghei, P. yoelii, P. vinckei* and *P. chabaudi* by starch gel electrophoresis. Parasitology, *76*:241–268, 1978.

12. Chappell, L.H.: Physiology of Parasites. New York, John Wiley & Sons, 1980.

13. Chichilla, M., and Frenkel, J.: Mediation of immunity to intracellular infection (*Toxoplasma* and *Besnoitia*) with somatic cells. Infect. Immun., *19*:999–1012, 1978.

14. Cox, F.E.G.: Towards schistosomiasis vaccines. Nature, *314*:402–403, 1985.

15. Damian, R.T.: The exploitation of host immune responses by parasites. J. Parasitol., *73*:1–13, 1987.

16. Damian, R.T.: Common antigens between adult *Schistosoma mansoni* and the laboratory mouse. J. Parasitol., *53*:60–64, 1967.

17. Davey, K.G.: Hormones, the environment, and development in nematodes. *In* Comparative Biochemistry of Parasites. Edited by H. van den Bossche. New York, Academic Press, 1972, pp. 81–94.

18. Davies, A.J.S., Hall, J.G., Targett, G.A.T., and Murray, M.: The biological significance of the immune response with special reference to parasites and cancer. J. Parasitol., *66*:705–721, 1980.

19. Dineen, J.K.: Immunological aspects of parasitism. Nature, *197*:268–269, 1963.

20. Doherty, P.C., and Mussenzweig, R.: Progress on *Theileria* vaccine. Nature, *316*:484–485, 1985.

21. Erasmus, D.A.: The Biology of Trematodes. New York, Crane, Russak & Co., 1972.

22. Fallis, A.M.: Ecology and Physiology of Parasites. Toronto, University of Toronto Press, 1971.

23. Fletcher, M., Woodruff, D.S., Loverde, P.T., and Asch, H.L.: Genetic differentiation between *Schistosoma mekongi* and *S. japonicum*: An electropho-

retic study. *In* The Mekong Schistosome. Edited by J.I. Bruce, Sornmani, H.L. Asche, and K.A. Crawford. Malacol. Rev., Supplement 2, Whitmore Lake, Michigan, 1980.

24. Gershwin, M.E., and Cooper, E.L. (eds.): Animal Models of Comparative and Developmental Aspects of Immunity and Disease. New York, Pergamon Press, 1978.

25. Godson, N.: Molecular approaches to malaria vaccines. Sci. Am., December, pp. 52–59, 1986.

26. Grzych, J.M., et al: An anti-idiotype vaccine against experimental schistosomiasis. Nature, 316:74–76, 1985.

27. Gutteridge, W.E., and Coombs, G.H.: Biochemistry of Parasitic Protozoa. Baltimore, University Park Press, 1977.

28. Harris, K.R.: The fine structure of encapsulation in *Biomphalaria glabrata*. Ann. N.Y. Acad. Sci., 226:446–464, 1975.

29. Harte, P.G., Rogers, N., and Targett, G.A.T.: Vaccination with purified microgamete antigens prevents transmission of rodent malaria. Nature, 316:258–259, 1985.

30. Herman, R.: The effects of interferon and its inducers on experimental protozoan parasitic infections. Trans. N.Y. Acad. Sci. Ser. II, 34:176–183, 1972.

31. Herrington, D., et al.: Safety and immunogenicity in man of a synthetic peptide vaccine against *Plasmodium falciparum* sporozoites. Nature, 328:257–259, 1987.

32. Heyneman, D.: Host-parasite resistance patterns—some implications from experimental studies with helminths. Ann. N.Y. Acad. Sci., 113:114–129, 1963.

33. Hoeijmaker, J.H.J., et al.: Novel expression-linked copies of the gene for variant surface antigens in trypanosomes. Nature, 284:78–80, 1980.

34. Houba, V.: Pathophysiology of the immune response to parasites. *In* Pathophysiology of Parasitic Infections. Edited by E. Soulsby. New York, Academic Press, 1976, pp. 221–232.

35. Kagan, I.G.: Serodiagnosis of parasitic diseases. *In* Manual of Clinical Microbiology. 3rd Ed. Edited by E.H. Lennett, et al. Washington, D.C., American Society of Microbiology, 1980.

36. Lackie, A.M.: Invertebrate Immunity. Parasitology, 80:393–412, 1980.

37. Levandowsky, M., Hutner, S.H., and Provasoli, L. (eds): Biochemistry and Physiology of Protozoa. Vol. 1 to 3. New York, Academic Press, 1979 (Vols. 1 and 2), 1980 (Vol. 3).

38. Lie, K.J., Heyneman, D., and Jeong, K.H.: Studies on resistance in snails. 7. Evidence of interference with the defense reactions in *Biomphalaria glabrata* by trematode larvae. J. Parasitol., 62:608–615, 1976.

39. Lie, J.K., Heyneman, D., and Lim, H.K.: Studies on resistance in snails: specific resistance induced by irradiated miracidia of *Echinostoma lindoense* in *Biomphalaria glabrata* snails. Int. J. Parasitol., 5:627–631, 1975.

40. Lo, C.T., Burch, J.B., and Schutte, C.H.J.: Infection of diploid *Bulinus* s.s. with *Schistosoma haematobium*. Malacol. Rev., 3:121–126, 1970.

41. LoVerde, P.T.: Chromosomes of two species of *Paragonimus*. Trans. Am. Microsc. Soc., 98:280–285, 1979.

42. LoVerde, P.T., and Fredericksen, D.W.: The chromosomes of *Cotylogaster occidentalis* and *Cotylaspis insignis* (Trematoda: Aspidogastrea) with evolutionary considerations. Proc. Helminthol. Soc. Wash., 45:158–161, 1978.

43. Lumsden, R.D.: Surface ultrastructure and cytochemistry of parasitic helminths. Exp. Parasitol., 37:267–340, 1975.

44. MacInnis, A.J. (ed.): Molecular paradigms for eradicating helminthic parasites. Proceedings of the Upjohn-UCLA Symposium, Alan Liss, 1987, p.575.

45. McDonald, W.W.: The influence of genetic and other factors on vector susceptibility to parasites. *In* Genetics of Insect Vectors of Disease. Edited by J.W. Wright and R. Pal. Amsterdam, Elsevier, 1967, pp. 567–584.

46. Maegraith, B.: Interdependence. Am. J. Trop. Med. Hyg., 26:344–355, 1977.

47. Mathews, P.M., and Mansour, T.E.: Applications of NMR to investigation of metabolism and pharmacology of parasites. *In* Molecular Paradigms for Eradicating Helminthic Parasites. Edited by A.J. MacInnis. New York, Alan R. Liss, 1987, pp. 472–492.

48. Maramorosch, K., and Shope, R.E. (eds): Invertebrate Immunity. New York, Academic Press, 1975.

49. Metcalf, P., et al.: Two variant surface glycoproteins of *Trypanosoma brucei* of different sequence classes have similar 6 Å resolution X-ray structures. Nature, 325:84–86, 1986.

50. Mitchell, G.F.: Injection versus infection: The cellular immunology of parasitism. Parasitology Today, 3:106–111, 1987.

51. Müller, M., Gutteridge, W.E., and Köhler, P. (eds.): Molecular and Biochemical Parasitology. Amsterdam, Elsevier/North-Holland Biomedical Press, 1980.

52. Murrell, K.D., Clark, S., Dean, D.A., and Vannier, W.E.: Influence of mouse strain on induction of resistance with irradiated *Schistosoma mansoni* cercariae. J. Parasitol., 65:829–831, 1979.

53. Newton, B.A.: Extranuclear DNA, with special reference to kinetoplast DNA. Acta Protozool., 1:9–21, 1974.

54. Pearson, T.W. (ed.): Parasite antigens. New York, Dekker, 1986, p. 413.

55. Peterson, D.S., Wrightsman, R.A., and Manning, J.E.: Cloning of a major surface-antigen gene from *Trypanosoma cruzi* and identification of a nonapeptide repeat. Nature, 322:566–568, 1986.

56. Pfefferkorn, E., and Pfefferkorn, L.: Genetic recombination with *Toxoplasma gondii*. J. Parasitol., 64(Suppl.):Abst. No. 44, 50, 1978.

57. Poinar, G.O., Jr.: Insect immunity to parasitic nematodes. *In* Contemporary Topics in Immunobiology—Invertebrate Immunology. Edited by E.L. Cooper. New York, Plenum Press, 1974, pp. 167–178.

58. Price, P.W.: Evolutionary Biology of Parasites. Princeton, Princeton University Press, 1980.

59. Richards, C.S.: Susceptibility of adult *Biomphalaria*

glabrata to *Schistosoma mansoni* infection. Am. J. Trop. Med. Hyg., *22*:748–756, 1973.

60. Richards, C.S., and Merritt, J.W., Jr.: Genetic factors in the susceptibility of juvenile *Biomphalaria glabrata* to *Schistosoma mansoni* infection. Am. J. Trop. Med. Hyg., *21*:425–434, 1972.

61. Schad, G.A.: Immunity, competition, and natural regulation of helminth populations. Am. Naturalist, *100*:359–364, 1966.

62. Sharma, S., and Godson, G.N.: Expression of the major surface antigen of *Plasmodium knowlesi* sporozoites in yeast. Science, *228*:879–882, 1985.

63. Sharma, S., Svec, P., Mitchell, G.H., and Godson, N.: Diversity of circumsporozoite antigen genes from two strains of the malarial parasite, *Plasmodium knowlesi*. Science 229:779–782, 1985.

64. Shorey, H.H.: Animal Communication by Pheromones. New York, Academic Press, 1976.

65. Short, R.B.: Sex and the single schistosome. J. Parasitol., *69*:4–22, 1983.

66. Skamene, E., Kongshavn, P.A.L., and Landy, M. (eds): Genetic control of natural resistance to infection and malignancy. New York, Academic Press, 1980, p. 597.

67. Slutzky, G.M.: The Biochemistry of Parasites. New York, Pergamon Press, 1981.

68. Smithers, R.S., McLaren, D.J., and Ramalho-Pinto, F.J.: Immunity to schistosomes: the target. Am. J. Trop. Med. Hyg., *26*(Suppl.):11–19, 1977.

69. Smithers, R.S., and Terry, F.J.: The immunology of schistosomiasis. Adv. Parasitol., *14*:399–422, 1976.

70. Taylor, A.E.R., and Muller, R. (eds.): Symposia of the British Society for Parasitology. Vol. XIV: Genetic Aspects of Host-Parasite Relationships. Oxford, Blackwell Scientific Publications, 1976.

71. Ulmer, M.J.: Site finding behaviour in helminths in intermediate and definitive hosts. *In* Ecology and Physiology of Parasites. Edited by A.M. Fallis. Toronto, University of Toronto Press, 1971, pp. 123–160.

72. Van den Bossche, H. (ed.): The Host-Invader Interplay. Third International Symposium on the Biochemistry of Parasites and Host-Parasite Relationships. Amsterdam, Elsevier/North-Holland Biomedical Press, 1980.

73. Wakelin, D.: Genetic control of susceptibility and resistance to parasitic infection. *In* Advances in Parasitology. Edited by W.H.R. Lumsden, R. Muller, and J.R. Baker. London, Academic Press, 1978, pp. 219–308.

74. Walls, K., and Schantz, P.M.: Immunodiagnosis of Parasitic Diseases. Vol. I. New York, Academic Press, 1986, p. 312.

75. Ward, P.F.V., and Huskisson, N.S.: The role of carbon dioxide in the metabolism of adult *Haemonchus contortus, in vitro*. Parasitology, *80*:73–82, 1980.

76. Wright, C.A., and Ross, G.C.: Hybrids between *Schistosoma haematobium* and *S. matthei* and their identification by isoelectric focusing of enzymes. Trans. R. Soc. Trop. Med. Hyg., *74*:326–332, 1980.

77. Yoshino, T.P.: Encapsulation response of the marine prosobranch *Cerithidea californica* to natural infections of *Renicola buchanani* sporocysts (Trematoda: Renicolidae). Int. J. Parasitol., *6*:423–431, 1976.

78. Young, J.F., et al.: Expression of *Plasmodium falciparum* circumsporozoite proteins in *Escherichia coli* for potential use in a human malaria vaccine. Science, *228*:958–962, 1985.

3

Introduction to the Protozoan Group; Phylum Zoomastigina

Protozoa abound in oceans, fresh water, soil, and the bodies of other organisms. They are generally microscopic, and although they consist of a single cell with one or more nuclei, they are amazingly complex in structure, physiology, and behavior. Their complexity is so great that they are sometimes called "acellular," to distinguish them from the individual cells that make up a metazoan animal or plant. Although some protozoan cells may be grouped together in a colony, each cell maintains its independent functions.

The classification of protozoa has been revised. It raises Protozoa and related forms to the level of a new kingdom, Protista.

CHARACTERISTICS OF PROTOZOA

No major morphologic features, outside of those common to cells of all plants and animals, can be listed for *all* the protozoa. Each subdivision of the subkingdom possesses its own distinctive morphology. The outer surface of protozoan parasites is an organelle for the passage of substances in and out of the organism, as with free-living forms, but it is also an area of reaction with the host. Antibody and antigens come together here, and surface factors are involved in movement on the host, attachment, and penetration into host cells.

Feeding Mechanisms. In addition to diffusion of nutrients through the cell membrane, three other methods of feeding have been described for the Protozoa: **phagocytosis, pinocytosis,** and **cytostomal feeding.** The term "endocyto-

sis" is currently used by many protozoologists to include phagocytosis (the uptake of solid material) and pinocytosis (the uptake of material in solution through small pinocytic vesicles). There are, however, functional, biochemical, and morphologic differences among the three methods. In sporozoa, the cytostome is also called a **"micropore"** or **"micropyle."** The tubular cytostome of trypanosomes functions in intracellular forms. In malarial parasites, the micropore actively ingests host cell cytoplasm.

"In many haemosporidians, the host cell cytoplasm is ingested in small vacuoles pinched off from the cytostomal cavity. . . . The cytostome of *Leucocytozoon*, one of the Haemosporina, appears to be a transient structure. . . . In this parasite pinocytotic vesicles that form along the wall of the cytostome take nutrient into the parasite. . . . Another method of nutrient intake by intracellular parasites is by pinocytosis along their plasmalemma. . . . The digestion of host cell cytoplasm occurs within food vacuoles of most intracellular parasites. . . . In the Haemosporina, one of the digestion products of hemoglobin is the electron-dense pigment, hemozoin."[1]

Excretion. This is accomplished primarily by diffusion through the cell membrane. Whether contractile vacuoles represent a primary excretory system can be questioned. These vacuoles, which are said to represent an osmoregulatory system, are absent from most marine and parasitic flagellates and amebas, but are present uniformly in marine and parasitic ciliates.

Undigested matter in ciliates is eliminated through a pore called a **cytopyge** or **cytoproct,** or through a temporary opening in the body wall.

Respiration. In parasitic species, respiration is either aerobic (e.g., the malarial parasite, *Plasmodium*) or anaerobic (e.g., the amebic dysentery parasite, *Entamoeba histolytica*). In general, however, parasites carry on aerobic or anaerobic fermentations rather than complete oxidation, even when oxygen is plentiful. These fermentations are more varied than those occurring in vertebrate tissues, in which lactic acid is formed almost exclusively. Most parasitic protozoa, as well as metazoan parasites, ferment carbohydrates to succinate. Lactate is the primary, if not the exclusive, product of the malarial parasite and is produced in lesser quantities in some other species.

The host intestine is not completely devoid of oxygen. The oxygen tension in the intestinal gases of a pig has been estimated to average about 30 mm Hg or about one-fifth that of air. It should not be assumed that the oxygen tension of any one habitat is a uniform phenomenon—it varies considerably. Apparently, there is an oxygen gradient in the intestine, because large numbers of protozoa may be found near the intestinal mucosa, with decreasing numbers toward the center. A similar gradient exists in tissues near small blood vessels. However, these distributions of parasites might be due to other factors, such as differences in the pH of the different areas. Moreover, bacteria in the intestine may alter the conditions near a parasite by using the available oxygen themselves. Respiratory metabolism is further considered with discussions of various groups of parasites. For extensive considerations, see Van den Bossche.[83]

Blood-dwelling forms of trypanosomes depend on a glycerophosphate oxidase system for their high oxidation demands. Glucose is broken down to pyruvic acid, but no further. In cultivation, however, respiration is more conventionally aerobic. The Krebs cycle operates in oxidative decarboxylation, and cytochrome pigments can be detected. Respiration of forms in the insect gut is similar to respiration of those in cultivation.

Reproduction. This may consists of a series of simple binary fissions, multiple fissions **(schizogony),** or elaborate and precise integrations of sexual and asexual reproductions. Schizogony is a form of asexual multiplication. Since no gametes are involved, the process is sometimes called **agamogony,** in contrast to gamete production, which may be called **gamogony.** In schizogony, the nucleus undergoes repeated division; each nucleus then becomes surrounded by a separate bit of cytoplasm, and the original cell membrane ruptures, liberating as many daughter cells as there were nuclei. These daughter cells are **merozoites.** The parent cell undergoing nuclear division is called a *schizont.* If a multinucleate cell divides into portions that are still multinucleate, the process is called **plasmotomy.** When a syncytium (many nuclei within one cell membrane) is produced, the process is called **nucleogony.**

Budding is another method of reproduction by some unicellular parasites. Essentially, the process is simply mitosis with unequal cellular division. Endogenous budding within the cell membrane occurs in other parasites, including the Myxozoa.

Sexual reproduction in various forms occurs in parasitic protozoa. If two cells unite and exchange nuclear material, the process is **conjugation** (common in ciliates). When they separate, each cell may be called an **exconjugant.** If sex cells **(gametes)** are produced, they unite by **syngamy** to form a **zygote,** the first cell of a new individual. Gametes may be similar in appearance, in which case they are known as **isogametes,** or they may be dissimilar in appearance, resembling the eggs and sperm of higher organisms, in which case they are called **anisogametes.** For example, in the body of a mosquito, the gametocytes of malarial parasites form gametes of different size and shape. The large variety (female) is a **macrogamete,** whereas the smaller one (male) is the **microgamete.**

Coprophilous Protozoa. These organisms normally live in feces outside the body. This group does not include the usual intestinal protozoa. Often, however, cysts of coprophilous protozoa get into the gut of animals or man and are mistaken for parasites. Thirty or more species of such soil, sewage, and fecal forms have been reported from various animals. They include flagellates, amebas, and ciliates. The word **coprozoic** is also used to describe these organisms, but it should be restricted to those species that normally do not live in feces outside the body, but can be cultivated in feces.

If feces from any mammal remain at room temperature or even in a refrigerator for a few days, motile amebas, flagellates, and ciliates of the coprophilous varieties may be found (Fig. 3–1). Motile coprophilous amebas can be dis-

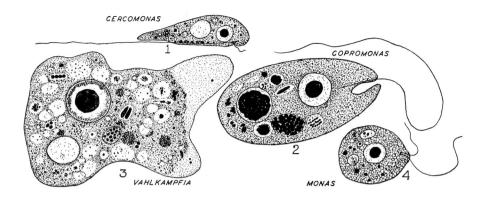

Fig. 3–1. Coprophilous protozoa from cattle feces. *1, Cercomonas* sp. ×5000. *2, Copromonas* sp. ×6000. *3, Vahlkampfia* sp. ×1600. *4, Monas communis* ×5000. (From Noble. Courtesy of the Journal of Parasitology.)

tinguished from most parasitic species by the presence in the former of a contractile vacuole and, usually, a large endosome in the nucleus, although much variation exists. The ciliates are usually clearly freshwater species.

For general references to parasitc protozoa see the following publications listed in the references at the end of this chapter. Chandra: Critical Reviews in Tropical Medicine, 1982.[9] Freyvogel: Environmental Management for Control of Parasitic Protozoan Diseases, 1886.[22] Honigberg: Mechanisms of Pathogenicity Among Protozoa, 1986.[30] Katz, et al.: Parasitic Diseases, 1982.[36] Kirkpatrick: Enteric Infections, 1984.[37] Lee, et al.: An Illustrated Guide to the Protozoa, 1985.[45] Lee: Protozoa as Indicators of Ecosystems, 1986.[44] Levine: Veterinary Protozoology, 1985.[47] Mettrick and Desser: Parasites—Their World and Ours, 1982.[58] Weber: Sexually Acquired Parasitic Infections in Homosexual Men, 1985.[88]

PHYLUM DINOFLAGELLATA

These free-swimming organisms, a few of which are parasitic, produce food reserves of starch and lipids and contain chromatophores. They are abundant in all oceans. The body possesses two grooves, one longitudinal **(girdle)** and one transverse **(sulcus).** A flagellum usually lies in each groove.

Members of the Dinoflagellida can occur in numerous invertebrates. *Blastodinium* lives in the gut of the crustacean, *Cyclops;* a parasite of proportionate size in man would be as large as his liver. Parasitic dinoflagellates may also be found on other crustacea, in annelids and salps (ascidians), and on the gills of some freshwater and marine fishes. These organisms can live within cells of siphonophores (colonial coelenterates) and in Protozoa (tintinnioid ciliates and Radiolaria). *Coccodinium,* for example, parasitizes other dinoflagellates, and *Duboscquella tintinnicola* is a large (100 μm) parasite in tintinnioid ciliates. *Paradinium poucheti* lives in the body cavity of copepods, whereas copepod eggs may be inhabited by *Chytriodinium parasiticum* and by *Trypanodinium ovicola.* The blue crab, *Callinectes sapdis,* along the Atlantic coast of the United States, harbors a pathogenic dinoflagellate, *Hematodinium* sp. The parasite is about the size of crab hemocytes and may occupy the vascular spaces of all tissues.

Order Rhizomastigida

The order Rhizomastigida shows affinities with amebas by possessing ameboid bodies that often produce pseudopodia. In general, they possess one to four flagella. Most are free-living. *Rhizomastix gracilis* is a species found in crane-fly larvae and in the salamander known as "axolotl." *Mastigina hylae* occurs in the intestines of many species of frogs; its nucleus is situated at the extreme anterior end of the body.

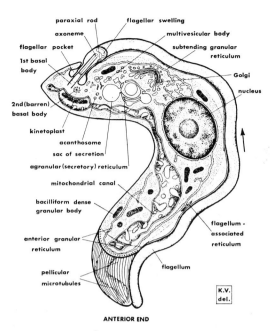

paraxial rod

axoneme

flagellar pocket

1st basal body

2nd(barren) basal body

kinetoplast

acanthosome

sac of secretion

agranular(secretory) reticulum

mitochondrial canal

bacilliform dense granular body

anterior granular reticulum

pellicular microtubules

flagellar swelling

multivesicular body

subtending granular reticulum

Golgi

nucleus

flagellum - associated reticulum

flagellum

K.V. del.

ANTERIOR END

Fig. 3–2. The fine structure of *Trypanosoma congolense* in its bloodstream phase as revealed by phase contrast observations and electron microscopy. (From Vickerman. Courtesy of the Journal of Protozoology.)

PHYLUM ZOOMASTIGINA

Order Kinetoplastida

FAMILY TRYPANOSOMATIDAE

Taxonomy. Trypanosomes are divided into two main groups: **Salivaria,** which develops infective forms in the salivary glands of the vector and infects new hosts by biting (e.g., *Trypanosoma rhodesiense* of African sleeping sickness, and *T. brucei brucei* that does not infect man); and **Stercoraria,** which develops infective forms in the hindgut of the vector. These forms leave the insect with its feces and infect new hosts by penetration through the skin or mucous membranes or by entry through lesions made by vector bites (e.g., *T. cruzi* of Chagas's disease).

The family consists of elongated leaf-like or sometimes rounded protozoa. The body is covered with the plasma membrane (also called the pellicle or plasmalemma) 10 μm thick. Beneath this membrane occur subpellicular microtubules, probably microskeletal in function (Fig. 3–2). The single flagellum emerges from the body at the anterior end. The attachment organ is a **hemidesmosome** formed within the flagellar

sheath. The flagellum is externally absent in certain attached stages and in the amastigote stage. All parasites in the family are internal, and most are apparently harmless commensals, usually found in arthropods or vertebrates. The flagellated stages range in length from 4 to 130 μm.

The body surface of *Trypanosoma brucei, T. rhodesiense,* and other blood forms possess long, slender cytoplasmic filaments named **filopodia** or **micronemata** that are visible only with an electron microscope. These minute appendages are up to 70 μm long. They arise from the anterior and posterior ends of the cell and also from the flagellum. Apparently, they are deciduous. Although their function is still in doubt, it has been suggested that they may possibly be released antigens; this concept is supported by the presence of considerable amounts of trypanosome cytoplasm within plasmalemma. Another possible function is environmental sampling, such as the detection of chemical gradients.

The flagellum arises from a **kinetosome** at the floor of an invagination that usually occurs at the anterior end of the body, but sometimes occurs near the posterior end or in between (see first basal body in Fig. 3–2). This invagination is called the **flagellar pocket,** and it may be shallow or long enough to reach almost to the posterior end of the body. A contractile vacuole, and occasionally more than one, opens into the flagellar pocket.

A cytostome has been described for several species (e.g., *Trypanosoma mega, T. conorhini, T. raiae., T. cruzi,* and *Crithidia fasciculata*). In *T. cruzi,* it begins as a funnel-shaped depression that continues as a narrow cylindrical tube passing deeply into the cell. Probably, in all species, the cytostome opens into the flagellar pocket or close to it in the anterior part of the flagellate.

The **kinetosome** (sometimes called a **blepharoplast, basal body,** or **centriole**) is a cylinder composed of nine equally spaced fibrils; it averages about 1.5 μm in overall diameter. The fibrils are connected to those of the axoneme within the flagellum. There is some evidence that the kinetosomes are self-duplicating, but there is more evidence that they arise anew each time the cell divides. They contain DNA, although its origin is obscure (possibly from the nucleus), and its function in the kinetosome is unknown.

Just posterior to the kinetosome, which often cannot be seen, lies the **kinetoplast,** a spherical, rod- or disc-shaped structure found mainly in members of the family Trypanosomatidae, but also in the Bodonidae and Cryptobiidae. It is usually deeply stained in prepared specimens. It is a modified mitochondrion and is made up of linear molecules, minicircles and maxicircles. The function of the former is obscure. The latter contains DNA (written kDNA to differentiate it from nuclear DNA), which undoubtedly codes for structures and functions related to mitochondria. In some members of this group, e.g., *Trypanosoma evansi*, the DNA component of the kinetoplast has the form of a dense compact mass that has lost its function but still replicates at division. These forms are **diskinetoplastic** trypanosomes. Either they have no cyclical stages in insect vectors or they are transmitted directly by the mouth parts of contaminated insects.

"Less than 20% of the kDNA is in the form of large (7 to 12 μm contour length) circular molecules which appear to correspond to the mitochondrial genome of other cells. The bulk of the kDNA takes the form of "minicircles" which are one tenth the size of the maxicircles and are interlinked by catenation with one another and with the maxicircles. The significance of the minicircle component of kinetoplast is quite unknown but its genetic role is now in some doubt."[84]

The presence and function of the kinetoplast is related to the life cycles of trypanosomes. It is large and active in blood forms of the Lewisi group (see discussion of genus Trypanosoma). This is the group in which the parasite develops in the hindgut of vector insects. The blood forms of the Brucei group (see also discussion of genus Trypanosoma) contain a small, more compact kinetoplast with less-pronounced DNA (Fig. 3–3).

Many blood forms are long and slender, but when they are taken into an insect, a pronounced modification of shape and changes in ultrastructure, antigenicity, and metabolic activity occur.

There are eight main body types in this family. They vary in size, body shape, location of the flagellum, location of the kinetoplast and other structures (Fig. 3–4). Each of these is described and defined below the figure. The life cycle of any one genus may include more than one of these configurations. These changes in form are associated with changes in ultrastructure and metabolic pathways and are associated with evasion of the host's defense mechanisms.

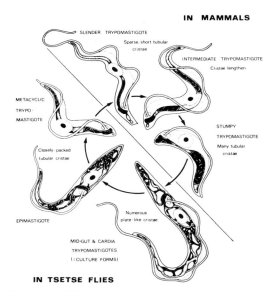

Fig. 3–3. Scheme showing changes in the form of the mitochondrion and its cristae during the life cycle of *Trypanosoma brucei*. (From Vickerman, K.: Morphological and physiological considerations of extracellular blood protozoa. *In* Ecology and Physiology of Parasites: a Symposium. Edited by A.M. Fallis. Toronto, University of Toronto Press, 1971.)

MAIN GENERA OF TRYPANOSOMATIDAE

Genus Leptomonas. These trypanosomatids sometimes reach 200 μm in length. They are pointed at the posterior end and are pointed or narrowly rounded at the anterior end. A reservior (flagellar pocket), whose depth does not exceed one-fifth of the body length, opens narrowly at the anterior end. The kinetoplast occurs below the base of the reservoir. The flagellum emerges from the bottom of the reservoir, passes out of the cytostome, and continues freely. There is no undulating membrane. The nucleus occurs in the middle third of the body length. The species of this genus are parasites of protozoa, nematodes, molluscs, and insects. There is only one host in the life cycle. A common species is *L. ctenocephalus* in the dog flea. The parasite forms amastigote cysts. Larval fleas become infected by eating flea feces contaminated with the cysts.

Genus Herpetomonas. Species of this genus are usually promastigotes or opistomastigotes but may be the intermediate form, paramastigotes, depending on the stages in its life cycle or temperature. The body is long and slender, pointed or truncate at the posterior end, and

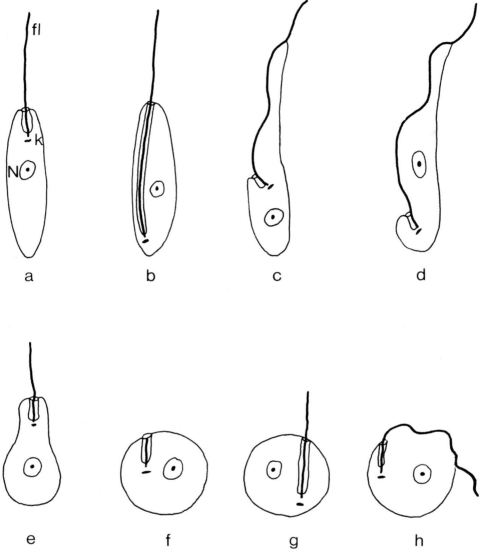

Fig. 3–4. Stages in the life cycles of trypanosomatid flagellates. *(a)* Promastigote: elongated form with antenuclear kinetoplast with flagellum arising near it and emerging from the anterior end of the body. *(b)* Opisthomastigote: elongated form with postnuclear kinetoplast; flagellum arising near kinetoplast and passing through the body and emerging from the anterior end. *(c)* Epimastigote: elongated form with juxtanuclear kinetoplast. *(d)* Trypomastigote: elongated form with postnuclear kinetoplast; flagellum arising near it and emerging from the side of the body to run along the surface to form a undulating membrane. *(e)* Choanomastigote: 'barley corn' form with antenuclear kinetoplast; the flagellum arises from a wide funnel-shaped reservoir and emerges anteriorly. *(f)* Amastigote: round, oval, or elongated form devoid of flagellum. *(g)* Paramastigote: form with kinetoplast close to nucleus; intermediate between Pro- and Opisthomastigote; flagellum emerges from anterior end and reservoir extends through body to point where flagellum and kinetoplast are close. *(h)* Sphaeromastigote: rounded form with a free flagellum, which represents a transitional stage between an amastigote and mastigote form. (From Molyneux and Ashford, courtesy of Taylor and Francis.[60])

pointed or rounded at the anterior end. The nucleus occurs in the middle or posterior third of the body length. The reservoir opens narrowly in the anterior end and extends a variable distance into the body. The kinetoplast is situated anywhere from a point one-fifth to about nine-tenths of the body length. The flagellum originates near the kinetoplast, passes through the reservoir, and emerges through the mouth of the reservoir. There is no undulating membrane. As in the leptomonads, development occurs only in one host. Cysts may be produced. These parasites are found in the gut, malpighian tubules or haemocoel of diptera, reduviid bugs, and hymenoptera. Sample species are *H. muscarum* in houseflies and *H. ampelophilae* in *Drosophila.*

Genus Crithidia. These are small (4 to 10 μm long) choanomastigote parasites of insects, with a body usually shorter and wider than in other genera. The anterior end is trucate. The body is often constricted in the anterior third, to produce a vase-like shape. The reservoir is wide, with its mouth (cytostome) occupying most of the truncate anterior end. The kinetoplast is large, lateral, and located posteriorly, sometimes near or besides the nucleus. The flagellum moves in a circular motion. These organisms are often clustered together or are attached like the pile of a carpet to the intestinal wall. They possess bacterial symbionts which supply essential vitamins or enzyme co-factors for their flagellate hosts. There are no intermediate hosts. Sample species: *C. fasciluata* in the gut of *Culex* and *Anopheles* mosquitoes.

Genus Blastocrithidia. Species of this genus are insect epimastigote flagellates (15 to 33 μm long) drawn to a long point anteriorly. The reservoir opens along the side. The kinetoplast is situated just posterior to the base of the reservoir and usually anterior to or beside the nucleus, but in occasional individuals, it occurs behind the nucleus. The flagellum arises near the kinetoplast and passes out through the reservoir and along an undulating membrane to the anterior end, whence it continues as a free flagellum. These parasites live in the mid-gut and hind-gut of their hosts. *Blastocrithidia familiaris* lives in the hemipteran, *Lygaeus pandurus.* According to Tieszen, et al.,[82] "Attachment of the parasites in the midgut and ileum occurs by interdigitation of expanded flagella over and between the microvilli." There is no attachment to the microvilla. In the rectum attachment is

to the cuticle of the gland cells. *Blastocrithidia* may form highly resistant and infective cysts which can be transmitted from one bug to another by coprophagy. Free-living stages of the parasite occur in water. Other species may be found in ticks and water striders.

Genus Trypanosoma. Elongate trypomastigote parasites (12 to 130 μm long) with a morphologically complex life cycle. In the form characteristic of the genus, the reservoir is considerably posterior to the nucleus. The kinetoplast is posterior to the base of the reservoir; a mass of electron-dense, anteroposterior fibers form a compact, sharply defined zone across the kinetoplast. The flagellum is attached to the body by an undulating membrane. The life cycle usually involves a vertebrate and an invertebrate host. If, in the invertebrate host, the flagellates are limited to the intestine they are called **stercorarian** trypanosomes. If they migrate to the insect salivary glands they are *salivarian* trypanosomes. A sample stercorarian species is *T. cruzi.*

Trypanosomes are usually found in the body fluids of vertebrates, especially in the plasma, and in the digestive tracts of arthropod or leech vectors. Trypanosomes, however, may occur in any body organ and some have a "preference" for certain organs such as the heart. The numbers of these parasites in one host may be enormous. For example, 20 million to 4 billion can be recovered from the blood of an animal 100 hours after infection. However, the relation between numbers and infectivity of trypanosomes varies during the normal development of these flagellates in the vector and in the mammal host. Host specificity also varies considerably. Some species, especially *Trypanosoma congolense,* occur in practically all domestic animals; others are found almost exclusively in one type of host, for example, *T. lewisi* in rats. Feeding apparently occurs by **pinocytosis** ("cell drinking"), which involves the passage of food down the cytostome and absorption through an enlargement or vacuole that forms at the end of the cytostome.

Differentiation of trypanosome species and subspecies involves a study of morphology, behavior, life cycles, and biochemical characteristics. The use of monoclonal antibodies and isoenzyme electrophoresis illustrates the modern techniques employed to solve difficult problems of identifying the parasites. An understanding of these characteristics also promotes improved diagnosis and treatment of the disease. Vickerman[86] stated that, as trypanosomes pass from one enviroment to another, changes in their gross and ultrastructural form are accompanied by changes in the pathways of energy metabolism associated with the mitochondrial system. Antigenic variation occurs in surface membranes, thus producing circulating variable antigenic types (VATs). This ability leads to evasion of the mammalian host's immune system and is a major factor in the success of these parasites.

Various investigators have suggested the possibility of sexual reproduction in trypanosomes in addition to longitudinal binary fission. Nyindo, et al.[67] gave evidence that this phenomomen occurs in in vitro cultivation of *T. brucei* from the insect vector.

The pathology of trypanosome infection varies from host to host and with different species of parasite. In general, at the site of infection, there may be edema due to lymph exudation and a cellular infiltration of lymphocytes associated with proliferation of histiocytes. In blood vessels, the parasite may cause hypoglycemia, an increase in numbers of lymphocytes and monocytes, a rise in serum IgM, and a drop in erythrocyte count. Anemia is characteristic of all trypanosome infections. Disease of the lymph glands and enlargement of the spleen are common. Mild myocarditis may occur, and the central nervous system may be involved.

Table 3–1 shows the classification of the major trypanosomes of man. We will not use the subgenus names nor the species from which subspecies names are derived; these are shown by letter only. Thus we list *T. (H.) lewisi* and *T. (T.) b. brucei* as *T. lewisi* and *T. brucei*.

For general references to trypanosomes see Baker: Perspectives in Trypanosomiasis Research, 1981[3]; Ginoux, et al.: Trypansomiasis[23], 1982; Hudson: The Biology of Trypanosomes, 1985[33]; Massamba, et al.: Distinction of African trypanosome species using nucleic acid hybridization, 1984[56]; Molyneux and Ashford: The Biology of Trypanosomes and Leishmania, Parasites of Man and Domestic Animals, 1983[60]; Newton, Trypanosomiasis, 1985.[65]

Trypanosoma lewisi (Fig. 3–5), a common blood parasite of rats, occurs throughout the world. Although it is generally nonpathogenic in adult animals, it can cause lethal infection in suckling rats. It is a slender flagellate, pointed at both ends and averaging about 25 μm long.

The life cycle is similar to that of *T. cruzi* but the intermediate host is the rat flea, *Ceratophyllus fasciatus.*

Trypanosoma theileri (Fig. 3–5) is a large blood parasite of cattle. Similar species occur in wild Bovidae and Cervidae. It reaches 70 μm long in the blood stage and up to 120 μm long in animals suffering from chronic infection. The life cycle is similar to that of *T. cruzi* (see next subject). Epimastigote and amastigote forms occur. As in *T. cruzi* reproduction occurs in tissues, but *T. theileri* may also divide in the bloodstream.

Ticks are vectors of this parasite. Transmission of the infection to cattle occurs when metacyclic forms enter broken skin or when mucous membranes of the mouth are contaminated. There is also evidence that transplacental transmission occurs. Parasitaemia is low and the infection is usually nonpathogenic.

Trypanosoma cruzi causes **Chagas' disease** (South American trypanosomiasis). It is found mainly in Central and South America, where it is a major cause of morbidity and mortality. In Chile alone it probably affects over one half million people. The parasites in the blood of man and animals are trypomastigotes. They are about 20 μm long and contain a large kinetoplast, which lies near the posterior end. These trypomastigotes do not multiply. Adult flagellates live in the blood and reticuloendothelial tissues of man, dogs, cats, monkeys, armadillos, opossums and other animals.

The acute phase of Chagas' disease starts with invasion of tissue cells, especially in heart muscle, smooth muscle, brain autonomic nervous system, and mononuclear phagocytes. In these cells the parasites multiply by equal binary fission forming amastigotes, epimastigotes, and spheromastigotes (Fig. 3–4). Invasion of host cells provides protection for the parasite from phagocytosis by macrophages. *T. cruzi* does not exhibit antigenic variations because the dividing stages occur within these cells. Host antibodies along with cellular immunity probably terminate the active phase of the disease (Vickerman[86]).

The various forms of *T. cruzi* in host cells reproduce rapidly, forming closely packed masses of **pseudocysts.** Rupture of pseudocysts liberates them into the blood. They may reenter other cells and be eaten by blood-sucking bugs belonging to the family Reduviidae. These insects pierce the skin of man and other verte-

Table 3–1. Classification of the Genus *Trypanosoma**

	Trypanosoma					
Section	Stercoraria			Salivaria		
Subgenus	*Megatrypanum*	*Herpetosoma*	*Schizotrypanum*	*Duttonella*	*Nannomonas*	*Trypanozoon*	*Pycnomonas*

Representative species	*T.(M.) theileri*	*T.(H.) lewisi*
		T.(H.) musculi
		T.(H.) rangeli
			T.(S.) cruzi	*T.(D.) uniforme*
				T.(D.) vivax
					T.(N.) congolense
					T.(N.) simiae
						T.(T.) brucei
						  T.(T.) b. brucei
						  T.(T.) b. rhodesiense
						  T.(T.) b. gambiense
						T.(T.) equiperdum
						T.(T.) evansi
							T.(P.) suis

*From Molyneux and Ashford,[60] courtesy of Taylor and Frances.

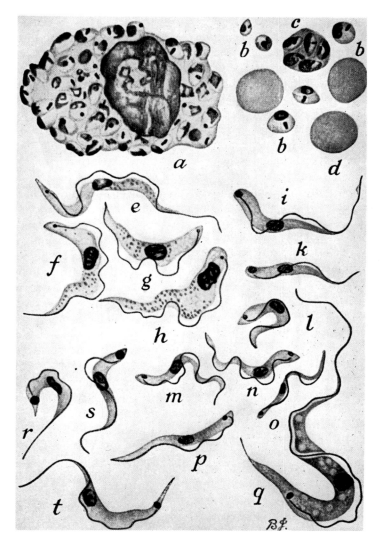

Fig. 3–5. *Leishmania* and *trypanosoma* from various animals. *a,b,c, Leishmania* in tissue smears; *a,* Macrophage packed with parasites; *b,* Parasites scattered by rupture of host cells; *c,* Detached portion of host cell with parasites; *d,* Erythrocytes; *e* to *t,* Trypanosomes; *e* to *f, T. evansi* or *T. equiperdum; e,* slender; *f,* intermediate; *g* to *h, T. brucei; g,* stumpy; *h,* posterior nuclear forms; *i, T. vivax; k, T. uniforme; l,m,n,o, T. simiae; p, T. congolense; q, T. theileri; r,s, T. cruzi; t, T. lewisi.* (From Richardson and Kendell, courtesy of Oliver and Boyd.)

brates and feed on blood which contains several types of trypansomes. Short or "stumpy" forms in the bug's midgut transform into small epimastigotes that multiply and maintain the infection for the rest of the bug's life. Other trypomastigotes transform into spheromastigotes. Longer epimastigotes pass into the rectum and become spheromastigotes which develop infective metacyclic forms. These do not divide. The cycle in the insect requires about 2 weeks. While feeding, the bugs defecate and urinate, thus depositing the infective metacyclic forms on the skin. Often the irritated host rubs the bitten area and creates ample opportunity for the parasites to be rubbed into the wound made by the bug.

Metacyclic forms may also penetrate mucous membranes, especially the ocular and, to a lesser degree, the oral. Infections may arise even from the use of contaminated syringes. Blood transfusion is an important method of spreading the disease. A fetus may receive the parasites through the placenta, and the infection may be transmitted to babies by mother's milk. Animals and man are sometimes infected by eating bugs or bug feces or by eating other infected animals. Infected bugs may infect "clean" bugs when they live close together.

During the chronic stage, 10 or more years after infection, symptoms of alimentary tract involvement may occur. Enlargement of the esophagus (megaesophagus) may be striking. Congenital megaesophagus has been reported. Enlargement of the esophagus and colon is associated with masses ("nests") of parasites in host tissues. Mortality is about 10%. The disease is lifelong and spontaneous cures do not occur. Digestive and/or cardiac manifestations are common, and sudden death is frequent. The infection has been described as the most important cause of myocarditis in the world. The incidence of infection may be as high as 50%, especially in children. In these young people, the infection usually subsides spontaneously; whereas adults with cardiac infections have a high ratio of morbidity and mortality.

Symptoms of Chagas' disease are so varied that a clear diagnostic picture is difficult to present. Often, however, the bug bites the area of the eye, especially in children. During the acute stage there occurs a unilateral swelling of the face (Romaña's sign), apparently due to the bite of the bug, not to the trypanosomes. The eyelid becomes puffy, and the eye is often closed. Both eyes and even the whole face and neck may become involved. As parasites invade body organs, enlargements of the spleen, lymph nodes, and liver occur, with headaches, fever, anemia, and prostration. The "mega" condition of enlargement of the esophagus or colon appears to be related to this disease. Urinary bladder muscles, striated muscles, and the nervous system may also be affected. Chronic myocarditis is the leading cause of death by heart failure in endemic areas. Intramuscular forms are mostly amastigote stages. The existence of distinct immunologic types may explain clinical diversity. The disease is diagnosed by several kinds of serologic test and by finding flagellates in the blood. Acute manifestations of the disease subside spontaneously in 2 to 3 months in 90 to 95% of cases.

Control of the infection centers mainly on the vectors. Pheromones, insect growth regulators, natural pathogens, and biologic control procedures are recommended. Mermithid nematodes and parasitic wasps reduce the vector populations. Education is important; householders should learn something about the life cycle of the parasites, keep their floors clean, spray piles of firewood, and improve house construction.

Trypanosoma cruzi infects more than 100 species of wild and domestic animals. In some parts of the southwest United States, wood-rat nests are inhabited by triatomid bugs that are parasitized with *T. cruzi*, and the disease in man has been reported a few times.

For zymodene characterization see Miles and Cibulskis,[59] 1986. For schizodeme analysis see Morel, et al.,[61] 1986. For monoclonal antibodies see Petry, et al.,[70] 1986. For cell biology see De Souza,[16] 1984. For the disease see Le Ray and Recacoeclea,[46] 1985. For a bibliography see Dvorak, et al.,[18] 1985.

Trypanosoma rangeli occurs in man, monkeys, opossums, dogs, cats, and anteaters in Central and South America. It is 25 to 37 μm in length and goes through the epimastigote and trypomastigote stages in man but may be as long as 80 μm in the triatomid bug, *Rhodnius prolixus*. Other triatomids also serve as vectors. Within a bug the flagellate develops in the gut, haemolymph, and salivary glands. Transmission of the small metacyclics occurs during the act of biting the vertebrate's host.

This parasite is not pathogenic to vertebrates but it is pathogenic to the bug by invading various tissues. It is a major factor in the control of populations of *Rhodnius prolixus*. Man may become infected with both *Trypanosoma rangeli* and *T. cruzi* at the same time. The two may be difficult to tell apart but may be differentiated by their different complement sensitivity.

Trypanosoma melophagium, a common blood parasite of sheep in England, is transmitted by the sheep ked (*Melophagus ovinus*), which is a wingless, bloodsucking fly. Probably up to 90% of British sheep are infected with this flagellate. A similar parasite, *T. theodori,* occurs in goats.

Trypanosoma vivax, an active blood parasite, occurs in practically all domestic animals in Africa, the West Indies, and parts of Central and South America. It causes one of the most important diseases of cattle in which anemia and thrombocytopenia are typical. Dogs and pigs are not easily infected. In Africa, the insect vector is the tsetse fly, but in other countries, transmisssion is by mechanical means, indicating that the parasites have been introduced to these countries and have been able to maintain themselves in spite of the lack of a suitable intermediate host (see Chap. 24, "Host Migrations"). These parasites average 22 μm in length, with a range of 20 to 26 μm. The posterior end of the body is characteristically wider than the anterior end.

Trypanosoma uniforme is shorter than *T. vivax* and occurs in cattle, sheep, goats and antelope. It is transmitted by tsetse flies.

Trypanosoma congolense is short flagellate that lacks a free flagellum. It measures about 13 μm in length with a range of 9 to 18 μm. There is considerable evidence that it is not just a simple species but is a group of widely diverse parasites. They are found in African game animals and horses, sheep, goats, camels and are especially serious in cattle. The disease is transmitted by tsetse flies in which the life cycle requires a minimum of 12 days involving both a midgut and proboscis stage. Infection occurs through the bite of the fly. Cattle suffer anemia and other symptoms common to other forms of trypanosomiasis in animals. The disease is called **nagana.** It is the same name given to the disease caused by *T. brucei.* Host blood serum lipids are altered. With some strains of the parasite the infection is almost always fatal. Pinder[71] studied the range of virulence in mice infected with *T. Congolense.* She stated that resistance was due to a "recessive trait controlled by a single autosomal gene (or gene cluster). In addition, sex-associated factors appear to confer higher resistance in females."

Trypanosoma simiae is a species similar to *T. congolense* but longer, averaging 14 to 24 μm. It is a common and virulent parasite of African monkeys and is also found in sheep, goats, pigs, and camels. It is the most important trypanosome of domestic swine, but apparently does not infect horses, cattle, or dogs. It is transmitted by the bite of tsetse flies and can be transferred mechanically by bloodsucking flies. The life cycle and infection rate are similar to those of *T. congolense.*

Some livestock and game animals are inherently tolerant to trypanosome infection. Trypanotolerance is probably a genetic factor controlled by a dominant gene.[62] This phenomenon explains why some breeds of animals survive in endemic areas, whereas others do not, thus proper selection of breeding stock can improve the herd.[17a]

Vertebrates are not the only hosts that have a genetic basis for trypanosome infection. Tsetse flies may inherit susceptibility to the parasite.[55]

Trypanosoma gambiense and *T. rhodesiense* cause **African sleeping sickness** (African trypanosomiasis of man). *T. gambiense* is responsible for the variety that is endemic in the west and central portions of Africa. In the blood of man *T. gambiense* may be long and slender, measuring about 25 × 2 μm with a flagellum,

or short and broad without a flagellum, or intermediate in shape. The parasite multiplies by longitudinal splitting, and it migrates through the body by way of the blood. Normal habitats are the blood plasma, cerebrospinal fluid, lymph nodes, and spleen. The disease occurs in men, women and children.

The vector of this flagellate is usually the tsetse fly, *Glossina palpalis,* which bites man and feeds on his blood. In the intestine of the fly, the parasite reproduces and forms both epimastigotes and trypomastigotes. After two weeks or more in the gut of the fly, the flagellates migrate to the salivary glands, where they become attached to the epithelium and develop into the infective stage. The ecologic relationships in human trypanosomiasis are summarized in Figure 3–6.

Congenital infection may occur when the placenta is damaged. The infection may also be transmitted through mother's milk. The parasites may also enter through the mucous membrane of the upper part of the alimentary canal. It is possible that Masai tribesmen become infected by drinking fresh blood. Laboratory workers have become infected through abraded skin. These unusual portals of entry are possible with both *T. gambiense* and *T. rhodesiense.* There is no evidence of natural *T. gambiense* infection in wild mammals, but such animals are possible reservoirs of infection. In guinea pigs terminally infected with *T. gambiense,* the level of glycogen and activity of glucose-6-phosphatase in the liver are decreased. This inhibition of enzyme function in dying guinea pigs is apparently the cause of hypoglycemia.[54]

The Gambian or chronic form of sleeping sickness primarily involves the nervous and lymphatic systems. There may also be invasion of the dermis with the production of a local chancre. After an incubation period of 1 or 2 weeks, fever, chills, headache, and loss of appetite usually occur, especially in non-natives. As time goes on, enlargement of the spleen, liver, and lymph nodes occurs, accompanied by weakness, skin eruptions, anemia, disturbed vision, and a reduced pulse rate. The infection leads to meningoencephalomyelitis and hemolysis. As the nervous system is invaded by the parasites, the symptoms include weakness, apathy, headache, and definite signs of "sleeping sickness." A patient readily falls asleep. Coma, emaciation, and often death complete the course of the

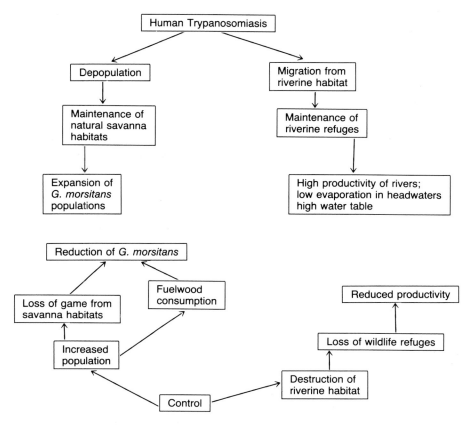

Fig. 3–6. A summary diagram of ecologic relationships in human trypanosomiasis. (From Molyneux and Ashford, courtesy of Taylor and Francis.[60])

disease, which may last for several years. The mortality rate is high.

Trypanosoma rhodesiense is closely related to *T. gambiense*. It is identical in appearance and has the same type of life cycle. Research strongly supports the concept that it is a set of variants of *T. brucei* rather than a subspecies as in Table 3–1.[80] The insect vectors are the tsetse flies, *Glossina morsitans*, *G. pallidipes*, and others. *T. rhodesiense* causes the Rhodesian or acute and more rapid type of human sleeping sickness, which usually results in death within a year. Because of its rapid course, Rhodesian trypanosomiasis rarely if ever causes the symptoms normally associated with sleeping sickness. Compared with Gambian sleeping sickness, there is less involvement of the central nervous and lymphatic systems, but cardiac involvement is more severe. Roelants and Williams[73] summarized the symptoms and pathology of African trypano-

somiasis caused by *T. gambiense* and/or *T. rhodesiense* as follows: irregular fever, diarrhea, hepatosplenomegally, kidney malfunction, oliguria, albuminuria, kidney malfunction, cardiomegaly, excessive sensitivity to touch, pain, various paralyses, nervous symptoms, delusions, epileptic crisis, and spectacular sleep disorders. In the absence of treatment patients die with extreme emaciation or general apathy from cardiovascular collapse. The disease is fatal in a few months to 2 years.

Eradication involves vector control, immunization (immunity can be conferred by infection in some cases), and treatment using suramin, pentamidine, or Berenil. These drugs gave no curative effect on late stages of infection. They are toxic to man. The incidence of infection is less than that with *T. gambiense,* and the parasite is restricted to a much more limited area, being almost confined to the high tablelands of southeast Africa. A malignant variety of Rhodesian trypanosomiasis has arisen among tourists who have visted East African wildlife reserves and national parks.

The greatest economic impact of African trypanosomiasis is its effect on domestic animals, mainly cattle. Millions of square kilometers of land would be available for raising cattle were it not for this disease. Some of the symptoms and disease occurring in man are also common in cattle, especially anemia and heart failure. Curative drugs used for cattle are isometamidium chloride and Berenil. There is evidence that some cattle are genetically resistant to trypanosomiasis. This trypanotolerance is probably due to a dominant gene. See Murry and Morrison[62] for a discussion of host susceptibility to African trypanotolerance.

Trypanosoma brucei is widely distributed in Africa in dogs, sheep, goats, horses, mules, camels, donkeys and many wild mammals. Cattle and pigs suffer mild infections, whereas laboratory animals may be severely infected. The morphology of the parasite is almost identical to that of *T. gambiense* and *T. rhodesiense*. It ranges from 12 to 42 μm in length and 2 to 3 μm width. Tsetse flies transmit the infection. In vertebrate hosts the parasite lives in blood and various tissues, especially in macrophages. For its developmental cycle see Figure 3–7. **Nagana** is the name for the disease in livestock. Wide variations in parasite size and shape give rise to the name **pleomorphism (polymorphism)**. As the parasites spread throughout the body, host reaction is shown by edema, anemia, nervous symptoms, fever, conjunctivitis, keratosis, blindness, paralysis, and, especially in horses, death.

Trypanosoma evansi (Fig. 3–5) is another serious parasite of horses, camels and many other domestic and wild mammals. It causes diseases known as **surra** and **mal de caderas**. Formerly *T. equinum* was thought to be a distinct species similar to *T. evansi*. It is now considered to be the same species. Some of these parasites possess a small kinetoplast; others lack this organ. *T. evansi* ranges in size from 15 to 36 μm. It can be found in dogs, in which the disease is often fatal, and in donkeys, cattle, and elephants. It is widespread throughout Africa, Central and South America, Russia, Burma, India, the Philippines, and Vietnam. The principal vectors are horse flies and stable flies in which there is no cyclic development. It is transmitted by bites with contaminated mouthparts. Ticks (*Ornithodoros*) and even vampire bats have been implicated. The pathology is similar to that of *T. brucei*. Anemia is the most common clinical sign of infection.

For a review of trypanosomiasis in domestic animals see Trypanosomiasis: A Veterinary Perspective by Lorne E. Stephen, 1986.[78a]

Trypanosoma equiperdum causes the disease **dourine** in horses (see Fig. 3–5). It is transmitted during copulation and thus is a veneral disease. The flagellate is morphologically identical to *T. evansi*, and its length is about 25 μm. It does not exhibit the usual changes in morphology during its life cycle. The parasite first affects the sexual organs of male and female horses and causes swelling and ulcers. Enormous numbers of parasites can be found in these ulcers. The nervous system may become involved, as evidenced by paralysis of the legs or parts of the face. The mortality rate is high, and there is no effective treatment. Dogs, mice, rats, and rabbits may become infected with *T. equiperdum*, but the parasites remain largely in the blood of these hosts. A complement fixation test has been developed to identify the disease in horses.

Trypanosomes of Other Vertebrates. Trypanosomes of birds have been reported primarily from domestic species, but undoubtedly many wild birds harbor as yet undescribed forms. *Trypanosoma avium* occurs in various birds; *T. gallinarum* is a parasite of chickens, and *T. hannai* may be found in pigeons. The names of other bird forms may be found in texts on protozoology.

Trypanosomes of fish are transmitted by leeches. Many fish trypanosomes, such as *Trypanosoma giganteum* in the ray and *T. percae* in perch, are unusually large (e.g., up to 130 μm long).

Trypanosoma mega lives in the toad, *Bufo regularis*. This parasite has been found to be a convenient and suitable flagellate for experimental studies. The taxonomy of anuran trypanosomes is in a much confused state. Many species have been described, but few with adequate care and detail. For a review of this neglected group, see Bardsley and Harmsen.[4]

Genus Leishmania. Trypomastigotes of this genus occur in two forms: round or oval amistogotes (2.5 to 5 μm) in vertebrate tissues, primarily within mononuclear phagocytes (macrophages), and promastigotes in the vector, a

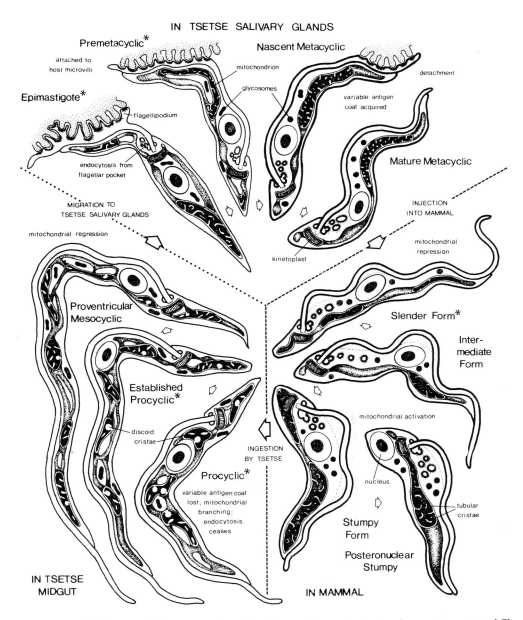

Fig. 3–7. Scientific diagram of *Trypanosoma brucei* developmental cycle. (From Vickerman, courtesy of Churchill Livingstone.[86])

phlebotomine fly. Vectors in the old world generally belong to the genus *Phlebotamas,* whereas those of the new world are usually *Lutzomyia.* The fine structure of the kinetoplast is like that of trypanosomes. The disease, **leishmaniasis,** is almost always a zoonosis. In addition to man it occurs in dogs, rodents, sloths, hyraxes, lizards, and other animals may carry one or more of the various species or strains of leishmania. Rarely the disease is transmitted directly from one mammal to another. This may occur by contact between a skin sore filled with amastigotes with a skin lesion on another mammal. In nonsylvatic areas or cities domestic dogs are a major source of infection. The geographic distribution, considering all species and varieties, includes Central and South America, Africa, the Mediterranian area, the Near East, India, the Soviet Union, and Asia.

Sand flies become infected by feeding on man or other animals, drawing blood infected with amistogotes into their intestines. In the midgut amastigotes transform into promastigotes which migrate to the salivary glands. When the insect again feeds on blood promastigotes are engulphed by, or penetrate into, the macrophages (Fig. 3–8). Penetration starts with attachment of flagellar tips. Lameller sheaths, veils or pseudopodia whorls progressively surround the parasite (Molyneux and Ashford).[60] This penetration is called "facilitated phagocytosis." It is possible however, that the parasite may actively contribute to the penetration process. Within the macrophage *Leishamania* transform again into amastigotes and rapidly multiply. Infection of these cells is closely related to genetic make-up. The internal susceptibility is apparently governed by single gene. Another gene may govern immunity and recovery. It is remarkable that parasites are able to enter and multiply within macrophages whose main function is to destroy invading organisms. Macrophages are havens for them. Their survival depends on getting inside this host cell within 10 or 15 minutes.

Leishmania causing either dermal or visceral disease may survive in the mammal host for many years without revealing themselves. There have been cases of patients moving away from areas of transmission years before the disease appeared. Where the parasites live during such periods of time is unknown.

Authorities disagree on the number of valid species and strains. WHO has listed 26 species

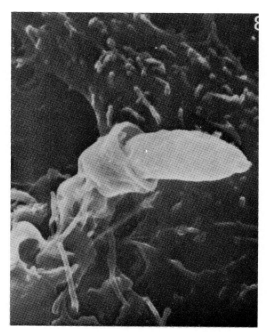

Fig. 3–8. Promastigotes of *Leishmania tropica* being engulfed by the lamellar sheath of a mouse macrophage. (From Zenian, A., Rowles, P., and Gringell, D.: Scanning electron-microscope studies of the uptake of *Leishmania* parasites by macrophages. J. Cell Sci., *39*:187–199, 1977.)

which include 16 subspecies or strains. Table 3–2 shows a simplified classification. Most of these flagellates are morphologically indistinguishable but with their morphologic, physiologic, biochemical and genetic differences, the various species and subspecies of *Leishmania* produce a broad spectrum of cutaneous, mucocutaneous, and visceral leishmaniasis. Differentiation of species and varieties is aided by clinical, epidemiologic, enzymologic, and other studies. Of the numerous species that have been described, we discuss the six most widely accepted. The first three are those that cause cutaneous leishmaniasis, an infection of reticuloendothelial cells of the skin and/or of the mucosae of the mouth and nose.

Leishmania tropica causes a serious cutaneous disease called **Oriental sore** or **cutaneous leishmaniasis** that occurs in tropic and subtropic regions of the world. Its distribution is, in general, similar to that of kala-azar, (see paragraph later in this section), but the specific areas of occurrence frequently do not coincide. It may be found in numerous parts of China,

Table 3–2.   Classification of the Genus *Leishmania**

Leishmania

Section	Hypopylaria (Parasites of reptiles)†	Peripylaria (parasites of reptiles* and mammals)	Suprapylaria (Parasites of mammals)
Representative species	"L. agamae" "L. ceramodactyli"	"L. adleri" L. braziliensis L. peruviana "L. tarentolae"	L. aethiopica L. donovani L. major L. mexicana L. tropica

*From Molyneux and Ashford,[60] courtesy of Taylor and Frances.
†Recent studies on reptile trypanosomes question the validity of lizard *Leishmania*

Asia Minor, southeast Russia, Greece, southern Europe, northern, western, and southwest Africa, and the New World. The transmitting agents include phlebotomine flies. Skin sores may vary from a small pimple to large ulcerated areas, usually on the face, arms, or legs. The parasites rarely invade adjacent mucocutaneous areas. Development from the bite of a sandfly to the appearance of skin sores requires a few weeks to a few months. The infection passes from rodent to rodent, the natural host.

Cutaneous leishmaniasis of the Old World resulting from infection with *Leishmania tropica* occurs in three major forms: (1) moist (or rural), caused by *L. tropica major;* a self-limited lesion begins at the site of the bite, develops rapidly, and subsides within about six months; the infection is largely transmitted out of doors; (2) dry (or urban) caused by *L. tropica minor;* the lesion grows slowly and may ulcerate, but heals with little scarring in about a year; (3) lupoid (or relapsing); the lesion enlarges slowly over a period of years, healing at the center while advancing at the periphery; this type occurs in the Middle East. The sore eventually heals, leaving a characteristic scar.

Leishmania mexicana occurs in some countries of Central and South America, principally Mexico and Guyana and, to a lesser extent, the Amazon region. It causea a cutaneous lesion and does not spread to mucous areas. Ear lesions are known as **chiclero ulcers.** In experimental animals, the infection may lead to visceralization, especially to involvement of the spleen and the liver. The vector seems to be *Lutzomyia flaviscutellata.* Several species of forest rodents are found infected in nature.

Leishmania mexicana is actually a group of parasites, three of which are not readily distinguished from one another. They differ in reservoir hosts, sandfly vector, geographic distribution, and clinical features. They are *L. mexicana mexicana, L. m. amazonensis, L. m. pifanoi,* and *L. enriettii.* The last is distinguished from the others by its unusually large amastigotes. It is found in laboratory guinea pigs.

Leishmania braziliensis is present in most parts of the tropic and subtropic regions of the New World, ranging at least from Panama to Argentina. The pathology, epidemiology, reservoir hosts, and species of sandflies involved in transmission vary considerably from one place to another.

The most severe form of this infection called

espundia, is endemic in the jungles of Brazil, Boliva, Peru, and some other South American countries. The clinical form frequently involves the mucous membranes of mouth, nose, and pharynx and may result in complete destruction of these tissues and associated cartilage. The lesions are resistant to treatment. In certain arid areas of western Peru, however, a benign form called **uta** is present. In these areas, involvement of the respiratory membranes has not been observed, and the skin lesions usually heal spontaneously. Between espundia and uta occur all the transitional forms of cutaneous leishmaniasis in the New World.

Throughout areas where *Leishmania braziliensis* is endemic, several mammals, both domestic and wild, have been found naturally infected. The list includes dogs, horses, wild rodents, and certain forest carnivores. The importance of such animals as reservoir hosts in relation to the disease in man is still to be determined. Among the several species of phlebotomine sandflies that could be incriminated as vectors are *Lutzomyia flaviscutellata, L. intermedius,* and *L. trapidoi.* Other species have been found naturally infected with promastigote flagellates, but further investigations are necessary in order to determine their relation, if any, to leishmaniasis.

Leishmania aethiopica is a cutaneous leishmanian parasite found in the highlands of Ethiopia and on Mount Elgon in Kenya. Infection is accompanied with serious diffuse lesions. Adults become immune and many children may be infected. Chronic blindness may occur. The parasite is transmitted by the sandfly *Phlebotomas longipes* and *P. pedifer.* Reservoir hosts are colonial rock hyraxes. Previous infections with the same species usually result in immunity that extends to other strains of *Leishmania*—cross immunity.

Leishmania major produces cutaneous leishmaniasis in man. It occurs in the semi-arid steppes of the U.S.S.R. where the gerbil (*Rhombomys opimus*), a colonial rodent, is a reservoir host. The disease is characterized by chronic lesions which usually heal and leave an immune host. Sometimes, however, they never heal completely and give rise to the disease **leishmaniasis recidivans.** The lesion is composed of a heavy lymphocyte infiltrate, giant cells, and rare epitheliod and histiocytic cells.

Leishmania donovani causes visceral leishmaniasis, a killing disease. Infection occurs chiefly in the spleen and liver, secondarily in bone marrow, intestinal villi and other areas. The disease is called **kala-azar** and **dum dum fever.** It occurs in east India, Assam, the Mediterranean area, southern Russia, north Africa, central Asia, northeast Brazil, Colombia, Argentina, Paraguay, El Salvador, Guatemala, and Mexico. It is a rural disease with reservoirs of infection in dogs, foxes, rodents, and other mammals. The strain in children is sometimes called *L. infantum.* Promastigotes may be found in flies (e.g., *Phlebotomus argentipes, P. orientalis*) three days after feeding. They occupy the anterior portion of the gut by the fourth or fifth day. On the seventh to ninth day after the fly has fed a second time, promastigotes are in the proboscis and become infective. The fly lives 14 to 15 days.

The first response to a bite is a small, non-ulcerated skin sore. Headache and weakness often accompany infection. Increase of blood volume, an increase of IgG, and a decrease of IgM may also occur. Infection of reticuloendothelial cells throughout the body is the basic factor. The disease is characterized by, "a lengthy incubation period, an insidious onset, and a chronic course attended by irregular fever, increasing enlargement of the spleen and of the liver, leucopenia, anaemia and progressive wasting. The mortality is high; death occurs in untreated cases in 2 months to 2 years." The most obvious physical signs are progressive enlargement of the spleen and, to a lesser extent, of the liver.

This disease occurs as three types or varieties, separated geographically. The first two types, Indian and Sudanese, are found mainly in human adults, rodents, and possibly lizards, but not in dogs. The third type, Mediterranean, is much more widespread, occurring in southern Europe, middle Asia, China, and Central and South America. In this type, dogs are important hosts, and jackals and other mammals have been found to be infected. Transmission of the disease may occur from vertebrate to vertebrate by direct contact, thus omitting the vector.

About two years after the acute stage of kala-azar occurs in the viscera, post-kala-azar leishmaniasis may appear. Manifestations range from depigmented areas of the skin to pronounced nodular lesions (Fig. 3–9).

Dogs also show pathogenic effects of infection. Loss of hair is a characteristic manifestation of severe visceral leishmaniasis in these an-imals. Ulcers on the lips and eyelids, weakness, and enlargement of the liver are common symptoms. Positive diagnosis relies on discovery of the parasite in infected tissues and in peripheral blood. Stray dogs should be eliminated, and insecticides should be used to control the sand-flies.

Visceral leishmaniasis may also appear in the new world. In Brazil the parasite is called *L. chagasi.* The vector is *Lutzomyia longipalpus.*

Probably many other species of *Leishmania* occur in animals but have not yet been reported. Those that have been described include *L. caninum* in dogs, *L. chamaelonis* in lizards, and *L. denticis* in the silverfish, *Dentex argyrozona.*

Endotrypanum is a digenetic genus of especial interest because species of this group are the only known kinetoplastids that live in red blood cells. Like *Leishmania* they are transmitted by sandflies, genus *Lutzomyia.* In the insect they occupy mainly the malpighian tubules. The mammal hosts are sloths in Central and South America. Details of the life cycle in these animals is unknown. The two known species are *Endotrypanum monterogei* and *E. schaudinni.*

For the biology of leishmania see Molyneux and Ashford;[60] also Chang, et al.;[10] for vector biology see Ward;[87] for ecology and control see Lainsons;[41] for clinical manifestation, diagnosis and treatment see Marsden and Jones;[55] for a bibliography see Heynaman, et al.;[27] for an analysis of isoenzyme data see Le Blancq, et al.,[42] 1986; for epidemiologic analysis see Wirth, et al.,[91] 1986; for vector information see Mutinga,[64] 1986; for the taxonomic status see Añez, et al.,[2] 1985; for direct non-insect vector transmission see Nuwayri-Salti, N. and Khansa, H.F.,[66] 1985; for a general reference see Chang and Bray,[11] 1985.

PHYSIOLOGY OF TRYPANOSOMATIDS

Dr. Seymour Hutner (in a personal communication) made some comments on biochemical parasitism that are worth repeating here:

"Perhaps accelerated by parasitic adaptation, trypanosomatid species have lost several biosynthetic systems, as evidenced by dependence on an impressive array of exogenous metabolites, supplied by the host in the case of uncomplicated parasitism, as in blood-dwelling stages, or, in the case of gut stages, by some combination of contributions from the accompanying gut microflora plus secretions of the host. Thus a trypanosome in the gut of a tsetse fly benefits directly from the metabolites rendered avail-

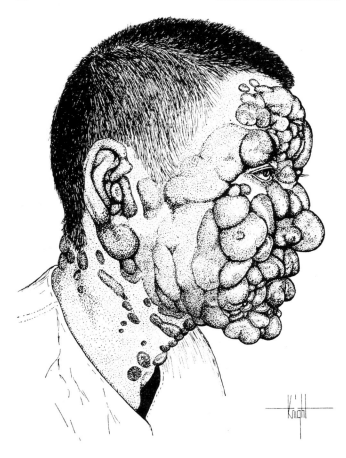

Fig. 3–9. A severe dermal, postvisceral case of kala-azar infection caused by *Leishmania donovani*. (Drawn from a photograph. Courtesy of Dr. Robert Kuntz.)

able by the fly's digestion of a blood meal plus contributions from the accompanying microflora of bacteria and fungi. All parasitic phagotrophs are more or less highly developed biochemical parasites, self-invited to the cellular smorgasbord provided by the unicellular prey or by metozoan host's cells. Where examined, trypanosomatids show ability to ingest particles. Therefore such a phagotroph feeding on a phagotroph may be regarded as biochemical hyperparasitism. For trypanosomatids it is as yet impossible to determine how many biosynthetic cripplings were due to parasitism, since the nutritional requirements of none of the free-living members of the Kinetoplastida have been worked out. This absence of biochemical scene-setting for the evolutionary drama of parasitism besets much of prasitology. Thus the nutritional requirements of no free-living platyhelminth are known aside from a little information obtained by radioactive tracer techniques. One of the great attractions of the Trypanosomatidae to biochemically minded parasitologists is the wealth of material pertinent to the biochemical adaptation required in passing between invertebrate vector and vertebrate host, and between cold-blooded and warm-blooded vertebrates.''

Carbohydrates are not stored in significant quantities within trypanosomatids. Therefore, most blood forms are dependent on exogenous glucose for their main energy source. Blood forms of the vivax, congolense, and brucei groups, as well as intracellular *Leishmania*, do not possess a functional tricarboxylic acid (TCA) cycle or the cytochrome type of terminal respiratory pathway. Energy (ATP) is produced when glucose is metabolized to pyruvate through glycolysis. The cytochrome system in blood trypanosomes has apparently been replaced by a particulate terminal L-α-glycerophosphate oxidase system in which NADH, reduced during glycolysis, is reoxidized to NAD+ in the presence of dihydroxyacetone phosphate. Although O_2 is used as a final electron acceptor, this system does not appear to be sensitive to cyanide. The absence of a functional TCA cycle and cytochrome system in blood trypanosomes is correlated with the presence of a mitochondrion that is thin and tubular with few cristae. In contrast, one-host trypanosomes, blood forms of the lewisi group (e.g., *T. cruzi*), the insect stages of all groups, and culture forms possess a functional TCA cycle and a terminal electron transport chain involving cytochromes. In culture forms, and presumably in insect stages as well, glucose is metabolized through both anaerobic and aerobic pathways to produce CO_2, succinate, acetate, and pyruvate as end products. Glucose is preferred as a respiratory substrate, but *T. brucei* also uses the amino acid proline. Mitochondria in these trypanosomes are considerably larger than those of the blood stages and possess numerous plate-like cristae.

Protein, nucleic acids, and lipids are obviously synthesized by trypanosomes, as indicated by their enormous reproductive capacities. The presence of transfer and ribosomal RNA systems, as well as the necessary cellular organelles, indicates that protein synthesis in the trypanosomes is typical of most eukaryotic cells.

Nucleic acid synthesis varies in different species. In general, most of the trypanosomatids are incapable of *de novo* purine synthesis and must rely on the host for these preformed nitrogenous bases. Presumably, host purines are made available for parasite incorporation as nucleosides or free bases. Purine salvage pathways may vary among parasite species. Pyrimidine synthesis has been described in some trypanosome species. *Crithidia fasciculata*, for example, uses carbonyl phosphate and aspartic acid for this purpose. Other trypanosomatids (e.g., *T. cruzi*) possess only a limited ability to synthesize pyrimidines. Pyrimidine requirements for these parasites most likely are met through salvage mechanisms.

Studies of nucleic acids and specific proteins may shed light on areas other than cell metabolism. For example, DNA analysis and isoenzyme electrophoretic techniques have been helpful in determining taxonomic and evolutionary relationships of various species and subspecies of *Leishmania*.

Lipid metabolistic pathways of trypanosomatids have received little study. The types of lipids they possess are, in general, similar to those in mammalian cells. However, considerable variation in lipid content has been noted among different trypanosome species or stages. Phosphatidylcholine predominates in most trypanosomes. Free fatty acid and acyl glycerol content is usually low, but phosphoglyceride and sphingolipid content is high, indicating that storage lipids predominate over structural lipids in most trypanosomes. Parasitic protozoa are incapable of free fatty acid synthesis and must rely on host reserves to satisfy this requirement. Fatty acids and other preformed lipids may serve as important sources of energy for blood *T. cruzi*. Knowledge of lipid metabolism can have some practical value. For example, chemotherapeutic activity of some arsenicals, against *T. brucei* and *T. congolense* may be related to interference with lipid and sterol synthesis.

Kinetoplast DNA (kDNA) has attracted considerable research interest. Although much is known about the molecular structure of the DNA itself, the actual role of this material in trypanosome genetics remains unclear. The mitochondria of dyskinetoplastic trypanosomes (those lacking kDNA) are apparently nonfunctional, and therefore these forms are unable to grow in culture or within the insect host. Blood stages, on the other hand, are able to function without kDNA because they rely primarily on extramitochondrial glycolysis to obtain their energy. The occurrence of dyskinetoplastic forms in nature, however, is rare.

Much research needs to be done. The statement of Hunter et al.[34] is of interest here:

"We think that no line of research on unicellular eukaryotes will surpass Trypanosomatidae for gaining insight into man's nature—even for comprehending the material basis of thought and emotion; here biopterin metabolism is especially pertinent. . . . Only malaria outdoes the Trypanosomatidae as causes of human suffering, death, and social disintegration. It is therefore prudent to keep updating the inventory of knowledge of these adversaries."

For detailed information on physiology and biochemistry of trypanosomes and other parasites, see Hill and Anderson,[28] von Brand,[6] and van den Bossche.[83]

IMMUNITY AND IMMUNOPATHOLOGY IN TRYPANOSOMATIDAE

Events of sleeping sickness begin with subcutaneous multiplication of salivarian trypanosomes inoculated by *Glossina*. The parasites then invade the circulatory system and tissues and

"increase logarithmically in the blood for one to three days after they are first detected in the bloodstream. They then seem to disappear from the circulation altogether, but their remarkable properties of recovery in mammalian hosts allow the next generation to develop in the blood; this population bypasses the earlier specific immune defense built up by the host by having antigens of a different constitution from those of the previous population. The interval between each parasitaemia wave in man may vary from one to eight days. This recurrent parasitaemia means that the host is exposed to a continuous sequence of infection with practically the same microorganisms and to different but closely related antigen-antibody reactions each time."[14a]

Vickerman[85] has provided evidence that the variable antigens for salivarian trypanosomes are located in the surface coat that is absent from the development cycle in the vector, but appears in the metacyclic forms that are introduced by the fly into the mammal.

Glycoproteins, which represent a major component of the trypanosome surface coat, are highly immunogenic. Specific antibody binding to surface glycoproteins induces a movement and concentration of the antigen to restricted areas of the parasite's surface (capping) and is followed by removal of these antigens from the surface membrane either by shedding or internalization.[7,78] Presumably, a new variant antigen appears in the surface membrane, thus allowing the parasite to avoid antibody destruction and to give rise to the relapse population. Whether specific antibody is functioning as an active inducer of antigenic change or merely acting as a selector of antigenic variants naturally occurring in a trypanosome population is at present unclear.[14] Immunoprophylactic control of human trypanosomiasis does not yet appear feasible. Numerous attempts at stimulating protective immunity against trypanosomes have been made. No effective vaccine exists for controlling African trypanosomiasis. Among domestic animals, cattle appear to develop some immunity against trypanosomes, but to date no technique has been developed to produce a clear, dependable, and effective artificial immunity to these parasites.

In addition to surface components of parasite origin, which constitute the variable antigens, host plasma proteins (nonspecific IgG, albumin, complement) also have been associated with trypanosome membrane coats.[17] In such cases, host proteins could be providing additional protection against trypanocidal antibody by masking parasite antigens that are immunoreactive or potentially immunogenic.

Trypanosoma lewisi infections in rats and *T. musculi* infections in mice are unique in that infected animals produce not only a trypanocidal antibody against rapidly dividing forms, but also a reproduction-inhibiting IgG called **ablastin** against "adult" parasite forms. Infections are self-limiting in rats, even though demonstrations have shown that blood forms of *T. lewisi* are capable of modifying surface antigen.[12]

Immunopathologic disorders in humans and in animals resulting from chronic African trypanosomiasis are commonly characterized by severe anemia, a general suppression of immune responsiveness, and histologic changes in lymphoid tissues. Enlargement of the spleen and the lymph nodes is the result of increased accumulation of macrophages and plasma cells in these organs as well as a disproportionate increase in B-lymphocyte-dependent areas. A concomitant depletion of splenic and nodal T lymphocytes also may be evident.[53] "Macrophages from the mononuclear phagocytic system (MPS) interact with *T. cruzi* either as host cells, in which the parasites readily multiply and differentiate, or as effector cells of the immune response which participate in the control of infection."[7]

Glycoproteins shed from surface membranes of blood trypanosomes (called **exoantigens**) ultimately are responsible for eliciting the production of antitrypanosome antibodies that, in some cases, can provide protective host immunity. Deposition of circulating exoantigen-antibody complexes on erythrocyte membranes, however, also may be responsible for blood cell destruction through autoimmune removal from the vascular system. Even in the absence of specific trypanosome antibody, exoantigen alone, binding to erythrocyte surfaces, is able to activate and to bind complement; this process results in hemolysis or removal of host cells.[63] These, and perhaps other immune mechanisms, are thought to be major contributors to host anemia associated with trypanosomiasis. Renal complications in the form of glomerulonephritis (inflammation of the kidney tubules) may also be the result of immune-complex deposition on tubule basement membranes. In the later stages of sleeping sickness, parasite invasion of the heart and brain choroid plexus may be associated with inflammatory tissue damage.

One of the most important immunopathologic disorders in trypanosomiasis is a suppression of immune reactivity, not only to the invading trypanosomes, but also to other disease-producing organisms. The question of how the parasite accomplishes this suppression is still unanswered. Mansfield summarized the current status of our understanding of trypanosome-induced immunosuppression by stating that obviously "none of the present theories concerning immunosuppression completely fit the established facts concerning immunologic function and pathology associated with the disease. A complete understanding of immunosuppression will come only after elucidation of the identity of affected target cells, the nature of suppression, and the molecular mechanism(s) by which the parasites induce suppression in vivo and in vitro.

The direct role of antibody in providing protective immunity against intracellular trypanosomes, *Trypanosoma cruzi* or *Leishmania* spp., is still poorly understood. Although trypanocidal antibodies are produced against the extracellular trypomastigotes of *T. cruzi*, cell-mediated immunity seems to be of primary importance in resisting the intracellular amastigote stages.[26,35] Experimental infections of *T. cruzi* in rodents that underwent thymectomy neonatally or that were serum-treated with antithymocyte antibody indicate that T lymphocytes are probably involved in mediating immune protection. The exact role of the T cell as a helper or effector or both has not yet been clarified. Granulocytes, for example, eosinophils and neutrophils, also may serve as accessory effector cells against amastigotes.[75] Milk transmission of *T. cruzi* antibodies from infected mice to their adopted offspring and human prenatal transfer of maternal antibody have been reported.[40]

Attempts to immunize experimental animals against acute infections of *T. cruzi* have been only partially successful. Mice vaccinated against Chagas' disease with parasite culture forms, hyperimmune serum, or killed antigens can induce a strong immune resistance, although in some studies, disappointingly low levels of immunity have been achieved using killed antigens or parasite extracts. The use of attenuated live vaccines, for example, irradiated trypomastigotes or amastigotes, has shown promise as an effective means of inducing acquired resistance to *T. cruzi* under controlled conditions.

Immunity to leishmanial infections appears to be of the cell-mediated type, involving sensitized cytotoxic lymphocytes and activated macrophages as primary effector cells.[72] A cooperative role for trypanolytic or opsonizing antibodies also has been suggested in in vitro experiments. Previous exposure to certain species or strains of *Leishmania* can provide immunity to reinfection. In humans, for example, immunity to cutaneous leishmaniasis is conferred upon the host following complete recovery from a primary infection with *L. tropica*. Artificial immunization against this disease, using living organisms, was practiced long before the etiologic agent was known. Natives of the Near East would inoculate material from an open skin lesion into an unexposed part of the body, thus preventing the future occurrence of a potentially disfiguring sore on the face. When dead and washed promastigotes of any species of *Leishmania* are inoculated into the skin of a person with Oriental sore, a delayed hypersensitivity reaction occurs. The reaction itself is no indication of immunity, but it is a mechanism for mobilizing immunologically competent cells to a site containing *Leishmania* antigens where the cells are effective when immunity is established. Protective immunity develops only after the complete elimination of the primary infection.

Primary lesions caused by cutaneous forms of leishmaniasis are usually indolent, with continuously extending boundaries. They generally heal with scarring. A second type of lesion is diffuse and represents a failure of cell-mediated immune (CMI) mechanisms. This failure of CMI function is directly tied to the failure of macrophages to kill internalized amastigotes, which, in turn, represents a key element in the overall immunopathology of disseminated visceral leishmaniasis. In kala-azar, "the unchecked proliferation of macrophages and their parasites leads to splenomegaly and hepatomegaly, and to extensive invasion of the bone marrow. Unless cell mediated immunity is permitted to develop, as with the aid of chemotherapy, the disease progresses to death."[93] The overcommitment of hemopoietic areas to macrophage proliferation also can cause an eventual reduction in red cell production and an anemic host condition.

For further details on trypanosomes and trypanosomiasis, see Taylor and Muller.[81]

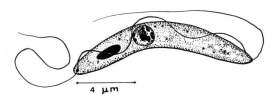

4 μm

Fig. 3–10. *Cryptobia stilbia* from the stomach of a deep-sea fish. The two kinetosomes probably indicate the start of mitosis. (From Noble. Courtesy of the Journal of Parasitology.)

FAMILY CRYPTOBIIDAE

Many members of this family of flagellates appear superfically to be like trypanosomes, but they all possess two anterior flagella instead of one. One of these flagella is usually trailing, while the other extends anteriorly. The trailing one is the outer margin of an undulating membrane in most species. Both flagella arise from a broad flagellar pocket. A cyostome-cytopharynx is present. A large and usually elongate kinetoplast is characteristic of the family. The genus *Cryptobia-Trypanoplasma* lives in the stomaches of marine fish and in the gonads of pulmonate snails, chaetognaths, siphonophores, and a few other animals. *Cryptobia borreli* in the blood of fish and *C. helicis* in the seminal vesicles of the snail, *Helix* are typical examples. The length of the body varies in different species from about 20 to 120 μm.

The kinetosome and kinetoplast are located anterior to the nucleus (Fig. 3–10). Transmissions is probably accomplished by ingestion of the flagellates, or, in snails, by way of spermatophores and fertilized eggs. Bower and Margolis[5] have shown that these haemoflagellates can escape from the coelom to the body surface of fish via ruptured areas. Contact with another fish provided the method of direct transmission of the parasite to a new vertebrate host. As the authors point out, the only criterion for distinguishing the genus *Cryptobia* from *Trypanoplasma* has been the direct versus indirect life history. Clear evidence for direct transmission among *Cryptobia* thus warrants placing *Trypanoplasma* as a junior synomym of the genus *Cryptobia*. Different varieties of fishes vary in susceptibility to the parasites. This variation is probably due to genetically transmitted resistance.

FAMILY BOBODIDAE

Ichthyobodo. Members of this genus have two short anterior flagella and two trailing flagella (all arising within a spiral longitudinal groove), and a cytostome that attaches to the skin of fish. *I. nector* causes an important disease of freshwater fish called "costiasis" because the parasite was formerly known as *"Costia necatrix"*—a now invalid name. The disease often causes a high rate of mortality in salmon, trout, and pond fish.

Order Retortamonadida

FAMILY RETORTAMONADIDAE

Chilomastix mesnili is the best-known member of the family. It is probably harmless and lives in the cecum and colon of man and other primates and the pig. The trophozoite, or motile form, is usually pear-shaped and averages about 12 μm in length. The spiral groove in the body is frequently not seen in the usual hematoxylin-stained preparations, and the three anterior flagella are also difficult to see. Probably, the most diagnostic feature is the location of the large nucleus near the anterior end of the body. The cyst is about 9 μm long, oval in shape, with a broad extension of the cyst wall at one end forming a spade-like or cone-shaped projection. The large nucleus, large cytostome, and various filaments can usually be seen in stained cysts. Other species of *Chilomastix* occur in the rectum of amphibians, fish, and many mammals (Fig. 3–11).

Retortamonas intestinalis is another harmless small flagellate of man (Fig. 3–12). It measures about 6 × 4 μm. Like *Chilomastix*, it possesses a cytostomal groove, but has only two anterior flagella, one extending anteriorly and the other passing backward through the cytosome and then trailing from the body. Other members of the genus have been found in insects (e.g., *Embadomonas* (= *Retortamonas*) *gryllotalpae* in the mole cricket).

Order Diplomonadida

FAMILY HEXAMITIDAE

At least 40 species names have been proposed for members of genus *Giardia*. *G. lamblia* has long been the preferred name for the species in man but now *intestinalis* or *duodenalis* is usually recommended. *G. cati* in cats and *G. canis* in dogs are common designations but *G. duodenalis*

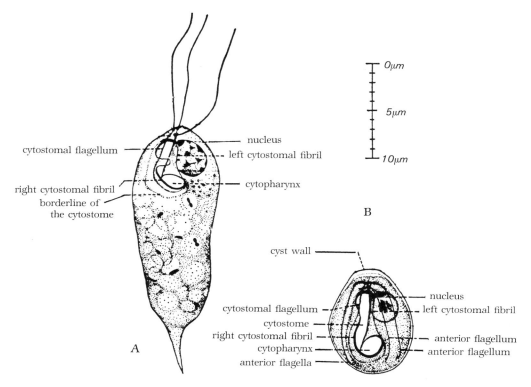

Fig. 3–11. *Chilomastix intestinalis* from the guinea pig. *A*, Trophic flagellate, ventral side, showing body structures. *B*, Cyst, ventral side. (From Nie. Courtesy of the Journal of Morphology.)

is also used for these and many other mammals. Some authorities believe that only two species infect mammals, the categories depending on whether the **median bodies** are smooth or rounded or shaped like the claw of a claw hammer. *G. muris* in mice, rats and hamsters belong to the first group, whereas all other mammals habor *G. intestinalis* with the claw-shaped median bodies. Confusion arises partly because these parasites are not host specific. They may be found in cattle, monkeys, sheep, squirrels,

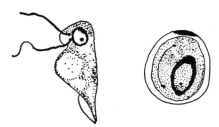

Fig. 3–12. *Retortamonas intestinalis* from man, showing a motile form and a variation of the cyst. (From Hogue. Courtesy of the Journal of Hygiene.)

bears, beavers, lizards, birds, and other animals.

Giardia intestinalis (Figs. 3–13 and 3–14) is a universal and common inhabitant of the duodenum, jejunum, and upper ileum. The parasites usually become attached to the surface epithelium. The percentage of infection varies in different countries and in different areas. In the United States the range is 1 to 20% but in most areas it is 4 to 10%. In some areas of the world almost all of the children are infected. Most infected people do not suffer from the disease.

Motile *Giardia intestinalis* range from 9 to 21 μm long, 5 to 15 μm wide and 2 to 4 μm thick, an average of about 7 × 14 μm. The pear-shaped body is unusual for protozoa because on its flat ventral side occurs an adhesive disc. The surface of this disc contains microtubules and ribbons which give it support and provide means for movement which occurs by grasping, contraction, or both. There are four pairs of flagella: anterior, posterior, ventral, and caudal. Each arises from a kinetosome and continues through the cytoplasm some distance before

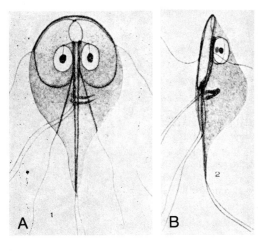

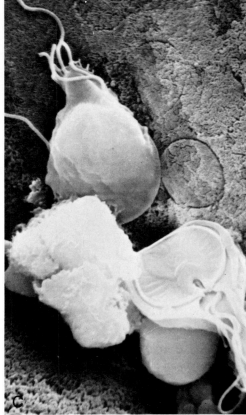

Fig. 3–13. *Giardia intestinalis,* motile form. (From Kofoid and Swezy. Courtesy of the University of California Publications in Zoology.)

emerging from the cell. The ventral pair lies in a ventral groove and assists the adhesive disc. The deeply staining fibers in the groove are known as axonemes or axostyles. There are two nuclei, each with an eccentric endosome. The median bodies lie behind the central disc. Their function is obscure but they may be associated with energy metabolism or simply may support the anterior portion of the parasite.

Cysts, usually 8 to 11 by 7 to 19 μm, contain two or four nuclei (Fig. 3–15). They lack the adhesive disc and exterior flagella. They also lack mitochondria, endoplasmic reticulum, Golgi apparatus, and lysosomes. The median bodies and intracytoplasmic portions of flagella stain deeply and make *Giardia* easily recognized. Cysts can survive weeks or months under moist conditions and can generally stand chlorinated water but are killed instantly in boiling water. Excystation occurs in the duodenum. Two trophozoites emerge from each ruptured cyst and can move about freely from place to

place. It is the cyst stage which is infective to another individual. Over 300 million cysts may be present in 1 mg of feces and an infection of only 10 cysts may cause the disease.

Transmission is by the anal-oral route, usually by drinking contaminated water. Contamination may be by animal or human feces carrying giardia cysts. Mountain streams or ponds thus become unfit for drinking because of defecation by careless hikers or by beavers, muskrats, or other animals. Municipal water supplies may be contaminated by broken sewer lines, inadequate chlorination or flooding. Dogs, cats and other pets are possible sources of infection. A rare method of transmission is through sexual contact. The disease in man is clearly zoonotic with many reservoirs of infection due to the fact that the parasite is not host specific.

Giardiasis is one of the top ten important human parasitic diseases. Two hundred million infections occur in the world each year, but there have been no deaths ascribed to this par-

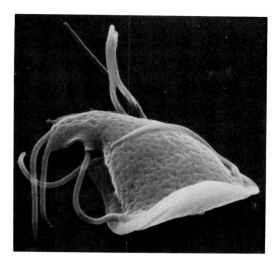

Fig. 3–14. SEM of *G. muris* trophozoite partially attached to the substratum. The anterior margin of the ventral disc was fixed in a state of dorsal flexion and the tail in a position of lateral flexion. Arrowheads, ventrolateral flange. Bar equals 1 μm. (From Erlandsen, S.L., and Feeley, D.E.: Trophozoite motility and mechanism of attachment. *In* Giardia and Giardiasis. Edited by S.L. Erlandsen and E.A. Meyer. New York, Plenum Press, 1984.)

asite. The disease shows great variability in pathologic conditions and clinical changes but is characterized by gastrointestinal disorders. Symptoms may be absent, minor or severe. Abdominal pains, nausea, diarrhea, cramps, vomiting, fever, bloating, chronic enteritis, headache, or malaise may occur. The most serious pathologic condition is malabsorption, due to toxic effects of the parasites attached to intestinal cells and to mechanical blockage caused by the enormous numbers of parasites (Fig. 3–16). In heavy natural infections of rats, mice and hamsters the parasites penetrate intestinal epithelia and submucosa, proceed through the lymphatic system and blood vessels to the lungs and other organs. If they get into the gallbladder, blockage of bile ducts may occur. Penetration in man, however, is rare and does not seem to be a factor in pathogenesis. Immunogenic activity is not pronounced but both humoral and cellular immune responses occur. Diagnosis is confirmed by finding cysts or motile forms in feces.

Since the infection often originates in the streams and ponds, it is sometimes called the hiker's or picnicker's disease. It is prevalent in crowded institutions such as some mental hos-

pitals and in areas of poor sanitation. The infection may persist for years and may be fatal. When people are hiking or fishing, their drinking water should be boiled. Treatment of water with various commercial purification compounds may be effective. Fortunately, giardiasis is easily and effectively treated with metronidazole or quinacrine hydrochloride. For a review of *Giardia* and giardiasis see Erlandsen and Feeley,[21] 1984; Kirkpatrick and Farrell,[38] 1982; Kirkpatrick in Green,[25] 1984.

Hexamita meleagridis causes infectious catarrhal enteritis in turkeys and other galliform birds. The disease is also known as **hexamitiasis** or hemaitosis and is characterized by diarrhea, nervousness, loss of weight, and death, especially in younger birds. The infectious organism, *H. meleagridis*, averages about 9 × 4 μm, is pear-shaped, and possesses eight flagella, two of which are attached to the posterior end of the animal, whereas six arise from the anterior end. Other members of the genus infect other birds. *Hexamita salmonis* (Fig. 3–17) is a species that develops in epithelial cells of the cecum of trout. *Hexamita intestinalis* infects frog intestines and blood.

Enteromonas hominis. About 4 × 7 μm, it has been reported from the intestine of man in warm countries of the world, but is not pathogenic. There are three anterior flagella and one that arises anteriorily, but adheres to the body as it passes backward, then emerges as a free flagellum posteriorly. A cytostome is said to be present. The cyst possesses four nuclei, two at each pole. This flagellate is possibly an ancestral form to the other diplomonads, e.g., *Hemita* and *Giardia*.

Order Trichomonadida

A diagnostic feature of this order is the parabasal body that is composed of a granule or one or more elongated bodies, often attached by fibrils to the kinetosomes of flagella. The parabasal body is probably homologous with the Golgi apparatus of metazoan cells. The **axostyle** (Fig. 3–18) is a filament or hyaline rod that passes length-wise through the body, often projecting from the posterior end. There may be a few or many flagella (typically four to six). A trailing flagellum is characteristic, and there is often an undulating membrane. Sexual stages are unknown. In addition to the two families described in the following section, there are the

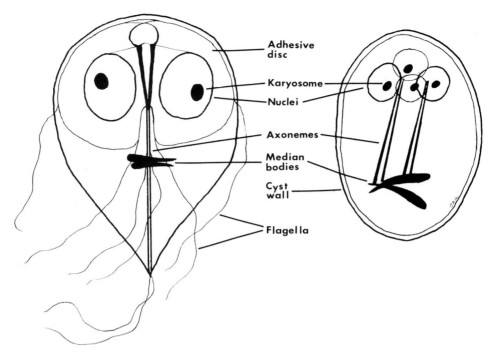

Fig. 3–15. Schematic diagram of a *Giardia* trophozoite (left) and a cyst showing morphologic features visible by light microscopy. (From Kirkpatrick and Farrell, courtesy Compend. Contin. Ed. Pract. Vet.[38])

Devescovinidae and Calonymphidae, both restricted to termites.

FAMILY MONOCERCOMONADIDAE

In this family are placed many flagellates that live in the digestive tracts of insects and vertebrates. There may be several flagella, but usually no undulating membranes.

Histomonas meleagridis (Fig. 3–19) ranges in diameter from 4 to 30 μm and it may possess up to four flagella or none. It usually lives in the liver of poultry and causes infectious enterohepatitis, but it may also infect the ceca, kidneys, or spleen. The bird's head characteristically turns a dark, almost black color—hence the common name of the disease, "blackhead." However, darkening does not always occur, or it may result from other diseases. No protective cyst is formed, so when the parasite is eliminated from the host's body with feces, it soon dies. The protozoa, however, may become enclosed within the eggshell of the poultry nematode cecal worm, *Heterakis gallinarum*, while still in the host intestine. The worm becomes an intermediate host of the protozoan. When the embryonated eggs of the worm are eaten

by poultry, the enclosed *Histomonas* are liberated. Earthworms may eat *Heterakis* eggs containing *Histomonas;* thus, a chicken can acquire two kinds of parasites by eating one earthworm.

In the invasive stage, *Histomonas meleagridis* is ameboid, and it feeds by phagocytosis. There follows a round, quiescent, vegetative stage, in which the parasite feeds by secreting proteolytic enzymes that diffuse from the parasite and digest host tissues. Digested products are taken into the protozoa by pinocytosis and probably also by diffusion. Neither of these two stages possesses mitochondria, and succinic dehydrogenase is absent. There is no cyst.

In the germinal zone of the nematode ovary, the parasite is extracellular. Here it feeds and multiplies. Moving down the ovary with oogonia, the histomonads penetrate developing oocytes in the growth zone of the ovary. The parasites feed and divide in the oocyte and in newly formed ova, where they appear to be similar to tissue-inhabiting stages, but are smaller. Mitochondria are not present. Apparently, the host is able to repair damage caused by the parasite to the reproductive system.

Blackhead is one of the most serious diseases

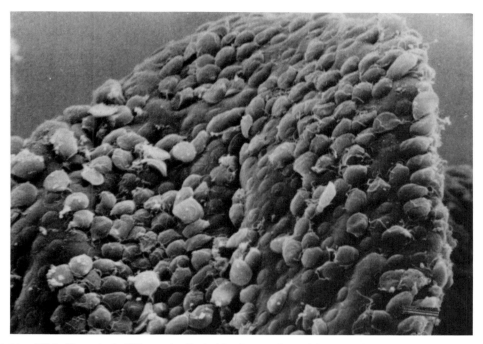

Fig. 3–16. SEM of intestinal villi in rat giardiasis. Massive numbers of *G. muris* trophozoites almost completely cover the surface of the villi. In areas free of attached trophozoites, circular dome-shaped lesions produced by adhesions of the ventral disc of *Giardia* trophozoites, can be seen in the MBV of the epithelial cells. The bar is equal to 10 µm. (From Erlandsen, S.L., and Feeley, D.E.: Trophozoite motility and mechanism of attachment. *In* Giardia and Giardiasis. Edited by S.L. Erlandsen and E.A. Meyer. New York, Plenum Press, 1984.)

of turkeys. It is less severe for chickens, which usually recover after a mild reaction.

Resistant chickens and turkeys may harbor both the protozoan, *Histomonas meleagridis,* and the worm, *Heterakis gallinarum,* at the same time. Such birds are reservoirs of infection for all new chicks. Chickens and turkeys obviously should not be raised together. Even raising young turkeys with older turkeys is apt to expose the young to serious protozoan and worm diseases.

Symptoms of histomoniais are similar to those of many other poultry diseases. The birds become listless, their wings droop, and their eyes are partly or completely closed much of the time. Young birds may die within two or three days, and older birds may last a week or two before death, or they may recover after a few days. Cleanliness and isolation are the keys to prevention. The nonpathogenic *Parahisto-monas wenrichi* is also transmitted by *Heterakis.*

Dientamoeba fragilis (Fig. 3–20). The histomonad affinities of *Dientamoeba* have long been recognized, but because of the presence of pseudopods and the absence of flagella, kinetosomes, and a pelta-axostyle complex, the genus was previously assigned to the order Amoebida. In 1974, Camp[8] placed it among the trichomonads in a new subfamily **Dientamoebinae,** after a careful comparison of its fine structure with those of trichomonads and of *Entamoeba* spp. An analysis[19] of antigenic relationships among *Trichomonas, Histomonas, Dientamoeba,* and *Entamoeba* lends strong support to this reassignment of affinities. *Dientamoeba fragilis* is characterized by two nuclei (arrested telophase). Only one-fifth to one-third of the population are mononucleate, and more than two nuclei are rare. Cysts have never been found. The parasite lives in the large intestine of man and is probably present in at least 20% of the population of most countries; an incidence of over 51%, however, may occur in local areas. It is found more frequently in younger than in older people, and more often in females than in males. The flagellate has been found in monkeys and was once reported from sheep. Examination techniques that are suited only for

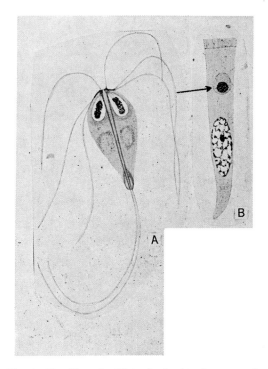

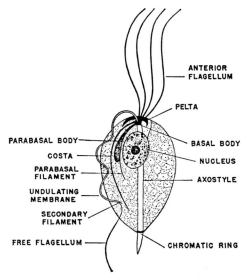

Fig. 3–18. Structures of *Trichomonas*. (From Levine. Courtesy of Burgess Publishing.)

Fig. 3–17. *Hexamita (Octomitus) salmonis,* a parasite flagellate of trout. *A,* Typical flagellate ×2360. *B,* Epithelial cell from the cecum of trout containing an early stage of the parasite (arrow). ×1230. (From Davis. Bulletin of the Bureau of Fisheries.) Dorsal and ventral aspects of *Giardia muris* trophozoites, lying on mouse jejunal mucosa. The exposed ventral surface of the lower trophozoite reveals the adhesive disc surrounded by the ventrolateral flange, and the paired ventral flagella. The upper trophozoite is adherent to the microvillous surface. Prior attachment sites appear as circular indentations formed by adhesive disc edges ×6700. (Reprinted by permission of the publisher from Owen, R.L., Nemanic, P.C., and Stevens, D.P.: Ultrastructural observations on giardiasis in a murine model. I. Intestinal distribution, attachment, and relationship to the immune system of *Giardia muris.* Gastroenterology *76*:757–769, 1979. Copyright 1979 by the American Gastroenterological Association.)

the identification of cysts would, of course, fail to reveal *Dientamoeba fragilis.* Because the parasite is often difficult to see, most surveys report a low incidence of infection.

The size of the organisms, ranges from 3 to 22 μm, with an average of about 9 μm. The cytoplasm, usually pale in appearance even in stained specimens, frequently contains small food vacuoles. Each nucleus contains a cluster of closely grouped granules, frequently four,

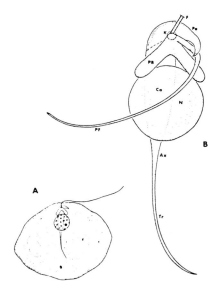

Fig. 3–19. *A, Histomonas meleagridis* (from Honigberg & Bennett, 1971, ×4270); *B, Histomonas meleagridis* (from Honigberg & Bennett, 1971). Diagram of mastigont system: f, flagellum; Pe, Pelta; PB, Parabasal body; K, Kinetosome; Ca, Capitulum; N, nucleus; PF, Parabasal Fibril; Ax, axostyle; Tr, Trunk of axostyle. (From Honigberg and Bennett. Courtesy of the J. Protozool.)

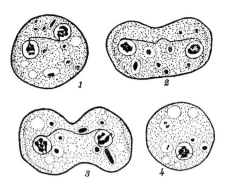

Fig. 3–20.   *Dientamoeba fragilis. 1,* Typical binucleate individual. *2,* and *3,* Stages in division. *4,* Uninucleate stage. (From Wenrich. Courtesy of the Journal of Morphology.)

which form the endosome. A fibril (*paradesmose*) may be seen connecting the two nuclei. There is practically no peripheral chromatin on the nuclear membrane.

The mode of transmission is uncertain, but it is probably carried in the eggs of an intestinal worm such as *Trichuris* or *Enterobius.* Usually there are few or no symptoms of infection with *Dientamoeba,* but the parasite may cause diarrhea, abdominal pains and pruritus, and loose stools.

FAMILY TRICHOMONADIDAE

This family is well known for its important parasites of man and of animals, both vertebrate and invertebrate. The flagellates are characterized by the presence of several anterior flagella, a **pelta** that lies at the anterior margin of the body, an undulating membrane, a deeply staining **costa** that extends along the base of the undulating membrane, and an axostyle. There is only one nucleus.

The name *Trichomonas* was first given to a genus that was believed to possess three anterior flagella. Unfortunately, confusion has arisen because species have been found with more than three flagella. The original group actually possesses four flagella, but the name *Trichomonas* remains. The following genera are now recognized: *Tritrichomonas,* with three anterior flagella (Fig. 3–21), *Trichomonas,* with four anterior flagella (see Fig. 3–18); and *Pentatrichomonas,* with five anterior flagella (Fig. 3–21).

Pentatrichomonas hominis inhabits the large intestine of humans, monkeys, dogs, cats, and rodents. Its size ranges from 6–14 × 4–6.5 μm.

There are usually five anterior flagella, one of which is directed posteriorly and is attached to the undulating membrane. At the anterior end there is a costa and a parabasal apparatus. True cysts have not been found but pseudocysts have been reported whose function is not clear. Transmission is by the fecal-oral route. The geographic distribution is worldwide, but the incidence of infection is heaviest in warm countries. Although it is frequently associated with pathogenic conditions, there is no evidence that the organisms are responsible for the symptoms.

Trichomonas tenax is smaller than *Pentatrichomonas hominis* and averages about 7 μm in length. It is similar to *P. hominis* in general appearance, but the undulating membrane extends about two-thirds the length of the body, and there is no trailing flagellum. The parabasal body is elongate and prominent. This flagellate lives in the human mouth and is a harmless commensal. One survey of 38 persons showed 15.7% infection. The parasites probably feed on bacteria and host food particles. Its habitat provides the basis for the name *T. buccalis,* by which it is sometimes called.

Trichomonas vaginalis (Fig. 3–22) is the largest of the three trichomonads of man. It ranges from 10 to 29 μm in length with an average of about 13 μm. The undulating membrane usually reaches to the middle of the body but it may be shorter. There is no trailing flagellum. The parabasal body with its parabasal filament is large. The axostyle characteristically curves slightly around the nucleus. Cysts are not produced.

Trichomoniasis is one of the most common sexually transmitted diseases. It is worldwide in distribution and affects an estimated 189 million people per year. *Trichomonas vaginalis* is found in 10 to 30% of women in the U.S.A. In tropical countries, healthy women of childbearing age show a rate of 15 to 40%. In men it varies from 4 to 15%. Its incidence in husbands or consorts of infected women, however, is probably 50%. Men, therefore, appear to be the cause of the persistence of trichomonal infection in women. Occasionally the flagellate is found in children, and there have been a few cases of infection from mother to baby during birth. The parasite may live in its host for years without manifesting itself. There is no tissue invasion. Clinical manifestations, however, are common. In women there is usually a pro-

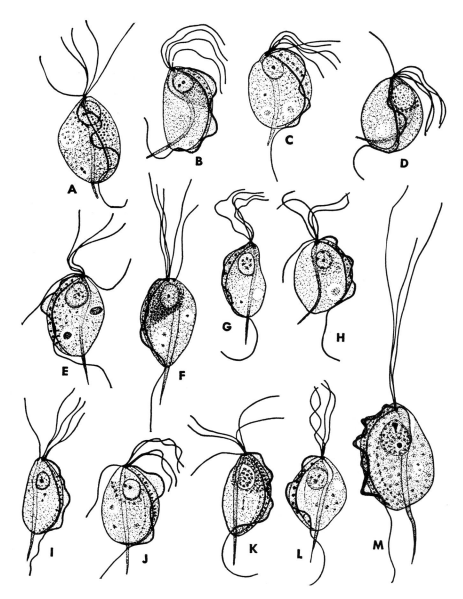

Fig. 3–21. *Pentatrichomonas hominis* from the intestine of man, *A* to *D,* compared with the same species from the monkey, *E* to *G. H* is from the cat, *I,* from the dog, *J* and *K* from the rat. *L,* is *Trichomonas gallinarum* from the pheasant. *M* is *Tritrichomonas fecalis.* (From Wenrich, courtesy of the J. Morphol.)

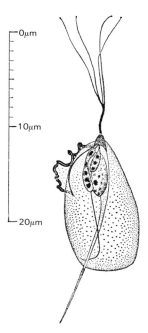

Fig. 3–22. *Trichomonas vaginalis.* The parabasal body is V-shaped, and the pelta appears as a dark-staining triangular membrane at the anterior end of the body. Note the double nature of the outer margin of the undulating membrane. (From Honigberg and King, courtesy of J. Parasitol.)

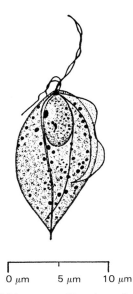

Fig. 3–23. *Trichomonas gallinae* from the intestine of a red-tailed hawk. (From Stabler, courtesy of J. Morphol.)

nounced vaginal discharge. Flagellates can be found in abundance in the leukorrheic fluid, which is sometimes purulent. Severe erosion of the vaginal mucosa may occur. The infection may be associated with itching, chafing, or a burning sensation. Acute trichinosis may accompany menstruation or pregnancy. In males the infliction may spread through the urethra to the seminal vesicles or prostate gland or other organs of the reproductive system. A thin discharge from the uretha is characteristic. Diagnosis is made by finding the flagellate in the discharge or secretions directly, or by finding them in cultures of the discharge.

Pathogenic strains of the parasite have been found within the cytoplasm of macrophages and epithelial cells. Damage may be due to direct contact with both the parasite and its toxic products. The flagellates were found to be 80 to 100% viable after long storage at −196°C.[32] For further information on *Trichomonas vaginalis,* see Su-Lin and Honigberg,[79] 1983 and Krieger,[39] et al., 1985.

Trichomonas gallinarum (see Fig. 3–21) causes avian trichomoniasis in chickens, tur-

keys, and other domestic birds. It is pear-shaped, like other members of the genus, and averages about 7×10 µm in size. The flagellate lives in the lower intestines of the birds and is the cause of diarrhea, loss of appetite and weight, ruffled feathers, and lesions in the intestinal wall. Cysts are not produced, and the method of transmission from bird to bird is unknown. *Trichomonas gallinae* (Fig. 3–23) is a similar parasite that infects the upper intestinal tract of poultry and other birds. In young pigeons, it is commonly fatal.

Tritrichomonas foetus (Fig. 3–24) reaches a length of 24 µm. It possesses a long undulating membrane, which is bordered on its outer margin by a flagellum that becomes free posteriorly. The axostyle is unusually wide, and it projects a short distance from the posterior end of the animal; the parabasal body is large. The parasites live in the preputial cavities of bulls and are thus readily transferred to the vaginas of cows during coitus. In the vagina, the trichomonads multiply for about three weeks. When the cow is in heat again, the parasites become reduced in numbers or disappear completely.

Fig. 3–24. *Tritrichomonas foetus,* a parasite that can cause serious disease in cattle × 1250. (From Wenrich and Emmerson, courtesy of J. Morphol.)

This cycle continues for 3 or 4 months, after which time the infection is usually lost. These flagellates may invade other reproductive organs and are apparently able to carry on their metabolic activites both aerobically (ineffectively) and anaerobically.

If the trichomonads enter into the uterus, as they easily do, they may cause temporary infertility or abortion. They may even infect the unborn young, appearing in the fetal membranes, the amniotic and allantoic fluids, and often in the stomach of the fetus. The parasites seem to have little effect on bulls. Some bulls apparently have a natural resistance to infection, and cows are able to develop some degree of immunity.

For trichomonads in general see Lee, J.J. 1985[43]: Order 5. Trichomonadida.

PHYSIOLOGY OF TRICHOMONADS

Trichomonads are generally anaerobic. They do, however, grow under low oxygen tensions. Mitochondria, cytochromes, and a functional TCA cycle are absent. Species of trichomonads vary in the number of carbohydrates they use, and different sugars exert different effects on parasite longevity and growth. Glucose, fructose, galactose, and maltose seem to be used effectively by all species, whereas pentoses and sugar alcohols do not support growth. Lactose, sucrose, trehalose, glycogen, and starch support growth poorly.

In the cytoplasm, metabolic energy is derived from the breakdown of glucose (commonly from endogenous glycogen stores) to pyruvate or succinate through the glycolytic pathway. Additional energy may be generated when pyruvate is converted to acetate by means of a coenzyme-A-linked reaction. This later conversion takes place within morphologically distinct organelles called hydrogenosomes, which contain the enzymes responsible for pyruvate metabolism. These organelles, because they are responsible for H_2 production in trichomonads, are distinct from mitochondria and peroxisomes.[48] They do not contain DNA. Flagyl, the commonly used and effective treatment for *T. vaginalis*, blocks H_2 production.

Aerobic respiration has been studied by several investigators. As usual, results have varied. For example, oxygen uptake by *Tritrichomonas foetus* was stimulated within 60 minutes by pure oxygen in the absence of glucose. In the presence of this sugar, the uptake was inhibited after 120 minutes. In contrast, the high oxygen tension was toxic to *Trichomonas vaginalis*. It increased endogenous respiration, but almost completely inhibited oxygen uptake with glucose. Energy metabolism seems to be centered around the intracellular glycogen reserves supplemented by any extracellular substrates. In general, carbohydrate use for respiration is similar to that for growth and anaerobic metabolism. Oxygen uptake and anaerobic fermentation in most trichomonads are inhibited by iodoacetate, arsenite, fluoride, and possibly azide. The presence of catalase is suggested by the lack of inhibitory action of arsenate, malonate, and hydrogen peroxide. Cyanide, in general, did not inhibit oxygen uptake, except for *Tritrichomonas foetus* and *Trichomonas vaginalis*. Some species, e.g., *Trichomonas suis*, were not inhibited by most of the compounds studied.

Nucleic acid metabolism in trichomonads has received scant attention. Apparently, there is little or no ability to synthesize purines anew, and therefore, parasites must depend on salvaging purine bases and nucleosides released into the medium by their hosts. Pyrimidines in trichomonads may be obtained either through direct synthesis or by salvage mechanisms.

Free amino acids may be taken up by trichomonads through direct absorption or mediated transport. Amino acids may be used in protein synthetic pathways, or they can be deaminated, converted to glucose, and used as an energy source. When carbohydrate reserves are depleted, cell proteins also may be broken down and used as an energy-producing carbon source.

Culture experiments have provided some information on lipid requirements. *Trichomonas vaginalis* is able to survive in culture media containing serum or preformed lipids (cholesterol and oleic and palmitic acids); this ability suggests that these parasites are able to metabolize sterols. Cholesterol represents the predominant sterol in trichomonads, but is probably derived from exogenous sources. *T. vaginalis* can use carbon derived from glucose and acetate to produce phospholipids, a major constituent of cell membranes. However, the rate of glucose uptake by parasites *in vitro* is low.[74]

Transformation of *Trichomonas gallinae* into pathogenic forms was described by Honigberg et al.[32] An avirulent strain, no longer infective for nonimmune pigeons, was exposed to a cell-free homogenate of a virulent strain. The former, when inoculated subcutaneously into mice, showed increased pathogenicity. DNase or RNase in the transformation medium eliminated the pathogenicity-enhancing effect. When the avirulent strain was exposed simultaneously to native DNA and to high-molecular-weight RNA from the virulent trichomonads, the effect was the same as when cell-free homogenates were used. DNase or RNase also eliminated the effect of the DNA-RNA mixtures. Omission of one or the other nucleic acid from the transmission medium negated their combined effect. The gift of pathogenicity was conferred only by the nucleic acids of the virulent strain. Avirulent strain homogenates and nucleic acids had no similar effect on their own trichomonad strain.

IMMUNITY OF TRICHOMONADS

Immunity conferred on the host as the result of infection by *T. vaginalis* is apparently temporary.

A complement-fixation test is positive in 80% of adult females and 40% adult male patients infected with *Trichomonas vaginalis*. Agglutination and agglomeration reactions vary, depending on the strains of *T. vaginalis* present. This variation is also true of the complement-fixation test, so a comparison of the two tests is valid only if the same strain of parasite is used. The intradermal or skin test is also used and is reliable for the diagnosis of genitourinary trichomoniasis. Protection against *T. vaginalis* is

achieved by injection of live and dead cultures. Heat-killed cultures have been effective in aborting the infection or in alleviating the symptoms of trichomoniasis when inoculated into the vaginal mucosa of infected women. See Woo for a review of studies on immunization.[92]

Order Hypermastigida

These protozoa possess many flagella, and they may also have many parabasal bodies and axostyles. There is a single nucleus and no cytostome. All the species are found in the alimentary tracts of termites, cockroaches, and wood roaches. Only one of at least eight families will be mentioned here. See Chapter 23 for these flagellates in termites.

FAMILY TRICHONYMPHIDAE

Members of this family possess a projection at the anterior end of the body. This "rostrum," which may be cone-shaped, is covered with long flagella except for its tip, while other flagella cover much of the rest of the body in longitudinal rows. The base, or more, of the body may be bare.

Trichonympha corbula is typical of the genus. It possesses circular bands of flagella, each band of a different length. The nucleus is surrounded by the narrow rods of the parabasal body, and the cytoplasm usually contains bits of wood that the animal has ingested. Several other species of this genus live in the intestines of termites and in wood roaches. The sexual cycle of *Trichonympha* has been thoroughly investigated and is partially illustrated in Figure 3–25. *Pseudotrichonympha grassii* in the termite, *Coptotermes formosanus*, and *Mixotricha*, *Deltotrichonympha*, and *Eucomonympha imla* from the wood roach are other examples of the family.

SUBPHYLUM OPALINATA

Order Opalinida

Members of this commensal group are called **opalinids** or **opalines**. There are about 375 species, found mainly in the large intestines of amphibians, but also in fishes, snakes, and in at least one lizard (*Varanus*). The parasites are flattened and are covered with rows of locomotory organelles called **cilia** by most investigators, **flagella** by a few and *unudulipoda* by a few. The fibrillar associates of the kinetosomes, however, are not like those of kinetosomes of ciliates.[50]

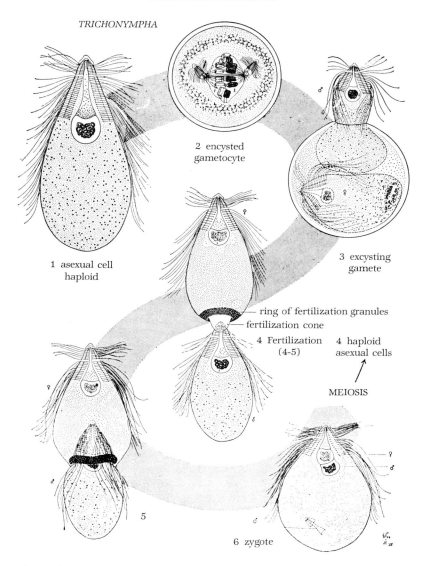

TRICHONYMPHA

2 encysted gametocyte

1 asexual cell haploid

3 excysting gamete

— ring of fertilization granules
— fertilization cone

4 Fertilization (4-5)

4 haploid asexual cells

MEIOSIS

5

6 zygote

Fig. 3–25. *Trichonympha* and some steps in its sexual cycle. (From Cleveland, courtesy of J. Protozool.)

Ingestion takes place by a modified pinocytosis. Usually, numerous monomorphic nuclei are scattered throughout the cytoplasm. Their geographic distribution is worldwide, but each has its own special distribution. The largest genus, *Opalina*, for example, can be found on any continent, except Australia and South America. Reproductive features plus the presence of two or more vesicular nuclei are the basis for placing the group with flagellates.

In the genus *Opalina* the reproductive cycle is controlled and synchronized with the repro-

ductive cycle of the host. The parasite cycle is complex (see Wesenberg[89] for details) and begins with tadpoles which have eaten cysts from anuran feces. A gamont emerges from each cyst, and it divides longitudinally until it forms uninucleate gametes. Meiosis is prezygotic, but apparently only a few gamonts undergo meiosis. Microgametes and macrogametes are formed and fuse to begin the cycle with a zygote. Asexual multiplication takes place in metamorphosing tadpoles and in adult anurans. Some of the multinuclear adults, in response to

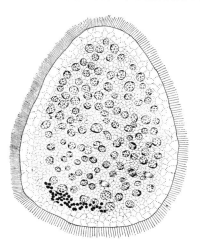

Fig. 3–26. *Opalina ranarum* × 230. This figure shows many nuclei, some of which are in the process of division. Most of the endospherules have been omitted. (From Metcalf, courtesy of the U.S. National Museum Bulletin.)

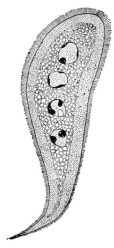

Fig. 3–28. *Cepedea lanceolata* × 850. (From Metcalf, after Bezzengerger, courtesy of the U.S National Museum Bulletin.)

host physiologic changes that precede and accompany the breeding season, begin a series of divisions and become successively smaller with fewer nuclei until they ultimately encyst. Encystation appears to be stimulated by breakdown products of androgenic steroid hormones excreted in the amphibian urine. There is evidence[9] that epinephrine induces sexual reproduction in *Opalina* in the toad, *Bufo*.

Protoopalina is considered to be a primitive genus. Its members are cylindrical or spindle-shaped and range in size from about 100 to over 300 μm long by about 20 to 70 μm wide. They possess two nuclei. *P. saturnalis* occurs in the marine fish, *Box boops,* while *P. intestinalis* and *P. mitotica* occur in the intestines of amphibia.

Opalina contains flattened, multinucleate species that are found in amphibia. The parasites range from 190–385 × 160–200 μm, with a thickness of 20 to 40 μm. Among the several known species are *O. ranarum* (Fig. 3–26), *O. hylaxena, O. oregonensis,* and *O. spiralis.*

Zelleriella contains individuals that usually have only two nuclei and whose bodies are more flattened than those of other genera. These parasites live in the intestines of amphibia. *Z. hirsuta* is about 130 × 60 × 22 μm and inhabits the gut of the toad, *Bufo cognatus. Z. opisthocarya* is itself parasitized by an ameba, *Entamoeba* sp. (Fig. 3–27).

Cepedea (Fig. 3–28) is also found in amphibians. The protozoa are cylindrical or pyriform in shape, and their range in size is about the

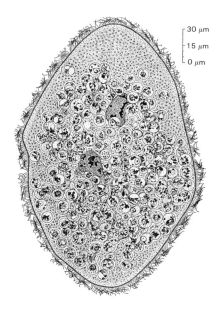

30 μm
15 μm
0 μm

Fig. 3–27. *Zelleriella opisthocarya,* an opalinid containing over 200 cysts of *Entamoeba.* (From Stabler and Chen, courtesy of the Biological Bulletin.)

same as that indicated for the genus *Opalina*. *C. cantabrigensis* is a representative species from the toad, *Bufo lentiginosus*.

REFERENCES

1. Aikawa, M., and Sterling, C.R.: Intracellular Parasitic Protozoa. New York, Academic Press, 1974.
2. Añez, N., Nieves, E., and Scorza, J.V.: [The taxonomic "status" of *Leishmania garnhami*, shown by the developmental pattern in its vector.] El "status" taxonómico de *Leishmania garnhami*, indicado por su patron de desarrollo en el vector. *Memórias do Instituto Oswaldo Cruz*, 80(1):113–119, 1985.
3. Baker, J.B. (ed): Perspectives in trypanosomiasis research. Proceedings of the Twenty-First Trypanosomiasis Seminar. Chichester, London, Research Studies Press, 1981.
4. Bardsley, J.E., and Harmsen, R.: The trypanosomes of Anura. Adv. Parasitol., *11*:1–7, 1973.
5. Bower, S.M., and Margolis, L.: Direct transmission of the haemoflagellate *Cryptobia salmositica* among Pacific Salmon (*Oncorhynchus* spp.) Can. J. Zool., *61*:1242–1250, 1983.
6. Brand, von T.: Biochemistry and Physiology of Endoparasites. Amsterdam, Elsevier/North-Holland Biomedical Press, 1979.
7. Brener, Z.: Immunity to *Trypanosoma cruzi. In* Advances in Parasitology. Edited by W.H.R. Lumdsen, R. Muller, and J.R. Baker. London, Academic Press, 1980, pp. 247–292.
8. Camp, R.R., Mattern, C.F.T., and Honigberg, B.M.: Study of *Dientamoeba fragilis* Jepps and Dobell. I. Electronmicroscopic observations of the binucleate stages. II. Taxonomic position and revision of the genus. J. Protozool., *21*:69–82, 1974.
9. Chandra, R.K.: Critical Reviews in Tropical Medicine. Vol. 1. New York, Plenum Press, 1982.
10. Chang, K.P., and Bray, R.S. (eds).: Leishmaniasis. Amsterdam, Elsevier B.V., 1986.
11. Chang, K.P., Fong, D., and Bray, R.S.: Biology of *Leishmania* and leishmaniasis. *In* Leishmaniasis. Edited by K.-P. Chang and R.S. Bray. New York, Elsevier, 1985, pp. 1–30. [from: Ruitenberg, E.J. and MacInnis, A.J. (Series eds.): Human Parasitic Diseases Vol. 1, 1985.]
12. Cherian, P.V., and Dusanic, D.G.: *Trypanosoma lewis*: ultrastructural observations of surface antigen movement induced by antibody. Exp. Parasitol., *44*:14–25, 1978.
13. Congresso International Sobre Doença De Chagas. Anais Abstracts. Instituto Oswaldo Cruz, Rio de Janeiro, 1979.
14. D'Alessandro, P.A.: The relation of agglutinins to antigenic variation of *Trypanosoma lewisi*. J. Protozool., *23*:256–261, 1976.
14a. De Raadt, P.: Immunity and antigenic variation: clinical observations suggestive of immune phenomena in African trypanosomes. Ciba Found. Symp., *20* (new series):199–224, 1974.
15. Desai, H.I., and Chandra, R.K.: Giardiasis. *In* Critical Reviews in Tropical Med. Vol. I. Edited

by R.K. Chandra. New York, Plenum Press, 1982, pp. 109–141.
16. De Souza, W.: Cell biology of *Trypanosoma cruzi*. Int. Rev. Cytol., *86*:197–283, 1984.
17. Diffley, P.: Comparative immunological analysis of host plasma proteins bound to bloodstream forms of *Trypanosoma brucei* subspecies. Infect. Immun., *21*:605–612, 1978.
17a. Dolan, R.B.: Genetics and trypanotolerance. Parasitology Today, 3:5, 1987.
18. Dvorak, J.A., Gibson, C.C., and Maekelt, A. (compilers): A Bibliography on Chagas' Disease. U.S. National Institutes of Health, Pan American Health Organization, and World Health Organization, 1985
19. Dwyer, D.M.: Analysis of the antigenic relationships among *Trichomonas, Histomonas, Dientamoeba* and *Entamoeba*. III. Immunoelectrophoresis technics. J. Protozool., *21*:139–145, 1974.
20. El Mofty, M.M., and Sadek, I.A.: The mechanism of action of adrenaline in the induction of sexual reproduction (encystation) in *Opalina sudafricana* parasitic in *Bufo regularis*. Int. J. Parasitol., 3:425–431, 1973.
21. Erlandsen, S.L., and Feely, D.E.: Trophozoite motility and mechanism of attachment. *In* Giardia and Giarsiasis. Edited by S.L. Erlandsen and E.A. Meyer. New York, Plenum Press, 1984.
22. Freyvogel, T.A.: Environmental management for the control of parasitic diseases. Insect Sci. Applic., 7(3):297–304, 1986.
23. Ginoux, P.Y., Frézil, J.L., and Alary, J.C.: Trypanosomiasis. Méd. Trop., 42(3):281–287, 1982.
24. Goodwin, L.G.: The African scene: mechanisms of pathogenesis in trypanosomiasis. Ciba Found. Symp., *20*:107–124, 1974.
25. Green, C.E.: Clinical Microbiology and Infectious Diseases of the Dog and Cat. Philadelphia, W.B. Saunders Co., 1984.
26. Hanson, W.L.: Immunology of American trypanosomiasis (Chagas' disease). *In* Immunology of Parasitic Infections. Edited by S. Cohen and E.H. Sadun. Oxford, Blackwell Scientific Publications, 1976, pp. 222–234.
27. Heyneman, D., Hoogstraal, H., and Djigounian, A.: Bibiliography of *Leishmania* and leishmanial diseases. Cairo U.S. Naval Med. Res. Unit No. 3. Special Publ. Vol. 1, 385 pp. Vol. 2, 303 pp., 1980.
28. Hill, G.C., and Anderson, W.A.: Electron transport systems and mitochondrial DNA in Trypanosomatidae: a review. Exp. Parasitol., *28*:356–380, 1970.
29. Honigberg, B.M.: Evolutionary and taxonomic relationships among the zoomastigophorea. *In* Parasites—Their World and Ours. Edited by D.F. Mettrick, and S.S. Desser. Amsterdam, Elsevier Biomedical Press, 1982, pp. 172–174.
30. Honigberg, B.M.: Mechanisms of pathogenicity among protozoa. Insect Sci. Applic., 7(3):363–378, 1986.
31. Honigberg, B.M., Farris, V.K., and Livingston, M.C.: Preservation of *Trichomonas vaginalis* and *Trichomonas gallinae* in liquid nitrogen. Prog. Protozool., *236*:199–200, 1965.
32. Honigberg, B.M., Livingston, M.C., and Stabler, R.M.: Pathogenicity transmission of *Trichomonas*

gallinae. I. Effects of homogenates and mixtures of DNA and RNA from a virulent strain on pathogenicity of an avirulent strain. J. Parasitol., 57:929–938, 1971.

33. Hudson, L.: The biology of trypanosomes. *In* Current Topics in Microbiology and Immunology. Edited by L. Hudson. Berlin, Springer-Verlag, 1985, 117:183.

34. Hunter, S.H., Baker, H., Frank, O., and Cox, D.: Nutrition and metabolism in Protozoa. *In* Biology of Nutrition. Edited by R.N. Fiennes. International Encyclopedia of Food and Nutrition. Vol. 18. Oxford, Pergamon Press, 1972, pp. 85–177.

35. Kagan, I.G.: American trypanosomiasis (Chagas' disease). *In* Immunological Investigation of Tropical Parasitic Diseases. Edited by V. Houba. New York, Churchill Livingstone, 1980, pp. 49–64.

36. Katz, M., Despommier, D.D., and Gwadz, R.W.: Parasitic Diseases. New York, Springer-Verlag, 1982.

37. Kirkpatrick, C.E.: Enteric protozoal infections. *In* Greene, C.E.: Clinical Microbiology and Infectious Diseases of the Dog and Cat, Philadelphia, W.B. Saunders Co., 1984.

38. Kirkpatrick, C.E., and Farrell, J.P.: Giardiasis. Compend. Ocont. Educ. Pract. Vet., 4:367–377, 1982.

39. Krieger, J.N., et al.: Geographic variation among isolates of *Trichomonas vaginalis;* demonstration of antigenic heterogeneity by using monoclonal antibodies and the indirect immunofluorescence technique. J. Infect. Dis., 152:979–980, 1985.

40. Lainson, R.: The American leishmaniases: some observations on their ecology and epidemiology. Trans. Roy. Soc. Trop. Med. Hyg., 77(5):569–596, 1983.

41. Lainson, R.: Our present knowledge of the ecology and control of leishmaniasis in the Amazon region of Brazil. Revis. Socied. Brasil. Med. Trop., 18:47–56, 1985.

42. Le Blancq, S.M., Cibulskis, R.E., and Peters, W.: *Leishmania* in the Old World: 5. Numerical analysis of isoenzyme data. Trans. Roy. Soc. Trop. Med. Hyg., 80:517–524, 1986.

43. Lee, J.J.: Order 5. Trichomonadida. *In* An Illustrated Guide to the Protozoa. Edited by J.J. Lee, S.H. Hutner, and E.C. Bovee. Lawrence, Kansas, Society of Protozoologists, 1985, pp. 119–127.

44. Lee, J.J.: Protozoa as indicators of ecosystems. Insect Sci. Applic., 7(3):349–353, 1986.

45. Lee, J.J., Hutner, S.H., and Bovee, E.C.: An Illustrated Guide to the Protozoa. Lawrence, Kansas, Society of Protozoologists. 1985.

46. Le Ray, D., and Recacoeclea, M.: Chagas' Disease. International Colloquium. Ann. Soc. Belge. Méd. Trop., 65, Supplement 1, 1985.

47. Levine, N.D.: Veterinary Protozoology. Dubuque, Iowa State University Press, 1985.

48. Lindmark, K.G., and Müller, M.: Hydrogenosome, a cytoplasmic organelle of the anaerobic flagellate, *Tritrichomonas foetus,* and its role in pyruvate metabolism. J. Biol. Chem., 248:7724–7728, 1973.

49. Lumsden, W.H.R.: Pathobiology of Trypanosomes. *In* Pathology of Parasite Diseases. Edited by Gaafar, et al.: Lafayette, Indiana, Purdue Univ. Press, 1971, p. 1–14.

50. Lynn, D.H., and Small, E.B.: Subphylum Opalinata. *In* Lee, J.J., et al. (eds): An Illustrated Guide to the Protozoa. Edited by Lee, J.J. et al. Soc. Proto. 1985, pp. 156–157.

51. McKelvey, J.J.: Man Against Tsetse: Struggle for Africa. Ithaca, N.Y. Cornell University Press, 1973.

52. Maegraith, B.: Adams & Maegraith: Clinical Tropical Diseases. 6th. Ed. Oxford, Blackwell Scientific Publications, 1976.

53. Mansfield, J.M.: Immunobiology of African trypanosomiasis. Cell. Immunol., 39:204–210, 1978.

54. Marciacq, Y., and Seed, J.R.: *Trypanosoma gambiense* in guinea pigs. J. Infect. Dis., 121:653–655, 1970.

55. Marsden, P.D., and Jones, T.C.: Clinical manifestation, diagnosis and treatment of leishmaniasis. Edited by K.-P. Chang and R.S. Bray. New York, Elsevier, 1985, pp. 183–198. [from: Ruitenberg, E.J., and MacInnis, A.J. (Series eds.): Human Parasitic Diseases Vol. 1. 1985.]

56. Massamba, N.N., and Williams, R.O.: Distinction of African trypanosome species using nucleic acid hybridization. Parasitol., 85:55–56, 1984.

57. Maudlin, I.: Inheritance of susceptibility to trypanosomes in tsetse flies. Parasitol. Today, 1(2):59–60, 1985.

58. Mettrick, D.F., and Desser, S.S.: Parasites—Their World and Ours. Amsterdam, Elsevier Biomedical Press, 1982.

59. Miles, M.A., and Cibulskis, R.E.: Zymodeme characterization of *Trypanosoma cruzi.* Parasitol. Today, 2(4):94–97, 1986.

60. Molyneux, D.H., and Ashford, R.W.: The Biology of Trypanosoma and Leishmania, Parasites of Man and Domestic Animals. New York, International Publications Service, Taylor & Francis Inc. 1983.

61. Morel, C.M., Deane, M.P., and Gonçalves, A.M.: The complexity of *Trypanosoma cruzi* populations revealed by schizodeme analysis. Parasitol. Today, 2(4):97–101, 1986.

62. Murray, M., Morrison, W.I., and Whitelaw, D.D.: Host susceptibility to African trypanotolerance. Adv. Parasitol., 21:1–68, 1982.

63. Muskoke, A.J., and Barbet, A.F.: Activation of complement by variant-specific surface antigen of *Trypanosoma brucei.* Nature, 270:438–440, 1977.

64. Mutinga, M.J.: Leishmaniases. Insect Sci. Applic., 7(3):421–427, 1986.

65. Newton, B.A. (ed.): Trypanosomiasis. Br. Med. Bull., 41(2):103–199, 1985.

66. Nuwayri-Salti, N., and Khansa, H.F.: Direct non-insect-vector transmission of *Leishmania* parasites in mice. Int. J. Parasitol., 15(5):497–500, 1985.

67. Nyindo, M., Chimtawi, M., and Owor, J.: *Trypanosoma brucei:* Evidence suggesting existence of sexual forms of parasites cultured from the tsetse, *Glossina morsitans morsitans.* Insect Science Applic., 1:171–175, 1981.

68. Olivier, M.C., Olivier, L.J., and Segal, D.B.: A Bibliography on Chagas' Disease (1909–1969). Index-Catalogue of Medical and Veterinary Zoology. Special Publication 2. Washington, D.C., U.S. Department of Agriculture, 1972.

69. Parr, C.W., and Godfrey, D.G.: The measurement

of enzyme ratios as a means of differentiating trypanosomes. Trans. R. Soc. Med. Hyg., 67:260–261, 1973.

70. Petry, K., Baltz, T., and Schottelius, J.: Differentiation of *Trypanosoma cruzi, T. cruzi marinkellei, T. dionisii* and *T. vespertilionis* by monoclonal antibodies. Acta Tropica, 43:5–13, 1986.

71. Pinder, M.: Trypanosoma congolense: Genetic control of resistance to infection in mice. Exp. Parasitol., 57, 185–194, 1984.

72. Preston, P.M., and Dumonde, D.C.: Immunology of clinical and experimental leishmaniasis. *In* Immunology of Parasitic Infections. Edited by S. Cohen and E.H. Sadun. Oxford, Blackwell Scientific Publications, 1976, pp. 167–202.

73. Roelants, G.E., and Williams, R.O.: African Trypanosomiasis. *In* Chandra, R.K. (Ed.): Critical Reviews in Tropical Med. Vol. I. Edited by R.K. Chandra. New York, Plenum Press, 1982, pp. 31–75.

74. Roitman, I., Heyworth, P.G., and Gutteridge, W.E.: Lipid synthesis by *Trichomonas vaginalis*. Ann. Trop. Med. Parasitol., 72:583–585, 1978.

75. Sanderson, C.J., and de Souza, W.: A morphological study of the interaction between *Trypanosoma cruzi* and rat eosinophils, neurotrophils and macrophages *in vitro*. J. Cell Sci., 37:275–286, 1979.

76. Sandon, H.: The species problem in the opalinids (Protozoa, Opalinata), with special reference to *Protoopalina*. Trans. Am. Microsc. Sci., 95:357–366, 1976.

77. Santos-Buch, C.A.: American trypanosomaisis: Chagas' disease. Int. Rev. Exp. Pathol., 19:63–100, 1979.

78. Schmunis, G.A., Szarfman, A., Langembach, T., and de Souza, W.: Induction of capping in blood-stage trypomastigotes of *Trypanosoma cruzi* by human anti-*Trypanosoma cruzi* antibodies. Infect. Immun., 20:567–569, 1978.

78a.Stephen, L.E.: Trypanosomiasis: A Veterinary Perspective. Oxford, Pergamon Press, 1986, 572 pp.

79. Su-Lin, K.-E., and Honigberg, B.M.: Antigenic analysis of *Trichomona vaginalis* strains by quantitative fluorescent antibody methods. Z. Parasitenkd., 69:161–181, 1983.

80. Tait, A., et al.: Enzyme variation in *T. brucei* ssp. II. Evidence for *T.b. rhodesiense* being a set of variants of *T.b. brucei*. Parasitol., 90:89–100, 1985.

81. Taylor, A.E.R., and Muller, R.: Pathogenic processes in parasitic infections. Symp. Br. Soc. Parasitol., 13, 1975.

82. Tieszen, K.L., Molyneux, D.H., and Abdel-Hafez, S.K.: Host-parasite relationships of *Blastocrithidia familiaris* in *Lygaeus pandurus* Scop. (Hemiptera: Lygaeidae). Parasitol., 92:1–12, 1986.

83. Van den Bossche, H. (ed.): Biochemistry of Parasites and Host-Parasite Relationships. Amsterdam, North-Holland Publishing, 1976.

84. Vickerman, K.: DNA throughout the single mitochondrion of a kinetoplastid flagellate: observations on the ultrastructure of *Cryptobia vaginalis* (Hesse, 1910). J. Protozool., 24:221–223, 1977.

85. Vickerman, K.: Morphological and physiological considerations of extracellular blood protozoa. *In* Ecology and Physiology of Parasites: a Symposium. Edited by A.M. Fallis, Toronto, University of Toronto Press, 1971, pp. 57–91.

86. Vickerman, K.: Developmental cycles and biology of pathogenic trypanosomes. Br. Med. Bull., 41(2):105–114, 1985.

87. Ward, R.D.: Vector biology and control. *In* Leishmaniasis. Edited by K.-P. Chang and R.S. Bray. New York, Elsevier, 1985, pp. 199–212. [from: Ruitenberg, E.J. and MacInnis, A.J. (Series eds.): Human Parasitic Diseases Vol. 1. 1985.]

88. Weber, J.: Sexually acquired parasitic infections in homosexual men. Parasitol. Today, 4:93, 1985.

89. Wessenberg, H.S.: Opalinata. *In* Parasitic Protozoa. Vol. 2. Edited by J. Kreier. New York, Academic Press, 1978, pp. 551–581.

90. Williams, P., and Coelho, M.D.: Taxonomy and transmission of *Leishmanaia*. *In* Advances in Parasitology. Edited by W.H.R. Lumsden, R. Muller, and J.R. Baker. London, Academic Press, 1978.

91. Wirth, D.F., et al.: Leishmaniasis and malaria: new tools for epidemiologic analysis. Science, 234:975–979, 1986.

92. Woo, P.T.K.: A review of studies on the immunization against the pathogenic protozoan diseases of man. Acta Trop., 31:1–27, 1974.

93. Zuckerman, A.: Current status of the immunology of blood and tissue protozoa. I. *Leishmania*. Exp. Parasitol., 38:370–400, 1975.

4

Phylum Sarcodina

PHYLUM SARCODINA

Class Lobosa

Order Amoebida

The Sarcodina are usually microscopic, and float or creep in fresh or salt water, although occasionally they may be sessile. The cytoplasm is commonly divided into an outer, homogeneous, clear ectoplasm and an inner, more granular endoplasm that makes up the bulk of the cytoplasm. The endoplasm contains fluid-filled vacuoles that may or may not contract, food vacuoles, mitochondria, other organelles, and the nucleus. Within the cytoplasm of young cysts of some species are **chromatoid bodies.** These structures are deeply stained with certain dyes and are convenient objects for taxonomic purposes. They develop in the motile stage of amebas and mature in the cyst, but disappear as the cyst ages. They are composed of masses of ribosomes and thus function in protein metabolism.

Pseudopodia are temporary projections of the body surface and cytoplasm that function for locomotion and feeding. They may be finger- or tongue-like (lobopodia), filamentous (rhizopodia), or stiff, narrow structures containing an axial rod (axopodia). Axopodia are semipermanent. Only lobopodia are commonly found among the parasitic sarcodina; they are thrust out and withdrawn in an unpredictable fashion, and the amebas may move rapidly or sluggishly.

Food capture occurs when a pseudopodium flows around a suitable object and engulfs it, that is, takes it into a temporary vacuole. Several variations of this method exist. Nutrients may also be absorbed directly through the cell membrane or by pinocytosis. Food is stored as glycogen in a vacuole in some species.

The nucleus contains an **endosome** (nucleolus) that may be large or small and may consist of a group of granules. It may be centrally located or eccentric in position. Chromatin material can be seen as small masses or minute granules on the inner nuclear membrane, and thus it appears as a distinct ring in optical section. Chromatin may also be scattered in the nucleus, or it may form a ring or "halo" around the endosome.

Reproduction usually occurs by binary fission, multiple fission, plasmotomy, or budding, but sexual processes are involved in some species. Cyst formation is common among many of the parasitic forms. The motile organism (vegetative or trophic stage, or **trophozoite** becomes rounded and secretes a resistant cyst wall around itself. Within this cyst, the nucleus may divide several times.

Parasites of this order are usually much smaller than the well-known, free-living *Amoeba proteus;* many average 10 μm in diameter, and many are usually seen only in the cyst stage. The motile stage moves by means of broad pseudopodia, and most of the species have only one nucleus, except in the cyst forms. Many species live in the digestive tracts of vertebrates and invertebrates.[1] See Bovee.[4]

FAMILY ENTAMOEBIDAE

The genus *Entamoeba* is usually found in the intestines of invertebrates and vertebrates. The vesicular nucleus possesses an endosome that is usually small and located at or near the nuclear center. Granules may occur on the nuclear membrane inner wall or around the endosome. Cyst nuclei are similar, but may number from one to eight. Chromatoid bodies are often present in young cysts.

The genus *Endamoeba* was originally created for *E. blattae* in cockroaches. There is no endosome, but numerous granules are located near the thick nuclear membrane. Cysts may contain many nuclei. Species of this genus are restricted to invertebrates.

Two methods of studying amebas (as well as other protozoa) involve axenic cultivation and cryogenic preservation. The former method is the maintenance of these organisms in the absence of all other cells. Axenic cultivation is essential for a full understanding of nutritional requirements, growth behavior, rates of multiplication, and metabolism of the parasites. Cryogenic preservation is freeze-preservation at ultra-low temperatures. Using liquid nitrogen as the refrigerant, amebas may be stored at −196°C for several years with little or no chemical or physical change. Long-term preservation eliminates most of the costly and time-consuming requirements of cultivation maintenance.[6]

AMEBAS IN INVERTEBRATES

Entamoeba sp. is morphologically similar to *E. ranarum* in frogs and lives in *Zelleriella* opalinids, which parasitize toads and frogs. (See p. 62). The motile or trophozoite stage of the amebas averages 8 μm in the diameter. These amebas live in pockets in the opalinids, sometimes in such large numbers that there is little room for anything else. Cysts averaging 9.4 μm occur in the opalinid host. Some of the amebas are parasitized by dot-like organisms that are probably *Sphaerita* (see Fig. 4–3). Thus the frog gut contains opalinids that harbor amebas that are host to *Sphaerita*.

Endamoeba blattae lives in the intestines of cockroaches and termites. The parasite varies from 10 to 150 μm in diameter, and it forms multinucleate cysts. Pseudopodia are usually broad, and striations often appear in the body. The organism is probably harmless to the insects.

Entamoeba phallusae has been found in the intestine of the ascidian, *Phallusia mamillata*, at Plymouth, England. It measures 15 to 30 μm in the motile stage; cysts averages 21 × 19 μm. The nucleus is prominent and eccentric. Small, unidentified amebas have also been found in the ascidian, *Clavellina lepadiformis*.

Entamoeba aulastomi lives in the gut of a horse leech, *Haemopsis sanguisuagae*. The cysts

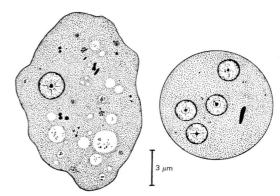

Fig. 4–1. *Entamoeba invadens,* motile and cyst forms.

contain four nuclei. *Endamoeba philippinensi* and *Endamoeba javanica* both may be found in the intestine of the wood-feeding roach, *Panesthia javanica*. The bee, *Apis mellifica*, is the host of *Endamoeba apis*, and crane fly larvae harbor *Endamoeba minchini*

Paramoeba periclosa is a pathogenic parasite of a few marine crabs and lobsters. Another species appears to be the causative agent of mass mortality of sea urchins.

AMEBAS IN FISH

Not more than nine species of amebas have been reported from fish, and some of these may be the same species. *Amoeba mucicola* may be found in the branchial mucus and on the skin of the marine fish, *Symphodus tinca*, whereas the mucus lining the stomach of the rainbow trout, *Salmo shasta*, contains *Schizamoeba salmonis*. *Entamoeba ctenopharyngodoni* occurs in the rectum of the freshwater fish, *Ctenopharyngodon idellus*, in China. The sunfish, *Mola mola*, harbors *Entamoeba molae* in its hindgut. This ameba is similar to *E. histolytica*, but the nucleus contains a subperipheral crescent of granules. A careful search would undoubtedly reveal other species of amebas in both freshwater and marine fish.

AMEBAS IN AMPHIBIA

Entamoeba ranarum lives in the large intestine of frogs. The motile stages usually vary in size from 10 to 50 μm in diameter. There is a small endosome in the nucleus, and the cysts usually posses 4 nuclei, although as many as 16 have been reported. It is possibly pathogenic. Similar amebas have been found in salamanders.

AMEBAS IN REPTILES

Entamoeba invadens (Fig. 4–1) is the best-known *Entamoeba* parasite from these hosts. It lives in the intestine, is similar in many respects to *E. histolytica* of man, and may be highly pathogenic. This ameba has become a favorite for *in vitro* studies of metabolism. A prerequisite for mass encystation of *E. invadens* within its host is the ingestion by the protozoon of minute particles of carbohydrate. The ameba is harmless in herbivorous turtles, as compared with its effects in experimental snakes. This difference is probably related to the abundance of particulate carbohydrate in the turtle's intestine and the absence of it in the gut of the snake. *E. invadens* has been found to be pathogenic to chicks. See Albach and Booden.[1]

Endolimax clevelandi lives in the intestines of the turtle, *Pseudemys floridana mobilensis*. The small trophic stage averages about 7 μm in diameter. A finely granular ectoplasm encloses a coarser endoplasm. Vaculoes may or may not be present, and some of them may contain bacteria or other food. The nucleus is vesicular in appearance with a large endosome. Cysts are often elliptical (an *Endolimax* trait), and they range from 4.5 to 10 μm in diameter. The mature cyst contains four nuclei, which are similar to those in the motile stages.

AMEBAS IN BIRDS

The freedom from amebas enjoyed by birds as compared with their burden of other parasites is difficult to explain. Probably a more thorough search of birds for intestinal amebas would be productive. The few amebic parasites reported from birds include *Entamoeba lagopodis* (= *E. gallinarum?*) and *Endolimax gregariniformis* (= *E. janisae*) in the cecum of the grouse, *Lagopus scoticus*; *Entamoeba anatis* in the duck; and *E. gallinarum* in various fowl.

AMEBAS IN MAMMALS

Considerable confusion exists with regard to the correct names for entamoebas of domestic animals. They also possess essentially the same morphologic features.[19]

Entamoeba bovis (Fig. 4–2) lives in the intestines of cattle and has a motile form that averages 5.3 × 7.5 μm, with a smoothly granular cytoplasm filled with vacuoles. The large nucleus has conspicuous peripheral chromatin and a large central endosome made up of compact granules. Cysts average 8.8 μm and possess a single nucleus; young forms also have irregular chromatoid bodies that stain deeply with hematoxylin. Glycogen vacuoles are common in the cysts. *E. histolytica* has also been reported from the intestine of cattle.

Entamoeba ovis lives in sheep intestines. Examination of host feces rarely discloses motile forms. Cysts average 7.2 μm in diameter and they contain chromatoid bodies of irregular shapes. The nucleus is often eccentric in position and is occasionally surrounded by a clear area. Peripheral chromatin is moderate in amount, and the endosome is composed of a large, compact mass of granules. Many variations of these characteristics exist. *E. caprae* has been reported from the stomach of sheep.

Entamoeba debliecki is found in the intestines of goats and pigs. The motile stage averages 13 × 16 μm, and it possesses a finely granular ectoplasm and a coarser endoplasm. The nucleus is large, usually pale, and the peripheral chromatin appears to be a homogeneous ring. The central endosome is large and often indistinct. Cysts range from 4 to 17 μm in diameter, averaging 8 μm. Variable chromatoid bodies occur in large cysts. The endosome varies from a large mass to a small one surrounded by a "halo" of granules. Three other species have been recorded from the goat intestine. They are *E. wenyoni*, *E. dilimani*, and *E. caprae*. The latter has also been found in the stomach.

Entamoeba intestinalis has been found in the cecum and colon of horses and is said to occur in other animals. Trophic forms are common, whereas the cysts are rare or absent. The nuclei are of the *E. coli* type, and the eccentric endosome is surrounded by a ring of granules. *E. gedoelsti* has been suggested as a better name for this ameba. *E. equi* (= *E. intestinalis?*) has also been found in the feces of horses.

Dogs harbor amebas, but the identification of species is uncertain. *Entamoeba venaticum* (= *E. histolytica? E. caudata?*) has been described from the large intestine, and *E. gingivalis* (= *E. canibuccalis?*) has been reported from the mouth. *E. histolytica* was identified in 8.4% of the dogs in Memphis, Tennessee. *E. coli* and *Endolimax nana* have also been reported from the intestine of the dog.

Cats apparently do not have their own amebas, but they can easily be infected with *Entamoeba histolytica* from man. For this reason, kittens are often used in experiments with this important human parasite.

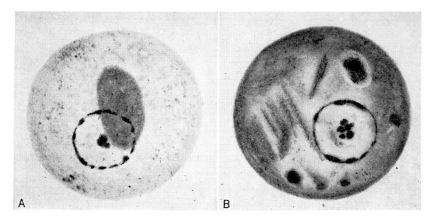

Fig. 4–2. *Entamoeba bovis* cysts from the intestine of cattle. (From Noble, E. Courtesy of the University of California Publications in Zoology.)

Rabbits may harbor *Entamoeba cuniculi* (= *E. muris*), which is similar to *E. coli* of man. *Endolimax* spp. also live in the cecum.

Guinea pigs are host to *Entamoeba cobayae* (= *E. caviae* and *E. muris?*) which resembles *E. coli* of man. This species occurs in the cecum, whereas the intestine harbors *Endolimax caviae*.

Rats and mice, like the two hosts above, possess an intestinal ameba that resembles *Entamoeba coli*. It is named *E. muris* (Fig. 4–3). Rats naturally infected with *E. histolytica* have been reported from England, Iraq, Indonesia, and the United States. *Endolimax ratti* occurs in the rat colon. It is morphologically indistinguishable from human *E. nana*.

Macaque monkeys may harbor several amebas that are morphologically identical to those of man. Those species reported from the large intestine are *Entamoeba histolytica* (= *E. nuttalli, E. duboscqi, E. cynomolgi?*); *E. coli* (= *E. legeri*); *E. chattoni*; *Endolimax nana* (= *E. cynomolgi*); and *Ioda-moeba buetschlii* (= *I. kueneni*).

AMEBAS IN MAN

In addition to the pathogenic soil amebas (see the end of the discussion of Sarcodina), the genera and species of amebas in man are *Entamoeba histolytica*, *E. hartmanni*, *E. coli*, *E. gingivalis*, *Endolimax nana*, and *Iodamoeba buetschlii*. As with the amebas of animals, the trophozoites of these species are usually uninucleate, and the cysts are commonly multinucleate. Most amebas of man are not pathogenic, but *E. histolytica* can cause one of the more serious parasitic diseases (amebiasis) of tropic and temperate countries. For an account of *Dientamoeba fragilis*, formerly considered to be an ameba, see Chapter 3.

A study of the character of DNA in various species can be an aid in species differentiation. Gelderman et al.[7] have shown that, although morphologically identical, the Laredo "strain" of *Entamoeba histolytica*, an isolate of *E. moshkovskii*, and the "classic" *E. histolytica* are sufficiently different in their DNA composition to

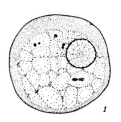 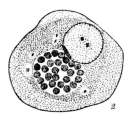 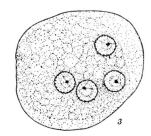

Fig. 4–3. *Entamoeba muris*, found in rats and mice. The figures show nuclear structure. *1*, Ameba with highly vacuolated endoplasm and endosome unstained (hemalum). *2*, With included parasite, *Sphaerita*. *3*, With four nuceli. (From Wenrich. Courtesy of the Journal of Morphology.)

warrant placing each in a separate species. For example, the Laredo amebas had the smallest amount of DNA per cell, the classic *E. histolytica* had approximately five times as much, and *E. moshkovskii* has approximately ten times as much. The amount of rapidly reassociating DNA (repeated sequences) also differed among the three amebas.

Entamoeba moshkovskii can easily be mistaken for *E. histolytica* when sewage-contaminated water is examined. The four-nucleate cysts of each species appear practically identical. It is easily cultivated at 24°C but poorly at 37°C, the temperature at which *E. histolytica* grows well. *E. moshkovskii* is not pathogenic and rarely occurs in humans.[23]

Entamoeba histolytica (Fig. 4–4), on a global scale[10] infects about 480 million people, and some 48 million suffer from invasive amoebiasis. "The latter, in the intestinal and extraintestinal forms, accounts for 40,000 to 110,000 deaths annually. In view of this considerable mortality and morbidity, more research is urgently needed so that more effective control strategies may be introduced."[27] Amoebiasis has been ranked third among parasitic causes of death, behind malaria and schistosomiasis. The prevalence in some endemic areas may be 20%. It is a major health problem in China, Mexico, much of Africa, eastern South America, all of southwest Asia, and the Indian subcontinent.[25]

The majority (about 85%) of people infected with *Entamoeba histolytica* are healthy carriers, but these parasites can invade the intestinal membranes and eventually destroy practically any tissues of the human body. Amebas have been found in almost all soft tissues and bone. Abscesses may occur in the skin (Fig. 4–5), liver, lungs, brain, uterus, vagina, and the gastric mucosa.[16a]

"The term 'invasive amoebiasis' reflects both parasite and host factors which result in pathologic lesions. **Invasive intestinal amebiasis** is characterized by the following markers: (1) clinical symptoms and signs of disease; (2) presence of haemtophagous trophozoites in the stool or rectal scrapings; (3) characteristic changes in the intestinal mucosa at endoscopic examination; (4) positive serologic tests for specific antibodies. Many gradations of pathologic change and clinical expression occur in invasive amoebiasis. This condition is usually characterized clinically either by acute amoebic dysentery with bloody, mucous stools, colicky pain, and rectal tenesmus, or by intermittent diarrhea, frequently with blood-stained feces. In general, there is no fever or other systemic manifestation and the symptoms disappear after a few days of treatment or even spontaneously."[27] The prevalence of intestinal amebiasis among male homosexuals has recently burgeoned in the United States and elsewhere. Amebiasis is basically the result of feces-to-mouth contamination, although the route to the mouth is usually indirect. See Turner's "Travellers' diarrhoea."[24]

The most common extraintestinal form of invasive amebiasis is *amebic liver abscess* which is 10 times more common in adults than in children, and with a higher frequency in males (3:1). The abscesses are generally located in the right lobe of the liver, and are easily recognized by CT scan. They may produce masses of necrotic material sufficient to replace close to 90% of the normal liver, or remain as pinpoint lesions. Many patients are cured with adequate treatment.[27]

Except for invasion of the liver, the extension of amebas from the intestine to other tissues via the blood is rare.

Cerebral amebiasis is considered a rarity, and is found only with the presence of amebic liver abscess. It occurs at any age and is more common in males than in females, causing convulsions, cranial nerve disturbances, paralysis of one or more parts of the body, or several other kinds of damage. Unless it is diagnosed early, it is almost always fatal.

Cutaneous amebiasis results from direct contact. The primary form usually occurs in genital regions and is a complication of intestinal amebiasis with dysentery. It is commonly secondary to homosexual activities. Other extraintestinal localities include lungs, the pericardium, kidneys, viscera, spleen, muscles, bone, esophagus, larynx, stomach, aorta, and more.

The adhesion of amebas to host cells is important in pathogenicity. This adherence precedes invasion of colon tissues and lysis of liver cells. "The capacity of *E. histolytica* to destroy tissues may involve a host of biochemical mechanisms, including secreted enzymes, cell-free 'cytotoxins,' and contact-dependent cytolytic mechanisms. It is not surprising that so successful a pathogen has evolved multiple mechanisms by which it invades tissues despite host defenses"[21] (Fig. 4–6).

Isolates of these parasites can be divided into

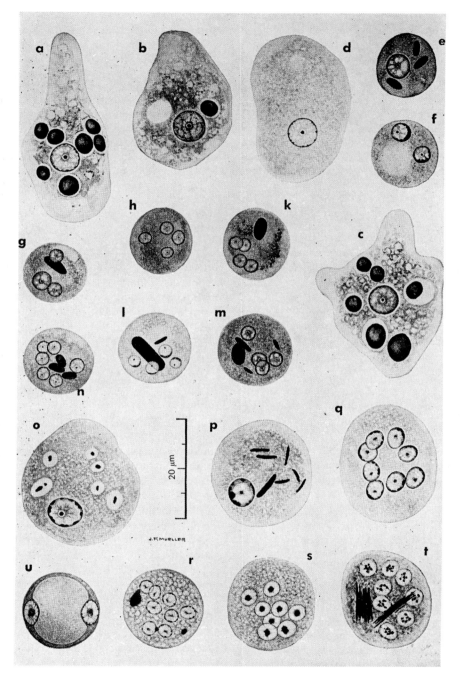

Fig. 4–4. *Entamoeba histolytica. a–c,* Trophozoites from a single case, showing ingested red blood cells. *d,* Trophozoite from another case showing delicate cytoplasm and typical nucleus. *e–n,* Various cysts, with and without chromatoid bodies, showing one, two, three, four, and six nuclei of varying character. Chromatoid bodies are commonly present in young cysts, absent in older cysts. In exceptional cases, eight nuclei are present.

    Entamoeba coli, o, Trophozoite with ingested bacteria, showing blunt pseudopod, typical nucleus. *p,* Cyst with chromatoids, single nucleus. *q–t,* Cysts with eight nuclei, showing varying conditions of cytoplasm, chromatoids, and nuclear structure. *u,* A common form of coli cyst containing a large central glycogen vacuole and two large nuclei. (From Mueller: The Story of Amebiasis, courtesy of Winthrop Laboratories.)

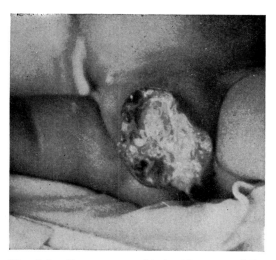

Fig. 4–5. Cutaneous amebiasis. (Courtesy of Dr. Francisco Biagi F., Ciudad Universitaria, Mexico.)

pathogenic and non-pathogenic zymodemes by an analysis of their isoenzyme patterns.[18a] "It can be concluded that the cytotoxic action of *Entamoeba histolytica* requires the activity of living trophozoites of pathogenic strains and the establishment of close contact with the target cells. Damage of mammalian cells by the amebas is achieved through a combination of mechanical and chemical means, the latter probably involving a membrane-bound toxin that produces osmotic damage in the plasma membrane of target cells. Lysis is followed by ingestion of the damaged cells by means of phagocytosis. Finally, degeneration of phagocytized cellular components within the phagocyte vacuole of the amebas complete the killing action of *E. histolytica*."[16]

Mature motile *Entamoeba histolytica* averages about 25 μm in diameter (range 10 to 60 μm). A surface coat (fuzzy coat or glycocalax) of variable thickness covers the trophozoite peripherally to the plasma membrane. Much of the cell surface has sucker-like circular openings (amebastomes) ranging from 2 to 6 μm in diameter. Internally the cytoplasm consists of a clear ectoplasm and a finely granulated endoplasm, that usually includes food vacuoles filled with bacteria or other organic material found in the intestine. Red blood cells are seen in pathogenic forms. Food vacuoles appear to be formed by invagination of the plasma membrane. Pinocytosis (see glossary) also takes place. Ribosomal elements are present as ribo-

nucleoprotein (RNP) bodies. In cyst stages these bodies become the crystalline structures known as chromatoid bodies. Microfilaments and macrotubules are also present. Trophozoites often have surface filopodia up to 40 μm in length. They may function in obtaining food, in the release of cytotoxic substances, in cytolysis, or in attachment to the substratum. There is no clear evidence for these functions. At the posterior end of many trophozoites engaged in active locomotion is the uropod, a refractile body to which foreign particles, cells, bacteria, and surface coat components become attached.

The vesicular nucleus is 3 to 5 μm in diameter, has a well-defined area of peripheral chromatin, containing RNA, but no nucleolus. The endosome may appear as a single granule or a closely packed cluster of minute granules, a site of DNA condensation.[16] Often a ring or halo appears to surround the endosome, and sometimes spoke-like lines radiate from the endosome to the nuclear membrane. Other granules, usually of uniform size, lie against the inner wall of the nuclear membrane which is perforated by numerous pores. Considerable variation occurs in these morphologic features. The nuclear membrane remains intact during nuclear division in *Entamoeba*, but the mitotic process is still little understood. "It is instructive to recall that several of the structural nuclear 'details' referred to in parasitology textbooks are, in fact, optical artifacts, since they are well below the limit of resolution of the light microscope."[16] See a review by McLaughlin & Aley.[17]

Movement is irregular and is associated with pseudopodia that are usually broad, but may be finger-like. The type and rapidity of movement varies, depending on the consistency of the surrounding medium, age of the parasite, temperature, strain or race of the ameba, stage of treatment of the host, and many other factors.

The amebas live in the lower small intestine and entire colon, and divide by binary fission. Lumen-dwelling forms engulf bacteria, and after a period of feeding and reproduction, vacuoles disappear and the amebas become rounded. Soon a cyst wall begins to form; these uninucleated stages are **precysts.** Glycogen vacuoles act as carbohydrate reserves. One or more deeply staining barshaped chromatoid bodies characteristically develop in the cytoplasm of the cyst. These bodies serve as a storage for ribosomes. Cysts (see Fig. 4–4) are about 8 to

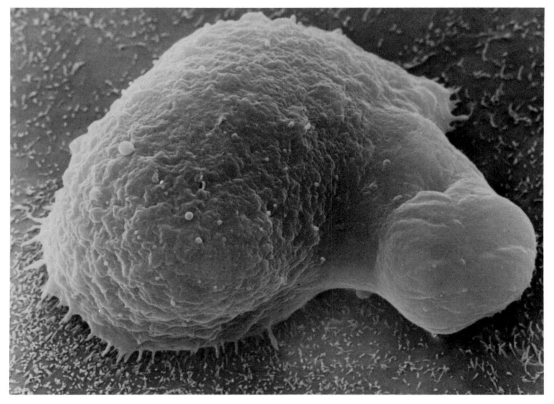

Fig. 4–6. Trophozoite of *Entamoeba histolytica* attached to a cultured epithelial cell. Note the smooth lobopodium at the right end and the wrinkled surface with numerous filopodia on the left. (From Gonzales-Robles and Martinez-Palomo, courtesy of the J. Protozool.)

20 μm in diameter (range 5 to 20 μm) and usually have four nuclei. **Metacystic stages** involve the division of the quadrinucleate organism into four small uninucleate stages, which subsequently grow and divide. Cysts pass out of the body with feces and become the infective stage. Forty-five million cysts may be discharged in the feces of one infected person in one day.

If cysts are kept moist, they can live for a few weeks to a few months, depending on the temperature. They are killed by drying. Cysts normally enter a new host in drinking water or food. They may be carried mechanically by such insects as flies and cockroaches or by people with unclean hands.

When eaten by a new host, a cyst is carried to the small or large intestine, where it escapes the confinement of the cyst wall. The process of escape is called **excystation.** It starts with increased activity of the ameba within its cyst. Soon a pseudopodium is seen to be applied to one point on the internal surface of the cyst wall. Shortly afterwards, the tip of the pseudopodium squeezes through a minute and previously invisible pore at this spot, and it appears like a tiny hernia on the outside.

When mature, the amebas usually remain in the lumen of the proximal part of the large intestine. Not only is the pH near the mucosa more stable than that of the rest of the lumen contents, but gases and organic materials in the paramucosal lumen (that portion of the lumen immediately adjacent to the mucosa) differ from those in the center of the lumen. These physiologic strata are obviously of importance in the physiology and distribution of intestinal parasites. They help to explain the different ecologic niches of such related species as *Entamoeba histolytica* and *E. coli*—the former appears to favor a closer contact with the mucosa.

Entamoeba histolytica can be cultivated and collected in suspension, and a sterile extract can be prepared that shows a high proteolytic activity against a casein or gelatin substrate.

The relationship between bacteria and amebas is important and little understood, but intestinal bacteria apparently play an essential and beneficial role in the growth and functions of this parasite in the alimentary tract. We find statements in the literature that bacteria provide nutrients through breakdown of extracellular materials. Other statements say that bacteria protect the host. It is important to remember that these parasites can cause lesions in the brain, bone and other organs where bacteria are absent. Germ-free adult guinea pigs with *Entamoeba histolytica* grown in the absence of other organisms (axenic cultivation) do not exhibit pathogenic symptoms in the ceca. Bacteria from the same axenic strain after 15 years of cultivation, however, were able to produce lesions and abscesses in adult hamsters, but only when the amebas were inoculated directly into the liver, peritoneal cavity, or skin.[8] The results of other experiments have differed, and differences in age or strain of animals, or other experimental procedures, may have been responsible.

Among the many unanswered questions are the following: In spite of the strong cellular immunity, why does amebic infection often become pathogenic, and sometimes fatal? "The balance between the commensal state and invasiveness in *E. histolytica* infection is a delicate one. It can be tipped to either side by many factors or a combination of factors. Some of these are the host's nutrition, habitat, overall health and immune state, the bacterial flora in his intestine, the lectin and toxin-producing capacity of the amoebae, and doubtless many others as well. The host-parasite relationship in amoebiasis is among the most complex ones in parasitic infections in general. This may be the reason why it took so many years to begin to understand the intricate relationship between the numerous factors determining the outcome of an amoebic infection. All the pieces of the puzzle must be put together before the outcome of an infection can be predicted."[18] Why do pathogenic and non-pathogenic strains have different isoenzyme patterns? Why is invasive amebiasis restricted to certain geographic areas? Are there different *genetic* types of amebas that cause amebiasis? Is it possible that what we now call *E. histolytica* is actually two different species?[16] Mirelman[18a] has emphasized the question

of what should be done in cases where trophozoites with non-pathogenic zymodemes are detected in feces of symptomatic carriers.

See Turner's "Travellers' diarrhoea"[24] and Reeves on *Entamoeba* metabolism.[22]

Entamoeba hartmanni is easily mistaken for *E. histolytica*, but the cyst diameter of the former is smaller (less than 10 μm), and it is not pathogenic. There are also enzymatic and antigenic differences. **E. polecki,** from several kinds of domestic animals, rarely infects humans, and generally is not pathogenic. Pigs to people transmission could be common where people and pigs live in close association in an unsanitary environment.

Prevention of the disease involves education, proper disposal of sewage, and avoidance of fecal contamination of food, water, and eating utensils.

Entamoeba coli (Fig. 4–4) is another non-pathogenic ameba that inhabits the human intestine. It has a worldwide distribution, and an overall prevalence of about 30%, making it the most common species of intestinal ameba in man. Ten to 50% of the population have been reported to be infected in various areas. The type and thoroughness of examination have influenced the accuracy of the reports.

The life cycle of this parasite differs from that of *Entamboeba histolytica* in that *E. coli* does not enter tissues. The trophozoite of *E. coli* ranges from 20 to 30 μm in diameter. It is more sluggish than *E. histolytica*, and the cytoplasm usually appears to be much more dense and crowded with food vacuoles. Although there is little clear ectoplasm, a thin, clear area is generally seen around the nucleus of the trophozoite in stained specimens. The nucleus possesses heavier peripheral chromatin than does *E. histolytica*, and the endosome is eccentrically placed. Young cysts are apt to have a large glycogen vacuole, and they may possess few nuclei; mature cysts possess eight nuclei, each being similar to the nucelus of the motile stage. The chromatoid bodies of the cysts are usually slim with pointed or irregular ends, thus appearing different from the cigar-shaped or bar-shaped bodies in *E. histolytica*. Unlike those of *E. histolytica*, the cysts of *E. coli* are not readily killed by drying; this resistance probably accounts for the high incidence of infection. *E. coli* cysts may possibly be airborne. Monkeys and apes also share this parasite with us; pigs apparently have it, but rats seem to be free from it.

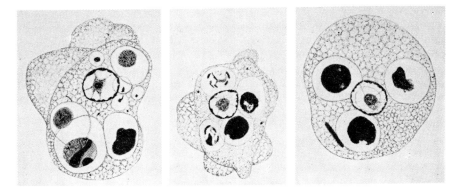

Fig. 4–7. *Entamoeba gingivalis* from monkey mouth. No cysts have been found. (Modified from Kofoid and Johnstone, courtesy of the University of California Publications in Zoology.)

Entamoeba gingivalis is a mouth inhabitant that usually is readily available for study in a classroom (Fig. 4–7). Only the trophic stage is known. The ameba lives in the gingival areas around the teeth of man, other primates, dogs, and cats. It can be gathered by gentle probing around the bases of the teeth with a toothpick. The incidence of infection is high, probably around 50%, but the organisms are more frequently associated with diseased conditions than with healthy mouths. There is no evidence, however, that the amebas cause disease. Related species occur in horses and pigs. Because there are no cysts and no intermediate hosts are involved, *E. gingivalis* represents the simplest kind of life cycle.

The size of *Entamoeba gingivalis* ranges from 5 to 35 μm in diameter, with an average of about 15 μm. In general, these amebas appear to be similar to *E. histolytica,* but usually they contain many more large food vacuoles that often enclose remnants of ingested host white blood cells. Chromatin of the nuclei of these host cells takes a deep stain, which often gives the cytoplasm of the ameba the apperance of containing several large, black nuclei. Sometimes, a definite ectoplasm may appear to be separated from the endoplasm by a deeply staining ring.

The nucleus is lined with beaded peripheral chromatin, and it contains an endosome that may consist of a single granule or, more commonly, of several closely grouped granules. Spoke-like fibrils often connect the endosome with the nuclear membrane.

Endolimax nana, an intestinal ameba of man, apes, monkeys, and pigs, averages about 10 μm in diameter in the motile stage (Fig. 4–8). It lives in the cecum and in the large intestine of 10 to 20% of the human population of most countries of the world. Although it is not pathogenic, its presence indicates that the host has been contaminated with fecal material from somebody else. Therefore, the presence of *Endolimax nana,* or of any other intestinal parasite whose method of transmission is directly from intestine to mouth, is an indicator that a pathogenic parasite may be present.

Trophozoites of *Endolimax nana* usually appear to be about half the size of those of *Entamoeba histolytica,* although there may be larger individuals. The cystoplasm often looks pale and vacuolated, and the pseudopodia are usually short and broad, with ends showing hyaline ectoplasm. Food vacuoles are often present. The single nucleus, usually not readily seen in unstained specimens, contains a large endosome that is often eccentric in position and may even lie against the nuclear membrane. Sometimes, minute threads appear to connect the endosome with this membrane. Because the nuclear membrane generally lacks the chromatin clusters or granules common to *E. coli* and *E. histolytica,* this membrane appears to be a fine line and is often barely discernible.

Cysts of *Endolimax nana* average 9 μm in diameter and are often so pale that they are difficult to see, even in stained material. Four nuclei are usually present, and they are similar to those of the motile stage, but only one or two nuclei may be present. In stained material, the cysts frequently appear to be no more than indistinct oval bodies containing four tiny, dark dots (the endosomes).

Iodamoeba buetschlii (= I. williamsi), an

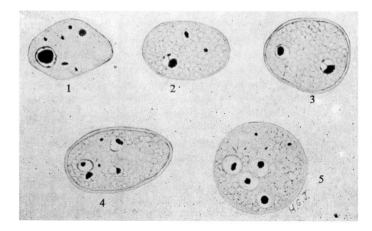

Fig. 4–8. *Endolimax nana,* a small intestinal ameba of man and monkeys. ×2500. *1,* Trophozoite with a few endoplasmic inclusions. *2, 3, 4, 5,* One-, two-, and four-nucleated cysts. Note the characteristic chromatoid bodies in *2. 5,* May be mistaken for a small cyst of *Entamoeba histolytica.* (From Anderson et al.: Amebiasis, courtesy of Charles C Thomas.)

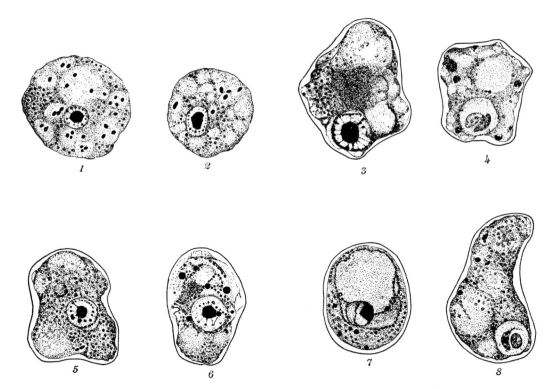

Fig. 4–9. *Iodamoeba buetschlii* from the intestine of man. All the figures on this plate were drawn by Mr. R.L. Brown from slides of a single infection.

1 to *3,* Trophozoites. *1,* Typical individual; nucleus with large, deeply stained endosome surrounded by a row of granules attached to it by radial fibrils. *2,* Small individual with larger nucleus; endosome elongated, granular ring not quite complete. *5,* Cyst of irregular contour; granules of periendosomal ring uneven in size.

6, Normal-sized cyst; note volutin-like granules in cytoplasm; one large intranuclear granule attached to endosome. *7,* Cyst with small nucleus; endosome a lateral plaque against the nuclear membrane, an unusual condition considered to be abnormal. *3,* Large cyst with periendosomal granules of nucleus arranged as a peripheral ring as in species of *Entamoeba. 4* and *8,* Cysts with lightly staining endosomes in nuclei; granules massed into crescents. (Modified from Wenrich, courtesy of the Proceedings of the American Philosophical Society.)

ameba in the large intestine (Fig. 4–9), gets its name from its large glycogen vacuole, which readily stains brown with iodine. The organism normally has a diameter of 9 to 14 μm in the motile stage in which the ectoplasm is not clearly differentiated from the endoplasm. The nucleus has little peripheral chromatin, and the large endosome is surrounded by a mass of refractile granules. In stained specimens, the endosome is dark, but the surrounding granules are usually lighter.

The cyst (about 9 μm long) is unusual because of its variable shapes. It, too, normally possesses a large glycogen vacuole. The single nucleus contains a deeply stainable endosome that usually lies against the nuclear membrane, and the granules may be clustered in the shape of a crescent near the endosome. There are many variations in the appearance of these cysts, and occasionally two nuclei are present. The ameba is probably nonpathogenic, but reports of pathogenicity exist. It has also been reported from the large intestine of monkeys, apes, and pigs, and is probably the same as *Iodamoeba suis*.

PATHOGENIC SOIL AMEBAS

Several kinds of amebas that live in soil and water can become facultative parasites in animals, including humans, and in invertebrates.[3,9] *Acanthamoeba* (Family Acanthamoebiidae) and *Hartmanella* (Family Hartmannellidae) have been reported from humans in many areas of the world.

Naegleria fowleri (Order Schizopyrenida, Family Vahlkampfiidae) has an ameboid stage (12 to 15 μm diameter) with short pseudopodia. Cysts are about 11 μm in diameter. When these amebas are suspended in water, within an hour they change into swimming flagellates, each with two flagella. For this change new RNA and protein molecules are synthesized. *N. fowleri* possess sucker-like amebostomes on the cell surface. These openings have been seen to engulf yeast cells, demonstrating a novel form of phagocytosis.[12] Mice infected with *N. fowleri* apparently are capable of infecting other mice. *N. fowleri* and *N. gruberi* have been found to be cytopathic for nine established mammalian cell cultures from several species. As few as a single *N. fowleri* per million neuroblastoma cells destroyed the mammalian target cells after 9 days.[14] Among the several other species, *N. australiensis* is highly pathogenic to mice; *N. lavaniensis* is non-pathogenic. There are some non-

pathogenic strains of *N. fowleri* that probably should be placed in the new species, *N. lavaniensis*. *N. australiensis italica* is a highly virulent strain.

Naegleria fowleri causes primary amebic meningoencephalitis (PAM). The disease is characterized by an acute hemorrhagic necrotizing meningoencephalitis, with high fever, nausea, neck rigidity, cough, and other symptoms usually associated with infections of the upper respiratory tract. The primary focus is the brain, where the olfactory lobes and cerebral cortex are most heavily damaged, owing to active phagocytosis of host tissues. Actively growing stages of *Naegleria* infect the brain by passing through the olfactory epithelium, through the cribriform plate, and along the olfactory nerve to the olfactory bulb. The parasites have been cultured from lungs, liver, spleen, and brains of human victims. Death usually occurs 3 to 7 days from onset of symptoms. The virulence factors relating to pathogenesis, and the determinative factors in host resistance, are not clear. Immunization attempts have been partially successful. There is no satisfactory treatment.

About 130 worldwide cases of PAM caused by *Naegleria fowleri* have been reported, and only three survivals were well documented. The victims were generally young adults and children, and the sources of infection were traced to swimming pools (including indoor pools filled with chlorinated water), warm ponds, and streams. Airborne infections by cysts in dust are possible. Virtually every type of water system has been implicated as a source of pathogenic soil amebas. For further considerations of soil amebas see Culbertson and Harper,[5] John,[11] Martinez,[15] and Warhurst.[26]

Acanthamoeba is a ubiquitous free-living genus, world-wide in distribution. It has been isolated from fresh water, well water, brackish water, soil, sewers, hot tubs, oceans, sewage, feces of domestic animals, wheat, barley, and London air. At least five species have produced a fatal, granulomatous encephalitis in predominantly chronically ill, debilitated, or immunosuppressed persons.[12a] These amebas have been implicated (rarely) in skin infections, pneumonitis, external otitis, and osteomyelitis in humans. *Acanthamoeba* has recently commanded wide publicity because of the devastating effects of its invasion of the human eye. In the past 3 years approximately 70 to 80 cases of Acanthamoeba keratitis have been reported. Most of

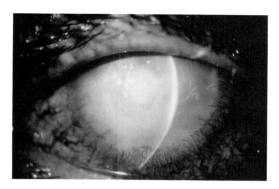

Fig. 4–10. The left eye of a 20-year-old male student with advanced *Acanthamoeba* keratitis. The patient is essentially left without any vision other than light perception. (Courtesy of Dr. Paul B. Donzis).

these were associated with contact lenses, soft lens wearers being more prone to infection than hard contact lens wearers. Damage to eyes (Fig. 4–10) includes collection of inflammatory cells around corneal nerves, nodular scleritis, severe iritis, necrotizing stromal suppuration, corneal perforation, and blindness. If diagnosed early, the disease appears to be curable. There have been approximately 12 cases of medical cures. Parasite cysts (10 to 26 μg) are highly resistant to the disinfectants and cleansing materials commonly used for contact lens. "Only boiling the lenses appears to be able to kill both *Acanthamoeba* trophozoite and cyst form." (Dr. Paul Donzis—personal communication.) This genus has also been found in invertebrates such as oysters, grasshoppers, and snails.

Many other free-living or coprophilous species have been described. *Flabellula calkinsi* and *Vahlkampfia patuxent* commonly live in the gut of oysters. The former species also invades other host organs. Coprophilous amebas are often found in the intestines of mammals. All these amebas, including the well-known genus *Amoeba*, possess large endosomes in the nucleus (the *limax* type), and all may be found in soil. *Sappinia diploidea*, sometimes occurring in great numbers in old ungulate feces, has two nuclei. See Chapter 26 for comments on the evolution of the Sarcodina.

IMMUNITY

Most animals and people apparently possess a natural resistance to amebic infection, as indicated by the large percentage of infected hosts without amebiasis. This protection against dis-ease is probably due to adjustment between parasite and host and does not necessarily depend on antigen-antibody responses. A large spectrum of specific host antibodies is induced by the amebas, especially when tissues are invaded. The existence of nonpathogenic strains of *E. histolytica* complicates the issue.[16,16a]

A protective role for parasite-induced antibodies is still questionable, but Meerovitch[18] has stated that work with experimental animals indicates the existence of acquired immunity. Cell-mediated immunity and the development of IgE and IgG antibody are involved. The amebic plasma-lemma is only weakly immunogenic, and antibodies are produced in response to amebic internal cytoplasmic antigens. In this manner, the most immune potential is diverted away from invading intact amebas toward those that have died and disintegrated in the tissues, and thus has no protective function. Aust-Kettis and Sundqvist[2] have suggested another way by which these parasites may escape immune recognition. They may alter their surface antigenic composition through either endocytosing or shedding membrane antibody-antigen complexes and reexpressing new surface antigens.

Serologic diagnosis of amebiasis has been successful using various techniques, especially the indirect hemagglutination test. Hemagglutination and precipitation tests yield a high percentage of positive results. However, because amebic antibodies can remain long after the parasites have disappeared, some problems have been encountered when serodiagnosing *active* amebic infections. Moreover, tests must be interpreted with regard to the background of antibodies ("noise"), which varies in different geographic areas. Skin test results have shown delayed hypersensitivity in human patients.

Fluorescent antibody tests are becoming increasingly popular. They involve coupling a fluorescent dye to an antibody, thereby making the antibody-antigen reaction visible. Relationships among species of *Entamoeba* have been clarified by such techniques. See Kagan[13] and Patterson et al.[20]

Blastocystis hominis, originally thought to be a yeast in the intestines of humans and other mammals, is a pathogenic sercodine parasite.

REFERENCES

1. Albach, R.A., and Booden, T.: Amoebae. *In* Parasitic Protozoa. Vol. 2. Edited by J. P. Kreier. New York, Academic Press, 1978, pp. 455–506.
2. Aust-Kettis, A., and Sundqvist, K.-G.: Dynamics of the interaction between *Entamoeba histolytica*

and components of the immune response. Scand. J. Immunol., 7:35–44, 1978.

3. Band, N.: The biology of small amoebae. Introductory remarks of the chairman. J. Protozool., 30:192–198, 1983.

4. Bovee, E.C.: Class Lobosea Carpenter, 1861. In An Illustrated Guide to the Protozoa. Edited by J.J. Lee, S.H. Hutner, and E.C. Bovee. Lawrence, Kansas, Soc. of Protozoologists, 1985, pp. 158–211.

5. Culbertson, C.G., and Harper, K.: Pathogenic free-living amoebae. J. Trop. Med. Hyg., 33(5):851–856, 1984.

6. Diamond, L.S., Meryman, H.T., and Kafig, E.: Preservation of parasitic protozoa in liquid nitrogen. In Culture Collections: Perspectives and Problems. Edited by S.M. Martin. Toronto, University of Toronto Press, 1963, pp. 189–192.

7. Geldermann, A.H., Keister, D.B., Bartgis, I.L., and Diamond, L.S.: Characterization of the deoxyribonucleic acid of representative strains of Entamoeba histolytica, E. histolytica-like amebae, and E. moshkovskii. J. Parasitol., 71:906–911, 1971.

8. Ghadirian, E., and Meerovitch, E.: Pathogenicity of axenically cultivated Entamoeba histolytica, strain 200:NIH, in the hamster. J. Parasitol., 65:768–771, 1979.

9. Griffin, J.L.: Pathogenic free-living amoebae. In Parasitic Protozoa. Vol. 2. Edited by J.P. Kreier. New York, Academic Press, 1978, pp. 507–549.

10. Guerrant, R.L.: The global problem of amebiasis: current status, research needs, and opportunities for progress. Rev. Inf. Dis., 8(2):218–227, 1986.

11. John, D.T.: Primary amebic meningoencephalitis and the biology of Naegleria fowleri. Ann. Rev. Microbiol., 36:101–123, 1982.

12. John, D.T., Cole, T.B., and Bruner, R.A.: Amebostomes of Naegleria fowleri. Report No. 17 at Ann. Meetings of Am. Soc. Parasitol., Aug. 1984.

12a. Jones, D.B.: Acanthamoeba—the ultimate opportunist? Am. J. Ophthalmol., 102(4):527–530, 1986.

13. Kagan, I.G.: Serodiagnosis of parasitic diseases. In Manual of Clinical Parasitology. 3rd Ed. Edited by E.H. Lennette, A. Balows, W.J. Hausler, and J.P. Truant. Washington, D.C., American Society for Microbiology, 1980, pp. 724–750.

14. Marciano-Cabral, F.M., Patterson, M., John, D.T., and Bradley, S.G.: Cytopathogenicity of Naegleria fowleri and Naegleria gruberi for estab-lished mammalian cell cultures. J. Parasitol., 68:1110–1116, 1982.

15. Martinez, A.J.: Free-living Amebas: Natural History, Prevention, Diagnosis, Pathology, and Treatment of Disease. Boca Raton, Florida, CRC Press, Inc., 1985.

16. Martínez-Palomo, A.: Biology of Entamoeba histolytica. In Amebiasis. Edited by A. Martinez-Palomo. New York, Elsevier, 1986.

16a. Martinez-Palomo, A.: The pathogenesis of amoebiasis. Parasitol. Today, 3:111–118, 1987.

17. McLaughlin, J., and Aley, S.: The biochemistry and functional morphology of the Entamoeba. J. Protozool., 32:221–240, 1985.

18. Meerovitch, E.: The jigsaw puzzle of host-parasite relations in amoebiasis begins to take shape. In Aspects of Parasitology. Edited by E. Meerovitch. Montreal, Institute of Parasitology, 1982, pp. 263–278.

18a. Mirelman, D.: Effect of culture and bacterial associates on the zymodemes of Entamoeba histolytica. Parasitol. Today, 3:37–40, 1987.

19. Noble, G.A., and Noble, E.R.: Entamoebae in farm mammals. J. Parasitol., 38:571–595, 1952.

20. Patterson, M., Healy, G.R., and Shabot, J.M.: Serologic testing for amoebiasis. Gastroenterology, 78:136–141, 1980.

21. Ravdin, J.I.: Pathogenesis of disease caused by Entamoeba histolytica: studies of adherence, secreted toxins, and contact-dependent cytolysis. Rev. Inf. Dis., 8(2):247–260, 1986.

22. Reeves, R.E.: Metabolism of Entamoeba histolytica Schaudinn, 1903. Adv. Parasitol., 23:106–142, 1984.

23. Scaglia, M., et al.: Entamoeba moshkovskii (Tshalaia, 1941): Morpho-biological characterization of new strains isolated from the environment, and a review of the literature. Ann. Parasitol. Hum. Comp., 58(5):413–422, 1983.

24. Turner, A.C.: Travellers' diarrhoea. Ann. Soc. Belg. Med. Trop., 59:109–115, 1979.

25. Walsh, J.A.: Problems in recognition and diagnosis of amebiasis: estimation of the global magnitude of mordibity and mortality. Rev. Inf. Dis., 8(2):228–238, 1986.

26. Warhurst, D.C.: Pathogenic free-living amoebae. Parasitol. Today, 1(1):24–28, 1985.

27. WHO Meeting: Amoebiasis and its control. Bull. WHO, 63(3):417–426. Prepared by A. Davis and Z.S. Pawlowski, 1985.

Phylum Apicomplexa

There are over 300 named genera and about 5000 named species in this phylum. We can mention only a small sample of this mass of parasite varieties. The phylum is characterized by complex apical organelles (Fig. 5–1) generally consisting of a conoid that, in gregarines, aids in attachment, but its function in the other parasites is unknown; by polar ring; by rhoptries that are probably secretory in function; and by subpellicular microtubules that may be related to locomotion. Cilia are absent, but the microgametes of some groups have flagella. Cysts are often present and sexual reproduction is by syngamy.

Host cell invasion by the Apicomplexa consists of three separate phases: (1) attachment of the parasite to the host cell; (2) invagination of the host cell plasmalemma, forming a parasitophorous vacuole; and (3) passage of the parasite into the vacuole. The mechanism of entry is still not completely understood. Several workers have proposed that parasite entrance results in a "capping" of the host-parasite membrane/membrane junction down the parasite body. In general, the taxonomic system used in this chapter is that proposed by Levine (1985).[87] See Canning[25] for a review of terminology, taxonomy and life cycles. This phylum is also called the **Sporozoa.**

Class Perkinsea

This class is characterized by flagellated 'zoospores' (sporozoites) with an anterior vacuole; a conoid that forms an incomplete truncated cone; lack of sexuality; and a homoxenous life cycle. A single species, *Perkinsus marinus.*

Class Sporozoea

The conoid, if present, forms a complete truncated cone. Reproduction is usually sexual and

asexual. Oocysts generally contain infective sporozoites that result from sporogony. Endodyogeny, a type of internal budding in which two daughter cells are formed inside the parent cell whose cell wall then disappears, is a major form of cell multiplication. Locomotion is by body flexion, gliding, or undulations of longitudinal ridges. Pseudopods, if present, are used only for feeding. These parasites are homoxenous or heteroxenous.

Syzygy (see p. 80) of gamonts (sporadins) usually occurs. Life cycles generally consist of gametogony and sporogony. Zygotes form oocysts within gametocysts. About 1500 species are described.

Subclass Gregarinia

The gregarines comprise a large group of parasites limited to invertebrates and lower chordates in which they occupy both tissue cells and body cavities. Mature gamonts (destined to form gametes) are extracellular. Zygotes form oocysts within gametocysts. About 1500 species have been described.

Gregarines within host cells are only a few micrometers in diameter, but species in body cavities may be much larger, the maximum being about 10 mm in length. In some species, the body is divided by a septum into two main parts: an anterior **protomerite** and a larger, posterior part, the **deutomerite,** which contains the nucleus. These types are **cephaline** gregarines. The **protomerite** sometimes possesses an anterior anchoring device, the **epimerite** (=mucron) (Fig. 5–2), formed from the conoid left in host tissue when the parasite breaks away. A chemicomechanical transduction is probably involved in the mechanism of gliding locomotion. Hosts become infected by swallowing spores. See Manwell,[94] Levine.[86]

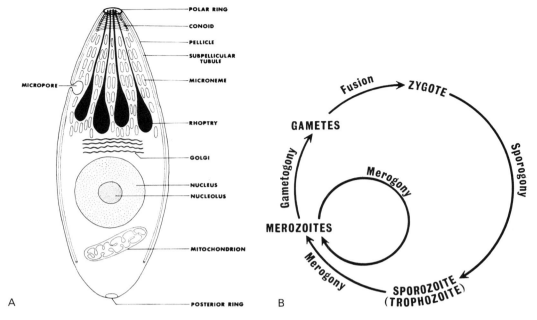

Fig. 5–1. *A,* Typical Apicomplexa species. *B,* Apicomplexa life cycle. (After Levine, 1974. Actualités Protozoologiques. Courtesy of the Université de Clermont.)

Order Archigregarinida

These parasites are found in the intestinal tracts of marine annelids, sipunculids, hemichordates, and ascidians. In their apparently primitive life cycles, the trophozoites undergo three multiplications: **merogony, gametogony, sporogony.** The nucleus divides a few or many times without division of the entire cell. When

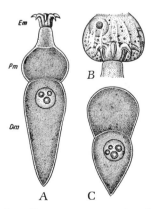

Fig. 5–2. *Corycella armata,* a gregarine showing the epimerite, *Em;* protomerite, *Pm;* and deutomerite, *Dm. A,* Entire parasite. *B,* Epimerite anchored in host cell. *C,*Gamont detached from epimerite. (From Grell: Protozoologie. Courtesy of Springer-Verlag.)

the cell does divide, there are as many daughter cells as there were nuclei. The trophozoites are called schizonts. When schizogony results in merozoites that repeat the cycle, the process is merogony. *Exoschizon* and *Selenidioides* are typical.

Order Neogregarinida

These gregarines are parasitic in malpighian tubules, intestines, hemocoel, and fat tissues of insects. They have a schizogonic phase (merogony) in their life cycles, and each spore forms eight **sporozoites.** *Mattesia* and *Schizocystis* are typical genera.

Order Eugregarinida

Sporozoites enter the host and develop into mature trophozoites without merogony. Trophozoites become gamonts that unite in pairs, forming a union called **syzygy.** Two (sometimes more) gamonts become united end to end. The anterior one is the primite, while the posterior one is the satellite. They then flatten against each other, secrete a cyst membrane around themselves, thus forming the **gametocyst.** The gamonts look alike, but they are physiologically different. The nucleus of each undergoes multiple divisions, and the resulting many nuclei move to the inside wall of the cyst. Then each one is budded off with a bit of cytoplasm and becomes a gamete. The gametes are morphologically the same, thus are called isogametes. Male and female gametes unite to form zygotes, each of which secretes a membrane around itself. Three divisions of a zygote (sporogony) result in 8 sporozoites.

The cephaline gregarines (Septatina) often do not enter host tissue cells, but the acephaline sporozoites are usually intracellular parasites. They often grow to a large size and protrude from the host cells. If they are in intestinal cells, they finally leave the gut lining and move freely in the intestinal cavity. Eugregarines are most commonly found in annelids, arthropods, and occasionally, in lower chordates.

Suborder Aseptatina

This suborder consists of nonseptate parasites living mostly in the coelom of the host.

Monocystis lumbrici (Fig. 5–3) lives in the seminal vesicles of the earthworm, *Lumbricus terrestris,* and in related worms. Gamonts are elongated, measuring about 200 by 65 μm. Infection of the worm occurs by ingestion of ma-

ture spores. In the gut of the worm, sporozoites are liberated; they penetrate the gut wall, migrate through the body, and enter the seminal vesicles. Here the young trophozoites enter tissue cells and increase in size. Older trophozoites become free from host tissue and migrate to the sperm funnel, where they become attached again to the host cells. Two of them join and secrete a wall around themselves, becoming a cyst. Within this cyst, gametes are formed and produce zygotes (sometimes called sporoblasts). Sporoblasts become spores, and the cycle is completed.

Aseptate gregarines have been found in many oligochaete annelids, sea cucumbers, nemerteans, insects, and many other invertebrates. *Lankesteria* in ascidians, *Selenidium* in the gut of polychaetes and sipunculids, and *Urospora* in annelids, sipunculids, nemertines, and echinoderms are common genera.

Suborder Septatina

This suborder is found primarily in the alimentary canal of arthropods and worms. The gamont is divided by a septum into a protomerite and a deutomerite. An epimerite is pres-

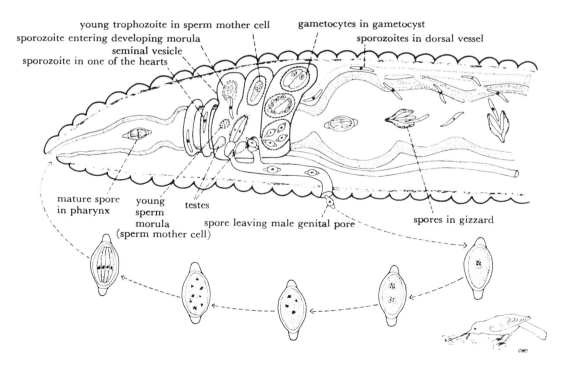

Fig. 5–3. Earthworm *(Lumbricus terrestris),* host of *Monocystis,* showing endogenous phase of life cycle of parasite. (From Olsen: Animal Parasites. Minneapolis, Burgess Publishing, 1967.)

ent. A typical life cycle is that of *Stylocephalus longicollis* (Fig. 5–4), a parasite of the beetle *Blaps mortisaga*. In some species, the gametocyst must pass out of the host body, it must dehisce, and the sporocyst must be eaten before infection can occur.

Gregarina garnhami occupies the intestinal ceca and midgut of the migratory locust *Schistocerca gregaria*. The protozoa sometimes destroy considerable areas of the cecal epithelium. Parasites may occur by the hundreds, in such masses that there is a barrier between food material in the gut lumen and the gut wall. Gregarines occur in both nymph and adult locusts.

Gregarina blattarum is a common species that lives in cockroaches (*Blatta orientalis* and others). Gamonts are enormous, reaching 110 μm; larger ones are easily seen with the naked eye. Cyst contents reach the outside through eight to ten sporoducts. The life cycle is basically similar to that of *Stylocephalus longicollis*.

Among other numerous genera are *Actinocephalus* and *Stylocephalus*, both in insects.

Subclass Coccidia

These intracellular parasites occur primarily in vertebrates. They enter epithelial cells and may cause considerable damage, often ending in the death of the host. Small gamonts are generally present. The conoid is not modified into an epimerite or mucron. Merogony, gametogony, and sporogony are characteristic of the life cycle. Only a few representative families are described here. See Long[91] for the biology of Coccidia.

Life cycles of coccidians generally include the following kinds of reproduction. Sporozoites enter host cells and multiply asexually (nuclear division) by schizogony. Resulting schizonts are released from host cells and usually are immobile. Motile schizonts enter other host cells and develop into merozoites by endodyogeny (see Glossary). Generations of merozoites grow within and destroy host cells. Merozoite formation is called merogony. Merozoites have all the morphology typical of Apicomplexa (Fig. 5–1A).[31] Some of them develop into gametocytes in host cells. This development begins the sexual gametogony phase of the life cycle. Gametogony ends with formation of female macrogametocytes and male microgametocytes. The latter leave host cells and enter cells containing macrogametocytes. Fertilization takes place, and the resulting zygote, forming a pro-

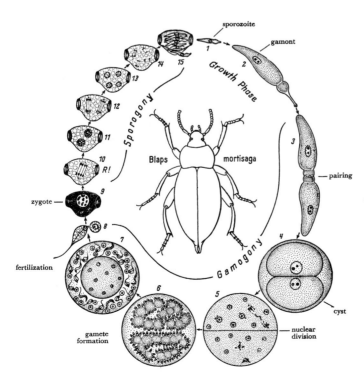

Fig. 5–4. The life cycle of the cephaline gregarine, *Stylocephalus longicollis*, a parasite of the intestine of the beetle, *Blaps mortisaga*. (From Grell, courtesy of Protozoologie.)

tective wall about itself, is the oocyst that begins the sporogony phase of the cycle. Sporocysts develop within oocysts, spores within sporocysts, and sporozoites within spores. Sporogony often takes place in another host (the "final" or "definitive" host). The oocyst, with its spores, is the infective stage acquired by a new host. See Figures 5–8 and 5–10. See Long on the biology of the coccidia.[91]

Order Eucocciida

This Order contains the **Coccidia**, a group consisting of more than 55 genera and about 2,000 species. A few are found in invertebrates, but most of them are intracellular parasites of vertebrates. Life cycles involve both asexual and sexual development. The zygotes are diploid; all other stages are haploid. Many are highly pathogenic. In most species the uninucleated oocyst passes from the host, and sporulation takes place outside the host. The taxonomy of coccidia needs much more study.

"A working hypothesis to account for the membrane changes during coccidial infection has been formulated. Briefly, as the infection progresses, gel-phase lipid accumulates and concomitantly the host cell becomes leaky. At this stage, exogenous trypsin could enter the infected cell and trigger the release of endogenous degradative enzymes that eventually lyse the infected cell from within. . . In coccidia-infected cells, the gel-phase lipid of the host cell membrane may create a structural alteration that maximizes the activities of various degradative enzymes. This sequence of events would ensure that schizonts are not released prematurely and that the host cell membrane becomes susceptible to lysis without the direct intervention of an elaborate mechanism that is governed by the parasite. This general view stresses that the induction and accumulation of gel-phase lipid plays an integral part in the infection cycle and is not merely the consequence of a pathogenic condition. At present, it is not known what causes the development of the gel-phase lipid."[114] These parasites are found in blood cells and epithelial cells of vertebrates and invertebrates.

Suborder Adeleina

FAMILY ADELEIDAE

Members of this family parasitize intestinal epithelium and associated glands of invertebrates. Occasionally, they occur in vertebrates.

Adelina deronis lives in peritoneal cells of the annelid worm, *Dero limnosa*. Oocysts are taken into the digestive tract of the worm, where its sporozoites are liberated. Sporozoites make their way to peritoneal cells, in which they become trophozoites. The trophozoites, in turn, may become attached end to end, thus forming a chain of individuals.

Other genera of the family include *Adelea, Klossia, Karyolysus,* and *Orcheobius*. Each species varies in the numbers of sporocysts and sporozoites that are produced from the oocysts and in the hosts and tissues that are parasitized. Sporogony of *Karyolysus lacertarum* takes place in a lizard, while schizogony and gametogony occurs in a mite (Fig. 5–5).

FAMILY HAEMOGREGARINIDAE

This group of coccidia possesses life cycles involving two hosts: a vertebrate host in which the parasite lives in the circulatory system, and an invertebrate host involving the digestive system. Oocysts are small, and there are no sporocysts. The zygote is an active ookinete.

Locomotion is accomplished by a gliding movement due to a series of "peristaltic" waves. Apparently, it involves a contractile element in the cytoplasm and differs from gregarine locomotion. See Manwell.[94]

Haemogregarina stepanowi begins its life cycle as a zygote (ookinete) in the gut of the leech, *Placobdella catenigera*. The zygote divides three times and forms eight sporozoites. When an infected leech takes blood from the turtle, *Emys orbicularis*, the sporozoites pass to the vertebrate host and enter red blood cells. There merogony occurs and produces merozoites that are liberated from the host cells and infect new red blood cells. Some of the merozoites eventually develop into gametocytes. If these cells are ingested by a leech, the gametocytes are liberated, mature to gametes, and produce zygotes.

Haemogregarines of fish may enter leukocytes as well as erythrocytes, and they can be separated into two broad categories: haemogregarines (without asexual multiplication in red blood cells) and schizohaemogregarines (with schizogony occurring in red blood cells). In spite of the large numbers of described species, the method of transmission from one fish to another is little understood, and evidence suggests that leeches are usually intermediate hosts,[96] but an isopod appears to be the intermediate host for *Haemogregania bigemina* in the

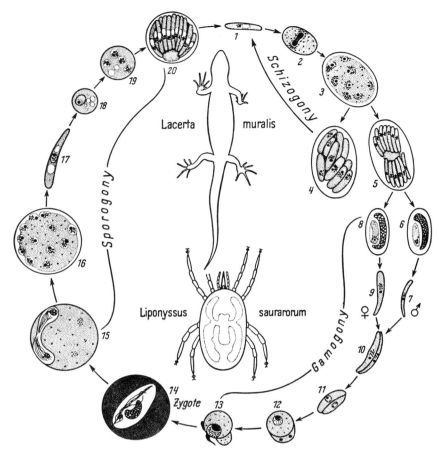

Fig. 5–5. The life of *Karyolysus lacertarum*. Sporogony occurs in a lizard, whereas schizogony and gamogony occur in a mite. (From Grell, courtesy of Protozoologie.)

marine *Blennius pholis*. Haemogregarines of reptiles are transmitted by the ingestion or possibly blood-sucking activity of infected arthropod vectors. Merozoites have been found in tissues of lungs, liver and spleen of lizards. Many species of *Hepatozoon* occur in the red blood cells or leukocytes of reptiles, birds, and mammals. In the vector, the oocyst is large and contains numerous sporocysts in which there are many sporozoites. The parasite is transmitted by flies, mosquitoes, lice, ticks, and mites, in which the sexual stages occur.

Suborder Eimeriina

Microgametes and microgamonts develop independently. A microgamont usually produces many microgametes. Sporozoites typically are enclosed in a sporocyst.

FAMILY LANKESTERELLIDAE

Oocysts with or without sporocysts; with eight or more sporozoites. Merogony, gametogony, and sporogony occur in the same vertebrate host. Sporozoites within red blood cells are transferred by mites, leeches, or mosquitoes.

Lankesterella occurs in amphibians and birds (Fig. 5–6). Schizogony, gametogony, and sporogony all take place in tissue cells (e.g., macrophage cells of spleen, liver, lungs, kidneys) of the same vertebrate host. Both **endodyogeny** and **endopolyogeny** occur (see glossary). Sporozoites, but not sporocysts, develop from oocysts, and they enter blood cells. Within these cells the parasites apparently ingest material from parasitophorous vacuoles by pinocytosis. These infected cells may be ingested by

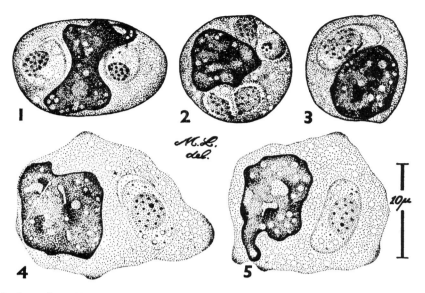

Fig. 5–6. *Lankesterella paddae* from several species of birds (family Zosteropidae, or Silver Eyes) from the South Pacific. *1* and *2*, Division stages, lymphocytes of *Gallirallus australis scotti*, New Zealand. *3*, Nondividing form, lymphocyte of *Zosterops lateralis lateralis*, New Zealand. *4* and *5*, Nondividing forms, monocytes of *Zosterops lateralis griseonota*, New Caledonia, and *Z. flavifrons flavifrons*, Futuna, respectively. (From Laird, courtesy of J. Parasitol.)

the transport hosts (leeches), in which the parasites do not develop. Infections in birds may be seriously pathogenic.

Schellackia in lizards and birds undergoes schizogony and gametogony in intestinal cells, and sporozoites enter erythrocytes or leuckocytes. Transmission is by mites. **Atoxoplasma** in bird leukocytes is transmitted by mites.

FAMILY AGGREGATIDAE

Schizogony takes place in cells of one host, and most of the gamogony and sporogony occur in another host. Marine annelids, crustaceans, and molluscs serve as hosts. Oocysts typically contain many sporozoites.

In the life cycle of **Aggregata eberthi** (Fig. 5–7), schizogony occurs in the crab, *Portunus depurator*, and both sporogony and gamogony in the cuttlefish, *Sepia officinalis*. Sporozoites develop into merozoites in the intestinal connective tissue cells of the crustacean host. When a cuttlefish eats an infected crab, the merozoites develop into gametocytes in the gut wall of the mollusc. Gametes are released, unite, and form zygotes. Sporoblasts, spores, and finally sporozoites develop from zygotes. Crabs eat the spores, and the sporozoites enter intestinal cells.

FAMILY CRYPTOSPORIIDAE

This family contains two genera: *Epieimeria*, a parasite of fish, and *Cryptosporidium*, occurring in a variety of vertebrates.[131] Both are most commonly found in the brush border of host epithelial cells, primarily of the intestine. Levine has considered only 4 of the 19 or 20 described species of *Cryptosporidium* to be valid. They are: *C. muris* in mammals, *C. meleagridis* in birds, *C. crotalis* in reptiles, and *C. nasorum* in fish. There is strong evidence that *C. parvum* from calves is also a valid species. *C. baileyi* has recently been described from chickens. There is a lack of host specificity of animal (including human) strains. In birds and humans the respiratory tract may also be infected, as well as the conjunctiva in birds. The life cycle of *Cryptosporidium* is similar to that of *Eimeria*, *Toxoplasma*, and other intestinal coccidians. Its oocyst has four sporozoites. Meronts are located outside the host cell proper. The genus *Cryptocystidium* has been used for species in fish.

The most prominent clinical sign of illness in livestock (especially calves) is diarrhea, which is usually profuse, yellow, and watery. Other results of infection are weight loss, depression, fever, dehydration, and death. Autoinfection may occur. Cryptosporidiosis appears to be a cosmopolitan disease. In humans, it is recognized as a zoonosis, and may cause severe dehydrating diarrhea and fever. There have been outbreaks among children in day-care centers, suggesting person-to-person transmission of the disease. Oocysts have been found in surface waters, some being used for recreational purposes, implying an important source of infection. "The presence of thin-walled autoinfective oocysts and recycling of type I meronts may explain why a small oral inoculum can produce an overwhelming infection in a suitable host and why immune deficient persons can have persistent, life-threatening cryptosporidiosis in the absence of repeated oral exposure to thick-walled oocysts."[42] For further information see Angus,[3] Anderson,[4] Kirkpatrick and Farrell,[79] Levine.[86]

FAMILY CALYPTOSPORIDAE

Calyptospora requires a crustacean intermediate host. Each sporocyst is surrounded by a membranous veil, and has an anterior apical opening. The genus *Goussia* is tentatively as-signed to this family. These coccidia occur in liver tissue of fish. *C. funduli*, a parasite of Cyp-prinodontidae, is transmitted by the shrimp, *Palaemonetes pugio.*[113]

FAMILY EIMERIIDAE

This family contains at least 15 genera and over 1300 described species, some of which have considerable economic importance. Eimerid sporozoan infections are known as coccidiasis, and the disease is known as coccidiosis. *Isospora* and *Eimeria*, have been found in the feces of chickens, turkeys, cattle, sheep, rabbits, fish, reptiles and many other animals, including man. The life cycle generally takes place in the intestine and is shown in Figure 5–8. It involves only one host. The developmental stages of *Eimeria*, and most other coccidia, are located in a parasitophorous vacuole which is a reservoir of nutrients for the parasite. During the process of invasion by the parasite the host cell plasmalemma is commonly invaginated and pinched off to form the vacuole around the parasite. Some investigators have reported that the plasmalemma is perforated as the parasite enters, and then a vacuole is formed by strands of the endoplasmic reticulum. Asexual devel-

LEGEND FOR PLATE I.

Stages of *Plasmodium vivax* in human erythrocytes from thin-film preparation.
1. Normal sized red cell with marginal ring form trophozoite.
2. Young signet ring form trophozoite in a macrocyte.
3. Slightly older ring form trophozoite in red cell showing basophilic stippling.
4. Polychromatophilic red cell containing young tertian parasite with pseudopodia.
5. Ring form trophozoite showing pigment in cytoplasm, in an enlarged cell containing Schüffner's stippling.*
6, 7. Tenuous medium trophozoite forms.
8. Three ameboid trophozoites with fused cytoplasm.
9, 11, 12, 13. Older ameboid trophozoites in process of development.
10. Two ameboid trophozoites in one cell.
14. Mature trophozoite.
15. Mature trophozoite with chromatin apparently in process of division.
16, 17, 18, 19. Schizonts showing progressive steps in division ("presegmenting schizonts").
20. Mature schizont.
21, 22. Developing gametocytes.
23. Mature microgametocyte.
24. Mature macrogametocyte.

*Schüffner's stippling does not appear in all cells containing the growing and older forms of *P. vivax* as would be indicated by these pictures, but it can be found with any stage from the young ring form onward.
(From Wilcox, A.: Manual for the Microscopical Diagnosis of Malaria in Man. National Institutes of Health, Bulletin No 180. Bethesda, MD, National Institutes of Health.)

PLATE I

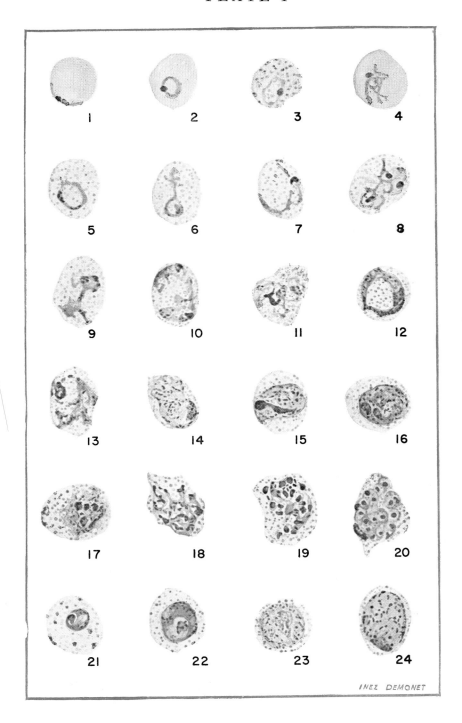

INEZ DEMONET

Stages of *Plasmodium malariae* in human erythrocytes from thin-film preparation.
1. Young ring form trophozoite of quartan malaria.
2, 3, 4. Young trophozoite forms of the parasite showing gradual increase of chromatin and cytoplasm.
5. Developing ring form trophozoite showing pigment granule.
6. Early band form trophozoite—elongated chromatin, some pigment apparent.
7, 8, 9, 10, 11, 12. Some forms which the developing trophozoite of quartan may take.
13, 14. Mature trophozoites—one a band form.
15, 16, 17, 18, 19. Phases in the development of the schizont ("presegmenting schizonts").
20. Mature schizont.
21. Immature microgametocyte.
22. Immature macrogametocyte.
23. Mature microgametocyte.
24. Mature macrogametocyte.

(From the Manual for the Microscopical Diagnosis of Malaria in Man. National Institutes of Health Bulletin No. 180. By Aimee Wilcox.)

PLATE II

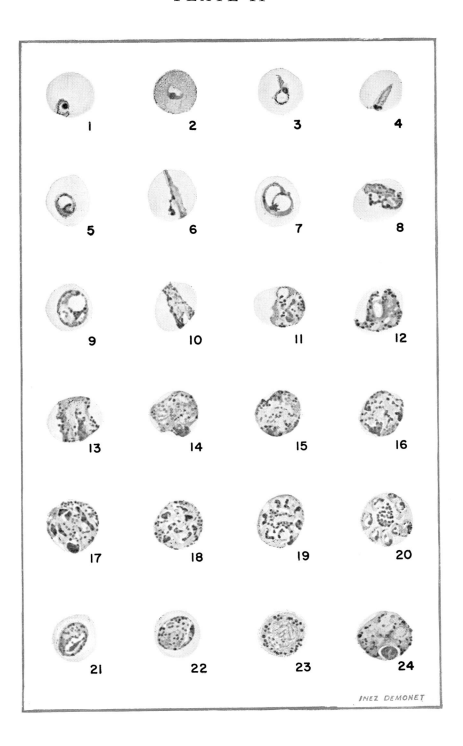

INEZ DEMONET

Stages of *Plasmodium falciparum* in human erythrocytes from thin-film preparation.

1. Very young ring form trophozoite.

2. Double infection of single cell with young trophozoites, one a "marginal form," the other, "signet ring" form.

3, 4. Young trophozoites showing double chromatin dots.

5, 6, 7. Developing trophozoite forms.

8. Three medium trophozoites in one cell.

9. Trophozoite showing pigment, in a cell containing Maurer's spots.

10, 11. Two trophozoites in each of two cells, showing variation of forms which parasites may assume.

12. Almost mature trophozoite showing haze of pigment throughout cytoplasm; Maurer's spots in the cell.

13. Estivo-autumnal "slender forms."

14. Mature trophozoite, showing clumped pigment.

15. Parasite in the process of initial chromatin division.

16, 17, 18, 19. Various phases of the development of the schizont ("presegmenting schizonts").

20. Mature schizont.

21, 22, 23, 24. Successive forms in the development of the gametocyte—usually not found in the peripheral circulation.

25. Immature macrogametocyte.

26. Mature macrogametocyte.

27. Immature microgametocyte.

28. Mature microgametocyte.

(From the Manual for the Microscopical Diagnosis of Malaria in Man. National Institutes of Health Bulletin No. 180. By Aimee Wilcox.)

PLATE III

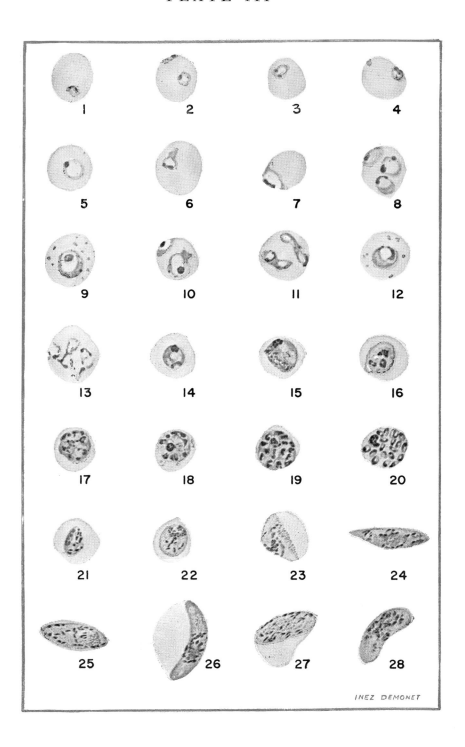

INEZ DEMONET

Legend for Plate IV.

Plasmodium ovale and *Plasmodium simium.* All preparations stained with Giemsa, except 13–14 (hematoxylin).

1–15. *Plasmodium ovale.*
1–6. Schizogony in erythrocytes.
7. Immature gametocyte.
8. Microgametocyte ($\times$ 1900).
9. Macrogametocyte ($\times$ 1900).
10. "Relapse" schizont.
11. Exflagellation.
12. Ookinete.

13, 14. Oocysts.
15. Sporozoite.
16–25. *Plasmodium simium.*
16–19. Ring forms ($\times$ 1900).
20–22. Schizonts ($\times$ 1900).
23. Immature gametocyte ($\times$ 1900).
24. Macrogametocyte ($\times$ 1900).
25. Microgametocyte ($\times$ 1900).

(From Garnham, P. C. C.: Malaria Parasites and Other Haemosporidia. Courtesy of Blackwell Scientific Publications.) (Figs. 1–7, 11–15, from James *et al.*, courtesy of Parasitology.)

PLATE IV

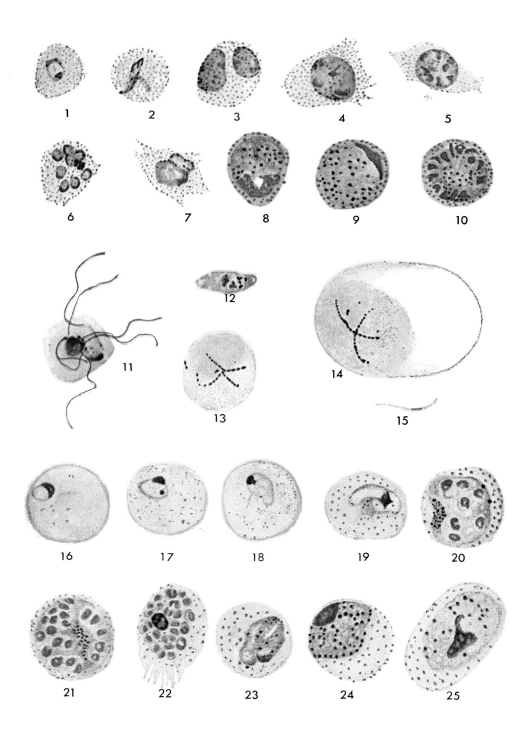

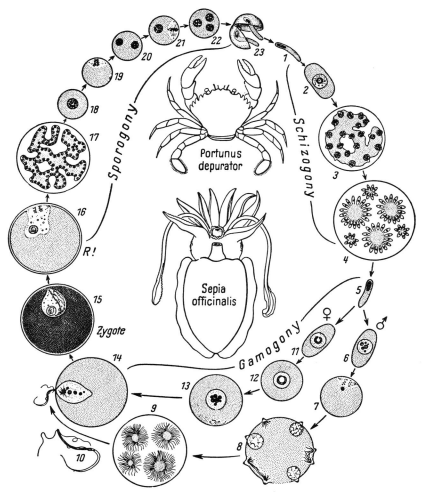

Fig. 5–7. *Aggregata eberthi* showing its life cycle, which include crabs, and cuttlefish. Schizogony occurs in the intestine of the crab, whereas sporogony and gamogony occur in the submucosal cells of the intestine of the cuttlefish. (From Grell, courtesy of Protozoologie.)

opment typically begins with endodyogeny in intestinal cells. The oocysts are shed unsporulated, and they sporulate within a few days upon exposure to oxygen and in a moist environment.

During the life cycles of *Eimeria necatrix* and *E. tenella* an obligatory third schizogony generation occurs. From merozoites of this generation the gametocytes are formed. Merozoites contain a nucleus with nucleolus, chonoid, rhoptries, micronemes, mitochondria, Golgi complex, and polysaccharide granules. Bile plays a role in excystation of sporozoites and in development of asexual stages. The life cycles of this family generally involve only one host.

The genus *Caryospora*[132] has a heteroxenous cycle involving reptiles and birds as primary hosts and mammals (e.g., rodents) as secondary hosts. *Eimeria funduli*, living in hepatic cells of the fish, *Fundulus grandis* and related fish, requires an intermediate host, the grass shrimp, *Palaemonetes pugio*.[51] Many other species of coccidians in poikilotherms probably require an intermediate host.

Much more needs to be known about immunity to coccidiosis. Younger animals become ill, whereas older hosts are generally clinically healthy. Premunition immunity apparently results from various coccidial infections. Older sheep and goats almost continually discharge

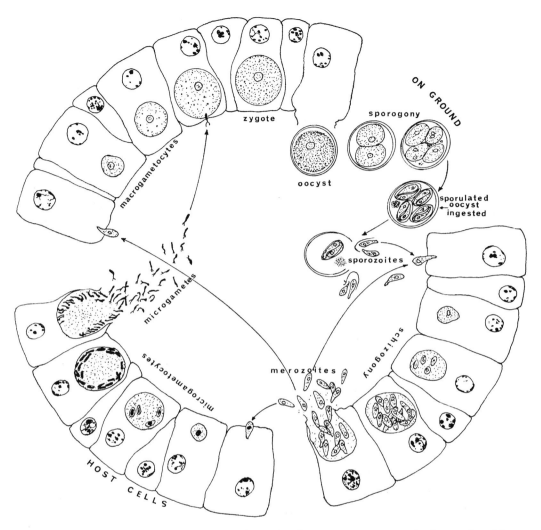

Fig. 5–8. Diagram of a generalized life cycle of *Eimeria*. The same pattern occurs in the species of *Isospora* that do not form tissue cysts. For details of sporozoite behavior see Lawn and Rose.[85]

oocysts; this phenomenon indicates a degree of premunition. During spontaneous outbreaks of infection, usually only lambs and kids become ill.

Eimeria is characterized by the presence of four sporocysts in each oocyst and two sporozoites in each spore. The many species (well over 1000) occur primarily in the intestinal cells of vertebrates (Fig. 5–9), but may also be found in epithelial cells of the liver, bile duct, or other organs. A species living in nonepithelial cells of fishes has been described. It requires a crustacean host to complete its cycle.[112] Some species infect invertebrates. *Eimeria schubergi*, for ex-

ample, inhabits the intestinal epithelium of the centipede, *Lithobius forficatus*. Cross-infection experiments indicate that *Eimeria* species are host-specific.

Coccidiosis is a self-limiting disease because, after a given number of schizont generations, all merozoites are transformed into gametocytes. Under such circumstances, the parasite's life cycle is completed. Thus, in coccidiosis, the severity of the disease depends largely upon the number of organisms that initiate infection. Effects of coccidiosis vary widely with the species of host, species of coccidia, age and resistance of host, numbers of host cells destroyed, size

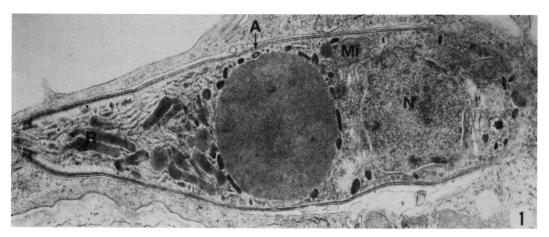

Fig. 5–9. T.E.M. micrograph of a sporozoite of *Eimeria vermiformis* in a villus tip in a mouse intestine; longitudinal section. Micronemes and rhoptries (R) fill the anterior third of the sporozoite. Posterior to these organelles is the anterior refractile body surrounded by amylopectin-like granules (A). A nucleus (N) is posterior to the refractile body. The mitochondria are small and have dense matrices (Mi). ×22,000. (From Adams and Todd, courtesy of J. Protozool.)

of infecting dose, degree of reinfection, and other factors. The following description of coccidiosis in calves, however, gives a general picture of the symptoms that might be seen in most animals with heavy infections.

"The symptoms of coccidiosis may include rough coat, weakness, listlessness, nervousness, poor appetite, diarrhea, and loss of weight or poor gains in weight. The general weakness may cause the calf to defecate without rising.

"When standing, the calf may attempt to defecate and not be able to pass feces; the intense straining results in an arched back, raised tail, and a "pumping" of the sides. Diarrhea may be watery or only slightly liquid, being quite unlike the "white scours" of calves less than three weeks old. Diarrhea caused by coccidiosis may contain many strands of gelatinous mucus and splotches or streaks of blood."

Coccidiosis is rare in horses, but cattle are infected by at least 15 species of *Eimeria*. Common parasites are *E. zurnii* and *E. bovis*, both highly pathogenic. Treatment hs not been completely successful. Cleanliness is the best preventive because the disease is carried from animal to animal by oocysts in manure.

At least ten species of *Eimeria* occur in sheep and goats. *E. arloingi*, *E. parva*, *E. faurei*, and *E. ninae-kohl-yakimova* are common examples. They infect tissues of the digestive tract and cause diarrhea, destruction of epithelial tissues, weakness, loss of weight, and death. As with calves,

clean, dry, uncrowded living conditions are strongly recommended.

Hogs may become infected with five species of *Eimeria*. *E. debliecki* and *E. scabra* (Fig. 5–10) are the most pathogenic. Dogs and cats are hosts to *E. canis* and *E. felina*. Some of the same coccidia occur in foxes. Rabbits may be infected with at least ten species of *Eimeria*. A well-known species is *E. stiedae*. In heavy infections of rabbits with *E. flavescens* the caecal crypts fail to maintain the integrity of the surface epithelium, "and by day 12 after infection there is widespread denudation of the caecal mucosa. Destruction of the crypt cells appear to be a crucial factor in the severity of the lesions. Death is apparently due to a combination of dehydration and tissue-invasion by bacteria."[59] Most wild mammals and birds have their own species of coccidia, and they are far too numerous even to mention here. Undoubtedly, many species are yet to be found and described.

Coccidia occur abundantly in domestic and wild birds.[92] The parasites cause a serious disease in chickens, geese, and turkeys, but apparently not in ducks. In chickens and turkeys, the parasites occur mainly in the small intestine, but they may also be found in other parts of the gut. They have a remarkable specificity for a particular region. Eight or nine species infect domestic poultry, and among the most common in chickens in the United States are *E. necatrix*, *E. acervulina*, *E. maxima*, and *E. tenella*. Turkeys

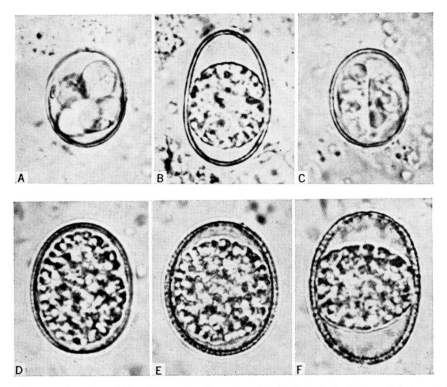

Fig. 5–10. Oocysts of species of *Eimeria* occurring in swine. *A* to *C, E. debliecki. A,* Sporulated cyst of smaller form with polar granule at one side of cyst. *B,* Cyst of larger form with rounded cytoplasm. Clear cyst wall. *C,* Mature cyst. *D,* to *F, Eimeria scabra. D,* Immature form with cytoplasm completely filling cyst. *E,* Cyst with cytoplasm beginning to contract. *F,* Cyst with cytoplasm almost entirely contracted, nearly "ball" stage. × 1500. (From Henry, courtesy of the University of California Publications in Zoology.)

are infected with *E. meleagridis, E. meleagrimitis,* and other species. Young birds are more susceptible than older ones, and immunity is built up against species harbored. A toxic component is often present in infected tissues. In a bird with heavy infection, the intestine may be enlarged and thickened, and villi are often damaged. The epithelial cells are sloughed off and are replaced about every two days. Droppings may be bloody, greenish, brownish, or watery. Birds may be obviously ill, and many of them die. One interesting result of infection is a rise in the blood sugar level of the bird. *E. necatrix* is probably the most pathogenic of the chicken coccidia. "Chickens dying from *Eimeria tenella* infection revealed four major physiological stresses before death: (1) hypothermia, (2) depletion of carbohydrate stores, (3) metabolic acidosis, and (4) renal tubule-cell dysfunction."[139]

Clean, uncrowded, dry living areas for poultry, as for mammals, are of great importance in prevention and control. For a detailed review

of the harm caused by coccidia in poultry, see Ryley.[120] For a review of host-parasite relationships in the alimentary tract of domestic birds, see Crompton and Nesheim.[40] For immune responses, see Rose.[118]

Isospora is characterized by the presence of two sporocysts in each oocyst and four sporozoites in each spore. Oocysts are discharged in feces. There are no intermediate hosts. Among several species, dogs and cats are infected with *I. canis* and *I. ohioensis,* and cats with *I. felis* and *I. rivolta. I. suis* in pigs causes diarrhea, lesions in the small intestine, enteritis, and villous atrophy (Fig. 5–11).[99] Cysts remain infectious for more than two years. Frogs may have *I. lieberkuhni. I lacazei* is found in the small intestine of sparrows, blackbirds, and other passerine birds. *Isospora belli* is common in many wild and domestic animals, and occasionally in man, where it may cause severe diarrhea and even death. If a freshly passed stool specimen is kept in a covered dish at room temperature for one

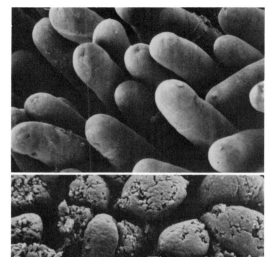

Fig. 5–11. 1. Uninfected villi of the small intestine of a pig. 2. Villi infected with *Isospora suis.* (about 210×). (From F.-R. Matuschka and K. Männer, courtesy of Gustav Fisher Verlag.)

to two days, the oocysts mature and contain the typical two spores, each enclosing four sporozoites. Human infection is not common.

Symptoms of this infection include abdominal pain, diarrhea, flatulence, abdominal cramps, nausea, loss of appetite, lassitude, loss of weight, and fever. After the initial infection, the incubation period lasts about a week; then, symptoms may or may not appear for about a month. During this time, oocysts are discharged. When oocysts are no longer found, the disease has run its course and reinfection apparently does not occur.

Caryospora is a facultative heteroxenous coccidium that infects intestinal epithelia of lizards, snakes and predatory birds. Rodents are secondary hosts that ingest sporulated oocysts.

FAMILY SARCOCYSTIDAE

The basic life-cycle pattern (Fig. 5–12) for all the tissue-cyst-forming coccidia is as follows. Oocysts, with their sporozoites, from feces of the final host are ingested by an intermediate host. The **zoites** multiply asexually (one to several generations) in various tissues and eventually form cysts. These tissue cysts contain infective zoites. Zoite multiplication in the intermediate host is by merogony. When the carnivorous host has ingested its infected prey, some of the tissue zoites, often preceded by generations of asexual multiplication, develop into gametes. This sexual development leads to oocyst formation, and sporogony leads to sporozoites.

This simplified outline of the life cycle does not emphasize the different functions of merozoites (Figs. 5–13 and 5–14) in the several phases of that cycle. To provide such emphasis, Frenkel[52] (Fig. 5–15) coined the term **tachyzoites** for the first, actively multiplying merozoites that develop within the intermediate host, irrespective of whether infection is from oocysts or tissue cysts. **Metrocytes** (noninfectious) and **bradyzoites** (infectious) are zoites that develop within tissue cysts. Other writers use **cystozoites** instead of bradyzoites. (See Markus.[96])

Toxoplasma gondii is worldwide in distribution and is most common in warm, moist areas. It has been reported from man, pigs, sheep, cattle, horses, dogs, cats, and other domestic animals, as well as rodents, wild carnivores, and birds. It is an important cause of abortion in sheep. Toxoplasmosis is now the second most common cause of diagnosed foetopathy of farm animals in the United Kingdom.[71]

Young animals (kittens and puppies) experimentally infected with cysts may show hepatitis, myocarditis, myositis, pneumonia, encephalitis, diarrhea, vomiting, or lymphadenopathy.

Herbivorous animals acquire the infection by ingesting oocysts in food or water contaminated by infective cat feces. Oocysts are able to survive in moist, shaded soil for one year. Man and carnivorous animals may acquire the infection by eating infected raw or poorly cooked meat, especially pig and sheep. Pigs are continually exposed to *T. gondii.* Humans habitually eat smoked, salted, and insufficiently cooked pork and pig products. Invertebrates such as flies, cockroaches, and earthworms may serve as mechanical disseminators. Entry into host cells apparently is accomplished either by passive uptake by phagocytosis or by active penetration in which an enzyme-like factor is involved (e.g., hyaluronidase or lysozyme). See Figure 5–12 for the life cycle involving a cat as the carnivorous final host, and containing the sporogony phase of the cycle, and Figure 5–15 for a comparison

TOXOPLASMA GONDII LIFE CYCLE

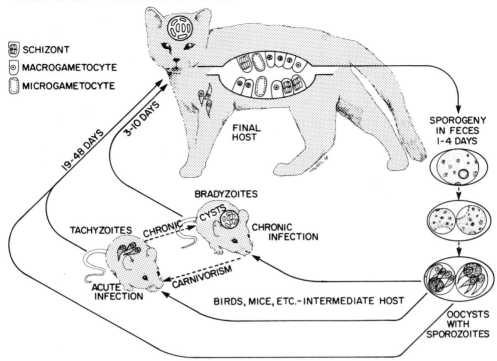

□ SCHIZONT
□ MACROGAMETOCYTE
□ MICROGAMETOCYTE

FINAL HOST

19-48 DAYS

3-10 DAYS

SPOROGENY IN FECES 1-4 DAYS

BRADYZOITES

TACHYZOITES   CHRONIC   CYSTS   CHRONIC INFECTION

ACUTE INFECTION   CARNIVORISM

BIRDS, MICE, ETC.-INTERMEDIATE HOST

OOCYSTS WITH SPOROZOITES

Fig. 5–12. *Toxoplasma gondii* life cycle, indicating that the ingestion of bradyzoites from cysts shortens the pre-patent period to 3 to 10 days, whereas after the ingestion of sporozoites from oocysts, it takes from 19 to 48 days to the shedding of oocysts. (From Frenkel, J.F.: Protozoan diseases of mammals and birds. *In* The Comparative Pathology of Zoo Animals. Edited by R.J. Montali and G. Migaki. Washington, D.C., Smithsonian Institution Press, 1980.)

Fig. 5–13. *Toxoplasma gondii* merozoites from host tissues. Cells average 3×6 μm.

of *Toxoplasma* with other cyst-forming coccidia. Figure 5–16 shows an attack on a macrophage.

Toxoplasma gondii, probably the most common infectious pathogenic animal parasite of man, infects about 13% of the world population. Most of the disease manifestations are self-limiting, and may be asymptomatic, and most cases result in chronic, persistent infection in which cysts persist in intermediate and definitive hosts (cats). Congenital toxoplasmosis, when symptomatic, most frequently involves the central nervous system and causes meningoencephalitis and retinitis. It is often fatal during the first few weeks of life. Surviving infants commonly suffer convulsions during early childhood, and severe ocular lesions which may lead to blindness; they may also be mentally deficient. Even some seemingly normal infants develop ocular lesions as children and young adults. Transplacental transmission of *Toxoplasma* probably occurs in a third of the cases of maternal toxo-

Toxoplasma gondii

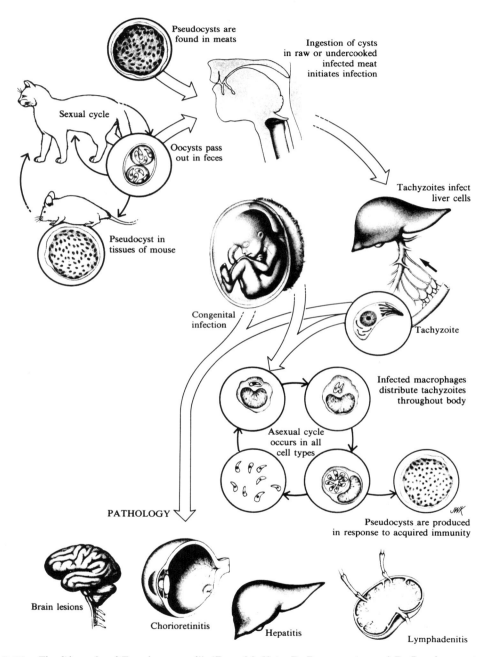

Pseudocysts are found in meats

Ingestion of cysts in raw or undercooked infected meat initiates infection

Sexual cycle

Oocysts pass out in feces

Tachyzoites infect liver cells

Pseudocyst in tissues of mouse

Congenital infection

Tachyzoite

Infected macrophages distribute tachyzoites throughout body

Asexual cycle occurs in all cell types

Pseudocysts are produced in response to acquired immunity

PATHOLOGY

Brain lesions

Chorioretinitis

Hepatitis

Lymphadenitis

Fig. 5–14.   The life cycle of *Toxoplasma gondii*. (From M. Katz, D. Despommier and R. Gwadz, courtesy of Springer-Verlag.)

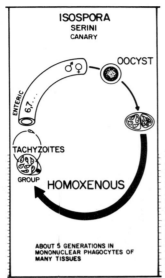

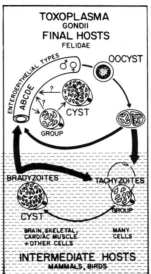

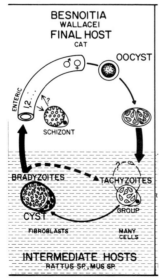

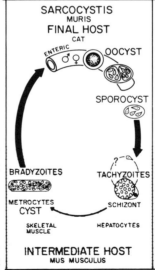

Fig. 5–15. Life-cycle patterns of four isosporid coccidia. (From Frenkel, J.K.: *Besnoitia wallacei* of cats and rodents: with a reclassification of other cyst-forming isosporid Coccidia. J. Parasitol., *63*:611, 1977.)

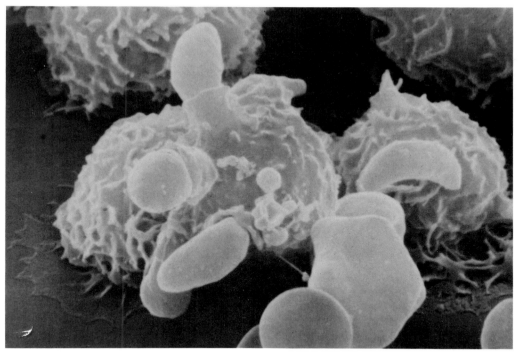

Fig. 5–16. A normal mouse macrophage simultaneously attacked by several toxoplasmas. Rudimentary sheath-like lamellipodia are formed around the invading parasites. Note the beginning stages of macrophage spreading with a concomitant reduction of surface ruffles. × about 7600. (From Chiappino, Nichols, and O'Connor, courtesy of J. Protozool.)

plasmosis acquired during human pregnancy.[90] Congenital infection may damage brain cells. New lesions cause by congenital toxoplasmosis can occur after 5 years of age. There are usually no observable symptoms if infection occurs in human adults.[66]

In women, infection prior to pregnancy does not result in fetal damage, but infection after the start of pregnancy may result in congenital infection of the baby. From 1 to 15% of infected babies die. The disease may be the cause of early abortion.

The principal symptoms in humans are lymphadenopathy (involving peripheral lymph nodes in the cervical, supraclavicular, and inguinal regions, and sometimes in the head and mesenteries) and fever. Parasites are found predominantly within tissues. Any nucleated cell may be parasitized by tachyzoites, but those of the central nervous system and muscles are preferred for the formation of bradyzoites in tissue cysts. Parasitemia, anemia, hepatitis, and hepatosplenomegaly may also occur, but many of these symptoms are observed in man only in

the acute and sometimes fatal form of the disease.

The main source of energy for *Toxoplasma* appears to be glucose. An active hexokinase has been demonstrated, and lactic, acetic, and another (unidentified) acid have been detected as metabolic products of glucose. *Toxoplasma* can oxidize reduced diphosphopyridine nucleotide, and it possesses a cytochrome system.

Specifically armed lymphocytes appear to be the effective mechanisms in immunity to the parasite. If a previous infection with *Toxoplasma* results in a high serologic test titer, the host may exhibit a strong protection against a challenge infection. The role of antibodies appears limited to lysis of extracellular tachyzoites. Antibodies are transferred via the placenta and with mother's milk to babies. Although partial immunity can be achieved through passive and active immunization, a safe and practical vaccine has not yet been developed. Maximum protection can only be obtained through contact with living parasites.

Prevention (especially for pregnant women)

includes eating only cooked meat, particularly pork and lamb. Raw, frozen meat, while containing fewer organisms, is not entirely safe. Pregnant women should not handle cats (Fig. 5–14). Hands should be washed after handling raw meat, cats, or contaminated soil. The Sabin and Feldman dye test is a definitive test for toxoplasmosis.

For reviews of *Toxoplasma* and toxoplasmosis, see Frenkel[53] (general), Jacobs[73] (general), Chinchilla and Frenkel[30] (immunity), and Dubey[44] (*Toxoplasma* and related genera).

Besnoitia cysts have been found in cattle, horses, reindeer, rodents, opossums and reptiles as intermediate hosts. Cats are the only known definitive hosts.[52] The life cycle of *Besnoitia* (Fig. 5–15) involves the predator-prey, two-host characteristic of the family. The parasite may infect the liver, heart, gut, mesenteries, blood vessels, lymphatic system, and skin, for example. *B. besnoiti* in South African cattle can cause both an inflammatory thickening of the skin that makes hides worthless and partial or total sterility in bulls due to involvement of the testes. This is because the organism prefers to form cysts in cooler areas of the body (skin, descended testes). The ovaries show similar slight involvement, as do other internal organs. In North America, cysts from deer, mice, and opossums inoculated into mice and hamsters can cause either subclinical infection or acute, fatal disease. See Figure 5–17 for a comparison of a *Besnoitia* merozoite with related genera.

Sarcocystis. This obligatory heteroxenous parasite produces muscle cysts (Fig. 5–18) in intermediate hosts such as domestic herbivores (cattle, sheep, pigs), opossums, lizards, and rodents. Definitive hosts are predators such as dogs, cats, owls, snakes,[98] and man. There are more than 100 named species. They shed infectious sporocysts in their feces from which herbivores become infected. Sarcocystosis can produce an acute, lethal disease in domestic animals and abortion in cattle. In the United States, 75 to 98% of cattle are infected. About 45 cases of infection in human muscle have been reported. People are infected by eating raw or lightly-cooked infected pork or beef, and the resulting intestinal infection is self-limiting. Experimental infections in man result in diarrhea, nausea, and abdominal pain.

Sarcocystis has adapted its life cycle (Fig. 5–19) completely to the predator-prey relationship existing between its two hosts. The prey host, for example, a cow, ingests sporulated sporocysts from feces of the predator, for example, a dog. Following infection with sporozoites, multiplication first takes place by schizogony and endodyogeny in endothelial cells of internal host organs, and, after some weeks, muscle cysts are formed. At first, the multiplying cells are not infectious, but in a few weeks, banana-shaped merozoites are developed. In the predator, merozoites change directly into macro- and microgametes and begin gametogony and sporogony below the epithelium of the gut. Contrary to other coccidia, no schizogonic multiplication occurs in the gut wall of the final host. Sporogony takes place in the gut and results in the formation of oocysts containing two sporocysts, each with four sporozoites. These sporulated sporocysts are eliminated with host feces.

In striated or cardiac muscle of the intermediate (prey) host, the parasites occur in elongated cysts that may be either so small that they cannot be seen without a lens or so large that they appear as short, white streaks. Cysts range in size from 25 μm to 5 cm. The motile merozoites, 6 to 20 μm in length, are crowded within compartments of the cyst.

The use of species names based on the scientific names of the intermediate and definitive hosts has been adopted by a number of writers. For example, instead of *Sarcocystis cruzi* whose definitive hosts are dogs, wolves, etc. and whose intermediate hosts are ox and bison, the new proposal is: *S. bovicanis*. Likewise, *S. ovifelis* and *S. bovihominis* and many others are being used. Levine[88] has pointed out that such usage is not acceptable because, although tempting, it is contrary to The International Code of Zoological Nomenclature. We shall follow Levine's advice in listing several more of the 122 species of this genus. *S. cuniculi* (definitive host—cat; intermediate host—rabbit), *S. gigantea* (definitive host—cat; intermediate host—sheep), *S. miescheriana* (definitive host—dog, fox, raccoon; intermediate host—pig), *S. falcatula* (definitive host—opossum; intermediate host—numerous species of birds). Members of this genus can use reptiles as definitive or intermediate hosts.

Frenkelia, Cystoisospora, Hammondia. The genus *Frenkelia* is closely related to *Sarcocystis*, but schizogony occurs in the liver and tissue cysts are formed in the brain of its intermediate host (voles). Predatory birds are the definitive hosts. *Isospora felis* and *I. rivolta*, contrary to other species of the genus, are heteroxenous,

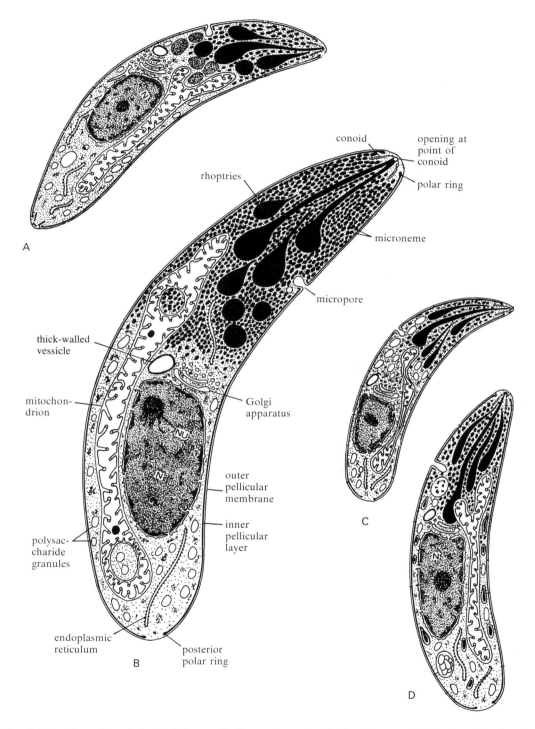

conoid

opening at
point of
conoid

rhoptries

polar ring

microneme

A

micropore

thick-walled
vessicle

mitochon-
drion

Golgi
apparatus

outer
pellicular
membrane

inner
pellicular
layer

polysac-
charide
granules

C

endoplasmic
reticulum

posterior
polar ring

B

D

Fig. 5–17. Merozoites of *A, Frenkelia* sp.; *B, Sarcocystis tenella; C, Toxoplasma gondii; D, Besnoitia jellisoni.* Schematic sections. N, nucleus; NU, nucleolus. (From Scholtyseck, Mehlhorn, and Müller, 1973, courtesy of Zeitschrift fur Parasitenkunde.)

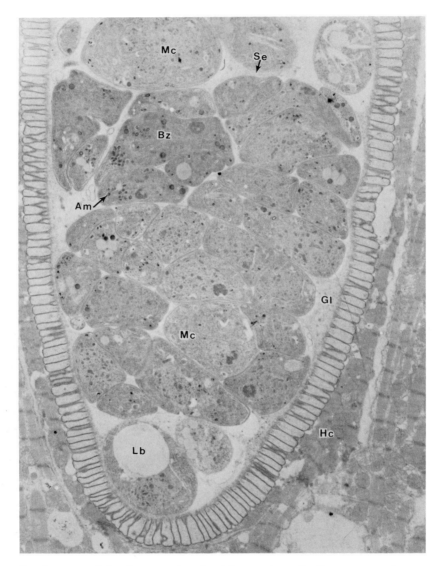

Fig. 5–18. Development of the sheep-canid cycle of *Sarcocystis tenella*. Transmission electron micrograph (58DPI) with fully formed sarcocyst wall; note indistinct granular septum (Se), bradyzoites (Bz), metrocytes (Mc), amylopectin (Am), and lipid bodies (Lb). (From Dubey, et al., courtesy of the National Research Council of Canada.)

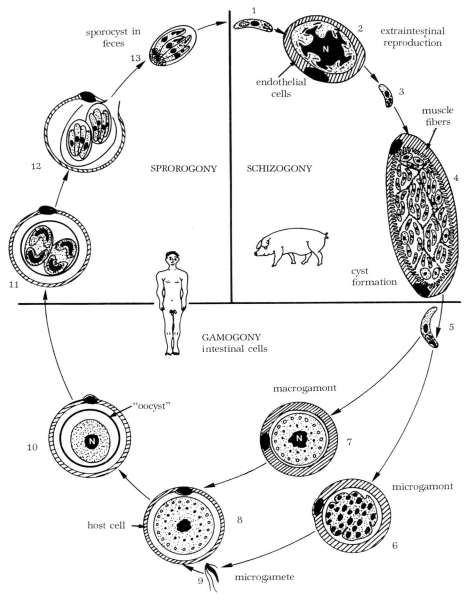

Fig. 5–19. A diagrammatic representation of the life cycle of *Sarcocystis suihominis*. *1*, Sporozoite. *2*, Within endothelial cells two generations of schizonts are formed giving rise to 50 to 90 merozoites (each) by simultaneous division of the giant nucleus. *3*, Merozoite; *4*, Cyst formation within muscle fibers with globoid metrocytes and elongate bradyzoites. *5*, After eating raw meat containing cysts, bradyzoites are set free within the intestine of man. *6, 7*, Micro- and macrogamonts develop in a parasitophorous vacuole within the cells of the lamina propria. *8, 9*, The stationary macrogamete *(8)* is fertilized by a motile microgamete *(9)*. *10*, The zygote is surrounded by a wall and becomes an "oocyst." *11*, Two sporocysts are formed in the interior of an oocyst while still in the host cell. *12*, The oocyst wall is broken and the two sporocysts are set free, containing four sporozoites. (From Frenkel, J.K., et al.: Sarcocystinae: nomina dubia and available names. Z. Parasitenkd., *58*:115, 1979.)

cyst-forming species and should perhaps be assigned to the genus *Cystoisospora*. The genus *Hammondia* forms cysts in skeletal muscle of intermediate hosts only, and is obligatorily heteroxenous.

For reviews of *Sarcocystis* and sarcocystosis, see Mehlhorn and Heydorn,[103] Markus,[95] and Dubey et al.[45]

Pneumocystis carinii is probably a yeast, possibly a fungus, but it is commonly included in protozoology and medical parasitology texts. This parasite is cosmopolitan in humans, domestic animals, and rodents. It is found only in the lungs, and causes an interstitial plasma cell pneumonia. The disease is particularly critical in immunosuppressed adults and children. Death is due to asphyxia.[135]

Suborder Haemosporina

Relationships between these parasites and the coccidia are shown by cellular fine structures and general similarities of life cycles. Sporozoites develop in blood-sucking insects that transmit parasites to vertebrate blood. Schizogony and merogony occur in red blood cells of the vertebrate body. Microgametes and macrogametes develop independently in blood cells, and the zygote, formed in the insect gut, is motile. Pigment (hematin) is formed from host hemoglobin by the malaria parasite. The conoid is generally absent.

Physiology. Although coccidia are able to carry on some metabolism anaerobically, sporulation is apparently strictly aerobic. The rate of metabolism is high in the oocyst during this period, but as soon as sporulation is completed, respiration and metabolism can barely be detected. Even over long periods of storage, however, polysaccharide reserves are depleted.

Excystation in the new host requires two separate stimuli. The first is carbon dioxide "triggering." Most, if not all, oocysts possess a micropyle. In the presence of carbon dioxide, the cap over the micropyle lifts, and permeability of the cyst wall in that area occurs. Body temperature is important. In birds, probably a mechanical rupture of cysts in the gizzard is the first step. Trypsin and bile are involved in the second stage of excystation. Each sporocyst possesses a plugged hole. The plug is called a **Stieda body** and is digested by trypsin. Bile enters through the hole, initiates motility of the sporozoites, and enables them to escape through the hole; this they can do rapidly. Initiation of excystation brings about increased respiration within the oocyst. During release from the sporocyst and subsequent penetration of host cells, sporozoites use most of their remaining amylopectin for energy requirements. As Ryley[120] has pointed out, the energy metabolism of *Eimeria tenella* sporozoites is in many respects similar to that of toxoplasms and malaria parasites, except that it has an appreciable endogenous respiration due to the presence of these amylopectin reserves. Evidence suggests that the developing schizont obtains its energy by glycolysis, whereas later in development, as merozoites prepare for an extracellular period, the aerobic Krebs cycle takes place. Developing macrogametes also change from a glycolytic to an oxidative metabolism. This switch in metabolism is mindful of trypanosomatids. Both cytostomal and pinocytic ingestion of nutrients have been described for macrogametes of *Eimeria acervulina,* with a transport of free ribosomes through tubules from the host cell to the parasite.[106]

FAMILY PLASMODIIDAE

This family contains protozoa that live in vertebrate tissues and blood cells and are transmitted by an insect vector. Schizogony occurs in vertebrate hosts, whereas sporogony occurs in insects. Zygotes are motile, and the sporozoites are naked. Opinion differs as to which genera should be placed within the family, and further research is needed on structure, physiology, and life cycles.

Plasmodium causes malaria. The sporogony phase of development almost always takes place in mosquitoes, but in some lizards it is transmitted by a sandfly (see Chap. 20). The genus *Plasmodium* has been separated into 11 subgenera, but several strains (subspecies?) exist within these groups, and their systematics needs more study. About 155 named species of *Plasmodium* parasitize a wide variety of vertebrates, especially mammals and birds.

MALARIA

Of all the diseases of mankind, malaria is one of the most widespread, best known, and most devastating. It has always been a disease associated with forests and with migrant populations such as armies and nomads. It has undoubtedly played a role in the history of civilization because large areas of the earth have been subjected to the ruinous effects of the illness and to debilitation and death caused by the parasites.

Masses of people in tropic and semitropic countries do not have the economic resources with which to combat the mosquito or to identify and treat the millions of patients. Nor, in many areas, do patients and their families possess enough social awareness of the problem to assist adequately the public health authorities. There is a shortage of well-trained specialists in epidemiology and entomology who understand malaria and have the resources needed for adequate eradication programs. Tourism pours hordes of new and susceptible people into and out of malarious regions. Although the disease is under control in many parts of the world, or at least the prospects for control are emerging, recent years have seen a worldwide resurgence of malaria. It is still a major disease in many tropical countries where the death toll is high, especially in children during the first 5 years of life. Estimates of fatalities vary considerably, but about 1.5 million people in this world die each year from malaria. "The fact remains that the over-all parasite reservoir throughout the world is still immense and, according to the WHO estimates, some 1620 million people inhabit areas of the world where the exposure to malaria infection is moderate to high, depending on the degree of endemicity."[21]

"Today, the concept of eradication is in abeyance and we find ourselves in many parts of the world hard pressed to contain malaria. Many species of *Anopheles* have developed resistance to insecticides, parasites have become resistant to drugs and the geographical distribution of resistant *Plasmodium falciparum* is spreading rapidly, compromising not only control in general but also the treatment of the acute illness. Malaria, thus, remains a disease of the greatest importance to civil and military authorities alike.

"We need new methods of mosquito control, preferably one to which the vectors cannot become resistant. We need new drugs for the treatment of drug-resistant malaria. We need new vaccines which can induce effective protective immunity towards sporozoites or asexual blood stage parasites or, alternatively, which can interrupt transmission by blocking the fertilization of macrogametes in the gut of mosquitoes."[102]

For a 1980 review of malariology based on a broad perspective, see Bruce-Chwatt.[20] For a history of hostilities among mosquitoes, malaria, and man, see Harrison.[64] For a short review of malaria, including treatment and general management, see Hall.[63] For a comprehensive coverage of malaria, see the numerous papers in the three volumes edited by Kreier.[80] For a symposium on malaria and the red cell see Evered and Whelan.[48] The WHO Expert Committee on Malaria has made a fine report, with emphasis on malaria control. (18th report; Technical Report Series 735. Geneva, 1986.)

Four species of *Plasmodium* are found in man. The species names listed below are followed by the several common names by which they are known.

Plasmodium vivax (vivax, simple, benign, or tertian malaria) is responsible for about 43% of cases of human malaria.

Plasmodium malariae (malariae or quartan malaria) causes about 7% of the world's malaria.

Plasmodium falciparum (falciparum, malignant, tropical, pernicious, or estivoautumnal malaria) is the most pathogenic of the malarias and is often fatal. It is responsible for about half the human cases of malaria.

These three malarias are worldwide in distribution, occurring largely in warmer countries (especially *P. falciparum*), but also in such winter-cold areas as Korea, Manchuria, and southern Russia (especially *P. vivax*).

Plasmodium ovale (ovale or ovale tertian malaria) is rare and is generally confined to tropical Africa and the islands of the western Pacific.

Table 5–1 compares important characteristics of these four species of *Plasmodium*, and Plates I, II, III, and IV show the stages of each species in human erythrocytes as they appear in stained blood smears. The mosquito vectors are discussed with other insects in Chapter 20.

The life cycle of *Plasmodium* (Fig. 5–20) starts with a zygote in the stomach of a female mosquito. The zygote changes into an **ookinete** (Fig. 5–21) over the next 10 to 25 hours by the formation of polar rings, the collar, microtubules of the cytoskeleton, and the inner two layers of pellicle membrane beneath the cell surface.[12] The ookinete migrates through the midgut wall, but does not penetrate the enveloping basal lamina. It then differentiates into an **oocyst** between the basement membrane and the basal lamina of the midgut. During this transformation the cytoskeleton, apical complex, and

Table 5-1. Comparative Characters of Plasmodia of Man (Stained Thin Smears)

Stage or Period	Plasmodium vivax	Plasmodium ovale	Plasmodium malariae	Plasmodium falciparum
Early trophozoite or ring	Relatively large; usually one prominent chromatin dot, sometimes two; often two rings in one cell	Compact; one chromatin dot; two rings in one cell uncommon	Compact; one chromatin dot; two rings in one cell rare	Small, delicate; two chromatin dots and two rings in red cell common; appliqué forms frequent
Late trophozoite	Large; markedly ameboid; abundant chromatin; prominent vacuole; pigment in fine rods	Small; compact; not ameboid; vacuole inconspicuous; pigment coarse	Smaller than vivax; compact; often band-shaped; not ameboid; vacuole inconspicuous; pigment coarse	Medium size; usually compact, rarely ameboid; vacuole inconspicuous; rare in peripheral blood after half grown; pigment granular
Young schizont or presegmenter	Large; somewhat ameboid; dividing chromatin masses numerous; pigment in fine rods	Medium size; compact; chromatin masses few; pigment coarse	Small; compact; chromatin masses few; pigment coarse	Small; compact; chromatin masses numerous; single pigment mass; rare in peripheral blood
Mature schizont or segmenter	Schizonts and merozoites large; pigment coalescent	Merozoites larger than in P. malariae; irregular rosette	Schizonts small, but merozoites large	Small merozoites; single pigment mass
Number of merozoites	12–24, usually 12–18	6–12, usually 8	6–12, usually 8	8–26, usually 8–18
Microgametocytes (usually smaller and less numerous than macrogametocytes)	Spherical; compact; no vacuole, undivided chromatin; diffuse coarse pigment; cytoplasm stains light blue	Similar to P. vivax but somewhat smaller; never abundant	Similar to P. vivax but smaller and less numerous	Crescents usually sausage-shaped; chromatin diffuse; pigment scattered large grains; nucleus rather large; cytoplasm stains pale blue

Macrogametocytes	Spherical, compact, larger than microgametocyte; smaller nucleus; pigment same; cytoplasm stains darker blue	Similar to *P. vivax* but somewhat smaller; never abundant	Similar to *P. vivax* but smaller and less numerous	Crescents often longer and more slender; chromatin central; pigment more compact; nucleus compact; cytoplasm stains darker blue
Pigment except in mature schizonts	Short, rather delicate rods, irregularly scattered; not much tendency to coalesce	Similar to but somewhat coarser than *P. vivax*; sometimes clumped or in lateral bands	Seen in young rings; granules rather than rods; tendency toward peripheral scatter	Pigment granular; early tendency to coalesce; typical single solid mass in mature trophozoite; coarse "rice grains" in crescents
Alterations in the infected red cell	Enlarged and decolorized; Schüffner's dots usually seen	Enlarged; decolorized; prominent Schüffner's dots appear early; infected cells may be oval-shaped with fimbriated ends	Cell may seem smaller; fine stippling (Ziemann's dots) occasionally seen	Normal size; Maurer's dots (or "clefts") common; Garnham's bodies occasionally seen
Length of asexual phase	48 hours	about 48 hours	72 hours	36–48 hours, usually 48
Prepatent period, minimal	8 days	9 days	14 days	5 days, average 8–12
Usual incubation period	8–31 days, average 14	11–16 days	28–37 days, average 30	7–27 days, average 12
Interval between parasite patency and gametocyte appearance	3–5 days	5 to 5 days; appearance irregular and numbers few	10–14 days; appearance irregular and numbers few	8 to 11 days
Developmental period in mosquito	30 days or more at 17.5°C; 16–17 days at 20°C; 10 days at 25–30°C	16 days at 25°C; 14 days at 27°C	30–35 days at 20°C; 25–28 days at 22–24°C	22–23 days at 20°C; 10–12 days at 27°C (80°F)

(From Beaver, Jung and Cupp: Clinical Parasitology, 9th Ed. Philadelphia, Lea & Febiger, 1983.)

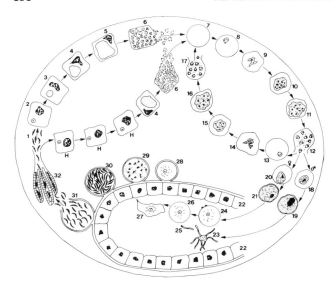

Fig. 5–20. The life-cycle of a primate malarial parasite. (From R. Bray and P. Garnham, courtesy of Churchill Livingstone.)

inner pellicular membrane are resorbed. Each oocyst undergoes numerous cytoplasmic subdivisions producing sporoblasts (Fig. 5–22) which yield many thousands of invasive sporozoites over a period of 7 to 12 days.[126] The weakly motile sporozoites vary (dependent upon species) from 9 to 16.5 μm in length, although aberrant forms up to 40 μm long have been described. Protective antibodies induce a capping reaction on the viable sporozoites that are carried throughout the mosquito within its hemocoele. Sporozoites invade the entire mosquito; many of them enter the salivary glands and thus are in a favorable position to infect the next host when the mosquito feeds on its blood. Parasites in mosquitoes cause pathologic changes that alter the behavior of the insect host.

In primate malaria the first attack on the vertebrate host by the parasite is the invasion of liver hepatocytes by sporozoites. The second attack is invasion of erythrocytes by merozoites formed in liver cells. The disease involves the macrophages and reticuloendothelial system of the entire host body. Thus malaria is a systemic disease rather than strictly a blood disease.[37] Any description of the first attack must center on the well-known phenomenon of malarial relapses, which have been reported to occur after weeks to many years after recovery. Much debate on the mechanisms responsible for relapses have been, and still are, taking place. The views of Krotoski and his collaborators[81–83] have been widely accepted and they are quoted below.

"According to the hypnozoite theory of malarial relapse, the infecting mosquito injects a population of varying numbers of sporozoites which differ according to their potential for entering the dormant state (as hypnozoites). Some (the majority in most temperate strains) do not enter the hypnozoite stage at all, but, shortly after infecting a hepatic parenchymal cell (hepatocyte), begin to undergo pre-erythrocytic schizogony; this process is completed within about 7 to 10 days to yield merozoites which, in turn, enter erythrocytes for the first time. Other sporozoites (the great majority in those strains capable of long-delayed primary parasitemias) do enter the dormant state, in which they persist as the uninucleate hypnozoites for varying, but intrinsically (genetically?) determined periods. Subsequently, upon completion of the latter, they are 'activated' by unknown (but, presumably, also parasite-endogenous) mechanisms, also to begin pre-erythrocytic schizogony, and, once again yielding merozoites some 7 to 10 days later for a totally new parasitemia and blood schizogonic cycle. Thus, each injected sporozoite acts totally independently of the others in producing merozoites, some after dormancy, some without, and the characteristic relapse patterns of different strains of *P. vivax* (and presumably *P. ovale, P. cynomolgi*, etc.) depend on the presence or absence of each type (no dormancy, short dormancy, intermediate dormancy, prolonged dormancy) or sporozoites in the population mixture injected." (Personal communication.)

Merozoites enter the erythrocytes (Fig. 5–23), and start the erythrocytic phase of the life cycle. Within the red blood cells, young plasmodia have a red nucleus and a ring-shaped, blue cytoplasm in typically stained smears. This appearance give rise to the name "signet ring" for

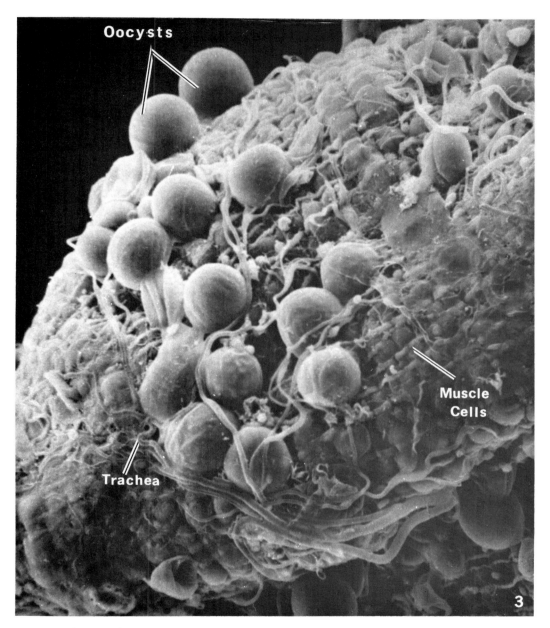

Fig. 5–21. Oocysts of *Plasmodium gallinaceum* on the haemocoel surface of the midgut of a female *Aedes aegypti*. ×300. (From Strome, C.P.A., and Beaudoin, R.L.: Exp. Parasitol., 36:131–142, 1974.)

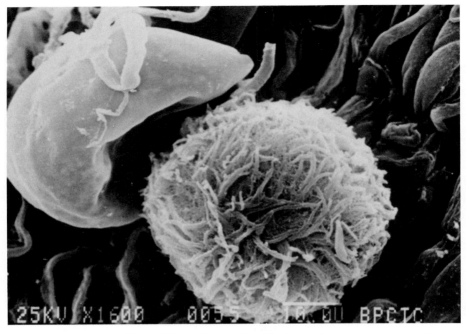

Fig. 5–22. An immature sporoblast of *Plasmodium yoelii yoelii* released from the oocyst, showing sporozoite buds formed on the surface of the sporoblast (6 days). ×2,166. (From P.H. Chen, et al., courtesy of J. Parasitol.)

the parasite at this stage. The ring configuration alters as the plasmodia begin to grow within the blood cells.

The plasmodium divides into merozoites, which break out of the red cell (or pass through an aperture in its surface), enter other erythrocytes, and repeat the multiplication process (Fig. 5–24).

A study[46] of the invasion of erythrocytes by *Plasmodium knowlesi* indicates that the anterior end of the merozoite must first make contact with the red cell. Then, the erythrocyte membrane becomes deformed (possibly a lesion is produced), and the parasite enters by a localized membrane invagination, resulting in an orifice through which sodium, water, and possibly calcium enter. The entrance of fluid through the orifice causes the merozoite to spin, and when the orifice closes, the spinning stops. For a review of the processes of red cell invasion by merozoites, see Bannister.[11] Schmidt[121] has challenged the hypnozoite theory, and he believes that the older cyclical theory is still valid. More research on this subject is clearly needed.

Some of the merozoites in the blood cells develop into sexual forms that grow into male microgametocytes or female macrogametocytes. When a mosquito feeds at this stage of the life cycle, the gametocytes are taken into the insect's stomach, where they mature into the microgametes and macrogametes. The microgametes are flagellate-like outgrowths from the microgametocytes. Their formation is called "exflagellation." Gametocyte emergence from red blood cells and exflagellation normally occur when the blood is exposed to air. The mosquito gut environment is not required. A decrease in blood CO_2 tension and a resulting rise of the pH are essential. The detached microgametes behave like the sperm cells of higher animals. Fertilization now takes place, and the zygote thus formed completes the life cycle.

The infected erythrocyte is not only altered functionally, but, with *Plasmodium falciparum* and *P. malariae*, small elevated knobs appear on its surface; and with *P. knowlesi*, *P. ovale*, and *P. vivax*, many small surface invaginations are seen with the electron microscope. The latter correspond to the Schüffner's dots that are often used as diagnostic features in the usual stained blood smears.[32] For further details of development within erythrocytes see Bannister and Sinden.[12]

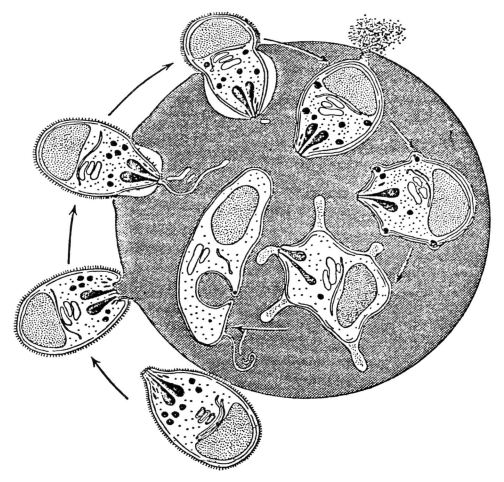

Fig. 5–23. The principal steps in invasion of a red blood cell by a *Plasmodium knowlesi* merozoite. Note that the parasitophorous vacuole begins as a cup formed by the invagination of the parasite and lined with the red blood cell membrane; also note that the merozoite coat is removed and discarded through a pore in the red blood cell membrane. The final stage in this figure is a feeding ring stage. (From Bannister, L.H., et al.: The invasion of red cells by *Plasmodium. In* Parasite Invasion. Symposium No. 15, British Society for Parasitology. Edited by A.E.R. Taylor and R. Muller. Oxford, Blackwell Scientific Publications, 1977.)

Physiology of Plasmodium

As was shown earlier in this chapter, intraerythrocytic plasmodia ingest the contents of host cells. In mature cells, the material is mainly hemoglobin; in reticulocytes and both erythrocytic stem cells, the material is mainly red cell organelles and hemoglobin precursors. A cytostome has been described for both erythrocytic and exoerythrocytic stages of several species of *Plasmodium* of reptiles, birds, and mammals. Many workers believe that, in mammalian malaria parasites, a combination of phagotrophy, pinocytosis, and cytosomal feeding commonly takes place. During endocytosis, droplets of host cytoplasm are engulfed through invaginations of the plasma membrane, with the subsequent formation of food vacuoles. Ingested hemoglobin and other host proteins are digested in these vacuoles by the parasite's proteases and provide a major source of free amino acids for plasmodial protein biosynthesis. Hemozoin, an insoluble by-product of hemoglobin degradation, can be seen as darkly pigmented deposits in host cell cytoplasm. Free amino acids also can be directly taken up by parasites from the external medium. Synthesis of proteins in *Plasmodium* is typical of other eukaryotic protozoans and does not require the use of ac-

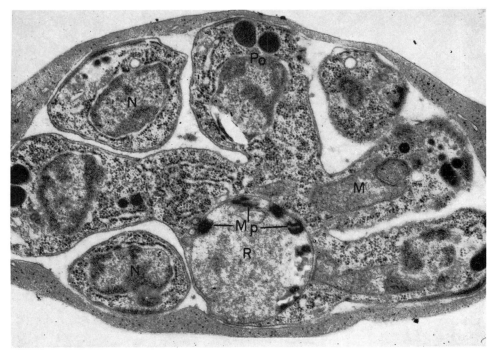

Fig. 5–24. A schizont of *Plasmodium cathemerium* with budding new merozoites. *Mp,* Malaria pigment; *N,* nucleus; *M,* mitochondrion; *Po,* paired organelles; *R,* residual body. (From Aikawa, Huff, and Sprinz, courtesy of Military Med.)

cessory host organelles. Precursors for lipd synthesis, for example, free fatty acids, cholesterol, or glycerol, in malarial parasites also must be derived from host cells.

Malaria parasites do not store carbohydrates as polysaccharides (for example, glycogen), and therefore a continual source of simple sugars, mainly glucose, must be present for energy production. Within host cells, plasmodia metabolize glucose by conventional anaerobic glycolysis, yielding adenosine triphosphate (ATP) and lactic acid. Under aerobic conditions, only avian species, which possess a functional tricarboxylic acid (TCA) cycle, are able to convert pyruvate to CO_2 and water for additional ATP formation. Coenzyme A, an essential cofactor in functional TCA cycles, must be provided exogenously because parasites are unable to synthesize this molecule. Reserves of host cell ATP may also serve as an alternate energy source for the parasite, as indicated in studies with *P. berghei.* Exoerythrocytic plasmodia possess a small cytostome, but presumably, these parasites obtain their nutrients mainly through their surface membranes, either by simple diffusion

or facilitated transport. Glucose also serves as the main energy source for exoerythrocytic stages.

Typical protozoan mitochondria are present in at least some *Plasmodia* species (e.g., *P. falciparum, P. lophurae*). In some forms (e.g., *P. berghei, P. coatneyi*), multilamellated or concentric-ring structures may have a function equivalent to that of classic mitochondria. The presence of cytochrome oxidase activity in mitochondria of both avian and mammalian malarias suggests an electron transport system using cytochromes. It is possible, however, that this enzyme may be serving other functions, for example, pyrimidine biosynthesis, and may not be associated with electron transport or oxidative phosphorylation.[61]

The formation of pentose sugars, for example, ribulose, for incorporation into DNA or RNA and other biochemical reactions, apparently is not accomplished through the pentose-phosphate shunt pathway because the first enzyme in this pathway, glucose-6-phosphate dehydrogenase (G6PD), is absent from parasite extracts. For this reason, it has been suggested that humans expressing G6PD deficiencies may be less susceptible to malarial infections. The sources of pentose sugars for parasite nucleic acid synthesis are not known, but it is probable that these sugars are derived from host cells. An external source of purines is also required by plasmodia and is thought to be provided by the adenine nucleotide pool of the host erythrocyte, primarily in the form of ATP. Parasites, however, are capble of *de novo* pyrimidine synthesis.

A milk diet suppresses an infection with some plasmodial species. The responsible factors are para-aminobenzoic acid (PABA), methionine, and presumably, additional amino acids. A pure milk diet lacks PABA, which is a necessary growth factor for these plasmodia. *Plasmodium* species apparently use PABA to produce various foliates, which act as essential cofactors. Sulfonamides and pyrimethamine block synthesis of folates in plasmodia, but drug resistance ensues rapidly.[80] Deficiencies in other dietary supplements, including vitamin A, the B vitamins, niacin, and ascorbic acid, have been shown to affect plasmodial growth and development adversely.

For further information on the physiology of *Plasmodium*, see von Brand,[16] Gutteridge and Coombs,[60] Sherman,[124] and Kreier.[80]

Immunology of Malaria

Human populations exposed over many years to frequent infections with *Plasmodium* acquire a high degree of protective immunity against the malarial agent. Immunity, however, is nonsterile (incomplete) and is characterized by the persistence of intermittent, low-level parasitemias. Infants born of immune mothers probably have received specific IgG antibodies from maternal serum across the placenta before birth. "Passively acquired immunity operates over the first few months of life and is probably responsible for the exceptional mildness of the primary malarial attack. . . . In hyperendemic areas, by the fifth year of life children possess sufficient immunity to protect them from most of the lethality and morbidity of malaria."[101]

Malarial parasites acquired through natural transmission elicit the production of antibodies comprising several immunoglobulin classes, although anti-*Plasmodium* activity appears to be associated primarily with the IgG type. The importance of humoral immunity in the control or elimination of the malarial parasite has been established through both in vitro parasite inhibition studies and passive transfer experiments with a variety of animals, including humans. Immunity acquired under natural conditions is usually species-specific, with effector mechanisms directed mainly against the asexual stages developing within host erythrocytes. Immune stimulation provided by sporozoite or exoerythrocytic stages is largely ineffective in preventing the establishment of malarial infections, although persons repeatedly exposed to sporozoites in hyperendemic areas can become partially immune. In a few persons, protection against malaria has been induced by inoculation of irradiated sporozoites of *P. falciparum*.[109]

The exact role of antibody in mediating protective immunity is still not clear. Results of in vitro studies indicate that antibody may act to inhibit erythrocyte invasion by extracellular merozoites.[115] The occurrence of inhibitory antibodies, characterized as a non-complement-dependent IgG or IgM, is correlated with clinical immunity developed in malaria-infested monkeys.[23] The cooperative interaction between antibody and various elements of the cellular immune system has also been recognized as an effective mechanism for elimination of the parasites. Antibody involvement in facilitating immune phagocytosis and in destroying infected erythrocytes by activated macrophages has been demonstrated by in vitro experiments.[123]

The role of thymus-derived lymphocytes (CMI) in mediating immunity to malaria is still poorly understood. It is generally accepted that T lymphocytes participate as helper cells in the formation of antimalarial antibodies. However, direct participation of these cells in an effector capacity is currently open to question. It has been suggested[3] that T lymphocytes synthesize and release mediators that activate macrophages. The latter, in turn, liberate mediators that produce degeneration of parasites within circulating erythrocytes. In experimental systems involving avian malaria, strong evidence suggests that CMI may be involved in the elimination of parasitic infections. Birds that are rendered agammaglobulinemic (unable to produce antibodies) are still able to develop effective immune resistance against various malaria species. Whether effector T-cells participate directly in mediating protective immunity in naturally acquired mammalian malaria has not yet been clearly demonstrated, although a requirement of T-cell reactivity is suggested in experiments involving artificial immunization of B-cell- and T-cell-deficient mice against *Plasmodium berghei.*[28]

Protective immunity in experimental animals has been achieved using both living and dead parasite preparations. In contrast to the slowly developing immunity acquired under natural conditions, artificially induced immunity develops quickly, persists for a short time, and is strictly species- (sometimes strain-) and stage-specific. Immune protection is the result of specific antibody activity and the cooperation of CMI components serving in helper and/or effector capacities. Antisporozoite immunity has been achieved against rodent *(Plasmodium berghei)*, monkey *(P. cynomolgi)*, and human *(P. falciparum, P. vivax)* malarias. Because of the strict stage-specificity of immunity produced by a sporozoite vaccine, protection in humans is of particular significance with *P. falciparum* infections, in which only a few sporozoites escaping immune destruction can lead to high multiplication rates in the liver, large numbers of erythrocytic schizonts, and a potentially fatal condition in the host.

Significant progress has been made in developing effective merozoite vaccines that produce a long-lived resistance, maintain their immunogenicity after disruption and storage, and are usually effective against different genetic strains or antigenic variants of a particular *Plasmodium* species.

"Blood stage merozoites constitute the most promising form of vaccine for potential use in a human trial. This is so in terms of its experimentally established immunogenicity of broad variant specificity against the stage of the parasite causing clinical illness. . . . Purified merozoite antigens might prove to be effective and in combination with vaccines active against e.e. [exoerythrocytic] stages or gametes would facilitate interruption of transportation. Results in animal models cannot be extrapolated to man and the ultimate evaluation of any vaccine chosen will depend upon controlled clinical trials in an exposed community."[33] (See Chapter 2 for genetic engineering of vaccines.)

Whole organisms or isolated antigenic components of malarial parasites, in addition to their usefulness as vaccines, are being employed in the serodiagnosis of human malarias. Laboratory and field trials indicate that the indirect fluorescent antibody test is probably the most effective method for detection of malaria-infected individuals, although passive hemagglutination, precipitating antibody tests, and enzyme-linked immunosorbent assays are also being used.[74] Serologic tests for malaria are not always specific, owing to the presence of antibodies that cross-react among different *Plasmodium* strains or stages. Antibodies are usually species-specific, but cross-reactivity between *Plasmodium* and *Babesia*[38] indicates that interspecific serologic barriers are not absolute.

Mosquito vectors possess varying degrees of innate resistance to plasmodia that may influence sporogony in various ways. Ookinete development may be arrested, sporogenesis may be aborted, or sporozoites may be nonfunctional. The mechanisms controlling parasite susceptibility or resistance are unknown. A strong genetic component, however, is indicated in mosquito breeding experiments.[136] The nutritional requirements of parasites or the "immune" reactivity of the host insect have been suggested as factors influencing plasmodial infectivity. Inactivation of gametocytes by vertebrate host immunoglobulins can also affect the oocyte's development.

Immunopathology

Parasitologists are generally convinced that most of the pathologic changes associated with malaria have an immunologic basis. Although this section deals primarily with the immunopathology of malarial infection, other pathophysiologic and clinical features not necessarily related to the immune responses are also mentioned.

The chronic character of malaria is probably due, at least in part, to the formation of a series of genetic variants of plasmodial strains in the host. Antibodies produced against one variant strain are not effective against another, and thus this other strain is able to evade immune destruction by the host. Antigenic variation has been shown to occur in simian and rodent, but not human, plasmodia.[17] It is presumed, however, that a similar escape mechanism must also be operative in chronic human infections.

If the host is able to survive the initially high parasitemia, immunity, in the form of variant-specific antibody production, reduces the parasite population to subpatent levels. As immunity declines, a new population of parasites, presumably representing a new antigenic type, is able to repopulate the host. This repopulation of the peripheral circulation with variant merozoites, in the continued presence of the original erythrocyctic infection, is called **recrudescence.**

Clinical signs and symptoms associated with the several human plasmodial infections are similar, but may differ with regard to intensity or complications. One characteristic response of human hosts is the paroxysm, which begins with chills followed by a gradually mounting fever that may reach 41°C (106°F). Profuse perspiration may last a few hours as the temperature subsides. The entire paroxysm lasts from 6 to 10 hours and occurs every third day in *Plasmodium vivax, P. falciparum,* and *P. ovale* infections. Hence the periodicity is tertian. Paroxysms associated with malariae malaria are quartan because they occur every 72 hours, on the fourth day. Chills and fever are thought to be triggered by the release of pyrogenic agents during the sporulation process.

Falciparum malaria is the most pathogenic of the malarias and is commonly fatal, especially in children and in the elderly. The incubation period is usually from 8 to 15 days and, as in vivax malaria paroxysms, is followed by chills with uncontrollable shivering, high fever, and finally, profuse sweating. Other symptoms may include headache, bone and muscle pains, malaise, anxiety, mental confusion, and even delirium. Severe anemia and leukopenia are common. The peripheral blood generally contains only parasite ring stages and gametocytes. Major causes of death in falciparum malaria are: severe renal, pulmonary and cerebral complications, preceded by anemia and immunosuppression.[140] Complicated attacks (pernicious malaria) are of five major types: (1) Cerebral malaria progresses gradually over a period of days, or it may appear suddenly, with rapid development of coma. The meninges are grossly congested, and hemorrhages occur in the cerebrum. Severe mental disturbances are common. Retinal hemorrhage is commonly a sign of cerebral malaria. (2) High fever (hyperpyrexia) may appear, characterized by a hot, dry skin, incontinence of urine and feces, mental disorientation, and delirium or coma. (3) Gastrointestinal syndromes include severe liver damage, nausea, vomiting, and diarrhea. Vascular collapse or acute hepatic failure may lead to death in severe cases. (4) Algid (chilly, cold) malaria is usually the result of disruption of the metabolite or salt balance in the body that manifests itself as medical shock. Symptoms include inelastic, pale skin covered with clammy sweat.

The skin feels cold, but the rectal temperature may reach 39° C. Other signs include shallow breathing, low blood pressure (both systolic and diastolic) and edema. On rare occasions, severe vomiting, convulsions, and gastrointestinal hemorrhage can occur. (5) **Blackwater fever** is the result of extensive intravascular hemolysis followed several hours later by the passage of brownish-colored urine due to hemoglobinuria. A catastrophic reduction in circulating red blood cells may occur, and although cardiac failure can result from this condition, renal failure accompanied by a great rise in blood urea is the most common cause of death.

The pathologic changes that accompany malarial infections are associated almost exclusively with the erythrocytic phase of plasmodial development. **Hemolytic anemia,** a hallmark of human malaria, is caused in part by the direct destruction of red blood cells by intraerythrocytic parasitism, but also, to a large degree, by immunopathologic mechanisms. One of the mechanisms, autoimmune hemolysis, is thought to be the result of antibody formation against normal red blood cell antigens or to host immunoglobulins (antiglobulins) adsorbed onto erythrocyte membranes.[93] The actual destruction of red blood cells probably involves complement either through direct activation of the lytic process by fixation to erythrocyte-associated antibody-antigen complexes, or indirectly, through enhancement of red blood cell removal by fixed macrophages (autoimmune phagocytosis). In the presence of complement, even circulating malaria antigen-antibody complexes nonspecifically adhering to uninfected red blood cells can trigger erythrocyte damage.

A severe renal disease *(malarial nephrotic syndrome),* occurring in a small percentage of chronic *Plasmodium malariae* infections, is believed to be a result of parasite immune complex deposition in glomerular capillary walls and renal tubules. Parasite immune complexes most commonly contain immunoglobulins of the IgG or IgM classes. Tissue damage is apparently mediated through the activation of complement by antibody-antigen (Ab-Ag) complexes associated with vascular basement membranes. In addition, lytic enzymes released from activated white blood cells (polymorphonuclear leukocytes) attracted to renal tissues by chemotactic complement components (C3a and C5a) may also contribute to renal membrane damage. Whether immune deposits in renal tissues originate as adherent circulating Ab-Ag complexes or whether they are the result of autoimmune reactions is still open to question. Transient nephritis, for example, proliferative glomerulonephritis, is associated with many malarial infections.

Nonspecific inflammatory responses constitute a major contributor to many of the pathophysiologic complications of falciparum malaria. The plugging of capillaries by red blood cells with accompanying circulatory stasis, commonly seen in cerebral and algid malaria, are believed to be the result of increased capillary wall permeability leading to local tissue edema and constriction of blood vessel volume. Pharmacologically active agents, for instance, kinins, released from parasite-stimulated inflammatory cells are responsible for permeability changes in capillary membranes. In addition, an increased "stickiness" of red blood cells that enhances the potential for capillary blockage may be the result of autoantibody production against infected and normal erythrocytes. Circulating hemagglutinins, detected in some malarias, may play a role in the disruption of microcirculation in various organs.

One of the characteristic signs of malaria infection is splenomegaly. Hyperactive malarious splenomegaly, apparently based on a disturbance in the T-lymphocyte control of the humoral responses to recurrent malaria, is common in many parts of the tropical and semitropical world. Mortality is high among those with gross disease.[39] This disorder is due primarily to an increase in the number of splenic macrophages (macrophage hyperplasia). The increase in macrophage content is apparently the result of an accumulation of circulating monocytes that are attracted to the spleens of infected hosts by a mononuclear cell chemotactic factor, probably of lymphocytic origin.[141]

Many cases have been reported in which concurrent malaria suppresses the host immune response to another parasite. Thus, malaria enhances Toxoplasma infections, in which T-cell control is believed to be important. Gastrointestinal bacterial and respiratory diseases are often unusually severe in children with acute falciparum malaria. Plasmodial infections have also been implicated as an etiologic factor in the expression of certain viral diseases. Because multiple infections by other disease agents are common occurrences in malaria-endemic regions, the phenomenon of **immunosuppression** is of critical importance. Houba characterized immunosuppression as an impairment of "humoral and/or cellular (or both) effectors of the immune response to antigenic stimulation in an otherwise normally developed immune competent system."[69] The mechanisms responsible for producing immunosuppressed states in malaria-infected hosts are not clearly understood. In acute malaria, immunosuppression is thought to be the result of a variety of complex and perhaps interacting factors. Large quantities of parasite antigen elaborated during infections may affect the immune system in several ways: (1) induction of macrophage dysfunctions by saturating Ag-binding sites, (2) production of immune paralysis in B- and/or T-cell populations, or (3) direct competition with other antigens. Distribution of antigen-processing functions in macrophages, breakdown of communication between T and B cells and the production of nonspecific (heterophile) antibodies have also been suggested as mechanisms contributing to immunosuppression.

In his 1983 presidential address before the Royal Society of Tropical Medicine and Hygiene, Sir Ian McGregor stated that recent studies "lend support to the long-recognized phenomenon that congenital infections are rare in highly endemic areas and where the majority of mothers are effectively immune. Extending this concept somewhat, they also suggest that of the malariostatic factors which may curtail parasite replication in the newborn infant, i.e., transplacentally acquired antibody, the presence of haemoglobin F within the erythrocyte, an aging red cell population and dietary deficiency of para-aminobenzoic acid, it is perhaps the inherited antibody factor which is predominantly responsible for the freedom from malaria that the young infant enjoys in the early weeks of life."[101]

For further information on the immunologic and immunopathologic aspects of malaria, see Speer and Silverman,[127] Voller,[133] and Weatherall and Abdalla.[137] For chemotherapy, see Bruce-Chwatt,[22] and Howells.[70]

Inherited Resistance

Ethnic differences in malaria remain largely unexplained. White people are more susceptible to malaria, and blacks have greater tolerance and more rapidly acquired clinical immunity. Blacks are less resistant to falciparum and malariae than to vivax malaria.

An erythrocyte surface antigen (Duffy blood group antigen) has been shown to serve as an attachment or internalization receptor for *Plasmodium vivax* and *P. knowlesi* and is necessary for the parasite's entry into host cells.[23,107] Many blacks lack the Duffy antigen; this lack could explain, at least in part, the foregoing racial differences.

An abnormal condition in man, **sickle cell anemia,** may provide protection against falciparum malaria. This condition is a molecular disease caused by the presence of a gene (Hb₁S) responsible for the production of an abnormal hemoglobin (hemoglobin S). Precipitation of hemoglobin S within erythrocytes destroys their oxygen-carrying abilities while causing the cells to assume a characteristic sickled shape. The homozygous pair results in the disease, which is generally fatal. The heterozygous individual is a healthy carrier of the trait, but for reasons not clear, is protected from malaria. Sickle cell anemia is most common in East African children between six weeks and six years of age, but it occurs in other parts of Africa and in some southern European and Asian populations. It also occurs in adults and in other parts of Africa and some southern European and Asian populations. Toxic heme (ferriprotoporphyrin IX) has been suggested as a cause of malaria parasite death in sickle cells.[111] Kidney damage and leg ulceration may be associated with sickle cell disease. "The possibility of vaccination and improvement in the effectiveness of nonimmunologic techniques, require much hardheaded thinking and provide an intellectual challenge of the highest order. There are no simple answers."[18]

Numerous hypotheses attempting to explain the relationship between the Hb₁S gene and its phenotypic expression include suggestions that parasites have difficulty in entering cells containing hemoglobin S, or that following successful entry, parasites' nutritional needs are not met. A more recent explanation is

"that sequestration of trophozoite-infected heterozygous cells in the venules leads to the sickling of the host cell, loss of erythrocytic potassium, and parasite death. The resulting attenuation of parasite multiplication would favor the survival of the sickle cell hemoglobin heterozygote and maintain the sickle cell hemoglobin gene at high frequencies in areas endemic for falciparum malaria."[55]

In the first five months of life, children with sickle cell trait (Hb₁S) are protected by maternal antibodies. For the next five years, before these children acquire immunity to malaria, they are at least partially resistant, probably because of preferential phagocytosis of parasitized red blood cells.[36] In a 1979 review of sickle cell disease in Africa, Molineaux et al. stated that "with rising standards of living and control of malaria, sickle cell anemia will become an immense medical, social and economic problem throughout the continent."[108]

Other genetic abnormalities thought to confer some degree of natural resistance to malaria are glucose-6-phosphate dehydrogenase enzymopenia,[14] hemoglobin C, and thalassemia. See Hill[67] for a discussion of the proposal that high frequencies of alpha thalassaemia are the result of selection by malaria.

Resistance of *P. falciparum* to antimalarial drugs has emerged as a major technical factor in the control of malaria. In Latin America, Asia, and Africa, *P. falciparum* is becoming resistant to all of the commonly used drugs except quinine.[43] Certain mosquitoes are becoming resistant to insecticides and developing the capacity to avoid sprayed surfaces.

Malaria in Animals Other Than Man

Reptilian Malaria. This disease is common in lizards of North and South America, Africa, the East Indies, the Pacific islands, and Australia, but much rarer in Europe and Asia. About half of the approximately 125 described species of *Plasmodium* are found in lizards. The parasite is occasionally also found in snakes. *Plasmodium* species develop in erythrocytes and sometimes in leukocytes. Actively ameboid asexual forms occur in peripheral blood. Gametocytes may persist in the circulation for many months, sometimes without accompanying asexual forms. The transmitting insect vectors of most of the 37 or more described species are unknown. Sandflies of the genus *Lutzomyia* transmit *Plasmodium mexicanum* to lizards of the genus *Sceloporus* in California. Pathogenicity among the species of reptilian *Plasmodium* is widely varied. Erythrocyte loss and anemia are characteristic. *P. mexicanum* is especcialy pathogenic, with endothelial schizogony and severe recrudescence. For a comprehensive review of reptilian malaria, see Ayala.[9]

Avian Malaria. The wide geographic range of bird malaria is easily explained by migratory habits of the hosts. Mosquitoes of the genera

Aedes and *Culex* are the most common vectors; several other genera are involved, but probably not *Anopheles,* except under experimental conditions. *Plasmodium gallinaceum* has been transmitted orally to chickens. The subgenus *Haemamoeba* includes the common *Plasmodium relictum,* infecting mostly passerine birds (especially sparrows); *P. cathemerium,* also in passerine birds (type host, the domestic sparrow); and *P. gallinaceum,* found in the jungle fowl and chickens. In other subgenera are a number of species, including *P. lophurae,* that infect chickens and other gallinaceous birds; and *P. elongatum,* which infects finches, sparrows, and other passerine birds.

In contrast to *Plasmodium* in mammals, the avian species always have a nucleolus and typical mitochondria.[1] Garnham[57] has listed the chief differences between avian and primate exoerythrocytic stages as follows:

"The exoerythrocytic stages of avian parasites are found in mesodermal tissues; those of the primate occur in the liver parenchyma. The primary cycle occupies two generations in the avian and one in the primate parasites. The avian EE parasites are infective on subinoculation, the primate are non-infective; the former can arise from the inoculation of blood stages, the latter arise only from sporozoites or their direct descendants in hepatic tissue. Avian exoerythrocytic schizonts produce a maximum of a thousand merozoites, primates produce a minimum of a thousand and normally many more."

Biochemical differences also occur between avian and mammalian plasmodial parasites. Glucose is catabolized in the avian form, *P. lophurae,* primarily by the Krebs citric acid cycle.[125] Other routes are used by most species in mammalian blood. Evidence[128] sugggests that complete aerobic respiration through the Krebs cycle depends on the presence of mitochondrial cristae. In general, erythrocytic stages of avian plasmodia possess cristate mitochondria, whereas similar stages in most mammals do not. Mammalian plasmodia (for example, *P. brasilianum*) that do possess cristate mitochondria probably use the Krebs cycle for at least part of their energy requirements.

Terzakis et al.[129] have studied the development of avian *Plasmodium gallinaceum* in the mosquito, *Aedes aegypti.* The researchers found that the oocyst cytoplasm rapidly segregates itself into sporoblasts, each with a number of nuclei, by a process of vacuole formation. Sporoblasts develop into sporozoites by a budding process (Fig. 5–25). Sporozoites, each with a single nucleus and a minute pore *(micropyle)* at one end, become distributed throughout the body of the mosquito when the enlarged oocyst bursts and are transferred to the vertebrate host through the insect bite.

Pathologic changes in avian malaria are primarily involved with dysfunction in circulation. Two characteristic symptoms are severe anemia and extreme water loss. The disease in *Plasmodium gallinaceum* is influenced by the site of secondary exoerythrocytic schizogony. Growth of tissue stages in capillary endothelium of various organs, especially in the brain, results in blockage of the vessels, and the host dies of "cerebral malaria." If the disease is caused by *P. elongatum,* parasites invade bone marrow, and the birds may die of aplastic anemia.

For a review of avian malaria, see Seed and Manwell.[122]

Malaria in Nonhuman Primates. Twenty species of malaria parasites have been described from nonhuman primates. There are a number of subspecies or varieties. *Plasmodium cynomolgi* and *P. inui* are among 13 species occurring in many Asian countries from Sri Lanka to the Philippines. Simian malaria in continental Africa is restricted to *Plasmodium gonderi* and *P. schwetzi;* and 2 species are found in the Malagasy Republic. In the New World (South America), *P. brasilianum* and *P. simium* (Plate IV) are found. Both these species, like *Plasmodium* species in birds and reptiles, possess mitochondria with prominent cristae and a nucleolus. Both ingest host cytoplasm almost exclusively through the cytostome. Moreover, in *P. brasilianum,* and in *P. malariae* and *P. falciparum,* knob-like processes occur on the surface of all parasitized erythrocytes. These processes are said to be the means by which parasitized cells are sequestered during development of the immune responses.

Among several *Plasmodium* species living in apes are *P. hylobati* in gibbons, *P. pitheci* in orangutans, and *P. schwetzi* in chimpanzees. The last species mentioned resembles *P. ovale* in man. *P. malariae* is also found in chimpanzees. Some of these species may have evolved in the following sequence: *P. malariae* in chimpanzees to *P. malariae* in man to *P. brasilianum* in South American monkeys.[56]

Human malarial parasites can be successfully introduced into other primates. *P. vivax, P. falciparum,* and *P. malariae* have been experimen-

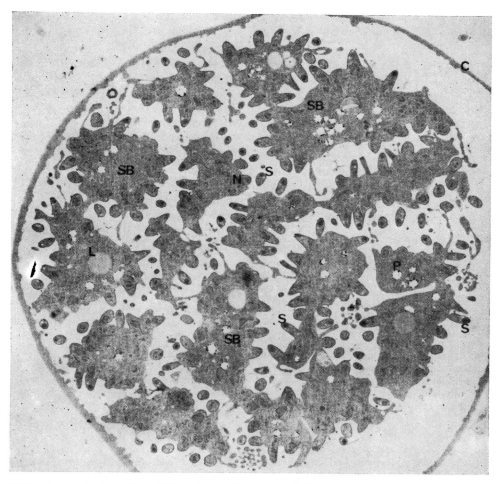

Fig. 5–25. Sporozoite formation of *Plasmodium gallinaceum.* Individual sporozoites *(S)* emerge from the sporoblast cytoplasm *(SB)* by budding N, Nuclei; *P,* pigment inclusions; *L,* lipid droplets. ×4500. (From Terzakis, Sprinz, and Ward, courtesy of Military Med.)

tally given to New World monkeys (for example, *Aotus),* and from the monkeys, mosquitoes have been infected with gametocytes; the resulting sporozoites can reinfect man. All four human species can easily be transferred to chimpanzees that have undergone splenectomy and to gibbons. The presence of hemoglobin A$_2$ in the blood of New World monkeys, chimpanzees, man, and probably gorillas (but not in Asian macaque monkeys) may be an important common factor in explaining this wide susceptibility to infection.[58]

Experimental infections in man can be achieved through the bites of mosquitoes in-fected with at least four species of *Plasmodium* from monkeys and apes, including *P. cynomolgi, P. brasilianum,* and *P. schwetzi.* Humans can also be infected by simian blood containing any one of five species of *Plasmodium,* including *P. inui, P. knowlesi,* and *P. schwetzi* (chimpanzee strain). A naturally acquired human infection of *P. simium* has been reported, and the first proof that simian malaria is a true zoonosis was published in 1965,[29] and involved a case of natural infection of man by *P. knowlesi,* a species normally found in monkeys in Malaysia. The Columbian owl monkey, *Aotus trivigatus,* is an excellent laboratory primate for experimental malarial studies, cultivation, and immunologic investigations. See Collins and Aikawa[34] for details of non-human primate malaria.

Simian malarial parasites, like those of man, have a complex of antigens.

"Immunity is conferred on one another by all homologous strains of simian malaria parasites. The various *P. cynomolgi* subspecies show no cross immunity either to each other or to different species. . . . The African parasite, *P. gonderi*, was able to infect monkeys immune to all oriental species tested, including *P. knowlesi*. . . . The three species, *P. knowlesi, P. coatneyi* and *P. fragile*, showed considerable cross immunity amongst themselves and against other oriental species, suggesting that they possess a broad spectrum of common antigens."[134]

Rodent Malaria. The generally accepted species of rodent *Plasmodium* are *P. berghei, P. yoelii, P. vinckei,* and *P. chabaudi.* Rodent malaria is found in murine rodents, flying squirrels, and porcupines and is easily transmitted to a number of different kinds of laboratory animals. It is absent from the New World. See Killick-Kendrick[77,78] for a review of taxonomy, zoogeography, and evolution. For further details of life cycles and morphology, see Landau and Boulard.[84]

A large part of our information about malaria has come from research on *Plasmodium berghei.* The erythrocytic forms have a strong predilection for reticulocytes or younger red blood cells. Growth and division synchrony can be abolished by experimental removal of host pineal body. The pineal also mediates the vascular capture and release of late growth forms of this parasite. The mechanism by which the pineal is activated and by which it carries out its part in regulating growth and division synchrony is unknown. It may be the junction point for a neural and chemical pathway. Evidence for this suggestion comes from the finding that the pineal hormone, ubiquinone, and the related compounds vitamin K and vitamin E, play a role in augmenting growth and division synchrony of *P. berghei* infections in mice.[10] Marked anemia and shock have been found to be the major causes of death in rats infected with *P. berghei.*

Bats are commonly infected with *Plasmodium.* See Bruce-Chwatt[19] for general biology, Sinden[125] for cell biology, Homewood[68] for biochemistry, Beale et al.[13] for genetics, Nussenzweig et al.[108] for immunology, and Carter and Diggs[27] for a general review of *Plasmodium* species in rodents.

Other Haemosporina

Only three of the several other genera of haemosporines will be mentioned. For a comprehensive review of them, see Fallis and Desser,[49] and for an account of their fine structure, see Aikawa and Sterling.[2] Schizogony does not occur in peripheral blood cells, but gametocytes are common in erythrocytes, and, for *Leucocytozoon,* in leukocytes. Sporogony takes place in insects other than mosquitoes. Vertebrate hosts are reptiles, birds, and mammals.

Hepatocystis is commonly found in arboreal mammals, especially in lower monkeys of the Old World tropics, in gibbons rarely, and in fruit bats. In African monkeys, *H. kochi* lives in parenchymal cells of the liver and causes pathologic changes. Glistening cysts (merocysts) appear on the liver's surface. Merocytes enter erythrocytes and develop into gametocytes. The host cells are pigmented. Biting midges of the family Ceratopogonidae are intermediate hosts. In *Culicoides adersi,* oocysts develop between the eyes and brain. Mature oocysts contain hundreds of sporozoites.

Haemoproteus occurs in domestic ducks and turkeys and in wild birds and reptiles. Gametocytes (Fig. 5–26) occur in host erythrocytes, and schizogony takes place in endothelial cells of blood vessels. Pigment is produced in infected red blood cells. Vectors are generally louse flies (Hippoboscidae). *H. columbae* is a familiar parasite of the pigeon, *Columba livia,* and other birds. The sexual phase occurs in the hippoboscid flies of the genera *Lynchia, Pseudolynchia,* and *Microlynchia.* Fertilization takes place in the fly's stomach, and zygotes enter the stomach wall. Asexual reproduction produces a cyst full of sporozoites. The cyst ruptures, and liberated sporozoites enter the bird host with the bite of the fly. Mosquitoes are also used as hosts.

Sporozoites now start the schizogonic cycle by entering endothelial cells of blood vessels of the lungs and other organs and dividing. Merozoites (or cytomeres) are produced and are released by rupture of infected exoerythrocytic host cells. These merozoites may enter other endothelial cells or red blood cells. In the latter, they become sausage-shaped and develop into gametocytes that mature to gametes in the fly's stomach.

Haemoproteus metchnikovi, a parasite of turtles, is transmitted by a tabanid fly, *Chrysops*

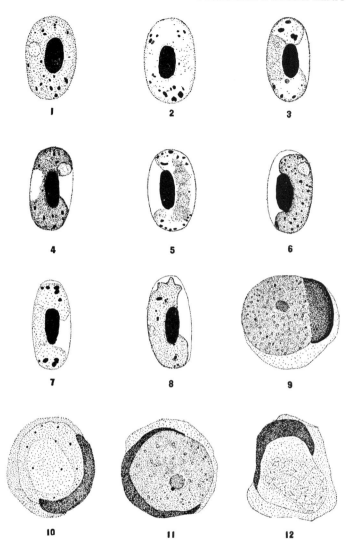

Fig. 5–26. Species of *Haemoproteus* and *Leucocytozoon* from the blood of various birds. *1* to *10* are × 1942; *11* and *12* are × 1750. *1* and *2*, *Haemoproteus archilochus* from the ruby-throated humming bird, showing a macro- and microgametocyte entirely enclosing the host cell nucleus. *3* and *4*, *Haemoproteus quiscalus* from the bronzed grackle, showing a micro- and macrogametocyte. *5* and *6*, *Haemoproteus* sp. from the Baltimore oriole, showing a micro- and macrogametocyte. *7* and *8*, *Haemoproteus* sp. from the blue jay, showing a micro- and macrogametocyte. *9* and *10*, *Leucocytozoon coccyzus* from the yellow-billed cuckoo, showing a macro- and microgametocyte. *11* and *12*, *Leucocytozoon sakharoffi* from the eastern crow, showing a macro- and microgametocyte. (From Coatney and West, courtesy of the American Midland Naturalist.)

callidus. See Atkinson, et al.,[78] for pre-erythrocytic development of *H. meleagridis* in turkeys.

Leucocytozoon (syn. *Akiba* is found in birds and is transmitted by blackflies (genus *Simulium*) or midges (genus *Culicoides*). The most marked pathologic feature is anemia. The infection may be highly pathogenic or fatal. Ovoid or fusiform parasites occur in erythrocytes and occasionally in leukocytes; and schizogony takes place in cells of the liver, kidney, and in reticuloendothelial cells throughout the body. Pigment is absent in all stages. More than 20 species of this genus exist. *L. simondi* in ducks (Fig. 5–27), *L. smithi* in turkeys, and *L. caulleryi* in chickens have all been reported to cause se-

vere losses in flocks. *L. simondi* is transmitted to a large extent in North America by *Simulium rugglesi*. The course of the disease in ducklings may be so rapid that the birds appear in good health one day and are dead the next. No treatment is effective.

Sporozoites are inoculated by an infected fly into a bird (Fig. 5–27), and then initiate schizogony in liver parenchymal cells. When hepatic schizonts (up to 45 μm) are mature, they release merozoites and syncytia. Merozoites penetrate erythrocytes and erythroblasts and become round gametocytes. Syncytia, with two or more nuclei, become phagocytized and are carried by the blood to various organs, espe-

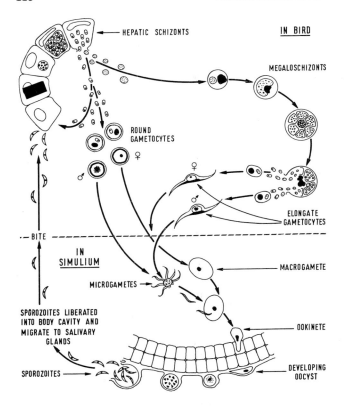

HEPATIC SCHIZONTS

IN BIRD

MEGALOSCHIZONTS

ROUND
GAMETOCYTES

♀
♂

ELONGATE
GAMETOCYTES

— BITE

IN
SIMULIUM

MACROGAMETE

MICROGAMETES

SPOROZOITES LIBERATED
INTO BODY CAVITY AND
MIGRATE TO SALIVARY
GLANDS

OOKINETE

SPOROZOITES

DEVELOPING
OOCYST

Fig. 5–27. *Leucocytozoon simondi* life cycle. (From Adam, K.M.G., et al.: Medical and Veterinary Protozoology. Edinburgh, Churchill Livingstone, 1971.)

cially spleen and lymph nodes, where they develop into large megaloschizonts (up to 200 μm diameter). The latter cause extreme hypertrophy of the host cell nucleus. Merozoites from megaloschizonts develop exclusively in leukocytes, predominantly in lymphocytes and monocytes whose nuclei become elongated and distorted, and they mature to form fusiform gametocytes. Pathogenicity is probably related to the occurrence of megaloschizonts. Many other species of *Leucocytozoon* do not form megaloschizonts.

Blackflies acquire the infection from adult birds and, on return for a second meal, feed preferentially on the downy young. Thus, the infection survives from year to year. Symptoms of infection include loss of appetite, lethargy, labored breathing, diarrhea, convulsions, weight loss, reduction of egg production, and occasionally death. The liver and the spleen may be markedly hypertrophied.

Sporozoites are formed in the insect vectors. Exflagellation, fertilization, ookinete formation from the zygote, penetration of the midgut epithelium by the ookinete, and transformation into oocysts follow a pattern of development similar to that exhibited by malarial parasites. Each oocyst forms about 50 sporozoites.

Birds surviving the infection continue to harbor parasites, but at low level. In spring, an increased parasitemia occurs, termed a relapse, possibly related to host reproductive activity. A state of premunition is initiated by the infection, and continual introduction of sporozoites into the birds is apparently necessary for protection against disease. A related genus, *Saurocytozoon*, occasionally occurs in South American lizards. See Fallis et al.[50] for a review of the family Leucocytozoidae.

Subclass Piroplasmia

Order Piroplasmida

Members of this order are pyriform, round, oval, rod-shaped, or ameboid parasites of various blood cells or histiocytes of vertebrates. Locomotion is by gliding or body flexion. Reproduction is by binary fission or schizogony; sexual stages have been described. The apical complex is reduced and, in the first two families discussed here, consists of a polar ring, rhoptries, and sometimes a conoid, micronemes, a micropore, and (in Babesiidae only) subpellicular tubules. Ticks are commonly used as intermediate hosts. For a detailed review of the piroplasmids, see Mehlhorn and Schein.[104]

Relationships among species in this order are not entirely clear. *Piroplasma, Babesiella,* and *Nuttallia* have been described from mammals, *Entopolypoides* from monkeys, *Aegyptianella* from birds, and *Babesiosoma* (= *Haemohormidium*) from amphibians and fish. Final determination concerning the relationships among these genera must await more detailed studies of immune reactions, electron microscope photographs, and life history investigations. Arthur[7] has discussed the ecology of ticks with reference to the transmission of piroplasms.

FAMILY BABESIIDAE

Babesia has caused dramatic loss of livestock and other animals. In 1981 it was estimated[117] that half a billion cattle throughout the world may be endangered by the disease known as *babesiosis, piroplasmosis,* or *cattle tick fever,* caused by one or more of the 17 species of *Babesia.* It is now under control in several countries.

In the erythrocytes of cattle and other ungulates, *B. bigemina* typically multiplies by binary fission, and two pear-shaped or roughly rounded trophic bodies are frequently observed. These trophozoites (2 to 5 μm long) ingest red blood cell cytoplasm by pinocytosis in some species and by a cytostome or tubular feeding structure in others. The pigment hemozoin is not formed. Vectors are ticks of the genus *Boophilus* (numerous species), *Haemaphysalis,* and *Rhipicephalus,* which feed on cattle blood and within which the sporogony phase of the parasite cycle takes place[119] (Fig. 5–28). Parasites ingested by ticks in one generation are infective to cattle during a parasitic stage of the next generation.[24]

"Irrespective of animal species infected, the major clinical signs are fever and intravascular hemolysis. The latter can induce anemia, hemoglobinemia, and, less frequently, hemoglobinuria . . . there is prominent evidence of bone marrow hyperplasia. A reduction in red blood cells greater than 50% may occur. . . Cases resulting in persistent anemia may manifest localized edema of the subcutis, ascites, and weight loss. Neurologic signs including delirium and convulsions have been observed in cattle . . . . and the dog. . . Renal disease characterized by uremia has been reported in dogs. . . Pregnant females may abort.

"In the spleen, lymph nodes and liver it is the histiocyte that undergoes remarkable hypertrophic and hyperplastic changes during babesiosis. This cellular response in the spleen is a major factor in the splenomegaly of babesial infections. . . . . It is the lymphocyte and monocyte-macrophage responses, clumping of erythrocytes, thrombosis, and vascular endothelial changes that focus attention on the role of vascular damage in these diseases. Generally speaking, however, the major causes for morphologic lesions encountered are anemic anoxia and red cell destruction."[66]

"The pathogenesis of *Babesia* infections can be categorized into two major pathways: (1) A primary pathway occurring early in infection when parasitemia is low, and in which the effects are through largely non-specific mechanisms mediated through vasoactive substances that interact with the coagulatory and complement cascades leading to circulatory collapse and blockade of the reticuloendothelial system. This pathway can occur in the presence of very low parasitemias and is therefore largely independent of parasite load. A major factor in the primary pathway is the enigmatic increase in osmotic fragility of the erythrocyte population; (2) the secondary pathway which depends on parasite load and the associated presence of soluble plasma antigen, specific antibody, immune complexes, and fibrin-fibrinogen complexes. This pathway triggers cell-mediated immunity and activation of the macrophage system and induces indiscriminate phagocytosis of normal and infected erythrocytes. Immune complexes, complement and auto-immunity are incriminated in this pathway on the basis of largely circumstantial evidence."[75]

Babesiosis is primarily a disease of older animals. Cattle may suffer fever, anoxia, loss of appetite, constipation or diarrhea, vomiting, bloody urine, and anemia. Experimental infections of cattle and dogs have revealed exoerythrocytic parasites in the lumen of capillaries of almost all internal organs, including the brain. The mortality rate is about 90%, and death may occur in one week. Treatment is of little value, but control is successful when the tick can be destroyed. Immunization, using living parasites, is a practical procedure.

Babesia in dogs enter red blood cells and cause an increase in temperature, pulse rate, and respiration rate. Dogs become weak and anemic and often die, but survivors generally acquire an immunity after infection. The disease is transmitted by the brown dog tick, *Rhipicephalus sanguineus. B. ovis* in sheep and goats is transmitted transovarially and alimentarily by the

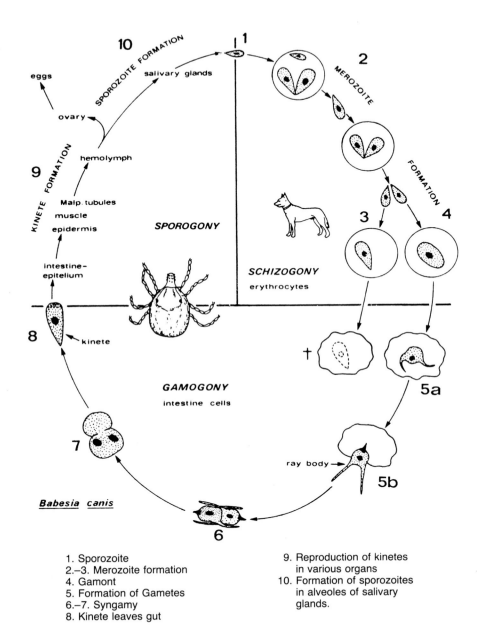

1. Sporozoite
2.–3. Merozoite formation
4. Gamont
5. Formation of Gametes
6.–7. Syngamy
8. Kinete leaves gut

9. Reproduction of kinetes
   in various organs
10. Formation of sporozoites
    in alveoles of salivary
    glands.

Fig. 5–28. Life cycle of *Babesia canis*. (From Mehlhorn and Schein, courtesy of Academic Press.)

tick *Rhipicephalus burse*. *B. bovis* is transmitted transovarially by the tick *Boophilus*.

Immunization is readily achieved with the administration of living parasites, and the innate immunity of *Bos indicus* (cattle) is well documented. Vaccination, however, is used with varying success, depending on age of host, previous exposure to parasites, and virulence of parasites.

Over 100 human cases of babesiosis have been reported as of 1980, mostly from North America. Out of 7 cases in Europe, 5 were fatal. All the patients reported from Europe had undergone splenectomy before becoming infected. These infections were generally caused by species of *Babesia* that normally infect cattle (for example, *B. bovis*), whereas human babesiosis in North America, usually caused by *B. microti*, a parasite of rodents, is much less pathogenic than the European species. Serologic tests in the United States and Mexico have revealed 54 presumptive cases; this finding probably indicates a high percentage of symptomless human infections. In stained human blood smears, the parasites have a remarkable resemblance to ring stages of *Plasmodium falciparum*, but without the pigment typical of the latter.[65]

Fluorescent antibody tests have shown some degree of relationship between antigens of *Babesia* and those of *Plasmodium*. Cross-immune reactions, however, may be due to nonspecific reactions independent of species-specific and genus-specific antigens. Tests have shown that common antigens might be shared by *Babesia bovis* (= *B. argentina*), *P. falciparum*, and *P. vivax*. There is thus a possibility that *B. bovis* of cattle might induce in man the production of antibodies capable of reacting with human malarial parasites. Other species occur in horses, sheep, goats, hogs, deer, camels, cats, rodents, mongoose, birds, and other animals. *B. bovis* can cause extreme skeletal muscle degeneration. It is transmitted transovarially by the tick *Boophilus bigemina*. Some other species cause a clogging of small capillaries of skin and brain with parasitized and non-parasitized erythrocytes. *B. ovis* in sheep and goats is transmitted transovarially and alimentarily by the tick, *Rhipicephalus burse*. *B. equi* differs from other species of *Babesia* in its formation of schizonts in lymphocytes of the horse, thus being similar to sporogony in *Theileria* (see below). See Rudzinska

et al.[119] for details of the invasion of *B. microti* into tick gut epithelial cells.

See Ristic and Levy[117] and Rudzinska et al.[119]

FAMILY THEILERIIDAE

See Young and Morzaria (1986)[142] for an excellent review of *Babesia* biology. There are areas of controversy in the life cycles of these parasites, and the genus apparently contains different groups of individuals that may require a modification of present classification. Immunization of cattle against theileriasis is successful.[41]

FAMILY DACTYLOSOMATIDAE

Theileria parva causes East Coast fever (theileriasis) in cattle, buffalo, and other mammals in Africa from Kenya to South Africa. The mortality in domestic cattle has exceeded 90%. Artificial immunization against the disease has not been successful, but the cattle that survive are immune to further infection. Symptoms include high fever, diarrhea, weakness, runny eyes, nasal discharge, swollen lymph nodes, and blood in feces. Schizogony takes place in the lymphatic system and in erythrocytes. Because red blood cells are not destroyed, there is no anemia or blood in the urine.

The principal vector is *Rhipicephalus appendiculatus*. Other genera of vectors are *Hyalomma*, *Amblyomma*, and *Haemaphysalis*. When a tick larva or nymph becomes infected, it is not infective to cattle until it reaches the next stage of development. The parasite's sexual cycle begins in the tick gut,[105] and sporogony occurs in the salivary glands. The modes of development of the kinete in tick cells is similar to that of *Babesia* in its tick cells. For an account of intraerythrocytic schizogony of *Theileria annulata* in cattle, see Conrad et al.[35] Among other species of *Theileria* that are parasites of ruminants are: *T. annulata* and *T. ovis*.

Dactylosoma is found in the blood of reptiles, amphibians and fish; schizogony and gametogony take place in red blood cells. Vectors are unknown. It has been transferred to the Eucoccidiida.

Members of this family are found in erythrocytes of fishes, amphibians, and reptiles. Vectors are unknown. Three genera are *Dactylosoma*, *Haemohormidium* (syn. *Babesiosoma*) and *Sauroplasma*. *Babesiosoma stableri* and *Rana* spp. is transmitted by a leech which is the only known vector for any member of this family.

Blastocystis hominis causes disease in humans.

REFERENCES

1. Aikawa, M.: The fine structure of the erythrocytic stages of three avian malarial parasites, *Plasmodium fallax, P. lophurae,* and *P. cathemerium.* Am. J. Trop. Med. Hyg., 15:449–471, 1966.
2. Aikawa, M., and Sterling, C.R.: Intracellular Parasitic Protozoa. New York, Academic Press, 1974.
3. Angus, K.W.: Cryptosporidiosis in man, domestic animals and birds: A review. J. Roy. Soc. Med., 76:62, 1983.
4. Anderson, B.C.: Location of cryptosporidia: Review of the literature. Am. J. Vet. Res., 45:1474–1477, 1984.
5. Allison, A.C., and Clark, I.A.: Specific and nonspecific immunity to haemoprotozoa. Am. J. Trop. Med. Hyg., 26:216–232, 1977.
6. Ash, L.R., and Orihel, C.: Parasites: A Guide to Laboratory Procedures and Identification. Chicago, Am. Soc. Clin. Pathol. Press, 1987.
7. Arthur, D.R.: The ecology of ticks with reference to the transmission of Protozoa. *In* Biology of Parasites. Edited by E.J.L. Soulsby. New York, Academic Press, 1966, pp. 61–84.
8. Atkinson, C.T., Greiner, E.C., and Forrester, D.J.: Pre-erythrocytic development and associated host responses to *Haemoproteus meleagridis* (Haemosporina: Haemoproteidae) in experimentally infected domestic turkeys. J. Protozool., 33(3):375–381, 1986.
9. Ayala, S.C.: Plasmodia of reptiles. *In* Parasitic Protozoa. Vol. 3. Edited by J.P. Kreier. New York, Academic Press, 1977, pp. 267–309.
10. Bahr, G.F.: Quantitative cytochemistry of malaria infected erythrocytes *(Plasmodium berghei, Plasmodium chabaudi* and *Plasmodium vinckei).* Milit. Med., 134:1013–1025, 1969.
11. Bannister, L.H.: The invasion of red cells by *Plasmodium. In* Parasitic Invasion. Symposium of the British Society for Parasitology. Vol. 15. Edited by A.E.R. Taylor and R. Muller. Oxford, Blackwell Scientific Publications, 1977, pp. 27–55.
12. Bannister, L.H., and Sinden, R.E.: New knowledge of parasite morphology. *In* Malaria. Edited by S. Cohen. British Med. Bull., 38(2):115–218, 1982.
13. Beale, G.H., Carter, R., and Walliker, D.: Genetics. *In* Rodent Malaria. Edited by R. Killick-Kendrick and W. Peters. London, Academic Press, 1978, pp. 213–235.
14. Bienzle, U., Guggenmoos-Holzmann, I., and Luzatto, L.: Malaraia and erythrocyte glucose-6-phosphate dehydrogenase variants in West Africa. J. Trop. Med. Hyg., 28:619–621, 1979.
15. Boufassa-Auzrout, S., Chermette, R., and Meissomier, E.: Cryptosporidiosis: a cosmopolitan disease in animals and in man. Paris, Office Internat. des Epizooties, Tech. Series No. 5, 1986.
16. Brand, von, T.: Biochemistry and Physiology of Endoparasites. Amsterdam, Elsevier North-Holland Biomedical Press, 1979.
17. Brown, K.N.: Antigenic variation in malaria. Adv. Exp. Med. Biol., 93:5–25, 1977.
18. Brown, K.N.: Host resistance to malaria. *In* Critical Reviews in Tropical Medicine. Vol. 1. Edited by R.K. Chandra. New York, Plenum Press, 1982, pp. 171–196.
19. Bruce-Chwatt, L.J.: Introduction. *In* Rodent Malaria. Edited by Killick-Kendrick and W. Peters. London, Academic Press, 1978, pp. xi–xxv.
20. Bruce-Chwatt, L.J.: Essential Malariology. 2nd Ed. London, Heinemann, 1985.
21. Bruce-Chwatt, L.J.: Imported malaria: an uninvited guest. Brit. Med. Bull., 38(2):179–185, 1982.
22. Bruce-Chwatt, L.J. (ed): Chemotherapy of Malaria, Revised 2nd Edition. World Health Organization. Geneva, Switzerland, 1986.
23. Butcher, G.A., and Cohen., S.: Antigenic variation and protective immunity to *Plasmodium knowlesi* malaria. Immunology, 23:503–521, 1972.
24. Callow, L.L.: Vaccination against bovine babesiosis. *In* Immunity to Blood Parasites of Animals and Man. Edited by L.H. Miller, J.A. Pino, and J.J. McKelvey, Jr. New York, Plenum Press, 1977, pp. 121–149.
25. Canning, E.U.: Terminology, taxonomy and life cycles of Apicomplexa. Insect Sci. Applic., 7(3):319–325, 1986.
26. Canning, E.U., and Lom, J.: The Microsporidia of Vertebrates. London, Academic Press, 1986.
27. Carter, R., and Diggs, C.: Plasmodia of rodents. *In* Parasitic Protozoa. Vol. 3. Edited by J.P. Kreier. New York, Academic Press, 1977, pp. 359–465.
28. Chen, D.H., Tigelaar, R.E., and Weinbaum, F.I.: Immunity to sporozoite-induced malaria infection in mice, I. The effect of immunization of T and B cell-deficient mice. J. Immunol., 118:1322–1327, 1977.
29. Chin, W., Contacos, P.G., Coatney, G.R., and Kimball, H.R.: A naturally acquired quotidian-type malaria in man transferable to monkeys. Science, 149:865, 1965.
30. Chinchilla, M., and Frenkel, J.: Mediation of immunity to intracellular infection *(Toxoplasma* and *Besnoitia)* with somatic cells. Infect. Immun., 19:999–1012, 1978.
31. Chobotar, B., Scholtyseck, E., Senaud, J., and Ernest, J.: A fine structural study of asexual stages of the marine coccidium *Eimeria ferrisi* Levine and Ivens 1965. Z. Parasitenkd., 45:291–306, 1975.
32. Ciba Foundation Symposium 94: Malaria and the Red Cell. London, Pitman, 1983.
33. Cohen, S.: Immunity to malaria. Proc. R. Soc. Lond. [Biol.], 203:323–345, 1979.
34. Collins, W.E., and Aikawa, M.: Plasmodia of nonhuman primates. *In* Parasitic Protozoa. Vol. 3. Edited by J.P. Kreier. New York, Academic Press, 1977, pp. 467–492.
35. Conrad, P.A., Kelly, B.G., and Brown, C.G.D.: Intraerythrocytic schizogony of *Theileria annulata.* Parasitol., 91:67–82, 1985.
36. Cornille-Brøgger, R., et al.: Abnormal haemoglobins in the Sudan savanna of Nigeria. II. Immunological response to malaria in normals and subjects with sickle cell trait. Ann. Trop. Med. Parasitol., 73:173–183, 1979.
37. Corradetti, A.: About the concept of malaria as a disease. Parassitologia, 24(2,3):101–103, 1982.

38. Cox, F.E.G.: Heterologous immunity between prioplasms and malaria parasites: the simultaneous elimination of *Plasmodium vinckei* and *Babesia microti* from the blood of double infected mice. Parasitology, 76:55–60, 1978.

39. Crane, G.G.: hyperactive malarious splenomegaly (tropical splenomegaly syndrome). Parasitol. Today, 2(1):4–9, 1986.

40. Crompton, D.W.T., and Nesheim, M.C.: Host-parasite relationships in the alimentary tract of domestic birds. *In* Advances in Parasitology. Edited by B. Dawes. London, Academic Press, 1976, pp. 96–175.

41. Cunningham, M.P.: Immunization of cattle against *Theileria parva*. *In* Immunity to Blood Parasites of Animals and Man. Edited by L.H. Miller, J.A. Pino, and J.J. McKelvey, Jr. New York, Plenum Press, 1977, pp. 189–207.

42. Current, W.L., and Reese, N.C.: A comparison of endogenous development of three isolates of *Cryptosporidium* in suckling mice. J. Protozool., 33:98–108, 1986.

43. Doberstyn, E.B.: Resistance of *Plasmodium falciparum*. Experientia, 40:1311–1317, 1984.

44. Dubey, J.P.: *Toxoplasma, Hammondia, Besnoitia, Sarcocystis,* and other tissue cyst-forming coccidia of man and animals. *In* Parasitic Protozoa. Vol. 3. Edited by J.P. Kreier. New York, Academic Press, 1977, pp. 101–237.

45. Dubey, J.P., Speer, C.A., Callis, G., and Blixt, J.A.: Development of the sheep-canid cycle of *Sarcocystis tenella*. Can. J. Zool., 60:2464–2477, 1982.

46. Dvorak, J., and Miller, L.H.: Invasion of erythrocytes by malaria merozoites. Science, 187:748–749, 1974.

47. Euzeby, J.: Les Parasitoses Humaines d'origine Animale, Characters Epidémiologigues. Paris, Flammarion Médecine-Sciences, 1984. (Ecole Vétérinaire, Lyon).

48. Evered, D., and Whelan, J. (eds.): Malaria and the red cell. Ciba Foundation Symposium 94. London, Pitman Books, 1983.

49. Fallis, A.M., and Desser, S.S.: On species of *Leucocytozoon, Haemoproteus,* and *Hepatocystis*. *In* Parasitic Protozoa. Vol. 3. Edited by J.P. Kreier. New York, Academic Press, 1977, pp. 239–266.

50. Fallis, A.M., Desser, S.S., and Khan, R.A.: On species of *Leucocytozoon.* Adv. Parasitol., 12:1–67, 1974.

51. Fournie, J.W., and Overstreet, R.M.: True intermediate hosts for *Eimeria funduli* (Apicomplexa) from estuarine fishes. J. Protozool., 30(4):672–675, 1983.

52. Frenkel, J.K.: *Besnoitia wallacei* of cats and rodents: with a reclassification of other cyst-forming isosporoid. Coccidia. J. Parasitol., 63:611–628, 1977.

53. Frenkel, J.K.: Toxoplasmosis. Pediat. Clin. N. Am., 32(4):917–932, 1985.

54. Frenkel, J.K.: Immunity in toxoplasmosis. Bull. Pan. Am. Health Org., 19(4):354–367, 1985.

55. Friedman, M., Roth, E.: Nagel, R., and Trager, W.: *Plasmodium falciparum*: physiological interactions with the human sickle cell. Exp. Parasitol., 47:73–80, 1979.

56. Garnham, P.C.C.: Recent research on malaria in mammals excluding man. Adv. Parasitol., 11:603–630, 1973.

57. Garnham, P.C.C.: Malaria Parasites and other Haemosporidia. Oxford, Blackwell Scientific Publications, 1966.

58. Geiman, Q.M., Siddiqui, W.A., Schnell, J.V.: Biological basis for susceptibility of *Aotus trivirgatus* to species of plasmodia from man. Milit. Med., 134:780–786, 1969.

59. Gregory, M.W., and Catchpole, J.: Coccidiosis in rabbits: the pathology of *Eimeria flavescens* infection. Int. J. Parasitol., 16(2):131–145, 1986.

60. Gutteridge, W.E., and Coombs, G.H.: Biochemistry of Parasitic Protozoa. Baltimore, University Park Press, 1977.

61. Gutteridge, W.E., Dave, D., and Richards, W.H.G.: Conversion of dihydroorotate to orotate in parasitic protozoa. Biochem. Biophys. Acta, 582:390–401, 1979.

62. Hadley, T.J., Klotz, F.W., and Miller, L.H.: Invasion of erythrocytes by malaria parasites; cellular and molecular overview. Ann. Rev. Microbiol., 40:451–477, 1986.

63. Hall, A.P.: Malaria. Roy. Soc. Health J., 100:57–61, 1980

64. Harrison, G.: Mosquitoes, Malaria and Man: A History of the Hostilities Since 1880. New York, E.P. Dutton, 1978.

65. Healy, G.R.: *Babesia* infections in man. Hosp. Pract., 14:107–116, 1979.

66. Hildebrandt, P.K.: The organ and vascular pathology of babesiosis. *In* Babesiosis. Edited by M. Ristic and J.P. Kreier. New York, Academic Press, 1981, pp. 459–473.

67. Hill, A.V.S.: Haemoglobinopathies and malaria: new approaches to an old hypothesis. Parasitol. Today, 3:83–85, 1987.

68. Homewood, C.A.: Biochemistry. *In* Rodent Malaria. Edited by R. Killick-Kendrick and W. Peters. London, Academic Press, 1978, pp. 169–211.

69. Houba, V.: Pathophysiology of the immune response to parasites. *In* Pathophysiology of Parasitic Infection. Edited by G.S.L. Soulsby. New York, Academic Press, 1976, pp. 221–232.

70. Howells, R.E.: Chemotherapy of malaria. Parasit. Today, 2(2):41–43, 1986.

71. Hughes, H.P.A.: Toxoplasmosis—a neglected disease. Parasitol. Today, 1(2):41–44, 1985.

72. Hutchison, W.M., and Hay, J.: The effect of congenital *Toxoplasma* infections on the motor performance of the mouse. *In* Parasitological Topics. Edited by E.U. Canning. Lawrence, Kansas, Society of Protozoologists, Allen Press, Inc., 1981, pp. 125–131.

73. Jacobs, M.R.: Toxoplasmosis *In* Medicine in a Tropical Environment. Proceedings of the International Symposium, South Africa, 1976. Cape Town, A.A. Balkema, 1977.

74. Kagan, I.G.: Serodiagnosis of parasitic diseases. *In* Manual of Clinical Parasitology. 3rd Ed. Edited by E.H. Lennett, A. Balows, W.J. Hausler, and J.P. Truant. Washington, D.C., American Society for Microbiology, 1980, pp. 724–750.

75. Kakoma, I., and Ristic, M.: Pathogenesis of ba-

besiosis. *In* Malaria and Babesiosis. Edited by Ristic, M., Ambroise-Thomas, P., and Kreier, J. Dordrecht, Martinus Nijhoff Publishers, 1984, pp. 85–93.

76. Khan, R.A.: Leeches as vectors of marine fish haematozoa. Fourth International Congress of Parasitology. Warsaw, Sect. C, 1978, p. 203.

77. Killick-Kendrick, R.: Taxonomy, zoogeography and evolution. *In* Rodent Malaria. Edited by R. Killick-Kendrick, and W. Peters. London, Academic Press, 1978, pp. 1–52.

78. Killick-Kendrick, R., and Peters, W. (eds): Rodent Malaria. London, Academic Press, 1978.

79. Kirkpatrick, C.E., and Farrell, J.P.: Cryptosporidiosis. Compend. Contin. Ed. Pract. Vet., 6:154–164, 1984.

80. Kreier, J.P. (ed.): Malaria. 3 Volumes. New York, Academic Press, 1980.

81. Krotoski, W.A.: Discovery of the hypnozoite and a new theory of malarial relapse. Trans. Roy. Soc. Trop. Med. Hyg., 79:1–11, 1985.

82. Krotoski, W.A., et al.: Observations on early and late post-sporozoite tissue stages in primate malaria. I. Discovery of a new latent form (hypnozoite), and failure of immunofluorescent technique to detect hepatic forms within the first 24 hours after infection. Am. J. Trop. Med. Hyg., 31(1):24–35, 1982.

83. Krotoski, W.A., et al.: Observations on early and late post-sporozoite tissue stages in primate malaria. II. The hypnozoite of *Plasmodium cynomolgi bastianellii* from 3 to 105 days after infection and detection of 36- to 40-hour pre-erythrocytic forms. Am. J. Trop. Med. Hyg., 31(2):211–225, 1982.

84. Landan, I., and Boulard, Y.: Life cycles and morphology. *In* Rodent Malaria. Edited by Killick-Kendrick and W. Peters. London, Academic Press, 1978.

85. Lawn, A.M., and Rose, M.E.: Mucosal transport of *Eimeria tennela* in the cecum of the chicken. J. Parasitol., 68:1117–1123, 1982.

86. Levine, N.D.: Taxonomy and review of the coccidian genus *Cryptosporidium* (Protozoa, Apicomplexa). J. Protozool., 31:94–98, 1984.

87. Levine, N.D.: Phylum II. Apicomplexa Levine, 1970. *In* An Illustrated Guide to Protozoa. Edited by J.J. Lee, S.H. Hutner, and E.C. Bovee. Lawrence, Kansas, Society of Protozoologists, 1985, pp. 322–374.

88. Levine, N.D.: The taxonomy of *Sarcocystis* (Protozoa, Apicomplexa) species. J. Parasit., 72:372–382, 1986.

89. Levine, N.D., and Baker, J.R.: The *Isospora-Toxoplasma-Sarcocystis* confusion. Parasitol. Today, 3:101–105, 1987.

90. Loke, Y.W.: Transmission of parasites across the placenta. Adv. Parasitol., 21:155–228, 1982.

91. Long, P.L. (ed.): The Biology of the Coccidia. Baltimore, Univ. Park Press, 1982.

92. Long, P.L., Boorman, K.N., and Freeman, B.M. (eds.): Avian Coccidiosis. Proceedings of the 13th Poultry Science Symposium, 1977. Edinburgh, British Poultry Science Ltd., 1978.

93. Lustig, H.J., Nussenzweig, V., and Nussenzweig, R.S.: Erythrocyte membrane-associated immunoglobulins during malaria infection of mice. J. Immunol., 119:210–216, 1977.

94. Manwell, R.D.: Gregarines and haemogregarines. *In* Parasitic Protozoa. Vol. 3. Edited by J.P. Kreier. New York, Academic Press, 1977, pp. 1–32.

95. Markus, M.B.: *Sarcocystis* and sarcocystosis in domestic animals and man. Adv. Vet. Sci. Comp. Med., 22:159–193, 1978.

96. Markus, M.B.: Terminology for invasive stages of Protozoa of the Subphylum Apicomplexa (Sporozoa). S. Afr. J. Sci., 74:105–106, 1978.

97. Martinez, A.J.: Free-Living Amebas: Natural History, Prevention, Diagnosis, Pathology, and Treatment of Disease. Boca Raton, Florida, CRC Press.

98. Matuschka, F.-R.: Life cycle of *Sarcocystis* between poikilothermic hosts. Lizards are intermediate hosts for *S. podarcicolubris* sp. nov, snakes function as definitive hosts. Z. Naturforsch, 36:1093–1095, 1981.

99. Matuschka, F.-R., and Heydorn, A.O.: Die Entwicklung von *Isospora suis* Beister und Murray 1934 (Sporozoa: Coccidia: Eimeriidae) im Schwein. Zool. Beiträge, 26:405–476, 1980.

100. McDonald, V., and Rose, E.M.: *Eimeria tenella* and *E. necatrix*: A third generation of schizogony is an obligatory part of the developmental cycle. J. Parasitol., 73:617–622, 1987.

101. McGregor, I.A.: Immunology of malaria infection and its possible consequences. Br. Med. Bull., 28:22–27, 1972.

102. McGregor, I.A.: Malaria—recollections and observations. Trans. Roy. Soc. Trop. Med. Hyg., 78:1–8, 1984.

103. Mehlhorn, H., and Heydorn, A.O.: The Sarcosporidia (Protozoa, Sporozoa): life cycle and fine structure. Adv. Parasitol., 16:43–92, 1978.

104. Mehlhorn, H., and Schein, E.: The piroplasms: life cycle and sexual stages. *In* Advances in Parasitology Volume 23. Edited by J.R. Baker and R. Muller. New York, Academic Press, 1984, pp. 37–103.

105. Mehlhorn, H., Schein, E., and Warnecke, M.: Electron-microscopic studies on *Theileria ovis* Rodhain, 1916: development of kinetes in the gut of the vector tick, *Rhipicephalus evertsi* Neumann, 1897, and their transformation within cells of the salivary glands. J. Protozool., 26:377–385, 1979.

106. Michael, E.: Structure and mode of function of the organelles associated with nutrition of the macrogametes of *Eimeria acervulina*. Z. Parasitenkd., 45:347–361, 1975.

107. Miller, L.H., and Carter, R.: Innate resistance in malaria. Exp. Parasitol., 40:132–146, 1976.

108. Molineaux, L., et al.: Abnormal haemoglobins in the Sudan savanna of Nigeria. III. Malaria, immunoglobulins and antimalarial antibodies in sickle cell disease. Ann. Trop. Med. Parasitol., 73:301–310, 1979.

109. Naval Medical Research Institute: Immunology of malaria. Bull. WHO, 57(Suppl.), 1979.

110. Nussenzweig, R.S., Cochrane, A.H., and Lustig, H.J.: Immunological responses. *In* Rodent Malaria. Edited by R. Killick-Kendrick and W.

Peters. London, Academic Press, 1978, pp. 243–307.

111. Orjih, A.U., Chevli, R., and Fitch, C.D.: Toxic heme in sickle cells: an explanation for death of malaria parasites. Am. J. Trop Med. Hyg., 34(2):223–227, 1985.

112. Overstreet, R.M.: Species of *Eimeria* in nonepithelial sites. J. Protozool., 28:258–260, 1981.

113. Overstreet, R., Hawkins, W., and Fourne, J.: The coccidian genus *Calyptospora* n.g. and family Calyptosporidae n. fam. (Apicomplexa), with members infecting primarily fishes. J. Protozool., 31:332–339, 1984.

114. Pasternak, J., and Fernando, M.A.: Host cell response to coccidian infection: an introspective survey. Parasitol., 88:555–563, 1984.

115. Quinn, T.C., and Wyler, D.J.: Mechanisms for action of hyperimmune serum in mediating protective immunity to rodent malaria (*Plasmodium berghei*). J. Immunol., 123:2245–2249, 1979.

116. Ristic, M., and Kreier, J.P. (eds.): Babesiosis. International Conference on Malaria and Babesiosis. Mexico City, 1979. New York, Academic Press, 1981.

117. Ristic, M., and Levy, M.G.: A new era of research toward solution of bovine babesiosis. *In* Babesiosis. Edited by M. Ristic and J.P. Kreier. New York, Academic Press, 1981, pp. 590–594.

118. Rose, W.E.: Immune responses of chickens to coccidia and coccidiosis. *In* Avian Coccidiosis. Edited by P.L. Long, K.N. Boorman, and B.M. Freeman. Edinburgh, British Poultry Science, Ltd., 1978, pp. 297–336.

119. Rudzinska, M., et al.: The sequence of developmental events of *Babesia microti* in the gut of *Ixodes dammini*. Protistologica, 20(4):649–663, 1984.

120. Ryley, J.F.: Why and how are Coccidia harmful to their hosts. *In* Pathogenic Processes in Parasitic Infections. Symposium of the British Society for Parasitology. Vol. 13. Edited by A.E.R. Taylor and R. Muller. Oxford, Blackwell Scientific Publications, 1975.

121. Schmidt, L.H.: Compatibility of relapse patterns of *Plasmodium cynomolgi* infections in rhesus monkeys with continuous cyclical development and hypnozoite concepts of relapse. Am. J. Trop. Med. Hyg., 35:1077–1099, 1986.

122. Seed, T.M., and Manwell, R.D.: Plasmodia of Birds. *In* Parasitic Protozoa. Vol. 3. Edited by J.P. Kreier. New York, Academic Press, 1977, pp. 311–357.

123. Shear, H.L., Nussenzweig, R.S., and Bianco, C.: Immune phagocytosis in murine malaria. J. Exp. Med., 149:1288–1298, 1979.

124. Sherman, I.W.: Biochemistry of *Plasmodium* (malarial parasites). Microbiol. Rev., 43:453–479, 1979.

125. Shuler, A.V.: ''Malaria''—Meeting the Global Challenge. Boston, Oelgeschlager, Gun & Hain, 1985.

126. Sinden, R.E.: The biology of *Plasmodium* in the mosquito. Experientia, 40:1330–1343, 1984.

127. Speer, C.A., and Silverman, P.H.: Recent advances in applied malaria immunology. Z. Parasitenkd., 60:3–17, 1979.

128. Sterling, C.R., Aikawa, M., and Nussenzweig, R.S.: Morphological divergence in a mammalian malarial parasite: the fine structure of *Plasmodium brasilianum*. *In* Basic Research in Malaria. Proc. Helminth. Soc. Wash., 39:109–129, 1972.

129. Terzakis, J.A., Sprinz, H., and Ward, R.A.: Sporoblast and sporozoite formation in *Plasmodium gallinaceum* infection of *Aedes aegypti*. Milit. Med., 131:847–1272, 1966.

130. Trager, W.: Living Together, the Biology of Animal Parasites. New York, Plenum Press, 1986.

131. Tzipori, S.: Cryptosporidiosis in animals and humans. Microbiol. Reviews, 47:84–96, 1983.

132. Upton, S.J., Current, W.L., and Barnard, S.M.: A review of the genus *Caryospora* Ieger, 1904 (Apicomplexa: Eimeriidae). System. Parasitol., 8:3–21, 1986.

133. Voller, A.: Aspects of the immunopathology of malaria. Symposium of the British Society of Parasitology. Vol. 13. Oxford, Blackwell Scientific Publications, 1975.

134. Voller, A., Garnham, P.C.C., and Targett, G.A.T.: Cross immunity in monkey malaria. J. Trop. Med. Hyg., 69:121–123, 1966.

135. Vossen, M.E.M.H., et al.: New aspects of the life cycle of *Pneumocystis carinii*. Z. Parasitenkd., 51:213–217, 1977.

136. Wakelin, D.: Genetic control of susceptibility and resistance to parasitic infection. Adv. Parasitol., 16:219–308, 1978.

137. Weatherall, D.J., and Abdalla, S.: Anaemia of *Plasmodium falciparum* malaria. Brit. Med. Bull., 38(2):147–151, 1982.

138. Weiss, L.: Hematopoietic tissue in malaria: Facilitation of erythrocytic recycling by bone marrow in *Plasmodium berghei*-infected mice. J. Parasitol., 69:307–318, 1983.

139. Witlock, D.R., Ruff, M.D., and Chute, M.B.: Physiological basis of *Eimeria tenella*-induced mortality in individual chickens. J. Parasitol., 67:65–69, 1981.

140. W.H.O. Malaria action programme. Severe and complicated malaria. Trans. Roy. Soc. Trop. Med. & Hyg., 80(Suppl.):1–50, 1986.

141. Wyler, D.J., and Gallin, J.I.: Spleen-derived mononuclear cell chemotactic factor in malaria infections: a possible mechanism for splenic macrophage accumulation. J. Immunol., 118:478–484, 1977.

142. Young, A.S., and Morzaria, S.P.: Biology of *Babesia*. Parasitol. Today, 2(8):211–219, 1986.

143. Young, L.S. (ed.): *Pneumocystis cariini* pneumonia. Pathogenesis, Diagnosis, Treatment. New York, Marcel Dekker, 1984.

6

Phyla Myxozoa, Microspora, and Ascetospora

PHYLUM MYXOZOA

This phylum is characterized by polycellular spores with a variable number of valves; one to six polar capsules, each with a coiled, hollow filament; a sporoplasm with one to many nuclei; and polycellular developmental stages in which spore formation takes place. Meiosis occurs at the end of sporogenesis, resulting in haploid, multicellular spores.[17]

Class Myxosporea

Order Bivalvulida

The parasites are primarily histozoic or celozoic in fishes, occasionally in amphibians and reptiles, rarely in annelids, and one in a trematode. The more primitive species are found in the gallbladder, urinary bladder, or uriniferous tubules of freshwater and marine fishes. Histozoic species are generally pathogenic in skin, muscles, gills, brain, liver, and other organs. More than 40 genera have been described. Figure 6–1 illustrates a few of them.

The life cycle of these parasites is presumed to be direct, without an intermediate host. The mature spore is the infective stage, and its sporoplasm contains two haploid nuclei which are considered to be gametes. The spore is the infective stage, and its sporoplasm, containing two nuclei, is generally thought to be haploid. These nuclei are considered to be gametes. When a spore is ingested by a host, its shell valves open, and the sporoplasm initiates development of the trophozoite by repeated divisions to form numerous generative cells. When two of these cells unite, the nondividing cell (wrapping cell or pericyte) surrounds the other and forms an envelope about it. The enveloped generative cell divides to form one or more spore-forming cells called *sporoblasts*. Sporoblasts giving rise to more than one spore— the procedure in most species—are called *pansporoblasts*. The earliest recognizable cluster of cells destined to become one spore consists of five cells. Two of these cells (valvogenic) form the two spore walls; two (capsulogenic) form the two capsules, each containing a polar filament; and the fifth cell forms the binucleate sporoplasm. In some species, two separate cells unite to form the sporoplasm. In addition to the two original generative cells that start pansporoblast formation, a third cell, a large, ameboid "lobocyte" with a possible excretory function, has been described by Grassé and Lavette.[6]

Myxosporan parasites in fishes commonly damage kidneys, muscles, and brain. Among the pathogenic species in salmonids are *Myxobolus* (formerly *Myxosoma*) *cerebralis* which infects cartilage that supports the central nervous system and causes "whirling disease";[8] and *Ceratomyxa shasta* which infects many organs and generally is fatal, especially in fry being raised in hatcheries. One example of a hyperparasite is that of *Fabespora* living in a digenetic trematode that lives in a fish.[14]

Order Multivalvulida

Spores have more than 2 valves: three families, one with 3 valves, in gallbladder of fish; one with 4 valves, in muscles and kidneys of fish; one with 6 valves, in muscles of fish.

The relationships of Myxozoa to other meta-

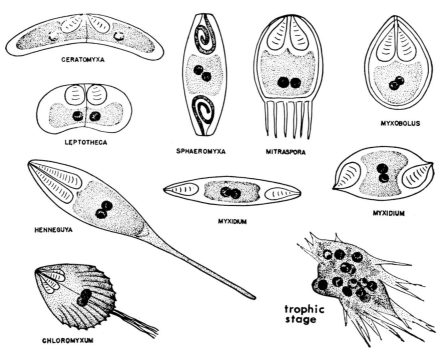

Fig. 6–1. Sample spores of the order Bivalvulida. At the lower right corner is a trophic stage with slim, pointed pseudopodia that are typical of several genera.

zoans are not definitely known, but similarities between myxosporan cnidocysts and nematocysts of coelenterates have convinced some workers that the Myxozoa have evolved from ancestral Cnidaria.[10] There is some validity for the belief of Grassé and Lavette[6] that the Myxozoa consists of a complex of three cellular kinds with distinct ends, and therefore "form one branch unconnected with both Protozoa and Metazoa." For a review of the Myxozoa, see Mitchell.[13] For a 1984 revised classification see Lom and Noble.[11]

Class Actinosporea

Order Actinomyxida

Each spore is surrounded by a membrane possessing three valves. Each of three polar capsules encloses a polar filament. In the mature stage, eight spores develop within the sporocyst. Five familes are known, occurring as intracellular parasites primarily in aquatic oligochaetes, and one genus in sipunculids. Two families are known: Tetractinomyxidae and Triactinomyxidae. All species occur in sipunculids and in tubificid annelids.

Tetractinomyxon intermedium, possessing spores 7 to 8 μm in diameter, is found in the coelom in the sipunculid, *Petalostoma minutum*. *Triactinomyxon legeri* live in the worm, *Tubifex tubifex*. Its spore contains 24 uninucleate sporoplasms. These organisms unite in pairs that are liberated as binucleate amebulae, each of which divides several times and forms a cyst *(pansporocyst)* containing cells of two sizes. These cells are the gametes. Each small cell fuses with a larger one (anisogamy) and forms a zygote, which develops into the spores that are transmitted from host to host.

Wolf and Markiw (1984)[21] provided evidence that Actinosporea and Myxosporea are alternating life forms of a single organism. The evidence was based on the production of whirling disease caused by *Myxosoma* (now *Myxobolus*) *cerebralis* in rainbow trout fry kept "in containers that had received tubificids from a fish hatchery where the disease was epizootic, or in containers that had received normal tubificids plus spores of *M. cerebralis*." The tubificids (annelid worms) are hosts to the "actinosporean" *Triactinomyxon*. This report immediately placed the Actinosporea in a taxonomic position of uncertainty. Hamilton and Canning[7] have provided a substantial challenge to these views. They added *M. cerebralis* spores to *Tubifex tubifex* colonized in sterile medium, and found no significant change in the prevalence of *Triactinomyxon dubin* (= *T. gyrosalmo*). Also, after the worms ingested spores of *M. cerebralis*, neither hatching of the spores nor further development within the worm was observed. In addition, field observations of distributions and occurrences of actinosporean species and *M. cerebralis* showed no obvious correlations. These results do not indicate that actinosporeans are involved with the life cycle of *M. cerebralis*.

PHYLUM MICROSPORA

There are two classes, Metchnikovellidea and Microsporididea. They occur in nearly all major animal groups. Fewer than 1000 species have been described, but they may be the largest group of parasitic animals.[2,16]

Microsporidans are intracellular parasites, mostly of arthropods (commonly in gut epithelium) and fish, often in skin and muscles. For species in vertebrates see Canning and Lom.[4] They are also found in mammals, birds (rarely), amphibians, reptiles, and in almost all invertebrate phyla, including trematodes and protozoa. Species reported from humans include *Nosema connori*, *Encephalitozoon brumpti*, *E. chagasi*, and *Enterocytozoon bieneusi*. Egg-shaped spores (Fig. 6–2) are generally about 3×5 μm. The outer layer of the 3-layered spore wall is proteinaceous, and the inside layer is chitinous.[19] A single polar filament is present. Transmission is by ingestion of spores. Intermediate hosts have not been found.

Infection of new hosts has generally been described as the result of the spore tubular filament serving as an "inoculating needle."[12] Great pressure builds up within the spore, causing eversion of the filament and expulsion of the sporoplasm through its lumen. Vavra[18] has pointed out that descriptions of eversion of the polar filament have been based on spores that germinate in an extracellular milieu. His observations appear to demonstrate that within the host cell the filament is not stable and is not a guide for the sporoplasm. Hatched microsporidia spores are almost never seen in tissues of living hosts. Using tissue culture methods, they demonstrated that spores of *Nosema heliothidis*

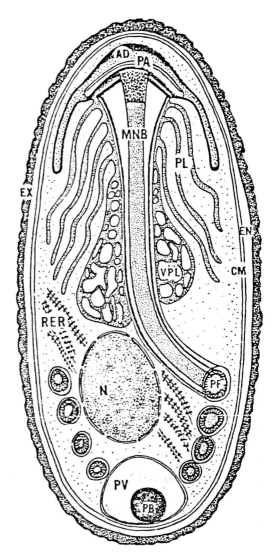

Fig. 6–2. Schematic representation of the fine structure of the microsporidian spore. AD, anchoring disc of the polar tube; CM, cytoplasmic membrane of the spore content; EN, endospore; EX, exospore; MNB, manubroid part of the filament; N, nucleus; PA, polar aperture; PB, posterior body; PF, polar tube; PL, lamellae of the lamellar polaroplast; PV, posterior vacuole; RER, endoplasmic reticulum densely populated with ribosomes; VPL, vesicular part of the polaroplast. (From Vávra, J.: Structure of the Microsporidia. *In* Comparative Pathology. Vol. I. Edited by C.A. Bulla and T.C. Cheng. New York, Plenum Press, 1977.)

do not hatch in their insect host cells, and that vegetative stages are responsible for cell transfer.

When the sporoplasm is released, it enters the gut epithelium and, by way of blood or the body cavity, is carried to various tissues. The sporoplasm then usually grows into a meront which begins a series of merogonic divisions (Figs. 6–3 and 6–4). Meiosis occurs in the sporoblastic sporonts. "Each sporont divides (sporogony) into two or more uninucleate or binucleate sporoblasts which transform (sporogenesis) into spores, thus completing the life cycle. These parasites produce a variety of lesions in their hosts and vary greatly in path-

ogenicity. In many invertebrates and some fish they may cause devastating mortalities. They are destructive parasites of honey bees, silkworm larvae, crabs, shrimps, fish and so on, and enemies of many undesirable insects species (mosquitoes, bollworms, corn earworms, grasshoppers, and so on)."[16] For example, *Nosema pyraustia* is a common parasite of the European corn borer, *Ostrinia nubilis;* and *N. epilanchae* is a highly virulent parasite of the larval and adult Mexican bean beetle, *Epilanchna varivestis,* as well as of bush beans and soy beans in the United States.

Microsporan cytopathologic features assume many patterns, and xenomas with proliferative

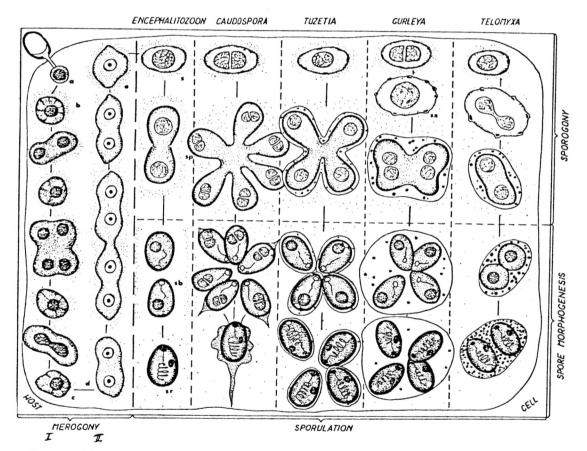

Fig. 6–3. Schematic representation of the microsporidian life cycle. *a,* Entry of the sporoplasm into the host cell through the polar tube of the spore; *b,* to *c,* stages of the first merogonial cycle; *d* to *e,* stages of the second merogonial cycle; s, sporont; sp, sporogonial plasmodium; sb, sporoblast; sr, spore. The zygote, which presumably occurs in some species, is sx. The course of sporulation in several microsporidan genera is presented to show various nuclear conditions and various cell membrane differentiations occurring during sporulation. The parasite-host cell interaction and the progressive lysis of the host cytoplasm during sporulation is also shown. (From Vávra, J.: Development of the Microsporidia. *In* Comparative Pathology. Vol. I. Edited by C.A. Bulla and T.C. Cheng. New York, Plenum Press, 1977.)

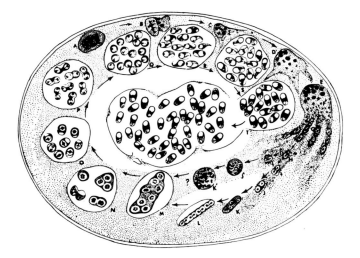

Fig. 6–4. Intracellular development of *Glugea*. *A,* Host cell nucleus that divides repeatedly by amitosis; the endoplasmic reticulum is continuous with the plasma membrane of plasmodial stages of the microsporidan parasites. *H* to *L* and *I* to *K,* Parasite plasmodia that expand and break up to form spores. *T,* Mature spores. (From Sprague and Vernick, courtesy of J. Protozool.)

inflammation and granuloma are common in various host organs. Differences in the reaction of invaded host cells are marked. An initial hypertrophy is usual, as occurs in infections of the fat body of crane fly larvae by *Thelohania tipulae*. *Amblyospora* is a pathogen of mosquitoes.[1] *Pleistophora*[3] is found principally in skeletal muscle bundles in numerous kinds of tropical freshwater fish. A frequent reaction by fish to microsporidan invasion is formation of host membranes around the parasite to produce a cyst, and massive invasion may destroy muscles and other organs. *Nosema apis* in honey bees is usually fatal. *Polydispyremia* occurs in black flies *(Simulium)*. The larval stages of digenic trematodes are often invaded by microsporeans, usually pathogenetic. *Nosema* has been reported from *Schistosoma* (experimentally). *N. leopocreadii* lives in the vitelline glands of a trematode.

Microsporida in mammals are generally considered to be *Encephalitozoon cuniculi,*[15] and lesions commonly occur in the brain and kidney tubules. Other sites of infection in mammals are liver, adrenal glands, optic nerves, retina, and myocardium. Routes of infection appear to be both oral and tracheal. Transplacental transmission can occur. Rodents and carnivores are frequently infected. Under conditions such as fish farming and breeding of rodents, the full microsporidan pathogenic potential is revealed. Infections in amphibians and reptiles generally do not produce external symptoms. See Weiser[20] for an account of Microspora in invertebrates, and Canning and Lom[4] for a discussion of these parasites in vertebrates.

Microsporidial spores in aquatic hosts appear to have little or no host specificity. There is a possibility that some species require a secondary host, although such a host has not been discovered. Females of the mosquito *Culex tarsalis* are able to suppress sporogony of *Amblyospora* (formally *Thelohania*) *californica,* and thereby maintain normal reproductive capacity. Infected male larval mosquitoes, however, usually die. Kellen et al.[9] have reported various other types of host-parasite relationships involving *Amblyospora* and 16 species of mosquitoes.

Although fewer than 1000 species of microsporeans have been described, they may actually be the largest group of parasitic animals.[16]

PHYLUM ASCETOSPORA

These organisms are primarily parasitic in aquatic invertebrates such as molluscs, annelids, tunicates, and arthropods. Multiplication is by a modified type of schizogony. Multicellular spores do not produce polar capsules or polar filaments. They are cytozoic, histozoic, and celozoic.[15]

At least two species of *Minchinia* proliferate in epithelial cells of gills and adjacent tissues of oysters and other invertebrates and cause extensive damage and frequent death of the hosts. This genus may belong to the Microspora. *Haplosporidium* is found in connective tissues and intestinal epithelia of annelids and in digestive glands of molluscs. *Urosporidium* is a hyperparasite in trematode microphallid metacercariae. Details of these and other genera are obscure. Animals become infected by ingesting spores that release small ameboid forms. These forms multiply repeatedly to form spores in various tissues.

Marteilia, pathogenic in European oysters, may be related to the Myxozoa. *Paramyxa*,[5] without a polar filament and with spores of multicellular origin, probably belongs with the Haplospora.[15] Details of these two genera are obscure, and their taxonomic positions are uncertain. See Desportes[5a] for an account of the class Paramyxea.

REFERENCES

1. Andreadis, T.G., and Hall, D.W.: Development, ultrastructure, and mode of transmission of *Amblyospora* sp. (Microspora) in the mosquito. J. Protozool., *26*:444–452, 1979.
2. Canning, E.U.: Microsporida. *In* Parasitic Protozoa, Vol. 4. Edited by J.P. Kreier. New York, Academic Press, 1977, pp. 155–196.
3. Canning, E.U., and Hazard, E.I.: Genus *Pleistophora* Gurley, 1893: An assemblage of at least three genera. J. Protozool., *29*:39–49, 1982.
4. Canning, E.U., and Lom, J., with a contribution by I. Dykova: The Microsporidia of Vertebrates. 1986.
5. Desportes, I.: Étude ultrastructurale de la sporulation de *Paramyxa paradoxa* Chatton (Paramyxida) parasite de l'annelide polychète *Poecilochaetus serpens*. Protistologica, *17*(3):365–386, 1981.
5a. Desportes, I.: The Paramyxea Levine 1979: an original example of evolution towards multicellularity. Origins of Life, *13*:343–352.
6. Grassé, P.-P., and Lavette, A.: Sur la reproduction des Myxosporidies et la valeur des cellules dites germinales. C.R. Acad. Sci. [D] (Paris), *286*:757–759, 1978.
7. Hamilton, A., and Canning, E.U.: Studies on the proposed role of *Tubifex tubifex* (Muller) as an in-termediate host in the life cycle of *Myxosoma cerebralis*. J. Fish Diseases, *10*:145–151, 1987.
8. Hoffman, G.: Whirling Disease in Trout. Fish Diseases Leaflet No. 47. Washington, D.C., U.S. Department of the Interior, 1976.
9. Kellen, W.R., Chapman, H.C., Clark, T.B., and Lindegren, J.E.: Host-parasite relationships of some *Thélohania* from mosquitoes (Nosematidae: Microsporidia). J. Inver. Path., *7*(2):161–166, 1965.
10. Lom, J.: Myxosporidian ultrastructure: an attempt at a synthesis. New York, Fifth International Congress of Protozoology. 1977, Abstract No. 271.
11. Lom, J., and Noble, E.R.: Revised classification of the class Myxosporea Bütschli, 1881. Folia Parasitologica, *31*:193–205, 1984.
12. Lom, J., and Vavra, J.: The mode of sporoplasm extrusion in microsporidian spores. Acta Protozool., *1*:81–89, 1963.
13. Mitchell, L.G.: Myxosporidia. *In* Parasitic Protozoa. Vol. 4. Edited by J. Kreier. New York, Academic Press, pp. 115–154, 1977.
14. Overstreet, R.M.: *Fabespora vermicola* sp. n. the first myxosporidian from a platyhelminth. J. Parasitol., *62*(5):680–684, 1976.
15. Sprague, V.: Classification of the Haplosporidia. Marine Fish. Rev., *41*:40–44, 1979.
16. Sprague, V.: Microspora. *In* Synopsis and Classification of Living Organisms. Vol. 1. Edited by S.P. Parker. New York, McGraw-Hill, 1982, pp. 589–594.
17. Uspenskaya, A.V.: New date on the life cycle and biology of Myxosporidia (Protozoa). Doklady Biol. Sci., *263*:118, 1982 (translated from Doklady Akad. Nauk SSSR, *262*:503–507, 1982).
18. Vávra, J.: Development of the Microsporidia. *In* Comparative Pathobiology. Vol. 1. Edited by L.A. Bulla and T.C. Cheng. New York, Plenum Press, 1977, pp. 87–110.
19. Vávra, J.: Structure of the Microsporidia. *In* Comparative Pathobiology. Vol. 1. Edited by L.A. Bulla and T.C. Cheng. New York, Plenum Press, 1977, pp. 1–86.
20. Weiser, J.: Microsporidia in invertebrates: host-parasite relations at the organismal level. *In* Comparative Pathobiology. Vol. 1. Edited by L.A. Bulla and T.C. Cheng. New York, Plenum Press, 1977, pp. 163–201.
21. Wolf, K., and Markiw, M.E.: Biology contravenes taxonomy in the Myxozoa: new discoveries show alternation of invertebrate and vertebrate hosts. Science, *225*:1449–1452, 1984.

7

Phylum Ciliophora

Ciliates move by means of hair-like projections of the cytoplasm called **cilia,** each of which is much like a little flagellum in appearance and structure.

Beneath the outer surface of the body in all ciliates is a system of granules and fibrils that together are known as the **infraciliature.** Two types of nuclei exist: the large **macronucleus,** usually single, and from one to many smaller **micronuclei.** Reproduction occurs by both asexual and sexual means. No fusion of independent gametes occurs, but two entire ciliates come together temporarily **(conjugation)** and exchange micronuclei. Most groups of ciliates contain at least a few parasitic species, and some groups are entirely parasitic. Ciliates may be found in or on molluscs, annelids, crustacea, echinoderms, and vertebrates. Only representative familes are discussed here. See Corliss[2] and Small et al.[7]

Class Kinetofragminophorea

Subclass Gymnostomata

Order Prostomatida

FAMILY BUETSCHLIIDAE

These ciliates have special interest for parasitologists because they live in the alimentary tract of herbivorous mammals. Their size varies considerably, but it averages about 35×55 µm. They are oval, barrel-shaped, or pear-shaped, are covered with cilia, and they possess a "mouth" or **cytostome** at the anterior end and an "anus" or **cytopyge** at the posterior end. Contractile vacuoles are common. If a drop of fluid from the cecum or colon of horses or from the rumen of cattle, camels, or other herbivores is placed on a slide and is examined immediately through a microscope, one might see swarming masses of these ciliates. For a discussion of the role these organisms play in the digestive tract of their hosts, see Chapter 23; see also Hungate.[5] *Buetschlia* is found in cattle, and among the many genera to be found in horses are *Blepharoconus, Bundleia,* and *Holophyroides.*

Subclass Vestibuliferia

Order Trichostomatida

The cytostome of ciliates in this order is usually situated at the base of an oral groove or pit, the wall of which bears dense cilia.

FAMILY BALANTIDIIDAE

The best-known member of the genus *Balantidium* is *B. coli.* Among the other species are *B. suis,* in the pig, *B. caviae,* in the cecum of the guinea pig, *B. duodeni,* in the gut of the frog, and *B. praenucleatum,* a large species, reaching 127 µm in length, found in the colon of the cockroach, *Blatta orientalis.*

Balantidium coli is the only ciliate parasite of man, although coprophilous species may be found occasionally. It is practically worldwide in distribution, lives in the large intestine, and is pathogenic, causing **balantidiasis.** This disease occurs generally in about 1% of the human population but the rate may be as high as 100% in pigs. *B. coli* may also be found in monkeys and dogs.

The motile form of *Balantidium coli* (Fig. 7–1) is rougly oval in shape and averages about 75×50 µm in size. The length ranges from 30 to 300 µm, and the breadth from 30 to 100 µm. Rows of cilia cover the body, and a peristomal region of longer cilia guards a cleft leading to the cytostome. The macronucleus is a sausage-shaped structure, and the small, dot-like micronucleus lies so close to the macronucleus that

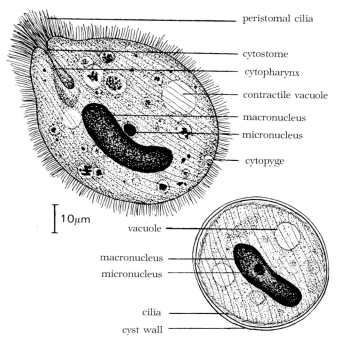

peristomal cilia

cytostome

cytopharynx

contractile vacuole

macronucleus

micronucleus

cytopyge

10μm

vacuole

macronucleus

micronucleus

cilia

cyst wall

Fig. 7–1. *Balantidium coli,* trophozoite and cyst.

it often cannot be seen. The cytoplasm may contain many food vacuoles and two contractile vacuoles, one large and one small. Asexual reproduction occurs by transverse fission. Sexual reproduction is by conjugation. A host diet rich in starch apparently favors the growth of this parasite.

The cyst of *Balantidium coli* is approximately round in outline and measures about 55 μm in diameter. Its large macronucleus and usually a vacuole can easily be seen. Living cysts are pale yellow or greenish in color.

When cysts are eaten, they excyst in the host intestine, and released ciliates begin to feed on cell fragments, starch grains, fecal material, and other organic matter. Often, the ciliates invade the mucosa and submucosa of the large intestine or cecum and cause severe ulcerative lesions that sometimes involve the entire length of the large intestine. Invasion is aided by hyaluronidase which it secretes, producing flask-like ulcers similar to those produced by *Entamoeba histolytica. B. coli* trophozoites may be found in the base of the ulcers and in the colon wall. A relation between balatidia and the whipworm, *Trichuris* often occurs. Worm eggs may be found in the ciliate food vacuoles. Symptoms of infection include diarrhea, abdominal pain,

dysentery, nausea, vomiting, weakness, and loss of weight. Diagnosis is confirmed by finding cysts in stool specimens, which are generally bloody and mucoid. See Zaman[9] for a review.

FAMILY ISOTRICHIDAE

This family is characterized by the possession of a cytostome at or near the apical end of the body and by uniform and complete surface ciliation. The ciliates are characteristically found in the stomachs of ungulate ruminants. There are three genera: *Isotricha, Dasytricha,* and *Protoisotricha. Isotricha prostoma* and *I. intestinalis,* both about 120 × 65 μm in size, and *Dasytricha ruminantium* are examples of species. One species of *Isotricha* may be found in the gut of a cockroach.

Order Entodiniomorphida

This order is especially well known as a rumen ciliate in cattle, sheep, and related hosts. The family Ophryoscolecidae has many species existing in the rumen of cattle and sheep. Representative genera are *Entodinium* (Fig. 7–2), *Diplodinium,* and *Ophryoscolex.* Cycloposthiidae are found more often in horses. The elephant, the rhinoceros, and even the gorilla and the

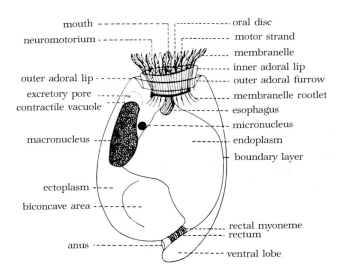

mouth — oral disc
neuromotorium — motor strand
— membranelle
— inner adoral lip
outer adoral lip — outer adoral furrow
excretory pore — membranelle rootlet
contractile vacuole — esophagus
— micronucleus
macronucleus — endoplasm
— boundary layer

ectoplasm —
biconcave area —

— rectal myoneme
— rectum
anus —
— ventral lobe

Fig. 7–2. *Entodinium biconcavum,* semidiagrammatic lateral view. (From Kofoid and MacLennan, courtesy of the University of California Press.)

chimpanzee may be hosts to members of the latter family. See Chapter 23 for a discussion of these and other ciliates in ruminants. See Hungate[5] and Coleman[1] for reviews of the rumen and its microbes.

Subclass Hypostomata

Order Chonotrichida

These commensal ciliates, which usually possess the general shape of a vase, are often stalked. They range in length from about 30 to over 100 μm and may be found attached to various aquatic animals, especially crustacea. Most of the described species are attached to marine animals, but one species, *Spirachona gemmipara,* occurs on the gill plates of freshwater gammarids (amphipods). Young amphipods are heavily infected. Marine genera include *Stylochona, Kentrochona,* and *Chilodochona.* A marine species, *Oenophorachona ectenolaemus,* is illustrated in Figure 7–3.

Order Rhynchodida

FAMILY ANCISTROCOMIDAE

The most interesting characteristic of this group is the presence of a tentacle that enables these ciliates to attach themselves to the body of the host and thereby ingest food. The attachment is made more secure by the action of thigmotactic cilia; hence the name of the order. Many genera have been identified, most species of which may be found attached to gills and palps of molluscs. A few species have been re-

ported from annelids. The ciliates are, in general, oval to elongate with a heavy covering of cilia of uniform size. The anterior tips are usually projected. These parasites are comparatively small, ranging from about 15 to 60 μm in greatest diameter.

Ancistrocoma myae is a transparent, pale green ciliate living in the excurrent siphon and pericardial cavity of the clam, *Mya arenaria.* It is 40 to 100 μm long.

Order Apostomatida

FAMILY FOETTINGERIIDAE

Foettingeria actinarum is a European ciliate that feeds on material within the gastrovascular cavity of a sea anemone. It leaves the host and

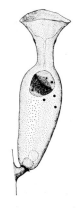

Fig. 7–3. Mature *Oenophorachona ectenolaemus,* marine chonotrichous ciliate. (From Matsudo and Mohr, courtesy of J. Parasitol.)

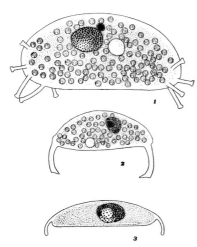

Fig. 7–4. Three species of *Allantosoma*, parasitic suctorea from the large intestine of the horse. ×853. *1, A. intestinalis; 2, A. bicorniger; 3, A. brevicorniger.* (From Hsiung, courtesy of Iowa State College J. Sci.)

encysts on some object in the sea while undergoing cell division. Products of this division are released and come to rest on crustacea (copepods, ostracods, and amphipods) as secondary hosts. Here they encyst. Sea anemones receive the ciliate when they feed on infected crustacea.

Chromidina lives on the renal epithelium and pancreas of squids. It may reach a length of 3 mm.

Terebrospira chattoni penetrates and feeds on the exoskeleton of a shrimp.

Synophrya causes tissue damage to gill lamellae of its marine decapod hosts.

Subclass Suctoria

Order Suctorida

Cilia are absent in adult stages, and most species are attached to the substrate by a stalk. *Allantosoma intestinalis* (Fig. 7–4), one of a few parasitic species, occurs in the large intestine and cecum of horses, in which the parasite becomes attached to other ciliates. *A. brevicorniger* may also attach to ciliates in the colon or cecum of horses, whereas *A. bicorniger* lives unattached in the colon of these hosts. The last two species are about 30 μm long.

Another suctorian, *Syphaerophyra sol,* may occasionally be found as parasitic in the ciliates *Stylonychia mytilus, Epistylis plicatilis,* and *Paramecium aurelia.* In host cytoplasm, the suctorian is rounded, but when it leaves the host, it rapidly produces radiating tentacles. In infected *P.*

aurelia, the cytoplasm becomes vacuolated and the nucleus disintegrates.

Feeding takes place by endocytosis and, apparently, by a process involving the inflow of host cytoplasm through a tubule within the feeding tentacle. "The hypothesis that suction in Suctoria is the force for the flow of cytoplasm from prey to predator seems very doubtful."[6]

Class Oligohymenophorea

Subclass Hymenostomatia

Order Hymenostomatida

FAMILY TETRAHYMENIDAE

Most members of this group are free-living, and the group is well known because it has been used extensively for physiologic and genetic studies. A few parasitic species show some degree of pathogenicity.[4] The genus *Colpidium* has a representative living in sea urchins. Various species of the genus *Tetrahymena* have been experimentally established in larval and adult insects, slugs, guppies, tadpoles, and embryo chicks. *T. corlissi* was established in guppies and tadpoles through artificially produced wounds. In experimental chicks, the circulatory system, body cavity, and muscle tissue may be invaded by two strains of these parasites, whereas the yolk sac or allantoic sac may be invaded by six strains.

Tetrahymena limacis lives in the renal organ of the European gray garden slug, *Deroceras reticulatum.* The ciliate can be grown axenically. *Tetrahymena* is also a normal endoparasite of mosquitoes, millipeds, and other animals. In insects, the ciliates live as facultative parasites in the hemolymph.

Glucoma is usually free-living, but it may occur as a facultative parasite of arthropods and in the central nervous system of fish and amphibians. The ciliate's tolerance of CO_2 favors the parasitic habit.

FAMILY ICHTHYOPHTHIRIIDAE

Ichthyophthirius multifiliis (Fig. 7–5), a worldwide ectoparasite, causes "white spot disease" or "ick" in many freshwater fishes, infecting skin, gills, eyes, and fins. Free-swimming **tomites** (swarmers) are the infective stage in the life cycle. They are 45 to 48 μm long and are shaped like a banana. They penetrate the epidermis by a boring action. Tomites possess food vacuoles, indicating that they feed before

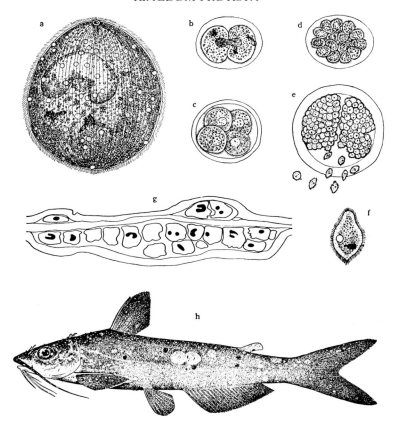

Fig. 7–5. *Ichthyophthirius multifiliis*, a parasite of fish skin. *a*, Free-swimming individual. ×75. (After Buetschli) *b* to *e*, Development within cyst. *f*, A young individual. ×400. (After Fouquet) *g*, Section through a fin of infected carp showing numerous parasites. ×10. (After Kudo) *h*, A heavily infected catfish, *Ameurirus albidus*. (After Stiles) (From Kudo, courtesy of Charles C Thomas.)

penetration. Inside host tissue they mature into the adult feeding stage—the trophozoite or **trophont.** Here they are visible as white spots which are 0.1 to 1.0 μm in diameter and contain 1 to 6 parasites. The infection may be highly pathogenic. It kills the fish by interfering with its osmotic regulation and gill function. Parasites leave the lesions and drop to the bottom of the stream, lake, or aquarium, secrete a gelatinous capsule, and reproduce by binary fission. Hundreds of new ciliates are formed in each capsule, and when released they are ready to attack new fish. The parasite is not host-specific.

With lowering temperatures the length of the life cycle slows. At 25°C it is 6 days, whereas at 4°C it requires more than 100 days. The disease is especially serious when fish are living under crowded conditions such as in fish hatcheries. Repeated heavy infections may cause massive

cellular necrosis. For development in gill epithelia see Ewing and Kocan.[3]

The crab, *Cancer magister* can be fatally infected by tissue-invading *Paranophrys*.

Order Scuticociliatida

FAMILY ENTORHIPIDIIDAE AND OTHER CILIATES OF SEA URCHINS

The members of this family are large ciliates, flattened laterally, with a lobe-like anterior end and a tapering or pointed posterior end. They live in the gut of sea urchins of the northern Pacific Ocean. Echinoids contain a number of other ciliates belonging to this order, some closely related to the Entorhipidiidae, including the genera *Cryptochilum, Biggaria, Entodiscus* (Fig. 7–6), *Madsenia,* and *Thyrophylax*. Species of the predominantly free-living genera *Ano-*

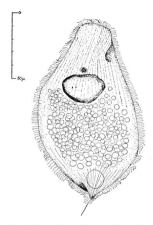

Fig. 7–7. *Trichodina parabranchicola,* a ciliate from gills of various intertidal zone fish. ×880. The scale line represents 10 μm. (From Laird. Courtesy of the Transactions of the Royal Society of New Zealand.)

Fig. 7–6. *Entodiscus borealis,* a ciliate from a sea urchin. (From Powers, courtesy of Biol. Bull.)

phyrs, Cyclidium, and *Pseudocohnilembus* also occur in sea urchins, as do ciliates belonging to other orders, including the trichostomes *(Plagiopyla, Lechriopyla),* the peritrichs *(Urceolaria),* the heterotrichs *(Metopus),* and the hypotrichs *(Euplotes).* Smaller species of the genera *Cryptochilum* and *Biggaria* occur in great numbers in the gut of algivorous echinoids. The geographic distributions of these ciliates generally follow the ranges of their hosts, so that the mixture of ciliate species found in echinoids of temperate regions is different from that found in tropical sea urchins.

Order Astomatida

As the name implies, this group is characterized by the absence of a mouth. The organisms are larger than those of some other groups, usually being between 250 and 350 μm in maximal length, but ranging fom 100 to 1,200 μm. Transverse fission and budding, common methods of reproduction, result in a chain of individuals. These parasites are found in fresh- and saltwater invertebrates, especially in oligochaete worms. An example is *Anoplophrya.*

Subclass Peritrichia

Order Peritrichida

The oral surface of these ciliates is flattened and forms a disc that has a counterclockwise spiral of one or more rows of cilia. Many of the free-living species are attached to a substrate by a stalk.

FAMILY SCYPHIDIIDAE

Ellobiophrya donacis is so modified that it has developed limb-like posterior projections that join around the trabecula of a molluscan gill and hold the ciliate in place. *Scyphida acanthoclini* similarly attaches itself to the gills of *Acanthoclinus quadridactylus,* and *Caliperia brevipes* is attached to the gills of skates (Rajidae).

FAMILY TRICHODINIDAE

Trichodina (Fig. 7–7) is found in or on fish, amphibians, molluscs, bryozoans, turbellarians, sponges, hydroids, various worms, crustaceans, and echinoderms. The aboral end is a flattened disc equipped with rings of cuticular teeth-like skeletal elements. Locomotor organelles consist of posterior membranelles; cirri and an undulating velum are present. Most species are ectoparasitic. One of the best-known species is *T. pediculus,* which "skates" on the surface of the coelenterate, *Hydra.* Economically important species live on various fish; one species is lethal to goldfish. *T. urinicola* inhabits the urinary bladder of amphibians. Another entozoic species, *Urceolaria urechi,* may be found in the intestine of the echiuroid marine worm, *Urechis caupo,* on the United States Pacific coast. *Trichodina truttae* lives on the skin of salmonid fishes in British Columbia. It is a large species with a body diameter of 114 to 179 μm. *T. oviducti* lives in the urogenital system of adult thorny skates, *Raja radiata,* and is transmitted venereally. Generally *Trichodina* is non-pathogenic but it may cause mild damage to host

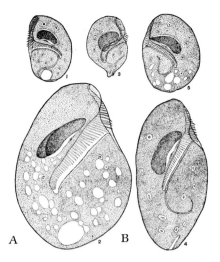

Fig. 7–8. *Nyctotherus cordiformis,* a ciliate found in frog and toad tadpoles. *A, Nyctotherus* from tadpole of *Rana clamitans.* Length, 340 μm; width, 230 μm. Note bilobed nucleus and anterior position of mouth and oblique pharynx. *B, Nyctotherus* from tadpole of *R. catesbeiana.* Length, 300 μm. Note narrowness, with macronucleus and pharynx almost longitudinal in position. (From Higgins. Courtesy of the Transactions of the American Microscopial Society.)

tissues. Members of this genus often feed on bacteria.

Class Polyhymenophorea

Subclass Spirotrichia

Cilia around the cytostome (adoral zone of membranelles) of this group are arranged in a spiral fashion, passing from the right side of the peristome into the cytopharynx.

Order Heterotrichida

FAMILY PLAGIOTOMIDAE

The peristome of these organisms contains an undulating membrane, and the entire body is ciliated. Parasitic forms may be found in many invertebrates and vertebrates. *Nyctotherus* is one of the best-known genera (Fig. 7–8). These ciliates are, in general, kidney-shaped; the indentation of the body contains a zone of cilia and leads to a cytostome that opens to a ciliated "esophagus." One species, *N. faba,* has been reported from the intestine of man. It is probably not pathogenic, nor are the species that live in animals noticeably harmful. *N. velox* can be found in the milliped, *Spirobolus marginatus; N. ovalis* inhabits the cockroach, *Blatta orientalis; N. parvus* and *N. cordiformis* live in amphibians. The last-named species can often be found in large numbers in the colons of frogs, toads, and tadpoles. This ciliate conjugates in the intestine of tadpoles at the time of host metamorphosis. *Plagiotoma lumbrici,* a related species, may be found in the coelom of earthworms, and *Metopus circumlabens* (Fig. 7–9) is an inhabitant of Bermuda sea urchins.

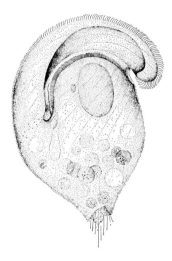

Fig. 7–9. *Metopus circumlabens,* a ciliate parasite in Bermuda sea urchins. (From Biggar. Courtesy of the Journal of Parasitology.)

REFERENCES

1. Colman, G.S.: Rumen Ciliate Protozoa. Parasitol., *18*:121–174, 1980.
2. Corliss, J.C.: The Ciliated Protozoa: Characterization, Classification and Guide to the Literature. 2nd Ed. New York, Pergamon Press, 1979.
3. Ewing, M.S., and Kocan, K.M: *Ichthyophthirius multifiliis* (Ciliophora) development in gill epithelium. J. Protozool., *33*(3):369–374, 1986.
4. Hoffman, G.L., et al.: A disease of freshwater fishes caused by *Tetrahymena corlissi* Thompson 1955 and a key for identification of holotrich ciliates of freshwater fishes. J. Parasitol., *61*:217–223, 1975.
5. Hungate, R.E.: The Rumen Protozoa. *In* Parasitic Protozoa. Vol. 2. Edited by J.P. Kreier. New York, Academic Press, 1978, pp. 655–695.
6. Rudzinska, M.A.: Do Suctoria really feed by suction? BioScience, *23*:87–94, 1973.

7. Small, E.B., and Lynn, D.H.: Phylum Ciliophora Doflein, 1901. *In* An Illustrated Guide to the Protozoa. Edited by J.J. Lee, S.H. Hutner, and E.C. Bovee. Lawrence, Kansas, Society for Protozoologists, 1985, pp. 393–575.

8. Ventura, M.T., and Paperna, I.: Histopathology of *Ichthyophthirius multifiliis* infections in fishes. J. Fish Biol., *27*:185–203, 1985.

9. Zaman, V.: *Balantidium coli. In* Parasitic Protozoa. Vol. 2. Edited by J.P. Kreier. New York, Academic Press, 1978, pp. 633–653.

8

Introduction to the Phylum and to Classes Turbellaria and Monogenea

Protozoans are single-celled eucaryotes. In Sections III to V we consider some multicellular eucaryotes with bodies that are generally longer than they are broad, commonly called "worms." Some, flat in cross section, are called **flatworms,** and are grouped in the **phylum Platyhelminthes.** Others, round in cross section, are termed **roundworms** and are grouped in the **phylum Nematoda** (Section V). The **"thorny-headed" worms,** characterized by a "crown of thorns," are also usually round in cross section, but their morphology differs sufficiently from flat- and roundworms to warrant grouping in a separate phylum, the **Acanthocephala** (Section IV). Many, but not all, platyhelminths and nematodes are parasitic. All Acanthocephala are parasitic. As you will learn, this is, indeed, a very wormy world.[13]

PHYLUM PLATYHELMINTHES

Members of this phylum are usually flattened dorsoventrally, lack true segmentation of the body, and are bilaterally symmetric; most possess an incomplete digestive tract (no anus), but cestodes have no gut. All Platyhelminthes lack a body cavity (are acoelomate) and are without special skeletal, circulatory, or respiratory systems; they have a flame cell (flame bulb) type of excretory system. The nervous system shows cephalization (i.e., specialization in the "head" region, defined as anterior) by a pair of anterior ganglia, and one to three pairs of longitudinal nerve cords connected by transverse commissures. Such networks are described as the "lad-der" type. Most are monoecious (hermaphroditic), but a few are dioecious (separate sexes). All reproduce sexually, and many also reproduce asexually during larval stages. Platyhelminths have some regenerative ability. They are not nucleotelic (i.e., do not have a set number of nuclei in adult forms) in contrast to nematodes, which are nucleotelic, do not reproduce asexually, and have no regenerative ability.

The major characteristics of the phylum Platyhelminthes were elegantly and simply defined by the late, great, invertebrate zoologist, Libbie H. Hyman: "They are acoelomote bilateria, without a definitive anus."[4]

CLASSIFICATION OF PHYLUM PLATYHELMINTHES

Because man-made classifications of organisms evolve more rapidly than the organisms themselves, there often is disagreement amongst authorities as to which grouping is most correct. The scheme shown here is at least useful, but it places the Monogenea closer to the Cestoidea in terms of evolutionary relationships for reasons given by Bychowsky[2] and by Llewellyn[6] (see also Chap. 26). It is only for simplicity and convenience that we discuss Turbellaria and Monogenea together in this chapter. See also Ehlers for phylogenetics of the phylum.[3]

Class Turbellaria
Class Trematoda
  Subclass—Aspidogastrea

Subclass—Digenea
Subclass—Didymozoidea
Class Monogenea
Class Cestoidea
    Subclass Cestodaria
    Subclass Cestoda

Class Turbellaria

These flatworms are mostly free-living. They usually have a ciliated epidermis, an undivided body, and a simple life cycle. The class is divided into ten orders, among which are over one hundred ectocommensal or parasitic species. Turbellarians may be found on or in echinoderms, crustaceans, molluscs, annelids, sipunculids, arachnids, coelenterates and other turbellarians, or occasionally as ectoparasites on turtles and snakes. In general, turbellarians show a high degree of host-type specificity.

The most primitive order, Acoela, has a few commensal species, living chiefly on echinoderms. The group known as rhabdocoels, formerly placed together under the order Rhabdocoela, comprises three orders, one of which, order Lecithophora (or Neorhabdocoela), is conspicuous for its large number of symbiotes. Almost all species of the family Umagillidae are associated with echinoderms; the remaining are found only in sipunculids. We shall mention a few of these parasites. See Jennings[5] for a detailed consideration of parasitism and commensalism in the Turbellaria; also see Riser and Morse.[12] The relationships between turbellarians and their hosts range from ectocommensal to endoparasitic. Ectocommensal rhabdocoels generally live on the gills or the body surface of freshwater crustaceans, or rarely, on gastropod molluscs or turtles. The diet of these parasites resembles that of free-living species—they are carnivorous and feed on protozoa, rotifers, nematodes, and crustacea. An example is *Ectocotyla paguri*, which lives on the outer surface of hermit crabs. Another, with tentacles and a posterior adhesive disc, is *Temnocephala*.

Endoparasitic species may live in body tissues or cavities. *Acholades asteris* is found in connective tissue of the tube feet of a sea star. *Syndesmis franciscana* (Fig. 8–1) lives in the intestine of a sea urchin and feeds entirely on ciliates also living in the host intestine. *Kronborgia*, in the body cavities of crustaceans, and *Oikiocolax plagiostomorum*, in the mesenchyme of another turbellarian, *Plagiostomum*, may cause castration of their hosts.

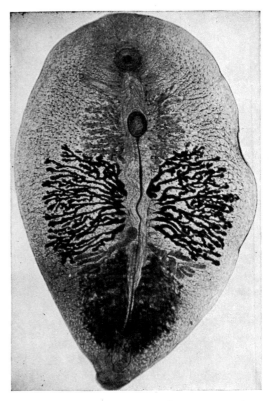

Fig. 8–1. *Syndesmis* sp., a turbellarian parasite from the intestine of the sea urchin. *Strongylocentrotus purpuratus.* (Preparation and photograph by Anthony T. Barnes.)

The life cycles of most turbellarian parasites are unknown. Those that have been studied, for example, some *Fecampia*, have a direct life cycle. *Fecampia erythrocephala* is pathogenic to its hosts, European marine crabs. The ciliated, immature parasite enters the hemocoel and undergoes profound modification, losing most of its viscera, but retaining its intestine. In the crustacean, it reaches sexual maturity; then it leaves its host, produces eggs, and dies.

Reproductive capacities may be increased in parasitic turbellarians. There is also a tendency toward reduction of the mouth, pharynx, or intestine and a resultant need for uptake of nutrients through the body surface. Reduced ciliation, lack of rhabdites (small rods in the epidermis or parenchyma secreted by gland cells), and lack of pigment in the epidermis may occur. A few species lose their eyes. As one

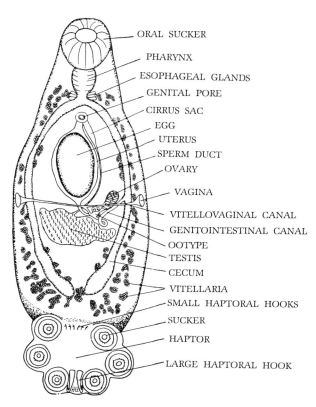

ORAL SUCKER
PHARYNX
ESOPHAGEAL GLANDS
GENITAL PORE
CIRRUS SAC
EGG
UTERUS
SPERM DUCT
OVARY
VAGINA
VITELLOVAGINAL CANAL
GENITOINTESTINAL CANAL
OOTYPE
TESTIS
CECUM
VITELLARIA
SMALL HAPTORAL HOOKS
SUCKER
HAPTOR
LARGE HAPTORAL HOOK

Fig. 8–2. *Polystomoidella oblongum,* a monogenetic trematode from the urinary bladder of turtles. (From Cable, courtesy of Burgess Publishing.)

would expect, many species have developed adhesive organs or attachment areas. Extreme modification of the parasite's structure is rare.

Class Monogenea

Although these worms are usually found as ectoparasites on lower vertebrates, especially fish, some inhabit gill chambers, mouth cavity, urinary bladder, cloaca, ureters, or body cavity of their hosts. About 435 species are known. Contrary to the browsers on the skin, gill dwellers are sedentary because their food, the host blood, is abundantly available. Life cycles of monogenetic trematodes do not involve more than one host. Adult parasites are attached to the host by a modification of the posterior end known as a **haptor** or, more accurately, **opisthaptor.** (The haptor at the anterior end is the **prohaptor.**) In general, an egg hatches, releasing a ciliated larva that swims to its host, attaches to the skin, gills, or elsewhere, and metamorphoses into the adult.

The morphology of monogenetic trematodes is fundamentally similar to that of digenetic trematodes; however, differences arise from the peculiar mode of life of the monogenetic trematodes (Fig. 8–2). These differences are sufficiently great to consider placing these worms with the cestodes.

Adult Monogenea range in length from less than 1 mm to 2 or 3 cm; the body outline varies from spindle-shaped to circular. The most striking feature is the opisthaptor. This organ usually has one or more suckers or cups used to attach the worm to the host. Simpler forms of Monogenea may not have a fully formed sucker, but only an expanded posterior end of the body. As the number of suckers increases, the posterior part of the body becomes larger and forms a disc. This disc may be divided radially by septa, and the whole organ, including septa and suckers, is activated by muscles. Added to this complex are hooks, also activated by muscles, which can be extended or withdrawn somewhat like cat claws. These organs form an adhesive structure or holdfast that helps the parasite to maintain its precarious perch on the host.

The anterior end also usually possesses one

or more suckers, ordinarily not as well developed as the oral sucker of digenetic trematodes. Sometimes, various other head organs, such as lappets, glandular areas, and extensions, exist instead of suckers. The function of these organs is primarily adhesive, probably to hold the mouth to the region of feeding, but it may also be sensory.

The mouth is normally located anteriorly, but it may be near the middle of the body. The shape of the opening ranges from slit-like to circular. It may be in, near, or removed from the anterior sucker or suckers. Most monogeneans feed principally on host blood, but they also eat mucus and epithelial cells.

The parasite's tegument is generally smooth and is pierced by a few other openings, such as excretory ducts and ducts from secretory glands. The outer layer of *Amphibdella flavolineata,*[9] a monogenean worm that lives in blood and on gill lamellae of electric rays, is a syncytial cytoplasmic epidermis bearing scattered microvilli and connecting with parenchymally situated "cell" bodies by means of conspicuous microtubule-lined cell processes. The epidermis of probably all monogeneans is secretory and contains numerous granular and vesicular bodies.[8]

These parasites (Fig. 8–3) lack a body cavity; the various organs lie in a sort of packing tissue, the **parenchyma.** Like the Digenea, the Monogenea carry on excretion with the aid of flame cells, tubules, and vesicles. The excretory pores are anterior and lateral, whereas in Digenea the pore is usually single and posterior. The digestive tract consists of the mouth, pharynx, esophagus, and gut. The gut may be simple or branched, sometimes with innumerable small blind pouches or **ceca.** There is no anus. The worms are hermaphroditic, and the reproductive systems are basically the same as those of digenetic trematodes (Chap. 9). There may be a single testis or many testes. Sperm pass through the vasa efferentia to the vas deferens and are assisted from the body by a copulatory organ.

Female structures are more complicated. The ovary is often branched or folded, and several organs provide nourishment and shell material to the egg and ensure its fertilization. Figure 8–4 presents a diagrammatic representation of the female genitalia and egg of a monogenetic trematode. Egg formation is essentially similar to that described for digenetic trematodes.

Although male and female genital pores may open either together or separately, they are usually closely associated on the ventral surface toward the anterior end of the animal. *Diclidophora merlangi* is a parasite of whiting, fish from the Northeast coast of Scotland. Blood is the primary food of this worm, but the parasite may supplement its diet with organic nutrient of low molecular weight absorbed directly from sea water by the tegument. The spined penis of one parasite attaches to another worm at a lateroventral position posterior to the genital openings. There is no vagina, so the sperm must be injected into the mate and travel between tissue cells to reach the seminal receptacle.[10] The term **oncomiracidia** refers to larvae of all monogenetic trematodes. Larvae find their host by chemical attraction.[7]

For a bibliography of monogenean literature, see Hargis et al.[1]

Subclass Monopisthocotylea

An oral sucker is lacking or is weakly developed in this group. The anterior end often possesses 2 or more lobes, formed by clusters of adhesive glands. The opisthaptor consists of a prominent disc armed with 2 or 3 pairs of large hooks (anchors) and up to 16 marginal hooklets.

Gyrodactylus, a viviparous member of the group (Fig. 8–5), serves as a representative genus. Only 0.5 to 0.8 mm long, it lives on the surface of fresh-water fish and frogs and infests host organs of locomotion and respiration, especially. The larvae develop within the uterus of the parasite and may contain clusters of embryonic cells. Within these larvae a second larva develops, followed by a third within the second, and a fourth within the third. This polyembryony, yielding four individuals from one zygote, leading to rapid amplifications of numbers of individuals, is perhaps reminiscent of asexual reproduction in digenetic trematodes, and might be considered when pondering the germ cell lineage in the Digenea (Chap. 9). The environmental or physiologic factors that control this polyembryony are unknown. This parasite is of concern to fish farmers, because populations of the parasite can increase rapidly on economically important hosts such as trout. Thus it provides academic as well as economic motives for study. The adult opisthaptor carries no suckers, but has a row of 16 small hooks along its edge and a large pair of hooks in the center.

Marine invertebrates may also be parasitized

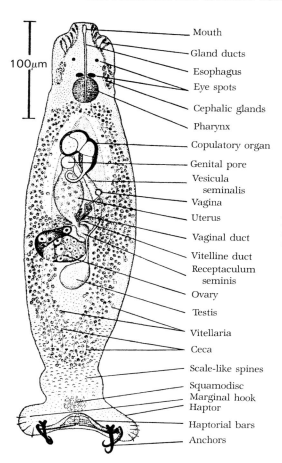

Mouth
Gland ducts
Esophagus
Eye spots
Cephalic glands
Pharynx
Copulatory organ
Genital pore
Vesicula
     seminalis
Vagina
Uterus
Vaginal duct
Vitelline duct
Receptaculum
     seminis
Ovary
Testis
Vitellaria
Ceca
Scale-like spines
Squamodisc
Marginal hook
Haptor
Haptorial bars
Anchors

100μm

Fig. 8–3. *Diplectanum melanesiensis* from the serranid fish, *Epinephelus merra*. Whole animal, ventral view. (From Laird, courtesy of the Can. J. Zool.)

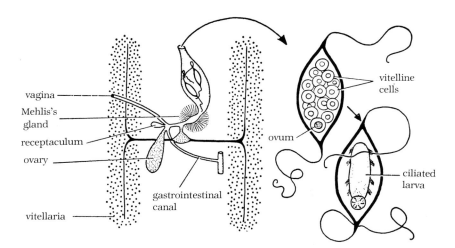

vitelline
cells

vagina
Mehlis's
gland
receptaculum
ovary

ovum

ciliated
larva

gastrointestinal
canal

vitellaria

Fig. 8–4. Diagrammatic representation of female genitalia and egg of a monogenetic trematode. (From Smyth and Clegg, courtesy of Exp. Parasitol.)

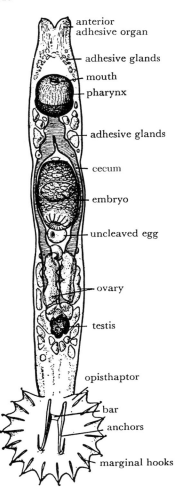

anterior
adhesive organ

adhesive glands

mouth

pharynx

adhesive glands

cecum

embryo

uncleaved egg

ovary

testis

opisthaptor

bar

anchors

marginal hooks

Fig. 8–5. *Gyrodactylus.* (From Hyman, after Mueller and Van Cleave, courtesy of McGraw-Hill.)

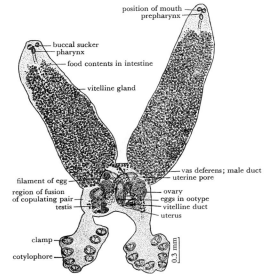

position of mouth
prepharynx

buccal sucker
pharynx

food contents in intestine

vitelline gland

vas deferens; male duct
uterine pore

filament of egg

region of fusion
of copulating pair

testis

ovary

eggs in ootype

vitelline duct

uterus

clamp

cotylophore

0.3 mm

Fig. 8–6. Two *Diplozoon ghanense* in permanent copulation. (From Thomas, courtesy of J. West African Sci. Assoc.)

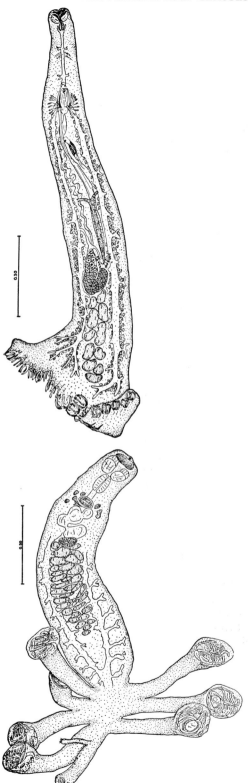

Fig. 8–7. *Axinoides raphidoma*, a monogenetic trematode from Gulf of Mexico fish. (From Hargis, courtesy of Proc. Helm. Soc. Wash.)

Fig. 8–8. *Choricotyle louisianensis*, from the gills of southern whiting. (From Hargis, courtesy of Trans. Am. Microscop. Soc.)

Fig. 8–9. Numerous *Allobivagina* sp. on gills of *Siganus luridus* (rabbitfish) from the Red Sea. Fish were kept in culture facilities (earthern ponds), where the monogeneid caused high mortality of the fish. (From Paperna et al., courtesy of Helgoländer Meeresunters. We thank Dr. Paperna for providing the original negative.)

by monogeneans belonging to this group. Squids are host to the genus *Isancistrum*, and the copepod *Caligus* may have clumps of *Udonella* attached to it. The latter parasite possesses two anterior suckers and a simple opisthaptor. The copepod is itself a parasite of marine fish.

Subclass Polyopisthocotylea

In this group, the mouth is surrounded by the prohaptor, which consists of one or two suckers or two pits. The opisthaptor may or may not be armed, but always has suckers or sucker-like bodies containing clamps. These parasites are almost exclusively blood feeders.

Polystoma integerrimum is endoparasitic in the urinary bladder of frogs and toads. Larval flukes usually attach themselves to the gills of tadpoles and remain in arrested growth until host metamorphosis takes place, when they migrate to the intestine and urinary bladder of the adult frog. The parasite's eggs are produced in the spring of the fourth year.[14] In some instances, the larva attaches itself to a young tadpole, matures as a neotenic (sexually mature) larva, and begins to lay eggs by the end of the fifth week. These worms are smaller than endoparasitic ones, and they die when their tadpole hosts metamorphose. This phenomenon of two possible generations, an ectoparasitic one

in young tadpoles and an endoparasitic one in older tadpoles and adult frogs, is reminiscent of the two possible life histories of the roundworm, *Strongyloides stercoralis* (Chap. 15).

Diplozoon ghanense (Fig. 8–6) lives on gills of fish *Alestes macrolepidotus*. Its greatest distinction is that, although it is hermaphroditic, a permanent union of the worms occurs. During the larval stage of this parasite, a small fleshy knob appears on the dorsal surface. Eventually, this knob becomes fitted into a ventral sucker of another larval worm. The two worms become securely fused together and cannot be separated. This unusual fusion of two separate individuals into one provides an interesting possibility for studying the concept of "self" in contrast to the old immunologic question of "host *vs* graft." Here we have an example of natural grafting, "parasite on parasite," in which the "graft" is not rejected, but indeed results in a sexual union. The gonads then begin to develop, and finally the vagina of one individual opens in the region of the uterus and vas deferens of the other. This arrangement is reciprocal, and cross-fertilization is made easy.

The family Microcotylidae is represented in Figures 8–7 and 8–8 by species from Gulf of Mexico fish. Some peculiarities of this group are well illustrated. Sucker-like clamps on the ends of four pairs of lateral peduncles occur in *Choricotyle louisianensis.*

For a description of monogenean evolution, see Chapter 26. Microcotylids can also cause damage to cultured fish in the crowded conditions of seawater ponds. Figure 8–9 shows an infestation of *Allobivagina* on gills of *Siganus luridus.*[11]

REFERENCES

1. Hargis, W.J., Lawler, A.R., Morales-Alamo, R., and Zwerner, D.E.: Bibliography of the Monogenetic Trematode Literature of the World. Special Scientific Report SS. Gloucester Point, VA, Virginia Institute of Marine Science, 1969.
2. Bychowsky, B.E.: Monogenetic Trematodes, their systematics and phylogeny. Acad. Nouk. SSSR. (English editor, W.J. Hargis; translator, P.C. Oustinoff). Washington, D.C., AIBS. 1957.
3. Ehlers, U.: Das phylogenetische system der Platyhelminthes. Stuttgart, Fischer, 1985.
4. Hyman, L.H.: The Invertebrates. Vol. 2. Platyhelminthes and Rhynchocoela. New York, McGraw-Hill, 1951.
5. Jennings, J.B.: Parasitism and commensalism in the Turbellaria. Adv. Parasitol., 9:1–32, 1971.
6. Llewellyn, J.: Monogenea. J. Parasitol., 56:493–504, 1970.
7. Kearn, G.C.: Experiments on host-finding and host-specificity in the monogean skin parasite *Entobdella soleae.* Parasitology, 57:585–605, 1967.
8. Lyons, K.M.: The epidermis and sense organs of the Monogenea and some related groups. Adv. Parasitol., 11:193–232, 1973.
9. Lyons, K.M.: Comparative electron microscopic studies on the epidermis of the blood living juvenile and gill living adult stages of *Amphibdella flavolineata* (Monogenea) from the electric ray *Torpedo nobiliana.* Parasitology, 63:181–190, 1971.
10. MacDonald, S., and Caley, J.: Sexual reproduction in the monogenean *Diclidophora merlangi*: tissue penetration by sperms. Z. Parasitenkd., 45:323–334, 1975.
11. Paperna, I., Diamant, A., and Overstreet, R.M.: Monogenean infestations and mortality in wild and cultured Red Sea fishes. Helgoländer Meeresunters, 37:445–462, 1984.
12. Riser, N.W., and Morse M.P. (eds.): Biology of the Turbellaria. Series in Invertebrate Biology. New York, McGraw-Hill, 1974.
13. Stoll, N.R.: This wormy world. J. Parasitol., 33:1–18, 1947.
14. Williams, J.B.: The dimorphism of *Polystoma integerrimum* (Frölich) Rudolphi and its bearing on relationships within the polystomatidae: part III. J. Helminthol., 35:181–202, 1961.

9

Class Trematoda, Subclasses Aspidogastrea and Digenea

These flatworms are all parasitic, some living on the surface of their hosts and some inside. The body is covered by a tegument that usually appears smooth, but may be spiny. Almost all species have one to several suckers. The life cycle of Aspidogastrea is direct, but that of Digenea involves two or more hosts, one of which is usually a snail.

Fish bear the heaviest burden of trematode parasites. Birds come next, having about three times as many kinds as occur in amphibians, reptiles, or mammals. A single host may harbor only one species, but hundreds of individuals.

The study of trematodes of fish is a wide field for investigators. Many of these parasites have yet to be described. Adult trematodes of numerous species occur in intestines or other organs of fish, and larval forms are commonly embedded in the skin, mesentery, muscles, liver, and other organs. The worms may or may not be injurious, depending on the numbers of individual worms and on the organ or organs infected.

Many of the species of Digenea discussed in Chapter 10 are serious pathogens of humans. Thus, in addition to taxonomic and ecologic studies on this interesting group, there has been an upswing in molecular and immunoparasitologic research which also offer new opportunities for students.

Subclass Aspidogastrea

This group is also called Aspidocotylea or Aspidobothria. This subclass is described by Rohde as follows:

"The Aspidogastrea are considered to be primitive, direct descendents of turbellarians which are not yet closely adapted to parasitism and have not yet incorporated the vertebrate host as a fixed component of their life-cycle. They are closely related to ancestors of the Digenea. Aspidogastrea and Digenea are both primarily parasites of molluscs."[35]

For an extensive account of Aspidogastrea, see Rohde.[34] For fine structure, see Rohde[33] and Halton.[18] The group is divided into two families, Aspidogastridae and Stichocotylidae. The aspidogastrids are characterized by an enormous circular or oval sucker occupying the greater part of the ventral surface (Fig. 9–1). The mouth (Fig. 9–2) lacks the typical sucker of higher forms. The illustration shows the mouth of an adult found in freshwater mussels. They are marine or freshwater parasites of the mantle and pericardial and renal cavities of clams and snails, of the gut of fish and turtles, and of the bile passages of fish. These parasites develop directly from free-swimming larvae, but details of their development are obscure. The stichocotylids, comprising the single genus *Stichocotyle*, are elongate slender worms, about 10 cm in length, that are parasitic in the bile passages or spiral valve of skates.

Subclass Digenea

These parasites are characterized by cuplike muscular suckers, usually without hooks or other accessory holdfast organs, by genital pores that normally open on the ventral surface between the suckers, and by a single, posterior excretory pore.

The classic shape of a digenetic trematode, or fluke, is that of a thick oval leaf. However, many variations exist, from those resembling a short

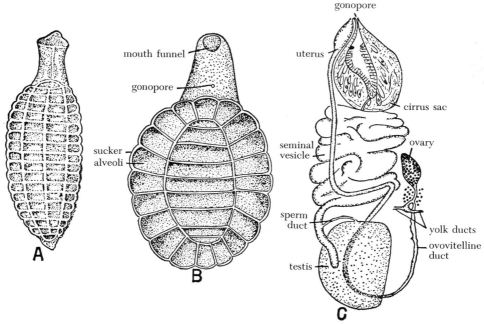

Fig. 9–1.   Types of Aspidogastrea. *A, Aspidogaster conchicola.* (After Monticelli.) *B, Cotylaspis.* (After Osborn.) *C,* Reproductive system of *Cotylaspis.* (After Stunkard.) (From Hyman, L.H.: The Invertebrates: Platyhelminthes and Rhynchocoela. Vol. II. New York, McGraw-Hill, 1951.)

piece of pencil, tapered at both ends, to narrow ribbons elongated to about 12 m (40 feet) or more (see Chap. 10). In contrast, some species *(Euryhelmis)* are wider than they are long. Although the presence of two suckers is typical, only one occurs on some flukes, and a few species have none. Normally, one sucker surrounds the mouth and is called the **oral** or **anterior sucker.** The other sucker is called the **acetabulum, posterior sucker,** or **ventral sucker. Monostomes** are flukes with one sucker; **distomes** possess two suckers.

Tegument

This metabolically active cellular unit has an outer layer, the **epicuticle,** which is anuclear, syncytial, and connected by narrow cytoplasmic tubes to the sunken nucleated regions of the tegument and, through the basal membrane and muscle layers of the body, to cytoplasmic masses lying in the parenchyma (Fig. 9–3). Microvillus-like and pinocytotic elements have been observed on the outer surface of larvae and adults. All these structural features suggest absorptive or protective functions. Absorption of glucose takes place through the tegument, thus assisting in the general nutritional activi-

ties of the worm. The tegument is also associated with respiratory and sensory functions. It is usually resistant to the action of pepsin and trypsin, probably because of the presence of acid mucopolysaccharides and polyphenols. This resistance is probably a main factor in protecting the worms from digestion by the host.

Histochemical analysis of the tegument of *Fasciola hepatica* and *Echinostoma revolutum* has demonstrated that it is formed by a lipoprotein complex. Characteristic granules in the cuticular layer consist of an acid mucopolysaccharide. Scales and spines of these trematodes are formed by a scleroprotein with a characteristically high content of cysteine and with a multitude of lipoid substances. For an account of the tegument of developmental and adult stages of trematodes, see Bils and Martin.[3] For tegumentary structures of various helminths, see Lee.[25] For tegumental ultrastructure of the blood fluke, *Schistosoma,* see Hockley.[20]

Muscle System

The bodies and parts of bodies of flatworms are often seen to expand, to contract, and to twist; this movement indicates the presence of

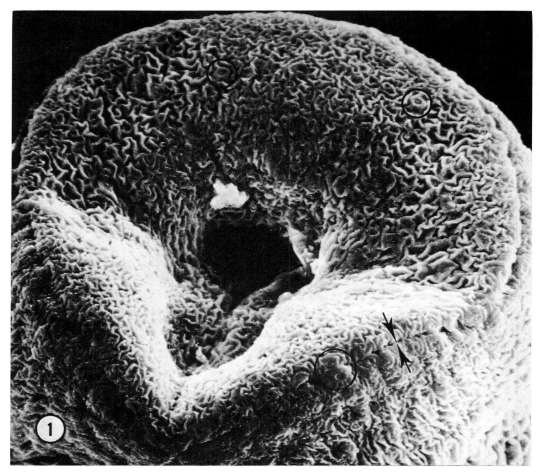

Fig. 9–2. *Aspidogaster conchicola.* Scanning electron micrograph of the oral region. Circles enclose uniciliate sense receptors within the mouth cavity and the margin of the lower lip. An infolding of tegument delimits the lips (arrows). ×1,200 (From Halton, D.W.: Ultrastructure of the alimentary tract of *Aspidogaster conchicola* (Trematoda: Aspidogastrea). J. Parasitol., *58*:455, 1972.)

muscles. These muscles, usually unstriated, lie in groups or layers primarily near the body surface as longitudinal or circular fibers. Other fibers are dorsoventral, while others occur in suckers or other organs.

Respiratory System

Some trematodes use oxygen in respiration when it is available. Intestinal forms are generally considered to be facultative anaerobes. The rate of oxygen consumption is controlled by several factors, especially temperature, body size, and oxygen tension. The respiration rate of some digenetic trematodes was found by Vernberg[39] to be directly proportional to the oxygen tension down to a partial pressure of 3%

or lower. Below this level, the parasite was able to regulate the rate of respiration. Obviously, the respiratory picture varies in a parasite, for example, a lung fluke, possessing life cycle stages in several locations in both vertebrate and invertebrate hosts (see the section on the physiology of trematodes in Chapter 10). Adult schistosomes, however, are homolactate fermentors.

Nervous System

Paired ganglia in the anterior end of the body serve as a brain, and from it lead various main nerve trunks, which, in turn, lead to branches innervating all parts of the body.

Harris and Cheng[19] described presumptive

DIGENEA

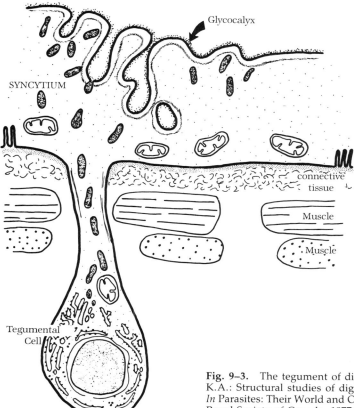

Fig. 9–3. The tegument of digenetic trematodes. (From Wright, K.A.: Structural studies of digestion in some helminth parasites. *In* Parasites: Their World and Ours. Edited by A.M. Fallis. Ottawa, Royal Society of Canada, 1977.)

neurosecretory cells associated with the cerebral ganglia and longitudinal nerve cords of the fluke, *Leucochloridiomorpha constantiae*. The investigators suggest that maturation may be directly or indirectly associated with a reduction in neurosecretion, as has been demonstrated in some annelids. Neurosecretory products contain a glycoprotein. Neurosecretory granules have also been found in the lancet fluke, *Dicrocoelium lanceatum*, the sheep liver fluke, *Fasciola hepatica*, and in other species.

Digestive System

The mouth leads to the pharynx, the esophagus, and the gut. The gut, also called the cecum, generally has two branches from the muscular pharynx, but it may have many. The epithelial lining of the cecum is the **gastrodermis.** In the rediae of *Cryptocotyle lingua*, many flexible ribbon-like folds extend from epithelial cells of the intestine into the lumen.[24] Nutritive

phagocytosis has been suggested. There is usually no anus, although in a few species (Echinostomatidae), an opening exists between the ceca and the excretory vesicle. Some flukes of fish have one or two anal pores that open to the outside.

The mode of feeding is suctorial, associated with the attachment process of the oral sucker and the muscular pharynx. Apparently, differences in the digestive processes are related to differences in the cells lining the gut of the parasite. This lining, the gastrodermis, occurs in a variety of cellular/syncytial organizations. A syncytium might increase the efficiency of the epithelium by elimination of leaky intercellular pathways for molecules. The gastrodermis is capable of both secretion and absorption. In some digeneans, for example, in *Schistosoma*, transtegumental feeding may also occur.

In the fluke *Cyathocotyle bushiensis*, parasitic in the cecum of ducks, the ventrally situated adhesive organ contains gland cells rich in protein and in RNA. These cells produce enzymes consisting of alkaline phosphatase, esterases of several kinds, and leucine aminopeptidase. The enzymes are effective in freeing the columnar epithelium of the host, and the cells thus freed become reduced to a granular material that is taken into the oral sucker of the fluke and is digested intracellularly within its cecum. The adhesive organ of the parasite thus serves for attachment and for extracorporeal digestion.[15]

Excretory System

Typical of all flatworms, this protonephridial system consists of flame cells (flame bulbs) connected by tubules uniting to form larger ducts that open either independently to the outside or join to form a urinary bladder that opens at or near the posterior end of the animal (Fig. 9–4). Flame cells and ducts function not only for excretion, but also for water regulation and,

possibly, to keep body fluids in motion. The ducts, or tubules, contain minute finger-like projections, which presumably aid reabsorption by increasing the internal surface area.

Excretory systems of certain cercariae and metacercariae contain concretions composed chiefly of calcium carbonate and a trace of phosphate. These concretions, or corpuscles, may be active in the fixation of carbon dioxide and in the buffering of acids.

Some metacercariae, for example, Strigeidae, possess a **paranephridial plexus** or "reserve bladder." This plexus of vessels ramifies to all organs and is connected to the protonephridial tubes. It contains calcareous corpuscles and fat globules. Possibly, it functions for storage of higher fatty acids.

"Packing tissue" of the body is a mass of parenchymal cells that function in secreting the interstitial material, in storing glycogen, and, in the absence of a circulatory system, in trans-

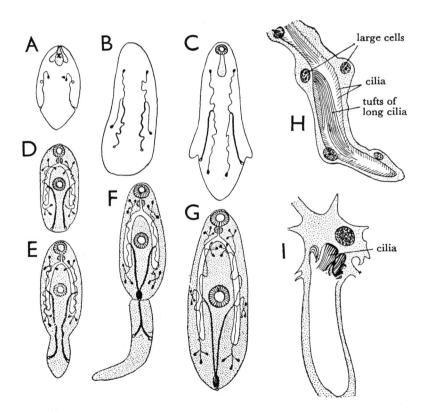

Fig. 9–4. The excretory system of Digenea. *A*, Miracidium. *B*, Sporocyst. *C*, Redia. *D, E, F*, Stages in development of the cercaria. *G*, Metacercaria. *H*, Tufts of long cilia and large cells forming the ciliated wall of the canal (not seen in the adult). *I*, Young-stage flame cell from *Dicrocoelium dendriticum*. (From Dawes. The Trematoda, courtesy of Cambridge University Press.)

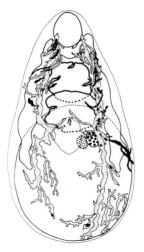

Fig. 9–5. *Paramphistomum bathycotyle.* Morphology of the lymphatic system; dorsal view. (From Lowe, C.Y.: Comparative studies of the lymphatic system of four species of amphistomes. Z. Parasitenkd., 27:169, 1966.)

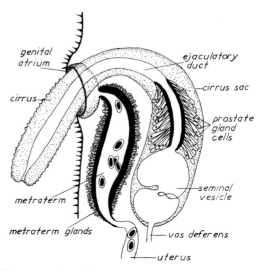

Fig. 9–6. Genital opening in a fluke, showing the point where eggs emerge from the body and an extended male organ, the cirrus, which transmits sperm to another worm.

porting metabolic products. A "lymphatic" system has been observed in several digenetic families and in the monogenetic family Sphyranuridae.[28] A tubular system distinct from the excretory system, the lymphatic system, consists of one pair of longitudinal canals branching repeatedly to supply the various tissues and organs (Fig. 9–5).

Reproductive System

The testes and ovary may be rounded or branched. Usually, two testes are present, but some digenetic flukes may have more than a hundred, for example, in some avian schistosomes. Most species in the family Monorchiidae possess a single testis. An enlarged detail of the distal end of the male reproductive system (Fig. 9–6) shows the apparatus used to transfer sperm to the female part of another worm or perhaps to the female part of the same worm. The **cirrus** is analogous to the penis of higher animals. In the female system, **Laurer's canal** may be a vestigial duct corresponding to the vaginal canal in tapeworms. In a few flukes, this canal has been shown to be involved in copulation. A sperm storage organ, a seminal receptacle or **receptaculum,** is also usually present (Fig. 9–7).

Vitellaria (vitelline glands) consist of yolk- and eggshell-producing cells. Most of this gland is probably associated with considerable protein demand for production of eggshells. The vitellaria may be dispersed (Fig. 9–8) or clumped; they are connected by minute ducts to larger vitelline ducts, which eventualy form main channels that come from each side of the organism. These channels meet near the midline and form the **vitelline reservoir**. The common vitelline duct coming from the reservoir joins the oviduct, and beyond this region it enlarges to become the **ootype**. The ootype is surrounded by a mass of minute gland cells known together as **Mehlis' gland** (Fig. 9–9). It has been suggested that secretions of Mehlis's gland may lubricate the uterus and may thus aid the passage of eggs along this tube, and also that the secretions might activate spermatozoa. This gland secretes a phospholipid that may cause the release of eggshell precursors from vitelline cells. It also produces free phenols from mucopolysaccharides and activates phenolases, thereby allowing oxidation of phenols to quinones involved in the "tanning" of eggshells. Undoubtedly, the function of the gland is similar in cestodes to its function in trematodes.

Diagrammatic representations of the genitalia of digenetic trematodes are shown in Figures 9–8 and 9–9. The following description of the method of shell formation is taken from Smyth and Clegg.[37] Mature eggs leave the ovary, pass

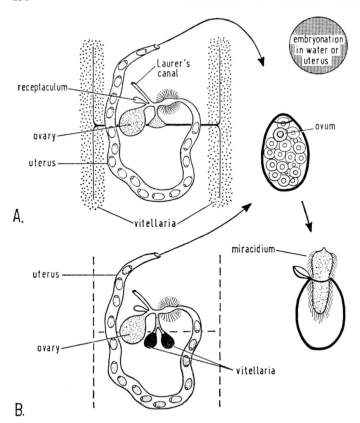

Fig. 9–7. Diagrammatic representation of female genitalia of digenetic trematodes. *A*, with extensive vitellaria; *B*, with condensed vitellaria. (From Smyth and Clegg, courtesy of Exp. Parasitol.)

down the short oviduct, and are fertilized by sperm that have been stored in the receptaculum or in Laurer's canal. The vitelline reservoir is filled with unusual cells that have been produced by the vitelline glands. These vitelline cells possess a nucleus, a mass of yolk material, and granules that are destined to produce the eggshell. These granules have been called **vitelline globules,** but are better named **shell globules.** As the egg moves through the ootype, a group of vitelline cells surrounds it, the number of vitelline cells being constant for each species of fluke. Shell globules are extruded from the vitelline cells and coalesce, thus forming a thin membrane. This membrane is the outer portion of the eggshell. More globules are released, and the shell is built up from within. Completion of the shell occurs in the lower part of the oviduct.

In *Syncoelium*, eggs may possess a thin membrane formed by Mehlis's gland and/or the vitelline gland, but the bulk of the shell is furnished by the uterus, which has large gland cells containing precursors to the eggshell. Vi-

telline cells of *Parastrigea* also contribute copious amounts of glycogen and RNA.[12]

Eggshells vary in their chemical nature.[29] Most of them are composed of a quinone-tanned protein and are hard and brownish in color. In others (amphistomes), a keratin type of protein is found, and shells are transparent and colorless.

Figure 9–10 illustrates eggs from representative trematodes, cestodes, and nematodes, and Table 9–1 is a key to most of these species.

Eggs pass through the uterus into a glandular **metraterm** and out of the body through the genital pore. The uterus, however, is more than a passive conduit for eggs, as was demonstrated by exposing adult *Philophthalmus megalurus* from the ocular sacs of birds to tritiated thymidine, glucose, tyrosine, and leucine.[32] These compounds entered and were incorporated in developing miracidia within eggs while they were passing through the uterus.

Life Cycles

In vertebrate hosts, adult flukes live in the digestive tract, in ducts associated with the alimentary canal, in the blood, lungs, gallbladder,

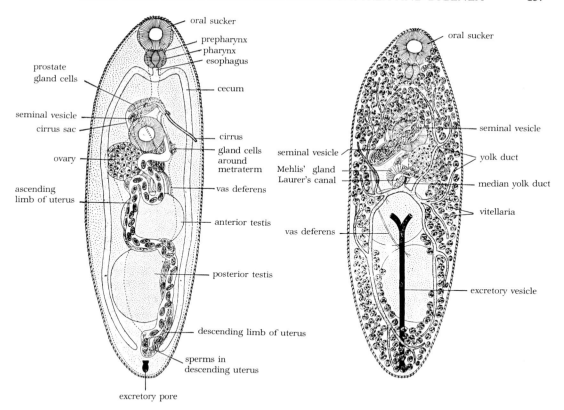

Fig. 9–8. *Plagiorchis (Multiglandularis) megalorchis,* showing the dispersed distribution of the vitellaria and vitelline ducts. (From Rees, courtesy of Parasitol.)

urinary bladder, and oviducts, or in almost any other body organ. Usually, the parasites are restricted to the paramucosal lumen and the mucosal and epithelial tissues.

Self-fertilization and cross-fertilization occur in trematodes. Self-fertilizing strains may be the result of adaptations to special environments where there are few snails, or where there is difficulty in contact with snails, for instance, in swift streams. Nollen[32] carried a self-fertilizing strain of *Philophthalmus megalurus,* an eye-dwelling parasite of birds, "through 3 successive life cycles with little or no deleterious effects when compared with cross-fertilizing strains." Parameters noted were growth, eggshell formation, and production of viable larval forms. "However, significant reductions in recovery rates of adults from monometacercarial infections in the second and third generations were found." In those species in which isolated adults apparently will not reach maturity in a normal manner, chemical factors may be in-

volved in attracting two worms. Michaels[31] postulated that tactile stimulation by the male through linear tegumental receptors is needed for normal egg laying in female *Schistosoma mansoni.*

The first cleavage of the zygote in an adult worm apparently produces a germinal cell and a somatic cell. The latter divides further to form the soma, and the germinal cell forms reproductive cells of the larval stages. This germinal lineage is preserved throughout the life cycle of trematodes. For review of germinal lineage see Tobler and MacInnis.[38a] Eggs usually leave the host through the intestine. If they enter water, they are either eaten by a snail in which they hatch, or they hatch in the water and become free-swimming, ciliated **miracidia** that penetrate the snail. Miracidia are equipped with penetration glands, an excretory system, germ cells, and in some cases, an eye spot (Fig. 9–11).

Miracidia freed from their capsules (eggshell) dart about randomly. They show great flexibility

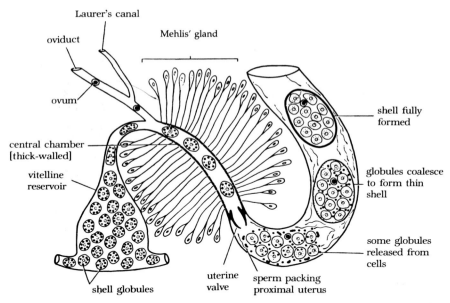

Fig. 9–9. Diagrammatic representation of eggshell formation in trematode, based on *Fasciola hepatica*. The process is essentially the same in tapeworms. (From Smyth and Clegg, courtesy of Exp. Parasitol.)

and continual peristaltic changes of form. They find a snail, clam, copepod or, rarely, and annelid host by chemotactic attraction or by some other means. When miracidia locate snails through chemical attractants, greater opportunity exists for contact with hosts in slowly flowing streams. Strong evidence suggests that at least some species perceive and are attracted by the mucus of the snail.[16,23,40] Many miracidia, however, reach their intermediate hosts purely by chance, although tropisms, for example, phototropism or geotropism, may guide them to areas of host concentration. Miracidia often demonstrate a high degree of selectivity for their host. Age of the mollusc is probably a controlling factor in the selection and entry by miracidia.

Because the link between a miracidium and its intermediate host may be one of the weakest in the life cycle of the parasite, and hence easiest to break, a substantial amount of information about it has accrued. The processes of host-finding by miracidia (and also by cercaria) have been separated into three aspects: (1) responses to the environment such as gravity, light and heat, which serve to bring the larval stage closer to the environment of the host; (2) response to the hosts' "active space"; the active space of a host is larger than the host itself, enhanced by diffusion of chemicals, currents, and other factors

emanating from the host, which serve to assist the larvae to "home in" on the host; (3) responses to the host itself. Further ecologic aspects and consequences of host-finding are described in other chapters.[28a]

Entry of miracidia into a snail host is either active or passive. If the egg is small, as in members of the fluke families Heterophyidae, Opisthorchiidae, Brachylaimidae, and Plagiorchiidae, it is usually eaten by the snail, whether aquatic or terrestrial, or sometimes by a bivalve mollusc. The miracidium emerges from the egg in the snail's intestine. Since snails readily eat feces, considerable opportunity exists for the passage of fluke eggs into snail intestines. If the egg is of medium size or is large, as in Schistosomatidae, Fasciolidae, and Troglotrematidae, it hatches in water or moist soil, and the miracidium must find the right species of snail. It penetrates through the body surface in the region of the snail's mantle, "foot," head, or tentacles.[8] According to some investigators, miracidia probe with their anterior end and attempt to bore into almost any object, including various species of unsuitable snails and other animals such as planarians. These parasites may keep trying to enter an unsuitable host or an unsuitable part of the snail host, such as the shell, until they die of exhaustion.[9]

A different description of the method of entry

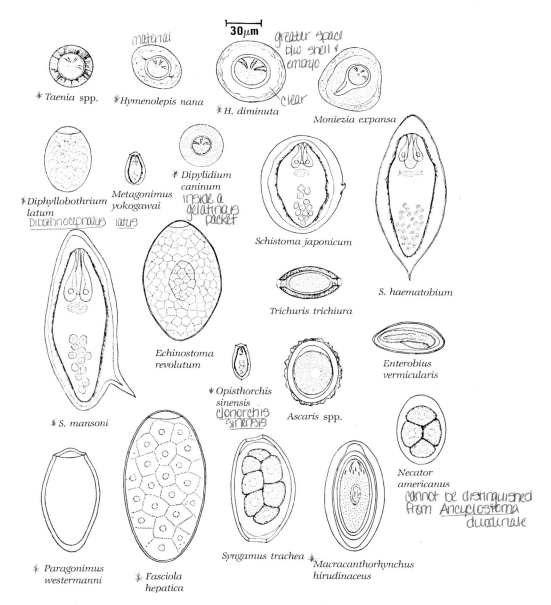

Fig. 9–10. Representative worm eggs of parasites of man and of animals. (From Schell, courtesy of John Wiley and Sons.)

Table 9–1.   Key to Helminth Eggs of Medical Importance

a. Egg nonoperculate, spherical or subspherical, containing a 6-hooked embryo .... b
    Egg other than above ....e
b. Eggs separate c
    Eggs in packets of 12 or more .... *Dipylidium caninum*
c. Outer surface of egg consists of a thick, radially striated capsule or embryophore .... *Taenia* spp.
    Outer surface of egg consists of thin shell, separated from inner embryophore by gelatinous matrix . .d
d. Filamentous strands occupy space between embryophore and outer shell .... *Hymenolepis nana*
    No filamentous strands between embryophore and outer shell .... *H. diminuta*
e. Egg operculate ....f
    Egg nonoperculate ....j
f. Egg less than 35 microns long .... *Opisthorchis* spp.
                                                                                          or
                                                                             Heterophyes heterophyes
                                                                                          or
                                                                             Metagonimus yokogawai
    Egg 38 microns or over .... g
g. Egg 38 to 45 microns in length .... *Dicrocoelium dendriticum*
    Egg over 60 microns in length .... h
h. Egg with shoulders into which operculum fits .... *Paragonimus westermanii*
    Eggs without opercular shoulders i
i. Egg more than 85 microns long .... *Fasciolopsis buski*
                                                                                          or
                                                                             Fasciola hepatica
                                                                                          or
                                                                             Echinostoma spp.
    Egg less than 75 microns long .... *Diphyllobothrium latum*
j. Egg 75 microns or more in length, spined .... k
    Egg less than 75 microns long, not spined .... m
k. Spine terminal .... *Schistosoma haematobium*
    Spine lateral l
l. Lateral spine inconspicuous (perhaps absent) .... *S. japonicum*
    Lateral spine prominent .... *S. mansoni*
m. Egg with thick, tuberculated capsule .... *Ascaris lumbricoides*
    Egg without thick, tuberculated capsule .... n
n. Egg barrel shaped, with polar plugs .... o
    Egg not barrel shaped, without polar plugs .... p
o. Shell nonstriated .... *Trichuris trichiura*
    Shell often striated .... *Capillaria* spp.
p. Egg flattened on one side .... *Enterobius vermicularis*
    Egg symmetrical .... q
q. Egg with large, blue-green globules at poles .... *Heterodera marioni*
    Egg without polar globules ....r
r. Egg bluntly rounded at ends, 56 to 76 microns long .... hookworm
    Egg pointed at one or both ends, 73 to 95 microns long .... *Trichostrongylus* spp.

(From Markell, E.K., and Voge, M.: Medical Parasitology. 4th Ed. Philadelphia, W.B. Saunders, 1976, pp. 29 and 30.)

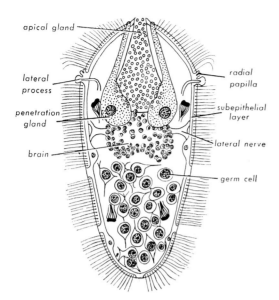

apical gland

lateral process

penetration gland

brain

radial papilla

subepithelial layer

lateral nerve

germ cell

Fig. 9–11. Miracidium of *Schistosomatium douthitti*, to show internal anatomy. (After Price.) (From Smyth. The Physiology of Trematodes. University Reviews in Biology. London, Oliver and Boyd, 1966.)

has been given by Dawes for *Fasciola gigantica*.[13] He stated that a preliminary attachment of the parasite to the host's body occurs by suctorial action. Attachment takes place at an unciliated region of the miracidium. The parasite produces a cytolytic substance that breaks down host tissue and makes a perforation in the skin. The cytolytic substance is produced by penetration glands. The miracidium usually sheds its cilia and thrusts itself into the snail's body. Subsequent to successful invasion, the ability of the parasite to reach a suitable site may be determined by both innate and acquired cellular immunity of the molluscs. Maintenance of trematodes in the snail demands the presence of suitable nutritional factors.

In the snail, the miracidium rapidly becomes a sac, the **sporocyst** (Fig. 9–12). Germinal cells lining the inner wall of the sporocyst develop into rediae (singular, **redia**), which mature and emerge from a birth pore or are liberated by rupture of the sporocyst. Each germ cell in a redia develops into a **cercaria** (Fig. 9–13). Cercariae also are released through a birth pore or by rupture of the redia. Sporocysts and rediae can be considered germinal sacs or cysts (Fig. 9–14), and the first division of the germinal cell in the germinal sac produces both a somatic and

a germinal cell. Generations in the snail, between miracidia and cercariae (Fig. 9–15), vary considerably. There may be two sporocyst stages followed by cercariae, omitting rediae, or one sporocyst followed by three redial stages, or no sporocyst, the miracidia giving rise directly to rediae.

Cercariae leave the snail host and usually develop into encysted **metacercariae** in a second intermediate host. Some metacercariae remain infective in their hosts for years. Bibby and Rees[2] demonstrated that developing metacercariae of *Diplostomum phoxini* in the brain of the minnow, *Phoxinus phoxinus*, feed at the expense of the host. In early stages, the parasites absorb nutrients through the body surface from surrounding brain tissue and cerebrospinal fluid. The metacercariae exhibit alkaline phosphatase activity that might be associated with active transport of glucose across the outer plasma membrane of the parasite's epidermis. Later, nutrients can also be taken through the mouth, and even small fragments of brain tissue may enter this opening. Glucose taken from the environment is converted into glycogen.

Metacercariae of several species of the genus *Diplostomulum*, however, inhabit the eyes of freshwater fishes. *D. spathaceum* occurs in the lens of trout and perch. Other species occupy the vitreous humor, choroid layer, retina, or sclera. An immature male *Schistosoma mansoni* was found in the cornea of a young boy. It had lodged in the anterior chamber of the eye and had probably entered it by the blood-vascular route. Experiments with rodents indicated that *S. mansoni* cercariae are unable to penetrate directly into the eye. Metacercariae have been reported from the manubrium and mesoglea of planktonic hydromedusa. Rarely, they become attached to aquatic vegetation. When eaten by the final host, they develop into adults.

In some flukes, four hosts are common. Cercariae from snails enter amphibians and become **mesocercariae,** which are unencysted and differ from the cercariae by being larger, by not having tails, and by having a more complex excretory system. In rodents that eat infected amphibians, the parasites become metacercariae that may be eaten by the final host, a mammal (see Chap. 10). Shoop and Corkum described yet another unusual aspect of the life cycle of these diplostomatid flukes. They reported that newborn cats and mice may become infected by transmammary passage of mesocercariae of *Alaria*

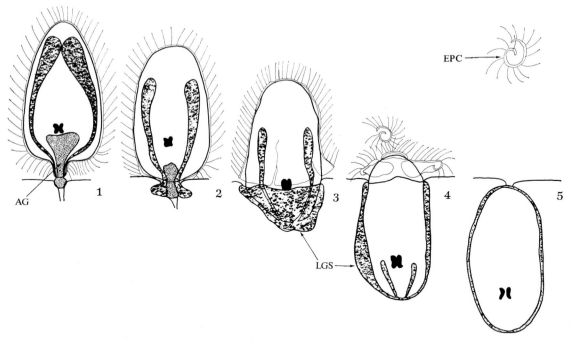

Fig. 9–12. The penetration process of a miracidium. *1,* Miracidium extruding adhesive material from apical gland (AG). *2,* Lateral gland secretion being extruded. *3,* Miracidium "burrowing" into lateral gland secretion (LSG) with epidermal plates sloughed off. *4,* Miracidium nearly enveloped by lateral gland secretion. Epidermal plate cells (EPC) sloughed off and moving away. *5,* Miracidium completely enclosed by lateral gland secretion, a sporocyst. (From Cannon. L.R.G.: The life cycles of *Bunodera sacculata* and *B. luciopercae* (Trematoda: Allocreadiidae) in Algonquin Park, Ontario. Can. J. Zool., 49:1417–1429, 1971.)

marcianae to the suckling young via the mother's milk!

Thus, we can divide the digenetic life cycle into three generations: the miracidium/mother-sporocyst generation, the daughter-sporocyst/redia generation, and the cercaria/metacercaria/adult generation.

Sporocysts may be short or branched, or they may resemble a long, irregularly coiled tube. A few sporocysts, for example, those of *Heronimus,* may pulsate. (See *Leucochloridium,* Chap. 10.) In some species, at least temporary contact with a snail is all that is necessary for a miracidium to metamorphose into a sporocyst. Snails apparently produce a substance that initiates the body transformation of the parasite. Once the process has started, it continues even if the developing sporocyst is removed from the snail. When infected with members of the family Ochetosomatidae, snails respond to the presence of sporocyst embryos by producing fibroblast cells that infiltrate other host cells surrounding the parasites, forming a protective membrane, the **paletot.** The fibroblast cells are **paletot cells.**

Ultrastructural studies of *Cryptocotyle lingua* rediae show ribbon-like folds extending from epithelial cells of its intestine into the lumen. Entrapped food globules in the folds suggest nutritive phagocytosis. The cytoplasmic folds, projections, or flaps of the tegument also suggest a nutritive function by phagocytosis or pinocytosis.[24] Rediae of other trematodes are known to ingest particulate matter as well as to absorb nutrients. Lie et al.[27] have described active ingestion of rediae of one species by rediae of another species in the same snail. Larval trematodes apparently use their host serum proteins, either directly or indirectly from the hepatopancreatic cells.

Daughter-sporocyst/redia generations are probably pedogenetic embryos. Rediae show the beginnings of adult characteristics, each having developed a sucker and an embryonic gut. Most of the internal tissue is germinative, and within the redia the cercariae are formed.

Mature cercariae possess two suckers, a

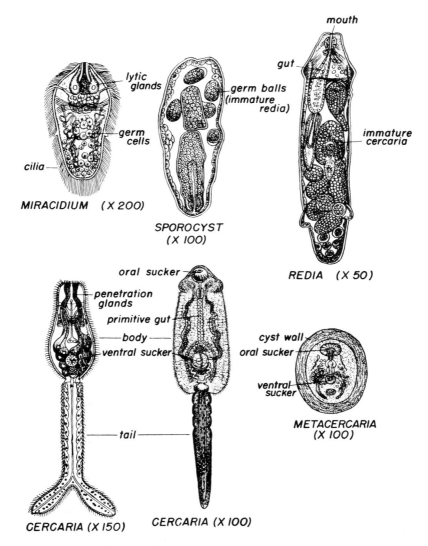

Fig. 9–13. Larval types of human flukes. (From the United States Naval Medical School Laboratory Manual.)

branched gut, and a tail (Fig. 9–16). They also contain several kinds of gland cells, including penetration and cystogenic cells. The cystogenic cells participate in the formation of the metacercarial cyst wall. Like the miracidia, cercariae may have eye spots or photoreceptors consisting of sensory and pigment-containing cells. The tail is considered homologous with the posterior region of the other generations of the life cycle because it is a secondary aquatic adaptation.

Reproduction by germinal sacs in trematodes may be diploid parthenogenesis.[6] Forty gener-

ations of rediae have been produced by transplantation from one snail to another. Apparently, a certain population density must be reached before the parthenogenetic rediae develop into cercariae. Clark[11] reviewed the nature of secondary multiplication in the Digenea and concluded that concepts of alternation of generations and polyembryony are not helpful and should be discarded. He also rejected the concept of parthenogenesis because ameiotic parthenogenesis cannot be distinguished from the "budding" of a single cell, owing to a lack of criteria. As Clark said,

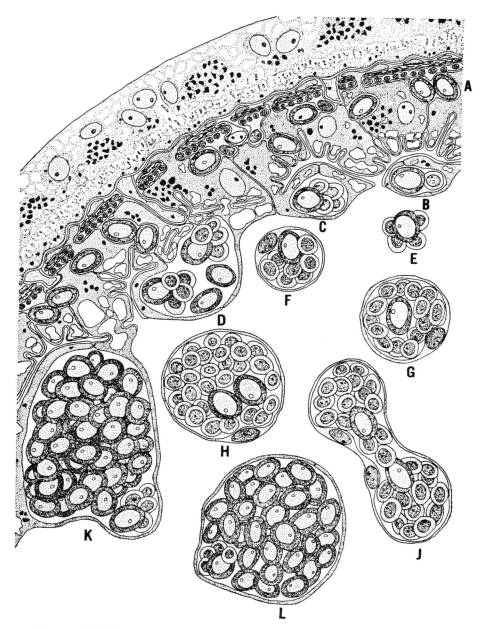

Fig. 9–14. Diagram of the body wall of a daughter sporocyst of *Cercaria bucephalopsis haimeana*, reconstructed from electron micrographs and transverse sections seen under the light microscope. *A*, Germinal cell in subtegument. *B*, Early germinal ball in germinal cyst, consisting of one germinal and one somatic cell. *C*, Germinal ball in germinal cyst, consisting of one germinal and 4 somatic cells. *D*, Germinal cyst containing 4 germinal cells, 2 of which have divided to produce a 5-celled germinal ball. *E*, Germinal ball in body cavity, consisting of one germinal and 4 somatic cells. *F*, Germinal ball in body cavity consisting of one germinal, 7 somatic, and one investing cell. *G*, Germinal ball in body cavity consisting of one germinal, 14 somatic and one investing cell. *H*, Germinal ball in body cavity consisting of 2 germinal, 28 somatic, and one investing cell. *J*, Late stage of division of advanced germinal ball into 2. *K*, Germinal mass enclosed in germinal cyst. *L*, Germinal mass free in body cavity. (From James and Bowers, courtesy of Parasitol.)

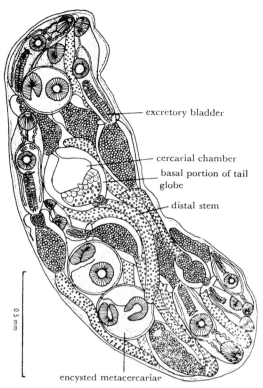

excretory bladder

cercarial chamber

basal portion of tail

globe

distal stem

0.5 mm

encysted metacercariae

Fig. 9–15. Sac-like daughter sporocyst of *Phyllodistomum simile;* the adult fluke is found in the urinary bladder of brown trout, *Salmo trutta.* Within the sporocyst can be seen cercariae and metacercariae. (From Thomas, courtesy of the Proceedings of the Royal Zoological Society, London.)

"If neither parthenogenesis nor sexual reproduction occur in the mollusc host, then there can be no alternation of generations, for all the stages are most simply explained as a sequence of larval forms linked by metamorphosis and regenerative multiplication."

Clark suggested budding and metamorphosis of larval stages as the explanation of digenean life histories. Bednarz[1] stated that each germ cell or "parathenogenic egg" gives rise only to one individual, and there is no trace of polyembryony. More recently, Whitfield and Evans[39a] have reviewed production of larval stages by trematodes and cestodes. They do not favor parthenogenesis as a process which might account for the phenomena. They further point out that the term polyembryony was defined as a production of more than one offspring from a single zygote. They also stress the fact that current information is inadequate to fully explain this reproductive process in platyhel-

minths. Thus we will use the phrase "sequential polyembryony" to describe the process of multiplication. It may well be that Platyhelminthes retain some "totipotent stemcells" which contribute to "multiplication" of larval stages and to regeneration. The interesting question for developmental biologists is whether a somatic cell, once segregated from the germ line, can be redifferentiated to become a germ line cell. As the student can see, experts differ. The final word has not yet been uttered. For semi-final words see Tobler and MacInnis.[38a]

The type and sequence of generations in the molluscan host are not fixed, but may depend on temperature, on the extent of differentiation that occurred before the parasite entered the mollusc, or on the intimacy of the parasite-host relationship. (See Chapter 23 for some effects of larval trematodes on the snail host.)

Serious pathogenic reactions may occur in snails infected by larval flukes. For example, friction between the worm surface and host cells causes damage to the snail's digestive glands. Cercariae may migrate through the body before leaving the snail. Various reports show that migration occurs by way of the blood, genital ducts, or digestive tract. Larvae may rupture the gut wall and enter the visceral hemocoel, the main venous vessels, the heart, and the mantle sinuses; then they may emerge through the inner surface of the mantle. In cyclocoelids and a few other groups, cercariae encyst in the rediae that produce them and never have a migratory phase. Once outside the snail, cercariae may penetrate their next host directly by burrowing into the skin. They do so more readily in those animals whose acellular elements of dermal connective tissue are least resistant to enzymatic action. Such susceptible hosts include young individuals, old individuals with youthful connective tissue, and individuals exposed to stresses.[26]

Some groups of cercariae possess an excretory epithelium that may immobilize wastes and may thus permit the encysted metacercariae to grow and to develop, even to functional maturity.[7] (See Chapter 10 for a description of the metacercarial cyst wall of the liver fluke, *Fasciola hepatica.*)

The final host devours the intermediate host or eats vegetation with metacercariae attached and so becomes infected. Excystation of the young fluke may be mediated by host digestive enzymes. Reducing and surface active agents

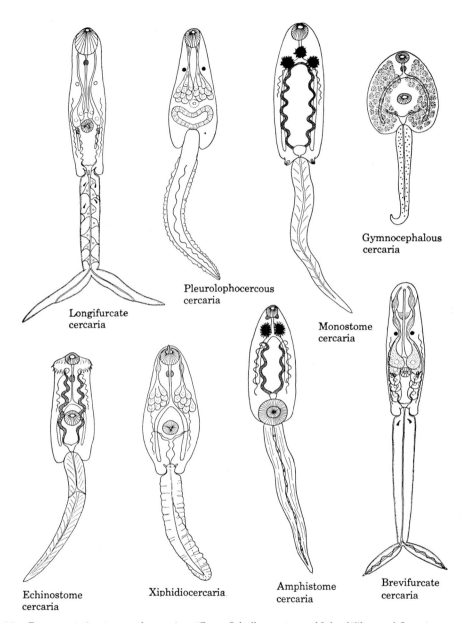

Longifurcate
cercaria

Pleurolophocercous
cercaria

Monostome
cercaria

Gymnocephalous
cercaria

Echinostome
cercaria

Xiphidiocercaria

Amphistome
cercaria

Brevifurcate
cercaria

Fig. 9–16. Representative types of cercariae. (From Schell, courtesy of John Wiley and Sons.)

and a carbon dioxide atmosphere may stimulate excystment. Released metacercariae reach the site of preference, mature, produce eggs, and thus complete the cycle.

Snails may serve as both primary and secondary intermediate hosts. For example, the echinostome, *Echinoparyphium dunni*, uses freshwater snails for its sporocyst, redial and cercarial stages; the emerged cercariae enter the same species of snail or another species and become metacercariae. (See *Gymnophallus*, Chap. 10.) Snails containing sporocysts can develop immunity to penetration by such cercaria. An example is the strigeid fluke of ducks, *Cotylurus*. Sexual maturity and egg production in metacercarial stages have been described by several authors. Stunkard[38] has suggested that this kind of progenesis seems to be a relict, the survival of an earlier developmental method. Adult *Parabrachylaema* (family Brachylaimidae) live in the kidney of *Euglandina rosea*, a terrestrial snail.

Certain members of the family Bucephalidae use a freshwater mussel, *Lampsilis siliquoidea*, as the intermediate host. *Cercaria tiogae* inhabits the freshwater unionid clam, *Alasmidonta varicosa*; and a related species, *C. catatonke*, has been seen emerging from the clam, *Strophitus undulatus quadriplicatus*. *Cercaria milfordensis* is a small, stout-bodied distome that lives in the gonad and digestive gland of the common mussel, *Mytilus edulis*. The sporocysts are orange and give the mantle lobes of the mussel an orange shade. *Phyllodistomum bufonis*, a fluke living in the urinary bladder of toads, *Bufo boreas*, produces fully embryonated eggs whose miracidia penetrate the clam, *Pisidium adamsi*. Mother and daughter sporocysts occur in the gill lamellae. Cercariae encyst in nymphs of dragonflies. Another species of this genus lives in the topminnow, *Fundulus sciadicus*, and also uses clams as intermediate hosts. The cercariae encyst within the daughter sporocysts. Minnows acquire the infection by eating infected clams.

Cephalopod molluscs may serve as second intermediate, paratenic, or final hosts of digenetic trematodes. Most larval forms in these molluscs, however, probably become adults in fishes.

The marine annelid worm, *Eupomatus dianthus*, serves as the first intermediate host of the fluke, *Cercaria loossi*, which is probably a fish blood fluke. In the Antarctic, available hosts would seem to be few and far between; the life of the larval trematode, *Cercaria hartmanae*, encourages this assumption because these larvae develop in other species of rediae that occur in the coelom of the marine annelid, *Lanicides vayssierei*, at Ross Island. *Allocreadium alloneotenicum* is found as a precociously mature adult fluke in the hemocoel of the caddis fly larvae belonging to the genus *Limnephilus*. Fluke eggs are liberated from disintegrating dead caddis fly larvae. Miracidia emerge and penetrate the bodies of the freshwater clam, *Pisidium abditum*. The life cycle within the clam involves one sporocyst stage, two rediae, and the cercariae. The cercariae leave the clam, penetrate caddis fly larvae, and mature in the hemocoel. Immature flukes have been found in larval, pupal, and adult mosquitoes.

"Cercariae" of members of the subfamily Gymnophallinae are ordinarily found in the snail, *Littorina saxtalis*. The parasite is either a cercaria that reproduces like a sporocyst or redia, or it is a sporocyst or redia that looks like a cercaria. Because of this peculiar combination of types, it is called a **parthenita.** The rounded organism appears to be a cercaria, but is filled with small, tailless cercariae.[22]

Trematode behavior has been studied almost entirely in free miracidia and cercariae, but in one species or another, all stages behave in a manner that seems to favor transmission, directly or indirectly, from one host species to the next in the life cycle.[5,28a]

Larval stages of trematodes, especially the early stages, are definitely host-specific, in contrast to many adult flukes, which may parasitize several species of vertebrate hosts. *Fasciola hepatica*, for example, may inhabit the livers of man, sheep, cattle, pigs, deer, rabbits, and other animals.

The effect of parasitism on the intermediate hosts is often not known. Trematodes, however, usually inhibit gonad growth or even cause castration. Thus, parasitized snails are unable to reproduce. Etges and Gresso[17] found that when the snail, *Biomphalaria glabrata*, is infected with larvae of the blood fluke, *Schistosoma mansoni*, the snail's egg laying is inhibited during the fourth week of infection, before the onset of cercarial emergence. Limited egg production is resumed well before the loss of infection, about 90 to 100 days after miracidial penetration. The rate of egg production by infected snails is about 10% of normal, but the eggs develop and hatch nearly normally. Normal and infected snails show a difference in thermal resistance; the latter succumb readily to increased temperature. Damage to the hepatopancreas often occurs. Rediae may ingest host cells and may cause sloughing of tissues, fat accumulation, and depletion of glycogen. Heavy infections often destroy the liver of the snail. See Cheng and Rifkin[10] for a review of the morphology, chemistry, and function of the different types of cells and tissues in molluscs that are involved in cellular reactions to the presence of helminth parasites. For phylogeny and taxonomy of marine species, see Cable.[4] For a study of medical and economic malacology, see Malek and Cheng.[30a] For a general treatment of trematodes, see Erasmus.[14,15] For snail-transmitted diseases, see Malek.[30] For general references, see Hyman[21] and Skrjabin.[36]

REFERENCES

1. Bednarz, S.: The developmental cycle of the germ cells in several representatives of Trematoda (= Digenea). Zool. Pol., 23:279–311, 1973.
2. Bibby, M.C., and Rees, G.: The uptake of radioactive glucose in vivo and in vitro by the metacercaria of *Diplostomum phoxini* (Faust) and its conversion to glycogen. Z. Parasitenkd., 37:187–197, 1971.
3. Bills, R.F., and Martin, W.E.: Fine structure and development of the trematode integument. Trans. Am. Microsc. Soc., 85:78–88, 1966.
4. Cable, R.M.: Phylogeny and taxonomy of trematodes with reference to marine species. *In* Symbiosis in the Sea. Edited by W.B. Vernberg. Columbia, University of South Carolina Press, 1974, pp. 173–193.
5. Cable, R.M.: Behavior of digenetic trematodes. Zool. J. Linn. Soc., *51 (Suppl. 1)*:1–18, 1972.
6. Cable, R.M.: Parthenogenesis in parasitic helminths. Am. Zoologist, 11:267–272, 1971.
7. Cable, R.M.: Thereby hangs a tail. J. Parasitol., 51:3–12, 1965.
8. Cannon, L.R.G.: The life cycles of *Bunodera sacculata* and *B. luciopercae (Trematoda: Allocreadiidae)* in Algonquin Park, Ontario. Can. J. Zool., 49:1417–1429, 1971.
9. Cheng, T.C.: The compatibility and incompatibility concept as related to trematodes and molluscs. Pacific Sci., 22:141–160, 1968.
10. Cheng, T.C., and Rifkin, E.: Cellular reactions in marine molluscs in response to helminth parasitism. *In* A Symposium on Diseases of Fishes and Shellfishes. Special Publication 5. Edited by S.F. Snieszko. Washington, D.C., American Fisheries Society, 1970, pp. 443–496.
11. Clark, W.C.: Interpretation of life history pattern in the Digenea. Int. J. Parasitol., 4:115–123, 1974.
12. Coil, W.H.: Observations on the histochemistry of *Parastrigea mexicanus* (Strigeidae: Digenea) with

emphasis on eggshell formation. Trans. Am. Microsc. Soc., *88*:127–135, 1969.
13. Dawes, B.: Penetration of *Fasciola gigantica* Cobbold, 1856, into snail hosts. Nature, *185*:51–53, 1960.
14. Erasmus, D.A.: The Biology of Trematodes. New York, Crane, Russak & Co., 1972.
15. Eramus, D.A., and Öhman, C.: The structure and function of the adhesive organ in strigeid trematodes. Ann. N.Y. Acad. Sci., *113*:7–35, 1963.
16. Etges, F.J., and Decker, C.L.: Chemosensitivity of the miracidium of *Schistosoma mansoni* to *Australorbis glabratus* and other snails. J. Parasitol., 49:114–116, 1963.
17. Etges, F.J., and Gresso, W.: Effect of *Schistosoma mansoni* upon fecundity in *Australorbis glabratus*. J. Parasitol., 51:757–760, 1965.
18. Halton, D.W.: Ultrastructure of the alimentary tract of *Aspidogaster conchicola* (Trematoda: Aspidogastrea). J. Parasitol., 58:455–467, 1972.
19. Harris, K.R., and Cheng, T.C.: Presumptive neurosecretion in *Leucochloridiomorpha constantiae* (Trematoda) and its possible role in governing maturation. Int. J. Parasitol., 2:361–367, 1972.
20. Hockley, D.J.: Ultrastructure of the tegument of Schistosoma. Adv. Parasitol., 11:233–305, 1973.
21. Hyman, L.H.: The Invertebrates: Platyhelminthes and Rhynchocoela. Vol. II. New York, McGraw-Hill, 1951.
22. James, B.L.: A new cercaria of the sub-family Gymnophallinae (Trematoda: Digenea) developing in a unique 'parthenita' in *Littorina saxatilis* (Olivi). Nature, *184*:181–182, 1960.
23. Kawashima, K., Tada, I., and Miyazaki, I.: Host preference of miracidia of *Paragonimus ohirai* Miyzaki, 1939, among three species of snails of the genus *Assiminea*. Kyushu. J. Med. Sci., *12*:99–106, 1961.
24. Krupa, P.L., Bal, A.K., and Cousineau, G.H.: Ultrastructure of the redia of *Cryptocotyle lingua*. J. Parasitol., 53:725–734, 1967.
25. Lee, D.E.: The structure of the helminth cuticle. Adv. Parasitol., 10:347–372, 1972.
26. Lewert, R.M., and Mandlowitz, S.: Innate immunity to *Schistosoma mansoni* relative to the state of connective tissue. Ann. N.Y. Acad. Sci., 113:54–62, 1963.
27. Lie, K.J., Basch, P.F., and Heyneman, D.: Direct and indirect antagonism between *Paryphostomum segregatum* and *Echinostoma paraensei* in the snail *Biomphalaria glabrata*. Z. Parasitenkd., 31:101–107, 1968.
28. Lowe, C.Y.: Comparative studies of the lymphatic system of four species of amphistomes. Z. Parasitenkd., 27:169–204, 1966.
28a. MacInnis, A.J.: How parasites find hosts: Some thoughts on the inception of host-parasite integration. *In* Ecological Aspects of Parasitology. Edited by C.R. Kennedy. Amsterdam, North-Holland Pub. Co., 1976, pp. 3–20.
29. Madhavi, R., and Rao, K.H.: *Orchispirium heterovitellatum*: chemical nature of the eggshell. Exp. Parasitol., 30:345–348, 1971.
30. Malek, E.A.: Snail-transmitted Parasitic Diseases. Vols. I and II. West Palm Beach, FL, CRC Press, 1980.

30a.Malek, E.A., and Cheng, T.C.: Medical and Economic Malacology. New York, Academic Press, 1974.

31. Michaels, R.: Mating of *Schistosoma mansoni* in vitro. Exp. Parasitol., *25*:58–71, 1969.

32. Nollen, P.M.: Uptake and incorporation of glucose, tyrosine, leucine, and thymidine by adult *Philophthalmus megalurus* (Cort, 1914) (Trematoda) as determined by autoradiography. J. Parasitol., *54*:295–304, 1968.

33. Rohde, K.: Fine structure of Monogenea, especially *Polystomoides* Ward. Adv. Parasitol., *13*:1–33, 1975.

34. Rohde, K.: The Aspidogastrea, especially *Multicotyl purvisi* Dawes, 1941. Adv. Parasitol., *10*:77–145, 1972.

35. Rohde, K.: Untersuchungen an *Multicotyle purvisi* Dawes, 1941 (Trematoda: Aspidogastrea) I. Entwicklung und Morphologie. Zool. Jb. Anat., *88*:138–187, 1971.

35a.Shoop, W.L., and Corkum, K.C.: Transmammary infection of newborn by larval trematodes. Science, *283*:1082–1083, 1984.

36. Skrjabin, K.I.: Trematodes of Animals and Man. Elements of Trematodology. (A continuing series of volumes in Russian.) Moscow, Akademiya Nauk SSSR, 1947–1962.

37. Smyth, J.D., and Clegg, J.A.: Eggshell formation in trematodes and cestodes. Exp. Parasitol., *8*:286–323, 1959.

38. Stunkard, H.W.: Progenetic maturity and phylogeny of digenetic trematodes. J. Parasitol., *45*:15, 1959.

38a.Tobler, H., and MacInnis, A.J.: Molecular and Developmental Biology of Parasitic Platyhelminthes and Nematoda. Amsterdam, Elsevier, in press.

39. Vernberg, W.B.: Respiration of digenetic trematodes. Ann. N.Y. Acad. Sci., *113*:261–271, 1963.

39a.Whitfield, P.J., and Evans, N.A.: Parthenogenesis and asexual multiplication among parasitic platyhelminths. Parasitology, *86*:121–160, 1983.

40. Wright, C.A.: Host-location by trematode miracidia. Ann. Trop. Med. Parasitol., *53*:288–292, 1959.

10

Class Trematoda, Subclass Digenea (Representative Families)

Cercariae possess a thin, nonepithelial excretory bladder and a forked or single tail.

Order Strigeatoida

FAMILY STRIGEIDAE

These intestinal flukes are found in reptiles, birds, and mammals. Usually, the anterior part of the body is flattened and concave, and the posterior part is more cylinder-shaped (Fig. 10–1). The concave portion has an adhesive organ or "holdfast;" the posterior portion contains most of the reproductive system. The parasite has two small suckers, and the overall length of the body ranges from a few millimeters to about 20 mm.

The excretory bladder enlarges and forms narrow tubules or a series of pockets or lacunae. These modifications, called the **reserve bladder system,** may extend over the dorsal and lateral regions of the body. The system functions in excretion and may also serve as a hydrostatic skeleton and a vehicle for the circulation of dissolved nutrients.

The life cycle of these flukes usually includes the miracidium, sporocyst, daughter-sporocyst, cercaria, and metacercaria.

Strigea elegans adults live in snowy owls. Miracidia enter snails and develop into sporocysts, rediae, and cercariae. These last emerge and become mesocercariae in frogs and toads. A third intermediate host, the garter snake, harbors the tetracotyle stage, which possesses a definite cyst wall. Frogs and toads may be only paratenic hosts.

Cotylurus flabelliformis may become a hyperparasite. Cercariae of this strigeid, a common fluke of ducks, may enter planorbid and physid snails, in which they develop into metacercariae only in those individuals that harbor sporocysts and rediae of other species of trematodes. They do not enter snails containing *C. flabelliformis* sporocysts. The strigeid cercariae enter other sporocysts and rediae in which they are apparently protected from immune reactions of the abnormal snail hosts and in which, as hyperparasites, they are able effectively to use the nourishment that the sporocysts and rediae have secured from the snail host. The normal environment for metacercariae of *C. flabelliformis* is the hermaphroditic gland of lymnaeid snails.

Uvulifer ambloplitis is one of several species of strigeoides that cause "black spot" disease in the skin of freshwater fishes in North America. The cercarial stage penetrates into the dermis and metamorphoses into a "neascus metacercaria." Subsequent melanization by the host results in the black spot. These metacercariae mature in fish-eating birds. The metacercariae are readily collected from minnows and other freshwater fish, and present the student with good practice and challenge in "keying out" the species. [42,73]

FAMILY DIPLOSTOMATIDAE

The body of these flukes is divided into a flat or spoon-shaped forebody containing the suckers and another attachment structure called **tribocytic organ,** and a cylindrical hindbody containing the reproductive organs. Intestinal ceca are long, testes tandem or opposite, cirrus sac absent, and the ovary is in front of the testes.

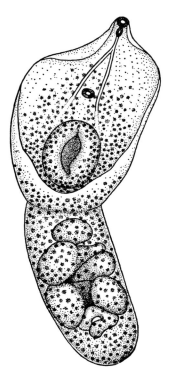

Fig. 10–1. *Neodiplostomum paraspathula,* ventral view. ×40. (From Noble, A.E., courtesy of J. Parasitol.)

The uterus is short, and abundant vitelline glands are variously located. The flukes are found in the intestines of birds and mammals.

Diplostomum phoxini possesses organs that have been called **pseudosuckers,** but are better known as **lappets.** There are two of them, one on each side, posterior and lateral to the oral sucker. They are musculoglandular areas that may be retracted or everted and vary in appearance, depending on the state of contraction. Electron microscopy reveals an outer covering of finger-like processes and an inner musculature. The mass of gland cells in each lappet discharges material to the exterior through the outer tegument layer.[31] In addition to a secretory function, the lappets may aid in the parasite's attachment to the host.

Alaria arisaemoides is another fluke that has more than one intermediate host. From the snail, cercariae enter tadpoles, frogs, or other vertebrates and become mesocercariae. The amphibians are eaten by raccoons, mice, rats, and other mammals in which the metacercarial stage develops. The final host eats the mice, rats, or raccoons. The parasites live as adults in the in-

testines of carnivores, such as dogs, cats, weasels, or minks.

Alaria americana is a related species, common in North America. Its mesocercarial stage may be fatal to humans.[35] Biologists and students who dissect recently killed frogs must be cautious, and frog legs must be cooked well-done before being eaten.

FAMILY SPIRORCHIIDAE

This family is composed of monoecious flukes that are distomes or monostomes inhabiting primarily the heart, larger arteries, and other blood vessels of turtles. Some parasites may occupy other tissues, such as the submucosa of the esophagus. The genus *Spirorchis* is characterized by a central row of testes; the genus *Haplorhynchus* contains two testes; and the genus *Vasotrema* has a single testis that is spirally coiled. Eggs make their way from the host's blood vessels to the intestine, pass out of the turtle's body, and hatch. Miracidia enter lymnaeid snails, develop into a first-generation sporocyst, then into another sporocyst, which produces the furcocercous (forked-tail) cercariae that leave the snail and penetrate the soft tissues of the next turtle host. Other genera are *Haplotrema, Learedium, Neospirorchis,* and *Amphiorchis.*

FAMILY SANGUINICOLIDAE

These small, delicate flukes live in the blood vessels and hearts of freshwater and marine fishes. Suckers and pharynx are absent and the intestinal ceca are X- or H-shaped. There may be only one testis or numerous testes. The ovary is in the posterior third of the body. An example is *Sanguinicola occidentalis,* which lives in North American wall-eyed pike (Fig. 10–2). Flukes classified under the name Aporocotylidae are included here with the Sanguinicolidae. See Smith.[75]

FAMILY SCHISTOSOMATIDAE (= BILHARZIIDAE)

Schistosomes are best known of all flukes because of the serious and major disease **(schistosomiasis** or **bilharziasis)** they cause in man. Animals are also hosts to these worms. *Schistosoma bovis* lives in the portal system and mesenteric veins of cattle and sheep. It occurs rarely in horses, antelopes, and baboons, and may be found in Africa, southern Asia, and southern Europe. The life cycle is essentially the same as

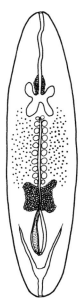

Fig. 10–2. *Sanguinicola,* a composite drawing. Note absence of suckers, the X-shaped ceca, and the column of paired testes, behind them the ovary.

that described for species in man. *S. intercalatum* may be a race of *S. bovis* that has become adapted to man. Snails belonging to the genus *Bulinus* serve as intermediate hosts for *S. bovis.* Other species parasitic in cattle, sheep, goats, carabao, antelope, and occasionally, in dogs and horses are *S. spindale* and *S. nasalis. S. spindale* also occurs in Africa and in southern Asia and parasitizes snails of the genera *Planorbis, Indoplanorbis,* and *Lymnea. S. incognitum* occurs in pigs in India. Schistosomiasis bovis in cattle and sheep generally causes few symptoms, but acute infections may be accompanied by anemia, hepatosplenomegaly, splenomegaly, increased mononuclear phagocytic activity, and increased erythrophagocytosis.

Heterobilharzia americana infects dogs in the United States. Its life cycle is similar to that of *Schistosoma mansoni,* described later in this chapter. As with other blood flukes, the most serious pathogenic response is the host reaction to masses of eggs that are produced and may disseminate to various parts of the body. Host tissue surrounding the eggs becomes infiltrated with epithelioid cells and leukocytes, producing a granular mass called a **granuloma.** The infection is common in several southern states and is associated with swamplands.

Other genera are *Ornithobilharzia, Bilharziella,*

Trichobilharzia and *Pseudobilharzia.* The relationships of these flukes are not well understood and deserve much more study. Sexes are separate, with the female often longer than the male and more slender. Females range from 3 to 15 mm (usually under 10 mm), but one species *(Gigantobilharzia acotyles),* which lives in the abdominal veins of the black-headed gull, may reach 165 mm.

Schistosomatium douthitti is a small blood fluke (1.9 to 6.3 mm long) that lives in the hepatic portal system of muskrats and various mice. It is found in North America and ranges as far north as Alaska. The ease of keeping this trematode in the laboratory in rodents has led to its use as an experimental animal.

Schistosoma rodhaini is a parasite of rodents and dogs; cats may be infected in the laboratory. It is normally transmitted by the snail, *Biomphalaria.* The fluke lives in the liver and mesenteries, and its eggs possess a terminal spine.

BLOOD FLUKES OF MAN

The disease schistosomiasis is estimated to affect 200 million people and exists in 72 countries around the world, from temperate zones to the tropics. Economic losses in Africa amount to hundreds of millions of dollars annually. Schistosomiasis can result in illness, disability, and death. Severe forms of the disease bring about enlarged liver and spleen, calcified bladder, deformity of the ureter, or malfunctioning of the kidneys.

The disease is of serious public health and socioeconomic importance. As Myrdal wrote, "Men and women were sick because they were poor; they became poorer because they were sick, and sicker because they were poorer."[67] Schistosomiasis is now ranked at least second to malaria in importance as a parasitic disease throughout the world.[92] Antigens from juveniles, adults, and eggs cause the formation of antibodies in the host. The resultant antibody-antigen reactions are responsible for the early clinical picture. Immunopathologic alterations in the human host become increasingly evident. Eggs containing living miracidia are the major pathogenic agents. Eggs from even a single worm can cause injury to various tissues, and a worm can lay eggs for up to 30 years.

Reactions similar to those in the vertebrate host may occur between sporocysts and their snail hosts. If the host recognizes the sporocyst as a foreign body, early larvae are surrounded by amebocytes and fibroblasts and are destroyed. If no cellular response is elicited, the host presumably does not recognize the parasite as foreign, possibly because the sporocyst's tegument surface may have a a molecular conformation identical to that of the snail.[6] For details of snail responses, see Chapter 23.

The general external anatomy of the three major schistosomes of man is illustrated in Figure 10–3, and a comparison of the two sexes is shown in Figure 10–4. Hermaphroditism has occasionally been reported.

Schistosoma japonicum occurs in Japan, the Philippines, Thailand, Laos, Kampuchea (Cambodia), Sarawak, Sulawesi (Celebes), and China. The overall prevalence of infection is from 10 to 25%. In Taiwan, the fluke is unusual in that it rarely infects man, but only various other mammals. In every other aspect, the parasite appears to be identical to *Schistosoma japonicum* found elsewhere in the Far East. The most common host of the Taiwan species is the rat, but flukes may also be found in shrews and bandicoots. Various laboratory reservoir hosts include pigs, dogs, cats, carabao, cattle, goats, horses, and rodents.

Male *Schistosoma japonicum* average 15 × 0.5 mm in size. An oral sucker lies at the anterior end, and an acetabulum is situated at the end of a short projection from the body. The body has a groove along much of its length and thus resembles a narrow boat. It was this groove that inspired the name "Schistosoma" or "split body." The female lies within this groove or fold, which is called the **gynecophoral canal.** Males apparently are required to help nourish and stimulate maturation of females. The tegument appears smooth to the naked eye. There is no muscular pharynx, and the intestinal ceca unite posteriorly to form a single cecum. Usually, seven testes are located near the anterior end of the body, and ducts from the testes join to form a seminal vesicle leading to a short duct; this duct, in turn, opens at the genital pore situated just posterior to the pedunculate acetabulum. No muscular cirrus exists.

Females average 20 × 0.3 mm, thus being both longer and narrower than the males. The gynecophoral canal is not usually long enough to enclose the female, so loops of its body generally can be seen extending from the canal. The ovary is situated in the posterior half of the body, and from it extends a long uterus that, in mature worms, is filled with 50 to 300 eggs. A genital pore opens just behind the ventral sucker. Numerous vitelline cells lie on either side of the long medium vitelline duct in the posterior quarter of the body.

These blood flukes live chiefly in the superior mesenteric veins, in which females extend their bodies from males or leave the males to lay eggs (Fig. 10–5) in small venules of the mesenteries

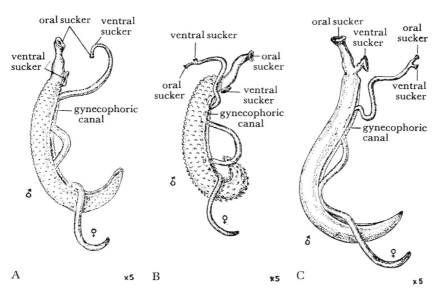

Fig. 10–3. Important schistosomes of man. *A, Schistosoma haematobium. B, S. mansoni. C, S. japonicum.* (From Belding: Textbook of Clinical Parasitology, courtesy of Appleton-Century-Crofts.)

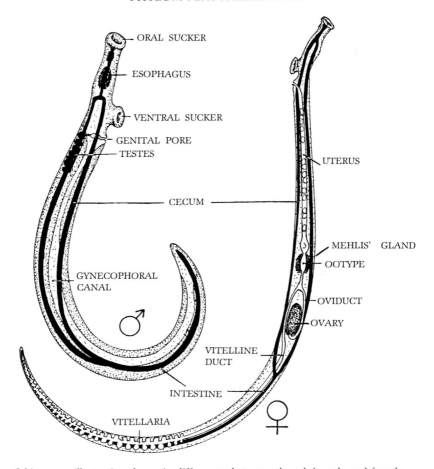

ORAL SUCKER

ESOPHAGUS

VENTRAL SUCKER

GENITAL PORE
TESTES

UTERUS

CECUM

MEHLIS' GLAND

OOTYPE

GYNECOPHORAL
CANAL

OVIDUCT

OVARY

VITELLINE
DUCT

INTESTINE

VITELLARIA

Fig. 10–4. *Schistosoma,* illustrating the main differences between the adult male and female.

of intestinal wall. The oval-to-rounded eggs require a few days to develop into mature miracidia within the eggshell. Eggs measure 90 × 60 μm and possess a minute lateral hook that is usually difficult to see. Masses of eggs cause pressure on the thin venule walls, which are weakened by secretions from the histolytic glands of the miracidia within the eggs. The walls rupture, and the eggs penetrate the intestinal lumen and thus pass to the outside. In a heavy infection, thousands of worms may be present in the blood vessels. See Figure 10–6 for the life cycle.

Hatching takes place in water. Although pH, salinity, temperature, and other environmental aspects are important, factors within the egg probably play a major role in the hatching process. Released miracidia can live for a few hours, during which time they actively swim about, as

though in search of their molluscan host. These hosts are various subspecies of the minute snail *Oncomelania hupensis* (Fig. 10–7). Miracidia penetrate the snail's foot or body with the aid of histolytic gland secretions; the process requires only a few minutes. Within the snail, the cilia are shed, and the miracidia become sporocysts. These sporocysts enlarge, and the germ cells within them develop into a second generation of sporocysts, which mature in the digestive gland of the snail. Germ cells within the second generation sporocysts give rise to cercariae, thus omitting the usual redial generation. At least 60 days, but usually 90, are required from miracidial penetration of the snail to emergence of cercariae. These cercariae possess forked tails, possibly to aid in swimming and penetration. These cercariae usually emerge during the early part of the night and are large enough (335

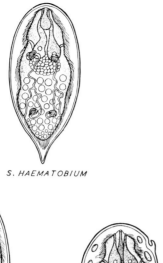

S. HAEMATOBIUM

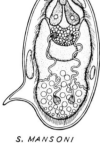

S. MANSONI

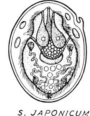

S. JAPONICUM

Fig. 10–5. Eggs of *Schistosoma*.

μm overall length) to be seen with the unaided eye. They are active swimmers, but may hang motionless from the surface of the water. One miracidium can eventually give rise to thousands of cercariae.

Penetration of man or other mammals usually occurs within 48 hours, and the life span of cercariae is limited to a few days. Penetration is probably similar to the following description of this process for *Schistosoma mansoni*. When cercariae come in contact with the skin of an appropriate host, they "loop" for variable periods of time by attaching themselves alternately with the oral and ventral suckers. When unattached, the oral end of the body constantly probes into every irregularity encountered. Points of entry include wrinkled areas, the bases of follicular eminences, points of scale attachment, distal hair-skin angles, orifices, follicular canals, and entry sites used by previous cercariae. Apparently, calcium and magnesium ions are essential for successful penetration,[49] which is strongly influenced by temperature, and stimulated by substances on the skin.[54]

Eventually, the cercariae become closely at-

tached by their oral suckers and assume a vertical or oblique position in relation to the surface. At this time, the body is elongated and slender, and the muscular oral sucker thrusts into the entry site. Penetrating cercariae release an enzyme from glands at their anterior ends. Several pairs of large glands are usually located on each side of the midline, and they empty through long, narrow ducts at the anterior end of the larva near the oral sucker. Although the exact nature of the secretion of the penetration glands is not known, some cercariae apparently produce hyaluronidase, an enzyme that hydrolizes hyaluronic acid, one of the principal substrates of connective tissue. Ramming motion and partial eversion of the sucker bring the openings of the ducts from the penetration glands in contact with the stratum corneum. Through these ducts, the gland content is secreted into the host tissue. Alternate contraction and elongation of the body and energetic movements of the tail accompany thrusts of the oral sucker.

Elongation of the body at the time the oral sucker initially probes into the entry site permits cercariae to push into small crevices. Thus attached and buried orally, the body is contracted and the diameter increased, resulting in an enlargement of the breach in the host tissue. This sequence of slenderizing the body while the oral tip is being thrust deeper, and then contracting the body and pouring out glandular secretions, seems to be the regular means of enlarging the entry pores for gradual penetration. Constant tail activity provides added forward thrust, but tail thrust is not necessary; tailless cercariae have been seen penetrating the host. Exploratory time for the cercariae is about 1.8 to 2.3 minutes, and entry time is about 3 to 7 minutes.

When a cercaria has penetrated the skin, it becomes a **schistosomule.** This post-penetration stage lasts until the schistosomule leaves the skin and develops into a young worm. The surface film of cercariae and schistosomules is important to parasitic functions. It is called the **glycocalyx.** See Stirewalt[78] for details.

Migration of *Schistosoma japonicum* through the body commences with entry into minute lymphatic or blood vessels, thence to the heart and circulatory system. In the portal system, the parasites develop to maturity. After several weeks, they migrate to superior mesenteric veins. Copulation takes place after females enter the gynecophoral canals of males. A month

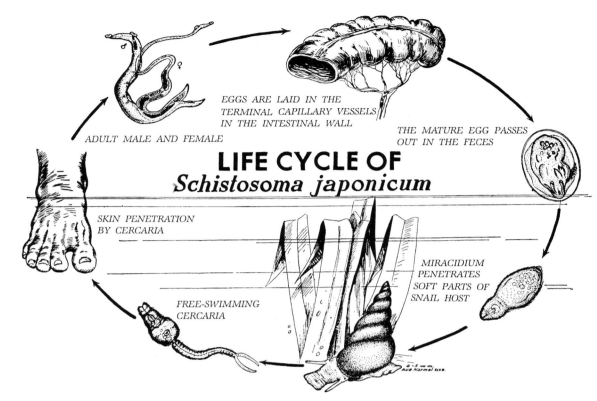

EGGS ARE LAID IN THE
TERMINAL CAPILLARY VESSELS
IN THE INTESTINAL WALL

ADULT MALE AND FEMALE

THE MATURE EGG PASSES
OUT IN THE FECES

LIFE CYCLE OF
Schistosoma japonicum

SKIN PENETRATION
BY CERCARIA

MIRACIDIUM
PENETRATES
SOFT PARTS OF
SNAIL HOST

FREE-SWIMMING
CERCARIA

Fig. 10–6.　Life cycle of *Schistosoma japonicum*. (Modified from Pesigan, courtesy of Santo Tomas J. Med.)

or more elapses from the time of penetration by cercariae to the appearance of schistome eggs in human feces. The life span of adult worms is usually between 5 and 30 years, and several thousand individuals may exist in one host.

Migrating worms usually cause little or no damage or symptoms, but occasionally serious reactions occur, such as pneumonia resulting from invasion of the lungs. Worms or eggs may be found in the ovary. The liver-inhabiting phase may be symptomless, or it may be marked by enlargement and tenderness of this organ and toxic reactions. Other symptoms include cough, fever, diarrhea, eosinophilia, enlargement of spleen, anemia, ascites, and an extreme wasting away, so that the arms, legs, and body in general become alarmingly thin while the abdomen is huge with fluid (Fig. 10–8). Ulceration and necrosis of tissues in the intestine are characteristic. **Schistosomiasis japonicum** is more severe and destructive than the disease caused by the other two common species of this parasite that infect man.

Symptoms and disorders are related to the number of worms, the tissues infected, and the sensitivity of the host. Although toxins, dead worms, and malnutrition may be involved in hepatosplenic schistosomiasis in experimental mice and rats, clear evidence suggests that the schistosome ovum is the primary factor in pathogenesis. This evidence was based on electron microscope study of liver parenchyma and schistosome pigment. Female worms do more damage than male worms because of the host reaction to the masses of worm eggs. For details of pathology, see Miyake.[64]

Prognosis is good in light cases when the patient does not become reinfected. It is grave in cases of long duration and in undernourished persons who often are exposed to the cercariae.

Diagnosis should be based on the finding of eggs in the feces or in biopsied tissue. Complement fixation reaction, dermal tests, and slide flocculation tests are of value.

Prevention and control involve educating people who use infected waters, developing

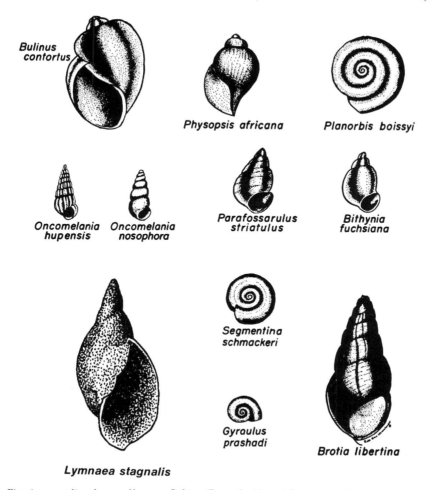

Fig. 10–7. First intermediate hosts of human flukes. (From the United States Naval Medical School Laboratory Guide, except *Lymnaea stagnalis,* which is drawn from a photograph by Malek.)

sanitary waste disposal measures, providing clean drinking water, and destroying snail vectors.

Schistosoma haematobium infects about 40 million people in Africa, Madagascar, Turkey, Cyprus, and southern Portugal. The most commonly and severely affected organ is the urinary bladder. "Bilharzial cystitis usually manifests itself in young persons between the ages of 10 and 30 years. Few cases occur in middle age and it is extremely rare in old age. . . . "[36] The disease may also affect the ureter, kidney, seminal vesicles, prostate, urethra, spermatic cord, testicle epididymis, and penis. For details, see Gazayerli et al.[36] and Maegraith.[56] In chimpanzees, these parasites have been reported from the urinary bladder, rectum, lungs, liver, appendix, and mesenteric veins.

These flukes are intermediate in size between the other 2 blood flukes of man, with males reaching 15 mm in length (averaging about 12 mm), and ranging from 0.8 to 1 mm in width. The body of the male is covered with small tegumentary tubercles, probably sensory in function. The slender, filiform females are about 2 cm long and are dark in color when they have been feeding on blood. Their morphology is similar to that of *S. japonicum,* and they have minute tubercles on the terminal portions of the body that usually project from the male gynecophoral canal. The female is able to extend her body in a quick, snake-like manner from the

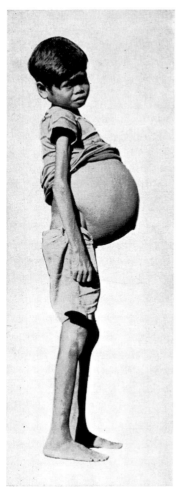

Fig. 10–8. An advanced case of schistosomiasis in a Filipino boy. (Courtesy of Dr. Robert E. Kuntz.)

canal into small veins, where her eggs are deposited, and then to withdraw quickly into the canal.

Eggs average 170×40 to $70 \mu m$ and are characterized by a pointed projection on the terminal end. Eggs penetrate the wall of the bladder or gut to the lumen and thence to the outside. A diurnal rhythm obtains in the appearance of eggs in host urine. The highest output occurs at midday or in the early afternoon. Probably, the miracidia within the egg secrete histolytic fluids that diffuse out of the egg and weaken the bladder wall sufficiently to permit the egg to be pushed through by various pressures on it. The spine may help.

The life cycle is similar to that of the other two flukes. The main intermediate hosts are snails belonging to the families Physidae and Bulinidae (Fig. 10–7). Soluble secretions that pass through eggshell micropores may aid in the release of miracidia. Cercariae develop directly within sporocysts.

In man, the infection is called **urinary schistosomiasis** or **vesical schistosomiasis.** As the urinary bladder gradually becomes infiltrated with eggs, ulceration and bleeding may occur. The eggs eventually become calcified. Symptoms usually do not appear for several months or more after infection.

Positive diagnosis is usually based on finding eggs in urine. Viable ova and miracidia may sometimes be found in cervical smears.

Cancer and its associated frequency with schistosomiasis have been noted by several investigators. A definite positive correlation seems to exist between bladder carcinoma and schistosomal infection in the Sudan.[58] More research is needed, however, to rule out other causes for the bladder cancer. It is known that early schistosomiasis may simulate carcinoma. A lesion of the penis, for example, simulated early carcinoma and almost resulted in autoamputation of the crown of the penis.[4] The lesion was caused by *S. haematobium.* Such tissue reactions, due to the presence of worms or their eggs outside the normal blood vascular circulation, are called *ectopic lesions.*

Schistosoma mansoni (Figs. 10–9, 10–10) is the smallest of the three. Males average 1 cm in length and females 1.6 cm. The tegumentary tubercles are larger than those of *S. haematobium.* The disease is called **schistosomiasis mansoni** or **intestinal bilharziasis.** It occurs in Arabia, Madagascar, Brazil, Venezuela, Puerto Rico, Surinam, and the Dominican Republic. Its African distribution is similar to that of *S. haematobium.* Wild rodents may be reservoirs of infection. See Marcial-Rojas for a general review.[61]

The presence or absence of the male is one of the most significant features affecting the functioning of the female of *S. mansoni.*[29] If a female develops in the absence of sexually mature males, the ovary is smaller and the ova lack cortical granules. It is possible that the female needs contact with the male to absorb from it the amino acids required for production of normal cortical granules. Another effect of the absence of the male is the inability of vitelline cells to mature. Again, the cause may be the absence

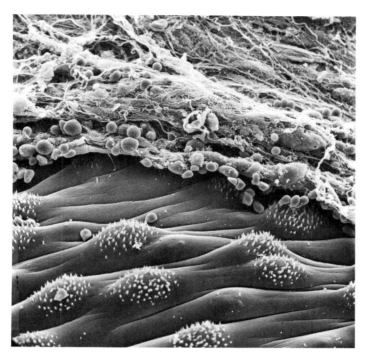

Fig. 10–9. *Schistosoma mansoni*. An adult male protrudes from a vein. ×800. The upper half of the electron micrograph shows the vein wall torn and pulled up to expose the worm. Rounded bodies among the connective tissue are red blood cells. Spinous tubercles occur on the worm's body surface. (Courtesy of Dr. Marietta Voge.)

of a metabolite required to trigger maturation of these cells. The uterus in this species is tegumentary in structure, but the vitelline duct and oviduct possess cilia and lamellae on their internal surfaces. The cellular nature of these surfaces suggest that they have a digestive function. One can assume that this function is common to other schistosomes.

Eggs measure 114 to 175 μm × 45 to 68 μm, and they possess a spine that projects from the shell on one side near the end of the egg. Miracidia enter several types of snails, but primarily those of the family Planorbidae. Responses of miracidia to chemical attractants were studied by using agar and starch gel pyramids impregnated with chemicals.[55] Short-chain fatty acids, some amino acids, and a sialic acid attracted the micracidia and stimulated them to attach and to attempt to enter the agar. The solvent action of various chemicals (e.g., acetone) removed the attracting substances from the snail *Biomphalaria glabrata*. Subsequent addition of butyric or glutamic acids to these snail tissues restored the capacity to attract.

Interestingly, Uhazy et al.[83] have shown that some of the same molecules which are released by snails into their environment and which stimulate miracidia, also serve to attract or trap snails. These authors proposed to use such molecules in conjunction with molluscicides to bring the snail to the poison.

Penetration by miracidia of *Schistosoma mansoni* was described by Wajdi,[87] as follows: (1) A miracidium is pressed against the snail by action of its cilia. (2) Adhesive glands in the miracidium produce a mucoid substance causing the miracidium to adhere to the snail. The substance also functions for lubrication. (3) An anterior papilla elongates and works its way through the snail's epithelium like a probe. (4) The penetration gland secretion apparently digests the tissue of the snail, thereby aiding the papilla in enlarging the area of damage through which the miracidium enters. Gerbils become infected with this fluke by eating infected snails.

Sporocysts of *Schistosoma mansoni* survive longest in snail tissues of the digestive gland, gonads, head-foot, tentacles, pseudobranch,

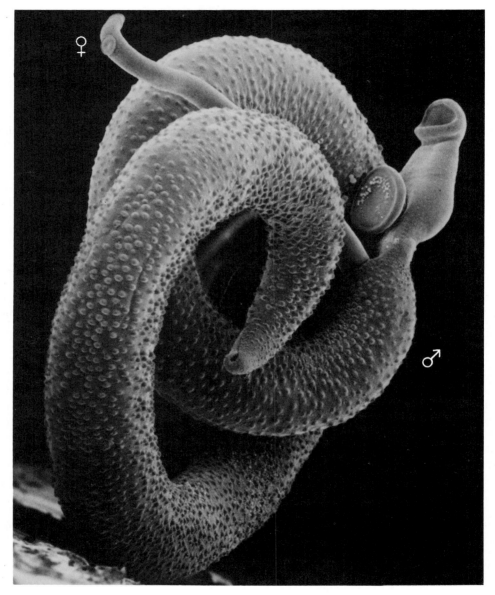

Fig. 10–10. *Schistosoma mansoni*. Scanning electron micrograph of male and female in copula. Note red blood cells in ventral sucker of male. (Courtesy of Dr. Ming M. Wong.)

and mantle collar. Cercariae survive longest in sinuses and veins, as well as in the tissues mentioned. Infected snails may show stunted growth and physiologic castration. Emergence of cercariae can be most frequently observed from the surface of the mantle collar and pseudobranch, and occasionally from the head-foot.

Jourdane et al.[45] reported that after cercariae are produced, a third generation of sporocysts that produce cercariae develops, then a fourth generation of sporocysts, and so on. This succession of generations of sporocysts appear to be a normal process, occurring only when snails remain infected longer than 40 days.

The schistosomule proceeds via the circulatory system to the lungs, then to the liver, and eventually to the inferior mesenteric veins of the sigmoidalrectal area.[11] Eggs penetrate the gut wall and pass from the body with feces. However, many eggs enter the hepatic portal system, are trapped in the liver, and are a major factor in pathology. The infection

"is self limiting, and although it might be clinically severe, it virtually never leads to death of the host. . . . In most instances the patient with intestinal schistosomiasis has not undergone the acute phase of the illness. The patient may present diffuse symptoms referable to the gastrointestinal tract. . . . The symptoms, which may be intermittent, include vague abdominal discomfort, watery or brownish diarrhea, anorexia, and dyspepsia. Whether these symptoms are due to schistosomiasis only cannot be definitely stated, because associated viral or bacterial agents may also give rise to these nonspecific gastrointestinal complaints. . . . Splenomegaly is present in practically all cases."[61]

A person with a worm burden of 50 mated pairs could be exposed to about 15,000 eggs per day for many years. Half the eggs might be retained in tissues with intense granulomatous reaction. These anti-egg granulomas are the main pathogenic responses of the chronic, hepatosplenic form of the disease. Warren[88] suggested that

"schistosomiasis leads to the development of an almost unique form of liver disease manifested by the signs and symptoms associated with portal hypertension unaccompanied by the stigmata of chronic hepatic parenchymal disease and marked derangement of liver function."

This hypertension is apparently due to the presence of fluke eggs in the tissues. Skin lesions of the external genitalia occur occasionally, but cutaneous lesions elsewhere are rare. Nevertheless, they have been found on the chest, the abdominal area, and the nose. For further details, see Marcial-Rojas.[61]

Boros[12] discusses immunoregulation of granuloma formation in mice as a model system.

A truly remarkable landmark in the treatment of flatworms, including many cestodes as well as trematodes, in humans and in veterinary practice, was the discovery of Praziquantel for use in chemotherapy.[40] It is effective against all species of schistosomes in humans, and many other flukes as well. It can be taken orally and possesses only minor side effects. For clinical applications see Beaver et al.[9] The mode of action has yet to be elucidated. Archer has recently reviewed chemotherapy.[1]

Schistosoma mekongi is a newly discovered species in man and dogs in Kampuchea (Cambodia). It is similar to *S. japonicum* but differs significantly in some structureal features and its biology.[13,86]

Schistosomal Dermatitis. Many species of flukes belonging to the family Schistosomatidae are parasitic in birds and in other animals. Snails harbor the developmental stages, and various species of cercariae may be found in large numbers in freshwater streams, ponds, or lakes or in the intertidal region of the seashore. These cercariae are not as host-specific as are the earlier stages, and they often try to penetrate the skin of vertebrates not normal to their life cycle. Studies on host recognition in schistosomes indicate that these cercariae apparently do not react to light, but to secretions from sebaceous glands that cause them to attach themselves to almost any nearby skin. In certain areas, bird cercariae may thus attempt to enter, or may actually enter, human skin. These parasites do not mature in the human body, but the penetration of the skin causes a dermatitis often accompanied by intense itching, a reaction known as "swimmer's itch." Occasionally, in particularly susceptible persons, the reaction is severe enough to cause prostration. This dermatitis should not be confused with another irritation, also called swimmer's itch, caused by certain blue-green algae.

At least one type of cercarial dermatitis results from the activity of cercariae from flukes that live in blood vessels of the noses of ducks. Eggs enter duck nostrils and can be found in large numbers in mucus. From there, they enter water and go through the usual stages in a snail.

Cercariae are released into water by the thousands, ready for unwary swimmers. Marine schistosome dermatitis occasionally occurs among bathers along the seashore. The cercariae of *Austrobilharzia variglandis* from the snail, *Nassarius obsoletus*, has been shown to be the culprit in San Francisco Bay, California.[39] Nasal schistosomiasis (snoring disease) may occur in buffaloes in India.

Accidental infection with cercariae, metacercariae, or even adult flukes of various species is not uncommon. Occasionally, metacercariae become attached to a person's throat and cause considerable irritation. The parasites are taken into the mouth with food or drinking water.

Stirewalt[79] has grouped the schistosomatids of man and animals into three large categories as follows:

"(a) The species causing human schistosomatiasis include: *Schistosoma haematobium, S. japonicum, S. mansoni, S. intercalatum, S. matthei, S. margrebowiei, S. rodhaini* and *S. bovis*. As we have seen, only the first three are of major public health importance. *S. intercalatum* is of concern in a few foci in Africa.

"(b) Schistosomes causing cercarial dermatitis and non-patent infections in subhuman primates without the development of adult parasites include: *S. spindale, Schistosomatium douthitti* and *Heterobilharzia americana*.

"It should be noted that some of the species listed in (a) and (c) are also responsible for cercarial dermatitis.

"(c) Schistosomes of veterinary significance and possibly of concern in meat production include species such as *S. nasalis, S. indicum, S. spindale* and *Orientobilharzia turkestanicum*. Spurious infections may be reported as a result of consumption of animals infected with the above-named species, and can lead to serious confusion when found in fecal samples of man in some areas."

Barrett[5] has provided an excellent overview of the biochemistry of these and other helminths.

See Warren and Newill[89] for a bibliography of this literature; Farley[33] for a review of the family Schistosomatidae, excluding schistosomes from mammals; Pan American Health Organization[70] for a guide for identification of snail hosts; Blanc and Nosny[10] for geography of schistosomiasis; Macdonald and Farooq[52] for a consideration of public health and economic importance; and Ansari,[2] Cheng,[16] Archer[1] and Thomas[80] for control, chemotherapy, and epidemiology.

FAMILY BRACHYLAIMIDAE

These small or medium-sized distomes are found in the intestines of many different groups of vertebrates, especially birds and mammals. Suckers are large, and the genital pore opens in the posterior part of the body. Testes are roughly tandem in position with the ovary between them. Genera include *Brachylaimus, Itygonimus,* and *Leucochloridium*. Unencysted metacercariae of some members of the family may damage their hosts. In slugs *(Limax)*, for example, these parasites destroy kidney tissue, and the results may be fatal.

Leucochloridium macrostomum is a European species found in the rectum of the crow, sparrow, shrike, jay, nightingale, and other birds. It is oval in outline, measures about 1.8×0.8 mm, and is about 0.45 mm thick. Eggs pass from bird intestines and are eaten by aquatic or terrestrial snails (for example, *Succinea*), in which the miracidia are liberated and penetrate host tissues. Sporocysts are formed and give rise to daughter sporocysts. These second-generation sporocysts make their way through the snail host, become branched, and enter the head and tentacles. There they become enlarged and colored, making the head and appendages of the snail appear to consist of elongated sacs ornamented with green, brown, or orange bands. The sacs pulsate with the rhythm of a human heart beat and are noticeable to birds, which peck at them. Inside the sacs are encysted cercariae that are infective to birds; because snail tissue ruptures easily, the sacs are readily released and eaten. Even when released from confinement, the sporocyst sacs pulsate. One wonders why it ever became necessary for a larval fluke to wave a colored flag at its next host and ask to be eaten.

FAMILY BUCEPHALIDAE (= GASTEROSTOMIDAE)

This family is characterized by the presence of the mouth near the middle of the body. The tiny adult flukes, less than 1 mm long, live in the gut of carnivorous fish. Miracidia enter oysters or other marine bivalves of freshwater clams and thus differ from the usual life cycle, which uses a snail as the first intermediate host. Cercariae develop within the sporocyst without rediae and then leave the mollusc, swim about in water, and become entangled in fins of small fish. Entanglement is aided by the presence of two long tails, which give the name "oxhead cercariae'' to this stage. These cercariae encyst under scales of the small fish that are eaten by larger carnivorous fish. Typical genera of these flukes are *Bucephalus* (Fig. 10–11), *Rhipidocotyle,* and *Prosorhynchus.*

FAMILY FELLODISTOMATIDAE

Sexually mature flukes belonging to the genus *Gymnophallus* inhabit the gallbladder, intestine, ceca, and bursa of Fabricius of shore birds and freshwater and marine fishes. Asexual generations of these parasites occur in bivalve molluscs, and they produce cercariae (from sporocysts) that belong to the Dichotoma group of furcocercous larvae.

On emerging from the first intermediate host, the cercariae attach themselves to the mantle or body wall of other bivalve or gastropod molluscs, or in ctenophores, echinoderms, clams, and crustacea, in which they develop as unencysted metacercariae that almost reach definitive size and are easily mistaken for adults. Metacercariae produce lesions on the mantles of their hosts, sometimes accompanied by deposition of nacreous material. Such pearly formations in the mantle of the *Mytilus edulis, Donax truncatus,* and other bivalves are not uncommon and are often reported as having been caused by a distome trematode. Cercariae of *Bacciger bacciger* are released from marine clams (e.g., *Donax*) and encyst in marine amphipods *(Tapes).* These are eaten by fish *(Atherina).*

Fellodistomum felis inhabits the gallbladder of catfish. It has an unusual life cycle in that the sporocyst and cercariae develop in clams, whereas the metacercariae develop in the brittle star, *Ophiura sarci.* The final host acquires the infection by eating brittle stars.

Order Echinostomida

FAMILY ECHINOSTOMATIDAE

This group of flukes is characterized by a collar of spines at their anterior end. The body is elongated and is covered with spines, and the two suckers are usually close together; the acetabulum is larger than the oral sucker.

Ducts associated with the reproductive system of echinostomes differ from those of other groups. Excess sperm and vitelline granules pass through Laurer's canal and are discharged at the dorsal surface of the worm. Sperm is stored in a uterine seminal vesicle, instead of the usual type of seminal vesicle. The **ovicapt**

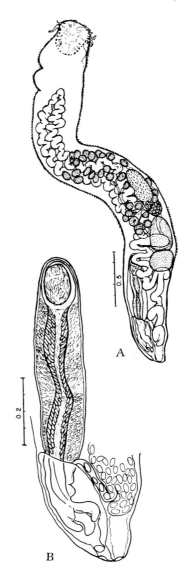

Fig. 10–11. *Bucephalus heterotentaculatus,* a gasterostome from marine fish on the Pacific coast of Mexico. The projected scale indicates millimeters in each figure. (*A,* Dorsal view of worm. *B,* Dorsal view of cirrus sac and genital atrium.) (From Bravo-Hollis and Sogandares-Bernal, courtesy of J. Parasitol.)

or **oocapt** occurs in echinostomes as well as in many other types of flukes. It is a muscular structure surrounding the base of the oviduct, and its contractions cause eggs to be sucked from the ovary into the oviduct. The Germans have a descriptive word for this structure, "Schluckapparat," or gulping apparatus. Laurer's canal, the ovicapt, and the part of the oviduct distal to the ovicapt are ciliated.

Echinostoma ilocanum (Fig. 10–12) occurs in the intestines of humans, dogs, cats, and rats in the Philippines. It averages about 5 × 1.5

mm in size and possesses a ring of spines around the anterior sucker. Spine-like scales occur on the tegument. Eggs (83 to 116 × 58 to 69 μm) hatch into free-swimming miracidia that penetrate the snail *(Gyraulus)*, and develop into sporocysts, rediae, another generation of rediae, and cercariae. Cercariae emerge from the snail and may either reenter the snail or enter almost any other snail or freshwater clam, in which metacercariae cysts are formed. If infected snails are used as food, parasites gain entry into the final host.

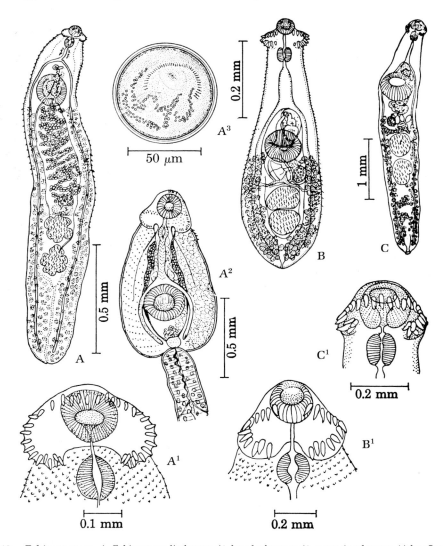

Fig. 10–12. Echinostomes. *A, Echinostoma lindoense; A¹,* head of same; *A²,* cercaria of same. (After Sandground and Bonne.) *A³,* encysted metacercaria of *E. ilocanum.*(After Tubangui and Pasco.) *B, Echinochasmus japonicus; B¹,* head of same. (After Yamaguti.) *C, Eurparyphium melis; C¹,* head of same. (After Beaver.) (From Chandler and Read: An Introduction to Parasitology, courtesy of John Wiley and Sons.)

Most other species of *Echinostoma* occur in birds. From 15 to 20 such species have been described from the intestines of such birds as the cormorant, grebe, owl, thrush, domestic duck, goose, pheasant, partridge, stork, crane, and hawk. Other genera of the family include *Echinoparyphium, Euparyphium, Echinochasmus, Patagifer, Himasthla, Petasiger,* and *Parorchis.*

Remarkable interrelationships between the larval trematodes *Cotylurus lutzi* (Strigeidae) and *Paryphostomum segregatum* (Echinostomatidae) were described by Basch as follows:

"Both trematodes use the snail *Biomphalaria glabrata* as first intermediate host. Rediae of *P. segregatum* prey upon and eliminate sporocysts of *C. lutzi* when individual snails are infected with miracidia of both species. Cercariae of *C. lutzi* enter *B. glabrata,* which also serve as second intermediate hosts for this trematode. When the snail harbors rediae of *P. segregatum,* the strigeid cercariae enter the rediae and develop as hyperparasites. When very young *P. segregatum* infections are exposed to cercariae of *C. lutzi,* no hyperparasites develop because advancing rediae consume the unencysted postcercarial larvae. Cercariae of *C. lutzi* normally do not encyst in snails harboring sporocysts of the same species. Such aversion is here termed the 'Winfield effect;' various authors have cited it as an example of immunity in snails. This effect was studied in snails in which sporocysts of *C. lutzi* had been eliminated by rediae of *P. segregatum.* Such snails, harboring only rediae of *P. segregatum,* were then exposed to cercariae of *C. lutzi* together with singly infected *P. segregatum* control snails, and the number of hyperparasites in the two groups were compared. No permanent immunity to cercarial penetration exists in snails cleared of sporocysts of *C. lutzi.*"[7]

Further complicated relationships involving *Echinostoma* have been reported by Lie et al.[50] Snails may defend themselves against trematode larvae by surrounding and destroying them with a mass of amebocytes. *Echinostoma lindoense* sporocysts that develop from irradiated miracidia are thus destroyed. If, however, normal sporocysts (nonirradiated) of *E. lindoense, Parayphostomum segregatum,* or *Schistosoma mansoni* are present, the survival time of the irradiated larvae is extended. Apparently, the protective ability of normal sporocysts is generously extended to their irradiated companions.

See Chapter 23 for other examples of antagonism and their possible use in biologic control. See also DeCoursey and Vernberg.[23]

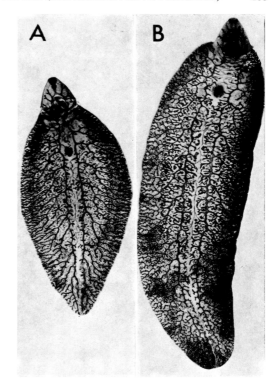

Fig. 10–13. *A, Fasciola hepatica,* and *B, F. gigantica.* ×2.5. (From Sahba, et al., courtesy of J. Parasitol.)

FAMILY FASCIOLIDAE

These broad, flat flukes are often seen in biology laboratories because of their large size and availability—some of them reach 100 mm in length. They may be covered with small scales; the two suckers are close together, and the anterior end is often drawn out into a cone-like projection. The genital pore is located just anterior to the ventral sucker.

Fasciola hepatica (Fig. 10–13A), the sheep liver fluke, measures 20 to 30 mm × 10 to 13 mm; as its common name indicates, it is found in the liver, gallbladder, or associated ducts. The parasite also occurs in cattle, horses, deer, goats, rabbits, pigs, dogs, squirrels, other animals and man, and its distribution is worldwide. The overall incidence of infection in man is less than 1%. A distinguishing feature of the adult fluke is the generous branching of most of the conspicuous organs of the body. In a poorly prepared specimen, it is difficult to identify the testes, ovaries, digestive canals, and uterus because of the complicated branching or coiling of all of them.

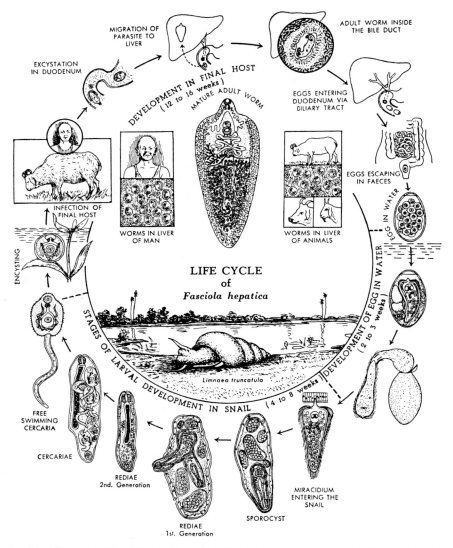

Fig. 10–14. The life cycle of *Fasciola hepatica*, a liver fluke of sheep, goats, and cattle, and occasionally, of man. (From Chatterjee, K.D.: Human Parasites and Parasitic Disease, courtesy of Dr. K.D. Chatterjee.)

The life cycle (Fig. 10–14) begins with fertilized eggs, whose shells do not possess quinone-tanned proteins; these eggs pass from the uterus of the worm into the gallbladder, bile, or hepatic ducts of the host. The eggs are about 140 × 75 μm, are operculated, are carried to the intestine, and leave the body in feces. If they enter water, the opercula fly open and the ciliated miracidia are liberated. Miracidia in eggs prepare for escape by producing an enzyme that digests the substance that holds the operculum to the shell. Factors within the host as well as external environmental conditions seem to ac-

tivate this enzyme. The substance binding the egg operculum to the shell probably consists primarily of a protein; thus, a proteolytic enzyme would seem to be involved. This enzyme apparently functions from inside the shell, because it has no influence on hatching when it is applied to the outside. Other explanations of the mechanism of hatching have been proposed. One of them postulates an initial response of the enclosed miracidium to light. Light activates the miracidium, which then alters the permeability of a viscous cushion that lies beneath the operculum. The cushion expands, and internal pressure is built up until the operculum ruptures and flies back, liberating both cushion and miracidium.[90]

Miracidia swim around for a few hours and then penetrate the body of a snail (for example, *Lymnea*). Within the snail, each miracidium develops into a sporocyst that produces first-generation rediae, which in turn produce second-generation rediae. During the hot summer, there is usually only one redial generation. Rediae give rise to cercariae that leave the snail, often at night, and swim to aquatic vegetation to encyst as metacercariae. One or two months are required for development from the miracidial to the metacercarial stage. Cyst walls of metacercariae of the liver fluke are composed of four major layers:[26] (1) an outer, tanned-protein layer with irregular meshwork containing cigar-shaped bodies; (2) a finely fibrous layer, composed mostly of mucopolysaccharide; (3) an inner layer, also mostly mucopolysaccharide; (4) a dense, compact layer composed of numerous protein sheets stabilized by disulfide linkages and formed from tightly wound scrolls. Within this cyst, the metacercaria develops into a juvenile fluke.

The vertebrate host acquires infection by ingesting the metacercariae with water plants or in drinking water. In the intestine of man, the parasite excysts and migrates through the gut wall and the body cavity to the liver, where it takes up residence in bile passages. Two months are required to reach maturity in the liver or gallbladder. Eggs pass down the bile ducts into the intestine and out to soil or water. The entire life cycle may require as many as 5 months for completion. For an excellent review of the relationships among various species of *Fasciola* and their molluscan hosts, see Kendall.[47]

The degree of pathogenicity of *Fasciola hepatica* to man depends on many factors, particularly the number of worms present and the organs infected. Mechanical and toxic damage are characteristic. The parasite occasionally infects the lungs, brain or other tissues. Pain in the region of the liver, abdominal pain, diarrhea, fever, and anemia are some of the usual symptoms. In cattle and sheep, the most important pathogenic processes are hepatic fibrosis and chronic inflammation of bile ducts. Growth rate, milk yield, and body weight may also be reduced.

In vitro cultures of *Fasciola hepatica* have shown a high rate of ammonia production. This phenomenon is the result of a diet of host blood and tissue. Addition of glucose to the saline solution in which the flukes are cultured decreases the amount of ammonia and increases lactic acid excreted.[65] These flukes, apparently, are not obligatory protein feeders, but they may use carbohydrate when it is available. Respiratory metabolism seems to be independent of oxygen. Oxygen may be used, however, for the tanning of eggshells with the aid of phenolase. Diagnosis of fascioliasis depends primarily on finding the eggs in stool specimens. For a review of fascioliasis, see Dawes and Hughes.[22]

Fasciola gigantica has a life cycle similar to that of *F. hepatica* and is common in cattle and other ruminants in Asia, Africa, and a few other areas. It is rare in man. *F. gigantica* is not as broad as *F. hepatica*, but it is longer, measuring 25 to 75 mm × 3 to 12 mm.

Fasciolopsis buski, a big relative of *Fasciola hepatica*, may reach 75 mm in length, but is usually smaller (Fig. 10–15). Its life cycle is similar to that of *F. hepatica*, but the adult lives in the small intestine of man and pigs. Most cases of infection have been found in China, but a low incidence exists in India. Man acquires the fluke by eating water chestnuts *(Eliocharis)*, the nuts of the red caltrop *(Trapa)*, or other water plants on which the metacercariae attach themselves.

Fascioloides magna, a large liver fluke, lives in cattle, sheep, moose, deer, and horses. It is an oval fluke that may reach 26 × 100 mm in size. The life cycle is similar to that of *Fasciola hepatica*.

Suborder Paramphistomata

FAMILY PARAMPHISTOMATIDAE (= AMPHISTOMIDAE)

The anterior end of these flukes possesses a mouth, but no sucker. The single sucker, or acetabulum, is located at the posterior end. The general shape of the body, which is often covered with papillae, is unlike the typical leaf-shape of other flukes. It frequently is rounded and sometimes looks more like a gourd or a pear with a hole at the top. Common genera are *Homalogaster, Gastrodiscus, Watsonius,* and *Gigantocotyle.*

Paramphistomes may be found in sheep, goats, cattle, and water buffalo throughout the world. Light infections are usually mild in effect, but massive infections of the small intestines with immature paramphistomes, such as

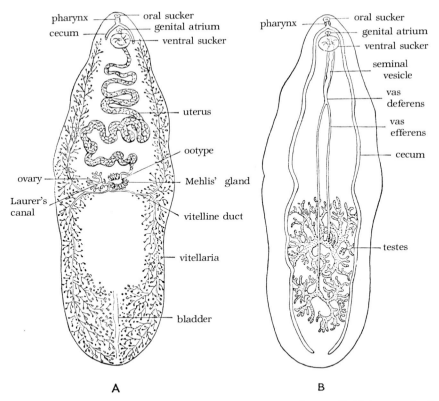

Fig. 10–15. *Fasciolopsis buski. A,* Female reproductive organs, ventral view. *B,* Male reproductive organs, ventral view. (Adapted from Odhner.) (From Belding: Textbook of Clinical Parasitology, courtesy of Appleton-Century-Crofts.)

occur in Africa, cause acute parasitic gastroenteritis with high morbidity and mortality rates. The disease is **paramphistomiasis.** For an excellent account, see Horak.[43]

Paramphistomum microbothriodes, a short, thick fluke, possesses a sucker around the genital pore. It lives in the rumen of cattle and other ruminants and may be found in South America and the United States, although the identity of the species in these two countries has been questioned. In appropriate snail hosts, the usual sequences of sporocyst, redia, and cercaria appear. Cercariae leave the snail and encyst as metacercariae on vegetation. No evidence exists that the adult flukes are pathogenic, but immature forms in the small intestine cause inflammation, edema, hemorrhage, and destruction of intestinal villi.

Superorder Epitheliocystidia

The cercaria possesses a thick-walled epithelial bladder and a single tail that is reduced in size or is lacking.

Order Plagiorchiida

FAMILY DICROCOELIIDAE

These flattened, translucent flukes occur in the gut, gallbladder, bile ducts, liver or pancreatic ducts of amphibians, reptiles, birds, and mammals. The ovary is situated behind the testes, and the uterus fills most of the posterior part of the body. Common genera are *Eurytrema, Dicrocoelium,* and *Brachycoelium.*

Eurytrema pancreaticum is a parasite (Fig. 10–16) that lives in pancreatic ducts of hogs, water buffaloes, and cattle in the Orient and of some camels and monkeys in the Old World. It has also been reported from the duodenum of sheep and goats in South America, and it was once found in man. The fluke measures about 12 × 7 mm. Cattle in Kuala Lumpur acquire the fluke by eating grasshoppers containing metacercariae.

Dicrocoelium dendriticum (= D. lanceolatum), an elongated, narrow fluke, measures

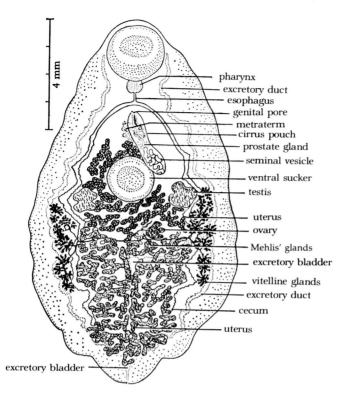

4 mm

pharynx
excretory duct
esophagus
genital pore
metraterm
cirrus pouch
prostate gland
seminal vesicle
ventral sucker
testis
uterus
ovary
Mehlis' glands
excretory bladder
vitelline glands
excretory duct
cecum
uterus

excretory bladder

Fig. 10–16. *Eurytrema pancreaticum.* (From Looss. *In* Chandler and Read: An Introduction to Parasitology, courtesy of John Wiley and Sons.)

about 8 × 2 mm and is known as the **lancet fluke** (Fig. 10–17). It occurs in the bile ducts of cattle, sheep, goats, horses, camels, rabbits, pigs, deer, elk, dogs, and occasionally man; distribution is worldwide. The fluke has a well-developed ventral sucker larger than the anterior sucker, unbranched ceca, slightly lobed testes that are almost tandem, no spines on the cuticle, and vitelline cells occupying the middle third of the body. Eggs of *D. dendriticum* must be eaten by an appropriate snail before they will hatch. The snails are land forms (*Cionella lubrica*), so a cycle of larval growth without swimming stages had to develop. There are two sporocyst stages, and the cercariae aggregate in masses called slime balls in the respiratory chamber of the snail. The slime balls are deposited on soil or grass as the snail crawls. Although sheep or other final hosts may possibly aquire infection by eating the slime balls or even the snails, ants (*Formica fusca*) serve as secondary hosts by eating these masses of cercariae, and the sheep acquire infection by eating the ants.

FAMILY MICROPHALLIDAE

This group resembles the Heterophyidae. They are small, some species being less than 1 mm long, although many are closer to 2 mm in length. The body is thick, pear-shaped, and spiny. The 2 branches of the intestines are unusually short, and there is a penis papilla. The usual host is a shore bird such as a sea gull, plover, godwit, sandpiper, or tern in which the parasites are found in the intestine.

Representative genera are *Microphallus*, *Maritrema*, *Levinseniella*, *Spelotrema*, and *Spelophallus*.

Levinseniella minuta is a tiny fluke averaging only about 0.16 mm long and 0.1 mm wide (Fig. 10–18). The normal vertebrate hosts are scaups or other diving ducks, but white mice easily become infected in the laboratory. Snails eat the fluke eggs, which hatch into miracidia and then develop into sporocysts. Cercariae emerge from the sporocysts and encyst, forming the metacercaria stage in the snail. Ducks become infected by eating infected snails.

FAMILY PLAGIORCHIIDAE

These flukes are found mainly in the intestines of frogs, snakes, and lizards, but they may

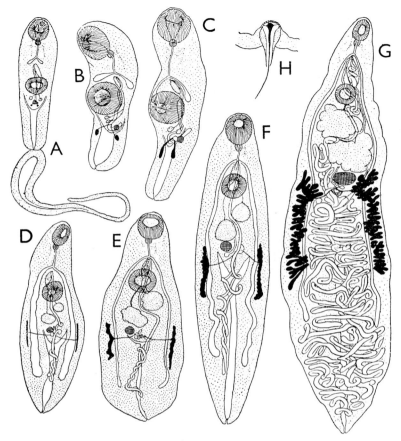

Fig. 10–17. Stages of *Dicrocoelium dendriticum* developing in the definitive host. *A, Cercaria vitrina*, removed from cyst. *B* to *F*, Juvenile flukes collected after infection: *B*, 8 days old; *C*, 9 days old; *D*, 12 days old; *E*, 14 days old, and *F*, 16 days old. *G*, Full-grown fluke. *H*, A touch papilla on the suckers of the adult. (After Neuhaus, 1939.) The magnifications of *ABC/DEF/G* stand in relation 5:1/2:4/1. (From Dawes: The Trematoda, courtesy of Cambridge University Press.)

occur in amphibians, birds, and bats (Fig. 10–19). The family is large and contains hundreds of species. The flukes are small or only moderately large and covered with spines; the two suckers are not far apart, and there is a Y-shaped urinary bladder. The parasites also live in the respiratory tract, urogenital tract, and sometimes the gallbladder. On leaving the snail, young worms (cercariae) penetrate aquatic insect larvae (Fig. 10–20), crayfish, or tadpoles. The vertebrate host aquires the infection by eating one of these second intermediate hosts.

Common genera are *Plagiorchis, Renifer, Haematoloechus* (= *Pneumoneces*, a common lung fluke of frogs), and *Prosthogonimus*. Most of the flukes are only a few millimeters long, but some may reach 20 mm or more in length. *Prostho-*

gonimus putschkowskii has been found in a hen's egg. This fluke usually occurs in the bird's oviduct.

FAMILY ALLOCREADIIDAE

Common intestinal flukes of marine and freshwater game fish such as trout and bass belong to this family. *Crepidostomum, Bunodera, Allocreadium*, and *Pharyngora* are the more common genera. Usually, they are small, often not longer than 1 or 2 mm. The cuticle is not spinous. In some genera (*Crepidostomum* and *Allocreadium*), the rediae are found in small bivalves, and the metacercariae occur in other bivalves, crayfish, amphipods, or mayfly nymphs. Other genera use snails as the first intermediate host and various other animals as

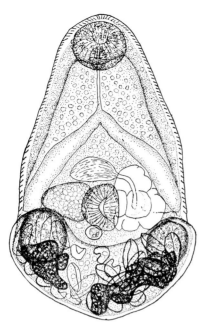

Fig. 10–18. *Levinseniella minuta.* Adult specimen, 0.17 mm long, from a white mouse. (From Stunkard, courtesy of J. Parasitol.)

the second intermediate host. *Helicometrina elongata* (Fig. 10–21), is about 4 mm long and lives in the small intestine of the tidepool fish, *Gobiesox meandricus*, in Bodega Bay, California. *Allocreadium neotenicum,*1.5 to 4.2 mm long, lives in the hemocoel of dytiscid beetles.

FAMILY TROGLOTREMATIDAE

Most of the genera belonging to this family are under 10 mm long, and some may be only 1 or 2 mm in length. In general, the group is composed of thick, spinous, or scaly flukes that have an oval outline and live in various organs and sinuses of birds and mammals.

The genus *Paragonimus* has special interest because it contains the common lung fluke of man, *Paragonimus westermani*. The following species of *Paragonimus* have been reported from various countries of North and South America: *kellicotti, mexicanus, peruvianus, caliensis,* and *rudis.* See Sogandares-Bernal and Seed[77] for details.

Species generally considered valid are: *Paragonimus westermani*, found in man, dog, cat, pig, and many other mammals; *P. kellicotti*, in cat, dog, pig, mink, goat, fox, and experimental animals; *P. ohirai*, in pig, dog, weasel, badger, wild boar, and experimental animals; *P. iloktsuensis*, in rat and dog; and possibly *P. compactus*. These lung flukes all have the same general characteristics and life cycle.

***Paragonimus westermani*,** a lung fluke, is one of the most important trematodes of man in some areas of the world, especially in the Far East. It has been found in Korea, Japan, mainland China, Taiwan, Philippines, Indonesia, Africa, India, some Pacific islands, Peru, and Ecuador. The incidence of infection ranges from 7 to 44%.

Anatomic details of an adult *Paragonimus westermani* are shown in Figure 10–22. No true cirrus or cirrus pouch is present. The fluke usually measures about $10 \times 5 \times 4$ mm in size. It is plump, reddish in color, and is covered with minute cuticular spines. It usually lives in the lungs, often paired with another fluke, and is enclosed within a host-produced cyst. Many other organs, however, may harbor the worms.

Eggs (Fig. 10–23), which average 87×50 μm, are brownish in color and sometimes are present in such numbers that they give the sputum a brown or rust-colored tinge. These operculated eggs are coughed up from the lungs and expectorated, or swallowed and leave the body in the feces. After two to several weeks of development in moist soil or water, the eggs hatch, and the miracidia swim about until a suitable snail appears. See Figure 10–24 for the life cycle.

Hatching is initiated by a sudden increase in activity of the larva inside the egg. The animal may, at first, lunge at the operculum, but it soon flattens its anterior end against this structure. A series of violent contractions then occurs, and the posterior portion of the body is extended until the operculum suddenly flies back as if on a hinge. Water passes through the exposed vitelline membrane and causes it to swell and to extend through the opercular opening. A bubble, which may become larger than the egg itself, is produced and forms a transparent sac into which the miracidium finally forces itself. The larva swims rapidly about inside this sac until the membrane is ruptured and the ciliated larva swims away.

Mature miracidia average 0.08×0.04 mm in size and are covered with four rows of ciliated ectodermal plates. *Semisulcospira, Brotea* and several other genera of snails are common first intermediate hosts. These snails are usually found in swift mountain streams, often away

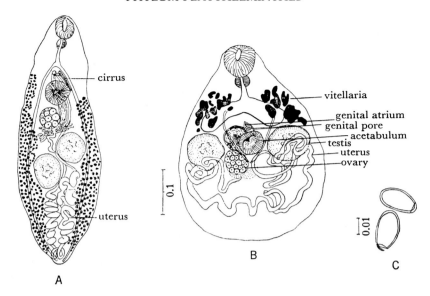

Fig. 10–19. *A, Plagiorchis vespertilionis; B, Acanthatrium jonesi; C,* Eggs. Both flukes are found in Korean bats. All figures drawn with the aid of a camera lucida. The projected scale indicates millimeters. (From Sogandares-Bernal, courtesy of J. Parasitol.)

from towns or villages. Infection of the snails probably depends primarily on animal reservoir hosts rather than on man. The incidence in snails is surprisingly low. Out of hundreds of snails examined in endemic areas in the Philippines, we have found only 2% to be infected. Age immunity apparently protects some snails. Miracidia shed their ciliated plates during the act of penetrating the snail. The sporocyst generation develops free in the lymphatic system, usually along the esophagus, stomach, or intestine. First-generation rediae emerge from the sporocyst about 4 weeks after infection. Second-generation rediae appear in the digestive gland 63 to 75 days after infection.

Microcercous cercariae emerge from the snail. They are ellipsoidal and possess a short tail with a spiny tip. The body is covered with small spines, and a prominent stylet lies in the dorsal side of the oral sucker. Two types of penetration glands are present, in addition to two irregular rows of mucoid glands on each side of the body. Cercariae of *Paragonimus westermani* in the laboratory do not readily escape from snails.

Cercariae enter a crayfish or a crab. Crabs belonging to the genera *Potomon, Eriocheir, Sesarma,* and *Parathelphusa* are the usual second intermediate hosts. Crabs probably ingest the cercariae. Metacercariae (0.34 to 0.48 mm) in the crustaceans may be found in the gills, heart,

muscles of the legs and body, liver, or reproductive organs. Infected crayfish usually belong to the genera *Cambaroides* or *Cambarus.* The time between infection of the crustacean and the appearance of mature metacercariae in its tissues may be several weeks to several months.

When a new host eats uncooked or poorly cooked crayfish or crabs, the metacercariae excyst in the intestine within an hour of ingestion. The minute worms are pinkish in color and possess a large, elongate, sac-like bladder. They penetrate the intestinal wall and usually go through the diaphragm to the pleural cavity. They may wander around the abdominal cavity for 20 days or more and may enter various organs in which they may reach sexual maturity. Among many organs that may be affected are the brain, intestine and other abdominal viscera, spermatic cord, epididymis, scrotum, vaginal wall, lumbar and gluteal muscles, spinal canal, subcutaneous tissues of the thorax and abdominal wall, and the extremities.[28] Even after penetrating the diaphragm, the worms may never reach the lungs. A heavy infection of the liver frequently is observed in experimental animals.

In animals, usually two worms exist in a lung cyst, but this phenomenon rarely occurs in man. It has been observed, in experimental animals, that worms probably locate one another

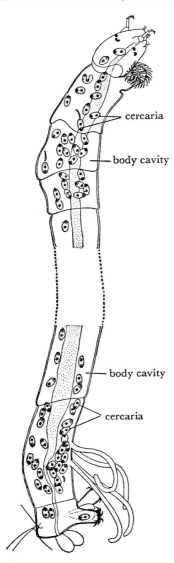

Fig. 10–20. Chironomid (midge) larva infected with *Plagiorchis megalorchis* cercaria. (From Rees, courtesy of Parasitology.)

by chemical attractants. Eggs appear in the sputum 2½ to 3 months after one eats infected crustacea.

Crabs are often prepared for cooking on chopping blocks, and the juice left on the block may contain metacercariae that adhere either to foods subsequently prepared on the same block or to the fingers of the cook. The juice of fresh crabs is used in various preparations in Korea and the Philippines. Crabs are often "pickled" or salted or dipped in wine or vinegar before

eating. These practices rarely kill all metacercariae. The encysted parasites can live for several days in a dead crab, one or 2 days in diluted wine, or 3 weeks in an ice chest at 10°C in 10% formalin.

Uncooked paratenic hosts, such as pigs and wild boar have been the source of infection in man.

Symptoms of infection include coughing, profuse expectoration, occasional appearance of blood in sputum, pain in the chest, brown sputum, and muscular weakness. Pathologic features involve bleeding spots where worms have penetrated tissues, leukocyte infiltration, tearing of muscles, scar tissue, host allergic reaction, and fibrous cyst formation around the worms. Cysts may contain blood-tinged material, purulent chocolate-colored fluid, worm eggs, living worms, dead worms, or no worms. Hemoglobin taken from the worms has been reported to be different from that in host red blood cells. Apparently, *Paragonimus* employs the heme moiety of the host directly in the formation of its own pigments. Migration of the worms to other tissues or organs may be troublesome. Cerebral involvement can be serious. Eggs distributed widely throughout the host cause inflammation, and secondary bacterial infection may occur.

Diagnosis consists primarily of finding eggs in sputum or feces, but the clinical picture also has value. Intradermal and complement fixation tests are useful.

Control should involve treatment of patients, elimination of reservoir hosts, disinfection of sputum and feces of patients, destruction of snails, crabs, and crayfish, prevention of human infection by cysts that become free from the arthropod host, and education against eating raw or poorly cooked or preserved crabs or crayfish.

Nanophyetus (= Troglotrema) salmincola is known by the curious name **salmon poisoning fluke.** Microbial pathogens are commonly transmitted to hosts by ectoparasites, but some years ago, parasitologists were surprised to discover that this trematode has taken the place of an arthropod vector of a rickettsial agent that causes a serious disease in dogs. The tiny fluke is oval in shape, about a millimeter in length, and lives in the intestine. Eggs are about 40 × 70 μm in size. Various snails (for example, *Goniobasis silicula*) serve as the first intermediate host within which cercariae similar to those of

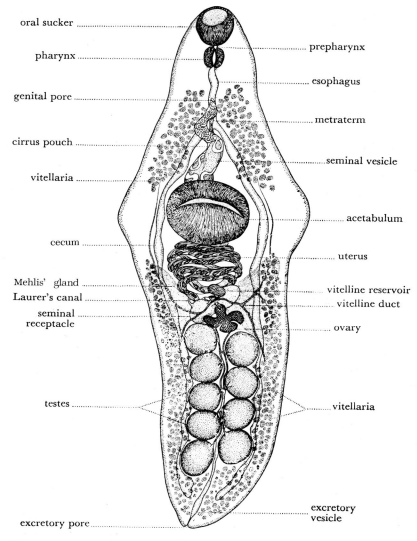

Fig. 10–21. *Helicometrina elongata* from a tidepool fish. ×42. Lateral bulges in the acetabular region are partially due to pressure during fixation. (From Noble and Park, courtesy of the Trans. Am. Microscop. Soc.)

Paragonimus westermani develop. These cercariae penetrate salmon, other salmonid and nonsalmonid fish, and salamanders (*Dicamptodon ensatus*), in which they enter kidneys. The parasites become infected with a rickettsia, *Neorickettsia helminthoeca*, which causes a serious disease in the natural final host that eats the salmon. Dogs, foxes, raccoons, and other fish-eating mammals harbor the fluke and may thus become infected with the rickettsia. It has also been found in man in Siberia. Fever, vomiting, and dysentery are some of the more severe symptoms. The mortality rate is high, but re-

covery confers immunity to the disease. For details, see Millemann and Knapp.[63]

Order Opisthorchiida

FAMILY OPISTHORCHIIDAE

The habitat of these lanceolate distomes is the bile passages of vertebrates. Suckers are small, the ovaries are located in front of the testes, and a genital pore is present just in front of the ventral sucker. The life cycle involves a snail and fish or amphibian intermediate hosts and the

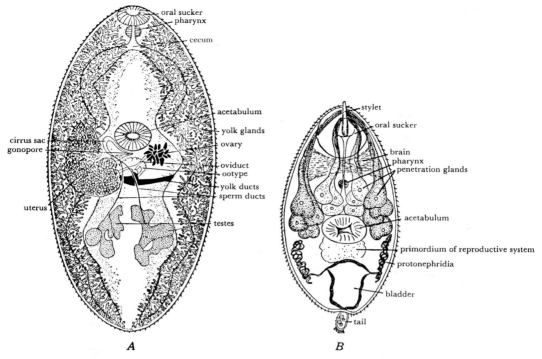

Fig. 10–22. *A, Paragonimus westermani,* adult. (After Chen.) *B,* Cercaria. (After Ameel.) (From Hyman: The Invertebrates, courtesy of McGraw-Hill.)

usual stages of miracidium, sporocyst, redia, cercaria, and metacercaria.

Opisthorchis (= Clonorchis) sinensis, the Chinese liver fluke, is the best-known member of the family. Its prevalence in China is evident from the name, but it is also found in Korea, Japan, and other Asian countries. The degree of infection in man reaches 15 to 70% in local areas and 100% in reservoir hosts such as cats, dogs, tigers, foxes, badgers, and mink. This fluke (Fig. 10–25) averages about 18 × 4 mm in size. Eggs (27 × 16 μm) are shaped like an electric light bulb with a distinct "door" or operculum on top. They are readily killed by desiccation, but can live for 6 months at 0°C if kept moist.

The life cycle (Fig. 10–26) includes a snail (for example, *Parafossarulus* or *Bulimus*) that eats the eggs. Miracidia hatch in the snail rectum or intestine and penetrate the walls of these organs and become sporocysts. On leaving the snail, cercariae swim to freshwater fish in which they penetrate the skin and encyst as metacercariae. At least 80 species (10 families) of fish, most belonging to the family Cyprinidae in China,

Korea, and Japan, have been incriminated as second intermediate hosts, but only 12 of them are chiefly responsible for human infection.

Man is infected by eating uncooked, infected fish. The normal route of this fluke to the liver appears to be by way of the bile duct. The incubation period in man is about 3 weeks. Although the presence of a few dozen worms constitutes an average infection, a patient having 21,000 worms has been reported.

Symptoms of infection are usually absent or mild. They may, however, include eosinophilia, nausea, epigastric pain, edema, diarrhea, vertigo, fluid in body tissues, and wasting of the body. The duration of infection, degree of liver damage, and the presence of metastases, and secondary infection determine the nature of the disorder. Infected organs include the liver, bile duct, gallbladder, pancreas, lymphatic glands, diaphragm, lungs, omentum, kidney, heart, bone, and brain. It is possible that the worms may be responsible for cancer of the liver.[28]

Prevention rests mainly on cooking all freshwater fish from endemic areas. Diagnosis is confirmed by finding eggs in fecal samples.

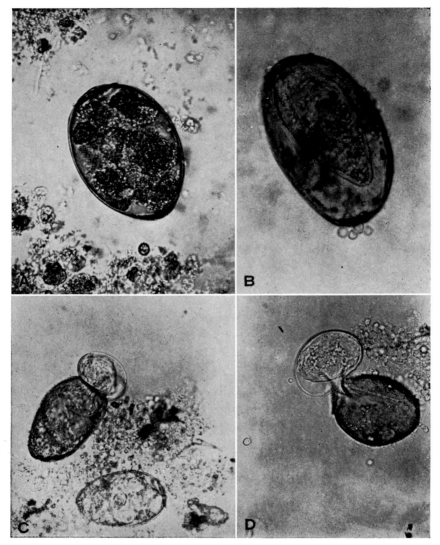

Fig. 10–23. Photographs showing hatching of the egg of the lung fluke *Paragonimus westermani*. (Courtesy of Dr. K.J. Lie.)

Opisthorchis felineus (= O. tenuicollis) is found in much of Europe and the Orient. It averages about 10 × 2.5 mm in size and lives in the liver of man, cats, dogs, foxes, and other mammals. The life cycle is similar to that of *O. sinensis;* eggs are eaten by the snail, *Bulimus.* Liberated cercariae possess penetration glands and large tails and are active swimmers. They become attached to various species of fish, drop their tails, penetrate the skin under scales, and become metacercariae in the skin or muscles. Man and other vertebrate hosts become infected by ingesting uncooked or partially cooked infected fish. The mature metacercariae in fish sometimes reach a diameter of 300 μm, so they can be seen as tiny spots by the naked eye. In the mammalian host, they excyst and make their way to the bile ducts of the liver, where they mature.

A few dozen of these worms in the human liver seem to cause no appreciable harm. When the number of parasites reaches the hundreds, however, there may be pain, congestion, liver enlargement, bile stones, and cirrhosis. For details of pathogenesis, see Gibson and Sun.[37]

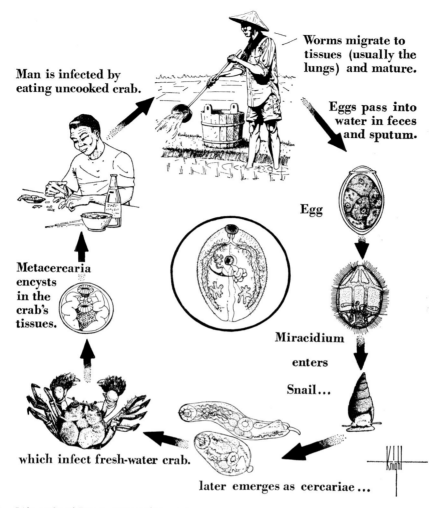

Man is infected by eating uncooked crab.

Worms migrate to tissues (usually the lungs) and mature.

Eggs pass into water in feces and sputum.

Egg

Metacercaria encysts in the crab's tissues.

Miracidium

enters

Snail...

which infect fresh-water crab.

later emerges as cercariae ...

Fig. 10–24.   Life cycle of *Paragonimus westermani.*

FAMILY HETEROPHYIDAE

These small intestinal flukes live in birds and mammals including dogs, cats, and man. The parasites are usually only 1 or 2 mm long, rarely reaching 3 mm. Some are distomes and some monostomes; the cuticle is scaly, at least on the anterior part of the body. The general body shape is ovoid or pyriform, and the ovary is located in front of the testes or testis. Metacercarial stages occur in fish or amphibians.

Heterophyes heterophyes (Fig. 10–27) averages only 1.3 × 0.35 mm; it may be pear-shaped, oval, or elongated. The outer surface is covered with spine-like scales. The midventral surface seems to possess two adjacent suckers, but careful examination shows that one of

them is the genital opening. This organ is retractile, and it may be called a genital sucker or a **gonotyl,** which may lie within the acetabulum. The oval eggs average about 28 × 15 μm; thus, they are smaller than many protozoan cysts. Eggs are eaten by snails.

Metacercariae occur in various fish that may be ingested by man. The adult parasite lives in the intestine and normally causes little harm, but it may produce pain and diarrhea. Eggs of this species and related forms may also reach the blood and cause mild to serious trouble in organs they enter. The fluke is especially prevalent throughout the Orient. A number of related species occasionally infect man, for example, *Cryptocotyle lingua, Haplorchis yokogawai,*

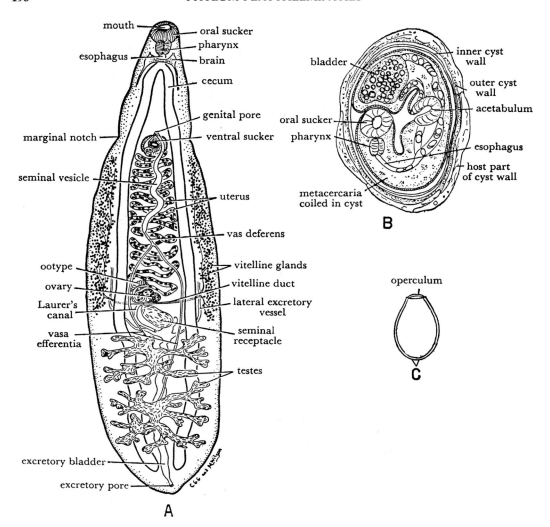

Fig. 10–25. *Opisthorchis sinensis. A,* Dorsal view of adult. (From Brown: Selected Invertebrate Types, courtesy of John Wiley and Sons.) *B,* Metacercaria. (After Komiya and Tajimi, 1940.) *C,* Capsule. (After Looss.) (*B* and *C* from Hyman: The Invertebrates. Vol. II, courtesy of McGraw-Hill.)

and *Heterophyes katsuradai. Cryptocotyle lingua* lives in dogs among the intestinal villi and in the lumen and causes considerable damage mainly because of mechanical factors. A heavy infestation causes the production of much mucus, sloughing of tissue, pressure atrophy, and necrosis.

Metagonimus yokogawai (Fig. 10–28) also lives in the small intestine of man, dogs, and cats, and it has been found in pelicans. It occurs in the Orient and a few other places. This fluke is only slightly larger than *Heterophyes heterophyes* and has a similar life cycle, with the metacercariae encysting in freshwater trout and

in other salmonid and cyprinoid fishes. A diagnostic character of adults is the eccentric position of the ventral sucker, which is located about halfway between the anterior and posterior ends of the body, but to the right of the midline. When man ingests the metacercariae, young worms grow to maturity in 2 to 3 weeks. Infection may produce colicky pains and diarrhea.

FAMILY HEMIURIDAE

Hemiurus is an elongated, cylindrical worm with a nonspinous tegument. It varies in length from a few to 15 mm. A characteristic feature

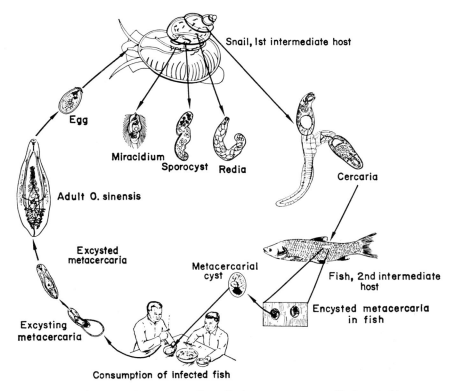

Fig. 10–26.   Life cycle of *Opisthorchis sinensis*. (From Yoshimura, courtesy of J. Parasitol.)

of the body is its division into an anterior **soma** and a posterior **ecsoma**. These parts may be telescoped together, the ecsoma being withdrawn into the soma. The dividing line between the thick-walled anterior part and the thin-walled posterior part is often clearly evident as a constriction (Fig. 10–29); hence, the worms are sometimes called "appendiculate." Often, the vitellaria consist of a few large bodies rather than many scattered particles. The flukes usually inhabit the gut, stomach, gallbladder, esophagus, or pharynx of marine fish, and they are worldwide in their distribution. Other genera include *Aphanurus, Sterrhurus,* and *Lecithochirium.*

Nonappendiculate genera include *Hysterolecitha* and *Derogenes.* The latter (Fig. 10–30) is the most widely distributed digenetic trematode of marine fish, being found in cod, flounder, sturgeon, sole, salmon, mackerel, halibut, and many others. The parasite is usually found in the stomach of its vertebrate host, but it may occur in the intestine. *Hysterolecitha trilocalis* is found on the gills of mudsuckers, *Gillichthys*

mirabilis, and unusual site "preference" for an adult digenean.

Halipegus is generally placed in a separate family (Halipegidae), and it commonly occurs in frogs and salamanders. Nonciliated miracidia are eaten by snails within which sporocysts, rediae, and cercariae are produced. Cercariae are eaten by insect larvae, copepods, or ostracods (Fig. 10–31).

Heinz[41] provides a review of many families and genera of flukes parasitizing humans in the Philippines.

FAMILY DIDYMOZOIDAE

The taxonomic position of this group is obscure. The worms have been placed in the family Hemiuridae and, recently, in a new Subclass, **Didymozoidea.** They usually live in pairs in cysts or cavities of their marine fish hosts and may occur in the body cavity, kidney, body surface, esophagus, gut, musculature, subcutaneous tissue, pharynx, or other places. The worms may have a broad posterior portion of the body *(Didymocystis),* they may be thread-

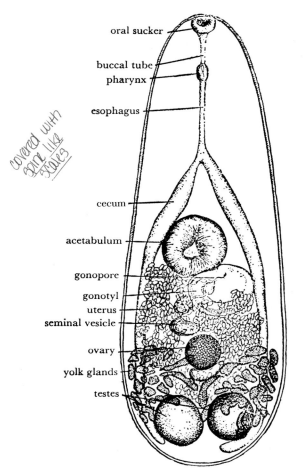

oral sucker

buccal tube
pharynx

esophagus

covered with spine like scales

cecum

acetabulum

gonopore
gonotyl
uterus
seminal vesicle

ovary

yolk glands

testes

Fig. 10–27. *Heterophyes heterophyes,* intestinal fluke of man. (After Witenberg, 1929.) (From Hyman: The Invertebrates. Vol. II, courtesy of McGraw-Hill.)

like (*Nematobothrium*), or they may have more bizarre shapes. The ventral sucker is absent in most members of this family; the pharynx is reduced or absent, and the life histories are not known (Fig. 10–32). Most of them are hermaphroditic, but in some the sexes are separate. In at least one cyst-dwelling species (*Kollikeria filicollis*), the sexes are not entirely separate, but the functional "males" and "females" are dimorphic.

Nematobibothroides histoidii may be found under the skin and between muscles of the sunfish, *Mola mola.* We have tried to remove the entire worm from the fish, but find it to be a difficult task. The parasite may be over 40 feet (12 m) long and follows a tortuous course, extending from one side of the host to the other,

winding in and out through fin supports, dipping down among muscles, and becoming entangled with other individuals.[68]

PHYSIOLOGY OF TREMATODES

Progress in our knowledge of the physiology and biochemistry of helminths has developed so rapidly during the last decades that it is impossible, within the confines of this book, to mention more than a few generalities and to take note of some items of special interest. See also Barrett.[5]

Carbohydrate metabolism in trematodes resembles that in cestodes. Simple sugars, mainly glucose, are acquired primarily through the external tegumental surface, presumably through membrane transport systems similar to those in cestodes. Larval stages[19] as well as adults use the tegument for glucose uptake. The digestion of complex carbohydrates within the worm's intestine may also occur, although this has not been demonstrated directly. The role of the gut epithelium in absorbing glucose and other simple sugars appears to be minimal. Glucose is stored as glycogen, which may be deposited throughout the body, mainly in the parenchyma, muscles, gonads, vitellaria, and tegument.

A report on nervous control over glycogen synthesis is pertinent here. Erlich[32] found that in *Fasciola hepatica,* glycogen was equally distributed in both halves of the worm. He cut animals in two with both suckers in the anterior half. When the suckerless halves were incubated in a suitable medium, they were unable to synthesize glycogen. When the anterior ends were incubated, they synthesized glycogen. When the anterior cone of each anterior half was removed, thereby removing the cerebral ganglia, these amputated portions, with two suckers, failed to synthesize glycogen. When intact anterior halves were incubated with eserine, which poisoned the cerebral ganglia, glycogen synthesis was inhibited. Erlich concluded:

"It appears that the presence of the anterior cone is indispensable for glycogen synthesis, and as its activity is inhibited by eserine, a specific inhibitor of acetylcholinesterase, it is probably the pharyngeal ganglion which controls the synthesis of glycogen in the liver fluke."[32]

Progress in our knowledge about the effects of regulatory molecules on trematodes reveals

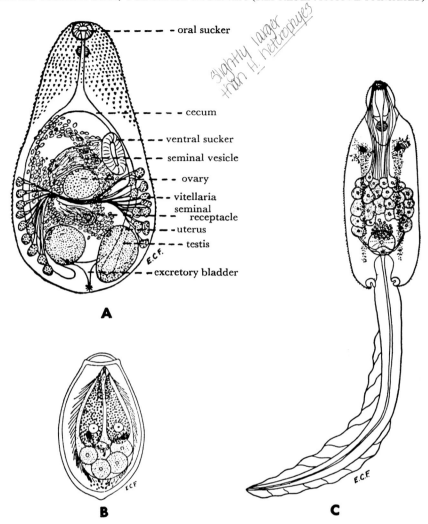

slightly larger than H. heterophyes

Fig. 10–28. *A,* Adult specimen of *Metagonimus yokogawai,* ventral view. ×36. (After Faust.). *B,* Egg of *M. yokogawai,* with mature miracidium. ×1300. (After Faust.) *C,* Cercaria of *M. yokogawai.* ×200. (Modified from Russell, P.F., Jung, R.C., and Cupp, W.C.: Clinical Parasitology. 9th Ed. Philadelphia, Lea & Febiger, 1984.)

the transition from classic biochemistry to the present impact of molecular biology. Nearly 30 years ago, in 1959, Tag Mansour reported the effects of the regulatory molecule, serotonin, on the carbohydrate metabolism of *Fasciola.* Subsequent studies have shown that serotonin also regulates metabolism and movement in schistosomes, and acts by way of receptors, transducers, and second messengers such as cyclic AMP.[59] Such systems can now be explored by the powerful tools of molecular biology.[60]

Adult trematodes are facultative anaerobes and acquire most of their energy through phos-phorylative glycolysis or fermentation, even in the presence of oxygen. Schistosomes, which are homolactic fermenters, rely almost entirely on the glycolytic pathway for energy production, with lactate being the sole end product of glucose catabolism. Even in the blood circulation, in which oxygen levels are high, blood flukes still use anaerobic respiration, although they can derive some of their energy needs (approximately 20%) from oxidative processes. Other flukes, for example, *Fasciola hepatica,* in addition to the regular glycolytic path, can produce the metabolic intermediate oxaloacetate

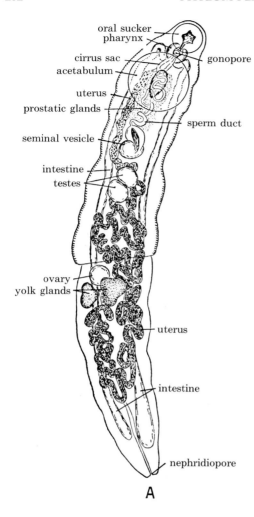

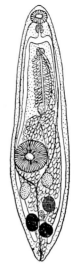

Fig. 10–30. *Derogenes varicus.* × 30. (From Dawes: The Trematoda, courtesy of Cambridge University Press.)

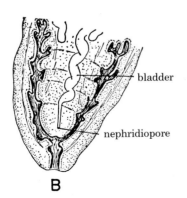

Fig. 10–29. *Hemiurus. A,* With rear end extended. *B,* Rear part, with end telescoped into the interior. (After Looss.) (From Hyman: The Invertebrates. Vol. II, courtesy of McGraw-Hill.)

through the enzyme-catalyzed fixation (phosphoenolpyruvate carboxykinase) of carbon dioxide to phosphoenolpyruvate. Malate, a product of oxaloacetate reduction, can then serve as a substrate for further metabolism within mitochondria that has also been shown to be an energy-yielding process. Acetate and propionate are the major metabolic end products in this pathway. Intestinal-dwelling echinostomes use glucose under both aerobic and anaerobic experimental conditions; however, in addition to the metabolic end products acetate, propionate, and lactate, large quantities of volatile fatty acids are produced.[72]

Sugar transport capacity in relation to **niche segregation** has been studied in a fascinating report by Uglem and colleagues.[82] They studied three species of *Proterometra* (Digenea: Azygiidae) which differ in their location within a host: (1) the lumen of the gut; (2) body cavity or tissue; or (3) external surface. The species living in the gut transported glucose by facilitated diffusion. The tissue-dwelling species used active transport. The ectoparasite dwelling on the external surface had no detectable system for transporting glucose across the tegument! In addition to demonstrating the evolutionary specializations of flukes, this paper provides a superb demonstration of the union of physiologic and ecologic research to advance our understanding of parasites.

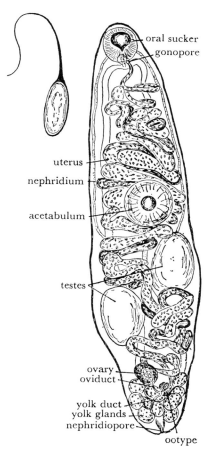

Fig. 10–31. *Halipegus,* nonappendiculate hemiurid from the oral cavity of frogs. Egg capsule is shown at upper left. (After Krull.) (From Hyman: The Invertebrates. Vol. II, courtesy of McGraw-Hill.)

Respiration is considered to be facultative anaerobic. Trematodes, however, are dependent on oxygen for the completion of the life cycle because eggs, miracidia, and cercariae are obligatory aerobes requiring oxygen. The requirement of oxygen for larval survival and the presence of a cytochrome system suggest that miracidia and cercariae may have a functional tricarboxylic acid (TCA) cycle. The overall importance of a TCA cycle in adult trematodes presently is not known, but it appears to be minimal, especially for helminths living in anaerobic environments such as the intestine or liver ducts.

Dicrocoelium dendriticum and others contain all the enzymes of the TCA cycle, but this cycle functions to a limited degree. The α-ketoglu-

tarate dehydrogenase complex is active at a low level compared to the other enzymes of the cycle thus creating a bottleneck. A comparable bottleneck is probably the reason that the pentose phosphate pathway is of minor importance in helminths. As one would expect, respiratory pathways vary from one stage in the life cycle to another. Evidence suggests the presence of a functional TCA cycle in adult *Fasciola hepatica,* but not in its miracidia.

Hemoglobin has been found in several species of trematodes. In *Fasciolopsis buski,* it occurs in the parenchyma, excretory channels, and around proximal uterine coils. In *Fasciola hepatica,* it is particularly abundant near the uterine coils and vitellaria. The hemoglobin differs from that of the host, and its role is obscure. *Dicrocoelium dendriticum* hemoglobin has a higher affinity for oxygen binding than vertebrate myoglobin and appears to be well adapted to an oxygen-carrying function. The minor role of oxygen in *Dicrocoelium* glucose metabolism, however, suggests that oxygen probably participates in other biochemical reactions besides energy-producing ones.[5,83]

Protein metabolism is conspicuously evident by the tremendous larval multiplication and fecundity of adult worms. Amino acids for larval protein synthesis undoubtedly are obtained from host hemolymph as free amino acids[38] or through the degradation of host proteins by cecal or extracorporeal digestion. Absorption of amino acids by sporocysts occurs through the tegumental surface, whereas in actively feeding rediae, amino acids probably are taken up by the gut epithelium as well as by the tegument. The tegument and cecal epithelium also serve as sites for amino acid absorption by adult trematodes. Glycine and proline are transported predominately through the tegument of *Schistosoma mansoni.*[3] Hemoglobin from ingested red blood cells, however, is broken down to small peptides and amino acids in the cecum of schistosomes that are presumably transported directly across the gut epithelium. An acidic thiol proteinase produced in the cecum degrades ingested hemoglobin.[27] Other trematode species probably possess similar kinds of enzyme systems for digestion of host proteins in the gut. Urea and ammonia represent the major by-products of protein metabolism in trematodes.

Lipid metabolism of trematodes is not well understood. Although worms can readily ab-

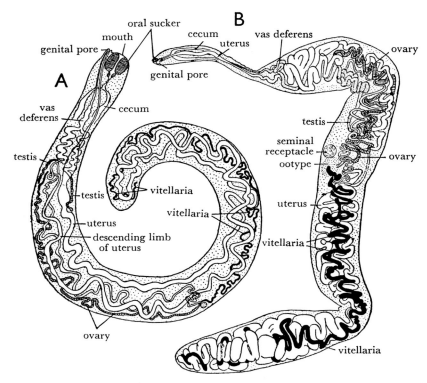

Fig. 10–32. Two species of *Didymozoon*. A, *D. scombri*. B, *D. faciale*. (From Dawes: The Trematoda, courtesy of Cambridge University Press.)

sorb host lipids, we are limited in our knowledge of their ability to synthesize them. Fatty acids of helminths are surprisingly similar, though not identical to those of their hosts. In studies of *Fasciola hepatica*[69] and *Schistosoma mansoni*,[62] it has been demonstrated that, although they are unable to initiate synthesis of sterols or fatty acids, these flukes can elongate preexisting fatty acid chains and can thereby produce specific lipids required for their metabolism. By employing radiolabeled precursors such as ^{32}P-Phosphate or ^{3}H-glycerol, *F. hepatica* also has been shown to have the capacity for initiating synthesis of many of its own phospholipids including lysolecithin, phosphatidylcholine, and others.[69] Nonspecific esters found in the tegument and intestinal, cecal epithelium of adult worms may function in catalyzing the hydrolysis of short-chain fatty acids. Lipid-like droplets in electron micrographs suggest that deposits of lipids are scattered throughout the worm's body. Such deposits may represent lipid waste products resulting from anaerobic fermentation or storage lipids such as sterols or

phospholipids. Little evidence suggests that lipids are used as energy sources.

Nucleic acid synthesis has only been investigated in a few trematode species, most notably the human blood fluke, *Schistosoma mansoni*. Schistosomes are unable to synthesize nucleotides initially and must derive these compounds either preformed, for example, in purines from digested host red blood cells, or from other precursor molecules such as adenine.[74]

Excretion is accomplished through the tegument, alimentary canal, and excretory system. The lining of the excretory system may also be involved in alkaline phosphatase activity, which may be associated with transport through the membranes. Excretory products include fatty acids, ammonia, urea, uric acid, and possibly amino acids. The excretory system has long been considered to be osmoregulatory, but recent studies indicate that this function has minimal priority in excretory metabolism. Doubt is also thrown on the concept that the system is protonephridial.[30]

The tegument and alimentary tract of trematodes both represent important organs in regulating parasite metabolic functions. As indicated previously, the tegument functions in the active transport of nutrients, for example, simple sugars or nucelotides, into the worm, as well as in the excretion of parasite-derived materials, including metabolic waste products or lytic secretions such as acid hydrolases.

Digestion in the alimentary tract occurs when trematodes feed on mucus, secretions, or host tissues or blood. Both intraluminal and extracellular digestion may take place. Microvilli lining the luminal surface of cecal epithelial cells increase the absorptive area of the gut allowing for the most efficient uptake of digested nutrients.

For further information, see Coles,[20] Barrett,[5] Bryant,[10] and Van den Bossche.[85]

IMMUNITY

Studies of immunity to trematode infections have been devoted primarily to schistosomes, and most of the work has centered on *Schistosoma mansoni*. Among mammals, there appears to be little innate immunity or resistance to *S. japonicum* or to *S. haematobium*, but hereditary or congenital resistance to *S. mansoni* is well established. Cross-immunity among schistosome species exists.

Mechanisms of innate resistance to schistosomal infections appear to be related to the ability of cercariae to penetrate and to migrate through the host's skin. Cercarial age, energy requirements, and host age are contributing factors to the survival of the parasite. Acquired immunity to all species of schistosomes generally occurs in animals and also in humans.[15] The conviction that man aquires an immunity (protective?) is based largely on epidemiologic surveys and extrapolation of results from studies of primates.

A primary schistosomal infection may last for only three or four weeks, as in rats, or may persist for several years, as in primates. If a rhesus monkey, in which a primary infection of *Schistosoma mansoni* is well established, is administered a second (challenge) infection, the latter is destroyed; resistance induced by the primary infection generally persists, but the adult worms continue to produce eggs. One hypothesis to account for this apparent dichotomy, that is, the existence of a strong protective

immunity in the presence of an active schistosomal infection, termed **concomitant immunity,** suggests that schistosomules of the primary infection acquire a coating of host material (blood group antigens, serum components) that protects the parasite from immune responses subsequently developed. Schistosomes secondarily introduced after the establishment of host immunity, however, are eliminated because they may not have enough time to invest themselves with a protective host coating. Other mechanisms by which schistosomes may be able to escape immune destruction include the sharing of host-like antigens, immunologic blockade, or modulation of surface antigens.[21] A successful method of vaccination, therefore, should be effective against the parasites upon entry, or should be able to neutralize the mechanism that enables older parasites to escape attack by the immune reaction.

The degree of acquired resistance to schistosomal infection is influenced by numerous variables, including host species or strain, host age, time duration between primary and secondary infections, infection route, and the presence of ova, to name a few. Therefore, the mechanisms of immune protection have not been precisely defined. Infected rodents, primates, and humans elaborate antibodies that, in the presence of complement, damage and kill schistosomules in culture. The appearance of these so-called lethal antibodies, however, does not positively correlate with the patterns of acquired immunity in infected animals; some doubt exists as to the involvement of these antibodies in mediating protection. Other in vitro studies have provided strong evidence that immunity to reinfection in rodents, in primates, and presumably, in man may involve a combination of humoral (opsonizing or cytophilic antibodies) and cellular (eosinophils, macrophages, basophils) mechanisms.[76] The time course of protective immunity in rats correlates well with the elaboration of opsonizing antibody and further supports a protective immune mechanism involving cell-mediated cytotoxicity that is antibody-dependent.

It is clear that antigens secreted by egg stages (soluble egg antigens) are directly responsible for eliciting a cell-mediated inflammatory response referred to as *granuloma formation*. This leading contributor to the pathologic features exhibited in schistosomiasis mansoni and japonicum is characterized by the accumulation of lymphocytyes, granulocytes, and fibroblasts around eggs trapped in host tissues. Although the predominant cell type involved in granulomatous reactions is the eosinophil, lymphocytes and macrophages may also be present. The inflammatory response is under thymic (T cell) regulation.[12,34] Investigations of the role of the schistosome egg in stimulating acquired resistance generally have revealed conflicting results.

New ways in which the immune response affects trematodes are still being discovered, and sometimes such results also contribute new basic information to immunology in general. An excellent example of this was the discovery by Capron's group,[44] that blood platelets in the presence of IgE kill schistosomes. This discovery implied another function, previously unknown, for platelets.

Antigenic heterogeneity of schistosomes, and parasitic Metazoa in general, undoubtedly contribute to the complexity of elicited immune responses. The use of dead vaccines that include worm homogenates, killed larvae, secretory antigens, and surface membrane components has been only marginally successful.[19] Similarly, injection of live vaccines, for example, irradiated cercariae or schistosomules, may produce a consistently high level of immune resistance in some animals, whereas in others, little or no protection is stimulated.[66]

However, the use of molecular techniques has revealed exciting possibilities for a vaccine, as described in the following paragraphs.

A potential vaccine for schistosomiasis mansoni now appears attainable based on the extraordinary accomplishments of André Capron's group at the Pasteur Institute in France. Where better for "chance to favor the prepared mind?" In their studies on the anti-idiotype approach to vaccine development (described in Chap. 2) they made the following serendipitous observation. One of their controls in immunizing rats to antibody-2 (Ab$_2$) coupled to Keyhole Limpet Hemocyanin (KLH) as an adjuvant, was to immunize with KLH alone. Much to their surprise (and probable delight!), antibodies to KLH alone bound to the same antigen they were studying from the surface of *S. mansoni*. Further, the rats immunized with only KLH were significantly protected from subsequent challenge with cercariae! Professor Capron immediately recalled experiments he had done 20 years previously showing cross-reactivity between antigens from *S. mansoni* and its intermediate host, the snail, *Biomphalaria glabrata*. They soon discovered that the antigen from this and other snails also possess the epitope which cross-reacts with the epitope on KLH.[25] Other analyses revealed that the antigen is glycosylated (contains sugar moieties), and such epitopes are essential for protective immunity. It will still take several years of difficult research before this and other cloned antigens have been adequately tested in model systems before trials can begin in humans. Studies such as these point out the importance and necessity for use of animals in research, for the betterment of mankind. The student should read the paper of Capron et al. in its entirety, as well as related papers in the volume edited by MacInnis.[53]

Another antigen recently cloned from *S. mansoni* by Alan Sher's group also may be useful in vaccination. The gene coding for this antigen has DNA sequences similar to that for the invertebrate "catch-muscle."[48] Perhaps it is only chance that the two molecules discovered as candidates for a vaccine have counterparts in free-living invertebrates, but such discoveries suggest that there may be other protective antigens awaiting researchers.

Control of schistosomiasis and other trematodes may also be accomplished by eliminating intermediate hosts. Molluscicides have proved effective in some areas, especially when applied in hyperendemic foci. However, bad experience has taught us that introduction of poisons into the environment can produce unwanted, residual effects for years. This does not imply that efforts to produce a safe molluscicide should cease. Rather, we need more efforts, such as those of Woodruff, aimed at control through genetic manipulation of populations of snails selected or "engineered" for resistance to parasites.[91] As with most problems of public health, eventual control of parasites requires an integrative approach: education, diagnosis, chemotherapy, vaccines, vector control, clean water supplies, safe disposal of sewage, continued surveillance, and of course, the money to accomplish all of this.

Diagnosis by serologic tests has been moderately effective, but problems with antigen cross-reactivity, standardization of methods, and sensitivity make it difficult to determine infections on an individual basis. On the other hand, serodiagnosis, when applied at the population level (termed **seroepidemiology**), can provide valuable information regarding the status of a disease in a given area and any changes in prevalence or intensity of the infection over time.[15,51] More than a dozen serologic tests have been used in the diagnosis of schistosomiasis with varying degrees of success. Enzyme-linked immunosorbent assay (ELISA), indirect immunofluorescent tests, and circumoval precipitin test (COP), in conjunction with highly purified antigens (membrane glycoproteins, soluble egg antigens), are currently being investigated for use in schistosomiasis screening programs.

Innate resistance to schistosomal infection has been demonstrated in various snail intermediate hosts. Host resistance appears to be under genetic control. The final state or host-parasite compatibility or incompatibility, however, probably depends on a combination of qualities inherent in both the infecting schistosome and the snail host.[6] Miracidia penetrating snails that exhibit innate resistance stimulate a hemocytic encapsulation reaction, whereby snail blood cells (hemocytes) rapidly accumulate around the parasite and form a multilayered capsule in which the larva is destroyed. The mechanisms of parasite recognition and destruction in refractory snails are presently unknown. Direct activation of hemocytes, or indirect cell activation through soluble hemolymph (plasma) factors, is presumed to mediate encapsulation reactions.[8] Hydrolytic enzymes released from encapsulating hemocytes may be involved in parasite killing.[18] Capsule formation represents the primary effector mechanism against not only larval trematodes, but also other parasitic Metazoa.[18]

Acquired (induced) resistance to trematodes in snails has also been reported. A specific resistance (protective immunity) to normal echinostomes, for example, *Echinostoma lindoense*, in a susceptible strain of *Biomphalaria glabrata* can be elicited by a primary challenge with irradiated echinostome miracidia.[50] Similar attempts to induce resistance in snails to schistosomal infection, using irradiated larvae or unisexual infections, have been unsuccessful. This failure suggests the possible existence of more than one type of activation or recognition mechanism in molluscs.

Immunologic research on trematodes other than schistosomes has been meager indeed. *Opisthorchis sinensis*, *Fasciola hepatica*, and *Paragonimus westermani* have received some inten-

sive study.[46] Circulating antibodies in laboratory animals infected with *Opisthorchis* have been detected by complement-fixation and hemagglutination tests. Precipitating antibodies have been detected in rabbits. The presence of protective immunity to *O. sinensis*, however, has not been firmly established. Experimental efforts to demonstrate aquired immunity to *Fasciola hepatica* have generally been disappointing, but some reports indicate that rats rapidly acquire immunity. Passive transfer of immunity to this liver fluke can be achieved in laboratory rats. Antibody apparently plays the major role, and cell-mediated immunity a secondary role.[71] Attempts at artificial immunization with worm tissues and excretion products, for example, have been largely unsuccessful, but normal mice become immune to *F. hepatica* when injected with peritoneal exudate cells from infected isologous donors.

Some protective immunity to *Paragonimus westermani* has been demonstrated. The prevalence of infection is much lower in children under 10 years old than in older children or adults. A host sex factor seems to be operative, because in Korea, 80% of the cases in children are boys, although this finding may simply reflect a difference in the eating habits or other behavior of boys. Circulating antibodies in human and experimental infections have frequently been demonstrated, but their effect on survival of worms is problematic. The role of cellular reactions is little understood.

REFERENCES

1. Archer, S.: The chemotherapy of schistosomiasis. Ann. Rev. Pharmacol. Toxicol., 25:485–508, 1985.
2. Ansari, N. (ed.): Epidemiology and Control of Schistosomiasis (Bilharziasis). Baltimore, University Park Press, 1973.
3. Asch, H.L., and Read, C.P.: Transtegumental absorption of amino acids by male *Schistosoma mansoni*. J. Parasitol., 61:378–379, 1975.
4. Badejo, O.A., Soyinka, F., and Laja, A.O.: Ectopic lesion of schistosomiasis of the penis simulating an early carcinoma. Acta Trop., 35:263–267, 1978.
5. Barrett, J.: Biochemistry of Parasitic Helminths. London, MacMillan, 1981.
6. Basch, P.F.: Intermediate host specificity in *Schistosoma mansoni*. Exp. Parasitol., 39:150–169, 1976.
7. Basch, P.F.: Relationships of some larval strigeids and echinostomes (Trematoda): hyperparasitism, antagonism, and "immunity" in the snail host. Exp. Parasitol., 27:193–216, 1970.
8. Bayne, C.J., Buckley, P.M., and DeWan, P.C.:

Schistosoma mansoni: cytotoxicity of hemocytes from susceptible snail hosts for sporocysts in plasma from resistant *Biomphalaria glabrata.* Exp. Parasitol., *50:*409–416, 1980.

9. Beaver, P.C., Jung, R.C., and Cupp, E.W.: Clinical Parasitology, 9th Ed. Philadelphia, Lea & Febiger, 1984.

10. Blanc, F., and Nosny, Y.: Nosographie des schistosomes. Ann. Soc. Belge Med. Trop., *47:*17–34, 1967.

11. Bloch, E.H.: *In vivo* microscopy of schistosomiasis. II. Migration of *Schistosoma mansoni* in the lungs, liver and intestine. Am. J. Trop. Med. Hyg., *29:*62–70, 1980.

12. Boros, D.L.: Immunoregulation of granuloma formation in murine Schistosomiasis mansoni. Ann. N.Y. Acad. Sci., *465:*313–323, 1986.

13. Bruce, J.I., and Sornmani, M. (eds.): The Mekong Schistosome. Whitmore Lake, Michigan, Malacol. Rev. Suppl. 2, 1980.

14. Bryant, C.: The regulation of respiratory metabolism in parasitic helminths. Adv. Parasitol., *16:*311–331, 1978.

15. Butterworth, A.E., and Hagan, P.: Immunity in human schistosomiasis. Parasitol. Today, *3:*11–16, 1987.

16. Cheng, T.C. (ed.): Molluscicides in Schistosomiasis Control. New York, Academic Press, 1974.

17. Cheng, T.C., and Garrabrant, T.A.: Acid phosphatase in granulocytic capsules formed in strains of *Biomphalaria glabrata* totally and partially resistant to *Schistosoma mansoni.* Int. J. Parasitol., *7:*467–472, 1977.

18. Cheng, T.C., and Rifkin, E.: Cellular reactions of marine molluscs in response to helminth parasitism. *In* A Symposium on Diseases of Fishes and Shellfishes. Edited by S.F. Snieszko. Washington, D.C., American Fisheries Society Publication No. 5, 1970, pp. 443–496.

19. Clegg, J.A., and Smith, M.A.: Prospects for the development of dead vaccines against helminths. Adv. Parasitol., *16:*165–217, 1978.

20. Coles, G.C.: Metabolism of schistosomes: A review. Int. J. Biochem., *4:*319–337, 1973.

21. Damian, R.T.: Molecular mimicry revisited. Parasitol. Today, *3:*263–266, 1987.

22. Dawes, B., and Hughes, D.L.: Fascioliasis: the invasive stages in mammals. Adv. Parasitol., *8:*259–274, 1970.

23. DeCoursey, P.J., and Vernberg, W.B.: Double infections of larval trematodes: competitive interactions. *In* Symbiosis in the Sea. Edited by W.B. Vernberg. Columbia, University of South Carolina Press, 1974, pp. 93–109.

24. DiConza, J.J., and Basch, P.F.:Incorporation of ^{3}H-thymidine and ^{14}C-glucose by *Schistosoma mansoni* daughter sporocysts in vitro. J. Parasitol., *60:*1045–1046, 1974.

25. Dissous, C., Grych, J.M., and Capron, A.: *Schistosoma mansoni* shares a protective oligosaccharide epitope with freshwater and marine snails. Nature, *323:*443–445, 1986.

26. Dixon, K.E., and Mercer, E.H.: The fine structure of the cyst wall of the metacercaria of *Fasciola hepatica.* Q. J. Microsc. Sci., *105:*385–389, 1964.

27. Dresden, M.H., and Deelder, A.M.: *Schistosoma*

mansoni: evidence that the "hemoglobinase" of adult worms is a thiol proteinase. Exp. Parasitol., *48:*190–197, 1979.

28. Edington, G.M., and Giles, H.M.: Pathology in the Tropics. 2nd Ed. London, Edward Arnold, 1976.

29. Erasmus, D.A.: A comparative study of the reproductive system of mature, immature and "unisexual" female *Schistosoma mansoni.* Parasitology, *67:*165–183, 1973.

30. Erasmus, D.A.: The Biology of Trematodes. New York, Crane, Russak & Co., 1972.

31. Erasmus, D.A.: The host-parasite interface of strigeoid trematodes. IX. A probe and transmission electron microscope study of the tegument of *Diplostomum phoxini* Faust, 1918. Parasitology, *61:*35–41, 1970.

32. Erlich, I.: On the regulation of glycogen synthesis in the liver fluke *Fasciola hepatica* L. Proc. Int. Congr. Parasitol. (Rome 1964), *1:*64–65, 1966.

33. Farley, J.: A review of the family Schistosomatidae: excluding the genus *Schistosoma* from mammals. J. Helminthol., *45:*289–320, 1971.

34. Fine, D.P., Buchanan, R.D., and Colley, D.C.: *Schistosoma mansoni* infections in mice depleted of thymus-dependent lymphocytes. I. Eosinophilia and immunologic responses to a schistosomal egg preparation. Am. J. Pathol., *71:*193–206, 1973.

35. Freeman, R.S., et al.: Fatal human infection with mesocercariae of the trematode *Alaria americana.* Am. J. Trop. Med. Hyg., *25:*803–807, 1976.

36. Gazayerli, M., Khalil, H.A., and Gazayerli, I.M.: Schistosomiasis hematobium (urogenic bilharziasis). *In* Pathology of Protozoal and Helminthic Diseases. Edited by R.A. Marcial-Rojas. Baltimore, Williams & Wilkins, 1971, pp. 434–449.

37. Gibson, J.B., and Sun, T.: Clonorchiasis. *In* Pathology of Protozoal and Helminthic Diseases. Edited by R.A. Marcial-Rojas. Baltimore, Williams & Wilkins, 1971, pp. 546–566.

38. Gilbertson, D.E., Etges, F.J., and Odle, J.D.: Free amino acids of *Australorbis glabratus* hemolymph: comparison of four geographic strains and effects of infection by *Schistosoma mansoni.* J. Parasitol., *53:*565–568, 1967.

39. Grodhaus, G., and Keh, B.: The marine, dermatitis-producing cercaria of *Austrobilharzia variglandis* in California (Tremata: Schistomatidae). J. Parasitol., *44:*633–638, 1958.

40. Groll, E.: Praziquantel. Adv. Pharmacol. Chemother., *20:*219–238, 1984.

41. Heinz, E.: Human Helminthiases in the Philippines. New York, Springer-Verlag, 1985, p. 315.

42. Hoffman, G.L.: Parasites of North American Freshwater Fishes. Berkeley, University of California Press, 1967.

43. Horak, I.J.: Paramphistomiasis of domestic ruminants. Adv. Parasitol., *9:*33–72, 1971.

44. Joseph, M., Auriault, C., Capron, A., and Viens, P.: A new function for platelets: IgE-dependent killing of schistosomes. Nature, *303:*310–312, 1983.

45. Jourdane, J., Theron, A., and Combes, C.: Demonstration of several sporocyst generations as a normal pattern of reproduction of *Schistosoma mansoni.* Acta Trop., *37:*177–182, 1980.

46. Kagan, I.G.: Other trematodes and cestodes. *In* Immunological Investigations of Tropical Parasitic Diseases. Edited by V. Houba. New York, Churchill Livingstone, 1980, pp. 148–156.

47. Kendall, S.B.: Relationships between the species of *Fasciola* and their molluscan hosts. Adv. Parasitol., 3:59–98, 1965.

48. Lanar, D., Pearce, E.J., James, S.L., and Sher, A.: Identification of paramyosin as the *Schistosoma* antigen recognized by intradermally vaccinated mice. Science, 234:593–596, 1986.

49. Lewert, R.M.: Schistosomes. *In* Immunity to Parasitic Animals. Edited by G.J. Jackson, R. Herman, and I. Singer. New York, Appleton-Century-Crofts, 1970, pp. 981–1008.

50. Lie, K.J., Heyneman, D., and Lim, H.K.: Studies on resistance in snails: specific resistance induced by irradiated miracidia of *Echinostoma lindoense* in *Biomphalaria glabrata* snails. Int. J. Parasitol., 5:627–631, 1975.

51. Lobel, H.O., and Kagan, I.G.: Seroepidemiology of parasitic diseases. Ann. Rev. Microbiol., 32:329–347, 1978.

52. MacDonald, G., and Farooq, M.: The public health and economic importance of schistosomiasis. Its assessment. *In* Epidemiology and Control of Schistosomiasis. Edited by N. Ansari. Baltimore, University Park Press, 1973, pp. 337–353.

53. MacInnis, A.J. (ed.): Molecular Paradigms for Eradicating Helminthic Parasites. UCLA Symposia Series, Vol. 60. New York, Alan Liss, Inc., 1987.

54. MacInnis, A.J.: Identification of chemicals triggering cercarial penetration responses of *Schistosoma mansoni*. Nature, 224:1221–1222, 1969.

55. MacInnis, A.J.: Responses of *Schistosoma mansoni* miracidia to chemical attractants. J. Parasitol., 51:731–746, 1965.

56. Maegraith, B.: Adams & Maegraith's Clinical Tropical Diseases. 6th Ed. Oxford, Blackwell Scientific Publications, 1976.

57. Malek, E.A.: Snail-transmitted Parasitic Diseases. Volumes I and II. Boca Raton, Florida, CRC Press, 1980.

58. Malik, M.O.A., Veress, B., Daoud, E.N., and El Hassan, A.M.: Pattern of bladder cancer in the Sudan and its relation to schistosomiasis: a study of 255 vesical carcinomas. J. Trop. Med. Hyg., 78:219–223, 1975.

59. Mansour, J.M., and Mansour, T.E.: GTP-binding proteins associated with serotonin-activated adenylate cyclase in *Fasciola hepatica*. Molec. Biochem. Parasitol., 21:139–149, 1986.

60. Mansour, T.E., Mansour, J.M., and Iltzsch, M.H.: Receptors, transducers, second messengers, and transmembrane signalling in parasitic worms. *In* Molecular Paradigms for Eradicating Parasitic Helminths. Edited by A.J. MacInnis. New York, Alan Liss, Inc. UCLA Symposia Series, 60:371–381, 1987.

61. Marcial-Rojas, R.A. (ed.): Pathology of Protozoal and Helminthic Diseases. Baltimore, Williams & Wilkins, 1971, pp. 373–413.

62. Meyer, F., Meyer, H., and Bueding, E.: Lipid metabolism in the parasitic and free-living flatworms, *Schistosoma mansoni* and *Dugesia doroto-cephala*. Biochim. Biophys. Acta, 210:257–266, 1970.

63. Millemann, R.E., and Knapp, S.E.: Biology of *Nanophyetus salmincola* and "salmon poisoning" disease. Adv. Parasitol., 8:1–41, 1970.

64. Miyake, M.: Schistosomiasis japonicum. *In* Pathology of Protozoal and Helminthic Diseases. Edited by R.A. Marcial-Rojas. Baltimore, Williams & Wilkins, 1971, pp. 414–433.

65. Moss, G.D.: The excretory metabolism of the endoparasitic digenean *Fasciola hepatica* and its relationship to its respiratory metabolism. Parasitology, 60:1–19, 1970.

66. Murrell, K.D., Clark, S., Dean, D.A., and Vannier, W.F.: Influence of mouse strain on induction of resistance with irradiated *Schistosoma mansoni* cercariae. J. Parasitol., 65:829–831, 1979.

67. Myrdal, G.: Economic aspects of health. WHO Chron., 6:203–218, 1952.

68. Noble, G.A.: Description of *Nematobibothrioides histoidii* (Noble, 1974) (Trematoda: Didymozoidae) and comparison with other genera. J. Parasitol., 61:224–227, 1975.

69. Oldenborg, V., Van Vugt, F., Van Golde, L.M.G., and Van den Bergh, S.G.: Synthesis of fatty acids and phospholipids in *Fasciola hepatica*. *In* Biochemistry of Parasites and Host-Parasite Relationships. Edited by H. Van den Bossche. Amsterdam, North-Holland Publishing Co., 1976, pp. 159–166.

70. Pan American Health Organization: A Guide for the Identification of the Snail Intermediate Hosts of Schistosomiasis in the Americas: Scientific Publication No. 168. Washington, D.C., Pan American Health Organization, World Health Organization, 1968.

71. Rajasekariah, G.R., and Howell, M.J.: *Fasciola hepatica* in rats: transfer of immunity by serum and cells from infected to *F. hepatica* naive animals. J. Parasitol., 65:481–487, 1979.

72. Schaefer, F.W., III, Saz, H.J., Weinstein, P.P., and Dunbar, G.A.: Aerobic and anaerobic fermentation of glucose by *Echinostoma liei*. J. Parasitol., 63:687–689, 1977.

73. Schell, S.C.: How to know the trematodes. Dubuque, Iowa, W.C. Brown Co. 1970.

74. Senft, A., Meich, P., Brown, P. and Senft, D.: Purine metabolism in *Schistosoma mansoni*. Int. J. Parasitol., 2:249–260, 1972.

75. Smith, J.W.: The bloodflukes (Digenea: Sanguinicolidae and Spirorchidae) of cold-blooded vertebrates and some comparison with the schistosomes. Helminthol. Abst. [Ser. A], 41:161–198, 1972.

76. Smithers, S.R., and Terry, R.J.: The immunology of schistosomiasis. Adv. Parasitol., 14:399–422, 1976.

77. Sogandares-Bernal, F., and Seed, J.R.: American paragonimiasis. Curr. Top. Comp. Pathobiol., 2:1–56, 1973.

78. Stirewalt, M.A.: *Schistosoma mansoni*: cercaria to schistosomule. Adv. Parasitol., 12:115–182, 1974.

79. Stirewalt, M.A.: Important features of the schistosomes. *In* Epidemiology and Control of Schistosomiasis (Bilharziasis). Edited by N. Ansari. Baltimore, University Park Press, 1973, pp. 17–31.

80. Thomas, J.D.: Schistosomiasis and the control of molluscan hosts of human schistosomes with particular reference to possible self-regulatory mechanisms. Adv. Parasitol., *11*:307–394, 1973.

81. Tuchschmid, P.E., Kunz, P.A., and Wilson, K.J.: Isolation and characterization of the hemoglobin from the lanceolate fluke *Dicrocoelium dendriticum*. Eur. J. Biochem., *88*:387–394, 1978.

82. Uglem, G.L., Lewis, M.C., and Larson, O.R.: Niche segregation and sugar transport capacity in digenean flukes. Parasitology, *91*:121–127, 1985.

83. Uhazy, L.S., Tanaka, R.D., and MacInnis, A.J.: *Schistosoma mansoni*: Identification of chemicals that attract or trap its snail vector, *Biomphalaria glabrata*. Science, *201*:924–926, 1978.

84. Van den Bossche, H., Thienpont, D., and Jansens, P.G. (eds.): Chemotherapy of Gastrointestinal Helminths. Berlin, Springer Verlag, 1986, p. 719.

85. Van den Bossche, H. (ed.): Comparative Biochemistry of Parasites. London, Academic Press, 1972.

86. Voge, M., Bruckner, D., and Bruce, J.I.: *Schistosoma mekongi* sp. n. from man and animals, compared with four geographic strains of *Schistosoma japonicum*. J. Parasitol., *64*:577–584, 1978.

87. Wajdi, N.: Penetration of the miracidia of *S. mansoni* into the snail host. J. Helminthol., *40*:235–244, 1966.

88. Warren, K.S.: Pathophysiology and pathogenesis of hepatosplenic schistosomiasis mansoni. *In* Clinical Tropical Medicine. Edited by K.M. Cahill. Baltimore, University Park Press, 1972, pp. 51–65.

89. Warren, K.S., and Newill, V.A.: *Schistosomiasis*: A Bibliography of the World's Literature from 1852–1962. Vol. I. Keyword Index. Vol. II. Author Index. Cleveland, The Press of Case Western Reserve Univ., 1967.

90. Wilson, R.A.: The hatching mechanism of the egg of *Fasciola hepatica* L. Parasitology, *58*:79–89, 1968.

91. Woodruff, D.S.: Genetic control of schistosomiasis: A technique based on the genetic manipulation of intermediate host snail population. *In* Parasitic and Related Diseases: Basic Mechanisms, Manifestations and Control. Edited by T.C. Cheng. Adv. Comp. Path., *8*:41–68, 1985.

92. Wright, W.H.: Schistosomiasis as a world problem. *In* Clinical Tropical Medicine. Vol. I. Edited by K.M. Cahill. Baltimore, University Park Press, 1972, pp. 72–83.

11

Class Cestoidea, Subclass Cestodaria

Class Cestoidea

General Considerations

All cestoidea are parasitic. They are platy-helminths possessing the general characteristics of the phylum given in Chapter 8. They differ from trematodes by the complete absence of an alimentary canal: no mouth, no gut, no anus. Hence, all nutrients are acquired through a specialized tegument. The Cestoda differ from trematodes in body plan, possessing "segments," called **proglottids,** each of which contains one or more sets of reproductive organs. The Cestodaria have only one proglottid, as do a few cestoda. The presence of multiple proglottids, similar to a roll of perforated tape (toilet paper may be a better analogy!), led to the common name, **tapeworms** (Fig. 11–1). Adults live in the intestines of vertebrates. Some adult tapeworms, and especially those whose larval stages parasitize man, pose serious problems of disease in humans.

The tapeworm head is a holdfast called a **scolex,** which is armed with hooks, suction organs ("suckers"), or both (Fig. 11–2). There are three major kinds of holdfasts:

1. **Bothria** (from the Greek word meaning "hole" or "trench") are usually slit-like grooves with weak suction powers, as in *Diphyllobothrium latum,* the fish tapeworm (Fig. 11–2).

2. **Phyllidia** (meaning "like a leaf") are ear-like or trumpet-like and have thin, flexible margins. The order Tetraphyllidea (Chap. 12), as the name indicates, possesses four phyllidia.

3. **Acetabula** are suction cups that are found, four in a circle, around the head of the order Cyclophyllidea (Chap. 12), which includes the common tapeworms of man.

Below the scolex is the neck (**zone of proliferation**) and the body (**strobila**). Whereas each

larval stage of a cestode is unquestionably an individual organism, the adults have sometimes been considered to be a "linear colony of highly specialized zooids." However, the adult cestode is considered to be an individual rather than a colony.

The proglottids, which number from 3 to 3000, become progressively more mature toward the posterior end of the tapeworm. Terminal segments, especially in the more primitive families of cestodes, may become detached at an early stage in their development to live and to mature independently in the intestine of their hosts. More commonly, however, the terminal gravid or ripe proglottids become little more than sacs filled with eggs.

The biology of cestodes has been reviewed recently by many authorities in the two volumes edited by Arme and Pappas.[1] For the student seeking more details than we provide about these fascinating worms we commend them to such books.

TEGUMENT

The tegument of cestodes (Fig. 11–3) is syncitial, largely proteinaceous, containing certain polysaccharides, glycoprotein, mitochondria, vacuoles, and membranes. The parasite-host interface has been defined as the "region of chemical juxtaposition of regulatory mechanisms of both host and parasite."[14]

High levels of protein synthetic activity occur in the subtegumental cells, where the nuclei reside (Fig. 11–3), and these proteins are presumably secreted into the tegumental matrix. The tegument is basically the same as that described for trematodes in Chapter 9, but the functions are consistent with the nutritional requirements of a parasite that does not have a

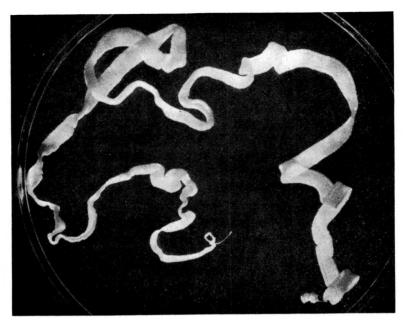

Fig. 11–1. Entire tapeworm, *Hymenolepis diminuta*. (Photograph by Zane Price.) (From Markell and Voge: Medical Parasitology, courtesy of W.B. Saunders.)

digestive tract. See Lee,[8] and Lumsden and Specian[11] for detailed structure.

REPRODUCTIVE SYSTEM

Each mature proglottid usually contains one set of female organs and one set of male organs. A few species, such as the dog tapeworm (Chap. 12), have duplicate sets of male and female organs in each proglottid (Chap. 12). Occasionally, the male organs mature, then degenerate before the female organs become functional. These organs are fundamentally similar to those of most flukes, but a vaginal canal opens to the outside. Scattered testes occur throughout each proglottid, and the vitellaria, containing shell globules and yolk material, are commonly grouped in one clump, but they may be dispersed along the lateral margins of the worm. Some proglottids in a mature strobila may be sexually undifferentiated. These and other reproductive organs are illustrated in Figures 11–4 and 11–5.

Self-fertilization involving the sexual structures of a single proglottid may occur. Hyman[5] felt that "self-fertilization by eversion of the cirrus into the vagina of the same proglottis is probably the most common method of impregnation in cestodes." She also mentioned mutual copulation between different proglottids of the same strobila and those of different strobila and other modes of sperm transfer, including hypodermic impregnation in those tapeworms that lack a vaginal opening, for example, the Acoleidae.

Williams and McVicar[22] briefly reviewed the literature on sperm transfer in cestodes, and they reported on their own observations on Tetraphyllidae found in sharks and rays. The investigators concluded,

"it thus seems possible that self-fertilization may often occur in many species and they may occasionally cross-fertilize, thus retaining heterozygosity. . . . It can be said that tapeworms of the order Tetraphyllidae show (i) self-insemination, (ii) protandry, (iii) protogyny, (iv) copulation between proglottides where there is no distinct evidence of either protandry or protogny and (v) copulation between dioecious individuals."

They wisely warn the reader that "transfer of sperm should not be taken as conclusive proof of fertilisation. . . . " These authors observed that those species of Tetraphyllidea that show self-copulation are restricted to sharks that are generally regarded as the most primitive elasmobranchs, and that the cross-fertilizing species have been found only in the more advanced rays (Rajiformes).

D. canium

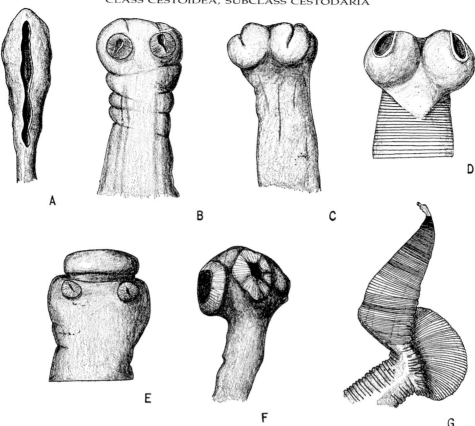

Fig. 11–2. Cestode scoleces. *A, Diphyllobothrium.* Note the bothria or grooves. *B, Mesocestoides.* Note the absence of hooks and rostellum. *C, Moniezia. D, Thysanosoma.* Note the absence of hooks and rostellum. *E, Raillietina.* Note the presence of rostellum. *F, Hymenolepis.* Note the retracted rostellum. *G, Fimbriaria.* Note the special holdfast organ. (From Whitlock: Illustrated Laboratory Outline of Veterinary Entomology and Helminthology, courtesy of Burgess Publishing.)

Hymenolepis nana was maintained in a self-fertilizing strain for 5 successive generations,[15] but several anomalies appeared: the frequency of cysticercoid abnormalities increased, fewer eggs developed into cysticercoids in insects, and fewer cysticercoids established themselves as adults in mice. No differences in size of adults or cysticercoids have been found between self- and cross-fertilizing strains. Prolonged self-fertilization of tapeworms may thus result in loss of viability. One study,[7] however, produced 14 self-fertilized generations without an apparent change in viability of either eggs or cysticercoids. A single cysticercoid of *Hymenolepis microstoma,* the mouse bile-duct tapeworm, was fed to a mouse. When the worm matured, gravid proglottids were fed to flour beetles, *Tribolium confusum.* One F_1 cysticercoid from the beetles was fed to each of 10 mice; the resulting mature worms furnished proglottids to feed to flour beetles, as described. One wonders whether the experiment could be conducted indefinitely.

Reciprocal fertilization between proglottids of different worms may take place when more than one tapeworm is present in the same region of the intestine. A single tapeworm may lay thousands of eggs a day, and one to two million eggs during its lifetime.

The organization of the reproductive organs may be used to divide cestodes into two groups. Group I includes those worms with vitellaria scattered throughout the peripheral region of the proglottid (Pseudophyllidea and Tetrarhynchidea), or those with vitellaria occurring in broad or narrow lateral masses (Tetraphyllidea, Proteocephalidea, and Lecanicephalidea). In some species (Pseudophyllidea), the eggs are

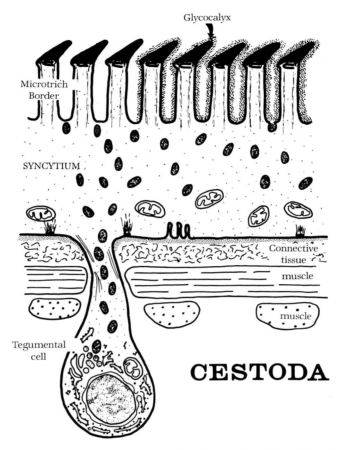

Fig. 11–3. The tegument of cestodes. (From Wright, K.A.: Structural studies of digestion in some helminth parasites. *In* Parasites: Their World and Ours. Edited by A.M. Fallis, Ottawa, Royal Society of Canada, 1977.)

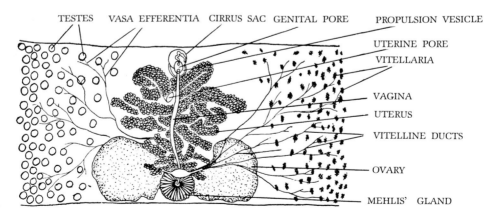

Fig. 11–4. A proglottid of the fish tapeworm *Diphyllobothrium latum.* Testes have been omitted from the right side and vitellaria from the left. (From Cable: An Illustrated Laboratory Manual of Parasitology, courtesy of Burgess Publishing.)

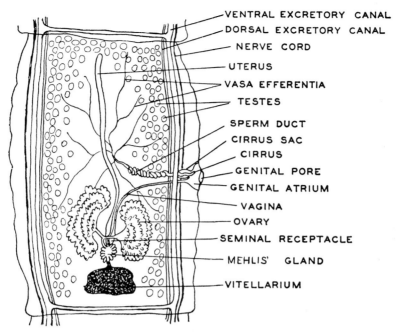

Fig. 11–5.   A mature proglottid of *Taenia pisiformis*, found in the dog and in other mammals. (From Cable: An Illustrated Laboratory Manual of Parasitology, courtesy of Burgess Publishing.)

thick-walled, oval, and operculated, as in flukes, whereas in others (Tetraphyllidea), the eggs are thin-walled, round, and nonoperculated.

Group II includes those cestodes with compact vitelline cells usually located in the midline of the proglottid, as typified by the Cyclophyllidea. Two small, median vitellaria occur in the Mesocestoididae, but some Anoplocephalidae do not have vitelline cells. Embryos of this group are usually protected by the **embryophore** (Figs. 11–6 and 11–7). Figures 11–8 and 11–9 illustrate these two groups.

The reproductive organs and eggshell formation of the Pseudophyllidea are closely related to the digenetic trematodes, as described in Chapter 9. Globules released from the vitelline cells coalesce around the mass of these cells, which surround the ovum. These globules form the eggshell (see Fig. 11–8). Cyclophyllidean cestodes have small vitellaria, and the ovum is surrounded by a few cells, occasionally only one. The ova of *Taenia* and species of other genera are surrounded by a thick embryophore.

Egg layers of a tapeworm (*Hymenolepis diminuta*, Chap. 12) were studied by Lethbridge[10] using histochemical tests. He suggested that ar-

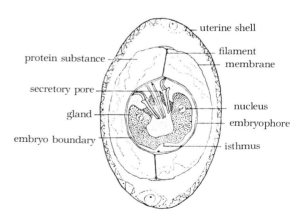

Fig. 11–6.   Ventral view of oncosphere (or embryo) of *Raillietina cesticillus*, enclosed in embryonic membranes. (From Reid, courtesy of Trans. Am. Microscop. Soc.)

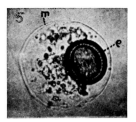

Fig. 11–7. *Taeniarhynchus saginatus*; a photomicrograph of an egg. m, External membrane of the egg. e, Embryophore containing oncosphere or six-hooked embryo. (From Gradwohl and Kouri: Clinical Laboratory Methods and Diagnosis, courtesy of C.V. Mosby.)

omatic and heterocyclic acids act as cross-linking agents in the shell protein. The subshell membrane contains lipids and functions to protect the enclosed embryo against fluctuations in pH and osmolarity in the external environment and infiltration of enzymes and toxic compounds. There is also a cytoplasmic layer that consists of a glucosamine-containing mucoprotein in a dehydrated or semidehydrated state in the intact egg. In a damaged egg, this cytoplasmic layer is rapidly dispersed by proteolytic enzymes. A sulfur-rich protein that is weakened by some proteolytic enzymes and is destroyed by others makes up the embryophore that encloses the hexacanth embryo.

Ova of *Echinococcus granulosus*, the hydatid worm (Chap. 12), are resistant to subfreezing temperatures. In one experiment,[3] eggs kept at $-50°C$ for 24 hours infected a mouse. Eggs kept at $-30°C$ had a much higher rate of viability and infectivity. Viability at subfreezing temperatures is generally related to the ability to survive the loss of a great proportion of intracellular water.

RESPIRATION

As in trematodes, respiration in cestodes is facultative anaerobic. Because free-living stages (coracidia) occur in some life cycles, genes for aerobic respiration are present in adults living in essentially anaerobic environments. For a discussion of respiration see the section on physiology in Chapter 12.

DIGESTIVE SYSTEM

Because cestodes possess no alimentary canal, there cannot be a system similar to that of higher animals. There is, of course, a system for obtaining nutrients. These enter through the tegument by diffusion or by active transport. The outer cytoplasmic surface is a syncytium and provides cytoplasmic/metabolic control of movement of all materials into or out of internal tissues. With the possible exception of phosphatases, these worms probably produce few enzymes for digestion outside the body. For metabolism of nutrients, see Chapter 12.

NERVOUS SYSTEM

This system consists of cerebral ganglia in the anterior end of the worm, from which extend two main lateral trunks that span the length of the body. Several other longitudinal nerves are present, and connecting all these are many branching nerves that extend to all parts of the body. A heavier concentration of nerves and neurons occurs in such organs as suckers and excretory vesicles (Fig. 11–10).

EXCRETORY SYSTEM

The excretory or osmoregulatory system is similar to that of flukes and consists of flame cells (flame bulbs) that connect with transverse and longitudinal collecting tubules (Figs. 11–5 and 11–11). These longitudinal vessels are located along each side of the body. Fluid flows anteriorly in the smaller, dorsal vessel and posteriorly in the larger, ventral vessel. These vessels generally open directly to the surface, but a caudal excretory vesicle and a caudal excretory pore might be present.

The exact function of the excretory vessels is in doubt, but it seems clear that, in addition to the usual role of ridding the body of certain metabolic wastes, these tubules help to maintain hydrostatic pressure. They also have a secretory function.

MUSCLE SYSTEM

Movements of the body and locomotion are accomplished by various sets of muscles. A thin, circular muscle layer and one or more longitudinal muscle layers occur close to the tegument. Within the parenchyma are well-developed transverse, longitudinal, dorsoventral, and occasionally, diagonal muscles. The scolex is especially well endowed with muscles associated with the suckers and hooks. Transmission of contraction waves is apparently inherent in the muscle fibers; detached, ripe segments often move actively.

Life Cycles

In a thorough review of cestode ontogeny, Freeman[4] stated that

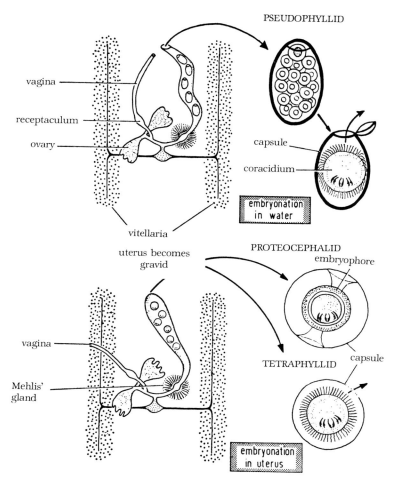

Fig. 11–8. Diagrammatic representation of genitalia and eggs of cestodes in Group I, with extensive vitellaria. (From Smyth and Clegg, courtesy of Exp. Parasitol.)

"cestodes may be more dependent for their development on various sites at various trophic levels than on hosts *per se*, a fact frequently over-looked when one thinks only in terms of one-host, two-host, or three-host cycles."

He suggested that developmental stages in cestode life cycles consist of: (a) an ovum (presumably requiring fertilization), (b) an **oncosphere** (a six-hooked larva often called a "hexacanth") that undergoes metamorphosis, (c) a **metacestode** that includes

"all growth phases between oncosphere and first evidence of sexuality, i.e. when it has (i) a fully differentiated scolex of adult size and (ii) a body showing proglottidation (= proglottisation), or first signs of approaching sexual maturation, and (d) the sexually reproducing adult in the enteron. . . ."

The cycle may require a few weeks or a few years. For a review of the biology of cyclophyllidean cestodes, see Lethbridge.[9]

Freeman's usage of the terms "oncosphere" and "metacestode" is not universally accepted, but we believe that his system of terminology is a sound and logical attempt to bring order out of the widely divergent and conflicting descriptions of life cycles in the literature. Following his concepts, the only stage appropriately called a "larva" is the oncosphere. Metacestode growth usually occurs outside the gut and in an invertebrate. One end of the metacestode may become modified to form a "cercomer" or tail. It is this end, presumably posterior, that was the functional anterior end (with hooks) of the oncosphere that moves with its posterior

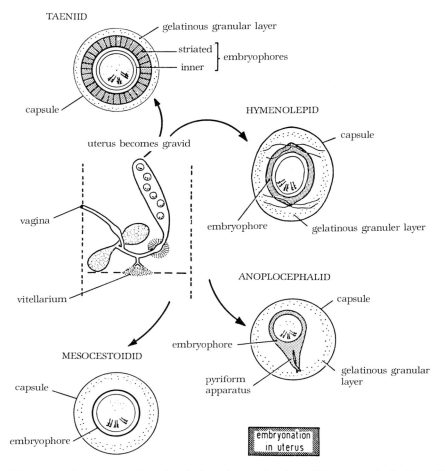

Fig. 11–9. Diagrammatic representation of genitalia and eggs of cyclophyllidean cestodes in Group II. (From Smyth and Clegg, courtesy of Exp. Parasitol.)

end forward. If the metacestode develops as a compact mass of cells (without a cavity), it is primitive, as in *Diphyllobothrium* and becomes a plerocercoid. Freeman has proposed the name **plenametacestode** to replace "plerocercoid." It is a mature, transient, fully differentiated pre-adult.

If the metacestode develops a primary cavity (or lacuna), its development is designated as **neoteric,** and it is apparently restricted to the Cyclophyllidea. When neoteric development results in a discard of part of the metacestode body wall as well as the cercomer and a regeneration of an excretory vesicle and pore, the result is a **cysticercoid,** as in *Hymenolepis.* When the posterior body becomes a large bladder, it is a **cysticercus** metacestode, as in the beef and pork tapeworms.

Freeman added numerous descriptive names to certain root terms in order to make the terminology more precise. For example, if the forebody of the cysticercus has begun segmentation, the metacestode is called a "strobilocysticercus." "Multicephalo-cysticercus" is more descriptive than the well-known "coenurus" (see Chap. 12). The term "acaudate bothrio-plerocercoid" gives a better mental picture than does "plerocercoid." We will not employ these descriptive terms in the following pages because they are not essential for an introductory understanding of cestode life cycles. For further details and divergent views on terminology, see Jarecka,[6] Mackiewicz,[12] Malmberg,[13] Slais,[16] and Voge.[19]

For the Cestoda order Caryophyllidea, in which adults consist of a single "proglottid"

junction of transverse commissures with
dorsoventral commissure

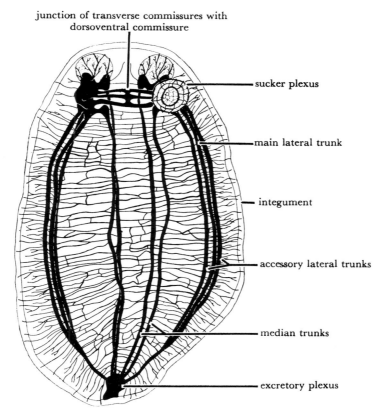

sucker plexus

main lateral trunk

integument

accessory lateral trunks

median trunks

excretory plexus

Fig. 11–10. The nervous system of *Mesocestoides* sp. based on histochemical and histologic preparations. (From Hart, courtesy of J. Parasitol.)

and one set of reproductive organs, Mackiewicz[12] has suggested the useful term **monopleroid,** defined as a "mature stage of cestodes, lacking a cercomer, internal or external proglottidization, and with a single set of reproductive organs." He also suggests that the term could be used for cestodarians.

The main types of cestode developmental stages are represented in the orders Pseudophyllidea and Cyclophyllidea (Fig. 11–12). In the Pseudophyllidea (for example, the broad fish tapeworm, *Diphyllobothrium latum,* p. 230), the larval stage in the egg hatches and becomes a free-swimming ciliated **coracidium** (see Fig. 12–3.) Copepods, for example, *Diaptomus,* eat the coracidia of *D. latum,* the cilia are shed, and the oncosphere migrates into the body cavity (hemocoel) and becomes a **procercoid** metacestode. The next host is a fish that eats the copepod. Salmon, trout, perch, pike, and other fish thus become infected. The procercoid

is freed in the intestine of the fish and makes its way into the muscles. Here it elongates into a **plerocercoid** metacestode (see Figs. 12–3 and 12–4). Because of an early mistake in identification, the plerocercoid metacestode was thought to be another type of worm and was called a **sparganum.** The name persists, and infection of a fish or other animal with plerocercoids is **sparganosis.** A plerocercoid may be minute or, in a few species, 5 to 10 cm in length. Fish-eating mammals, including man, are the final hosts. Plerocercoids develop into mature tapeworms in the intestine, and eggs appear a few weeks after infection. (See Fig. 12–3.) *D. latum* plerocercoids shed their hind bodies before starting to develop adult strobila. Shedding occurs piece by piece during a two-day period while the worm is migrating to the anterior third of the small intestine. In contrast, *D. dendriticum* and *D. ditremum* develop into adults without shedding the metacestode body.

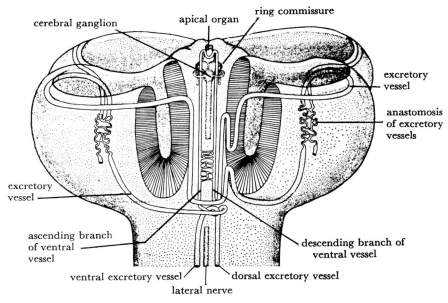

Fig. 11–11. The scolex of *Tetrabothrius affinis;* reconstruction of the left half of the excretory system with part of the bothridia removed. (From Rees, courtesy of Parasitology.)

The order Cyclophyllidea, which includes most of the important cestodes of man, displays three main types of immature tapeworms (Fig. 11–12): oncosphere, cysticercoid, and cysticercus. Hatching of oncospheres of the sheep tapeworm, *Moniezia expansa,* requires mechanical breakage of the eggshell and subshell membrane and enzyme digestion of the pyriform apparatus.[2] The three pairs of hooks may become vigorously active when the embryo hatches and may enable it to penetrate host tissues. In cysticercoids, the scolex is retracted.

These metacestodes may or may not have a solid tail; often, the entire body appears to be solid. This stage is frequently found in an arthropoid intermediate host. A cysticercus (bladderworm) has a rounded, fluid-filled cyst or bladder into which the scolex is invaginated. A **coenurus** is a larger bladder worm that has many invaginated scoleces attached to its inner, or germinative, layer. **Hydatid cysts** (Figs. 11–12, 12–29) are of two types, unilocular or multilocular, as shown.

For pathogenesis of larval cestodes in mammals, see Smyth and Heath.[18] For immunity, see Wassom, et al.[21] For a comprehensive treatment of tapeworms, see Wardle, et al.[20]

Subclass Cestodaria

In this group of worms there is no scolex, and the body is not segmented. These cestodes are therefore sometimes called **monozoic,** in contrast to the **polyzoic** forms that have proglottids. The two orders of Cestodaria described here are also characterized by oncospheres that have ten hooks, in contrast to the six hooks possessed by practically all the orders of Cestoda. The general appearance of the Cestodaria resembles that of a monogenean trematode rather than a tapeworm. The absence of a digestive tract, the presence of parenchymal muscles similar to those of tapeworms, and the developmental stages resembling tapeworms seem to justify placing the group with cestodes. The worms are found in the intestines and body cavities of fish. The cestodarians have generally been considered to be the most primitive of the cestoides, but considerable evidence suggests that they are progenetic plerocercoid metacestodes. See Chapter 26 for a discussion of their phylogeny.

Order Amphilinidea

Amphilina foliacea lives in the body cavity of sturgeons (*Acipenser*). It is an oval, flat worm without a scolex or digestive tract and with a protrusile proboscis. It is hermaphroditic with ovaries scattered throughout the body. A long, loosely coiled uterus, a vagina, and male and female openings, both at the posterior end of

PSEUDOPHYLLIDEA

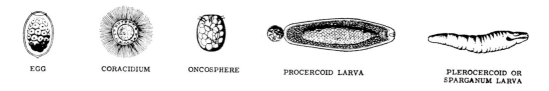

EGG	CORACIDIUM	ONCOSPHERE	PROCERCOID LARVA	PLEROCERCOID OR SPARGANUM LARVA

CYCLOPHYLLIDEA

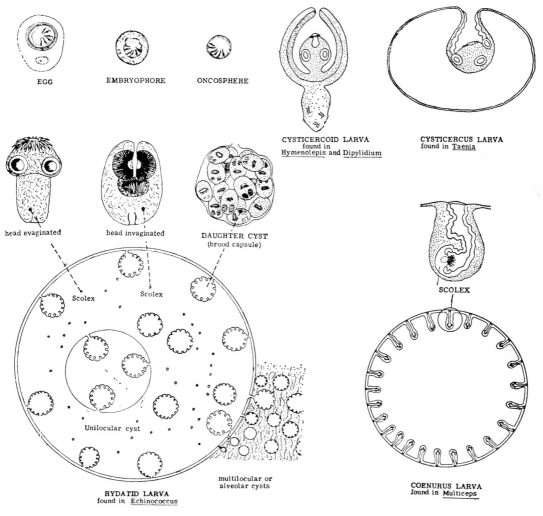

EGG   EMBRYOPHORE   ONCOSPHERE

CYSTICERCOID LARVA
found in
Hymenolepis and Dipylidium

CYSTICERCUS LARVA
found in Taenia

head evaginated   head invaginated   DAUGHTER CYST
(brood capsule)

Scolex   Scolex

SCOLEX

Unilocular cyst

multilocular or
alveolar cysts

HYDATID LARVA
found in Echinococcus

COENURUS LARVA
found in Multiceps

Fig. 11–12. Tapeworms of man, immature stages. (Courtesy of the Navy Medical School, National Navy Medical Center.)

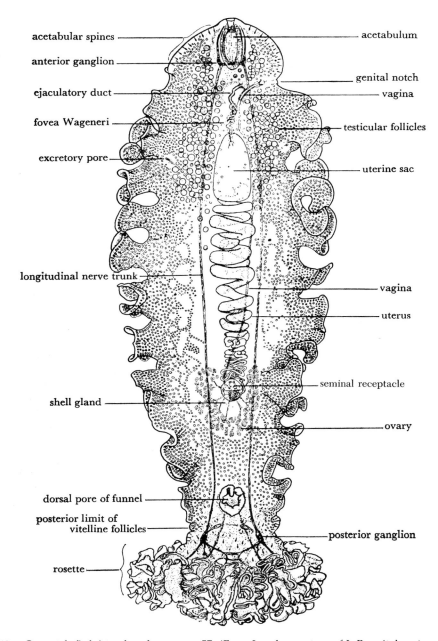

Fig. 11–13. *Gyrocotyle fimbriata*, dorsal aspect. ×57. (From Lynch, courtesy of J. Parasitology.)

the body, are present. Adults range in length from a few to about 40 mm.

The life history of *Amphilina foliacea* starts with the egg, which develops into a ciliated larva while still in the uterus of the mother worm. The adult worm presumably penetrates the abdominal wall of the host before expelling eggs. The larva is called a **lycophore** or **decacanth,** and it does not emerge from the egg until it is eaten by the second host, an amphipod. In the crustacean host, the larva makes its way into the body spaces and develops into a procercoid, then into a plerocercoid stage. The latter is infective to sturgeons, which eat amphipods. Sexually mature worms contain larval hooklets.

Order Gyrocotylidea

These elongated flatworms are nonsegmented, and they live in the spiral intestine of primitive fishes.

Gyrocotyle urna. This and *G. fimbriata* are representative species. The genus consists of worms with bodies composed of one segment flattened dorsoventrally. At the anterior end is a muscular sucker, the **acetabulum.** The posterior end is funnel-shaped, with an anterior dorsal pore and a wide posterior attachment organ. The borders of this organ are thin and are folded in a complex manner, forming the rosette. The lateral borders of the body, which has spines on its surface, are thin and undulant or ruffled. The vaginal pore is dorsal in position, the male genital pore and the uterine pore are ventral. All three pores are found in the anterior fourth of the body.

Little is known of the life cycle of these parasites. A ten-hooked ciliated larva emerges from the egg and may enter the host tissues directly without using an intermediate host.

Gyrocotyle fimbriata lives in the intestine of the ratfish, *Hydrolagus,* which occurs along the western coast of the United States, and *Chimaera* in the Atlantic. The parasites average 32 mm in length, but range from 13 to 63 mm. The general anatomy is as described for *G. urna.* Other details may be seen in Figure 11–13.

What probably represents the first attempt to apply molecular techniques to the taxonomy of helminthic parasites was performed on these interesting cestodarians. Simmons et al.[17] provided some physical characterizations of the DNA from four species of gyrocotylideans. They also demonstrated that DNA could be extracted from specimens preserved in ethyl alcohol, could be made radioactive by tritiating it in vitro, then subsequently successfully used for hybridization (heteroduplex formation) of DNA strands to determine relatedness of species. Because it is now much easier to prepare radioactive DNA and RNA by enzymatic methods, it would be interesting to perform similar experiments amongst the various, platyhelminth taxons.

REFERENCES

1. Arme, C., and Pappas, P.W. (eds.): Biology of the Eucestoda. Vol. I. and II. New York, Academic Press, 1983.
2. Caley, J.: *In vitro* hatching of the tapeworm *Moniezia expansa* (Cestoda: Anoplocephalidae) and some properties of the egg membranes. Z. Parasitenkd., 45:335–346, 1975.
3. Colli, C.W., and Williams, J.F.: Influence of temperature on the infectivity of eggs of *Echinococcus granulosus* in laboratory rodents. J. Parasitol., 53:422–426, 1972.
4. Freeman, R.: Ontogeny of cestodes and its bearing on their phylogeny and systematics. Adv. Parasitol., 11:481–557, 1973.
5. Hyman, L.H.: The Invertebrates: Platyhelminthes and Rhynchocoela. The Acoelomate Bilateria. Vol. II. New York, McGraw-Hill, 1951.
6. Jarecka, L.: Phylogeny and evolution of life cycles of Cestoda from fresh water and terrestrial vertebrates. J. Parasitol., 56:169–170, 1970.
7. Jones, A.W., et al.: Prolonged selfing in *Hymenolepis microstoma* (Cestoda). Exp. Parasitol., 29:223–229, 1971.
8. Lee, D.E.: The structure of the helminth cuticle. Adv. Parasitol., 10:347–372, 1972.
9. Lethbridge, R.C.: The biology of the oncospheres of cyclophyllidean cestodes. Helminthological Abstracts, Series A. Vol. 49. Slough, U.K., Commonwealth Agriculture Bureaux, 1980, pp. 49–72.
10. Lethbridge, R.C.: The chemical composition and some properties of the egg layers in *Hymenolepis diminuta* eggs. Parasitology, 63:275–288, 1971.
11. Lumsden, R.D., and Specian, R.D.: The morphology, histology, and fine structure of the cyclophyllidean tapeworm, *Hymenolepis diminuta. In* Biology of the Rat Tapeworm, *Hymenolepis Diminuta.* Edited by H.P. Arai. New York, Academic Press, 1980, pp. 157–280.
12. Mackiewicz, J.: Caryophyllidea (Cestoidea): a review. Exp. Parasitol., 31:417–512, 1972.
13. Malmberg, G.: On the procercoid protonephridial system of three *Diphyllobothrium* species (Cestoda, Pseudophyllidea) and Janicki's cercomer theory. Zool. Scripta, 1:43–56, 1971.
14. Read, C.P., Rothman, A., and Simmons, J.E.: Studies on membrane transport, with special reference to parasite-host integration. Ann. N.Y. Acad. Sci., 113:154–205, 1963.
15. Rogers, W.A., and Ulmer, M.J.: Effects on continued selfing on *Hymenolepis nana* (Cestoda). Proc. Iowa Acad. Sci., 69:557–571, 1962.

16. Šlais, J.: Functional morphology of cestode larvae. Adv. Parasitol., *11*:396–480, 1973.

17. Simmons, J.E., et al.: Characterization and hybridization of DNAs of Gyrocotylidean parasites of chimaeroid fishes. Int. J. Parasitol., *2*:273–278, 1972.

18. Smyth, J.D., and Heath, D.D.: Pathogenesis of larval cestodes in mammals. Helminth. Abst., *39*(Ser. A):1–24, 1970.

19. Voge, M.: Systematics of cestodes—present and future. *In* Problems in Systematics of Parasites. Edited by G.D. Schmidt. Baltimore, University Park Press, 1969, pp. 49–72.

20. Wardle, R.A., McLeod, J.A., and Radinovsky, S.: Advances in the Zoology of Tapeworms, 1950–1970. Minneapolis, University of Minnesota Press, 1974.

21. Wassom, D.L., DeWitt, C.W., and Grundmann, A.W.: Immunity to *Hymenolepis citelli* by *Peromyscus maniculatus:* genetic control and ecological implications. J. Parasitol., *60*:47–52, 1974.

22. Williams, H.H., and McVicar, A.: Sperm transfer in Tetraphyllidea (Platyhelminthes: Cestoda). Nytt Mag. Zool., *16*:61–71, 1968.

12

Class Cestoidea, Subclass Cestoda

The members of this subclass (sometimes called Eucestoda) are tapeworms which have a scolex with holdfasts, and hooks may or may not be present; the strobila consists of three to many proglottids; the oncosphere (hexacanth) has six hooklets. Twenty-one orders of cestoda have been described, as well as hundreds of species.[51] Undoubtedly, many species remain to be described. Schmidt has provided useful texts for identification of tapeworms.[40] The update of "Zoology of Tapeworms" by Wardle, McLeod, and Radinovsky[51] provides further access to this field.

Among the 21 orders of cestodes, the majority of the common tapeworms parasitizing humans reside in two orders, the Pseudophyllidea and Cyclophyllidea. Humans may serve as hosts for adult tapeworms, for metacestodes (larval stages), or for both. As travel throughout the world becomes easier and more common and "ethnic eating" more uncommon, the likelihood of encountering rare tapeworms increases. Likewise, the role that cestodes play in the big, ecological picture is difficult to assess or address. Do they influence populations of hosts? The interested student may consult the chapters by Kennedy, and by Esh in Arme and Pappas, Volume I for further details.[3] The comments by Rauch[34] are pertinent to all who study parasites, particularly cestodes, especially *Echinococcus.*

Order Tetraphyllidea

These tapeworms are commonly found in the intestines of elasmobranchs, for example, sharks and rays; they are characterized by four phyllidia on the scolex. These "suction" organs of attachment are usually broad and leaf- or trumpet-like, and they may be simple or complex. The worms are moderate in size, usually not exceeding 10 cm in length and possessing at most only a few hundred proglottids. The ovary is bilobed, and each lobe is constricted horizontally; the vagina lies dorsal to the uterine sac, and vitellaria occur as two marginal bands. The order contains two families: Phyllobothriidae and Oncobothriidae. Life cycles of the various genera have not been completely delineated, but they are basically similar to those of the pseudophyllids discussed later in this chapter.

Order Lecanicephalidea

Like the tetraphyllids, these tapeworms live in the intestines of elasmobranchs. The two groups of worms are similar in many respects, but the scolex of the lecanicephalids consists of two main parts in tandem. The anterior portion may be bulb-like (many possess tentacles or suckers), and the posterior part may bulge like a cushion and may bear four suckers. The order is composed of small tapeworms possessing few proglottids and more or less cylindrical bodies. Complete life cycles have not yet been described. The oyster, *Crassostrea virginica,* may be the natural intermediate host of metacestode stages of members of the genus *Tylocephalum.*

Order Proteocephalidea

Proteocephalidae is the only family, but it is a large group with many well-known species. The worms inhabit the intestines of amphibians, reptiles, and fish. The genus *Proteocephalus* occurs in many freshwater teleosts throughout the world. Figure 12–1 illustrates larval stages of *P. parallacticus. Lintoniella adhaerens* has been found in hammerhead sharks. The worms, in general, are only a few centimeters long; mature proglottids are longer than they are broad. The scolex is varied, but usually has four simple

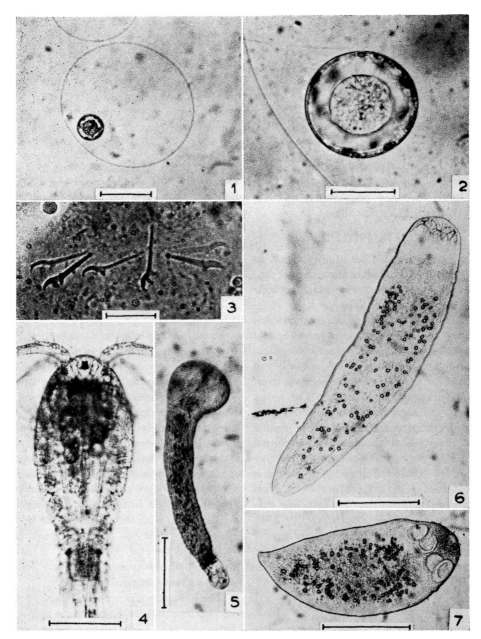

Fig. 12–1. Photomicrographs of the cestode *Proteocephalus parallacticus* from lake trout. *1*, The egg. *2*, Greater magnification of the egg. *3*, Embryonic hooks. *4*, *Cyclops bicuspidatus* with subspherical metacestodes in situ. *5*, Cercomer stage. *6*, Unfixed plerocercoid showing general morphology. *7*, Same plerocercoid as in *6* during fixation with 10% formol-saline solution. (Value of scale for each figure is *1, 4, 5* = 0.10 mm; *2* = 0.03 mm; *3* = 0.01 mm; *6, 7* = 0.02 mm.) (From Freeman. Can. J. Zool., *42*:393, 1964. Reproduced by permission of the National Research Council of Canada.)

suckers flush with the surface of the body and near the anterior tip. The scolex may or may not extend beyond the suckers. This extension sometimes possesses hooks. Vitellaria occur in two marginal bands.

The life cycle of proteocephalids involves an oncosphere that develops into a procercoid in the body cavity of a copepod *(Cyclops)*. Copepods are eaten by fish or amphibians, in which adult worms develop. Plerocercoids of *Proteocephalus ambloplitis* have been reported to migrate from the parenteral cavity of bass through the gut wall and into the gut lumen. Apparently, they do this in response to an increase in temperature and with the aid of histolytic secretions. Vertebrates may also serve as intermediate hosts when they are eaten by larger fish, amphibians, or reptiles. In these cases, the plerocercoid metacestodes usually inhabit the liver or other organs of the first vertebrate host, and, when eaten, remain in the intestine of the second vertebrate host.

Order Diphyllidea

Echinobothrium is the only genus, and it contains few species. Adults occur in the intestines of elasmobranch fishes, and developmental stages inhabit marine molluscs and crustaceans. The group is characterized by a scolex that possesses large hooks at its anterior end and two large, boat-like bothridia, each formed by a fusion of two of these sucker-like attachment organs. The long "neck" of the worm is spiny, and the entire worm is small, having fewer than 20 proglottids.

Order Tetrarhynchidea (= *Trypanorhyncha*)

The scolex of these tapeworms has four long tubes, within each of which lies a slender tentacle armed with rows of hooks (Fig. 12–2). These tentacles can readily be extended from the scolex and withdrawn into the tubes. In addition, the scolex possesses two to four phyllidia that are not usually well developed. The entire holdfast end of the worm is long and cylindrical. Vitellaria are usually distributed in a sleeve-like layer in the cortex of each proglottid. Testes extend into the region behind the ovary. The vagina and its opening are ventral to the cirrus pouch; the sperm duct does not cross the vagina before entering the cirrus pouch. Embryonation is completed when eggs have been expelled from the proglottid.

These tapeworms, usually under 100 mm in length, may be only a few millimeters long. They normally inhabit elasmobranchs, especially the spiral valve. The life cycle involves two intermediate hosts, the first a copepod and the second a teleost fish.

Order Pseudophyllidea

An important characteristic of this group is the presence of two **bothria** on the scolex instead of suction cups. The bothria usually are not specialized. The bothrium may be a short, longitudinal, slit-like groove in some species and a wider depression in others. In either case, the bothria possess weak holdfast properties. In some groups of worms, bothria margins join to form a tube. The length of worms varies from a few millimeters to 25 m or more. Some forms are **monozoic,** that is, possessing a body without segmentation; but most of them are **polyzoic,** a term referring to the familiar divisions of the body (strobila) into proglottids.

Usually, only one set of reproductive organs occurs in each proglottid, but a few species have two sets. The genital opening is often on the midventral surface and even on the middorsal, but it may also occur laterally. The ovary is bilobed, vitellaria are numerous and scattered, and the uterus opens to the outside on the ventral surface.

FAMILY HAPLOBOTHRIDAE

Haplobothrium globuliforme has four retractile, spined "tentacles" that suggest a relationship with the tetrarhynchs, but the anatomy of its gravid segments places it with the pseudophyllids. The life cycle of this cestode includes a coracidium that is eaten by *Cyclops,* within which it develops into a procercoid. A bony fish (for example, the bullhead, *Ictalurus nebulosus*) eats the crustacean, and the procercoid is liberated and develops into a plerocercoid in the liver of the fish. A ganoid fish, *Amia calva,* eats the bullhead and thus becomes infected.

FAMILY DIPHYLLOBOTHRIIDAE

Diphyllobothrium latum is an important parasite of man, but most of the several genera of this family are parasites of marine mammals. *D. latum*, the fish tapeworm or broad tapeworm (also known as *Dibothriocephalus latus*), has a scolex that is almond-shaped, measuring 2 to 3 × 0.7 to 1 mm, with deep dorsal and ventral

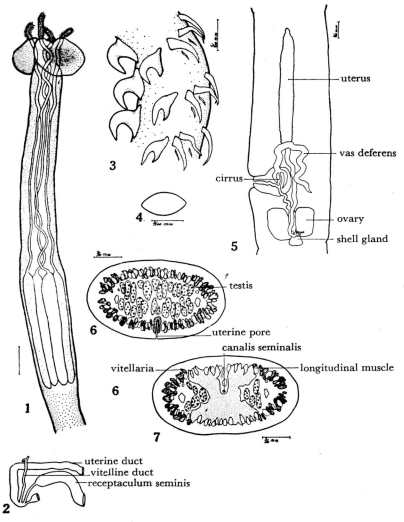

Fig. 12–2. *Tentacularia* found in elasmobranch fish. (From Hart, courtesy of the Transactions of the American Microscopical Society.)

grooves. The anterior 20% of the body is composed of small, immature segments, while the rest consists of mature and gravid segments. The entire worm may be 3 to 10 m or more in length. Most of the segments are wider than long, but most of the posterior gravid segments are longer than wide. Testes are numerous, small, rounded bodies situated in the lateral folds or the dorsal side of the proglottid. The vas deferens is much convoluted and proceeds anteriorly from the midplane at the beginning of the posterior third of the proglottid. It enlarges into a seminal vesicle and ends in a cirrus that is median in position, approximately between the first and second third of the body.

The ovary is symmetrically bilobed and is located in the posterior third of the proglottid with Mehlis's gland (see Chap. 9) between the lobes. Vitelline cells occupy the same lateral areas as do the testes, but the former are ventral to the latter. The uterus is in the form of a rosette and occupies the middle field of the proglottid. The vagina is a narrow, coiled tube, its coils interspersed with those of the uterus.

Diphyllobothrium latum eggs are broadly ovoid, yellowish to golden brown, operculated,

55 to 76 × 41 to 56 μm, and nonembryonated when voided with feces of the host.

The life cycle (Fig. 12–3) is described in Chapter 11. *D. latum* plerocercoids shed their entire bodies before starting to develop adult strobila. The shedding occurs piece by piece during a two-day period while the worm is migrating to the anterior third of the small intestine. In contrast, *D. dendriticum* and *D. ditremum* develop into adults without shedding the larval body. Adult worms may be found in the small intestine of man, pigs, dogs, cats, or many other mammals. Some of these parasites may have both birds and mammals as definitive hosts. Intermediate hosts are freshwater copepods and fish (Fig. 12–4).

Symptoms of infection are often absent. Digestive discomfort, anemia, abdominal pains, nervous disorders, or enteritis may occur in man. If a mature tapeworm is situated in the proximal part of the small intestine, the worm may make vitamin B_{12} inaccessible to the host and may incorporate most of the vitamin into its own body, thereby depleting the amount of available B_{12} essential for the formation of erythrocytes. Tapeworm anemia is the result. Prevention consists of thoroughly cooking fish before eating, or of freezing them at −10°C for 24 hours. Reservoir hosts such as bears should be prevented from eating garbage, and untreated sewage should be prevented from flowing into freshwater lakes and rivers.

Sparganosis (Chap. 11) may occur in man from infection by the plerocercoid larvae of three species of *Diphyllobothrium* from marine fishes in Japan. The normal hosts for these larvae are fish, frogs, snakes, or amphibious mammals. Southeast Pacific islanders occasionally used crushed fresh frogs as a poultice and have been known to acquire plerocercoids from this practice.

Spirometra mansonoides lives as an adult in dogs and cats, while its sparganum may be found in various vertebrates including man. The life cycle (Fig. 12–5) is similar to that of *Diphyllobothrium latum*. The number of plerocercoids in a single frog may range from one to 25, located mainly in the hind legs and abdomen. Oral transfer to cats and dogs results in the establishment of adult tapeworms in the intestine.

See von Bonsdorff[7] for a monograph on diphyllobothriasis in man.

In a remarkable series of observations and experiments, Justus Mueller serendipitously discovered that spargana of *S. mansonoides* cause mice and rats to grow abnormally large![29] Subsequent studies by Mueller and others revealed that the sparagana produce a growth factor which partially mimics the action of the host's pituitary growth hormone. Even more interesting are the recent results of Phares and Cox[32] suggesting that *S. mansonoides* "acquired" a gene for growth hormone from humans. Further data are required before this hypothesis can be substantiated or rejected. But if it is true, a whole new area of molecular parasitology will open. We may be not only what we eat, but also composed of genes exchanged with our parasites. Perhaps the results of Phares' group represent only partial sequence similarities in the DNA concerned, or are related to phenomena such as "onc" genes. However, in an era when transgenic mice are routinely produced in many laboratories, it is easy to postulate even the one rare event of acquisition of a gene from the host by a parasite, or vice versa. Are transposons involved?[47] Such new fields of study are now available to the curious.

Order Caryophyllidea

FAMILY CARYOPHYLLAEIDAE

Adults of these little worms live in freshwater teleost fishes, and all use species of tubificid or naidid oligochaetes as intermediate hosts. In annelid worms, they may be larval forms that have become sexually mature (pedogenetic). *Archigetes* (Fig. 12–6), a well-known genus, is apparently capable of maturing in tubificids as well as in fish, and some species do not require a fish host.[23]

Developmental as well as physiologic and ecologic factors may have a profound influence on the timing and duration of seasonal cycles. Genital organs are developed in the metacestode stages of *Caryophyllaeus laticeps*. The worms may spend six months or longer in the tubificid and only two months in fish. One of the factors, therefore, that could contribute to the observed periodicity is seasonal variation in the availability of infective larval stages.

Order Spathebothridea

These small worms were formerly included with the order Pseudophyllidea. They do not have true bothria or suckers. There is no external segmentation, but some internal proglotti-

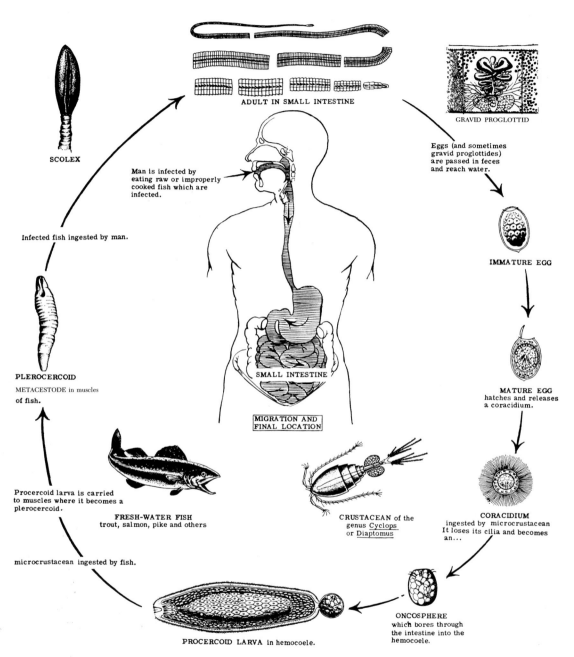

SCOLEX

ADULT IN SMALL INTESTINE

GRAVID PROGLOTTID

Man is infected by eating raw or improperly cooked fish which are infected.

Eggs (and sometimes gravid proglottides) are passed in feces and reach water.

Infected fish ingested by man.

IMMATURE EGG

PLEROCERCOID

METACESTODE in muscles of fish.

MATURE EGG hatches and releases a coracidium.

SMALL INTESTINE

MIGRATION AND FINAL LOCATION

Procercoid larva is carried to muscles where it becomes a plerocercoid.

FRESH-WATER FISH trout, salmon, pike and others

CRUSTACEAN of the genus Cyclops or Diaptomus

CORACIDIUM ingested by microcrustacean It loses its cilia and becomes an...

microcrustacean ingested by fish.

PROCERCOID LARVA in hemocoele.

ONCOSPHERE which bores through the intestine into the hemocoele.

Fig. 12–3. Life cycle of *Diphyllobothrium latum*. (Courtesy of the Naval Medical School, National Naval Medical Center.)

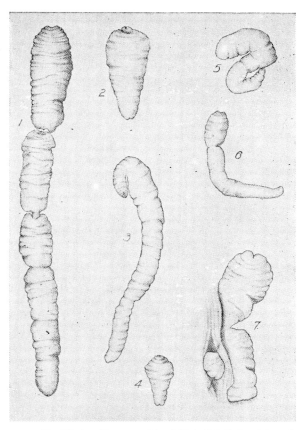

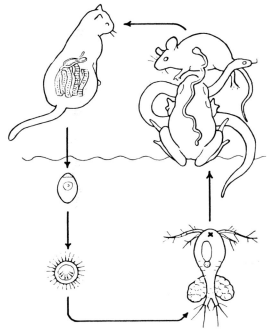

Fig. 12–5. Life cycle of *Spirometra mansonoides*, a pseudophyllidean tapeworm. The oval egg hatches and releases a coracidium, which is eaten by *Cyclops* sp. in which the procercoid develops. Various vertebrates, with the exception of fishes, become hosts to the procercoid stage by eating *Cyclops*. The definitive host, usually a cat, becomes infected by eating an intermediate vertebrate host. (From Mueller, J.F.: The biology of *Spirometra*. J. Parasitol., *60*:3–14, 1974.)

Fig. 12–4. Plerocercoids of *Diphyllobothrium latum*. *1*, Old preserved plerocercoid of *D. latum* showing deep constrictions. ×5. *2*, *3*, *4*, *5*, and *6*, Young plerocercoids of *D. latum* preserved while in flesh to show normal positions. ×11¼. *7*, The same, with part of the flesh of the fish still in position. × 11¼. (From Vergeer, courtesy of J. Infect. Dis.)

dation exists, and the uterus opens between the male and female apertures. Medullary testes occur in two oval lateral bands; the ovary is rosettiform or bilobed. The operculated eggs have thick shells. Adults have been described as neotenic procercoids, and they are found in the more ancient groups of fish.

Order Cyclophyllidea

Most of the important tapeworms of man and domestic animals belong to this order. The order is also well represented among adult tapeworms of birds, but not so well among those of amphibians and reptiles. Like the pseudophyllids, members of this order have a wide range in length. Some of them are only a few millimeters long, while others may reach a length of 30 meters. An important characteristic of the group is the presence of four well-developed suckers on the scolex. The anterior tip of the scolex usually projects as a *rostellum*, which may or may not bear hooks and may be retractable (Fig. 12–7). Proglottids are usually flattened, and the genital apertures are located marginally on one or both sides. The ovary is typically bilobed or fan-shaped, and the testes consist of scattered granules. The yolk gland is normally compact and lies posterior to the ovary. The gravid uterus may be branched or sac-like and contains the embryos enclosed by embryonic membranes. In some species, the uterus forms egg capsules containing one or several eggs. Almost all the species are hermaphroditic, but a few are dioecious. In some species, the sexes are completely separate (for example, *Dioecocestus*, a parasite of grebes),[10] whereas in others, the male organs disappear before female organs become functional, or the male system occupies

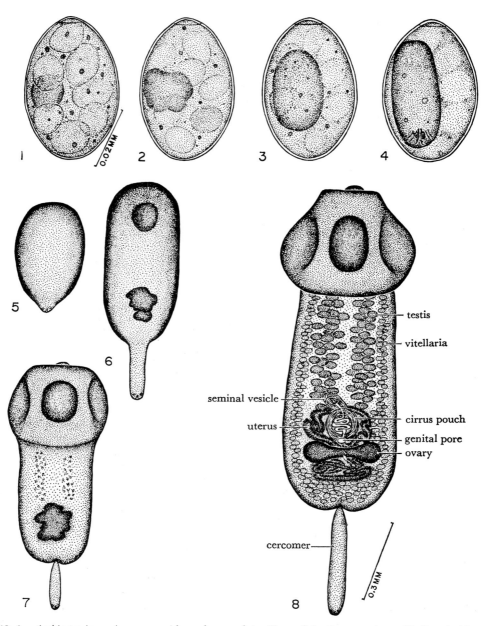

Fig. 12–6. *Archigetes iowensis,* procercoid nearly complete. (From Calentine, courtesy of J. Parasitol.)

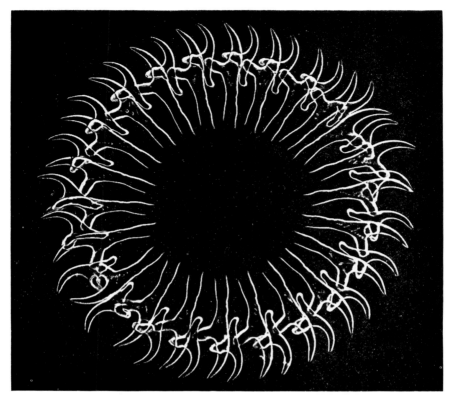

Fig. 12–7. Hooks on the scolex of the tapeworm *Taenia pisiformis*. (Courtesy of the American Institute of Biological Sciences. Drawing by D.W.C. Marquardt.)

one part of the strobila and the female organs another part (Fig. 12–8). Males of *Shipleya inermis* in birds are small and delicate and occupy an anterior position in the small intestine. Behind the male lies the more robust female, which may be hermaphroditic. Male and female worms usually occur in single pairs in nodules in the intestinal mucosa. The pairing assures cross-fertilization (see Coil[11]).

Arthropods, annelids, molluscs, and vertebrates serve as intermediate hosts, and amphibians (rarely), reptiles, birds, and mammals harbor the adult tapeworms.

FAMILY DAVAINEIDAE

The scolex of these small to moderately large tapeworms has hooks, suckers, and a cushion-shaped rostellum. An important species is *Raillietina (= Skrjabinia) tetragona*, a common tapeworm of domestic fowl. This worm may reach 25 cm in length. Larval stages occur in ants or in maggots of the housefly.

Raillietina bonini live in the pigeon. Eggs from the feces of a bird are eaten by a snail or slug in which the cysticercoid develops. Pigeons become infected by eating molluscs (Fig. 12–9).

Raillietina cesticillus is probably the best-known member of the family because it is a common tapeworm of poultry. Chickens, pheasants, guinea fowl, and wild birds are often infected. The adult worms may reach 130 mm in length and are about 2 mm wide. The four suckers are small, and 400 to 500 minute hooks encircle the scolex. Eggs are passed from the host with feces and are eaten by intermediate hosts, which may be one of several species of beetles or even the housefly. Birds become infected by eating insects containing cysticercoids. Figure 12–10 illustrates the stages in the life cycle. Figure 11–6 shows the oncosphere within its embryonic membrane. This tapeworm has been grown in bacteria-free (gnotobiotic) chicks without showing abnormal growth.

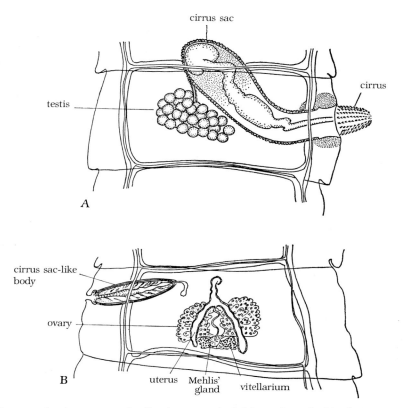

Fig. 12–8. The reproductive system of a dioecious cestode *Infula* sp. from the North American dowitcher, *Limnodrumus griseus hendersoni* Rowan. A, Proglottid of a male worm. B, Proglottid of a female worm. (From Burt: Platyhelminthes and Parasitism. London, St. Paul's House, 1970.)

Raillietina loeweni is a common parasite of the hare, *Lepus californicus melanotis*. The intermediate host is the harvest ant belonging to the genus *Pheidole*, which the hare inadvertently ingests while eating vegetation.

Davainea meleagridis, a small member of the family, is found in the turkey, *Meleagris gallopavo*. Mature specimens are only 5 mm long by 950 μm wide. The scolex is about 165 μm wide. When one remembers that the lower limit of vision with the unaided eye is about 100 μm, one can appreciate the difficulty encountered in hunting for such tiny tapeworms (Fig. 12–11).

FAMILY DILEPIDIDAE

Dipylidium caninum is worldwide in distribution and is common in the small intestines of dogs, cats, and other carnivores, but is rare in man. The few infections that do occur in man are usually in children. The tapeworm averages about 30 cm (10 to 70 cm) in length and can be recognized by the elongated, almond-shaped mature proglottids (Fig. 12–12). The rhomboidal scolex possesses an anterior projection, the rostellum, which is armed with several transverse rows of hooks and can be retracted into a rostellar sac. Below the rostellum are 4 prominent suckers. A mature proglottid (Fig. 12–12) contains 2 sets of reproductive organs with an opening on each side of the body. The uterus develops as a network of canals or cavities. Eggs (Figs. 12–13 and 12–14), 24 to 40 μm in diameter, occur in oval packets containing 5 to 20 eggs each. Ripe proglottids containing these packets, or balls of eggs, break loose from the strobila and look like active little worms about the size and shape of a pumpkin seed. When they reach the outside, they rupture, and the eggs may be ingested by larvae of the fleas, *Ctenocephalides canis* and *C. felis*, or of the human flea, *Pulex irritans*, or adult biting lice, *Trichodectes canis*. Within these insects, the eggs hatch, and larvae migrate to the body cavity. By the time the insect has reached maturity, the tape-

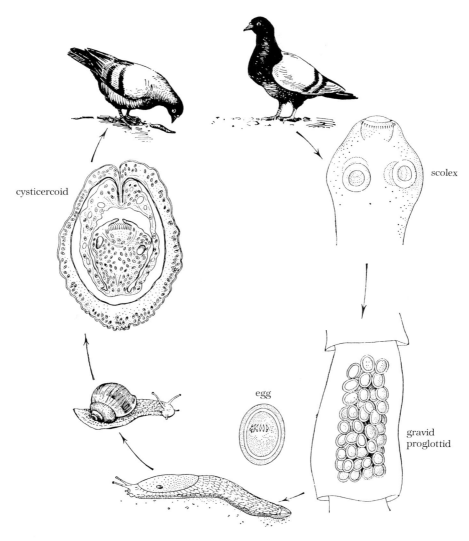

cysticercoid

scolex

egg

gravid
proglottid

Fig. 12–9. *Raillietina bonini,* a tapeworm of pigeons. The illustration shows two possible intermediate hosts. (From Joyeux and Baer. *In* Grassé: Traité de Zoologie, courtesy of Masson Editeurs.)

worm has developed into a cysticercoid stage or acanthacetabulo-pleurocercoid metacestode. Dogs and cats get the fleas or bits of them into their mouths; if swallowed, the parasite is carried to the intestine, where the cysticercoid stage matures into an adult tapeworm (Fig. 12–15). Children apparently get infected fleas or parts of them under their fingernails and become infected by putting their fingers into their mouths.

Symptoms in children are absent to mild. Diagnosis involves finding egg packets or entire proglottids in feces. Dogs and cats and their sleeping quarters should be kept as clean as possible, their bodies should be "wormed" often, and they should be treated frequently with insecticides.

FAMILY HYMENOLEPIDIDAE

Hymenolepis nana, (*Vampirolepis nana* is used by some authors) a common cosmopolitan species, is found in man as well as in rats and mice. Because the parasite averages only about 32 mm (usually 25 to 40 mm) in length, it is called the dwarf tapeworm (Fig. 12–16). The rostellum of *H. nana* is retractable, like that of *Dipylidium caninum,* but it possesses a single ring of hooks

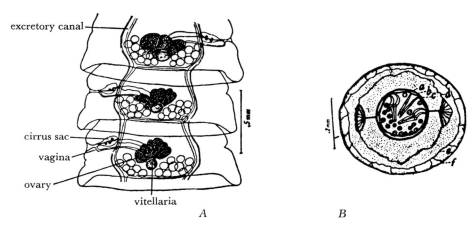

excretory canal

cirrus sac

vagina

ovary

vitellaria

A

B

Fig. 12–10. *Raillietina cesticillus.* A, Mature segments. B, Oncosphere. *(a)* Surface layer of oncosphere; *(b, c, d, e, f)* The five membranes of the oncosphere. (From Lapage: Veterinary Parasitology, courtesy of Oliver and Boyd.)

(Fig. 12–17). Proglottids are wider than long, and they contain one set of reproductive organs (Fig. 12–18).

The life cycle is unusual for tapeworms (Fig. 12–19). Eggs (Fig. 9–10) are 30 to 50 μm in diameter. They reach the outside and may be ingested by grain beetles, fleas, or other insects in which the oncospheres develop into tailed cysticercoids (Fig. 12–20). Often, these eggs are swallowed directly by humans or mice, and when oncospheres are liberated, the eggs develop into tailless cysticercoids in the intestinal villi. Thus, the intermediate host in the life cycle can be eliminated. Factors that cause larval release from eggs include bile salts, trypsin, and probably, pepsin. Cysticercoids mature, drop into the lumen of the small intestine, and develop into adult tapeworms in one to two weeks. The life span is short, and worms are eliminated.

The incidence of infection in man ranges from less than 1 to 28%. Worm numbers in one host may be high—7000 were taken from one human patient. Symptoms of infection, if any, are usually mild, but toxic reactions such as nervous disorders, sleeplessness, diarrhea, and intestinal pain may occur. Diagnosis is best made by finding eggs in stool specimens. Preventive measures include personal cleanliness, destruction of rats and mice, and a well-balanced diet to promote resistance to infection. Because the direct cycle of *H. nana* in humans may result in continuous "autoinfection," this parasite is of special concern to those receiving immunosuppressant drugs, for example, transplant recip-

ients, sufferers of colitis, as well as to victims of immunodeficiency diseases such as AIDS.

Hymenolepis diminuta (Fig. 11–1), the common species in rats, occasionally infects man. The incidence of infection in man is usually less than 1%, but in favorable localities (for example, a few areas of India) it may run as high as 6%. The maximal number of worms recorded from one man is 19. Worms average 45 cm in length, thus being considerably larger than *H. nana.* Size, however, is partly a function of the age of the rat host, but the basis for this relationship is not well understood (see Chap. 23). This tapeworm was established as a model system in the laboratory of Asa Chandler at Rice University more than 50 years ago. Because it is relatively easy to maintain in beetles and rats in laboratories, it has served as the "*Escherichia coli*" of cestodology. An entire volume has been devoted to reviews on the biology, biochemistry, and physiology of this relatively harmless but useful tapeworm.[4]

The tegument surface is composed of microvilli (**microtriches**, singular **microthrix**), similar to the microvilli constituting the brush border of mammalian intestinal cells, which provide a digestive-absorptive and secretive surface. In addition to transport functions, this border possesses hydrolases acting on phosphate esters and probably on monoglycerides. The phosphoesterase is not part of the mechanism involved in the mediated transport of hexose sugars in *Hymenolepis.* It may have a digestive function. Microtriches are also believed to aid both in maintaining the position of worms in

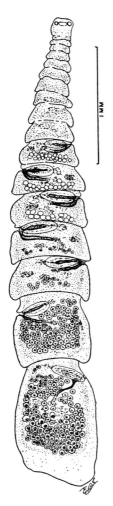

Fig. 12–11. *Davainea meleagridis,* a tapeworm from the turkey. Whole specimen, whole mount. (From Jones, after Fuhrmann. Proceedings of the Helminthological Society of Washington.)

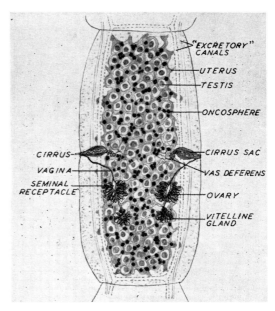

Fig. 12–12. *Dipylidium caninum,* mature proglottid. The uterus forms a network around the testes.

the gut and in migrational movements of these parasites.

Eggs of *Hymenolepis diminuta* (40 to 50 μm), unlike those of *H. nana,* have a sculptured shell and lack polar filaments.

"The results of the histochemical tests indicate that the outer envelope of *H. diminuta* oncospheres is built of protein with properties of 'soft' keratin, bound with minute quantities of polysaccharide. The underlying thin coat is composed of proteins containing a large proportion of basic amino acids. It is adhered to the next underlying inner envelope through a mucopolysaccharide-protein complex with strong swelling capacity in water. The embryophore is composed of pro-

teins, containing minute quantities of phosphoric acid residues."[27]

Adults of *H. diminuta,* in contrast to *H. nana,* have an indefinite life span and do not show aging. *H. diminuta* adults were maintained in rat hosts for 14 years by 13 successive surgical transplantations.[35] The worm obviously has a life span potentially longer than that of its host.

The life cycle requires an intermediate host in which cysticercoids develop. Many kinds of insects serve as this host, for example, grain beetles, earwigs, fleas, flies, dung beetles, and cockroaches. The intermediate layer of the cysticercoid body wall is a dynamic area containing a number of enzymes. The physiology and pattern of development of this metacestode is essentially similar to that of the indirect cycle of *H. nana.*

Hymenolepis diminuta in rats exhibits an anterior migration, in the small intestine, between 12 midnight and 6 A.M., and a posterior migration between 12 noon and 6 P.M..[45] See the discussion of site selection in Chapter 24. This circadian migratory behavior appears to be correlated with the periods of host feeding[36] (rats normally feed at night). Apparently, the migration is more likely to be a response to gut secretions stimulated by the entry of food than a response to the food itself. Investigations of

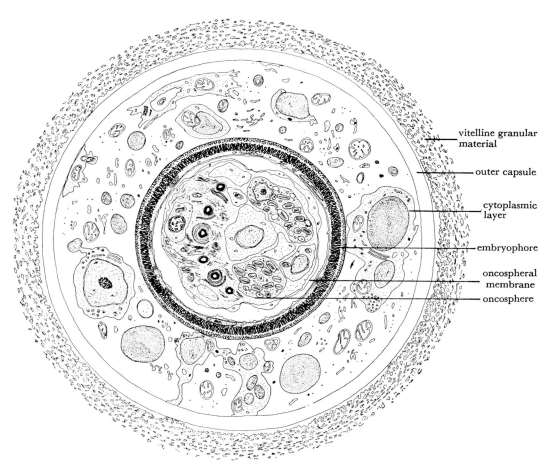

vitelline granular
material

outer capsule

cytoplasmic
layer

embryophore

oncospheral
membrane

oncosphere

Fig. 12–13. Schematic drawing of the mature egg of *Dipylidium caninum,* based on electron microscope photographs. (From Pence, courtesy of J. Parasitol.)

population distributions of other helminths in the vertebrate gut should include the possibility of similar behavior.

Symptoms of infection are mild or absent. In one study,[20] no effect on the host was measurable, and a proposal was made that *H. diminuta* and other cestodes should be regarded as endocommensals. Control of the tapeworm involves keeping rats and insects away from stored fruits and grains and being careful not to eat insect-contaminated food.

FAMILY TAENIIDAE

Taenia solium, the pork tapeworm (Fig. 12–21), as the common name indicates, is one that humans may acquire from eating uncooked pork. The incidence of infection in humans varies with the locality from less than 1% to about

8%, with a worldwide figure of 2 to 3%. The figures may be inaccurate, however, because this species may be confused with the beef tapeworm, *Taeniarhynchus saginatus* (formerly called *Taenia saginata*) (see discussion later in this chapter).

The adult pork tapeworm lives in the small intestine, ranges in length from 2 to 7 m, and has a scolex with a rounded rostellum. This structure is armed with large hooks that alternate with small hooks, and give the appearance of a double ring. Four prominent, round suckers are present. Microscopic, spine-like projections cover the body surface. The ovary possesses an accessory third lobe that is difficult to see in usual stained and mounted specimens. The gravid proglottids (Fig. 12–22) are longer than wide and contain a single set of repro-

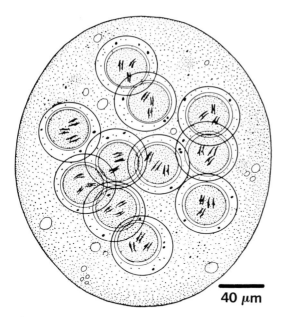

Fig. 12–14. *Dipylidium caninum,* packet of eggs.

ductive organs. The parasites are common in Europe, but are rare in parts of the Orient, the United States, and England. They are most common in the Middle East, Kenya, Ethiopia, and Mexico. Humans are the only definitive host. Camels, dogs, monkeys, and humans in addition to pigs may serve as intermediate hosts.

The life cycle of *Taenia solium* (Fig. 12–23) starts with a thick-walled, round egg averaging about 38 μm in diameter. It contains the characteristic embryo with its 3 pairs of hooks. Usually, the eggs remain in proglottids that become isolated from the rest of the strobila and pass from the body of the host. Pigs, man, dogs, or other animals ingest these eggs in contaminated food. Oncospheres are liberated in the small intestine, make their way through the gut wall to blood vessels, and are carried to all parts of the body. In various organs, especially muscles, larval worms leave the blood and develop into cysticerci or bladder worms (Figs. 12–24 and 12–25). Symptoms of cysticerci in humans might not appear for several years after infection, if at all. Organs affected are, in decreasing order of frequency, the cerebrum (Fig. 12–25), the meninges, the skeletal muscles, the cerebellum, and the heart. With central nervous system infection, the clinical symptoms vary widely and may include visual and psychic dis-

turbances, epileptiform fits, personality changes, and motor and sensory paralysis. Cysts may finally disintegrate, but usually, especially in muscles, they become calcified. A study of cysticercosis in swine and cattle revealed the location of the parasites to be within the lymphatic capillaries of muscles.[42] Muscles of hogs may sometimes become so filled with these parasites that the meat is called **measly pork.** Fluid within cysticerci is composed largely of host blood plasma. Humans usually acquire adult tapeworms by eating uncooked or poorly cooked infected pork. Bladder worms become evaginated in the small intestine, hooks and suckers enable them to become attached to the gut wall, and the worms develop to maturity. Humans can be infected directly by ingesting eggs in feces-contaminated food or water. Autoinfection can occur by putting fingers, similarly contaminated, into the mouth.

For the morphology and pathogenicity of bladder worms, see Slais.[41]

Symptoms of adult tapeworm infection may be absent, or there may be mild general reactions. Rarely, there is diarrhea, loss of weight, nervous symptoms, and even perforation of the intestinal tract. Reactions in man depend on the extent of infection and the location of the bladderworms. Cysticercosis of the brain obviously produces symptoms different from cysticercosis of the forearm muscles. Severe pathologic reactions may not occur until cysticerci degenerate and release toxins. Diagnosis of intestinal infection is based on finding proglottids or eggs in stool specimens and is confirmed by finding the scolex. If the scolex is not recovered, 4 to 6 months are required to be sure the entire worm is no longer present. Treatment for cysticercosis, other than surgical excision, has not been possible, although relatively new anthelmintics such as praziquantel may be helpful.[48] Prevention consists of thoroughly cooking all pork before eating it. Proper sewage disposal is important.

As is apparent from this discussion, bladder worms can be more serious to the host than can the adult worms. Some of these parasites were first discovered as larval stages and were thought to be adults of new species or genera, and so were given new names. It was later found that they were stages in the life cycles of other parasites, so two sets of names arose for the organisms. For example, *Cysticercus tenuicollis* is the thin-necked bladder worm of domestic ruminants, but a study of its complete life cycle showed it to be the larval stage of

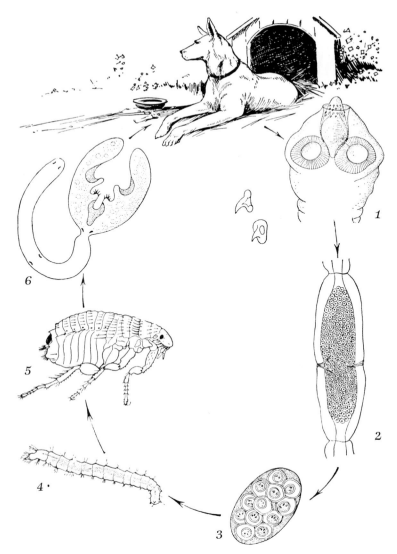

Fig. 12–15. Life cycle of *Dipylidium caninum*. *1*, Scolex with partially retracted rostellum. *2*, A gravid proglottid. *3*, An egg packet containing 12 eggs. *4*, A larval flea that eats the eggs. *5*, An adult flea. *6*, The infective cysticercoid stage found in the flea. *7*, A dog, the definitive host. (From Joyeux and Baer. *In* Grassé: Traité de Zoologie, courtesy of Masson Editeurs.)

Taenia hydatigena of dogs. *Cysticercus ovis*, causing sheep "measles," is the larva of *Taenia ovis* of dogs.

The larval stage of the tapeworm, *Taenia solium*, is only one of several types of bladder worms of pigs. These larvae have also been reported from other domestic mammals. This parasite usually measures about 5 × 10 mm when fully mature and infective to man. Sometimes, the cysticerci are so numerous as to occupy more than half the total volume of a piece of

flesh (Fig. 12–24). Worms are characteristically located in the connective tissue of striated muscles, but they may be found in any organ or tissue of the body.

Various authorities disagree as to the site of "preference." *Cysticercus bovis* in cattle are found most frequently in the masseter muscle, the tongue, and the muscles of the foreleg and heart. A tissue reaction may occur in these muscles. Belding[6] lists possible infected muscle tissue in the following decreasing order of pref-

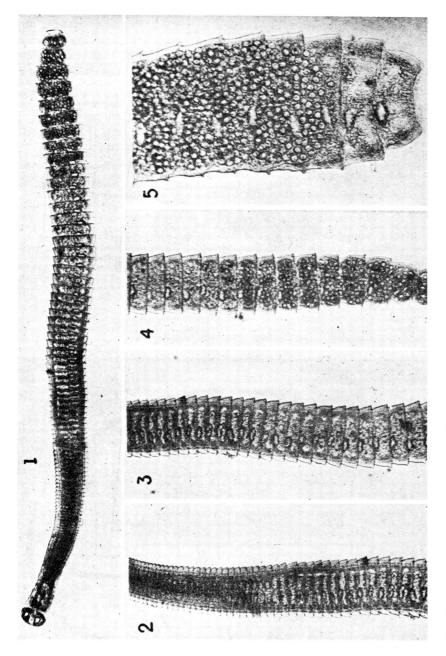

Fig. 12–16. *Hymenolepis nana.* (Originals of Kouri). *1,* Complete specimen. ×26. *2, 3,* and *4,* Anterior, center, and posterior thirds of the parasite, respectively. ×40. *5,* Posterior fourth of the parasite. ×100. The gravid segments are filled with eggs. Most of the caudal segments are completely emptied. The third from last has partially lost its ova. (From Gradwohl and Kouri: Clinical Laboratory Methods and Diagnosis, courtesy of C.V. Mosby.)

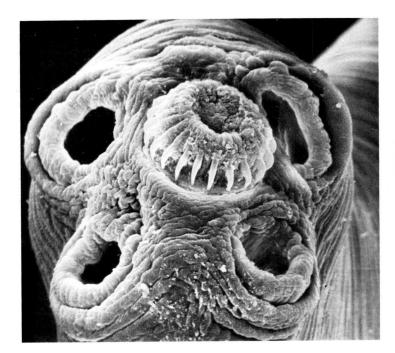

Fig. 12–17. *Hymenolepis nana.* Scolex showing four suckers and rostellum with its row of hooks. (Courtesy of Holm and Schulz. Bayer AG Leverkusen.)

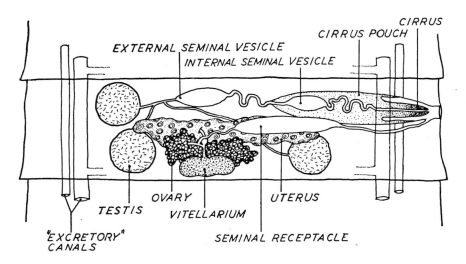

Fig. 12–18. *Hymenolepis nana,* mature proglottid.

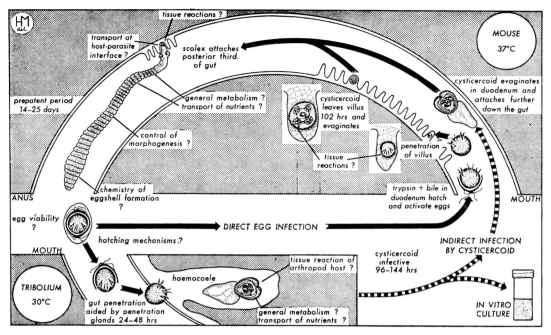

Fig. 12–19. Life cycle of the rodent cestode, *Hymenolepis nana,* and the physiologic problems associated with it. (From Smyth: The Physiology of Cestodes. University Reviews in Biology, courtesy of Oliver and Boyd.)

erence: tongue, neck, shoulder, intercostal, abdominal, psoas, femoral, and posterior vertebral. No definite symptoms of bladder worm infection appear in animals, and there is no known practical method for removal of the parasite from swine.

Taeniarhynchus saginatus, the cosmopolitan "beef tapeworm," is also called the "unarmed tapeworm" because the scolex does not possess hooks (Fig. 12–26). It is more prevalent in man than is *Taenia solium.* It is longer than the pork species, usually measuring from 5 to 10 m, and it lives in the small intestines. One extreme specimen reached 25 m in length, about 3 times as long as the entire human intestine. The entire strobila possesses from 1000 to 2000 proglottids. Table 12–1 summarizes major differences between the 2 species.

The life cycle is essentially similar to that of *Taenia solium* and starts with an almost identical egg (Fig. 12–27). Ova are expelled only from detached proglottids that migrate to the perianal area. Intermediate hosts are cattle, buffalo, or other ungulates. Heavy infections cause the "measly" condition to occur primarily in jaw muscles and in the heart. Humans are probably the only definitive host, acquiring the infection

by swallowing the cysticerci in uncooked, infected beef.

Eggs of various helminths, for example, *Taeniarhynchus saginatus,* can be carried by flies both internally and externally.[16] Flies, such as the housefly, *Musca domestica,* ingest only the smaller eggs. These eggs remain infective after passage through the fly gut. Other invertebrates, for example, beetles and earthworms, may also carry helminth eggs. Symptoms and treatment are the same as for the pork species. For a review of taeniasis and cysticercosis, see Pawlowski and Schultz[31] and Márques-Monter.[26]

Taenia pisiformis (about 500 mm long) possesses a life cycle much like that of *T. solium.* The intermediate hosts are rabbits, rats, squirrels, or some other rodent that might be eaten by canines. Oncospheres usually are found in the liver of intermediate hosts, whereas the infective stage, the cysticercus, inhabits the peritoneal cavity. The rodents are eaten by dogs, cats, wolves, foxes, and other carnivores that thus become definitive hosts.

Taenia taeniaeformis occurs in domestic and wild cats. Adult worms apparently do little damage, but the metacestode may cause con-

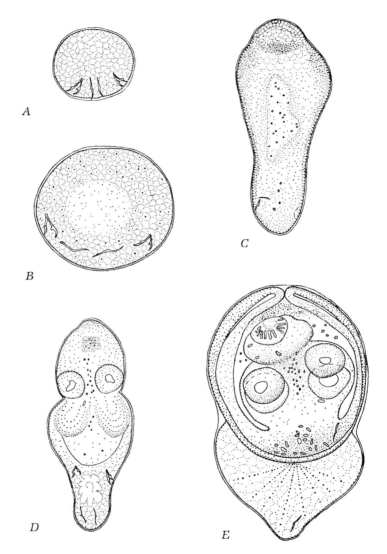

Fig. 12–20. Five stages in the growth of *Hymenolepis nana*. *A*, Stage 1. Solid oncosphere, showing paired hooks and external membrane. *B*, Stage 2. Cavity and dispersal of oncosphere hooks. *C*, Stage 3. Two body divisions, elongation of cavity and anterior zones of sucker and rostellum primordia. *D*, Late stage 3. Process of withdrawal, showing separation of "neck" tissue, which will become layer immediately enveloping scolex. *E*, Stage 4. Fully formed cysticercoid. Withdrawn scolex, enlarged rostellum with partly developed hooks, and clearly demarcated suckers. All drawings freehand; relative size indicated by oncosphere hooks (10 to 12 μm). (From Voge and Heyneman, courtesy of the University of California Publications in Zoology.)

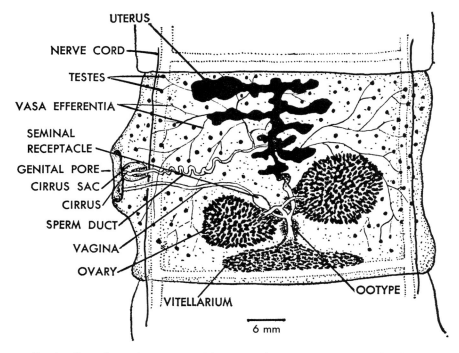

UTERUS
NERVE CORD
TESTES
VASA EFFERENTIA
SEMINAL
RECEPTACLE
GENITAL PORE
CIRRUS SAC
CIRRUS
SPERM DUCT
VAGINA
OVARY
VITELLARIUM
OOTYPE

6 mm

Fig. 12–21. *Taenia solium,* the pork tapeworm. Mature proglottid. (From Noble and Noble: Animal Parasitology Laboratory Manual, Lea & Febiger.)

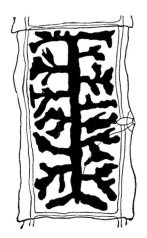
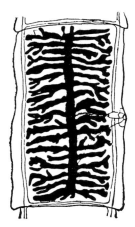

Fig. 12–22. *Taenia solium* left and *Taeniarhynchus saginatus* right. Gravid proglottids showing typical uterine patterns. Considerable variation in numbers of uterine branches occurs in both worms. (From Roudabush: An Aid to the Diagnosis of Helminths Parasitic in Humans, courtesy of Wards Natural Science Establishment.)

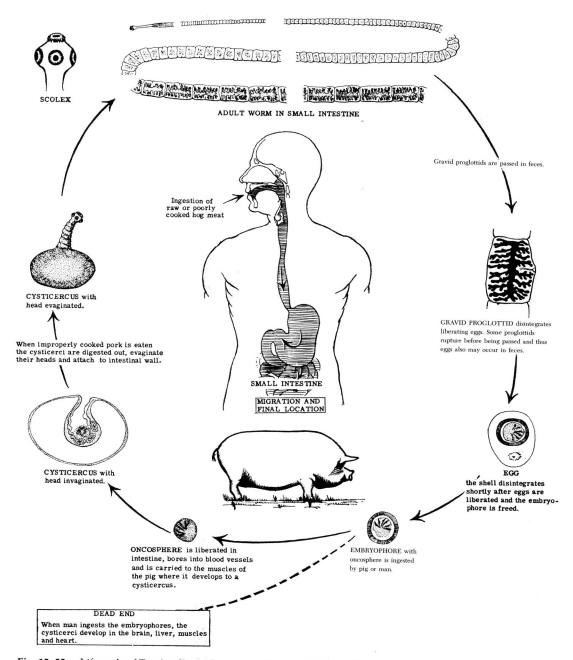

SCOLEX

ADULT WORM IN SMALL INTESTINE

Gravid proglottids are passed in feces.

Ingestion of
raw or poorly
cooked hog meat

CYSTICERCUS with
head evaginated.

When improperly cooked pork is eaten
the cysticerci are digested out, evaginate
their heads and attach to intestinal wall.

GRAVID PROGLOTTID disintegrates
liberating eggs. Some proglottids
rupture before being passed and thus
eggs also may occur in feces.

SMALL INTESTINE

MIGRATION AND
FINAL LOCATION

CYSTICERCUS with
head invaginated.

EGG
the shell disintegrates
shortly after eggs are
liberated and the embryo-
phore is freed.

ONCOSPHERE is liberated in
intestine, bores into blood vessels
and is carried to the muscles of
the pig where it develops to a
cysticercus.

EMBRYOPHORE with
oncosphere is ingested
by pig or man.

DEAD END

When man ingests the embryophores, the
cysticerci develop in the brain, liver, muscles
and heart.

Fig. 12–23. Life cycle of *Taenia solium*. (Courtesy of the Naval Medical School, National Naval Medical Center.)

Fig. 12–24. *Cysticercus fasciolaris (= Cysticercus tae-niaeformis)* in an experimental rat liver. (Courtesy of Dr. Robert E. Kuntz.)

siderable harm to the rat or mouse host in which it is normally found. It infects the rodent liver and there becomes encapsulated. A serious cancer-like growth apparently may rise from this encapsulation.

Taenia hydatigena in sheep produces resistance to the liver fluke, *Fasciola hepatica*. Infection of sheep with the larval stage, called *Cysticercus tenuicollis,* 12 weeks before *F. hepatica* challenge generates a high level of protection against the trematode. Sheep do not normally acquire significant immunity to this fluke.

Multiceps multiceps is normally found in dogs and wolves. The worm is similar to *Taenia solium* in appearance, and some investigators have placed it in the genus *Taenia*. When the eggs are eaten by ruminant animals, such as sheep, cattle, and horses, or related wild mammals, the metacestodes develop into the bladder stage, each of which contains many scoleces (Fig. 12–28). This stage is a **coenurus,** which resembles a brood capsule of *Echinococcus granulosus* (see Fig. 12–33). Although it may develop in almost any tissue, it often occurs in the brain, causing giddiness, thus the common name "gid worm" or "gid tapeworm" (sometimes called *Coenurus cerebralis*). This worm is most prevalent in sheep, but has rarely occurred in man, probably owing to accidental ingestion of eggs from dog feces. Prognosis in man is grave, but praziquantel may help. Dogs acquire the adult tapeworm by eating infected parts of sheep.

Echinococcus is a genus of taeniids that includes tapeworms whose metacestode stages are amongst the most serious parasites of hu-

mans. The metacestodes are called *hydatids,* and the infections in humans are properly termed **hydatidosis** or hydatid disease. The term echinococcosis should be reserved for infections with the adult worm in Carnivores, but not all authors follow this convention. Hydatidosis is considered to be one of ten "great neglected diseases" of humans, because there is as yet no sure treatment and no vaccine, and the disease is often fatal.

There are four species in the genus, but two of the species are responsible for nearly all of the disease in humans. These two species, *E. granulosus* and *E. multilocularis,* produce hydatids of different types and morphology. *Echinococcus granulosus* forms a "unilocular" hydatid (a cyst with "one chamber" or locule), whereas *E. multilocularis* forms a hydatid with "many chambers," and is often called "alveolar hydatidosis," implying many chambers or holes. Figure 12–29 illustrates the difference. Because these descriptive terms are not perfect (e.g., a unilocular cyst of *E. granulosus* may contain daughter cysts!), and the shape of an *E. granulosus* cyst need not be round, resulting from the confines of surrounding tissue, or rupture and outgrowth of a "bud" on a unilocular cyst, care must be taken when attempting to diagnose hydatids to species. Gross morphology can be deceiving.

These parasites are **Zoonoses,** and the hosts and geographic distribution are somewhat different in the two major species. *Echinococcus granulosus* is more widespread throughout the world, whereas *E. multilocularis* is primarily holarctic in distribution. For convenience of discussion, as well as application of control measures, the epidemiology of *E. granulosus* may be divided into two major types: *pastoral* and *sylvatic,* implying "domesticated" and "in the wild," respectively. The definitive hosts are carnivores, and predation is an essential element of the life cycles, except for accidental infections, the worst of which, occur in humans. The *pastoral* cycles include intermediate hosts such as domestic sheep, cattle, horses, camels, and pigs, with dogs as the most important definitive host, but other wild canines may be included. The *sylvatic* cycle includes game animals such as deer, moose, elk, caribou, reindeer, and the wolf as the predator and definitive host. In other areas of the world the cycle may include dingos and kangaroos, jackal and deer, or man and hyena. The epidemiology of *E. multilocularis*

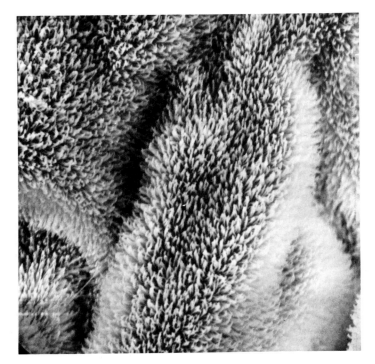

Fig. 12–25. Cysticercus of *Taenia solium* from the human brain. ×4000. The irregular body surface is covered with minute spines. (Electron micrograph courtesy of Dr. Marietta Voge.)

is mostly sylvatic and primarily involves microtine rodents such as voles or mice as intermediate hosts and foxes as the primary definitive host, but dogs, wolves, coyotes, and rarely, cats, may be involved. A description of the two major species follows.

Echinococcus granulosus causes one of the most serious larval tapeworm infections in man. It is called the **hydatid worm** (Fig. 12–30) because it forms hydatid cysts in various organs. Normal hosts for the adult parasite are dogs in which hundreds of the worms may occur in the small intestine. Wolves and jackals also harbor the adult worm, and in some areas foxes are

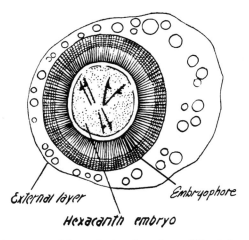

External layer   *Embryophore*

Hexacanth embryo

Fig. 12–27. Schematic drawing of egg of *Taeniarhynchus saginatus*. (From Gradwohl and Kouri: Clinical Laboratory Methods and Diagnosis, courtesy of C.V. Mosby.)

Fig. 12–26. *Taeniarhynchus saginatus,* scolex. The beef tapeworm. The line on the right represents 0.22 mm.

Table 12–1.   Differential Diagnosis

Character	*Taenia solium*	*Taeniarhynchus saginatus*
Rostellum	Present, armed	Absent
Testes	375–575, confluent posterior to vitellarium	880–1200, not confluent posterior to vitellarium
Cirrus pouch	Extends to excretory vessels	Does not extend to excretory vessels
Ovary	3 lobes	2 lobes
Vaginal sphincter	Absent	Present
Ova	Spherical	Oval
Gravid segments	Do not leave host spontaneously	Leave host spontaneously
Uterine branches	Less numerous	Numerous

The most reliable and easily assessed criteria for the differentiation of the 2 species in man are:
1. The presence or absence of an armed rostellum
2. The number of ovarian lobes
3. The presence or absence of a vaginal sphincter
(Adapted from Verster. Courtesy of Zeitschrift fur Parasitenkunde.)

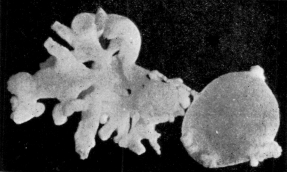

Fig. 12–28.   Two types of coenurus of *Multiceps.* (From Russell, P.F., Jung, R.C. and Cupp, E.W.: Clinical Parasitology. 9th Ed. Philadelphia, Lea & Febiger, 1984.)

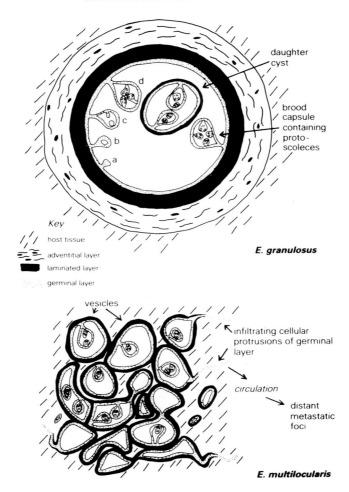

Fig. 12–29. Comparison of unilocular hydatid cyst of *E. granulosus* with multilocular (vesicular) hydatid of *E. multilocularis*; a,b,c,d represent stages in the development of a brood capsule, eventually forming a daughter cyst. (From R.C.A. Thompson, Ed. The Biology of *Echinococcus*. Courtesy of George Allen and Unwin, Ltd.)

probably infected. The parasite is especially common in sheep-raising areas such as Australia, parts of South America, and the Middle East. The entire worm consists of 3 segments, as well as the scolex, and it is only 3 to 5 mm long. The scolex has a retractable rostellum armed with a double circle of hooks and 4 suckers. Only a single ripe proglottid occurs at any one time and it, although tiny, resembles that of *Taenia solium*. The tegument is covered by minute spines (Fig. 12–31) and possesses sensory endings (Fig. 12–32). Apparently, adult worms do little damage to their hosts.

The fine structure of the nervous system of *Echinococcus granulosus* was described by Morseth[28] as follows: "Lateral nerve trunks are composed of unmyelinated fibers without a cellular sheath. Mitochondria and vesicles of several sizes occur in the fibers. Clear vesicles accumulate on one side of synaptic junctions, whereas dense vesicles, possibly neurosecretory, occur at a presumed neuromuscular junction and in nerve processes."

Self-insemination appears to be the normal process of sperm transfer in these worms. The important part of the life cycle (Fig. 12–33), so far as its pathogenicity is concerned, is the metacestode, which may occur in man, cattle, sheep, camels, horses, moose, deer, pigs, and rabbits, for example. These hosts ingest *Echinococcus granulosus* eggs, which are almost identical to those of *Taenia*. The oncospheres are

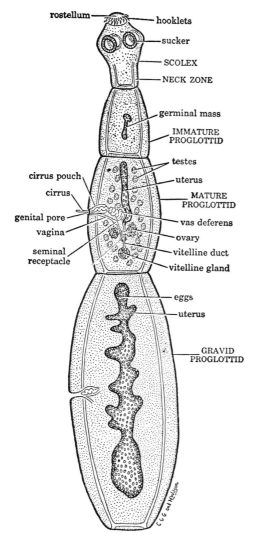

rostellum
hooklets
sucker
SCOLEX
NECK ZONE
germinal mass
IMMATURE PROGLOTTID
testes
cirrus pouch
uterus
cirrus
MATURE PROGLOTTID
genital pore
vas deferens
vagina
ovary
seminal receptacle
vitelline duct
vitelline gland
eggs
uterus
GRAVID PROGLOTTID

Fig. 12–30. *Echinococcus granulosus,* surface view of whole worm. (From Brown: Selected Invertebrate Types, courtesy of John Wiley and Sons.)

liberated in the intestine and, as in other species, enter mesenteric veins and make their way to various organs, especially lung and liver. The spleen, kidneys, heart, brain, or even bone may be infected. In these organs, the oncosphere develops into a spherical cyst or hydatid that may grow to a diameter of 15 cm (6 inches).

The size and shape of the cyst may be limited by the organ or the space in which it develops. The cyst's inner lining is germinative and gives rise to scoleces, brood capsules, and daughter cysts (Fig. 12–34). Brood capsules and scoleces

may become free from their attachments and may form a loose mass on the floor of the cyst; this mass is known as **hydatid sand.** The cyst is filled with **hydatid fluid.** The fluid in sheep hydatids is a mixture of host serum components and parasite antigens. Host IgG molecules have been found in hydatid cysts from several species of mammals. As many as 2 million scoleces may be present in a large cyst and may contain 2 liters of fluid. Old, enormous cysts have been reported to contain over 15 liters of fluid. These fluid-filled cysts are **unilocular** and may persist for years. They may develop within bone, thus forming osseous hydatid cysts. Figure 12–35 shows a cyst-like mass larger than a fist, taken from the abdomen of a woman in New York. The small (0.5 to 2.5 cm in diameter) echinococcal cysts it contained are illustrated in Figure 12–36.

"The hydatid cyst grows at a rate of 1 mm per month, or approximately 1 cm per year. Clinical symptoms do not usually develop until the cyst is at least 10 cm in diameter. Physical signs of a tumor mass do not usually develop until the cyst reaches a diameter of 20 cm. Symptoms of hydatidosis are due to two factors: localization with mechanical effects brought about by the space-occupying growing cyst, and the generalized toxic and/or allergic reactions secondary to absorption of the toxic products of the parasite."[33]

Heavy infections may result in severe tissue damage and death. See Pool and Marcial-Rojas,[33] and Thompson[46] for details.

In heavily endemic areas, 50% of dogs are infected with adult worms, and up to 90% of sheep and cattle and 100% of camels may be infected with hydatid cysts. The incidence of hydatid disease (echinococcosis or hydatidosis) in man may be as high as 20% in a few areas in South America, but it is usually much lower.

Hydatid disease is caused by at least nine different biologic strains of *Echinococcus granulosus* and several other species of *Echinococcus.* Although clinical manifestations and host responses differ, serologic and other methods for diagnosis of the disease are effective.

Damage by any cyst is, of course, related to its size and location. Simple cysts seem to do little harm to animals, although pressure on surrounding organs may be of consequence. Rupture of a cyst in man may cause allergic reactions such as rash or wheals, redness of the skin, fever, shortness of breath, cyanosis, vomiting, diarrhea, circulatory shock, and sudden death. Intravenous injection of sheep hydatid cyst

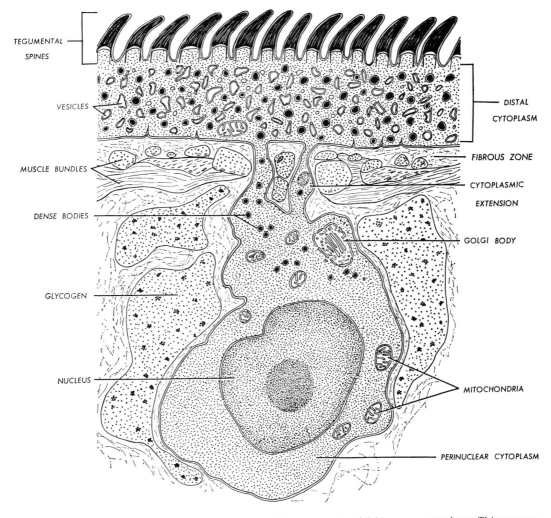

TEGUMENTAL SPINES

VESICLES

MUSCLE BUNDLES

DENSE BODIES

GLYCOGEN

NUCLEUS

DISTAL CYTOPLASM

FIBROUS ZONE

CYTOPLASMIC EXTENSION

GOLGI BODY

MITOCHONDRIA

PERINUCLEAR CYTOPLASM

Fig. 12–31. Cellular arrangement of the tegument of the protoscolex of *Echinococcus granulosus*. This arrangement is the same in the adult form. (From Morseth, courtesy of J. Parasitol.)

fluid into animals causes similar cardiovascular responses. Enlarging cysts may destroy bone or may impair the normal functioning of other organs. Migrating cysts can be serious. Although the adult worms do not live long, the cysts may remain alive for many years. Host tissue reactions, toxemia, eosinophilia, pressure effects, obstruction of blood vessels, and other factors indicate the presence of parasites. Work with *Echinococcus multilocularis* in mice[1] suggests a combination of humoral and cell-mediated immunity factors in host resistance to echinococcosis.

Treatment for infection by cysts of *E. granulosus* is surgical excision, and even this procedure is often unsuccessful or impractical. Recently, trials with benzimidazoles suggest that chemotherapy may eventually be possible, if enough of the right drug can be delivered to the parasite.[12] Prevention consists mainly of keeping dogs free from infection and avoiding accidental ingestion of tapeworm eggs. Obviously, the best way to prevent dogs from becoming parasitized by those worms with intermediate hosts is to prevent dogs from eating the intermediate hosts. To prevent man from becoming infected with the cysts, personal hygiene is important. Keep dogs clean; do not pet infected dogs; never allow a dog to "kiss" your face.

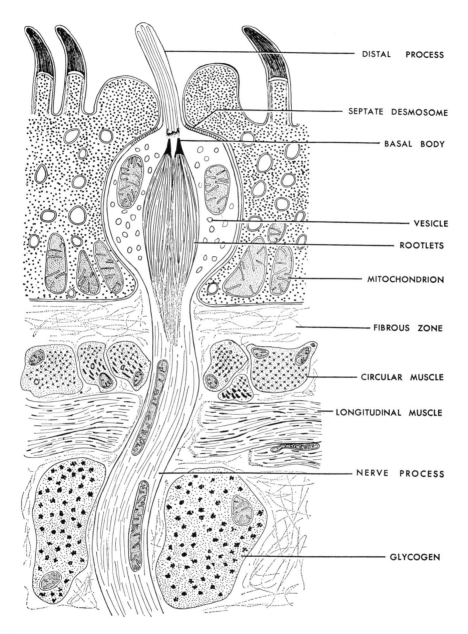

Fig. 12–32. Longitudinal section through a sensory ending in the tegument of *Echinococcus granulosus*. (From Morseth, courtesy of J. Parasitol.)

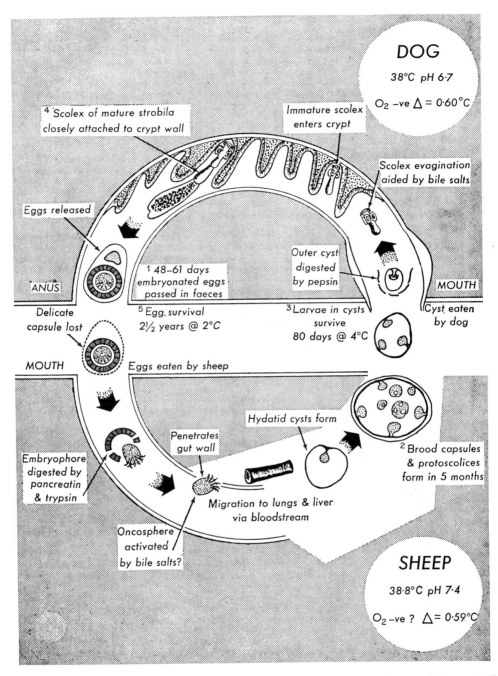

Fig. 12–33. The life cycle of *Echinococcus granulosus* and some of the physiologic factors relating to it. (From Smyth, courtesy of Academic Press.)

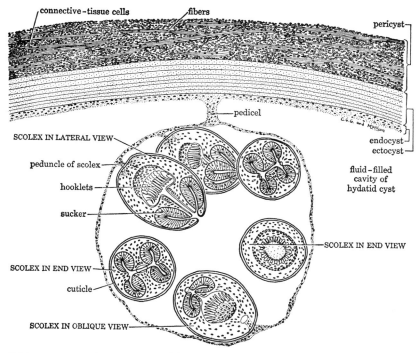

Fig. 12–34. *Echinococcus granulosus,* portion of the hydatid. (From Brown: Selected Invertebrate Types, courtesy of John Wiley and Sons.)

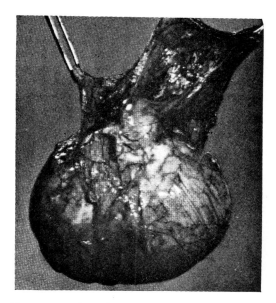

Fig. 12–35. An *Echinococcus* cyst larger than a fist. (From Bohrod, courtesy of Med. Radiol. Photography.)

Fig. 12–36. Small cysts taken from the mass illustrated in Figure 12–35. (From Bohrod, courtesy of Med. Radiol. Photography.)

Echinococcus multilocularis produces an even more serious condition because the cyst becomes irregular, filled with connective tissue and gelatinous masses, and grows like a malignant tumor with metastasis. This type is a **multilocular** or **alveolar** cyst. The tapeworm, found chiefly in Europe and northern Asia, looks much like a small *E. granulosus*. In length it ranges from 1.4 to 3.4 mm. The testes lie posterior to the cirrus sac, and the uterus does not possess lateral pouches. Although dogs or cats may harbor the adult worm, the principal host is the fox. Intermediate hosts are usually rodents in which the liver may become infected. In humans, unless the alveolar hydatid is small, confined, and without metastasis, surgical excision is most often unsuccessful. But early diagnosis, surgery where appropriate, coupled with aggressive, monitored chemotherapy may extend the life of the victim.[12] In humans, the germinal membrane may not produce scoleces, making histologic diagnosis difficult, especially differentiation from tumors. The germinal membrane of the alveolar hydatid is a deadly but fascinating tissue, worthy of further research. Ali-Khan et al.[2] have made some useful observations on the role of these germinal cells in metastasis. What could be useful, would be adaptation of these cells (or tissue) to in vitro culture, where they could be used to assess effects of drugs, antibodies, and the like.

For a review of the biology of hydatid organisms, see Smyth,[43] and Thompson.[46]

FAMILY ANOPLOCEPHALIDAE

Members of this family possess scoleces without hooks. Female reproductive organs may be single or double. The parasites live in intestines of birds, herbivorous mammals, and primates. The cysticercoid stage occurs in oribatid mites and insects. A few genera, for example, *Avitellina*, lack vitellariae.

Adult worms that live in intestines of domestic and wild ruminants include the genera *Anoplocephala*, *Moniezia*, and *Thysanosoma*. Rabbits, hares, and other rodents may harbor *Andrya* or *Cittotaenia*. Other genera may be found in birds, and *Bertiella* is a parasite of monkeys, apes, and occasionally, man.

Anoplocephala perfoliata, *A. magna*, and *Paranoplocephala mamillana* are cosmopolitan tapeworms in horses and other equine animals. Heavy infections may result in perforation of the intestine and death of the host animal. Intermediate hosts are oribatid mites.

Moniezia expansa (Fig. 12–37) is a common tapeworm of sheep. Worm eggs in host feces contaminate pastures. Minute, free-living oribatid mites creep over the soil, usually in the evening, and probably ingest the worm eggs accidentally with their natural food. Within two to five months, the young tapeworms have developed into the cysticercoid stages in the mites.

Sheep eat the mites, which often cling to forage grass. Cysticercoids are released in the sheep small intestines, where they mature in about 30 days. Adult *M. expansa* reach a length of 600 cm (20 feet). The mature proglottids are wider than long.

Thysanosoma actinioides is found in domestic and wild ruminants and develops in psocid lice (order *Psocoptera*). These small insects may be found on pasture vegetation.

FAMILY LINSTOWIIDAE

These worms are small to medium in size. The scolex is unarmed. They primarily inhabit insectivorous animals, and the metacestodes occur in beetles. Because of the unarmed scolex and other features, the group is sometimes placed as a subfamily of the Anoplocephalidae. An important genus is *Inermicapsifer*. *I. madagascariensis* is unusual in that a few cases of human infections have been reported, although this species normally parasitizes rats. The adult is about 40 cm in length. The life cycle is unknown. *Oochoristica* occurs in reptiles and mammals. *O. ratti*, in rats and mice, uses various insects as intermediate hosts.

FAMILY MESOCESTOIDIDAE

Members of this family are found as adults in birds and mammals. They are small to medium in size. The holdfast has four prominent suckers, but no rostellum. The genital aperture is median and on the ventral surface of the body. There is no uterine pore. The uterus may be present, or it may be replaced by a **parauterine organ.** This organ, which contains the eggs, is formed by parauterine cells arising in medullary tissue between anterior margins of the ovarian lobes. The cells form a reticular syncytium, and the organ is apparently involved with the developmental process of embryos. The cysticercoid stage develops in invertebrates. When rodents, lizards, or other vertebrates ingest these invertebrates, the cysticercoids may develop

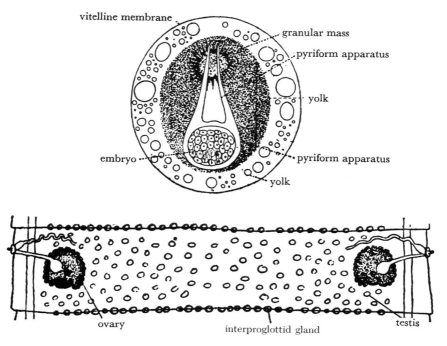

Fig. 12–37. *Moniezia expansa.* Proglottid (after Mönnig) and egg (after Railliet). (From Lapage: Veterinary Parasitology, courtesy of Oliver and Boyd.)

into tetrathyridia. A tetrathyridium is flat, broad anteriorly, and tapering posteriorly. The invaginated portion, including the scolex, may appear snake-like. Active forms may be motile and may assume various shapes. Vertebrates containing these modified cysticerci are second intermediate hosts.

PHYSIOLOGY OF CESTODES

The physiology and biochemistry of cestodes are similar to those of trematodes. There is a linear gradient in metabolic rate along the strobila, so that observations made on proglottids at one point in the body may differ from those on other proglottids.

Numerous species of cestodes and some trematodes contain large amounts of inorganic substances located in noncellular **calcareous corpuscles,** which are scattered throughout the parenchyma, especially in larval stages. This material is composed of concentric lamellae containing an organic base that includes DNA, RNA, glycogen, protein, a hyaluronic-type polysaccharide, and alkaline phosphate. Inorganic components are mainly calcium, magne-sium, phosphorus, carbon dioxide, and trace elements such as aluminum, cadmium, iron, magnesium, and sodium. The function of the calcareous corpuscles is obscure, but they probably provide a buffering action that protects the parasite from the harmful effects of host secretions. They may also be reserve centers for inorganic ions required especially for the high level of metabolic activity in the parasite's early larval stages.[44]

Carbohydrate metabolism in adult tapeworms is predominantly anaerobic. Glucose and galactose are the only carbohydrates known to be metabolized by cestodes, although some species absorb other mono- and disaccharides. Sugars are moved across the tegumental membrane into the body by an active transport mechanism. Glucose represents the major source of fuel for metabolic energy and is stored in the form of glycogen.

Seasonal changes in the behavior of tapeworms and in their carbohydrate content may be correlated with the feeding behavior of the host. The imposition of the dynamics of host gut physiology on the specific carbohydrate requirement of cestodes may be of importance in the distribution of the parasites and in the selection of species of parasites that appear and prosper in a particular host.

Respiration is primarily anaerobic, as noted, and energy is produced primarily through the Embden-Meyerhof glycolytic pathway. Instead of using the classic series of reactions yielding lactic acid as the final excretory product, however, the metabolic intermediary, phosphoenolpyruvate (PEP), is converted to oxaloacetate (via the CO_2-fixing enzyme PEP carboxykinase), which in turn is reduced to malate. The fixation of carbon dioxide is accomplished by the formation of high-energy triphosphates (GTP, ITP). In the absence of CO_2, glucose uptake and glycogen synthesis are inhibited. Malate can also enter mitochondria, in which it is further converted to fumarate and, finally, reduced to succinic acid, which is excreted. Reoxidation of nicotinamide adenine dinucleotide (NADH) during the fumarate-to-succinate reduction also results in the production of adenosine triphosphate.

We have noted that oxygen is present in the host gut, and under appropriate conditions, some cestodes may be able to consume this oxygen. For example, *Moniezia expansa* and *Taenia hydatigena*, are capable of aerobic respiration, as shown by the presence of mitochondrial cytochrome a_3, which is functional in oxidative phosphorylation reactions.[5,9] Because these cestodes have more than one terminal oxidase (a_3, and O), they can modify their respiratory chain systems to adapt to the environment. Nonheme iron proteins affect electron transport and are universal among plants and animals; hence the proteins occur, unsurprisingly, in cestodes. For a critical review of anaerobic energetics in cestodes, see the papers by Fioravanti, and by Scheibel in the volume edited by MacInnis.[22] Barrett provides a useful overview of the older literature.[5]

Protein metabolism in cestodes is based on the uptake of preformed amino acids by active transport across the tegumental membrane or on larger protein units through pinocytotic mechanisms. Extracorporeal digestion may possibly occur, as in *Echinococcus granulosus* or *Hymenolepis diminuta*, in which hydrolytic enzymes secreted by the parasite into its external environment aid in the breakdown of complex host molecules to simpler transportable units. Proteins are used for energy and in the production of numerous types of body tissues. The nitrogenous end products of protein catabolism include urea, uric acid, and ammonia.

Cestodes studied thus far have been found to be incapable of initiating synthesis of purines and pyrimidines.[5,17] Therefore, they must rely on exogenous nucleoside (or free base) sources for the formation of ribo- and deoxyribonucleic acids. Active transport loci for purine and pyrimidine nucleosides have been identified in the outer tegumental membrane of various *Hymenolepis* spp.,[3,30] and at least in *H. diminuta*, pyrimidine nucleoside transport appears to be NA^+-dependent.[24]

Lipid metabolism of tapeworms needs more extensive study. The tegument is permeable to various fatty acids. Body tissues may contain large amounts of lipids with highly unsaturated fatty acids. At least some cestodes (*Spirometra mansonoides*, *Hymenolepis diminuta*) have lost their ability to initiate synthesis of long-chain fatty acids and sterols, including cholesterol. Presumably, these substances must be obtained from the host.

Synthesis of saturated fatty acids apparently does not require oxygen, but oxygen is essential for the production of double bonds of unsaturated fatty acids. One can conclude that the unsaturated fatty acids are of host origin. Cerebrosides (glycolipids) are widely distributed in cestodes. The unsaponifiable lipid fraction of these parasites contains sterols, with cholesterol predominating. Although cestodes freely absorb fatty acids and deposit much of them in tissues as glycerides, nothing indicates that they use glycerides or fatty acids.

For futher reading, see Barrett,[5] Bryant,[8] Van den Bosshe,[48,49] and Arme and Pappas.[3]

IMMUNITY

Developmental stages of *Hymenolepis nana*, *Hydatigera taeniaeformis*, several species of *Taenia*, and *Echinococcus* are commonly studied by immunologists because these parasites make intimate contact with host tissues. A high degree of acquired resistance is common for these species, and it may approach absolute immunity. Artificial immunity to tapeworms is frequently induced. For example, a parenteral injection of eggs or activated oncospheres of *Taenia hydatigena*, *Echinococcus granulosus*, or *Taenia ovis* into lambs at three months of age immunizes the host against homologous metacestodes.[15]

Immunization of dogs, using secretory antigens derived from adult *E. granulosus* cultivated in vitro, has been successful.[18] Sheep hydatid fluid is a mixture of parasite antigen and components of sheep serum, including IgG. For methods of serodiagnosis of hydatid disease, see Kagan,[21] Schantz and Kagan,[39] Walls and Schantz,[54] and Chapter 8 in Thompson.[46]

"The persistence of strong infection immunity long after the removal or loss of the parasites is uncommon among helminths, but it apparently occurs in some cestodes. . . . Worms residing in the intestinal lumen are not immunologically inert; they abundantly secrete antigens. . . . In general, observations with living material have indicated that the antigens most functional in eliciting protective immunity are associated with the more actively metabolizing stages, including active oncospheres."[53]

For example, larval stages of *Taenia pisiformis*, collected during in vitro cultivation, can be used to vaccinate rabbits.[38] Limited protective immunity may possibly be passively transferred to offspring from infected mothers.

The development of immunologic tolerance to cestode antigens has been demonstrated. Although passive immunization, using antisera as a vaccine, can be achieved in laboratory animals, vaccination against important human tapeworm infections has not been successful.

The rat tapeworms, *Hymenolepis nana* and *H. diminuta*, have been used extensively for immunologic investigations. The host apparently reacts to the first infection by changing the intestinal epithelium in some manner or by producing some substance in the intestine that renders most oncospheres from a second infection unable to penetrate the intestinal epithelium. Even if the parasites are successful, they soon die; this phenomenom indicates a second wall of defense on the part of the host.

Possibly, some parasites are able to maintain themselves within their host by changing their antigenic structure from time to time and thus escape the host antibodies that would otherwise destroy them. Immunity to *Hymenolepis nana* in mice occurs after egg infection, but not directly after infection with cysticercoids. This difference is understandable, because egg infection leads to the tissue phase of the parasite, whereas cysticercoid (indirect) infection leads to the adult phase in the gut.

If rabbits are immunized by injections of homogenates of adult *H. nana* and then are bled and the serum collected, this serum is sensitized against the worms. An adult *H. nana* placed in such serum immediately becomes unusually active and withdraws its rostellum. In a few hours, a precipitation is found on the worm, beginning with the anterior end and progressing posteriorly along the strobila. See Hopkins and Stallard for a study of the rejection of *H. citelli* by mice.[19]

In dual infections of mice or rats with *Hymenolepis nana* and *H. diminuta*, the *H. nana* are much more successful in maintaining themselves. The immunogenic tissue phase of the *H. nana* elicits an inhibitory reaction to infectivity and growth of the less immunogenic *H. diminuta*. The host reaction that develops against *H. nana* is just as strong against *H. diminuta*. Large numbers of common antigens seem to exist among helminths. Antibodies penetrate the lumen of the host gut from sloughed and disintegrated mucosa cells and also by passing through cell membranes.

Circulating antibodies to *Hymenolepis diminuta* have been identified in experimentally infected rats. Destrobilation and rejection of adult worms in rats and mice are common. These observations and others provide strong evidence that immunologic systems can limit the success of adult cestodes in gut lumen.[54,55] Acquired immunity to cysticercosis appears to take place in domesticated ruminants. Considerable success has been achieved in immunizing calves and lambs against cestodes, but to date, no immunologic mechanism appears to be capable of destroying cestodes already established in host tissues. An antibody-mediated protective mechanism is a major feature of immunity to *Taenia* spp. in experimental animals.

It is possible that complementary genetic systems have developed in some animal parasites and their hosts that involve the specific reaction of one gene in the parasite with one gene in the host, giving a common phenotype of resistance or susceptibility. Such a genetic relation has developed between the flax rust and its host. The genetics of immunity to animal parasites is almost an untouched field, but Wassom et al.[52] have shown that "acquired resistance to *Hymenolepis citelli* in *Peromyscus maniculatus*, the white-footed deer mouse, is . . . controlled by a single autosomal dominant gene R (recessive gene designated r)." They "postulate that *H. citelli* is propagated in such foci by genetically nonimmune (r/r) hosts comprising about 25% of the host population."

Little work has been done on the immunology of taenid parasites, but considerable evidence suggests that living metacestodes can stimulate protective immune responses. Cell-mediated immunity plays a protective role against adult *Taenia taeniaeformis*, and humoral immunity is also involved.

"Cross immunity between heterologus species of tapeworms may play an important role in regulating infections with larval cestodes under some circumstances. Man is apparently highly resistant to infection with *Echinococcus granulosus*, i.e. the rate of infection relative to the chances of exposure appears to be very low. It is interesting to speculate on the role that chance ingestion of the usually more readily available eggs of heterologus parasites such as *Taenia hydatigena*, *T. ovis*, *T. pisiformis* or *Multiceps serialis* may play in preventing *E. granulosus* infection of humans."[37]

Immunoprophylaxis in experimental *Echinococcus multilocularis* infections has been achieved.

For further reviews on immunity, see Gemmell,[13] Gemmell and Johnstone,[14] and Williams.[54]

If your curiosity about the biology of cestodes is not yet satisfied, the two volumes edited by Arme and Pappas will serve as a good starting point.[3] Mettrick et al. provide an overview of putative neurotransmitters in cestodes.[25] Van den Bossche et al. review chemotherapy.[48]

REFERENCES

1. Ali-Khan, Z.: Host-parasite relationship in Echinococcosis. II. Cyst weight, hematologic alterations, and gross changes in the spleen and lymph nodes of C57L mice against graded doses of *Echinococcus multilocularis* cysts. J. Parasitol., 60:236–242, 1974.
2. Ali-Khan, Z., Siboo, R., Gomersall, M., and Faucher, M.: Cystolytic events and the possible role of germinal cells in metastasis in chronic alveolar hydatidosis. An. Trop. Med. and Parasitol., 77:497–512, 1983.
3. Arme, C., and Pappas, P.W. (eds.): Biology of the Eucestoda. Vol. I. and II. New York, Academic Press, 1983.
4. Arai, H.P. (ed.): Biology of the Tapeworm *Hymenolepis diminuta*. New York, Academic Press, 1980.
5. Barrett, J.: Biochemistry of Parasitic Helminths. London, MacMillan Publishers LTD, 1981, p. 308.
6. Belding, D.L.: Textbook of Clinical Parasitology. 2nd Ed. New York, Appleton-Century-Crofts, 1965.
7. Bonsdorff, Bertel von: Diphyllobothriasis in Man. London, Academic Press, 1977.
8. Bryant, C.: The regulation of respiratory metabolism in parasitic helminths. Adv. Parasitol., 16:311–331, 1978.
9. Cheah, K.S.: Cytochromes in *Ascaris* and *Moniezia*. In Comparative Biochemistry of Parasites. Edited by H. Van den Bossche. New York, Academic Press, 1972, pp. 417–432.
10. Coil, W.H.: Studies on the biology of the tapeworm *Dioecocestus acotylus* with emphasis on the Oogenotop. Z. Parasitenkd., 33:314–328, 1970.
11. Coil, W.H.: Studies on the biology of the tapeworm *Shipleya inermis* Furhmann, 1908. Z. Parasitenkd., 35:40–54, 1970.
12. Eckert, J.: Prospects for the treatment of the metacestode stage of *Echinococcus*. In The Biology of *Echinococcus* and Hydatid Disease. Edited by R.C.A. Thompson. London, George Allen and Unwin, 1986, pp. 250–284.
13. Gemmell, M.A.: Immunology and regulation of cestode zoonoses. In Immunology of Parasitic Infections. Edited by S. Cohen and E. Sadun. Oxford, Blackwell Scientific Publications, 1976, pp. 333–358.
14. Gemmell, M.A., and Johnstone, P.D.: Experimental epidemiology of hydatidosis and cysticercosis. Adv. Parasitol., 14:312–369, 1977.
15. Gemmel, M.A., and MacNamara, F.N.: Immune response to tissue parasites. II. Cestodes. In Immunity to Animal Parasites. Edited by G.J.L. Soulsby. New York, Academic Press, 1972, pp. 235–272.
16. Goddeeris, B.: The role of insects in dispersing eggs of tapeworms, in particular *Taeniarhynchus saginatus*. 1. Review of the literature. Ann. Soc. Belg. Med. Trop., 60:195–201, 1980.
17. Heath, R.L.: Biosynthesis de novo of purines and pyrimidines in *Mesocestoides* (cestoda). Int. J. Parasitol., 56:98–102, 1970.
18. Herd, R.P., Chappel, J., and Biddell, D.: Immunization of dogs against *Echinococcus granulosus* using worm secretory antigens. Int. J. Parasitol., 5:395–399, 1975.
19. Hopkins, C.A., and Stallard, H.E.: Immunity to intestinal tapeworms: the rejection of *Hymenolepis citelli* by mice. Parasitology, 69:63–76, 1974.
20. Insler, G.D., and Roberts, L.S.: *Hymenolepis diminuta*: parasite or commensal? Am. Soc. Parasitol., 49th Ann. Meet. Abst. No. 78, p. 36, 1974.
21. Kagan, I.G.: Serodiagnosis of hydatid disease. In Immunology of Parasitic Infections. Edited by S. Cohen and E.A. Sadun. Oxford, Blackwell Scientific Publications, 1976, pp. 130–142.
22. MacInnis, A.J. (ed.): Molecular Paradigms for Eradicating Helminthic Parasites. UCLA Symposia Series. Vol. 60. New York, Alan Liss, 1987.
23. Mackiewicz, J.: Caryophyllidea (Cestoidea): a review. Exp. Parasitol., 31:417–512, 1972.
24. McCraken, R.O., Lumsden, R.D., and Page, C.R., III: Sodium-sensitive nucleoside transport by *Hymenolepis diminuta*. J. Parasitol., 61:999–1005, 1975.
25. Mettrick, D.F., Sukkedo, M.V.K., and Sukkedo, S.C.: Acetylcholine—a neurotransmitter in parasitic Platyhelminthes? In Molecular Paradigms for Eradicating Helminthic Parasites. UCLA Symposia Series. Vol. 60, pp. 421–423. Edited by A.J. MacInnis, New York, Alan Liss, 1987.

26. Márques-Monter, H.: Cysticercosis. *In* Pathology of Protozoal and Helminthic Diseases. Edited by R.A. Marcial-Rojas. Baltimore, Williams & Wilkins, 1971, pp. 592–617.

27. Moczón, T.: Histochemistry of oncospheral envelopes of *Hymenolepis diminuta* (Rudolphi, 1819) (Cestoda, Hymenolepididae). Acta Parasitol. Pol., *20*:517–531, 1972.

28. Morseth, D.J.: Observations on the fine structure of the nervous system of *Echinococcus granulosus*. J. Parasitol., *53*:492–500, 1967.

29. Mueller, J.F.: The biology of *Spirometra*. J. Parasitol., *60*:3–14, 1974.

30. Page, C.R., III, and MacInnis, A.J.: Characterization of nucleoside transport in hymenolepidid cestodes. J. Parasitol., *61*:281–290, 1975.

31. Pawlowski, Z., and Schultz, M.G.: Taeniasis and cysticercosis *(Taenia saginata)*. Adv. Parasitol., *10*:269–343, 1972.

32. Phares, C.K., and Cox, G.S.: Molecular hybridization and immunological data support the hypothesis that the tapeworm, *Spirometra mansonoides* has acquired a human growth hormone gene. *In* Molecular Paradigms for Eradicating Helminthic Parasites. UCLA Symposia Series. Vol. 60, pp. 391–405. Edited by A.J. MacInnis. New York, Alan Liss, 1987.

33. Pool, J.B., and Marcial-Rojas, R.A.: Echinococcosis. *In* Pathology of Protozoal and Helminthic Diseases. Edited by R.A. Marcial-Rojas. Baltimore, Williams & Wilkins, 1971, pp. 635–657.

34. Rausch, R.L.: Parasitology: Retrospect and Prospect. J. Parasitol., *71*:139–151, 1985.

35. Read, C.P.: Longevity of the tapeworm, *Hymenolepis diminuta*. J. Parasitol., *53*:1055–1056, 1967.

36. Read, C.P., and Kilejian, A.Z.: Circadian migratory behavior of a cestode symbiote in the rat host. J. Parasitol., *55*:574–578, 1969.

37. Rickard, M.D., and Coman, B.J.: Studies on the fate of *Taenia hydatigena* and *Taenia ovis* larvae in rabbits, and cross immunity with *Taenia pisiformis* larvae. Int. J. Parasitol., *7*:257–267, 1977.

38. Rickard, M.D., and Outteridge, P.M.: Antibody and cell-mediated immunity in rabbits infected with the larval stages of *Taenia pisiformis*. Z. Parasitenkd., *44*:187–201, 1974.

39. Schantz, P.M., and Kagan, I.G.: Echinococcosis (hydatidosis). *In* Immunological Investigation of Tropical Diseases. Edited by V. Houba. New York, Churchill Livingstone, 1980, pp. 104–129.

40. Schmidt, G.D.: How to know the tapeworms. Dubuque, IA, W.C. Brown Co., 1970. (See also: Schmidt, G.D.: Handbook of Tapeworm Identification. Boca Raton, Florida, CRC Press, 1985.)

41. Slais, J.: The Morphology and Pathogenicity of the Bladder Worms. The Hague, Dr. W. Junk N. V. Pub., 1970.

42. Slais, J.: The location of the parasites in muscle cysticercosis. Folia Parasitol., *14*:217–224, 1967.

43. Smyth, J.D.: The biology of the hydatid organisms. Adv. Parasitol., *7*:327–345, 1969.

44. Smyth, J.D: The Physiology of Cestodes. San Francisco, W.H. Freeman & Co., 1969.

45. Tanaka, R.D., and MacInnis, A.J.: An explanation of the apparent reversal of the circadian migration by *Hymenolepis diminuta* (Cestoda) in rat. J. Parasitol., *61*:271–280, 1975.

46. Thompson, R.C.A. (ed.): The Biology of *Echinococcus* and Hydatid Disease. London, George Allen and Unwin, 1986.

47. Tobler, H., and MacInnis, A.J.: Molecular and Developmental Biology of Platyhelminths and Nematoda. Amsterdam, Elsevier, in press.

48. Van den Bossche, H., Thienpont, D., and Janssens, P.G. (eds): Chemotherapy of Gastrointestinal Helminths. Berlin, Springer Verlag, 1985.

49. Van den Bossche, H. (ed.): Comparative Biochemistry of Parasites. London, Academic Press, 1972.

50. Walls, K., and Schantz, P.M.: Immunodiagnosis of Parasitic Diseases. Vol. I. New York, Academic Press, 1986, p. 312.

51. Wardle, R.A., McLeod, J.A., and Radinovsky, S.: Advances in the Zoology of Tapeworms, 1950–1970. Minneapolis, University of Minnesota Press, 1974.

52. Wassom, D.L., DeWitt, C.W., and Grundmann, A.W.: Immunity to *Hymenolepis citelli* by *Peromyscus maniculatus:* genetic control and ecological implications. J. Parasitol., *60*:47–52, 1974.

53. Weinmann, C.J.: Cestodes and Acanthocephala. *In* Immunity to Parasitic Animals. Vol. 2. Edited by G.J. Jackson, R. Herman, and I. Singer. New York, Appleton-Century-Crofts, 1970, pp. 1021–1059.

54. Williams, J.F.: Cestode infections. *In* The Immunology of Parasitic Infections. Edited by S. Cohen and K.S. Warren. Oxford, Blackwell, 1982, p. 676.

55. Williams, J.F.: Recent advances in the immunology of cestode infections. J. Parasitol., *65*:337–349, 1979.

13

Phylum Acanthocephala

GENERAL CONSIDERATIONS

The phylum Acanthocephala[8] is composed of thorny-headed worms, so called because of the many thorn-like hooks that occur on the proboscis (**acanth** means a thorn and **cephala** the head). These cosmopolitan worms are all endoparasitic as adults in the digestive tracts of terrestrial and aquatic vertebrates ranging from fish to man, but are found especially in fish. They are whitish or yellow parasites with wrinkled or smooth bodies that range in length from 1 mm to over 1 m. Larval stages are found in invertebrates. Hundreds of young stages may be found in a single intermediate host, and the vertebrate host may harbor thousands of adult worms. Damage to the intestinal epithelium and proliferation of host connective tissue are the most pronounced histopathologic effects. We shall mention only a few of the many species, and two of the four classes of this phylum.

The proboscis and associated structures are called the **presoma.** The rest of the body is the **trunk.** Small spines occur on the trunk in some genera, for example, *Corynosoma strumosum* in seals. The proboscis, the neck, and the trunk may become modified as accessory anchoring devices, such as a bulb-like inflation of the anterior end and a general covering of cuticular spines. In adults, the body cavity is a fluid-filled space between the viscera and the body wall and is a **pseudocoel** rather than a true coelom. Neither a digestive tract nor a true circulatory system is present in either adult or larval worms. Sexes are separate, females are almost always larger than males, and the posterior aperture, or **gonopore**, is the only body opening.

In general, adult tissues have lost cellular identity, and few nuclei are present in the entire body.

The tegument is a syncytium with many functions, which include protection, inactivation of the host's digestive enzymes by charge effects, osmoregulation, concentration of nutrients, facilitation of assimilation by pinocytosis, ion accumulation, and the facilitation of ion transport. The outer surface coat is a carbohydrate-rich **glycocalyx** (epicuticle). Closely packed pores at the tegument surface lead to pore canals that branch and anastomose. The lacunar canal system, a "circulatory system," is probably made up of the spaces between tegumentary fibers, normally packed with lipid. The pseudocoel's contents are accessible to the lacunar system. Under the tegument lies an infolded plasma membrane and below that, a **basement lamina.** The tegument contains smooth endoplasmic reticulum, mitochondria, lipid, Golgi clusters, glycogen deposits, free ribosomes, and various kinds of vesicles. Giant nuclei in the tegument are helpful in species identification, based on their location, number, and size. The outer body muscles are largely circular, and the inner muscles are a multinuclear syncytium. The physiology of acanthocephalids is essentially the same as that of cestodes.

The proboscis is globular or cylindrical in shape, bearing rows of recurved hooks or spines that serve to attach the worm to the gut of its host. In some species, the proboscis of the adult and that of the infective larva are identical in appearance; in others, the adult proboscis may be modified. In some worms, the proboscis is permanently anchored in host tissue, as in

263

Filicollis in birds, *Polymorphus* in ducks, and *Pomphorhynchus bulbocolli* in fish. The proboscis can usually be withdrawn into a muscular **proboscis receptacle.** This sac extends into the body cavity from the neck region. Fluid in the proboscis and sac is independent of fluid in the body cavity. These two separate hydraulic systems permit trunk movement without loosening the hold of the proboscis on the intestinal wall. Hooks on the proboscis vary according to size, shape, number, and arrangement and thus have considerable taxonomic value. See Figure 13–1 for general anatomy.

Lemnisci are paired organs, usually elongated and pendulous, that extend into the body cavity from the neck region. The central canal of each lemniscus is continuous with the lacunar system. The lemniscus apparently serves as a fluid reservoir when the proboscis is invaginated. Histochemical evidence suggests that the lemnisci may also serve a metabolic function, especially for fat metabolism. They are often surrounded by muscles at their basal portion.

Ligament sacs extend from the proboscis sheath or from the adjacent body wall and form tubes that surround the reproductive organs. They do not commonly persist in adults, and only one may be present. A **ligament strand** is attached to the gonads and extends the length of the ligament sacs.

Protonephridia serve as excretory organs. They consist of flame bulbs and collecting tubules found in some members of the class Archiacanthocephala. Flame bulbs are grouped into two masses attached to the reproductive organs and empty by way of a canal or bladder into the sperm duct or uterus.

The nervous system consists mainly of a ganglion in the proboscis sheath and of nerves that connect the ganglion to other organs and tissues of the body. In addition, a pair of genital ganglia, with nerves, is present in the male. **Sense organs** are found in the proboscis and in the penis and male bursa.

Male reproductive organs consist of a pair of testes, one behind the other, and a common sperm duct formed by the union of a duct from each testis. The common duct leads through the penis to the outside. There is usually a cluster of large **cement glands** that empty into the common sperm duct, sometimes by way of a cement gland reservoir. The penis projects into an eversible bursa. This cup-like terminal structure with a thick, domed cap is used to hold the female during copulation.

Female reproductive organs consist of an ovary fragmented into **ovarian balls** that lie in the ligament sac or are free in the pseudocoel. A **uterine bell,** at the end of the ligament sac, is a funnel or cup-like structure that receives ova from the ovarian balls. At the base of the uterine bell are the **selector apparatus** and bell pockets. The former permits only mature eggs to enter the uterus. The reproductive tube continues as an elongated uterus. A vagina connects the uterus to the terminal opening, or gonopore.

During copulation, the male discharges sperm into the vagina. Then secretions from the cement glands block the exit of the vagina and thus prevent the escape of sperm. In some species of worms, the cement gland secretions form a cap over the entire terminal end of the female.

''Homosexual rape results in the male victim having the genital region sealed off with cement and effectively removed from the reproductive population. . . . We interpret the evolution of the cement gland and sexual behavior as the result of sexual selection.''[1]

Fertilized eggs go through early embryologic development in the ligament sac or pseudocoel. When eggs emerge from the gonopore, they contain a hooked larva called the **acanthor** (Fig. 13–2). A host must eat the eggs before the embryos can hatch. See Figure 13–3 for representative eggs.

Eggs of *Moniliformis dubius* (see discussion later in this chapter), an acanthocephalan of rats with larval stages in the cockroach, hatched experimentally in certain electrolytes, provided the molarity of the solution was greater than 0.2 and the pH greater than 7.5. The addition of CO_2 lowered the required pH level and increased the percentage of eggs that hatched. The embryo released small amounts of chitinase, which possibly assisted in the decomposition of a chitinous membrane that surrounds the embryo. Activation of egg hatching was probably ionic rather than osmotic. The optimal hatching temperature seemed to be between 15 and 37°C. Eggs hatched in stimulating fluids at 20°C after storage in distilled water at 5°C for at least 4 weeks.

Little is known about life-cycle details of most species, but undoubtedly they all follow the same general plan. Each female usually has only

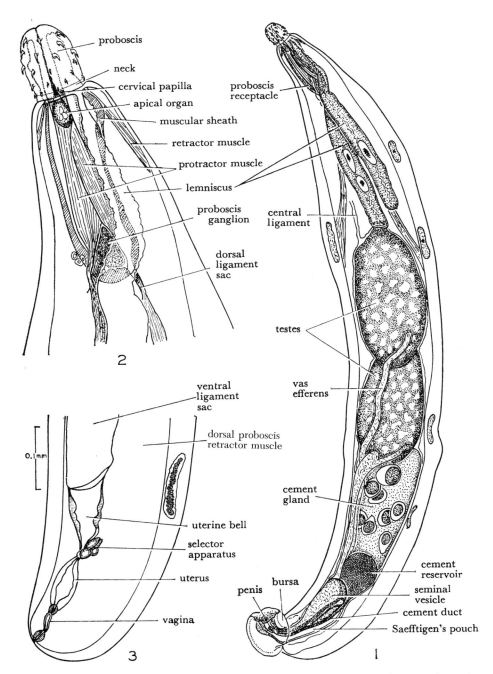

Fig. 13–1. Adults of *Paulisentis fractus. 1*, Mature male, lateral view. *2*, Same, proboscis and anterior trunk region enlarged. *3*, Posterior portion of female, lateral view. (From Cable and Dill, courtesy of J. Parasitol.)

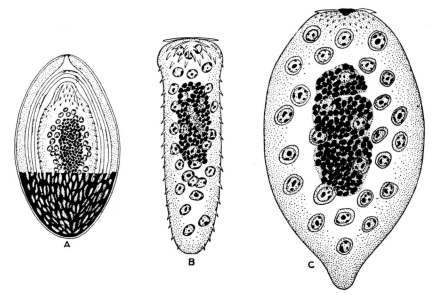

Fig. 13–2. Shelled embryo and acanthor stages of *Macracanthorhynchus hirudinaceus* (Pallas). Magnification approximately 500 diameters. *A,* "Egg" or shelled embryo. Surface appearance shown only on lower pole; on remainder of drawing, membranes are shown in optical section. *B,* Acanthor, stage I, from lumen of midgut of beetle larva. Note the larval rostellar hooks and small body spines. *C,* Acanthor, stage II, after penetrating the wall of the midgut of the beetle larva, about 5 to 20 days after artificial infection. (Redrawn from Kates.) (From Van Cleave, courtesy of J. Parasitol.)

one reproductive cycle. At the time of copulation, all eggs are subject to fertilization, and the female may become distended with stored embryonated eggs, which she can discharge selectively over a long period of time. Eggs are normally spindle-shaped, often resemble diatoms, and are commonly eaten by aquatic insects and crustaceans.

Intermediate hosts for acanthocephalids of fish are usually benthic amphipods (*Gammarus, Pontoporeia*), isopods (*Asellus*), or ostracods (*Cypris*). Snails and aquatic insects sometimes act as either transport hosts or as secondary intermediate hosts. Some species of acanthocephalids may require a second and even a third intermediate host, but usually only one is involved. No superimposed multiplicative cycle occurs similar to that in flukes, and there is no asexual or parthenogenetic development. One adult arises from each zygote. No free-living stage occurs.

The following is a life cycle involving one invertebrate host: An egg eaten by an arthropod hatches into an **acanthor,** develops into an **acan-**

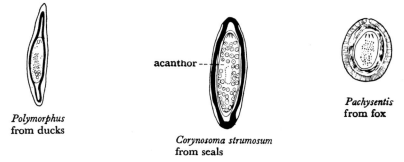

Polymorphus
from ducks

acanthor - -

Corynosoma strumosum
from seals

Pachysentis
from fox

Fig. 13–3. Representative eggs of Acanthocephala. (From Olsen: Animal Parasites. Minneapolis, Burgess Publishing, 1967.)

thella, becomes a **juvenile** (which may progress to **cystacanth**), and is eaten by the final, vertebrate host, in which it becomes an adult. Transport hosts can be either vertebrates or invertebrates. Within the invertebrate host, the acanthor is liberated from the egg, bores through the gut wall, and develops into the acanthella. In several weeks or more, the acanthella becomes a juvenile, which is generally called a cystacanth (the second infective stage), differing from an adult primarily by being sexually immature. The term "juvenile" should be used only when the larva resembles the adult. The vertebrate host becomes infected by eating the arthropod intermediate host or a transport host. Sexual development occurs in the final host. (See Figure 13–5.)

Because host specificity for the intermediate host is more limited than that for vertebrate hosts, the distribution of acanthocephalids is largely determined by invertebrate hosts. For example, some birds acquire *Corynosoma* by eating fish. This parasite occurs in aquatic birds and marine mammals, and the fish is presumably a transport host.

In the life cycle of *Leptorhynchoides thecatus*, a parasite of the rock bass *Ambloplites rupestris*, an amphipod is the intermediate host, but if the fish swallows an amphipod containing a juvenile worm that has not completed its full development, the parasite does not mature in the intestine of the final host. Instead, it penetrates the gut wall and becomes encysted in the mesenteries.

The biochemistry of the Acanthocephala needs much more study, and only a few generalizations can be made at the present time. Adult *Moniliformis dubius*, and probably other members of the phylum, depend on host dietary carbohydrate for energy needs. Glycogen constitutes the principal carbohydrate reserve. However, trehalose, a disaccharide (two glucose molecules), also serves as a reserve store of energy. Glycolysis is carried out primarily by the Embden-Meyerhof scheme, but acanthocephalans should be classified as facultative anaerobes. The integument carries out pinocytosis and can absorb at least some triglycerides, amino acids, nucleotides, and sugars. *M. dubius* liberates aminopeptidase from its surface membrane and thus can hydrolyze peptides in the ambient medium.[7] Lipids accumulate in the worm tissues in considerable amounts, but are

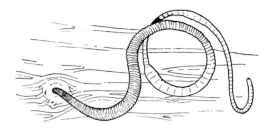

Fig. 13–4. *Macracanthorhynchus hirudinaceus* adult attached to the gut wall of a pig.

not apparently used as energy sources. For the biology of Acanthocephala, see Nicholas.[5]

Class Archiacanthocephala

FAMILY OLIGACANTHORHYNCHIDAE

Macracanthorhynchus hirudinaceus (Figs. 13–4, 13–5) is rare in man, but is a common species in hogs, in which it occurs in the small intestine attached to the gut wall. It is usually large; the females range from 20 to 65 cm in length, whereas the males range from 5 to 10 cm. The largest worms may be about the width of an ordinary pencil. The animal is flattened in normal life, but it soon becomes cylindric when collected and preserved. Irregular transverse folds give the worm a wrinkled appearance. The small, protrusible proboscis is armed with hooks in 6 spiral rows. The body tapers from just behind the proboscis to the posterior end, at which there is a bell-like bursa. Testes are located in the anterior half of the male worm.

Female worms produce eggs (67 to 110 × 40 to 65 μm) containing mature acanthors. Eggs leave the pig's intestine with feces and become widely scattered. These eggs, which are resistant to adverse environmental conditions, remain viable for years. Birds may carry the eggs on their feet or may even eat them with contaminated food and pass them unharmed through their own bodies. The usual invertebrate host is a larval beetle that eats the eggs.

Cockchafer beetles (*Melolontha vulgaris*), rose chafer beetles (*Cetonia aurata*), water beetles, (*Tropisternus collaris*), and various species of dung beetles are the common hosts. Eggs hatch in the beetle gut, and the acanthor uses its hooks to bore through the intestinal wall to the hemocoel, in which it develops into the acanthella that gradually assumes most of the adult characteristics. Acanthors of *M. hirudinaceus*

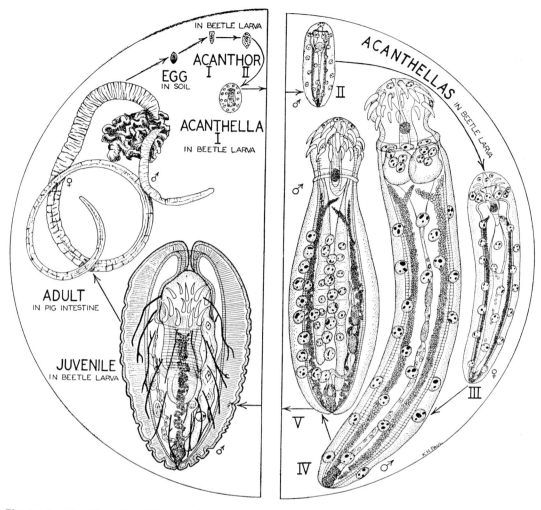

Fig. 13–5. The life cycle of *Macracanthorhynchus hirudinaceus* (Pallas), the thorny-headed worm of swine. Drawings of adult worms are about natural size; all other stages are at uniform magnification of approximately 38 diameters. The elongated, branching, dark bodies in the juvenile are the nuclei that become the giant nuclei of the subcuticula in the adult. (After Kates, 1943) (From Van Cleave, courtesy of J. Parasitol.)

may induce extensive temporary changes in the histologic structure of the gut in the area of penetration. Some of the effects observed in a cockroach are destruction of epithelial cells, aggregation of hemocytes, and hyperplasia of adjacent tissues. Acanthors are soon melanized.

By the end of 6 weeks to 3 months, the juvenile (cystacanth) stage is reached. The beetle may be eaten by a second intermediate host, but usually it is devoured by the final host. This juvenile is thus the infective stage for pigs or man. The life span of the adult in the pig is usually less than a year. Although 400 acanthors have been found in one naturally infected beetle, usually fewer than 30 worms are present in a pig. Occasionally, dogs and other mammals that might eat beetles are parasitized by this worm.

Ulcerations, necrotic areas, anemia, and even penetration of the vertebrate's intestinal wall with subsequent peritonitis may occur in serious infections. Control consists mainly of sanitary measures in raising pigs, such as rearing them on concrete floors and promptly removing feces.

FAMILY MONILIFORMIDAE

Moniliformis moniliformis (probably a combination of several species) is a cosmopolitan acanthocephalan that lives in the small intestine of rats, mice, dogs, and cats. The worms have a beaded appearance because of annular thickenings. Males range from 4 to 13 cm in length, whereas females are 10 to 30 cm long. The proboscis possesses 12 to 15 rows of hooks. Eggs, 85 to 120 × 40 to 50 μm, are eaten by beetles (*Blaps gigas, Tenebrio molitor, Calandra orizae*) or by cockroaches (*Blatta orientalis, Blattella germanica*). The life history closely resembles that of *Macracanthorhynchus hirudinaceus*. Juveniles appear in 4 to 6 weeks, and as many as 100 may inhabit a single cockroach. The infection rate in rats varies, but may run as high as 20% in *Rattus alexandrinus*. Frogs, toads, and lizards have been suggested as potential transport (paratenic) hosts. Growth of the worm ceases when the rat is placed on a diet devoid of carbohydrate.

Moniliformis dubius is also common in rats (Fig. 13–6) and has been reported from man. The life cycle is similar to that of *M. moniliformis*.

Acanthors within the cockroach become surrounded by a protective membrane. This capsule is made of host tissue.[4] The initial response of the host is a gathering of hemocytes around the parasite. Hemocytes are then repulsed as the acanthor alters its tegument and produces a microvilli-type structure for absorption of nutrients.[6] Thus, encapsulation involves a combination of host and parasite defense mechanisms.

Successful establishment of the larval stages of this worm in the cockroach, *Periplaneta americana*, apparently depends on their ability to inhibit the phenoloxidase system of the insect. This inhibition may be associated with polyanionic mucins occurring in the trilaminate envelope surrounding each developing acanthocephalan larva or cystacanth.[2]

Order Palaeacanthocephala

FAMILY POLYMORPHIDAE

Polymorphus minutus is a small bright-orange acanthocephalan that lives in the small

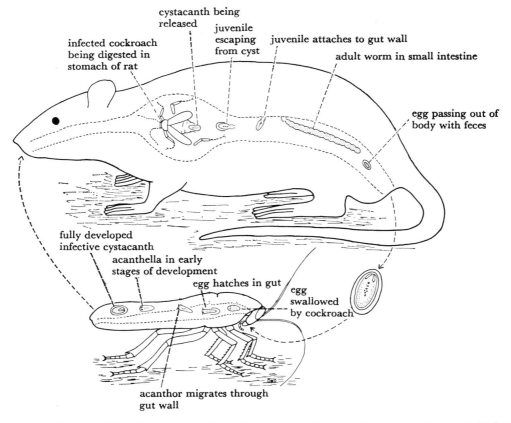

Fig. 13–6. Life cycle of *Moniliformis dubius*. (From Olsen: Animal Parasites. Minneapolis, Burgess Publishing, 1967.)

intestine of chickens, ducks, and possibly geese and other water birds. The male is about 3 mm long, and the female grows up to 10 mm. It can cause serious disease in domestic and wild fowl. The intermediate host is the amphipod, *Gammarus pulex*.

A closely related species, *Filicollis sphaerocephalus*, is also found in aquatic birds. The invertebrate host is a crustacean. *Filicollis anatis* is another small species that inhabits the intestines of ducks, geese, and other water birds. The male is 6 to 8 mm long and the female, 10 to 25 mm long. *Asellus aquaticus*, a freshwater isopod, is the intermediate host.

Affinities of this phylum to other worms are not clear. Developmental stages are more similar to those of flatworms than to nematodes, but the group is probably closest to the cestodes, although distinctly separate. About 500 species are known and probably many more have not been described.

For futher information on the Acanthocephala see Crompton.[3] For evolution, see Chapter 26. Yamaguti provides useful keys to genera.[9]

REFERENCES

1. Abele, L.G., and Gilchrist, S.: Homosexual rape and sexual selection in acanthocephalan worms. Science, *197*:81–83, 1977.

2. Brennan, B.M., and Cheng, T.C.: Resistance of *Moniliformis dubius* to the defense reactions of the American cockroach, *Periplaneta americana*. J. Invertebr. Pathol., *26*:65–73, 1975.

3. Crompton, D.W.T.: An Ecological Approach to Acanthocephalan Physiology. Cambridge, Cambridge University Press, 1970.

4. Mercer, E.H., and Nicholas, W.L.: The ultrastructure of the capsule of the larval stages of *Moniliformis dubius* (Acanthocephala) in the cockroach *Periplaneta americana*. Parasitology, *57*:169–174, 1967.

5. Nicholas, W.L.: The biology of the Acanthocephala. Adv. Parasitol., *11*:671–706, 1973.

6. Rotherham, S., and Crompton, D.W.T.: Observations on the early relationship between *Moniliformis dubius* (Acanthocephala) and the haemocytes of the intermediate host, *Periplaneta americana*. Parasitology, *64*:15–21, 1972.

7. Uglem, G.L., Pappas, P.W., and Read, C.P.: Surface aminopeptidase in *Moniliformis dubius* and its relation to amino acid uptake. Parasitology, *67*:185–195, 1973.

8. Van Cleave, H.: Expanding horizons in the recognition of a phylum. J. Parasitol., *34*:1–20, 1948.

9. Yamaguti, S.: Systema Helminthum, Vol. 5. Acanthocephala. New York, Wiley Interscience, 1963.

14

Introduction

The phylum Nematoda contains an almost unbelievable number of worms that are free-living in water and soil and an impressive number of species that are parasitic in plants and animals. Biology students frequently find thread-like worms coiled in the muscles, connective tissues, or other organs of laboratory-dissected animals. Fish are especially remarkable for their burdens of worms. Hundreds of individual worms and several species often inhabit one host. Insects and other invertebrates have their own roundworm parasites and also serve as intermediate hosts. Even fossil nematodes are known. The worm *Heydonius antiguus* was found projecting from the anus of the beetle *Hesthesis immortua* in Rhine lignite. Various other species have been found in amber.

Many nematodes have yet to be described, and the parasite-host relationships are just beginning to be understood. Parasitic nematodes often cause important pathologic changes in their hosts, and it is among these pathogenic species that relationships with the host have been intensively studied and are most completely understood. Some parasitic nematodes posses their own parasites. Thus, *Aphelenchoides parietinus*, a nematode parasitic in plants, may be infected with a microsporidian. Nematology is certainly an open field for anyone interested in investigating parasites.

Nematodes, called "roundworms," are active, slender worms, usually pointed at both ends. They possess a mouth and an anus, thus having a complete digestive tract. The body cavity is a pseudocoelom. Figs. 14–1, 14–2, and 14–3 give a general picture of some of the organ systems of these worms.

For detailed descriptions of nematode structures, see Bird[4] and Croll.[11] For a short account of nematode immunology, see Chapter 17. For keys to nematode parasites of invertebrates and vertebrates, see Poinar[28] and Anderson et al.[1] respectively.

CUTICLE AND HYPODERMIS

The cuticle covers the external surface and lines the buccal cavity, esophagus, vagina, excretory pore, cloaca, and rectum. It is usually smooth, but may be covered with spines or have longitudinal and transverse ridges. It serves as a tough, flexible, protective covering, resistant to host digestive enzymes and, especially in adult worms, is generally impervious except to water and some small ions in aqueous solution. Cuticular absorption of physiologically insignificant quantities of glucose occurs in the intestinal worm, *Ascaris*. Some substances are excreted through the cuticle (for example, sodium and ammonium ions and urea).[8] Worms that lack a functional gut (for example, mermithids in insects—see Chapter 17), may aquire nutrients through a specially modified cuticle.

The cuticle also functions as part of the hydrostatic skeleton of nematodes. Its role is to resist the high internal pressure of the pseudocoelomic fluid. By keeping the fluid contained, the nematode's form is maintained and anchorage for the muscles is provided.

The cuticle consists basically of three layers: a surface **cortical layer,** a middle **matrix layer,** and a **basal layer** (Fig. 14–4). They are usually subdivided and strengthened by fibrils often compacted to form fibers. The subdivisions vary

271

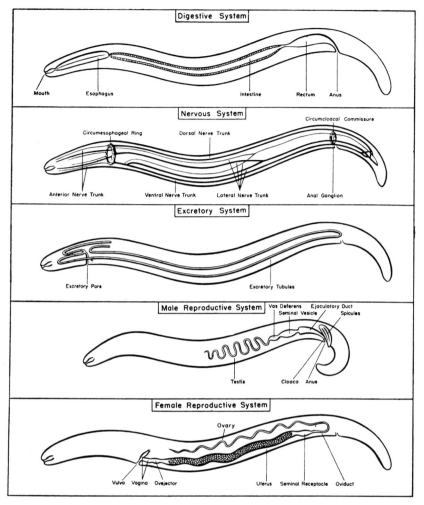

Fig. 14–1. Major organ systems of roundworms. (From the United States Navy Medical School Laboratory Guide.)

in number and in structure among the numerous species of nematodes. The external part of the cortical layer is a cuticular plasmalemma, or epicuticle, which in many species is invested with a fuzzy coating, a glycocalyx (Fig. 14–5). The cuticular surface is antigenic and the glycocalyx may contain antigen-antibody complexes. The epicuticle contains lipids, proteins and traces of carbohydrate. The internal cortical layer contains enzymes and RNA and is metabolically active. The matrix layer may be filled with fluid and may contain structures such as filaments, rods, and plates. This layer is made up of proteins that resemble collagens. Inglis[20] considered the rods of this layer to be canals

and that modifications of the cuticle, based on this system of canals, result from the need for a cuticle that is flexible longitudinally and strong radially. Radial strength may also be achieved by the presence of transverse annules, alternately rigid and flexible, by spiral fibers (as in *Ascaris*; see Chap. 16), or by some other kind of strongly patterned modification. Cuticular canals have been demonstrated in a few species, for example, *Ascaris*, but they have not been seen in most nematodes. The rods may have a skeletal function. The basal layer contains fibers, and it also resembles collagens chemically.

Twenty amino acids have been found in the cuticle of several species of nematodes, as well

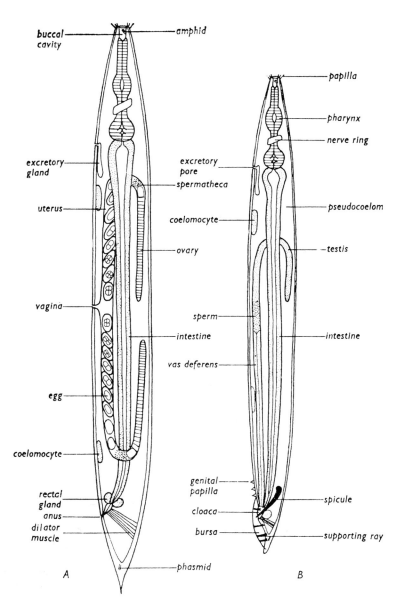

Fig. 14–2. General morphology of a nematode (hypothetical). *A*, Female. *B*, Male. Lateral view. (From Lee, D.L.: The Physiology of Nematodes. San Francisco, W.H. Freeman, 1965.)

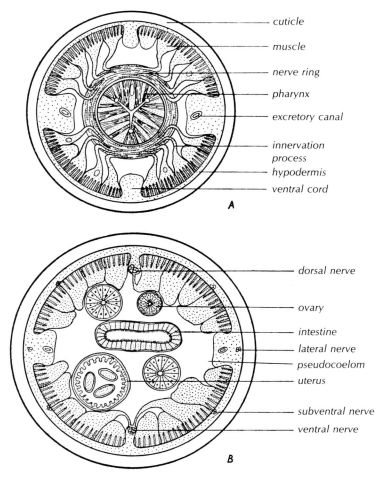

cuticle
muscle
nerve ring
pharynx
excretory canal
innervation process
hypodermis
ventral cord

A

dorsal nerve
ovary
intestine
lateral nerve
pseudocoelom
uterus
subventral nerve
ventral nerve

B

Fig. 14–3. Transverse sections of a nematode. *A,* Pharyngeal region. *B,* Middle region. (From Lee, D.L.: The Physiology of Nematodes. San Francisco, W.H. Freeman, 1965.)

as small amounts of carbohydrates and lipids. RNA, ascorbic acid, adenosine triphosphate, hemoglobin, acid phosphatases, and other enzymes have been demonstrated in the inner cortex of *Ascaris.* The presence of these substances indicates that this complex membrane is not an inert covering, but is in a state of constant metabolic activity[4] and is involved in the synthesis of proteins. The cuticle plays an important role in osmoregulation and in ion regulation, and several writers have suggested a nutritive function in species lacking an intestine.

Laterally the cuticle may project from the body as a fin or a wing-like flange. In the neck region these structures are known as **cervical alae** (singular, **ala**), whereas those on the tail are **caudal alae.** The latter may be supported by

fleshy papillae and are used in copulation. The cuticular surface also bears several sets of papillae. These include circumoral labial papillae, cervical papillae **(deirids),** and various genital and caudal papillae, most of which have a tactile function.

The **hypodermis** lies beneath the basal layer of the cuticle and is responsible for its synthesis. The hypodermis is not sharply delimited from the body musculature to which it is attached (Fig. 14–6). It aids in regulating body wall permeability. It is a thin layer with four longitudinal thickenings or cords containing the hypodermal nuclei, and it may be cellular or syncytial. Once the larval forms have developed, the number of nuclei present in the hypodermis remains constant. This metabolically active layer is rich

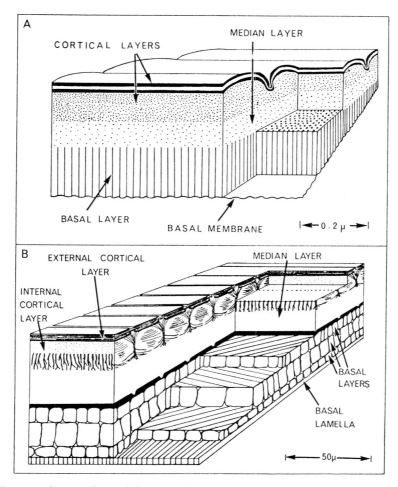

Fig. 14–4. Diagrams of nematode cuticle layers. *A*, Typical infective larva. *B*, Adult *Ascaris lumbricoides*. (From Bird, A.: The Structure of Nematodes. New York, Academic Press, 1971.)

in glycogen, lipids, enzymes, mitochondria, and Golgi bodies. For details of cuticle ultrastructure, see Bird[6] and Lee.[21] For details of ultrastructure and cytochemistry, see Lumsden.[24]

MUSCULAR SYSTEM

Nematodes have two kinds of muscles: (1) somatic (unspecialized), which consist of a single layer lying next to the hypodermis; and (2) specialized, which have a variety of functions, depending on their location, for example, spicular muscles to protrude the spicules in males. Body wall muscles are placed longitudinally and are responsible for the sinuous movements of the worms. The heavily fibrous zone of each myofiber is attached to the hypodermis, but the

other less fibrous end of the muscle cell (seen clearly in cross section, Figs. 14–3, 14–6) is connected to either the dorsal or ventral nerve cord, from which it receives its motor stimulation. Noncontractile parts of somatic muscles serve for glycogen storage. Between the muscle layer and the digestive tract is the body cavity known as a pseudocoelom, filled with fluid under high pressure. This pressurized fluid-filled cavity, with the body wall that provides containment, functions as a **hydrostatic skeleton.**

"In any muscle system some antagonising mechanism must be available. That it is largely hydrostatic in nematodes, based on the fluid contents of the pseudo-coelom, cannot be seriously questioned. The nematode body is cylindrical, the muscles are arranged longitudinally and the antagonisation of the

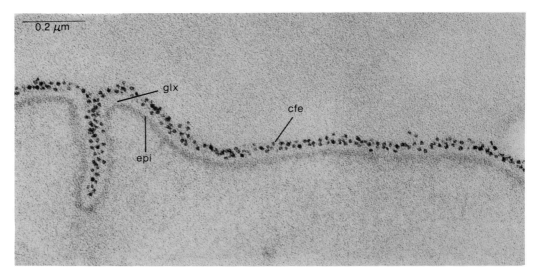

Fig. 14–5. Longitudinal section through the cuticle of an adult *Caenorhabditis briggsae* surface-labeled with cationized ferritin, showing the ferritin particles (cfe) on the external surface of the glycocalyx (glx), with the underlying epicuticle (epi), the dark band. (From Himmelhoch and Zuckerman. Exp. Parasitol. 43:161–168, 1978.)

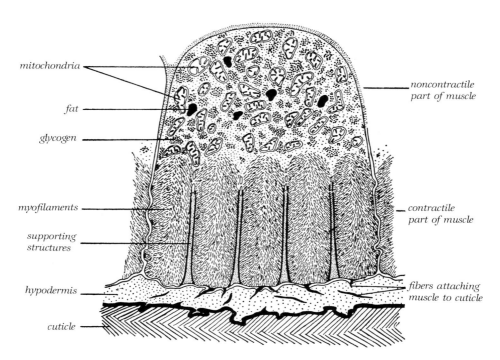

Fig. 14–6. Transverse section through a muscle of the body wall of *Nippostrongylus brasiliensis* toward the end of the muscle. Drawn from several electron micrographs. (From Lee: The Physiology of Nematodes, University Reviews in Biology, Oliver and Boyd.)

muscle system is dependent on the action of the internal pressure: volume. Thus, when the muscles contract, so as to straighten the body, they will elongate the contracted muscles of the opposite side of the body and there will be a concomitant increase in the internal pressure. However, the antagonising effect is only required to act on antero-posteriorly arranged muscle units but the pressure generated will act in all directions, radially as well as longitudinally. If there is no method of compensating for, or overcoming, this, the body would not straighten but would simply tend to swell and become, or tend to become, spherical. The cuticle must, therefore, be able . . . to stretch radially, that is it must be anisometric."[20]

The aforementioned concept may be too rigid because many nematodes are able to pull back the anterior part of the body, thus foreshortening and widening the body. Locomotion involves a series of backwardly moving waves, generated by the muscles and formed in a dorsoventral plane. A simplified model of forces and wave action was outlined by Croll[13] as follows:

"The combined effect of the hydrostatic pressure of the body fluids, the fibres in the elastic cuticle, and the longitudinal muscle bands, all in a flexible cylinder, provides a mechanism for propagating waves. Muscular contractions allow such changes in the overall shape of the nematode as determined by the cuticular structure and internal hydrostatic pressure."

DIGESTIVE SYSTEM

The alimentary tract of nematodes is a simple tube made of cells arranged in a single layer. The mouth leads to a buccal capsule (not always present), thence to a muscular esophagus (= pharynx) that empties into the intestine. Minute projections called *microvilli* line the inner surface of the intestine. The anus is located near the posterior tip of the worm. Intestinal cells generally are rich in mitochondria, Golgi complexes, ribosomes, glycogen, protein bodies, lipid or fat bodies, and endoplasmic reticulum. Gland cells within the oral and esophageal regions function in protein and mucopolysaccharide synthesis, and their products are shed into the digestive tract or directly to the exterior.

Food of parasitic nematodes consists of host tissues (*Nippostrongylus*), blood and tissues (hookworms) or various secretions and intestinal contents (*Ascaris*). A dog hookworm (*Ancylostoma*) sucks the ends of a group of intestinal villi into its buccal cavity, in which the action of digestive juices and the shearing of the teeth form a bolus of tissue that is separated from the host gut and is partially digested. The tissue is then drawn into the esophagus with any ingested blood and forced into the intestine by the pumping action of the muscular esophageal bulb. Additional digestion occurs in the intestine, but most of the red blood cells pass through the intestine intact.

Intestinal microvilli apparently absorb food and have a secretory function. The presence of numerous secretory granules and ribosomes in intestinal cell cytoplasm of such nematodes as *Nippostrongylus brasiliensis* indicates the synthesis and secretion of digestive enzymes. The intestine may also function as an excretory organ. For a review of nutrition in intestinal nematodes, see Lee.[22] See Chapter 17 for a discussion of physiology.

NERVOUS SYSTEM

The nervous system consists basically of a ring of nerve fibers and associated ganglia encircling the esophagus (see Fig. 14–1) and another nerve ring around the posterior region of the intestine. Longitudinal nerve cords, generally 4 main ones, connect the rings and extend to the extremities of the body. There are about 250 nerve cells in most nematodes, and these cells appear to have a neurosecretory function. Nematodes possess a sensitive receptor system and a high degree of nervous coordination. Sense organs (sensilla) include papillae, amphids, phasmids and pigment spots, and ocelli in some freeliving forms.

The **sensilla** of the nematode head include: (1) two bilaterally placed **amphids,** (2) two sets of six, small, **labial papillae** arranged in inner and outer rings, and (3) four, bilaterally situated **cephalic papillae** (Figs. 14–7 and 14–8). All of these sensilla have a basic tripartite structure consisting of one or more neurons, a sheath cell that provides a channel for the ciliated neuronal tips, and a socket cell that anchors the sheath cell in the cuticle (Fig. 14–9). In some nematodes the sheath cell also has a secretory function. In adult hookworms, for example, it is particularly large and secretes acetyl cholinesterase and, perhaps an anticoagulant as well.

The paired amphids (Fig. 14–10) are chemosensory organs that probably play a role in host recognition among parasitic species. Each am-

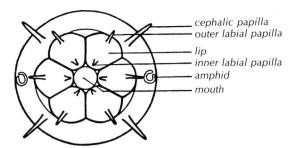

cephalic papilla
outer labial papilla
lip
inner labial papilla
amphid
mouth

Fig. 14–7. En face view of a nematode head showing the positions of the mouth, lips, amphids, and papillae. (From Jones; Plant Nematology. MAFF Technical Bulletin No. 7, courtesy of Her Majesty's Stationery Office.)

phidial channel opens through a pore on the lateral surface of the head. The ciliated tips of some of the amphidial neurons lie in this channel, and therefore, are exposed to stimuli from the outside. The tips of the remaining neurons enter more or less complex pockets in the sheath cell. Mutant nematodes with structurally abnormal amphids show altered behavior in a chemical gradient, providing evidence that the amphids are chemoreceptors.

In some species, the inner labial sensilla also open to the exterior, and, therefore, they too

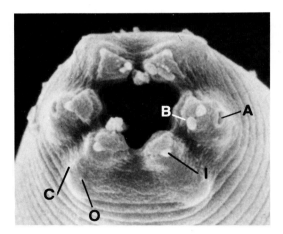

Fig. 14–8. Scanning electron micrograph of an en face view of *Caenorhabditis elegans* showing the triangular mouth opening surrounded by six lips. Each lip has a large internal labial papilla (I) and a minute outer labial papilla (O). The laterally situated slit-like openings (A) are the amphids. Four small bilaterally arranged, cephalic papillae are closely associated with the sublaterally placed outer labial papillae. Bacteria (B) are seen adhering to some of the lips. (Courtesy of Donald L. Riddle.)

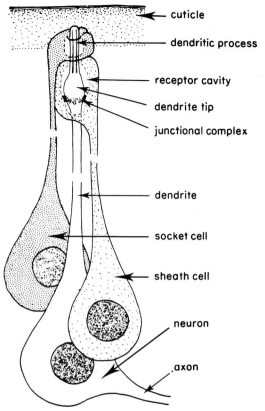

cuticle
dendritic process
receptor cavity
dendrite tip
junctional complex
dendrite
socket cell
sheath cell
neuron
axon

Fig. 14–9. Diagram of a cephalic sensillum of a nematode. The tip of the dendritic process ends in the cuticle indicating that the sensillum illustrated is a mechanoreceptor. In chemoreceptors, the receptor cavity opens through the cuticle to the exterior. (From Wright, K.A.: Nematode sense organs. *In* Nematodes as Biological Models, Volume 2. Edited by B.M. Zuckerman. New York, Academic Press, 1980, pp. 237–296.)

may have a chemosensory function (Fig. 14–11). The tips of the neurons of the remaining anterior sensilla end in the cuticle and probably function as mechanoreceptors (Fig. 14–12).

Phasmids are small paired sensilla, opening through the cuticle of the tail, with the same general structure found in the sensilla of the anterior end (Fig. 14–9). Because they open to the exterior, they too are thought to be chemoreceptors. Phasmids occur on the tail of a large group of nematodes, the Secernentea (= Phasmidia) and this distinguishes them from another group, the Andenophorea (= Aphasmidia).

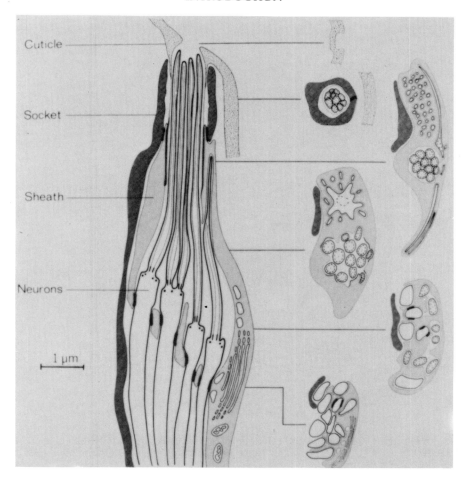

Cuticle

Socket

Sheath

Neurons

1 µm

Fig. 14–10. Diagram of an amphid of *Caenorhabditis elegans* in longitudinal and transverse section. The scale of the transverse sections is half that shown for the longitudinal section. (From Ward, Thomson, White, and Brenner: Electron reconstruction of the anterior sensory anatomy of the nematode *Caenorhabditis elegans*. J. Comp. Neurol., *160*:313–338, 1975.)

EXCRETORY SYSTEM

Considerable morphologic variation occurs among excretory systems. The class Secernentea possess longitudinal ducts opening to the outside through a midventral pore that is associated with ventral gland cells and an excretory duct. The longitudinal ducts are often connected by a cross tube, giving the whole apparatus an "H" shape.

The class Adenophorea usually has a single excretory gland that opens by way of a short duct through a pore on the midventral line. Ammonia is the major end product of nitrogen metabolism in the infective stages of the parasites, but urea and uric acid are frequently present.

See Rogers[30] for a review of nitrogenous components and their metabolism.

This system also has a secretory function. The so-called excretory cells produce and store enzymes (peptidases) that are activated and released through the excretory duct into the space between the old and the new cuticles at molting. These enzymes lyse the old cuticle in a specific weak zone, causing the anterior end of the worm's cuticular sheath to fracture and be shed (Fig. 14–13).

RESPIRATION AND CIRCULATION

Nematodes do not have blood vessels, but body fluids that contain hemoglobin bathe the

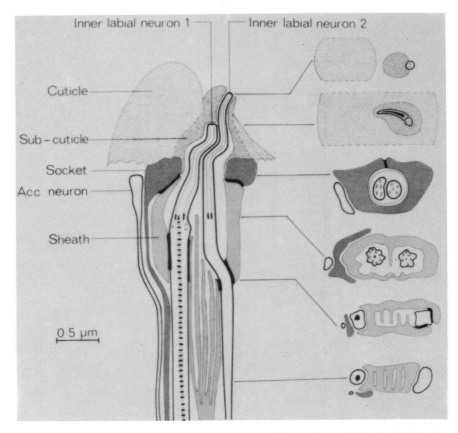

Fig. 14–11. Diagram of an inner labial sensillum from *Caenorhabditis elegans* in longitudinal and transverse sections. (From Ward, Thomson, White, and Brenner: Electron reconstruction of the anterior sensory anatomy of the nematode *Caenorhabditis elegans*. J. Comp. Neurol., *160*:313–338, 1975.)

various organs. Hemoglobins may also be found in the tissues of the body wall. These pigments have an affinity for oxygen in some, but perhaps not all, nematodes. The requirement for oxygen varies among worms. However, even those usually considered to be completely anaerobic may require minute amounts of free oxygen.

The hemoglobin in body fluids may be a by-product of digestion and may play no role in respiration, but hemoglobin in the body wall probably accepts oxygen from the surrounding medium. The common stomach worm of ruminants, *Haemonchus contortus*, and the cecal worm of fowl, *Heterakis gallinarum*, use molecular oxygen readily under laboratory conditions, and presumably, their tissue hemoglobin plays the same role in vivo. See Chapter 17 for a discussion of physiology.

REPRODUCTIVE SYSTEM

Nematodes are usually dioecious and commonly exhibit sexual dimorphism. This condition, in which one sex differs in size or shape from the other, reaches an extreme in *Trichosomoides crassicauda*, a parasite of the urinary bladder of rats. The male worm of this species is a tiny creature compared with the female, and lives parasitically in her uterus.

Parthenogenesis, often mitotic with polyploidy, is found in the order Tylenchida, which predominantly parasitizes plants, and in the order Rhabditida, which has species infecting amphibians, reptiles, and mammals. In the former order, various types of parthenogenesis supplement or replace amphimixis. The rhabditoids occurring in the lungs of amphibians and reptiles are generally conceded to be gynomorphic

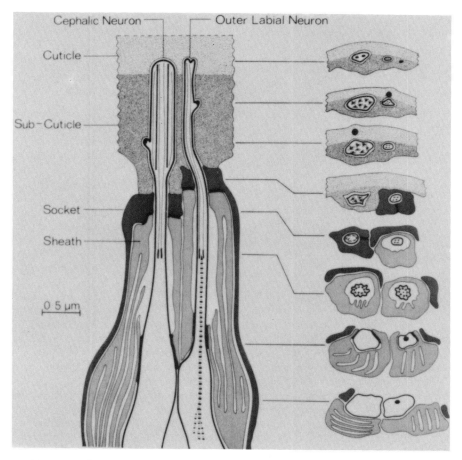

Fig. 14–12.   Diagrams of the outer labial and cephalic sensilla from *C. elegans* in longitudinal and transverse sections. (From Ward, Thomson, White, and Brenner: Electron reconstruction of the anterior sensory anatomy of the nematode *Caenorhabditis elegans*. J. Comp. Neurol., *160*:313–338, 1975.)

hermaphrodites. Such a development possibly arose in situations in which the opportunity for males and females to meet was limited. Female *Strongyloides* living in the intestinal wall of mammals are generally thought to be parthenogenetic (see Chap. 15). *S. ratti* has a diploid chromosome number of 6 in females and 5 in males. *S. papillosus*, a parasite of sheep, apparently has a diploid chromosome number of 4 in free-living adults of both sexes. In some species of *Strongyloides*, the sperm enters the oocyte, but takes no further part in development, a situation called "pseudogamy."

The female reproductive system consists of one or two coiled tubules uniting to form a vagina that opens through a vulva. The vulva is usually located on the anterior portion of the body. The distal ends of the tubes form the ova-ries, the portions next to them are the oviducts, and the remainder are the uteri. The anterior glandular portion of the uterus has a high metabolic and synthetic activity. Lipids tend to be abundant in reproductive organs of male and female nematodes.

Eggs (Fig. 14–14) vary considerably in appearance. The eggshell consists of three major layers. The outer or *vitelline layer* is submicroscopic and possibly originates from the oolemma. It is covered with a thin uterine coat. The next, *chitinous layer* is the most distinct and contains various amounts of chitin. The innermost lipid layer is the last to be laid down, and is presumably responsible for the shell's impermeability. Eggshell proteins contain about 35% proline. See Bird[5] and Wharton.[35]

Male worms possess reproductive organs that

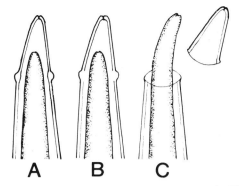

Fig. 14–13. Steps in the molting of a nematode. The diagram shows a larva with a detached cuticular sheath (A), that has been lysed internally at an anterior, circumferential fracture zone (B), and that has cast off the tip of the sheath and is emerging (C). (Redrawn from Lee and Atkinson.)

are also modifications of long coiled tubes. The worms usually have only one testis, which is the distal end of a tube that continues as the vas deferens and joins the lower end of the gut at the cloaca. Before this junction, the vas deferens enlarges and forms a seminal vesicle, a storage sac for sperm. The sperm of some species resemble amebas, those of others are spindle-like. The terminal end of the male organs may be called the ejaculatory duct. Transfer of sperm to the female worm is aided by a pair of **spicules** (Fig. 14–1) in many species of roundworms. These long, hardened structures arise from primordia in the rectal epithelium. They are composed largely of protein and may be thrust out through the cloaca and may serve the additional function of tactile sensory organs, or receptors for pheromones emitted by female worms.

Another male structure, the **gubernaculum,** is a sclerotized thickening of the cuticle of the spicule pouch; it lies on the dorsal side of the cloaca and probably helps to guide the spicules as they are thrust out. A larger, less well-developed organ, the **telemon,** lies on the ventral and lateral walls of the cloaca. It too develops as a thickening of the cuticle of the cloacal lining and also helps to direct the spicules in copulation. The posterior end of the male of some species is flared and curved in such a manner as to suggest a hood. This structure, the **copulatory bursa** (see Figures 15–5 and 15–11), helps the male hold itself to the female during copulation. The walls of the bursa may be supported by finger-like rays. The number and arrangement of these rays serve as diagnostic features in species identification.

LIFE CYCLES

Life cycles take a variety of forms among the thousands of species of nematodes that infect animals. Early developmental stages within the egg are basically the same for all nematodes and are well represented in Figure 14–15, which illustrates embryogenesis in the egg of *Contracaecum aduncum*, a parasite of marine fish. It has been suggested that larvae within eggs of the hookworm, *Necator americanus*, pump fluid into the intestine by means of the esophagus. This "feeding" behavior results in osmotic changes that flush enzymes into the egg and cause membrane changes, an influx of water, and distention of the egg. Rotary movements of the worm's stoma against the egg membranes release the pressure and cause a break in the egg shell through which the larva passes.[12]

One of the simplest life cycles is exhibited by *Trichuris* (= *Trichocephalus*) *trichiura*, the common whipworm of man (described in Chapter 17), and by similar species in many other mammals. Eggs of whipworms pass out of the body with feces and develop into embryos within a few weeks. They are then infective to a new host and gain entrance by being ingested. Embryonated eggs may remain viable for many months if they remain in moist areas. When ingested, they pass to the cecum where they hatch and, in about four weeks, the young worms mature. As is characteristic of nematodes, the sexes are separate. Mating occurs as soon as the adult stage is reached, and not long thereafter the females start producing eggs.

In many species, the eggs hatch in the external environment and the larvae, living in moisture films, feed on the fecal or soil microflora and fauna until they reach the infective third larval stage (L_3). At this stage, feeding ceases, growth and development are interrupted, and the larvae move up onto soil surface particles or low vegetation as far as the moisture films extend. There they remain, surviving on stored food, until a host ingests them, or, in some skin-penetrating species, a host makes contact with them. The free-living, microbiverous first and second stage larvae (L_1 and L_2) of these nematodes have a short, *rhabditiform* esophagus with

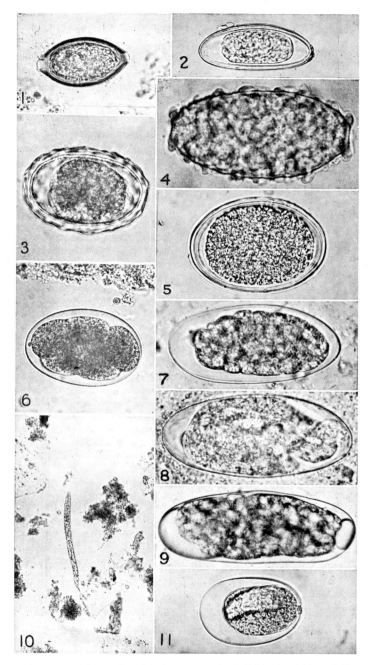

Fig. 14–14. Nematode eggs and larvae found in human stools. *1, Trichuris trichiura. 2, Enterobius vermicularis. 3, Ascaris lumbricoides*, fertilized egg. *4, Ascaris*, unfertilized eggs. *5, Ascaris*, decorticated egg. *6,* Hookworm. *7, Trichostrongylus orientalis*, immature egg. *8, Trichostrongylus*, embryonated egg. *9, Heterodera marioni. 10, Strongyloides stercoralis*, rhabditiform larva. *11, Strongyloides* egg (rarely seen in stool). All figures ×500 except *10* (×75) (From Hunter, Frye, and Swartzwelder: A Manual of Tropical Medicine. Philadelphia, W.B. Saunders.)

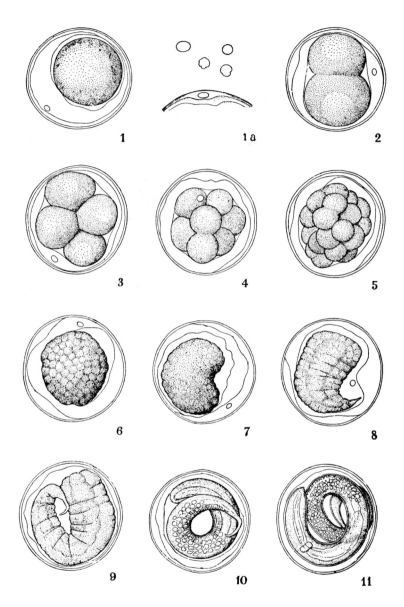

Fig. 14–15. Development within the egg of *Contracaecum aduncum,* a nematode of fish. (From St. Markowski. Courtesy of the Bulletin de l'Académie Polonaise des Sciences.)

a definite, posteriorly placed bulb, whereas the non-feeding, infective third-stage larva (L₃) has a less distinctly bulbed *strongyliform* or *filariform* esophagus. The types of freeliving larvae have been named on the basis of these structural differences (Fig. 14–16).

Skin-penetrating infective larvae are generally thought to migrate to the intestines via the blood, heart, lungs and respiratory tree, pharynx, esophagus and stomach. This migratory pathway is well established for hookworms[3] but, for some other species for which it has been accepted as fact, recent studies have raised serious questions.[36] Even species that infect orally undergo complex migrations through the host before returning to the gastrointestinal tract to reside as adult worms. For instance, the larvae of *Strongylus vulgaris*, the most important helminth of horses, migrate from the intestine under the intima of the mesenteric arteries, causing serious pathologic changes manifesting as colic. After several months of development in the arteries, the parasites return to the colon, where they mature, attach to the mucosa and feed on the blood and tissue fragments.

Other species have *indirect* life cycles involving intermediate hosts. These hosts are often ingested by definitive hosts, but in some host-parasite systems, the intermediate host is a biting arthropod (e.g., tick, mosquito, or flea) which actively carries L₃ to the definitive host, thus serving as a vector. This mode of transmission is particularly characteristic of the *fila-*rial worms, most of which are transmitted by haematophagous arthropods (see Chapter 16). *Wuchereria bancrofti*, the causative agent of *elephantiasis* in man, is one of the best-known examples of the group. Depending on the particular host-parasite association, the filariae develop to the infective stage in the muscles, fat body, Malpighian tubes or hemocoel. Wherever they develop, the juvenile worms (see Fig. 16–17) are usually specific to the particular type of tissue selected. These larval worms pass through morphologically distinct stages. The stage taken up from the blood, the microfilaria (Fig. 14–17), is a slender, motile embryo that sometimes retains the vitelline membrane as a sheath. In the insect it is transformed into a more complex but shorter "sausage stage" larva which elongates in later stages to beome an infective larva adapted to transmission by an arthropod's bite (see Fig. 16–17).

The widely held view that larvae complete a strictly delineated itinerary through a sequence of organ systems during their migration is strongly supported by some authors (e.g., Ulmer[33]) but challenged by others. Croll,[13] for example, has provided evidence that the clumped aggregations of *Nippostrongylus brasiliensis* (Chap. 15) in the host's small intestine are directly correlated to the dispersion of host food boli. Habitat selection of this worm, and probably of other species, apparently results at least in part to random forces involving such

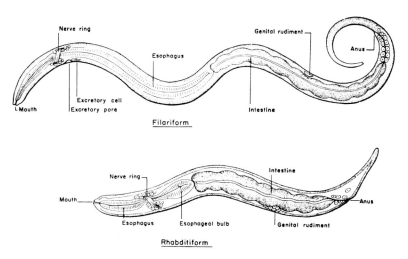

Fig. 14–16. Free-living larvae of intestinal roundworms. *A*, Filariform; *B*, Rhabditiform. (From the United States Navy Medical School Laboratory Guide.)

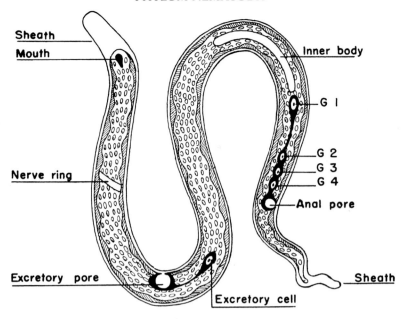

Fig. 14–17. An embryonic nematode (microfilaria) showing the sheath and other structures. G1 to G4, genital cells. (From the United States Navy Medical School Laboratory Guide.)

factors as the "plumbing and engineering of the host's organs."[13]

A number of studies on sex attractants indicate that pheromones produced by each sex in many species attract the opposite sex. One sex is frequently more strongly attracted than the other. Chemical attractants are also involved in the recognition of both suitable hosts and sites within hosts. (See Chapter 15.)

INFECTIVE STAGE

The infective stage, whether a larva or an egg, is a bridge between free-living and parasitic modes of life. The organism does not feed or grow and thus is different from other stages. It must be able to "recognize" the host and to withstand marked environmental changes (e.g., temperature, osmotic pressure, pH). Its food reserves are carbohydrate or fat, and although the need for oxygen varies greatly, oxygen is consumed when available at atmospheric pressure. The mechanism of osmoregulation of infective stages is unknown, but permeability of eggshells is known to be low.

Among animal parasitic species, the embryonated egg is often the infective stage; in-variably, therefore, these species enter their hosts via the mouth. Other nematodes produce eggs that hatch in the external environment. The larvae that emerge feed, grow, molt twice, and come to the surface where they wait for an opportunity to enter a host. The infective larvae are often environmentally resistant, non-feeding, dispersal stages with sealed mouths, fully ensheathed in the uncast loose-fitting cuticle of the preceding stage. Parasites of grazing animals are generally infective via the mouth, while those of other animals may infect orally, or, in some cases, by skin penetration.

Almost invariably, when there are a series of freeliving stages, these include only three larval stages (L_{1-3}) with L_3 being infective. Species of the genus *Strongyloides* are an exception to this rule because a facultative, freeliving adult stage may occur. These nematodes are highly sensitive to environmental clues and, when conditions in the host become unfavorable (e.g., immune resistance increases) or external ecologic conditions become particularly favorable, one or more freeliving generations may occur (Chap. 15).

A complex of environmental, host- and parasite-specific factors affect the success of the infective process, whether this involves hatching of eggs in the gastrointestinal tract, penetration of skin, or invasion of the mucosa. These factors include, for example, stimuli provided by the host (CO_2, warmth), age and motility of the larvae, the thickness of the water films in which the larvae occur, and many others. Exactly how the parasites are stimulated to contact, achieve attachment and penetrate the host is incompletely known. However, recent work suggests that entry into and survival in the skin involves stimulation of parasite penetration by fatty acids (e.g., linoleic acid) and triggering of a series of biochemical changes with the ultimate production of prostaglandins, some of which are immunodepressive. It has been suggested that this facilitates the survival of parasites that have entered the host (Fusco, Salevsky, and Delbrook).[18]

Under certain conditions some species remain dormant in the subcutaneous or other tissues for a prolonged period after invasion of the host. Many factors influence this interruption of migration and development. Parasites may have a genetically or environmentally programmed, interrupted developmental period. Alternatively, this may be density dependent or influenced by innate or acquired host resistance.

Many parasitic roundworms, on entering an arthropod host, become surrounded by chitin or other special membranes or become impregnated with a brown deposit forming a "capsule" around them. Studies have shown that this encapsulation or melanin deposit is a defense mechanism on the part of the host. Sometimes, the reaction occurs only when a parasite is introduced into an abnormal host.

MOLTING

Four molts occur during the life cycles of nematodes. Shedding of the cuticle occurs by deposition of a new cuticle and **exsheathment** (= **ecdysis**) of the old. Molting appears to be under hormonal control. Evidence[7] suggests that cuticle formation is regulated by messenger RNA specific for molting. The initial stimulus for each molt varies with the species. For *Haemonchus contortus*, *Trichostrongylus colubriformis*, and the like, the first parasitic molt is stimulated by the host's undissolved gaseous carbon dioxide and/or undissociated carbonic acid. The latter species and *Dictyocaulus viviparus* require pepsin, whereas *Haemonchus* larvae are indifferent to

pepsin. The triggering stimulus for *Nematodirus battus* is hydrochloric acid. The cuticle may be shed in one piece, or the anterior end may be shed separately as a cap.

The process of exsheathment in *Phocanema* (= *Porrocaecum*), a larval parasite of muscles of codfish and an adult parasite of the gut of seals, and probably in other species of nematodes, is under neurosecretory control. On the basis of reports from several workers, including their own observations, Davey and Summerville[16] presented the following hypothesis for ecdysis in *P. decipiens*.

"Environmental conditions present in the digestive tract of seals . . . lead to the activation of neurosecretory cells in the ganglia associated with the nerve ring. One or more hormones released from these neurosecretory cells act upon the excretory cell, leading to an increase in the activity of aminopeptidase and other enzymes and to the subsequent secretion of these enzymes into the space between the two cuticles. These enzymes affect the cuticle in such a way as to weaken it in a circumferential ring at about the level of the nerve ring. The cuticle breaks at this circumferential weakening, and the last stage wriggles free."

An increase of water in the excretory cell activates the enzymes that are stored in a crystalline form.

RESTING STAGES

Most parasites require the ability to interrupt development and to enter a resting state such as **diapause** or **quiescence**. Diapause is arrested development resulting from a response to a seasonal change, for example, temperature, in the environment of an infective nematode larva, and it is related to a failure to resume development following a molt. It may also occur in the egg. To date, a diapausing stage in nematodes has been clearly established for only a few species. Quiescence is inhibited or interrupted development resulting from such host-associated factors as dehydration, lack of oxygen, high salt concentration in plants, or the action of immunity in animals.[17]

Developmental arrest in nematodes frequently has more than a single function. It can be regulatory. For example, "storage of larvae in the host in a quiescent form limits large oscillations in parasite abundance or biomass. . . . Large oscillations place the host, the parasite, or both, in jeopardy, and strong selection for mechanisms to dampen these oscillations would be expected."[31]

The significance of arrested development has been interpreted in two different ways: (1) It performs

"a regulatory function, tending to maintain a constant burden of adult worms. . . . On this basis, arrested development is seen as a consequence of host resistance and effects of the weight of infection and of the presence of adult worms are regarded as central phenomena. The renewed development of arrested larvae is believed to be controlled by a sensitive feedback mechanism which allows larvae to develop in sufficient numbers to replace adult worms that are lost."

(2) Arrested development fulfills

"the function of synchronizing the life history of the parasite with that of its host or with seasonal changes in the outside environment. Punctuality (which may be defined as being in the right place at the right time) is the essence of successful parasitism and the most effective aid to punctuality is the ability to mark time. . . . As a means of synchronization, arrested development implies a response to signals either to induce development to be halted or to cause it to be resumed, or both."[26]

Two patterns of synchronization are discernible: one with the seasons and the other with reproduction of the host.

ENDOCRINE RELATIONSHIPS

Nematode life cycles are closely synchronized with those of the hosts. For example, the development of some tylenchoids is blocked during the overwintering of the insect host. Numerous studies describe the correlation between molting of an insect and the life history of its parasites (see the account of flagellate sexual development in wood roaches in Chapter 23). A report[29] that α-ecdysone, the molting hormone of insects, resembles the ecdysial hormone of the nematode parasite *Phocanema depressum* would seem to have wide and important implications. However, Davey and Hominick[15] have made a careful and detailed review of "the evidence for a possible endocrine basis for the remarkable synchrony often observed between the development of entomophilic nematodes and that of their hosts . . ." and they concluded that "there is no rigorous

evidence which points to such an association." Davey[14] has stated that

"a hypothesis which involves the neurosecretory system as the mediator of the effects of JH [juvenile hormone] and steroids will explain all of the observations which have so far come forward. Thus, the nematode is responding to these exogenous hormones via its own endocrine system. The occurrence of JH and ecdysone are regarded simply as part of the normal pattern of distribution of these compounds, which are ubiquitous."

The function of neurosecretory cells and the shedding of the cuticle has already been described. Functions of other neurosecretory cells (for example, primary sense cells in the lips of *Ascaris*, neurosecretory vesicles in a specialized area of the cuticle near the excretory pore of *Haemonchus contortus*) are little understood.

See Davey[14] for a review of hormones in nematodes, and Poinar.[27]

PARATENESIS

Nematode larvae commonly migrate through an unnatural or inadequate host in which the larvae do not develop into adults. For example, larvae of the cat and dog hookworm, *Ancylostoma brasiliensis*, may wander into the skin of man, causing "creeping eruption" characterized by visible tracks with red, painful, and swollen advancing ends, associated with intense itching. This kind of infection is called **cutaneous larva migrans**. A similar larval infection in internal organs, as occurs with larval *Toxocara canis* in man, is called **visceral larva migrans**.

Visceral larva migrans, as defined by Sprent,[32] is

"a biologic phenomenon to which all animals are subject . . . [comprising] . . . the invasion of, and migration through, any of the tissue of the animal body by nematode larvae, whether they be derived from nematodes which are natural parasites of the affected host, or from parasites which normally mature in other hosts."

Beaver,[2] however, recommends the restriction of the definition "to include only the prolonged migration and long persistence of larvae whose behavior clearly reflects that which occurs in a norml intermediate or paratenic host." Examples in man include larval migrations of *Toxocara canis, Gnathostoma spinigerum,* and certain species of the hookworm *Ancylostoma*. Beaver introduced the term **paratenesis** to

"denote the passage of infective-stage larvae without essential development through a series of transport (paratenic) hosts to the final host with the transport host serving at the same time to maintain the infective-stage larvae from one season of transmission to the next."

Much more work needs to be done on larval morphology, behavior, physiology, and immunity.

For epidemiology and control of some nematode infections of grazing animals, see Michel.[25] For general references to the nematodes, see Levine,[23] Bird, Chitwood and Chitwood[9] Chitwood,[10] and Lee.[22] For speciation, see Inglis.[19] For keys to nematode parasites of vertebrates and invertebrates, see Anderson et al.,[1] and Poinar,[28] respectively. For parasites of humans, see Warren and Mahmoud.[34]

REFERENCES

1. Anderson, R.C., Chabaud, A.G., and Willmott, S. (eds.): C.I.H. Keys to the Nematode Parasites of Vertebrates. No. 1. St. Albans, England, Commonwealth Institute of Helminthology, 1974.
2. Beaver, P.C.: The nature of visceral larva migrans. J. Parasitol., 55:3–12, 1969.
3. Behnke, J.M., Paul, V., and Rajasekariah, G.R.: An improved technique for experimental infections with skin penetrating nematode larvae (*Necator americanus*). Int. J. Parasitol., 16:461–464, 1986.
4. Bird, A.F.: The Structure of Nematodes. New York, Academic Press, 1971.
5. Bird, A.F.: The development and organization of skeletal structures in nematodes. *In* The Organisation of Nematodes. Edited by N.A. Croll. New York, Academic Press, 1976, pp. 107–137.
6. Bird, A.F.: The nematode cuticle and its surface. *In* Nematodes as Biological Models. Vol. 2. Edited by B.M. Zuckerman. New York, Academic Press, 1980, pp. 213–234.
7. Bonner, T.P., Evans, K., and Kline, L.: Cuticle formation in parasitic nematodes: RNA biosynthesis and control of molting. Int. J. Parasitol., 6:473–477, 1976.
8. Brand, von T.: Biochemistry and Physiology of Endoparasites. Amsterdam, Elsevier/North-Holland Biochemical Press, 1979.
9. Chitwood, B.G., and Chitwood, M.B.: Introduction to Nematology. Baltimore, University Park Press, 1974.
10. Chitwood, M.B.: The systematics of biology of some parasitic nematodes. *In* Chemical Zoology. Vol. III. Edited by M. Florkin and B.J. Scheer. New York, Academic Press, 1969, pp. 233–244.
11. Croll, N.A. (ed.): The Organisation of Nematodes. New York, Academic Press, 1976.
12. Croll, N.A.: *Necator americanus*: activity patterns in the egg and the mechanism of hatching. Exp. Parasitol., 35:80–85, 1974.
13. Croll, N.A.: The Behavior of Nematodes: Their Activity, Senses and Responses. New York, St. Martin's Press, 1970.
14. Davey, K.G.: Horomones in nematodes. *In* The Organisation of Nematodes. Edited by N.A. Croll. New York, Academic Press, 1976, pp. 273–291.
15. Davey, K.G., and Hominick, W.M.: Endocrine relationships between nematodes and their insect hosts—a review. Exp. Parasitol., 33:212–225, 1973.
16. Davey, K.G., and Summerville, R.I.: Molting in a parasitic nematode, *Phocanema decipiens*—VII. The mode of action of the ecdysial hormone. Int. J. Parasitol., 4:241–259, 1974.
17. Evans, A., and Perry, R.: Survival strategies in nematodes. *In* The Organisation of Nematodes. Edited by N.A. Croll. New York, Academic Press, 1976, pp. 383–424.
18. Fusco, A.C., Salevsky, and Delbrook: *Schistosoma mansoni*: production of cercarial eicoconoids as correlates of penetration and transformation. J. Parasitol., 72:397–404, 1986.
19. Inglis, W.G.: Speciation in parasitic nematodes. Adv. Parasitol., 9:185–223, 1971.
20. Inglis, W.G.: the structure of the nematode cuticle. Proc. Zool. Soc. Lond., 143:465–502, 1964.
21. Lee, D.L.: The structure of the helminth cuticle. Adv. Parasitol., 10:347–379, 1972.
22. Lee, D.L., and Atkinson, H.J.: Physiology of Nematodes. New York, Columbia University Press, 1977.
23. Levine, N.D.: Nematode Parasites of Domestic Animals and of Man. 2nd Ed. Minneapolis, Burgess, 1980.
24. Lumsden, R.D.: Surface ultrastructure and cytochemistry of parasitic helminths. Exp. Parasitol., 37:267–339, 1975.
25. Michel, J.F.: The epidemiology and control of some nematode infections in grazing animals. Adv. Parasitol., 14:355–397, 1976.
26. Michel, J.F.: Arrested development of nematodes and some related phenomena. Adv. Parasitol., 12:279–366, 1974.
27. Poinar, G.O.: Entomogenous Nematodes. A Manual and Host-List for Insect-Nematode Associations. Leiden, E.J. Brill, 1975.
28. Poinar, G.O., Jr.: C.I.H. Keys to the Groups and Genera of Nematode Parasites of Invertebrates. St. Albans, England, Commonwealth Institute of Helminthology, 1974.
29. Rajulu, G.S., Kulasekarapandian, S., and Krishnan, N.: Nature of the hormone from a nematode *Phocanema depressum* Baylis. Curr. Sci., 41:67–68, 1972.
30. Rogers, W.P.: Nitrogenous components and their metabolism: Acanthocephala and Nematoda. *In* Chemical Zoology. Edited by M. Florkin and B.J. Scheer. Vol. III. New York, Academic Press, 1969, pp. 379–428.
31. Schad, G.A.: The role of arrested development in the regulation of nematode populations. *In* Regulation of Parasite Populations. Edited by G.W

Esch. New York, Academic Press, 1977, pp. 111–162.

32. Sprent, J.F.A.: Visceral larva migrans. Aust. J. Sci., 25:344–354, 1963.

33. Ulmer, M.J.: Site-finding behavior in helminths in intermediate and definitive hosts. *In* Ecology and Physiology of Parasites. Edited by M. Fallis. Toronto, University of Toronto Press, 1971.

34. Warren, K.S., and Mahmoud, A.A.I.: Tropical and Geographic Medicine. New York, McGraw-Hill, 1984.

35. Wharton, D.: Nematode egg-shells. Parasitology, *81*:447–463, 1980.

36. Wilson, P.A.G.: Roundworm juvenile migration in mammals: The pathways of skin penetration reconsidered. *In* Aspects of Parasitology: A Festschrift Dedicated to the 50th Anniversary of the Institute of Parasitology, McGill University, 1932–1982. Edited by E. Meerovitch. Montreal, McGill University Press, 1982.

37. Wright, K.A.: Nematode sense organs. *In* Nematodes as Biological Models. Vol. 2. Edited by B.M. Zuckerman. New York, Academic Press, 1980, pp. 237–296.

15

Class Secernentea (= Phasmidia), Orders Tylenchida, Rhabditida, Strongylida

Order Tylenchida

Most of the important plant parasitic nematodes belong to the order Tylenchida; others are in the order Dorylaimida. These nematodes are commonly included only in courses in plant pathology. A voluminous literature exists, including books and journals devoted exclusively to nematode-plant relationships, with particular emphasis on plants of agricultural importance. We can do no more than present a brief introduction to the subject. In this presentation, only the genera *Xiphinema* and *Trichodorus* have been selected from the order Dorylaimida (class Adenophorea). All other genera belong to the order Tylenchida.

The nematode parasites of plants are often called "eelworms." They do enormous damage to cultivated plants. In the United States alone, the cost has been estimated to be over $1.5 billion annually. This loss is due to direct damage done by the worms and by other organisms that enter the damaged areas. Nematodes may transmit plant viruses.

In structure, plant nematodes are much like the familiar ascarid worms of vertebrate digestive tracts, but they are very much smaller. The smallest are only about 400 μm long, and few reach a length much over 3 to 4 mm. One of the most characteristic structures in these worms is a stylet, or spear, in the anterior end. This structure apparently helps the parasites to penetrate plant tissues. Life cycle stages consist of eggs, larvae, and adults, but the details vary, depending on the species of nematode, species of plant host, and temperature and other environmental factors. Various species of Tylenchida reproduce by parthenogenesis as well as by amphimixis. In general, female worms lay eggs that hatch either in soil or in the host plant. If host plants are not available, eggs frequently do not hatch, but remain dormant for years. Even larval stages of some forms can remain alive for a surprisingly long time. The larvae of the wheat nematode *Anguina tritici*, for example, can live in galls for 20 years. In any event, once nematode larvae enter a plant, they begin to feed on plant tissues. All crop and ornamental plants are probably attacked by nematode parasites.

Some nematodes are ectoparasitic on parts of plants that appear above the ground. *Aphelenchoides besseyi*, for example, feeds on leaves or on the developing buds. This parasite may also be endoparasitic, feeding on tissues within the stems and leaves of the strawberry plant. *A. ritzema-bosi* may even be ectoparasitic on some parts of gooseberry or blackberry plants and endoparasitic on other parts. Apparently, the host plant determines the nature of parasitism. Gall formation is one of the responses to endoparasitic activity. Galls may be formed in stem tissues, in a leaf, or even in flowering tissues. *Anguina* spp. produce galls in various flowering and other parts of plants. *Ditylenchus dipsaci* is a stem and bulb nematode that lives in the stems of wheat, alfalfa, and potato and in the leaves of onion, daffodil, garlic, and other plants. Intercellular lamellae of the host plant break down, and the tissues become loose and spongy. Secondary infection may be produced by bacteria or fungi. Dried *D. dipsaci* may live for as long as 21 years.

Underground parts of plants may be attacked by either ectoparasites or endoparasites or by both. The ectoparasitic worms simply lay their eggs on or near roots, and when young worms hatch, they start to feed on the tender parts of roots, such as the tips. Plant roots swell, and the plant becomes stunted or dies. Many species penetrate the underground parts of plants (Fig. 15–1).

Meloidogyne is one of many genera that cause gall formation. Galls have given rise to the name "root-knot" nematodes for the group. Thousands of plants in a small area may be attacked, resulting in wilting or death. Damage is done to many farm and garden plants including potatoes, peas, cereals, tomatoes, sugar beets, beans, clover, watermelon, and decorative plants. Unfortunately, the presence of parasites is not usually known until they occur in such large numbers that serious damage is well under way. *Meloidogyne marioni* (Fig. 15–2) is an example of one species that causes great damage among cultivated plants.

For a detailed account of nematodes parasitic in plants, see Zuckerman et al. 1971.[26] For host-parasite relationships, see Webster.[23]

Order Rhabditida
SUPERFAMILY RHABDIASOIDEA

The evolution of a parasitic mode of life among roundworms might well have started with groups like the rhabditoids. Many of these worms live in decaying flesh, dung, decomposing plant material, or similar substances from which transfer to the intestine of an animal or the tissues of a plant was probably easy. Members of this large order possess head sensory organs in the form of papillae, and the amphids are reduced to small pockets.

Strongyloides stercoralis is a relatively common, although focally distributed, nematode parasitic in humans, primates, and dogs. In humans, it is most prevalent in warm, moist parts of the world with inadequate sanitation. However, foci of high prevalence occur elsewhere, for example in some mountain valleys of Kentucky, U.S.A., where at least 3% of the rural children are infected. The parasite is common in puppies from pet shops or breeding kennels having poor hygienic conditions.

In dogs, and probably in humans as well, the infection is usually self-limiting, with the host becoming resistant to reinfection. However, approximately one-third of infected hosts fail to eliminate their parasites completely and harbor chronic infections, some of which persist in humans for more than 35 years.[8] This extraordinary persistence is attributed to this parasite's rare ability to reproduce within the host by autogenous, internal reinfection (*autoinfection*). Autoinfection is also the basis for two particularly severe forms of strongyloidiasis, namely

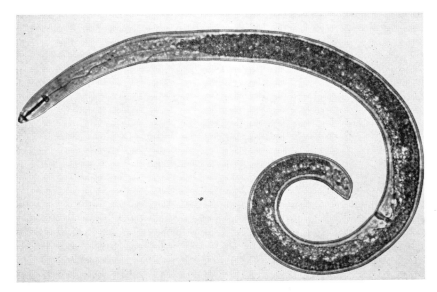

Fig. 15–1.   Mature female of *Helicotylenchus dihystera,* a plant pathogenic nematode. (Courtesy of W.T. Mai, Cornell University.)

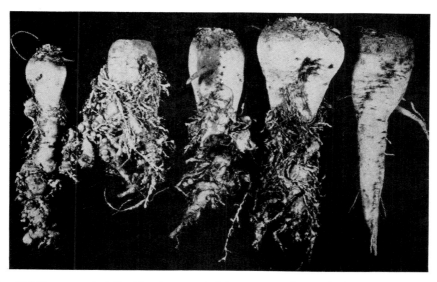

Fig. 15–2. *Meloidogyne marioni.* Root-knot nematode injury to sugar beets. The beet on the right is from soil treated with a nematocide. (From Thomason and Lear. Courtesy of California Agriculture. University of California, Division of Agricultural Sciences.)

hyperinfection and disseminated infection. These occur in immunodepressed individuals, including humans receiving organ transplants or patients with intercurrent immunosuppressive disease, such as acquired immunodeficiency syndrome (AIDS).

The parasitic adult is a small, thin female (Fig. 15–3), measuring 2 to 2.5 mm in length, that burrows into the mucosa of the small intestine. The common name of any species of this genus, "threadworm," aptly describes the parasite's form and relationship to the mucosa through which it is "threaded." The eggs are produced by mitotic parthenogenesis and embryonate in the mucosa. They hatch, liberating a rhabditiform larva (L_1), either into the mucosa per se or into the intestinal lumen (Fig. 15–3). Usually the rhabditiform larvae are voided in the feces and they initiate the free-living phase of the life history. Occasionally, some rhabditiform larvae develop so rapidly that they attain the infective filariform, third-stage internally. These precocious larvae invade the tissues of the infected individual in the large intestine. This is called *autoinfection.*

Rhabditiform larvae that pass to the outside in feces can develop further by one of two pathways. In the simplest case, female rhabditiform larvae develop directly into infective, skin-penetrating, third-stage larvae (Fig. 15–3, Direct Cycle). These larvae are thought to migrate to the intestine via the blood, lungs, airways, mouth, esophagus, and stomach. Recently, this almost universally accepted migratory pathway has been challenged by Wilson,[24] who argues that the concept is based on a poorly designed, over-interpreted experiment involving too few experimental animals. In this connection, it is of interest that a related species, *Strongyloides ratti,* a particularly well-studied experimental model, does not appear to migrate via the circulation, lung, and airways. Instead it migrates subcutaneously to the head and then apparently moves to the intestine via the mouth, esophagus, and stomach.

When rhabditiform larvae passed in the feces develop directly to third-stage larvae, the life cycle is said to be *direct.* It is also called the *homogonic* cycle. This contrasts with the indirect or *heterogonic* cycle in which, after a total of 4 larval stages, adult females and males develop, giving rise to a free-living generation (Fig. 15–3 [not all larval stages are shown]). The inclusion of a free-living generation in a life cycle occurs in several species of *Strongyloides,* but is virtually unknown in other groups of parasitic helminths.

The factors that determine whether *Strongyloides stercoralis* larvae develop precociously and enter an autoinfective cycle in the host, or

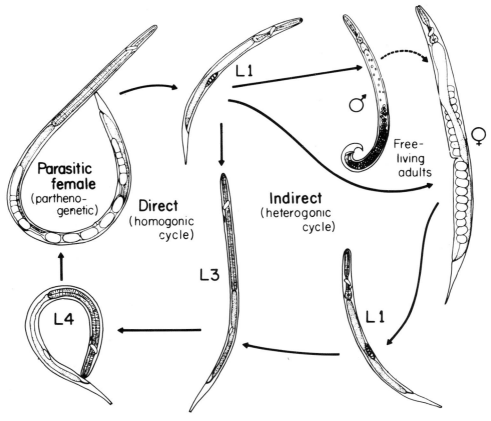

Fig. 15–3. Diagrammatic representations of the morphologic stages occurring in the direct (homogonic) and indirect (heterogonic) life cycles of nematodes of the genus *Strongyloides*. The rhabditiform (L_1) and filariform (L_3) larval stages are shown. The L_2 (not shown) is larger than the L_1, but has the same rhabditiform structure. Note the marked dimorphism shown by the parasitic and free-living adult females. (From Georgi: Strongyloidiasis. Courtesy of *1982 CRC Handbook Series in Zoonoses.*)

whether they follow the homogonic or heterogonic pathway in the external environment, are poorly understood. The occurrence of autoinfection appears to be governed, at least in part, by an increased intestinal transit time and a compromised immunologic responsiveness. The more frequent occurrence of hyperinfection in immunodepressed hosts, as compared to fully immunocompetent hosts, suggests that the parasite's developmental rate and survival during migration are immunosensitive.

With reference to external development, few studies have involved *Strongyloides stercoralis*, but investigations of *Strongyloides ransomi* and *Strongyloides papillosus* indicate that external or internal environmental factors determine the pathway that is taken. In a masterful synthesis of the evidence available to date, Georgi[7] suggests the following control mechanisms based largely on the work of Triantaphyllou and Moncol.[12,22] Females, embedded in the intestinal epithelium, produce eggs, that are genotypically female, by mitotic parthenogenesis. Internal environmental factors (host species, age and immunologic status or density and age of the parasite infection), acting during oogenesis or embryogenesis, affect gene expression or the parasite's hormone balance so that the rhabditiform larvae develop as *either* potential males or females. In the external environment the potential females develop to either infective filariform larvae following the direct (homogonic) cycle, or, alternatively, grow and molt several times in the soil to produce free-living adult female worms, following the indirect (heterogonic) cycle. The environmental factors that determine which pathway potential females follow include pH, pO_2, pCO_3, temperature, abundance of nutrients, and the consistency of the medium in which the larvae are developing. Georgi's interpretation of these events are presented in Figure 15–4. Perhaps, as in *dauer* larva formation in *Caenorhabditis elegans* cultures, pheromones are also involved.

In the case of *Strongyloides stercoralis*, as has already been mentioned, the rhabditiform larvae may even develop to infectivity precociously in the host's intestines. This moves the direct life cycle from the environment into the large bowel. Ecologically, this may be a minor change, merely substituting a warmer, anaerobic, internal fecal habitat for one that is cooler, more aerobic and external.

The free-living females reproduce by meiotic parthenogenesis. Although the males inseminate females, and each egg must be invaded by a sperm for development to be initiated, the egg and sperm pronuclei never fuse. Instead, the nuclei of the egg's second maturation division recombine to form a nucleus having a diploid chromosome number. The larvae hatching from these eggs eventually develop into infective, filariform larvae, and hence, the parasitic adult worm population consists solely of females.

It has been proposed recently that strongyloidiasis in humans should be considered a spectral disease with asymptomatic carriers at one end of the spectrum and persons with fatal hyperinfective or disseminated infection at the other. In newly acquired heavy infections, abdominal pain and diarrhea occur; in long standing chronic infections there are bouts of pain and diarrhea, interspersed with periods of asymptomatic infection. A linear skin reaction, *larva currens* (racing larva) emanating from the anal region and accompanied by intense itching is characteristic of chronic infections. Autoinfection can cause the infection to develop into an overwhelming strongyloidiasis with profuse diarrhea, severe abdominal pain, and respiratory involvement. Secondary bacterial invasion of the blood (septicemia) and even of the lungs and brain occur. These severe forms of the disease are often fatal.

Strongyloidiasis also occurs in other animals, including reptiles, birds, and mammals. Among the more important parasites of animals are *S. ratti*, *S. westeri*, *S. ransomi*, and *S. fulleborni*, parasites of rats, horses, swine, and primates, respectively. The latter is transmissible to humans. *S. westeri*, *S. ransomi* and *S. fulleborni* are stored in the tissues of resistant adult hosts. They are released from dormancy during lactation and transmitted in mother's milk to nursing young, in which heavy infections cause diarrheal disease.

SUPERFAMILY RHABDITOIDEA

Rhabditis coarctata must be carried as ensheathed larvae on the surface of dung beetles to fresh dung before they can mature. *R. ocypodis* is carried on the gills of crabs. An immune response by earthworms to *R. pellio* has been described.[13]

Neoaplectana glaseri is generally free-living, but its larva often penetrates and destroys the larvae and adults of certain beetles. *N. glaseri* has been used as a biologic control agent against the Japanese beetle.

This group of nematodes includes free-living species as well as parasites of plants and animals. The animal parasites occur mainly in invertebrates and include several species useful for the biologic control of insects. A particularly important rhabditoid nematode is the free-living species, *Caenorhabditis elegans.* This is the first metazoan to be described completely at the cellular level. Not only has each cell been placed in the structural plan of the worm, but also the lineage of each has been traced back to the early cell divisions of the ovum. Given this detailed developmental and anatomic background, the nematode has become an important model for molecular, cellular, and developmental biologists. It is also beginning to attract the attention of parasitologists, who will undoubtedly find it a useful model for investigations ultimately concerned with nematode parasitism.

Order Strongylida

This order contains many of the most important nematodes of humans, other mammals, and birds. Members of the group are characterized by the presence in the male of a posteriorly situated copulatory bursa (Fig. 15–5B, C). This funnel-shaped extension of the cuticle, which holds the female during copulation, is supported by fleshy, finger-like rays, that frequently exhibit a species-specific pattern. Females have a simple, conical tail (Fig. 15–5D). The mouth (Fig. 15–5A) is *not* surrounded by the three conspicuous lips that characterize the ascarids (see Fig. 16–3), nor does the esophagus have a distinct, spherical bulb (see Fig. 16–8)

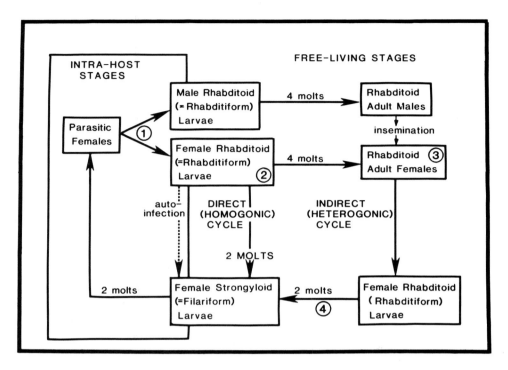

Fig. 15–4. Direct (homogonic) and indirect (heterogonic) life cycles of *Strongyloides*.

1. Parasitic females produce only genotypically female eggs by mitotic (apomictic) parthenogenesis, but both male and female rhabditoid larvae are produced through the agency of extrinsic intrahost factors (duration and density of infection; species, age, and immunologic status of host) operating during oogenesis or embryogenesis. These extrinsic intrahost factors are believed to act by modifying gene activity or by modifying sex hormone balance in the developing embryos. For example, in sheep infected with *S. papillosus*, most of the eggs develop into infective larvae and the free-living generation is sterile because of a rarity of males, but when rabbits are artificially infected with *S. papillosus* of ovine origin, almost as many males as females appear in cultures and large numbers of heterogonic infective strongyloid larvae are produced. In addition, in the reported case of human strongyloidiasis contracted from dogs,[18] the human fecal cultures consisted, after 24 to 36 hours' incubation, predominantly of homogonic strongyloid larvae, whereas the canine cultures underwent almost exclusive heterogonic development with infective strongyloid larvae first appearing at 4 to 5 days.

2. Female rhabditoid larvae may develop into either free-living rhabditoid females or into strongyloid infective larvae; the relative proportions depend on extrahost (cultural) conditions, such as pH, pO_2, pCO_3, consistency of substrate, temperature, and abundance of food materials. In man, *S. stercoralis* rhabditoid larvae may complete the necessary two molts to the infective strongyloid stage before leaving the host's digestive tract; this leads to autoinfection.

3. Each egg of the free-living female must be penetrated by a sperm for development to occur but, at least in *S. papillosus* and *S. ransomi*, male and female pronuclei do not fuse (pseudofertilization). Instead, the products of the second oval maturation division recombine to form the egg nucleus with a diploid chromosome number. Thus, these rhabditoid females reproduce by meiotic parthenogenesis.

4. Rhabditoid larvae of the free-living generation develop only into strongyloid larvae (and thus, in turn, into strongyloid parasitic females). (From Georgi: Strongyloidiasis. Courtesy of *1982 CRC Handbook Series in Zoonoses*.)

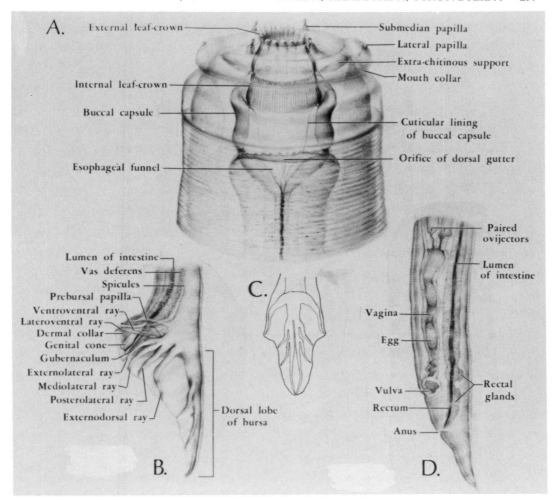

Fig. 15–5. Morphology of a strongylid nematode *Cyathostomum coronatum*. A, Dorsoventral three dimensional representation of the anterior end. B, Lateral view of posterior end of a male showing the rays of the left side of the bursa. These rays are duplicated on the right side of the nematode. C, Ventral view of the dorsal lobe of the bursa showing its bilateral symmetry. D, Lateral view of the posterior end of the female. In the species shown, the vulva is located near the posterior end. In some other strongylids, the vulva may be situated near midbody. (From Lichtenfels. Courtesy of Proceedings of the Helminthological Society of Washington.)

that occurs in the oxyurids (pinworms). One or two leaf crowns occur in the oral openings of many (but not all) species of strongylid nematodes (Fig. 15–5A). Virtually all species lay characteristic thin-shelled eggs that have undergone only a few cell divisions before passage from the host (see Fig. 14–14, 6). It is convenient to refer to these ova as "strongyle-type eggs."

The free-living life cycles of most members of this group are similar, with unembryonated eggs passing in the feces, full embryonation, hatching, and early larval development ($L_1 \rightarrow L_3$) occurring in the fecal deposit, and ensheathed,

non-feeding, infective larvae (L_3) dispersing from the feces into the soil or onto herbage. Larvae on herbage ascend as far as moisture films extend, where dispersal ceases and host-finding behavior comes into play. This may involve quiescence on vegetation, or accumulation in spindle-like aggregates on soil particles that reduces the risk of desiccation to the individual. Depending on the species, these larvae are infective by ingestion, skin penetration, or both.

Some species have intermediate hosts, in which case infection occurs at an earlier stage

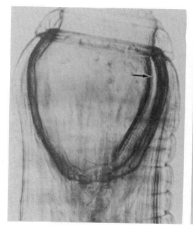  

Fig. 15–6. Lateral views of the anterior ends of the large strongyles (*Strongylus* spp.) of the horse. Left to right: *Strongylus edentatus* lacking teeth in the base of the buccal capsule, *S. equinus* with pointed teeth, and *S. vulgaris* with rounded teeth in the buccal capsule. The *dorsal gutter* (arrow) forms a channel along the inner dorsal surface of the buccal capsule of each species. It extends from the esophagus to the mouth opening, and, presumably, conducts the flow of esophageal enzymes to the attachment site of the worm. (From Lichtenfels. Courtesy of Proceedings of the Helminthological Society of Washington.)

and the L_3 remains the infective form for the definitive host.

The superfamily Mestastrongyloidea includes many species that use intermediate hosts. It also differs from other superfamilies of the Strongylida structurally and biologically as described when the group is discussed in greater detail (p. 313).

SUPERFAMILY STRONGYLOIDEA

The Greek term "strongylos" means round or compact; it is used in many scientific names other than for roundworms. However, the name "strongyle" has become a popular term to denote many roundworms of large domestic animals. Actually, strongyles belong primarily to the subfamilies Strongylinae and Cyathostominae. The name is confusing because neither *Strongyloides stercoralis* nor *Rhabditis strongyloides* are strongyles. Strongyles are found in horses and other equids, and in elephants, rhinoceroses, rodents, swine, and other mammals. Species common to horses (Fig. 15–6) are *Strongylus equinus, S. vulgaris, S. edentatus,* and *Triodontophorus* spp. The larger strongyles measure up to 4 cm in length; they are stout, grayish worms found anchored to gut mucosa.

A ring of fence-like projections known as a leaf crown (Figs. 15–5A and 15–6) or **corona radiata** surrounds the mouth of these worms. Unlike hookworms, they do not have teeth or cutting plates in the mouth opening, but some strongyles do have teeth in the base of the mouth capsule (Fig. 15–6).

This group contains many important parasitic worms of domestic animals and man. Many of them ingest blood by sucking a portion of the intestinal mucosa into their mouths; they keep the blood from clotting by secreting an anticoagulant. The host's intestinal lumen is essentially anaerobic, but parasites obtain oxygen from the ingested blood cells. The presence of fresh blood in worms gives some of them a bright red color.

Strongylus edentatus (Fig. 15–6) has a characteristic life cycle. Eggs reach the ground with host feces. In one or two days, the first-stage rhabditiform larva hatches and develops into the second-stage rhabditiform larva. Within a few days, the elongated, ensheathed, infective larva has developed. It cannot feed, but may live for several months, sometimes throughout the winter. The host ingests the infective larva with vegetation. The larva penetrates the intestinal wall and enters the connective tissue under the abdominal peritoneum, where it causes hemorrhagic nodules and increases in size for about three months. It then migrates back to the gut wall, where it produces other nodules, matures, enters the intestinal lumen, and attaches to the mucosa. See Lichtenfels[10] for an account of helminths of domestic equids, including keys to the species.

SUPERFAMILY SYNGAMOIDEA

Syngamus trachea, the gapeworm (Fig. 15–7), is a red nematode that lives in the tracheas of various species of poultry and game birds. It derives its common name from the fact that heavily infected birds apparently try to rid themselves of the annoyance caused by this worm by gasping, coughing, swallowing, or stretching the neck. Male worms average about 4 mm in length and the females about 17 mm. Male and female worms are fused in permanent copula.

Eggs are coughed up by the host, swallowed, and reach the outside with feces. In moist soil, the eggs become infective in a week or two, depending on the temperature, and may be ingested by a chicken or some other bird, or the eggs may hatch in the soil and the larvae may be ingested. The worm is versatile in its life history: larvae may enter an earthworm, slug, or snail, or the egg may be eaten by one of these intermediate hosts. Thus, chickens or other birds can become infected by eating snails, slugs, or earthworms.

Within susceptible birds, the larvae penetrate the intestinal wall, enter the circulation, and are carried to the lungs, where they leave the circulation by penetrating into the alveoli. They then move up the airways to their predilection site in the trachea. Within 2 weeks the larvae are mature; they live 23 to 92 days in chickens and 48 to 126 days in turkeys. Birds may suffocate if the worms are present in large numbers. General cleanliness in bird management is important, and care should be taken not to raise birds in areas contaminated from previous infections.

Stephanurus dentatus, the common kidney worm of pigs (Fig. 15–8), may also inhabit the liver, lungs, spleen, muscles, spinal canal, and body cavities. Male worms are 30 to 45 mm long. They appear black and white because some internal organs can be seen from the outside.

Female worms in kidneys produce oval eggs that average a little more than 100 μm in length; thus, they can just be seen without a microscope. Eggs pass down the ureter, into the uri-

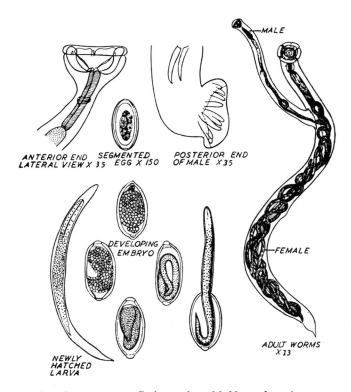

ANTERIOR END SEGMENTED   POSTERIOR END
LATERAL VIEW X 35 EGG X 150   OF MALE X 35

MALE

DEVELOPING
EMBRYO

FEMALE

NEWLY
HATCHED
LARVA

ADULT WORMS
X 13

Fig. 15–7. *Syngamus trachea,* the gapeworm. Redrawn from M. Neveu Lemaire.

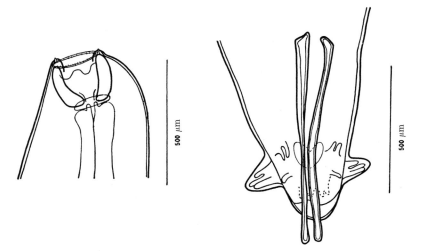

Fig. 15–8. *Stephanurus dentatus,* anterior and posterior end of male. (From Morgan and Hawkins: Veterinary Helminthology, courtesy of Burgess Publishing.)

nary bladder, and to the outside with urine. A heavily infected pig may pass a million eggs a day.

In moist, shaded soil, eggs hatch into tiny, free-living larvae that grow and molt twice within the next three to five days. Young worms are then infective to pigs when ingested with contaminated food. Larvae also enter their hosts through cuts or sores on the skin, or they may even penetrate the unbroken skin. Within the body, larvae are carried by blood and lymph vessels to the liver, where they may remain for several months. Then the parasites make their way to kidneys, in which they become mature. The entire life cycle requires 6 to 8 months.

Larvae and adults cause abscesses, hemorrhages, adhesions, loss of weight, and death. The economic loss in the United States due to this worm alone amounts to millions of dollars annually. Good management practices help to reduce infection. Keeping the area around hogs dry and exposed to sunlight prevents the development of enormous numbers of eggs and larvae.

SUPERFAMILY ANCYLOSTOMATOIDEA

Hundreds of millions of people have hookworms, and the daily loss of blood due to this infection has been estimated as the equivalent of the total volume of blood of 1,500,000 persons.[21] The hookworms of humans are avid bloodsuckers. They produce minute intestinal ulcers with their teeth, digestive enzymes, and

the sucking action of their muscular esophagi. They feed on the intestinal mucosa, ingesting lysed epithelial cells, tissue fluid, and blood. However, only a few of the ingested erythrocytes are digested; most are expelled from the anus periodically and explosively in red plumes of undigested blood. Serum appears to be needed for the normal functioning of the worm's esophagus and for the incorporation of glucose into the parasite's glycogen. Untreated heavy infection results in anemia and hypoalbuminemia. Patients become pallid, weak, short of breath, and tire rapidly at work or play. Children may show several years of physical, sexual, and/or mental retardation.

Hookworms (Fig. 15–9) have a buccal capsule containing ventral teeth or cutting plates. The male bursa is conspicuous. The name "hookworm" is said to derive from the position of the anterior end, which is bent backward (dorsally). It is also purported to derive from the hook-like appearance of the bursal rays. Figure 15–9 shows diagnostic features of the two species common in humans.

The two most important hookworms of humans are *Ancylostoma duodenale* and *Necator americanus*. Their geographic distributions may once have been disjunct, with the former having the more northerly distribution. However, migrations of human populations particularly during the 18th and 19th centuries have obscured the original pattern. It is thought that *Necator americanus* (a misnomer) was brought into the New World by infected West Africans at the time of the slave trade.

Many other species of hookworms occur in mammals. These include *Ancylostoma caninum*, an important cause of anemia and even death in dogs, and *Ancylostoma braziliense*, a parasite of cats and dogs that can cause severe dermatitis in humans. While the various species differ with respect to the details of their structure, life histories, and pathogenic potential, they show many similarities.

Ancylostoma duodenale (Fig. 15–9) is found in the small intestine of millions of people, chiefly in southern Europe, Africa, the Middle East, India and the Orient. The male worm is 8 to 11 mm × 0.4 to 0.5 mm in size. The pos-

terior end of its body is flared and forms a bursa supported by fleshy rays with a characteristic pattern. A pair of long spicules that remain separate distally is characteristic of this species. The females average 10 to 13 mm × 0.6 mm in size. The posterior end of the body tapers to a blunt point. The vulva is located at a point about two-thirds the length of the body from the anterior end. Eggs are ovoidal, thin-shelled, and measure 56 to 60 μm × 34 to 40 μm. When eggs are found during fecal examinations, they are usually already in the early stages of segmentation (Fig. 14–14).

The daily output of eggs from a single female worm ranges from 10,000 to about 25,000. At temperatures ranging from 25 to 35° C, larvae

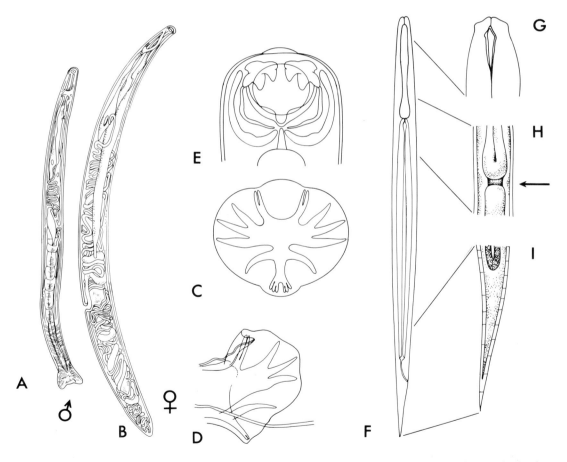

Fig. 15–9. Differential diagnostic features of adult and larval hookworms of man. A–I. *Ancylostoma duodenale:* A,B, *Robust* adult male and female *A. duodenale.* C, Bursa of male (ventral view of opened bursa). D, Bursa, lateral view. E, En face view of adult worm showing 2 pairs of large *teeth* in oral opening. F, Filariform third-stage larva showing major landmarks and proportions. G, *Lightly* chitinized oral spear. H, *Transition* area (arrow) between end of esophagus and intestine. I, Larval tail showing widely spaced, *inconspicuous* striations.

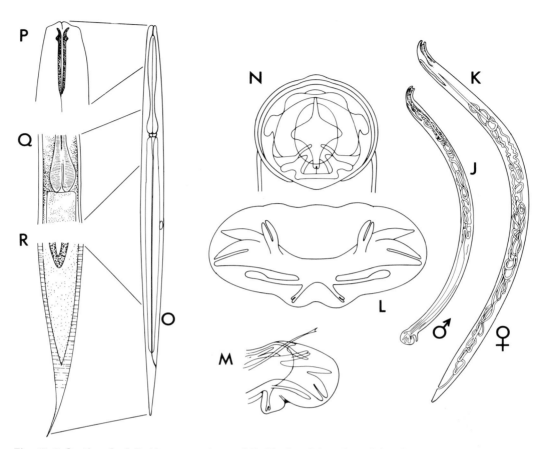

Fig. 15–9 Continued. J–R. *Necator americanus*: J,K, *Slender* adult male and female *N. americanus*. L, Bursa of male (ventral view of opened bursa). M, Bursa, lateral view. N, En face view of adult worm showing *cutting plates* in oral opening. O, Filariform third-stage larva showing major landmarks and proportions. P, *Heavily* chitinized oral spear. Q, Junction between esophagus and intestine *without* transition area. R, Larval tail showing closely spaced *prominent* striations. (Modified from Chandler: Hookworm Disease, courtesy of Macmillan and Co.)

hatch from eggs in 24 hours. Free oxygen is essential for hatching and for further development. Larvae grow rapidly, molt twice, and in about a week become ensheathed, non-feeding, slender, filariform larvae (Fig. 15–10).

These infective larvae crawl to a high point on a soil particle, vegetation, or bit of rock, so long as it is moist, and wait for a new host to come along. Alternate wetting and drying of the soil hastens the death of infective larvae, since it causes them to migrate up and down frequently, exhausting their food reserves; hence frequent rains, with dry weather in between, reduce the number of larvae in the soil.

When a filariform larva comes into contact with the skin of a new host, it burrows into the skin, undergoes lung migration, and finally settles in the small intestine as a tissue-feeding fourth stage larva equipped with a provisional buccal capsule. It molts this capsule along with the larval cuticle to become an adult. Adult worms mature and eggs begin to appear in the feces about 8 weeks after infection. If the infective larva enters via the mouth and is swallowed, it does not migrate via the circulation and lungs. Instead, it enters the intestinal mucosa, develops for a short period of time, returns to the lumen, and the fourth-stage larvae attach to the intestinal epithelium as after lung migration. The remainder of the life cycle continues as already described.

In parts of the world where free-living de-

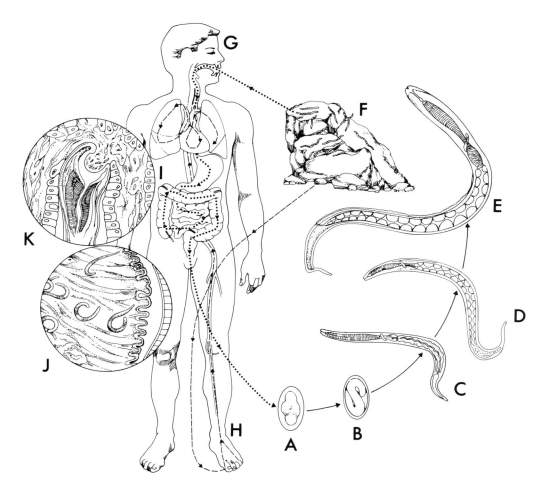

Fig. 15–10.   The life cycles of the two major hookworms of man, *Ancylostoma duodenale* and *Necator americanus*. A, Egg leaving body in feces. B, Embryonated egg containing larva. C, First stage (L$_1$), microbiverous, rhabditiform larva. D, Microbiverous, rhabditiform L$_2$. E, Ensheathed, non-feeding, strongyliform (= filariform) infective L$_3$. F, Several L$_3$ positioned for penetration on soil particles. G, Oral infection. *A. duodenale*, but not *N. americanus*, is infective via the mouth. H, Percutaneous infection. Both species are infective via the skin. I, The arrows extending from the foot and continuing to the heart, lung, respiratory tree, pharynx, and gastrointestinal tract trace the internal migration of the developing larvae following percutaneous infection. J, Adult worms attached to the mucosa of the small intestine. K, A longitudinal section through the buccal capsule of a hookworm showing a plug of the intestinal mucosa in the buccal capsule of the worm. (Modified from The Microscopic Diagnosis of Tropical Disease, courtesy of Bayer; also from Chandler: Hookworm Disease, courtesy of Macmillan and Co.)

velopment and larval survival are seasonal, at least some of the worms entering the human host become arrested in their development. Many of these larvae resume growth and mature subsequently, so that an increase in the egg output of the infected population is associated with the rainy season favorable for the hookworm's freeliving stages.

Ancylostoma duodenale larvae will also become dormant in the muscles of some food animal species, suggesting that meat-borne infection is possible.[17] Arrested development in the tissues of the normal definitive host is a correlate of lactogenic infection (transmammary transmission) in several host-parasite associations. This, along with the observation that *A. duodenale* parasitizes young children, including suckling infants, suggests that it, too, may be transmitted by this route.[2,16]

Necator americanus is a common parasite of humans in the moist tropics and subtropics. Where its geographic range is coextensive with that of *Ancylostoma duodenale*, it is often the dominant species. Its present range covers Central and South America, including the Caribbean Islands, sub-Saharan Africa, much of India, southeast Asia and parts of the Orient.

Necator americanus differs from *Ancylostoma duodenale* in a number of structural characteristics (Fig. 15–9); it is thinner, its anterior end is more distinctly bent, and the anterior edge* of its mouth capsule is armed with a pair of simple, curved, cutting plates, rather than teeth. The two spicules of the male are fused distally, whereas those of *Ancylostoma duodenale* are separate. Other distinguishing details are illustrated in Figure 15–9. The strongyle-type eggs measure 70×38 μm. The life history of *Necator americanus* (Fig. 15–10), while generally similar to that of *Ancylostoma duodenale*, differs in the following important details:

(1) lower fecundity (10,000 eggs per day), that may be more a matter of sensitivity to density than an inherent difference in reproductive potential, (2) obligate skin penetration, (3) absence of seasonal arrested larval development within the host, (4) 5-year life span under natural conditions (apparently much longer in persons protected from re-infection), and (5) rare occurrence in infants.

Because, as mentioned previously, *Ancylo-*

stoma duodenale* may be found at high density in nursing infants, its transmission in mother's milk seems probable. In contrast, no evidence of neonatal *N. americanus* infection has been found, even when specifically sought, in areas of Africa where hookworm infection is prevalent, but only *Necator americanus* occurs.[2,16]

The symptoms of hookworm infection are easily understood within the framework of the life history and the feeding behavior of the adult worms.[2,16] Larval worms, when they penetrate the skin, cause hookworm dermatitis ("ground itch"), particularly in previously infected, sensitized individuals. Minute, red lesions mark the points of larval penetration. With time, these macules become small, itchy blisters. In naturally acquired relatively light infections, the signs of larval migration through the lungs are usually absent or so trivial that they cause the patient only slight discomfort. However, in heavy infections, simultaneous movement of parasites through the respiratory system may cause varying degrees of tissue damage and is associated with coughing and laryngeal irritation, which may result in hoarseness or even a temporary loss of voice.

The adult worms feeding on the mucosal epithelium cause the blood loss that is the ultimate cause of the anemia called "hookworm disease." This disease is attributable to chronic blood loss; when the host's iron reserves become exhausted and adequate hemoglobin levels can no longer be maintained, a simple iron-deficiency-type anemia develops, characterized by small (microcytic), pale (hypochromic) erythrocytes. The small size and pale color of the cells reflect the iron deficiency.

The bloodsucking activity, along with bleeding from lesions after the worms move, leads to the loss of plasma proteins (hypoproteinemia). Additionally, there is protein loss due to the damage to the intestinal epithelium. Protein-losing intestinal disease is discussed more fully on page 308, but it is this protein loss that is the basic cause of the diarrhea and edema that sometimes occurs in hookworm diseases. Heavily infected individuals often have puffy faces; this is particularly characteristic of severely infected children.

Where it is prevalent, hookworm disease has an important socioeconomic impact. Children may show several years' worth of physical, sexual, and even mental retardation, and adults, lacking physical stamina, are easily exhausted

---

*The anterior edge is actually the ventral edge because the worm is dorsally bent.

by physical labor. This led sensationalist newspapers to print headlines calling hookworms the "laziness germ." In the early part of the century, hookworm disease was common in rural areas of the southern United States. It was the regionally depressing effect of widespread, heavy hookworm parasitism, along with the availability of an effective chemotherapeutic agent, that prompted the Rockefeller Foundation to select hookworms from among other infectious agents for its initial program of disease control to promote public health. This program supported hookworm treatment and sanitation campaigns, as well as research and health education, in the United States and eventually throughout the world. Hookworm control has been a success in much of southern Europe, Japan, Taiwan, Korea, and the United States. It has been less successful elsewhere, where defecation in the open is still common, or where uncomposted human feces are used for fertilizer. Successful control involves health education, the building, use, and maintenance of sanitary latrines, the wearing of shoes, and anthelmintic treatment.

Field investigations conducted in endemic areas have shown that reinfection after treatment occurs rapidly.[18] This observation led to the advocacy of repeated mass-chemotherapy for community-wide hookworm control, because it would, in turn, reduce the adult worm and free-living larval populations and lead to the reduction of the prevalence and intensity of infection. However, since hookworms are aggregated in their distributions in populations of hosts, it is probable that repeated treatment of only the most heavily infected persons would achieve a similar level of control more economically. This is particularly so because most of the heavily infected individuals are predisposed to heavy infection.[1,18]

Hookworm Infection in Other Species. Hookworms are pathogenic, economically important, intestinal parasites of ruminants, carnivores, and sometimes of other groups of mammals as well. There are two subfamilies, the Necatorinae, having curved, oral, cutting plates, and the Ancylostominae, having one to three pairs of teeth at the anterior edge of the mouth capsule. The Necatorinae occur in ruminants (*Bunostomum* spp.) or carnivores (*Uncinaria* spp.), whereas the Ancylostominae are mainly parasites of carnivores. All species are "plug feeders" in the sense that they suck a plug of intestinal cells into the buccal capsule and they abrade, lyse, and ingest, but unlike both species occurring in humans, some are *not* bloodsucking worms.

Species of *Bunostomum* do suck blood as both fourth-stage larvae and adults. Because even as larvae these parasites are relatively large, hematophagous parasites, serious blood loss can occur before eggs appear in the feces, making prepatent disease possible, and diagnosis difficult. Species of *Bunostomum* have a wide geographic distribution but are particularly important parasites of young livestock in warm, moist parts of the world.

In contrast, the geographic distribution of two other species of Necatorinae, *Uncinaria stenocephala* of canids, and *Uncinaria lucasi* of fur seals, extends into the Arctic. *Uncinaria stenocephala* is *not* an active bloodsucker, and therefore disease caused by this species results only from damage to the intestinal mucosa and takes the form of protein-losing enteropathy. In contrast, *Uncinaria lucasi* sucks blood, causing anemia and death in the pups of Northern Fur Seals. It is especially interesting that adult seals do not harbor adult worms. Instead, they carry dormant larvae in their tissue. In females, these larvae resume development and migrate to the mammary glands of the nursing mother. Pups are infected by ingesting larvae in milk, and become the source of hookworm eggs that contaminate the breeding beaches. Only they carry patent infections; if *adult* seals are invaded they store the larvae in their subcutaneous fat, and take them to sea during the winter months in an arrested state of development. Larvae that remain on the beach die during the cold, arctic fall and winter. Thus, overwintering depends on survival of arrested parasitic larvae in the tissues of the adult seals.

The transmammary route is also important for transmission of *Ancylostoma caninum*, a particularly damaging parasite of puppies and young dogs. Heavily infected nursing pups may die of severe blood loss within the first few weeks of life. Many young, less heavily infected dogs develop hookworm disease, with anemia, diarrhea, weight loss, and occasionally death. With age and periodic reinfection, dogs become resistant to the disease, but they may remain carriers of light intestinal infections and harbor dormant larvae in their body musculature, the intestinal mucosa, or both. It is interesting that the muscle larvae are, in fact, intracellular parasites. In the female the larvae resume development around the time of parturition and go to the mammary tissues. Experimentally, injections of the hormones progesterone and oestradiol can initiate resumption of larval development with an attendant appearance of larvae in samples of milk.

The observation that mature dogs exposed to periodic infection limit their worm burdens and become resistant to the disease led to the production of a vaccine. The vaccine, constituted of irradiated infective larvae, protected pups against the disease and against the accumulation of arrested larvae in the tissues, but it did *not* completely prevent infection with adult worms. A few of these established, but they were stunted and nonpathogenic. However, the presence of worms, as indicated by the occurrence of hookworm eggs in the feces, was unacceptable to many dog owners and veterinary practitioners. Thus, although the vaccine was a scientific success, it became a commercial failure and is therefore no longer available.

SUPERFAMILY TRICHOSTRONGYLOIDEA

Trichostrongyles are small-mouthed, slender, bursate nematodes infecting mainly the gastrointestinal tract and occurring in all vertebrate groups except fishes. They are the most important helminth parasites of cattle, sheep, and other ruminants.

The life cycles of the numerous species of trichostrongyles are similar and need not be described in detail in the discussions of each species included later in the text. Generally speaking, adult worms lay "strongyle-type" eggs into the gastrointestinal lumen. The eggs pass out in feces, and given sufficient moisture and warmth, embryonate and hatch in the feces. Rhabditiform larvae (L_1 and L_2) feed on the fecal microflora and develop to the infective third-stage (L_3). This larva, morphologically in the third-stage, fails to cast off the loose cuticle of the previous stage, and is, therefore, an ensheathed, non-feeding form. It disperses from the fecal mass onto herbage, where, while moisture is adequate, it survives until it invades the host. Infection is usually via the oral route, because most trichostrongyles are parasitic in grazing herbivores or omnivores. After ingestion, larvae exsheath in the gastrointestinal tract just proximally to the parasite's predilection site. Thus, in ruminants, parasites of the abomasum exsheath in the rumen, those of the du-

odenum in the abomasum and those of the large intestine in the ileum. Exsheathment is initiated in response to appropriate environmental stimuli, "physiologic triggers," which, for species that exsheath in the rumen, are a high concentration of dissolved carbon dioxide and an appropriate body temperature. The rate of exsheathment is enhanced by other conditions characteristic of the rumen; a pH near neutrality, a low redox potential, and the presence of reducing agents.

The trichostrongyles are of great importance to veterinary medicine. More complete accounts of these parasites may be found in the veterinary parasitology textbooks by Georgi,[6] Levine,[9] and Soulsby.[19]

Haemonchus contortus, the twisted stomach worm, or barber's pole worm (Fig. 15–11), an abomasal parasite of sheep, goats, deer, and other ruminants, has a worldwide distribution. It measures 10 to 20 mm and 18 to 30 mm in length, males and females, respectively. The common names describe the female, the white ovaries of which are twisted around its blood-filled, red intestine. Males appear more uniformly red. The parasite produces strongyle-type eggs and undergoes a characteristic trichostrongyle-type life cycle. As it is the most important nematode parasite of sheep, aspects of its pathogenesis, the immune response to its presence, and its interaction with other parasites are known in interesting detail.

Haemonchus contortus is remarkable among trichostrongyles because both fourth-stage larvae and adult worms are specialized blood-suckers having an oral lancet (Fig. 15–11) that slits the abomasal blood vessels. The adult worms ingest about 0.05 ml of blood per day. Additionally, blood and plasma proteins are lost from the remaining lesions when the worms move.

In experimental ovine haemonchosis, anemia develops in three phases that illustrate the underlying pathophysiologic mechanisms. All pathologic features of haemonchosis can be produced by the periodic mechanical bleeding of sheep, showing that blood loss per se, and not the release of toxins by the parasite, is all that is necessary to account for the disease.

The bloodsucking activities of the worms lead to the loss of erthyrocytes and plasma proteins. Only part of the iron that enters the intestine during lysis of the red blood cells and breakdown of the hemoglobin is reabsorbed. Early in

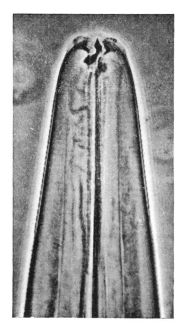
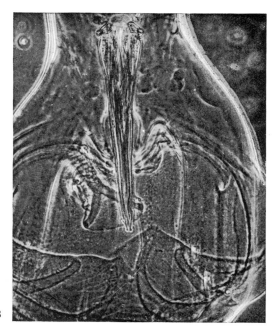

A   B

Fig. 15–11. *A,* Lateral view of the anterior end of *Haemonchus contortus.* × 400. *B,* Spicules and bursa of *Haemonchus.* Note barb on spicules and asymmetrical dorsal lobe of the bursa. × 138. (From Whitlock: Diagnosis of Veterinary Parasitism. Lea & Febiger.)

the course of a heavy infection, the volume of the erythrocytes in given amount of blood (the packed cell volume [PCV]) decreases. However, part of the lost iron is reabsorbed and, unless the host is already iron deficient, it has reserves to replenish the hemoglobin supply. Then homeostatic mechanisms of the host swing into action and the erythropoietic system compensates for the erythrocyte loss. Thus, after an initial lag phase during which the PCV decreases, it levels off and may even increase again. During this second, compensated phase of the infection, a continuous net loss of iron occurs and, if a high level of infection is maintained, iron reserves gradually become depleted, the PCV begins to fall again and the host, if not already anemic, becomes so. The clinical signs of haemonchosis include, in addition to anemia, hypoproteinemia (due to the loss of plasma proteins), edema, and loss of appetite and weight. Death may occur in heavily infected animals under 6 months of age that cannot muster an immune response.

The immune response to *H. contortus* in sheep has been under investigation since 1929, when the great American parasitologist, Norman Stoll,[20] described a phenomenon called "self

cure." This occurs when an immunologically competent sheep has been exposed to a number of sensitizing infections with *H. contortus* and, then, after reinfection, responds dramatically, eliminating the entire existing population of *H. contortus.* The immunologic response to *H. contortus* is discussed in greater detail in Chapter 16.

Stoll, in fact, described self-cure *and* "protection," but unfortunately protection does not always occur. Indeed, efforts to immunize sheep against *H. contortus* have proved disappointing. Parasitologists have found that sheep less than 6 months of age are immunologically immature with reference to this parasite and do not respond to vaccination. Furthermore, there is great individual and breed variation with regard to the ability to respond. Individual unresponsiveness, a common phenomenon in host/parasite associations, was first described many years ago (see Chap. 16), but its implications for vaccination against helminth parasites have only recently been widely appreciated. Under some conditions, it is uneconomical to vaccinate a flock of sheep because many individuals remain resistant carriers and some continue to be fully susceptible. These problems have led to an increased interest in immunogenetics and resistant breeds. Some breeds from highly endemic tropical areas show strong resistance to infection with the parasite. For instance, under experimental conditions in East Africa, native Red Masai sheep have much higher anti-larval responses and become much more resistant to *H. contortus* infection than imported Merino sheep.

Haemonchus placei, a species closely related to *H. contortus* occurs mainly in cattle. It has been the subject of numerous experimental investigations, some of which have general importance for parasitology (see also p. 469). One particularly interesting investigation provides evidence that a population of adult worms can act in a density dependent manner to inhibit the development of incoming larvae. A corollary of this concept, for species that store inhibited larvae in tissues, suggests that the sudden loss of a population of adult worms will trigger the massive, simultaneous development of arrested larvae. This is precisely what happened in an investigation involving the chemotherapeutic termination of *H. placei* infections in young cattle, in which anthelmintic treatment precipitated an outbreak of severe haemonchosis.[19]

Ostertagia ostertagi (Fig. 15–12) is the most important nematode parasitizing cattle in temperate areas of the world. It occurs in the abomasum, and, in the young animals, heavy infections interfere with digestion, causing diarrhea, edema, anorexia, and weight loss. The edema, which occurs characteristically in the submandibular space, is commonly known as bottle jaw.

Infected animals pass strongyle-type eggs in their feces. Under favorable conditions, larvae emerge from the eggs and develop to the infectivity in about 2 weeks. Free-living development is favored by cool (55 to 75°F) moist, climatic conditions. The L_3 is positively phototactic and migrates from the feces onto the adjacent vegetation. In damp weather, hordes of larvae (Fig. 15–13) may accumulate in drops of water on herbage. Because hot, dry weather that causes fecal deposits to crust over is inimical to larval dispersal, spring and fall conditions, rather than those of summer, are associated with the acquisition of heavy infections.

After ingestion by cattle, the infective larvae exsheath, enter the gastric glands, develop to

young adulthood, emerge from the glands, and mature in the abomasal lumen. Parasitic development involves a tenfold increase in the size of the worm, causing marked distension of the gastric glands. When the relatively large worms emerge, they rupture the glands and damage the surrounding capillaries as well. The parietal and zymogen cells of normal gastric glands produce hydrochloric acid and pepsinogen, respectively, but damaged parietal cells produce little acid. As a result, pepsinogen is not converted to pepsin, and, in turn, proteins are not digested. Furthermore, the abomasal pH rises and, therefore, bacterial growth is no longer inhibited. The damaged epithelium allows serum proteins that have escaped from the damaged capillaries to leak into the abomasal lumen. The pathogenesis of ostertagiasis is summarized in Figure 15–14.

The presence of undigested food, serum proteins, and an abnormally large bacterial population alters the osmotic balance and fluid moves into the intestinal lumen, causing diarrhea. Meanwhile, the loss of serum proteins (hypoalbuminemia) changes the osmotic balance in the blood, and fluids accumulate in the tissues, particularly in the dependent (i.e., lower) parts of the body. In cattle, the submandibular space is a particularly characteristic site for such accumulation.

The faulty tight junctions of the abomasal epithelium also allow the accumulating pepsinogen to move into the blood. An increased plasma pepsinogen level is a characteristic abnormality occurring in ostertagiasis and can be used to distinguish this disease from others that produce diarrhea in cattle.

A particularly interesting form of ostertagiasis is characterized by a sudden onset of severe disease at a time when, enigmatically, pasture populations of larvae are at a low. This unseasonable form of the disease occurs when hordes of arrested larvae that had remained quiescent in the gastric glands suddenly resume development and emerge from the gastric mucosa, causing simultaneous massive damage. Large accumulations of larvae are, therefore, quite rightly compared to a time bomb set to explode when least expected. The mechanism that causes the apparent spontaneous release of the larvae from dormancy remains unknown.

Although other genera of intestinal trichostrongyles, including *Cooperia*, *Nematodirus*, and *Trichostrongylus* (Fig. 15–15), are significant par-

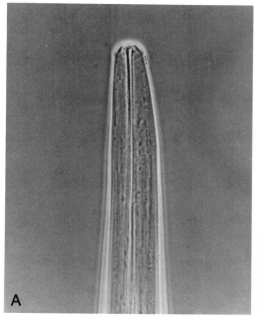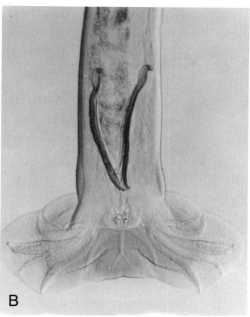

Fig. 15–12. *Ostertagia ostertagi.* A, Anterior end. B, Bursa and spicules of the male. The bursae has been spread out and flattened. (From Georgi: Parasitology for Veterinarians, courtesy of W.B. Saunders.)

asites of wild and domestic ruminants, they are not as important as *Haemonchus* and *Ostertagia* economically, and, therefore, are less well studied. On the other hand, some trichostrongyles parasitize smaller animals, and are valuable experimental models. Investigations using *Obeliscoides cuniculi*, a parasite of rabbits, have contributed importantly to understanding the interaction of genetic and environmental factors in the causation of developmental arrest in parasitic nematodes. Similarly, the classic investigations of Michel,[11] that first demonstrated individual immunologic unresponsiveness to a nematode, involved *Trichostrongylus retortaeformis*, another parasite of rabbits (see Chapter 17: Immunity to Nematodes). Still other species of trichostrongyles (e.g., *T. colubriformis*), although normally parasites of ruminants, will infect small laboratory animals (guinea pigs, hamsters) and these too have served as valuable experimental models.

Among the trichostrongyles occurring in ruminants, *Nematodirus* merits further comment. Unlike the other genera that produce "strongyle-type" eggs, species of *Nematodirus* produce particularly large (150 to 250 μm long), thicker-shelled eggs (Fig. 15–16). Their large size is probably related to the fact that the stage

that hatches is *not* the first stage larva, but a larger, more highly developed, infective third-stage. Therefore, the egg itself has to contain sufficient nutrients to support larval development through two stages.

It is especially interesting ecologically and epidemiologically that, in summer and fall, only a small proportion of L_3-containing eggs of *Nematodirus battus* are infective. Most eggs of this species require prolonged exposure to cold, winter conditions before the larvae are infective. Hatching occurs in the external environment during the following spring when the soil temperature rises. The hatched infective larvae crawl up onto herbage where they can be ingested by the young, susceptible lambs of the year.

This diapause phenomenon is important for the parasite's survival, since there is an annual rapid, density-dependent expulsion of adult worms and previously infected animals show strong resistance. Obviously, then, this species needs a resting stage that will permit it to survive the winter outside the host until new lambs become available in the spring. In many trichostrongyles, developmentally arrested, tissue-dwelling parasitic larvae serve this purpose, but in *Nematodirus battus*, larval diapause within the egg serves instead.

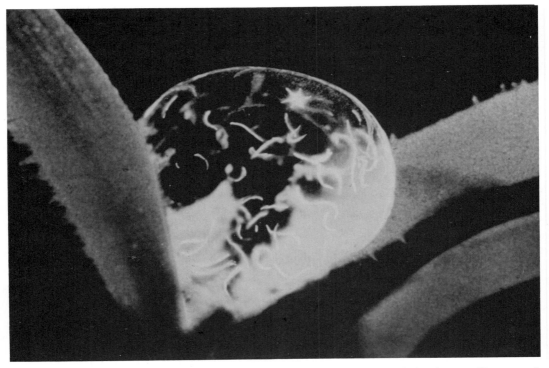

Fig. 15–13. Infective larvae of *Ostertagia ostertagi* in a droplet of water on a blade of grass. (Courtesy of American Hoechst Corporation.)

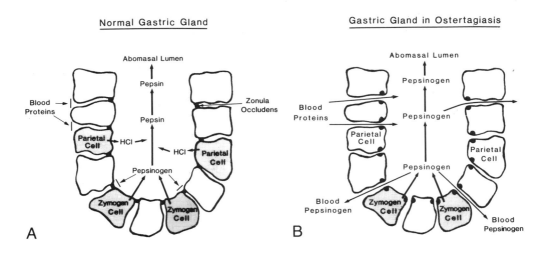

Fig. 15–14. Pathogenesis of *Ostertagia ostertagi* infection. A, A normal gastric gland of the abomasum of a bovine. B, A parasite-damaged gland. See text for explanation. (Courtesy of American Hoechst Corporation.)

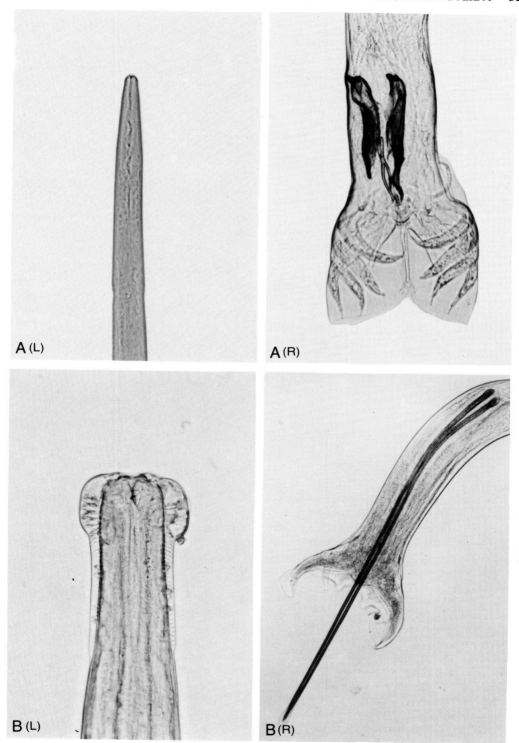

Fig. 15–15. Trichostrongyle nematodes. A, *Trichostrongylus.* Anterior end (left); bursa and spicules (right). B, *Nematodirus.* Anterior end (left); bursa and spicules (right). (A, from Georgi: Parasitology for Veterinarians, courtesy of W.B. Saunders. B, from Whitlock: The Diagnosis of Veterinary Parasitisms, courtesy of Lea & Febiger.)

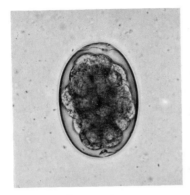

Strongyle

Fig. 15–16. Eggs of trichostrongyle nematodes, showing the normal thin-shelled, blunt-ended, strongyle-type egg along with the larger eggs of *Nematodirus* spp. (From Georgi, J.R.: Parasitology for Veterinarians. 4th Ed. Philadelphia, W.B. Saunders, 1985. From Fig. 12–23, p. 245.)

Nematodirus sp.

Nematodes of the genus **Dictyocaulus** are unusual among the trichostrongyles because they parasitize the lungs. One species, *Dictyocaulus viviparus* (Fig. 15–17), occurs in cattle, bison, and various species of cervids (deer). The adult worms live in the airways and lay thin-shelled eggs that usually hatch internally so that larvae, rather then eggs, pass in the feces. The free-living life history is like that of other trichostrongyles except that the L$_3$ is not sufficiently motile to move from the fecal deposit onto herbage. However, the parasite has evolved a fascinating behavioral mechanism to disperse from the feces when it attains infectivity. At this stage, the larvae migrate onto the sporangia of fungi that grow on the fecal mass, and when the sporangia explode, launching their spores into the surrounding environment, they also disperse the larvae explosively.

   D. viviparus is the causative agent of parasitic bronchitis in bovines. The larvae that arrive in the lungs are sufficiently small to reside in the bronchioles, where the worms and the exudate produced in response to their presence obstruct air flow. This is associated with coughing and abnormally rapid breathing. As the worms grow, mature and become gravid, eggs begin to appear in the feces (i.e., the patent phase of the infection), and the adult worms move into the larger spaces provided by the bronchi and trachea, damaging the epithelium severely. Blockage becomes attributable to the large adults, a foamy exudate produced by the host, and eggs and larvae as well. Because some eggs and larvae may be aspirated into the bronchioles and even into the alveoli, much of the respiratory tree, including its finest ramifications, may be blocked. During this patent phase, affected animals cough, have difficulty breathing, lose condition rapidly, and, if severely affected, may

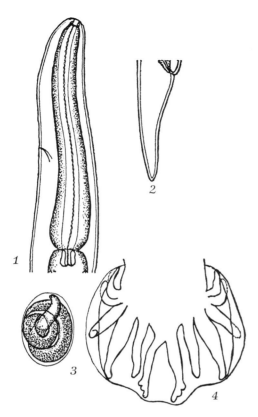

Fig. 15–17. *Dictyocaulus viviparus,* a lungworm of cattle and other mammals. 1, Anterior end. 2, Posterior end of female. 3, Egg with larva. 4, Bursa of male. (After Skrjabin, et al.)

die. Survivors eventually expel the adult worms. The host then enters the post-patent phase of the infection, and, ultimately, much of the damage is repaired. The animals are now largely resistant to reinfection, i.e., if challenged by a heavy infection, they tolerate few worms that cause only a mild, transient infection.

This resistance suggests that cattle can be protected from parasitic bronchitis by immunization (i.e., protected from the disease, although not from infection), and, in fact, the only commercially successful vaccine against a helminthic disease is one that protects against bovine parasitic bronchitis. This live vaccine is constituted of irradiated, infective larvae. Two doses of 1,000 larvae each, given to young calves, provide a transient protection that wanes in the absence of further stimulation. However, under field conditions, periodic reinfection occurs, maintaining a high level of resistance.

SUPERFAMILY HELIGMOSOMATOIDEA

Nippostrongylus brasiliensis, a small, reddish worm, is a common parasite of the small intestine of wild rats. The worm has been used in numerous experimental studies of parasitic nematodes, including investigations of their physiology, immunology and behavior. The male is 3 to 4 mm in length; the female 4 to 6 mm. In both sexes, the head is enlarged. The bursa is asymmetrical in shape, and the spicules are long and slender. Eggs pass from the intestine of the host, and the young worms within them develop into the infective stage in one day under favorable environmental conditions. The life cycle resembles that of the hookworm, *Ancylostoma duodenale,* and the adult stage is reached in about a week. On entering the host body, the larvae are carried by blood and lymph to the heart and thence to the lungs. By this time, they have developed to the fourth-stage larva and are carried up the bronchi to the pharynx and down the digestive tract to the intestine, where a final molt ushers them into adulthood. They feed primarily on host cells and blood, and eggs appear in the host's feces by the sixth day after infection.

Heavy infections are self-limiting, most worms being expelled in 2 to 3 weeks. Meanwhile, however, the feeding activity and the irritation due to the worm's movements can cause extensive damage to the intestinal mucosa. The normally delicate intestinal villi fuse to form larger, blunt-tipped units, or, in severe cases, the normal surface architecture is so radically changed that it becomes a flat sheet. Under these circumstances, the intestinal surface area is reduced and local malabsorption occurs. In simple, experimental host-parasite systems involving only one species of parasite, this can be compensated for by increased absorption in normal areas of intestinal epithelium lower in the intestine. However, natural host-parasite associations often consist of several parasitic species distributed sequentially along the gut. In this case, distal compensation may be impossible, and malabsorption may occur.

A single *Nippostrongylus brasiliensis* male is strongly attracted to a sex pheromone liberated from females, but when several males are jointly exposed to the pheromone, the response decreases. This inhibitory trend is reversed when the male:female ratio is increased above 10:35. This shift in the response to pheromone is possibly the result of male homosexual interference (thigmokinesis?) or the effect of multiple pheromones.[3] Apparently, pheromones produced by male worms attract only females, whereas the latter emit substances that attract both sexes. Females of *N. brasiliensis* are attracted to lipid fractions of excretory and secretory products of both males and females, whereas males attract only females.[14]

Changes in behavior and longitudinal movement in the host gut are correlated with the feeding regimen of the host, but, as with the tapeworm *Hymenolepis* (see Chap. 12), such movement appears to be either a response to gut secretions or a byproduct to host-food interactions, rather than a response to the food itself.

SUPERFAMILY METASTRONGYLOIDEA

The buccal capsule of the metastrongyles is absent or rudimentary; the bursa is reduced or absent; the vulva is situated near the anus; and the worms usually require an intermediate host. Typically, metastrongyle eggs hatch internally so that the larvae are passed in the feces. These are easily recognized by their characteristic tails. Most nematode larvae have smooth, uniformly tapering, conical tails of varying, species-specific length. The larval tails of the metastrongyles are frequently irregular in outline with terminal knobs or spines and sometimes with a distinct subterminal bend or kink. Most metastrongyles are "lungworms," and inhabit the respiratory (or circulatory) systems of carnivores, ungulates, rodents, and primates throughout the world. In North America, species of the genus *Protostrongylus* are important pathogens of wild Bighorn sheep and have probably been responsible for the failure of these animals to survive in abundance in some areas of the Rocky Mountains. For a bibliography of lung nematodes of mammals, see Forrester et al.[5] For a review of lungworms of sheep, see Rose.[15]

Metastrongylus elongatus, M. salmi, and *M. pudendotectus* are all thread-like, white worms that may reach 60 mm in length, although some are much shorter. They live in air passages of the lungs of pigs, where they lay their eggs. These eggs are coughed up, are swallowed, and are passed from the host with the feces.

Various species of earthworms belonging to the genera *Helodrilus, Lumbricus,* and *Diplocardia* eat the eggs, which hatch in the intestine. Larval

parasitic worms leave the gut lumen and enter the walls of the esophagus, crop, gizzard, and intestine. After a period of growth, the parasites enter the circulatory system, and in about a month, they are infective to swine that eat the earthworms. Larvae are released into the pig's intestine where they penetrate the walls and, through the lymphatic and blood vessels, reach the lungs. Three to four weeks after being eaten by pigs, the larvae have matured in the lungs and start producing eggs.

The interdependence of these congeners with reference to their establishment in a host is of particular ecologic interest. *Metastronglyus elongatus* can establish itself in monospecific infections and, in natural infections, occurs alone. Invariably, the other two species are found occurring with *M. elongatus,* and experimental investigations have shown that, when they are administered to pigs singly, the infections will not establish. This interspecific interaction contrasts with most others that have been described by parasitologists in which congeners have interacted antagonistically.

Infected pigs suffer from malnutrition, lung hemorrhage, difficult breathing, and coughing. The parasites may kill young pigs, and they may also carry the virus of swine influenza.

SUPERFAMILY PROTOSTRONGYLOIDEA

Angiostrongylus cantonensis normally lives in the blood vessels of rodents. It may cause cerebral angiostrongylosis (eosinophilic meningoencephalitis) in man. Infection occurs by eating raw prawns, land crabs, molluscs, or raw vegetables over which infected molluscs (slugs or snails) have crawled. Infective larvae are deposited in the mucus tracks of active gastropods.

These invertebrates serve as intermediate hosts of the parasite, which occurs in western and eastern Asia, Australia, the islands of the Pacific, Madagascar, Sri Lanka, Sarawak, Egypt, and the southern United States. The larvae have also been found in frogs and terrestrial planarians and have been experimentally established in pigs and calves. *A. costaricensis* has a life cycle involving terrestrial molluscs and wild rodents. The parasite causes abdominal angiostrongylosis in humans in Costa Rica. One case was reported in the United States. For a review of angiostrongylosis in eastern Asia and in Australia, see Cross.[4]

Parelaphostrongylus tenuis, the meningeal

worm of the white-tailed deer (*Odocoileus virginianus*), is a metastrongylid lungworm taxonomically, possessing the morphologic and life characteristics of this group. However, as its common name indicates, the adult worms reside in the central nervous system. This raises an interesting question: By what route does the parasite's dispersal stage, the first-stage larva, escape from the host? The parasite is also of special ecologic interest because it is a well-studied example of a pathogenic organism that affects the interspecific interactions of its hosts. As the range of its natural definitive host, the white-tailed deer, expanded, and the parasite was introduced into new geographic areas and host species, the new, more susceptible host species experienced an increased mortality with an attendant decrease in range. This three-way interaction is discussed more fully in Chapter 24.

Mature males and females measure about 1.6 to 5.5 and 3 to 7.7 cm in length, respectively. In natural hosts, they live in the subdural space closely associated with the venous sinuses of the brain, into which the female worms release their eggs. The eggs are then carried to the lungs, where they become trapped in the alveolar capillary beds, embryonate, and hatch. The first-stage larvae move into the alveoli and up the bronchial tree, aided by the directional beat of the ciliated respiratory epithelium. When they reach the pharynx, the larvae are swallowed, and eventually leave the host in the feces.

Early larval development occurs in terrestrial gastropods (snails or slugs). When a gastropod moves across fecal pellets deposited by an infected deer, the larvae actively invade the mollusc's foot. The developing larvae molt twice within the gastropod before attaining infectivity as third-stage larvae. Deer and other ruminants become infected when they feed on low-growing vegetation. The third-stage larvae emerge from the tissues of the ingested mollusc in the abomasum, penetrate into the peritoneal cavity, and then invade the spinal cord via a peripheral nerve. The growing larvae migrate anteriorly at first in the neural parenchyma, and subsequently in the subdural space. They are already immature adults by the time they reach the brain. Within about 3 months after infection, the worms have settled in the subdural spaces, matured, copulated, and produced eggs.

In white-tailed deer, disease is rare, but, in other wild ruminants (moose, elk, caribou) and domestic livestock severe neurologic disease is common. Weakness, ataxia (loss of coordination), circling, blindness and behavioral changes, including depression or fearlessness, are manifestations of infection in abnormal hosts. Abnormal hosts are severely affected because the parasites do not leave the neural parenchyma to enter the subdural space while small, but continue to migrate and develop in the parenchyma, even after having attained the large, more destructive, size of the immature adult. Furthermore, after arrival and maturation in the subdural space, they may reinvade the neural parenchyma and lay their eggs.

Little can be done to prevent or treat the infection in free-ranging wild animals. In zoos, it is important that white-tailed deer and other more susceptible species do not share a common enclosure. With reference to wildlife management, white-tailed deer should be quarantined for several months before they are used to stock parasite-free areas, so that their feces may be examined for the presence of lungworm larvae. Infected animals should not be used to stock new areas. Under conditions where it is possible to administer treatment, high doses of ivermectin and albendazole are effective.

REFERENCES

1. Anderson, R.M.: The role of mathematical models in helminth population biology. *In* Parasitology—Quo Vadit? Proceedings of the Sixth International Congress of Parasitology. Canberra, Australian Academy of Science, 1986.
2. Banwell, J.G., and Schad, G.A.: Hookworm. Clin. Gastroenterol., 7:129–156, 1978.
3. Bone, L.W., and Shorey, H.H.: Interactive influences of male- and female-produced pheromones on male attraction to female *Nippostrongylus brasiliensis*. J. Parasitol., 63:845–848, 1977.
4. Cross, J.H. (ed.): Studies on Angiostrongyliasis in Eastern Asia and Australia. Taipei, United States Naval Medical Research Unit No. 2, 1979.
5. Forrester, D.J., Forrester, G.M., and Senger, C.M.: A contribution toward a bibliography on the lung nematodes of mammals. J. Helminthol., 40(Suppl.):1–122, 1966.
6. Georgi, J.R.: Parasitology for Veterinarians, 3rd Ed. Philadelphia, W.B. Saunders, 1985.
7. Georgi, J.R.: Strongyloidiasis. *In* CRC Handbook Series in Zoonoses: Parasitic Zoonoses, Section C, Volume II, 1982, pp. 257–267.
8. Grove, D.: Strongyloidiasis. *In* Tropical and Geographic Medicine. Edited by K.S. Warren and A.A.F. Mahmoud, New York, McGraw-Hill, pp. 373–379, 1984.

9. Levine, N.D.: Veterinary Parasitology. Minneapolis, Burgess Publishing, 1978.

10. Lichtenfels, J.R.: Helminths of domestic equids. Illustrated keys to genera and species with emphasis on North American forms. Proc. Helminthol. Soc. Wash., *42*:1–91, 1975.

11. Michel, J.F.: 'Self-cure' in infections of *Trichostrongylus retortaeformis* and its causation. Nature, London, *69*:881, 1952.

12. Moncol, D.J., and Triantaphyllou, A.C.: *Strongyloides ransomi*: factors influencing the in vitro development of the free-living generation. J. Parasitol., *64*:220–225, 1978.

13. Poinar, G., and Hess, T.: Immune responses in the earthworm, *Aporrectodea trapezoides* (Annelida), against *Rhabditis pellio* (Nematoda). *In* Comparative Pathobiology. Vol. 3. Edited by L. Bulla and T. Cheng. New York, Plenum Press, 1977, pp. 69–84.

14. Roberts, T.M., and Thorson, R.E.: Chemical attraction between adults of *Nippostrongylus brasiliensis*: Characterization of the substance which attracts females. J. Parasitol., *63*:849–853, 1977.

15. Rose, J.H.: Lungworms of the domestic pig and sheep. Adv. Parasitol., *11*:559–599, 1973.

16. Schad, G.A., and Banwell, J.G.: Hookworms. *In* Tropical and Geographic Medicine. Edited by K.S. Warren and A.A. Mahmoud. New York, McGraw-Hill, 1984, pp. 359–372.

17. Schad, G.A., et al.: Paratenesis in *Ancylostoma duodenale* suggests possible meatborne human infection. Trans. Roy. Soc. Trop. Med. Hyg., *78*:203–204, 1984.

18. Schad, G.A., and Anderson, R.M.: Predisposition to hookworm infection in humans. Science, *228*:1537–1540, 1985.

19. Soulsby, E.J.L.: Helminths, Arthropods, and Protozoa of Domesticated Animals (Mönnig). Baltimore, Williams & Wilkins, 1968.

20. Stoll, N.R.: Studies with the strongyloid nematode *Haemonchus contortus*: I. Acquired resistance under natural infectious conditions. Am. J. Hyg., *10*:384–418, 1929.

21. Stoll, N.R.: On endemic hookworm, where do we stand today? Exp. Parasitol., *12*:241–252, 1962.

22. Triantaphyllou, A.C., and Moncol, D.J.: Cytology, reproduction, and sex determination of *Strongyloides ransomi* and *S. papillosus*. J. Parasitol., *63*:961–973, 1977.

23. Webster, J.M.: Aspects of the host-parasite relationship of plant parasitic nematodes. Adv. Parasitol., *13*:225–250, 1975.

24. Wilson, P.A.G.: Roundworm juvenile migration in mammals: The pathways of skin penetration reconsidered. *In* Aspects of Parasitology: a Festschrift Dedicated to the 50th Anniversary of the Institute of Parasitology, McGill University, 1932–1982. Edited by E. Meerovitch. Montreal, McGill University Press, 1982.

25. Wood, W.B. (ed.): The Nematode *Caenorhabditis elegans*. Cold Spring Harbor Monograph 15. Cold Spring Harbor Laboratory Press, Cold Spring Harbor, New York, 1988.

26. Zuckerman, B.M., Mai, W.F., and Rohde, R.A. (eds.): Nematodes. Vol. I. Morphology, Anatomy, Taxonomy, and Ecology. Vol. II. Cytogenetics, Host-Parasite Interactions, and Physiology. New York, Academic Press, 1971.

16

Class Secernentea, Orders Ascarida, Spirurida

Order Ascarida

The presence of three prominent lips around the mouth is an important characteristic of this group of worms (Fig. 16–1). Usually no pronounced posterior bulb is present in the esophagus, but in many species the muscular esophagus is followed by a posterior **ventriculus.** The latter may be short, and an anterior **intestinal cecum** and/or a posterior **ventricular appendix** may arise from its posterior edge (Fig. 16–2). The tail of the male is usually coiled. Life cycles of the ascaridoids generally involve one or more intermediate hosts, but those of species parasitizing man and domestic animals are often direct. Members of this order are primarily parasites of vertebrates, and some species have considerable economic importance.

SUPERFAMILY ASCARIDOIDEA

Toxocara canis and *Toxascaris leonina* are common intestinal roundworms of dogs. The latter also occurs in cats. Adult male worms range in length from 7 to 9 cm, whereas females are 10 to 17 cm long. *T. canis* is longer in average length than *T. leonina*. Neither species makes use of an intermediate host but both may use a paratenic host. The two are easily distinguished by examining their eggs. *Toxascaris* eggs (Fig. 16–3) are colorless and have almost smooth shells and prominent vitelline membranes. Eggs of *Toxocara* are light brown and have a thicker, pitted, proteinaceous coat and less prominent vitelline membranes.

Both species are infective to young animals by ingestion of embryonated eggs containing second stage (L_2) larvae. *Toxascaris* larvae enter the wall of the intestine, where they continue their development; then they return to the lumen of the gut and mature. In young, highly susceptible dogs, *Toxocara* larvae penetrate the gut wall and, by the hepatic circulation, arrive in the liver; then they migrate to the alveoli of the lungs, trachea, mouth, esophagus, stomach and small intestine, where they mature. *Toxocara* females can produce 200,000 eggs per day. Adults probably live for about 4 months.

In addition to the direct life cycle described above, *Toxocara canis* has several life history options, including *prenatal* infection. Some authorities believe that virtually every puppy is infected at birth, and, because *Toxocara canis* is a health risk for children, they insist that all puppies should be treated in the first few weeks of life, even before the worms mature and the parasite's eggs appear in the feces, making it possible to determine the parasite's presence.

Prenatal, canine *Toxocara canis* infection, like lactogenic hookworm infection of suckling pups, depends upon the storage of arrested larvae in the tissues of resistant, mature female hosts. During pregnancy, stored larvae in a *Toxocara canis*-infected female are reactivated and move to the fetal liver, where they again become dormant. Shortly after the birth of an infected pup, these larvae resume development and migrate to the small intestine, using the same tracheal route as larvae that hatch from eggs in the intestine. It is interesting, and somewhat enigmatic, that prenatal infection is also important for the transmission of *Toxocara vitulorum*, a parasite of calves, whereas *Toxocara cati*, a very common parasite of cats, is not transmitted *in utero*.

Additionally, *Toxocara canis* uses paratenic

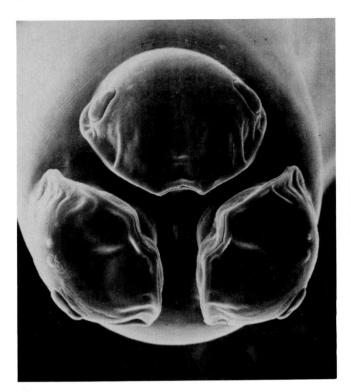

Fig. 16–1. *Ascaris suum:* scanning electron micrograph of the mouth opening with its three, large surrounding lips. (Courtesy of Dr. R. Lichtenfels, U.S. Department of Agriculture.)

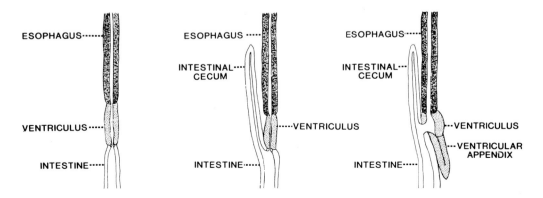

Fig. 16–2.   Variation in the structure of the esophagus among ascarid nematodes, showing esophagi with a posterior ventriculus, an intestinal cecum and a ventricular appendix. (Redrawn from Binford: Pathology of Tropical and Extraordinary Diseases. Courtesy of Armed Forces Institute of Pathology.)

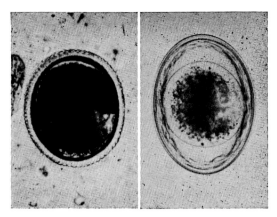

Fig. 16–3. Eggs of *Toxocara cati* (left) and *Toxascaris leonina* (right). *T. leonina* may be found in dogs as well as in cats; the one illustrated came from a tiger. It is 89 μm long. (From Georgi: Parasitology for Veterinarians, courtesy of W.B. Saunders.)

hosts. When for instance, a mouse ingests an embryonated egg, it hatches, but the larva fails to develop. Instead it migrates to the internal organs, including the liver, brain and even the eyes. Migration in mice is highly pathogenic, undoubtedly increasing their vulnerability to predation, and thus, the likelihood of transmission. Unfortunately, *Toxocara canis* larvae also survive in humans and migrate just as they do in mice and other mammals. When larval helminths migrate, without development, in the skin or internal organs of an abnormal host, the diseases they produce are called cutaneous and visceral *larva migrans,* respectively. In human *Toxocara canis* infection, hepatic, ocular, and even cerebral *larva migrans* occurs.[19] The latter may result in blindness.

Human toxocariasis is most prevalent in children, particularly children with a history of playing in soil or sand contaminated with dog feces. Soils and sandboxes in public parks, where dog owners walk their pets, are frequently heavily contaminated with *Toxocara canis* eggs.[11] The health risks associated with this form of environmental pollution contributed importantly to the passage of "Scoop Laws" that require dog owners to remove their pets' feces from the public domain in a sanitary manner.

The larvae of an ascarid of raccoons, *Baylisascaris procyonis*, are relatively large and neurotropic, i.e., they tend to invade the central nervous system, including the brain. They have

been associated causally with serious, often fatal, neurologic disease in free-ranging and captive species of wildlife, in domestic animals, and even in man.[22] Veterinary public health officials strongly discourage the adoption of "orphaned" raccoons for various reasons, including the risk of this particularly dangerous form of *larva migrans.* If, nevertheless, raccoons are kept as pets, they must be treated periodically to remove these nematodes before the nematodes mature and lay eggs.

Ascaris lumbricoides (Fig. 16–4), the largest nematode parasitizing the human intestine, is widely distributed in those parts of the world where poor socioeconomic conditions favor the occurrence of fecally transmitted parasites. Infection is extraordinarily prevalent, with 1.27 billion individuals, or approximately 25% of the world's population, harboring the parasite! In some parts of the world, where people normally defecate in the open, whole villages may be infected. Fortunately, morbidity and mortality attributable to ascariasis is low, but because infection is so prevalent, the actual number of fatal cases, estimated at 20,000 per year, is substantial. The annual global morbidity, due mainly to malnutrition and pulmonary disease, has been estimated at 1 million cases! This species, or more frequently its close relative *Ascaris suum,* has been used in numerous morphologic, immunologic, physiologic and biochemical studies, but as Pawlowski and Arfaa[29] have pointed out, much remains unknown about these important parasites.

Adult female worms measure 20 to 50 cm in length and 3 to 6 mm in diameter; males measure 15 to 30 cm × 2.4 mm. Thus, females are about as thick as a pencil, and end in a simple, conical tail. In contrast, the tail of the male is coiled ventrally and has two longitudinal rows of ventrolateral papillae that extend anteriorly from the tip of the tail to beyond the cloacal opening. The papillae probably aid the male in locating and embracing the female during copulation. A pair of short, curved, distally pointed spicules lie in the end of the ejaculatory duct and are protrusible through the cloacal opening (Fig. 16–5). The vulva is situated on the ventral surface near the junction of the anterior and middle third of the female worm. This area may be constricted into a "genital girdle," which is said to appear when the supply of sperm stored by the female is exhausted.

Three prominent lips, one dorsal and two

Fig. 16–4. *Ascaris lumbricoides,* adults. (Courtesy of Dow Chemical Co.)

sublateral, surround the mouth (Fig. 16–1). The mouth cavity leads to the muscular esophagus, which opens into the intestine. The intestine extends almost to the posterior end of the body, where it joins the rectum. The latter opens to the exterior directly through the anus in females, and into the cloaca in males.

The cuticle contains fibrous protein as well as small amounts of carbohydrates and lipids. The three basic cuticular layers (see Chap. 14) have been modified in *Ascaris* to form nine layers with an arrangement of spiral fibers. The outer cortex is not digested by pepsin or trypsin, but can be digested by ficin and papain. Evidence indicates that the cuticle is not permeable to amino acids and sugars. The gut, therefore,

must play an important role in absorption. For a description of growth of the cuticle and fine structure of the body wall, see Watson.[37]

The mechanisms by which *Ascaris* digests foodstuffs is not thoroughly understood. Amylase, lipase, esterase, and protease have been found in the intestine of this worm, but the exact role these enzymes play is still uncertain. Glucose and fructose are rapidly absorbed and seem to be quickly metabolized. The inner surface of the intestine is covered with microvilli, which increase its area and possibly have a secretory function. Evidence indicates that *Ascaris* in pigs ingest large numbers of cells that have been sloughed off from the host intestinal epithelium and lymphocytes that have found their way into the lumen. The gut probably functions as an excretory organ as well as in nutrition.

Ascaris possesses two distinct types of hemoglobin: (1) that which comprises about 2% of the hemolymph, and (2) that which occurs in the body wall of the worm. The functions of these hemoglobins are obscure. Their respiratory function is questionable because, under anaerobic conditions, the body-wall hemoglobin is deoxygenated in 6 hours, but the hemolymph hemoglobin remains unchanged. In general, the composition of the hemolymph is similar to that of the host intestinal fluids, as far as total ions, solids and ash are concerned. More specifically, 4.9% of the hemolymph of *Ascaris* is composed of proteins, of which 2.8% are albumens and 2.1% are globulins. Magnesium,

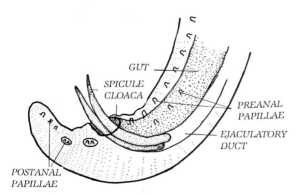

Fig. 16–5. *Ascaris lumbricoides,* posterior end of a male worm. The spicules emerge from the ventrally situated cloacal opening.

calcium, and potassium are kept at regular concentrations in this fluid, but sodium passes freely through the cuticle. The worm behaves like an osmometer. Most of the lipids of *Ascaris* are deposited in the reproductive tissues, especially the ovaries and uterine eggs; undoubtedly, these lipids are catabolized during embryonation. Starving worms apparently are unable to use these lipids.

The life cycle of *Ascaris lumbricoides* begins with the fertilized egg (see Fig. 14–14) within the body of the female. These eggs, 45 to 75 × 35 to 50 μm, are ovoidal with a thick, transparent shell and an outer, coarsely mamillated, albuminous covering. They are not embryonated when voided with the host feces. Unfertilized eggs are longer, narrower, more elliptical, and usually possess an irregular albuminous coating. Each female may lay as many as 200,000 eggs per day. Eggs are laid in the small intestine of the host and they contain refringent inclusion bodies, which contribute directly to the formation of the vitelline membrane and possibly to the chitinous shell. The eggshell is resistant to desiccation and to chemicals. Eggs in a 10% formalin solution remain viable for years, containing active larvae.

Growing and molting of the embryo occur within the egg, but the eggs do not hatch in the soil. Full embryonation requires two to three weeks. When embryos have developed to second-stage larvae (Fig. 16–6), the eggs are infective to man; when ingested, they reach the small intestine, where they hatch. Hatching is a complicated process that has been described as follows.[3] Dissolved intestinal gaseous carbon dioxide at low redox potentials, undissociated carbonic acid, and pH are all involved in producing a stimulus that causes the eggs to produce a hatching fluid, which contains the enzymes esterase, chitinase, lipase, and protease. Esterase alters the vitelline membrane of the eggs and thus permits the other enzymes to hydrolyze the hard shell. The inner membrane forms a bulge at this point, and the larva escapes by stretching and finally bursting through this bulge. Occasionally, the entire shell is hydrolyzed, leaving only the thin vitelline membrane surrounding the larva, which then ruptures the membrane.

Although larvae hatch in the same part of the intestinal tract where the adult worms will establish themselves, the larvae do not remain at this site; instead, they burrow into the intestinal wall, enter the circulation, and follow the same circuitous migratory route traveled by hookworm larvae. They go through the heart to the lungs, where they spend a week or more. Then they migrate to air spaces, up the trachea to the mouth, and are swallowed. The third molt takes place either in the lungs or the intestine. Back in the small intestine where they started, they now remain, molt once more, and become mature. The tour through the body is thought to be a phylogenetic reminiscence reflecting the occurrence of an intermediate host in which the ancestral *Ascaris* migrated. Indeed, primitive ascarids still have life cycles involving two intermediate hosts.

The odds against any one egg reaching maturity are indicated by the vast numbers of eggs produced by one female. As mentioned previously, one worm may produce as many as 200,000 eggs per day. A worm probably does not live for more than a year, but if it should live that long and should lay eggs at that rate, it would produce 73,000,000 eggs during its lifetime. If many adult worms are present at one time, we can easily understand why even an unconcentrated fecal smear always contains eggs.

Pathogenesis in ascariasis varies with the stage of parasites and the location in the host.

Larval worms migrating to the small intestine normally pass through the liver and lungs. In human ascariasis, passage through the liver is usually asymptomatic. The severity of pulmonary ascariasis depends upon the number of ascarid larvae migrating and on sensitization due to previous infection with the parasite. Symptoms vary from a slight cough in a primary, low-grade infection to a severe pneumonia with violent coughing, shortness of breath, wheezing and a high eosinophilia in subsequent infection.

In its intestinal phase, the manifestations of infection depend on the number of worms and on the nutritional status of the host. A few adult worms in an adequately nourished host may not cause noticeable disease. However, adult ascarids are notorious wanderers, and if the intestinal environment is disturbed, the worms may leave the small intestine and emerge spontaneously from the mouth, nose, or anus. Psychologically, this can be extremely stressful for the unsuspecting patient. The worms may also enter and obstruct the bile or pancreatic duct,

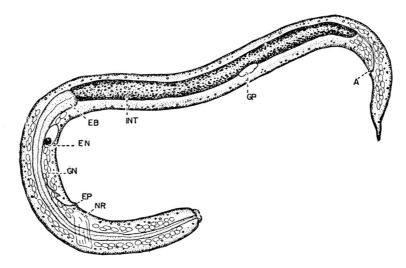

Fig. 16–6. *Ascaris lumbricoides,* second-stage larva. ×684. A, anus; EB, esophageal bulb; EN, nucleus of excretory cell; EP, Excretory pore; GN, ganglionic nucleus; GP, genital primordium; INT, intestine; NR, nerve ring. (From Nichols, courtesy of the Journal of Parasitology.)

and on rare occasions, even invade the upper respiratory system.

When the host is already malnourished and adult worms are numerous in the intestine, they may interfere with the absorption of nutrients, thereby exacerbating the existing nutritional deficiency.[9,33] Heavy intestinal worm burdens sometimes cause acute abdominal problems, including perforation of the intestinal wall, resulting in peritonitis or blockage of the intestinal lumen. Allergy to *Ascaris lumbricoides* and its relatives is common. This commonly manifests as facial edema, conjunctivitis and occasionally, severe asthma. These allergic responses may occur in persons showing no other evidence of infection.

Ascaris suum, a sibling species of *Ascaris lumbricoides,* is a common parasite of swine that will infect man, but only with great difficulty. This has been demonstrated experimentally and is supported by the observation that areas of high *Ascaris* prevalence in swine and humans do not coincide. The two species differ in minute structural details, such as the height of the labial denticles, as well as in the nature of the disease they cause.[2]

The ready availability of *A. suum* at slaughterhouses, its large size, and the fact that, except for its size, it is reasonably representative of nematodes, has made it the nematode commonly dissected by students in the zoology lab-

oratory. It is also frequently used as a laboratory model by experimental parasitologists.

This globally-distributed parasite of swine remains economically important, but its prevalence is decreasing wherever swine are raised indoors on concrete or steel grids and/or are given periodic treatment with modern anthelmintics. In swine raised with less concern about the control of parasitism, *Ascaris suum* can be a major cause of poor health and inadequate growth.

The most important pathologic changes are initiated when the larvae migrate through the lungs of young, non-resistant animals. In a primary infection the larvae cause small hemorrhages at the penetration points where they enter the alveoli. They also cause the loss of alveolar epithelial cells. In addition to this largely mechanical damage, more serious immunopathologic damage occurs in subsequent infections of sensitized hosts. This manifests as pulmonary edema, hemorrhage, and emphysema. Affected swine cough and have difficulty breathing; heavily infected individuals may die. Animals that recover are resistant to further disease.

Anorexia (lack of appetite) is another effect of parasitism; therefore, infected animals fail to maintain satisfactory rates of growth. This is exacerbated by the presence, in the small intestine, of adult worms that interfere with digestion and absorption of nutrients. Less apparent clinically, but significant economically, is the damage done by the larvae during migration through the liver. In swine sensitized by a previous infection, there is a severe response to the migrating larvae, with cellular infiltration of the affected areas, leaving them discolored and scarred after healing occurs.[3] These areas are called "milk spots," and if they are found during meat inspection at the slaughterhouse, the affected livers are condemned as unfit for human consumption.

Parascaris equorum is a related, large parasite of horses. Females measure up to 50 cm in length, and are about as thick as a small finger. Young, poorly maintained horses may harbor heavy infections; such animals may expel a bucket of worms after treatment! As in human and porcine ascariasis, heavily infected animals do not maintain adequate rates of growth. Very heavy infections can be fatal if the worms occlude the intestine and cause it to rupture.

This species was important in the early history of biology. Its availability and fecundity (each female containing thousands of eggs) made it the organism of choice for the great German cytologists, Boveri and Goldschmidt, who first described the cell cycle and meiosis using the ova of this and other, closely related ascarids.

Ascaridia galli is a common large parasite of the small intestine of various domestic and wild birds. It has a worldwide distribution. Males average about 50 mm long, and females may be longer than 100 mm, although they usually measure about 90 mm in length. These worms possess a preanal sucker. Eggs do not hatch in the soil, but when infective and eaten by a suitable host, they hatch in the intestine. During part of the growth period, the juveniles burrow into the intestinal mucosa for a few days, but sometimes this burrowing does not occur. Occasionally, young worms wander elsewhere in the host body, even penetrating the oviduct and becoming enclosed within the eggshell.

SUPERFAMILY HETEROCHEILOIDEA

Nematode larvae of the family Anisakidae are commonly found in the muscles, body cavity, mesenteries, and peritoneum of marine fishes and in invertebrates, as well. Larvae in the viscera of fish can migrate into the flesh on capture and death of fish. These larvae are generally at the third stage of development. Adult worms of the genus *Anisakis* usually occur in the stomach and intestines of marine mammals. Adults of other genera of anisakids are found in the intestines of amphibians, reptiles, birds, and mammals. Marine anisakids use one or two intermediate hosts, the first most commonly is a crustacean, especially an euphausid. Invertebrate hosts for larvae of *Contracaecum* are shrimps, copepods, arrow worms, mysids, cephalopods, and medusae.

Over 1200 human cases of anisakiasis (herringworm disease) have been reported, mostly from Japan,[34] but also from Europe and North America.[12] Lesions in the host's stomach are common. Symptoms may be similar to those of peptic ulcers or tumors, and they include nausea, vomiting, bloody diarrhea, and fever. The disease is acquired by eating raw, pickled, smoked, or salted fish such as herring, haddock, cod, mackerel, and bonito. Larval anisakids also are found in squid along the North American Atlantic coast. Among other genera whose larvae have been causally associated with anisakiasis are the codworms *Phocanema* (= *Terranova*) and *Porrocaecum*. Cooking fish or freezing it for 24 hours eliminates the danger of infection.[25]

SUPERFAMILY HETERAKOIDEA

Heterakis. The remarkable association between the protozoan flagellate, *Histomonas meleagridis*, which causes blackhead disease in poultry, and the cecal nematode, *Heterakis gallinarum* (Fig. 16–7), has been mentioned in Chapter 3. Male worms average about 10 mm in length, while females are usually about 13 mm long. Eggs pass from the host with feces and are eaten by earthworms, which serve as vectors. When the earthworms are swallowed by chickens, turkeys, and presumably ducks, geese, guinea fowl, quail, or other birds, the eggs hatch within a few hours, and the freed larvae burrow into the cecal mucosa for a day or two. The larvae then return to the cavities of the ceca and mature in three to four weeks. The distribution of this worm is probably worldwide.

SUPERFAMILY OXYUROIDEA

Oxyuroids, commonly known as pinworms, are characterized by an esophagus with a posterior bulb. Males are usually much smaller than females and have one or two equal spicules. They are found in both vertebrates and inver-

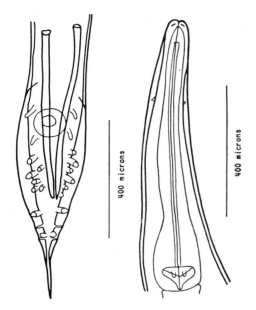

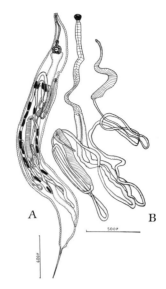

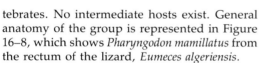

Fig. 16–7. *Heterakis gallinarum,* posterior and anterior ends of a male. (From Morgan and Hawkins: Veterinary Helminthology, courtesy of Burgess Publishing.)

Fig. 16–8. *Pharyngodon mamillatus* from the rectum of a lizard. *A,* A young female. *B,* The dissected genital structures. (From Chabaud and Golvan, courtesy of the Archives de l'Institut Pasteur du Maroc.)

tebrates. No intermediate hosts exist. General anatomy of the group is represented in Figure 16–8, which shows *Pharyngodon mamillatus* from the rectum of the lizard, *Eumeces algeriensis.*

Oxyuris equi is the common pinworm of the cecum and colon of horses. Males average about 11 mm long, whereas the females may be over 100 mm in length (40 to 150 mm). The adult worms are probably nonpathogenic during growth and maturation in the cecum. However, when fully gravid, the female worms emerge from the anus and crawl across the perianal area of the infected horse, depositing eggs which stick to the skin in an allergenic adhesive. The irritation of the crawling worms and itchiness elicited by the adhesive secretion cause infected horses to rub the base of the tail and neighboring areas against fences, stable furnishings, and other rough objects. Thus, affected horses tend to have an unkempt, rat-tailed appearance and if their skin is damaged, are subject to secondary bacterial infections. Sheep, rabbits, and other vertebrates may become infected with other species of pinworms.

Enterobius vermicularis is the cosmopolitan pinworm or seatworm of humans (Fig. 16–9). The adult female is most abundant in the cecum

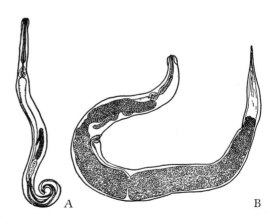

Fig. 16–9. *Enterobius vermicularis. A,* Male, ×23, showing characteristic coiling of the hindbody. *B,* Internal structures of the female, ×23, are obscured by the mass of eggs. (From Beaver, courtesy of Am. J. Clin. Path.)

and appendix. This short, whitish worm is shaped like a narrow spindle. A pair of lateral cephalic alae is situated at the anterior end. The mouth is surrounded by three lips or labia, and its cavity leads to an esophagus with an extra or prebulbar swelling and a distinct posterior bulb.

Males range from 2 to 5 $\times$ 0.1 to 0.2 mm in size. Each has a curved posterior end with a pair of small caudal alae supported anteriorly and posteriorly by pairs of papillae, and only a single spicule. Females range from 8 to 13 $\times$ 0.5 mm in size, each with a long tapering tail. The anus lies at the junction of the middle and posterior thirds of the body. The vulva is situated anteriorly in front of the junction of the anterior and middle thirds of the body. Eggs (Fig. 16–10) are flattened on the ventral side, measure 50 to 60 $\times$ 20 to 30 μm, have thick shells, and are embryonated when laid.[20]

The eggs are not deposited in the host's intestine, but remain in the body of the female worm until she crawls through the host's anus, usually at night. An infected person can sometimes feel the worms crawling on the perianal skin. The female worm contains about 11,000 eggs, and the pressure of these eggs sometimes causes her to burst and to scatter them on the body of the host or on the bedclothes. If the worm does not burst, she rapidly discharges her eggs and dies. Eggs are scattered about on clothes, bedclothes, hands, the bodies of in-

fected persons, and even in the dust of the room. In classrooms, up to 300 eggs have been found per 30 cm[2] in dust. Eggs embryonate fully and are infective within a few hours.

In women and girls, a pinworm may, on rare occasions, crawl into the genital opening and cause inflammation and irritation. Infected persons may experience intense itching in the anal region or only a mild tickling sensation. Children can become irritable and might not sleep well. The general symptoms are anal or vaginal pruritus, sleeplessness, irritability, abdominal pain, constipation, or diarrhea. Frequently symptoms are absent. Lesions in the appendix are not uncommon, and the worms may cause appendicitis.

People become infected by inhaling contaminated air, by sucking fingers, or by ingesting contaminated food or drink. When fully embryonated eggs are swallowed, they hatch in the small intestine; the larvae migrate to the cecum, appendix, colon, or ileum, where they mature in about a month.

Humans are generally thought to be the only hosts, but these worms have been reported in a few other primates, for example, chimpanzee and gibbon. Children are more commonly infected than are adults, presumably because they are less fastidious in matters of personal hygiene.

Diagnosis is usually made by a perianal swab, using cellophane tape, sticky side out, over the end of a tongue depressor, microscope slide, or

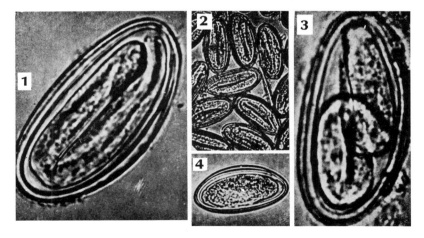

Fig. 16–10.   Eggs of *Enterobius vermicularis*. (Original photomicrographs of Kouri.) The eggs are asymmetric, double-walled, and colorless, and they contain a tadpole-shaped or worm-shaped embryo. *1, 2, and 3*, Eggs with vermiform embryo. (The embryo emerges through the convex part, located in the cephalic pole of the eggs—the adherence zone.) *4*, Nonembryonated egg. (After Gradwohl and Kouri: Clinical Laboratory Methods and Diagnosis, courtesy of C.V. Mosby.)

other suitable object. The tape is examined microscopically for the presence of the characteristic egg (Fig. 16–10). Prevention includes frequent laundering of nightclothes and bedding, keeping hands and fingernails clean, frequent bathing, and keeping rooms as dust-free as possible.[30]

SUPERFAMILY SYPHACIOIDEA

Pinworms of the genera *Passalurus, Aspicularis,* and *Syphacia* are cosmopolitan parasites, the former of rabbits, and the latter two of rats, mice, and other rodents (Fig. 16–11). Because they occur in hosts that are easily maintained in the laboratory, they are commonly used as representative pinworms for various kinds of experimentation and for evaluating new chemotherapeutic agents. *Syphacia* spp. have been reported in humans as rare, relatively harmless parasites.

Herbivorous tortoises harbor an abundant and complex pinworm fauna, which has been of interest in ecological parasitology. The ecological relevance of these nematodes is discussed in detail in Chapter 23.

Order Spirurida

SUPERFAMILY SPIRUROIDEA

These slender worms of moderate size usually possess two lateral lips surrounding the mouth, but four or six small labia may be present. The buccal capsule is cuticularized, and the esophagus does not have a bulb. In the female, the vulva is usually located near the middle of the body. The males of most species have two unequal spicules.

These worms are parasites of the digestive tract, particularly the upper part (including the mouth, esophagus and stomach), the respiratory system, eyes, nasal cavities, and sinus sacs of vertebrates. The life cycle involves one or two intermediate hosts, the first of which is invariably an arthropod, a beetle, grasshopper, fly, cockroach, or crustacean.

Gongylonema pulchrum (Fig. 16–12) is a parasite of sheep, goats, pigs, horses, cattle, and other mammals in which it lies embedded in a raised sinuous tunnel in the mucosa or submucosa of the esophagus and oral cavity. It is worldwide in distribution. Males may reach 62 mm in length, and females may be as long as 145 mm. The adult nematodes are recognized easily by the presence of raised, ovoid, cuticular blisters (Fig. 16–12). Eggs pass from the host in feces and are eaten by dung beetles or cockroaches, in which the larval worms develop. About a month after ingestion, these larvae become encysted in the insect and are infective to a new vertebrate host that eats the insect. Man is occasionally infected.

Gongylonema neoplasticum occurs in the stomach of rats, where it causes neoplastic (i.e., cancerous) changes in adjacent tissues.

SUPERFAMILY THELAZIOIDEA

Thelazia callipaeda, the oriental eye worm, and *T. californiensis* are two of many species normally found on the eyes of vertebrate animals and occasionally on the eyes of man. They are 5 to 20 mm long and usually lie on the surface of the eyeball (Fig. 16–13). When disturbed, they make their way under the eyelids or into the conjunctival sac. Morphological characteristics of the genus include a mouth without definite lips, a short buccal cavity, and usually numerous prenanal papillae. Males may or may not have alae, and the spicules are usually unequal. Members of the genus are ovoviviparous. Muscid flies ingest the larvae while feeding on lachrymal secretions, serve as hosts for larval development, and transmit the infection when they feed on the secretions of another eye. Most infections in domesticated animals are asymptomatic; in man, excessive lachrymal secretion and conjunctivitis may occur. Worms are removed from the eye easily and present only a mild, transient problem for the infected person. Related genera of nematodes occur on eyes, in nasal chambers, and in mouths of mammals and birds, in air sacs of birds, and in intestines of fish.

Spirocerca lupi (Fig. 16–14), an extraordinarily interesting parasite of wild and domestic canids, is distributed widely in warm, temperate parts of the world. The blood-red adult males and females measure 30 to 54 and 54 to 80 mm in length, respectively. They live mainly in large esophageal cysts, occurring rarely in the stomach and aorta. The minute, colorless, parallel-sided, embryonated eggs are deposited in the cystic spaces, pass into the esophageal lumen and leave the body with the host's feces. The eggs are ingested by dung beetles, in which larval development occurs. The larvae (L_3) in the beetles are infective to their definitive canine hosts, but since the beetles are too small to be usual prey for most canines, insectivorous transport hosts usually occur in the life cycle.

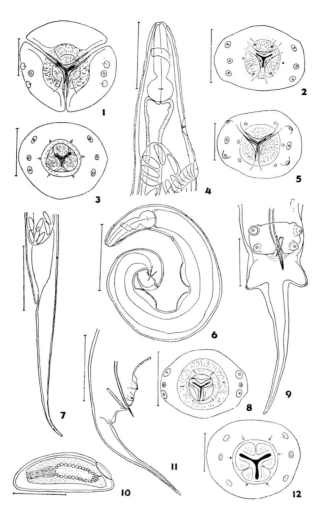

Fig. 16–11. *Syphacia*, several species from the intestines of various mammals showing especially the oral region. (Scales for *1*, *5*, and *12* represent 25 μm; for *2* and *8*, 30 μm; for *3*, *9*, and *10*, 50 μm; for *4* and *6*, 300 μm; for *7*, 500 μm; for *11*, 100 μm) *1* En face view of *S. thompsoni* from *Glaucomys sabrinus macrotis*, in Wisconsin. *2*, En face view of *S. obvelata* from *Microtus* in Alaska. *3*, En face view of *S. citelli*. *4*. Anterior end of *S. arctica*, female. *5*, En face view of *S. arctica*. *6*, Male, *S. arctica*. *7*, Tail of female, *S. arctica*. *8*, En face view, *S. peromysci* from *Peromyscus maniculatus bairdii* in Wisconsin. *9*, Ventral view, posterior extremity of *S. arctica*. *10*, Egg of *S. arctica*. *11*, Side view, posterior extremity of *S. arctica*. *12*, En face view, *S. eutamii* from (type locality and host) *Eutamias minimus*, Grand Marias, Minn. (From Tiner and Rausch, courtesy of Natural History Miscellanea.)

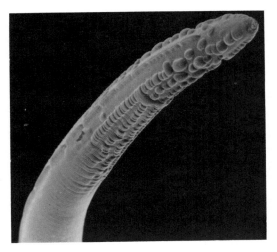

Fig. 16–12. *Gongylonema pulchrum:* scanning electron micrograph of the anterior end showing blister-like, cuticular bosses. (Courtesy of Dr. M.M. Wong.)

These include lizards, birds, and small mammals, in the viscera of which the larvae encyst. If a parasitized transport host falls prey to a suitable carnivore, the larva is released by digestion, penetrates the stomach, enters the aorta, and migrates anteriorly. After several months, it leaves the aorta, moves into the esophagus, and matures.

Spirocerca lupi is one of few metazoan parasites associated with neoplastic (cancerous) changes in its host. In predisposed breeds of dogs, *S. lupi* causes large, malignant, esophageal tumors. Indirectly, it causes the growth of bony spurs on the thoracic vertebrae lying just dorsad to its route of migration in the aorta. The mechanisms underlying these and other pathologic changes caused by the parasite remain unknown.

Gnathostoma spinigerum, a short, stout worm, is normally a parasite of cats and of other vertebrates, and only accidentally of man. The worms are worldwide in distribution, but are found mainly in the tropics. In normal hosts, they live in stomach tumors; thus, eggs get to the outside with feces. Eggs must enter water, in which they hatch into free-swimming larvae that are eaten by small crustacea such as *Cyclops,* which in turn are eaten by invertebrates or vertebrates (fish, frogs, water snakes), which are then eaten by cats, wild felines, and dogs. The parasites must develop in each of these hosts to become infective to the next host. Adults are 3 to 5 cm long and about 8 mm wide. They possess distinct, enlarged, spiny heads, and the spines continue down half the body.

In man, the worms are often found in an immature condition in skin tumors, although they may become lodged in almost any organ. Major foci of human infection are Thailand and Japan. People become infected by eating parasitized raw, marinated, or fermented freshwater fish.

SUPERFAMILY DRACUNCULOIDEA

Nematodes belonging to this group do not possess definite lips, but six conspicuous labial papillae are present. Eight external papillae occur in a circle surrounding the inner labial papillae. The esophagus has a muscular portion and a posterior, broader glandular portion.

Dracunculus medinensis, the Guinea worm

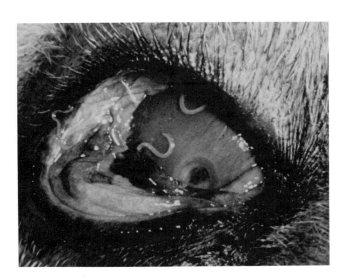

Fig. 16–13. *Thelazia* sp. on the eye of a dog. (Original photograph by Dr. E. Wagner. Courtesy of H. Zaiman: A Pictorial Presentation of Parasites, an unpublished slide collection.)

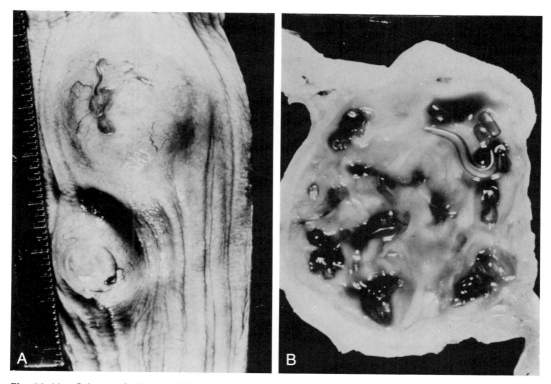

Fig. 16–14. *Spirocerca lupi.* A, Adult worm protruding from an esophageal granuloma of a dog; B, Section of an esophageal granuloma containing adult worms. (From Bailey, courtesy of Annals New York Academy of Sciences.)

(Fig. 16–15), occurs in many parts of Africa and Asia. It has been estimated that up to 40 million people are infected annually, about 6 million of these occurring in India alone. Areas of abundant rainfall are freer from the worm than are drier locations. Its distribution is affected less by climate than by poverty, the parasite occurring in areas where the public water supply consists of open wells. Adult females measure 75 to 120 cm in length (up to 4 feet), with a diameter of about 1.25 mm. Males average only about 25 mm in length. The adult worms live in the subcutaneous connective tissue of humans.

The life cycle begins with the development of young worms within the body of the female parasite. By the time the young are ready to emerge from the uterus, the female has produced a blister in the host's skin, into which a portion of the worm projects. When the host's skin comes into contact with water, during bathing or washing clothes, the blister bursts releasing myriads of young worms from the uterus into the water. Larvae are approximately

600 μm long. As they swim, some of them may be eaten by the small freshwater crustacean, *Cyclops* (Fig. 18–5). Within this crustacean vector, the young Guinea worms migrate to the hemocoel, undergo one or two molts, and become infective in 10 to 20 days. Definitive hosts (man and possibly dogs) acquire the infection by ingesting *Cyclops* in drinking water. In these mammals, the larvae leave the *Cyclops* while it is being digested and penetrate the host's intestinal wall. They migrate to connective tissue, where they mature in 3 to 4 months. After fertilization, the males die and are absorbed by the host. When females become gravid, about a year after infection, they migrate to the skin, which they pierce and thus provide the locus for the development of a small ulcer.[26]

Symptoms of dracunculiasis are absent until a skin sore begins to form. Then, nausea, diarrhea, giddiness, skin rash, itching, or asthma may occur. The sore is usually on feet, legs, or arms, but occasionally is on other parts of the body, the location corresponding to the parts that most often get wet. A reddish pimple de-

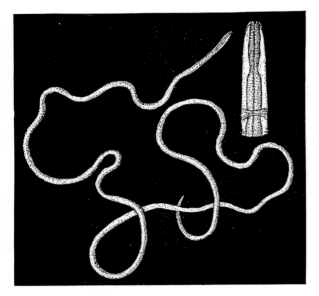

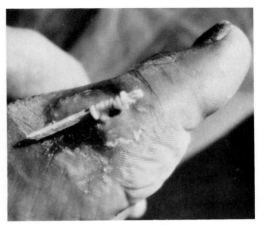

Fig. 16–16. The nematode, *Dracunculus medinensis,* being removed from the foot by periodic, gradual winding on a matchstick. (Courtesy of the Armed Forces Institute of Pathology, AFIP Neg. No. 67-1563-6.)

Fig. 16–15. *Dracunculus medinensis,* the Guinea worm. A female specimen is shown with the enlarged view of the anterior end.

velops, enlarges, and forms a blister that eventually breaks. The sore is usually not more than 5 mm in diameter, but it can become larger, and sometimes it becomes secondarily infected with bacteria.

The traditional procedure for ridding the patient of these worms is to remove them surgically. Modern drugs that provide symptomatic relief and reduce inflammation make worm extraction easier. Africans and Asians remove the worms by slowly winding them on small sticks (Fig. 16–16), but this feat must be done gradually, a few centimeters a day, or the worm will break, causing severe inflammation. Bacterial infection of the ulcer or tunnel containing the worm is a frequent complication of dracunculiasis.

Control measures include keeping people with the sores from contaminating wells, laundry, and bathing waters, and encouraging *Cyclops*-eating fish to become established in streams or ponds used by people. Even when infected crustaceans are present, human infections can be prevented by straining drinking water through a filter or even a tightly woven cloth.

The World Health Organization and other health-oriented national and international agencies have designated the '80s as the decade during which safe drinking water will become available to *all* people.[27] As part of this program, dracunculiasis has been targeted for eradication. If the program succeeds, *Dracunculus medinensis* will be the first helminth of humans to be eradicated globally.

SUPERFAMILY FILARIOIDEA

These long, thin, tapering worms have a simple mouth without lips and a long esophagus divided into an anterior muscular part and a posterior glandular part that lacks a distal bulb. They are often called filariae or filarial worms, the adults of which are viviparous, releasing vermiform embryos, microfilariae (Fig. 14–17) into the blood, lymph, or adjacent tissues. From here, they are taken up by hematophagus arthropod vectors, in which larval development occurs. Some of the species are important pathogens of man. Because most filarial parasites are host-specific and are well adapted only to their normal definitive hosts, research with commonly available small laboratory animals has been limited.

Wuchereria bancrofti and *Brugia malayi* are closely related, lymphatic-dwelling, filarial nematodes that are among the most important metazoan parasites of humans. They are included among the "Big Six," *the* six infectious diseases ranked by the World Health Organization as playing predominant roles in inhibiting the economic development of the Third World. These mosquito-transmitted filariae are widely distributed in the moist tropical and subtropical parts of the world, with foci of infection remaining even in such highly developed nations as Japan.

The adult parasites occur in the lymph vessels and nodes. *W. bancrofti* is the larger of the two species, but grossly they are indistinguishable, white, thread-like nematodes. The males, smaller than the females, have coiled tails and two spicules of unequal length and dissimilar shape. The vulva is located near the anterior end of the female, which, when mature, is filled with embryos, the microfilariae. The latter leave the female and circulate in the blood.

The microfilariae of *W. bancrofti* are usually nocturnally periodic, that is, they begin to appear in the peripheral circulation in the evening, become maximally abundant sometime after midnight, and essentially disappear again by morning. However, in some areas, an aperiodic form occurs, that is, there is *no* daily peak, and microfilarial density remains more or less constant throughout each 24-hour period. The microfilariae of *B. malayi* are nocturnally periodic in regions where the species is transmitted by anopheline mosquitoes, but are subperiodic where transmitted by mosquitoes of the genus *Mansonia* (see Mosquitoes, Chapter 20). Subperiodic forms show a diurnal rhythm with maximal microfilaremia occurring at night, but in contrast to the periodic forms, the microfilariae maintain an easily detectable level of abundance during the day.

It has been suggested that, as an adaptation for transmission, microfilarial behavior has evolved so that peak microfilaremia corresponds to the time of the vector's peak biting activity. Alternatively, the rhythms may be primarily adaptations that hold the parasites in the most suitable part of the host for the maximal amount of time daily, releasing them only briefly at times when conditions in the peripheral circulation are least adverse for microfilarial survival.

Although the mechanisms that control periodicity remain to be understood completely, Hawking[16,17] has provided an explanation for nocturnal periodicity that is supported by abundant experimental data. He proposed that this periodicity depends alternately on (1) accumulation of the microfilariae in the lung capillaries, usually by day (active phase), and (2) approximately even distribution throughout all the circulating blood, usually by night (passive phase). This periodicity is adjusted to the established circadian cycle of the host (sleeping and waking habits) instead of day and night as such. The microfilariae have their own endogenous rhythm, synchronized with the cycle of the host. The periodicity appears to be controlled by the magnitude of the difference in oxygen tension between venous and arterial blood during day and night.[16,17,39] Microfilariae of *W. bancrofti* (Fig. 14–17) disappear from the peripheral circulation during the day because they accumulate in the small arterioles of the lungs. If an experimental host breathes oxygen at night, when the microfilariae are normally in the peripheral blood, the parasites accumulate in the lungs.

Whatever the cause of dispersal of young worms to the peripheral circulation, if they are there when a person is bitten by a mosquito (*Culex, Anopheles, Aëdes, Psorophora, Mansonia*) appropriate to the species of the worm, these microfilariae are taken into the insect's stomach, in which they begin a period of growth and migration that usually requires about 2 weeks. During this time, the larvae molt twice and eventually change to an elongated, infective stage (Fig. 16–17). At first they become lodged in thoracic muscles; but later, near the end of their residence in the insect, the infective larvae travel to the proboscis, where they are in a position to penetrate the skin or wound when the insect bites another person (or the original host).

Unlike the life cycles of other parasites (i.e., trematodes and some protozoa) that require intermediate hosts, those of the filariae (and of nematodes generally) do *not* include a stage that multiplies in the intermediate host. Each microfilaria that is ingested and is able to develop and survive gives rise to only *one* infective larva. This has important implications for the epidemiology and control of parasites. Development is intracellular within muscle fibers, but tissue reaction is rarely observed. A light infection probably does the mosquito no harm, but a heavy infection is injurious and increases the death rate among mosquitoes. *Brugia* larvae in thoracic flight muscles of *Aëdes egypti* apparently feed on host muscle mitochondria during the last third of their intramuscular development phase. Mitochondria have been found attached to the worm's surface, lodged in its esophagus and partly disintegrated in the intestine.[4]

Wuchereria and *Brugia* larvae travel from the point of entry into human skin to the lymphatic

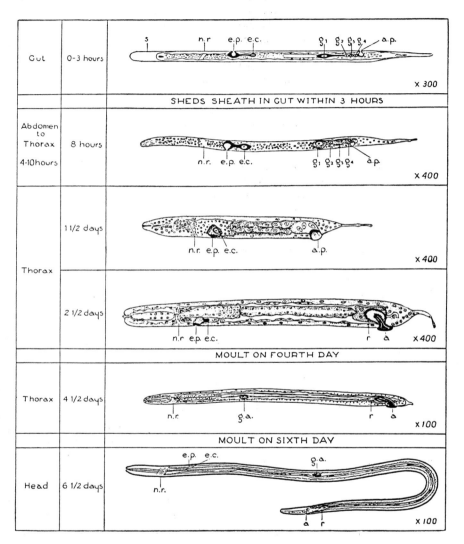

Fig. 16–17. Developmental stages of *Brugia malayi* in the mosquito. Column 1, site in the mosquito, Column 2, time required for development under tropical conditions, Column 3, larval stage: a, anus; ap, anal pore; ec, excretory cell; ep, excretory pore; g1, g2, g3, g4, so-called genital cells; gp, genital primordium; nr, nerve ring; r, rectum; s, sheath. (From Belding, D.L.: Textbook of Parasitology. 3rd Ed. New York, Appleton-Century-Crofts, 1965. Redrawn from Feng, 1936.)

vessels, where they grow and molt twice before becoming adults.

The pathologic changes occurring in lymphatic filariasis are attributable to the presence of adult worms in the delicate lymph vessels. Their presence results in dilation and thickening of the affected vessel and the formation of growths that extend into the vessel's lumen. The pathologic changes, along with the worm's presence, lead to inadequate lymph flow and damage to the intraluminal valves. Ultimately, abnormal back pressure, lymphstasis or both occur, causing lymphedematous changes of the affected leg, arm, breast, or genital organs. Death of adult worms elicits a strong immunologic response that exacerbates the damage done by the living worms. Under endemic conditions, individuals that develop a strong immune response and are subjected to prolonged, regular re-exposure, eventually develop *elephantiasis*, the horribly disfiguring enlargement of affected organs with thickening and wrinkling of the skin (Fig. 16–18).[24,28]

Not all infected individuals become symptomatic. In fact, lymphatic filariasis is a spectral disease, the manifestations of which depend on the timing of first exposure, the frequency of exposure, and variations in the degree of the host's response, which probably depends on genetic factors. Thus, infection in persons born in endemic areas ranges from infected but asymptomatic, through infected and showing annual temporary bouts of disease (filarial fevers), to those with permanent lymphedema and perhaps elephantiasis. These pathologic states differ with respect to the occurrence of microfilaremia and the nature and degree of immune responsiveness. Infections are generally more severe in immigrants.[28]

Tropical pulmonary eosinophilia or eosinophilic lung is another form of human filariasis, but the disease has a microfilarial rather than an adult-worm etiology. Patients with tropical eosinophilia are hyperresponsive and destroy the microfilariae as rapidly as they are produced. This is accompanied by an asthma-like pulmonary disease with a high eosinophilia. Microfilarial destruction apparently occurs in the lung, and, hence, microfilariae are absent from the peripheral blood. This absence makes parasitologic diagnosis difficult, but immunologic tests are positive for the presence of specific antibodies. Indeed, cells and sera from this class of patient are so reactive that they support

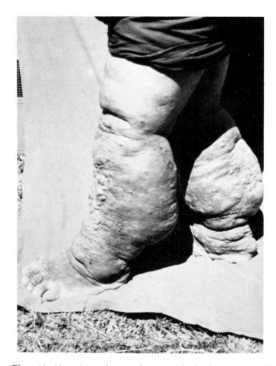

Fig. 16–18. An advanced case of elephantiasis of both legs. (Courtesy of Dr. Walter Vermeulen, Western Samoa.) In a personal communication, Dr. Vermeulen stated, "At present, this type of fantastic deformity is quite rare but we do estimate that in the age group of forty years plus, we find an incidence of elephantiasis of 2 to 3%."

the concept that eosinophilic lung is caused by the patient's hypersensitivity to the presence of microfilariae.

For many years diagnosis was made by microscopic demonstration of microfilariae in blood smears. Because only a few drops of blood are examined in a smear, many light infections can be missed. A modern, more sensitive method involves lysing the cells in a blood sample and forcing it through a filter that retains the microfilariae. This filter can then be cleared and examined microscopically for microfilariae. Charts presenting the morphologic features of microfilariae useful in differential diagnosis are found in textbooks of human parasitology and tropical medicine.[36] These direct parasitologic methods can be supplemented by a number of new immunodiagnostic methods,[1,7,15] but these are still regarded as research tools rather than established diagnostic tests.[23]

Vaccination to prevent filarial infection is not yet possible, but it is the goal of much research activity. Presently, the best form of individual prevention is avoidance of mosquito bites through the use of repellants, protective clothing, house sprays, netting, and screening. Community-wide prevention and control involves chemotherapy, mosquito control or both (see Chapter 20 for a discussion of mosquito control). Diethylcarbamazine (DEC), an anthelmintic with little if any adulticidal activity, is a potent microfilaricide and can therefore be used to interrupt transmission.[23] In fact, since it is also effective against molting larvae, it should be useful as a chemoprophylactic agent as well. In some parts of the world, DEC has been added to household staples (e.g., table salt) to provide continuous, low-level, community-wide prophylaxis.[8] By this and other methods, control has been achieved locally; but generally speaking, control efforts have not had permanent, ameliorative effects. Too often, administrative enthusiasm and adequate funding cannot be sustained for a sufficient period of time to have a permanent impact.

Onchocerca volvulus, when included with *Wuchereria bancrofti* and *Brugia malayi,* completes the trio of filarial worms that have a major depressing effect on socioeconomic progress in the Third World. *O. volvulus* is transmitted by black flies *Simulium* spp., and because of the habitat requirements of the aquatic immature stages of these vectors, the parasite occurs only near areas with free-flowing rivers and streams (Fig. 16–19). On a larger, global scale, the parasite is widely distributed in sub-Saharan Africa, in mountainous areas of the Arabian peninsula and in Central and South America. It has been suggested recently that the parasite is establishing new foci of infection in widespread areas of the New World tropics.[21] It causes a variety of pathologic changes in heavily infected humans, the most important of which occur in the eye.

Until recently, ocular onchocerciasis, popularly known as River Blindness (Fig. 16–20), was an increasing problem in East Africa, where its prevalence had become so great in certain riverine localities that entire villages were abandoned and land valuable for cultivation was lost. This serious economic problem and the attendant emotion-charged social situation attracted international attention, and led to the formation of the Onchocerciasis Control Program (OCP) under the auspices of the World Health Organization. The program has received broad-based support from the United States Agency of International Development (USAID), the World Bank, the World Health Organization and other major national and international agencies. The resulting complex of research and control activities is one of the great success stories of biomedical science, public health, and international collaboration.

Onchocerca volvulus is a parasite of humans, in which its adult stages live mainly in the subcutaneous connective tissue. The adult males and females are thread-like nematodes measuring 3 to 6 and 23 to 70 cm in length, respectively. The cuticle is raised in annulations that are much more prominent on the females than the males. Worms of both sexes occur together inseparably tangled in a nodule. The females deposit unsheathed microfiliariae (210 to 320 μm, in length) into the nodule (Fig. 16–19), from where the microfilariae migrate through the host's tissues, including the eyes. Ultimately, they become concentrated in the skin and are taken up by black flies (*Simulium* sp.) when they feed on an infected person. In the fly, the microfilariae invade the thoracic muscle and develop to the infective stage (L_3) in about a week. The L_3 migrate to the mouthparts of the fly and are transmitted when the fly resumes feeding. Related species (Fig. 16–21) with similar life histories occur in horses and cattle and have been used as experimental models, because *O. volvulus* does not develop in laboratory animals.

The subcutaneous nodules tend to be distributed around the neck, shoulders and head in infections occurring in Central America, whereas in Africa they occur over bony protuberances on the trunk and limbs (Fig. 16–19). Impaired vision occurs in infected individuals with high microfilaria counts. It is thought, therefore, that ocular onchocerciasis is caused by invasion of the eye by the microfilariae per se, although the exact mechanism by which they damage the eye remains unknown. Other important pathologic manifestations include inflammation and enlargement of the lymph nodes, particularly those of the groin and, in longstanding infections, thickening, premature wrinkling, and depigmentation of the skin. Parasitologic diagnosis depends on demonstrating the characteristic, unsheathed microfilaria in sediments from skin biopsies soaked in saline solution. The presence of infection can also be established by demonstrating the adult worms and microfilariae in excised nodules.[6,13,14,35]

Fig. 16–19. Epidemiology of onchocerciasis. (From Hunter, Frye, and Swartzwelder: A Manual of Tropical Medicine, courtesy of W.B. Saunders.)

Fig. 16–20.   Chain of men blinded by onchocerciasis being led by young boys not yet blinded by the disease. (Courtesy of the Armed Forces Institute of Pathology, AFIP Neg. No. 68-4062.)

No satisfactory drugs have been available for the treatment of onchocerciasis, but recently a benzimidazole (mebendendazole) and a fungicidal antibiotic, ivermectin, have shown great chemotherapeutic promise. Excision of nodules (nodulectomy) is a method of control that has been practiced in Central America for at least 50 years; it is thought to have limited the occurrence of blindness. In the absence of untreated control groups, the alleged beneficial effects are not supported by conclusive evidence.

Control at the community level has depended on larvacidal chemicals. In the 1950s, the parasite was eradicated from a relatively small, isolated area in East Africa by applying DDT to the small mountain streams that constituted the breeding habitats of its vector.[32] In the Volta River Basin the problem is many orders of magnitude greater. The black fly vector occurs over vast areas of savannah and forest, rather than in small, isolated mountain systems, and the breeding habitats include major rivers. Nevertheless, transmission had been reduced dramatically by larvacidal treatment of breeding habitats in an area covering 75,000 km^2, so that in 80% of this area there is now little risk of blindness. Actually, the larger sizes of the breeding habitats proved to be an advantage to

the Onchocerciasis Control Program, because it made them amenable to aerial spraying. This program has been remarkable for its conscious effort to minimize undesired adverse ecologic side effects; the safest effective larvacides were used at concentrations considered the minimum necessary to achieve control of the vector. The program is, in fact, using bacteria (*Bacillus thuringiensis*) and nematodes (mermithids) as biologic control agents where local ecologic conditions make this approach feasible.[32,36]

Loa loa (Fig. 16–22), the eye worm of man and primates, occurs locally in the forested areas of equatorial Africa. Its precise range is determined by the distribution of its intermediate hosts, certain deer- and horseflies of the genus *Chrysops*, that prefer densely shaded habitats. There are two physiologic races of the parasite. One, occurring in humans and transmitted by day-biting flies, has a diurnal microfilarial periodicity with peak densities occurring in peripheral blood about noon. The other, occurring in baboons (drills) and other primates, is nocturnally periodic and is transmitted by night-biting or crepuscular vectors. The parasite's behavioral responses that result in this periodicity are genetically determined and inherited as a simple, single gene trait in a classical Mendelian fashion.[10]

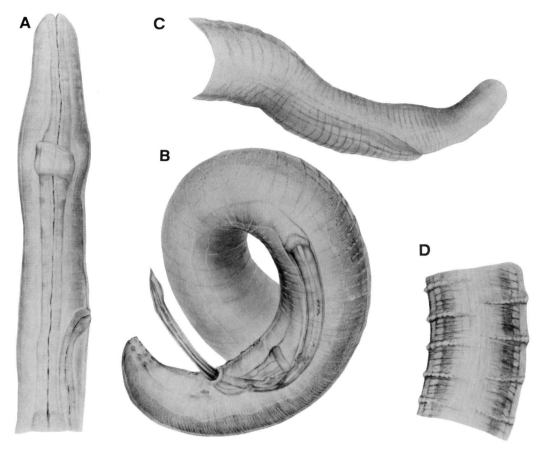

Fig. 16–21. *Onchocerca cervicalis* from the cervical ligament of a horse. A, anterior end of the female; B, coiled posterior end of the male; C, posterior end of the female; D, segment of the body showing the prominent cuticular annulations. (Courtesy of Dr. R. Lichtenfels, U.S. Department of Agriculture.)

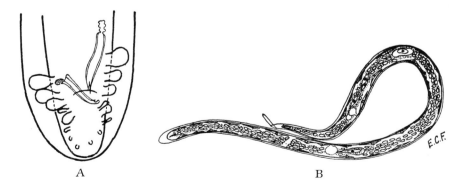

Fig. 16–22. *Loa loa.* A, Posterior end of male, ventral view, showing caudal alae, papillae and copulatory spicules. ×180. (From Faust. After Yorke and Maplestone. Nematode Parasites of Vertebrates, courtesy of J. and A. Churchill, Ltd.) B, Microfilaria of *Loa loa*. ×666. (From Faust, After Fülleborn. Arch. Schiffs Tropen-Hygiene, courtesy of Johann Ambrosius Barth.)

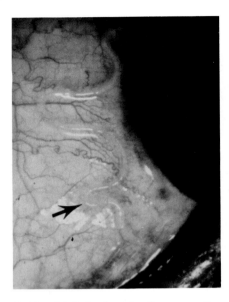

Fig. 16–23. Adult *Loa loa* migrating across the eye. (Courtesy of the Armed Forces Institute of Pathology, AFIP Neg. No. 73-6654.)

The adult worms, males measuring 30 to 34 and females 50 to 70 mm in length, are widely distributed in the subcutaneous tissues of their hosts. They are prone to wander and can move quite rapidly (up to 2.5 cm per minute). In wandering, they occasionally traverse the eye (Fig. 16–23), which for the unsuspecting patient may be a psychologically traumatic, first indication of infection. Adult females produce sheathed microfilariae that enter the circulation and accumulate in the small arteries of the lungs. They disperse to the peripheral circulation during the day or night, depending on the biting habits of the specific vector. After the fly ingests the microfilariae they move to its fat body, where in about 10 days they attain the infective stage. This stage is surprisingly large (2 mm long), and when the fly takes its next blood meal, it emerges by rupturing the membranes of the fly's proboscis. The infective larvae enter the subcutaneous tissues of the definitive host, where after molting twice they remain throughout their adult lives. The prepatent period is 4 to 6 months.

Although the parasite traversing the eye may be the infection's most dramatic manifestation, this is not the most common sign of loaiasis. This dubious distinction belongs to transient swellings (Calabar swellings) lasting 1 to 3 days,

measuring up to 3 cm in diameter. They occur anywhere on the body, but most commonly on the extremities. These swellings, named after the Calabar Coast of Africa, are probably local immunologic responses to the presence of the parasite in the underlying subcutaneous tissue. Pathologic changes also occur in the kidneys, due to the deposition of antigen-antibody complexes, and in the central nervous system if it is invaded by microfilariae. In mild cases this invasion may cause irritability, depression, and headache; in more heavily infected individuals encephalitis, coma, and death may occur. The latter may be associated with the sudden destruction of large numbers of microfilariae in the brain after anthelmintic treatment.[28]

Again, as in lymphatic filariasis, native-born individuals respond much less vigorously to the presence of the parasite than immigrants. The former may harbor asymptomatic infections that go unrecognized until the worm wanders across the eye. These patients usually have a microfilaremia, and can transmit the infection. In contrast, foreigners often react strongly, destroy the microfilariae as they are produced, and therefore do not transmit the infection. They often have a high blood eosinophilia and are more severely debilitated by the repeated appearance of painful Calabar swellings.[28]

Dirofilaria immitis, the "heartworm" (Fig. 16–24), lives in the pulmonary artery and heart (typically the right ventricle) of dogs and in some wild carnivores, such as the fox or wolf. It is transmitted by mosquitoes and is widely distributed in tropical and temperate parts of the world having sufficient rainfall.

After ingestion by a mosquito, the microfilariae migrate from the stomach to the Malpighian tubules. Here they initially develop as small intracellular sausage-stage larvae (Fig. 16–17) and later as extracellular, destructive parasites that ingest the cells of these tubules. Eventually they migrate to the labium via the abdominal, thoracic and cephalic cavities, developing further as they go. They attain the infective stage in the labium. The time required to complete development is temperature-dependent, and may be as short as 8 to 10 days in tropics or as long as 15 to 17 days in temperate areas. The infective larvae (L3) burst from the mosquito's proboscis when it feeds, and actively invade the skin via the opening made by the mosquito. Development in the definitive host occurs in the connective tissue, and in-

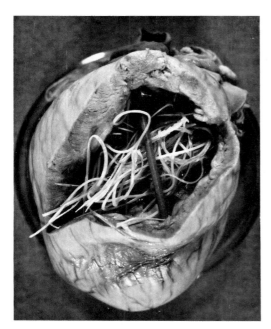

Fig. 16–24. The ventricle of a dog's heart cut open (sides held apart with a pencil) to show a heavy infection with *Dirofilaria immitis*. The atria and pulmonary artery were also packed with worms. (Courtesy of Mr. Michael Masson.)

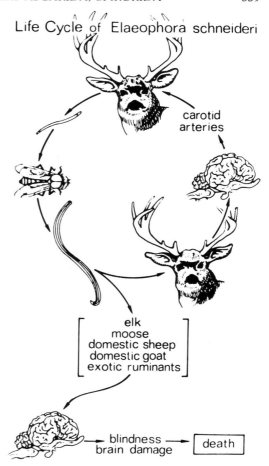

Fig. 16–25. Life cycle of *Elaeophora schneideri*. (From Hibler, courtesy of the Colorado Division of Wildlife.)

cludes 2 molts. Adult males and females, measuring about 14 cm and 27 cm in length, respectively, live in the heart and pulmonary artery (Fig. 16–24). The infection becomes patent in 6 to 7 months.

The adult worms in the chamber of the heart and adjoining vessels obstruct blood flow and interfere with the functioning of the heart valves. Furthermore, the movement of the adult worms in the pulmonary artery causes a roughening of the endothelium with the formation of thrombi (clots), impeding blood flow in this important vessel. When adult worms die, they may be carried to the lung, where they block the branches of the pulmonary artery. Dogs with heartworm disease tire quickly, cough and fail to thrive.[5]

In areas where this parasite occurs, dogs should be examined for the presence of circulating microfilariae annually before the transmission season begins. If none are found, they should be given daily prophylactic treatment (diethylcarbamazine) for the length of the mosquito season and for 2 months thereafter. If microfilariae are found, a program of chemother-

apy designed to kill the adult worms and the microfilariae should be initiated. This must be done under the direction of a competent veterinarian, since as noted above, dead worms may be carried to the lungs with possible serious consequences for the treated animals.

Immature *Dirofilaria* of various species may infect man as conjunctival or subcutaneous parasites. Occasionally, adult *Dirofilaria* have been reported from man, an abnormal host, but no evidence suggests that viable microfilariae are produced by female parasites.

Elaeophora schneideri (Fig. 16–25) is a filarial parasite of white-tailed deer (*Odocoileus virginianus*) and the mule deer (*Odocoileus hemionus*). In these, its normal hosts, the parasite occurs mainly in the carotid arteries and produces little disorder. In domestic sheep the parasite is associated with a form of filarial dermatitis commonly known as "sorehead." In elk and other abnormal hosts, the parasites occur in arteries supplying blood to the brain and eye; infection, therefore, is associated with behavioral abnormalities (e.g., circling and ataxia [incoordination]) and blindness. In these animals the infection is often fatal.

The adults are large worms, measuring up to 12 cm in length, and often occur in pairs in the arteries. Microfilariae are released into the arteries, but are usually found in the capillaries of the face, where their tabanid (horsefly) vectors tend to feed.[18] The developmental period in the fly is about 2 weeks. The infective larvae occur in the mouth parts of large species of horseflies. Because the infective larvae are extraordinarily large, small horseflies succumb to infection and are, therefore, unsuitable hosts.

The geographic distribution of the parasite in deer includes large areas of North America, whereas in sheep and elk, the parasite is found in high mountainous areas of the western United States. In elk, this may be determined largely by the restriction of those hosts to remote areas, but a similar explanation will not serve for sheep. Perhaps the distributional anomaly is determined by the feeding preference of the parasite's vectors.

REFERENCES

1. Ambroise-Thomas, P.: Immunological diagnosis of human filariases; present possibilities, difficulties and limitations (a review). Acta Trop., 31:108–128, 1974.
2. Ansel, M., and Thibaut, M.: Value of the specific distinction between *Ascaris lumbricoïdes* Linné 1758 and *Ascaris suum* Goeze 1782. Int. J. Parasitol., 3:317–319, 1973.
3. Areán, V.M., and Crandall, C.A.: Ascariasis. *In* Pathology of Protozoal and Helminthic Diseases. Edited by R.A. Marcial-Rojas. Baltimore, Williams & Wilkins, 1971, pp. 769–807.
4. Beckett, E.B, and Boothroyd, B.: Mode of nutrition of the larvae of the filarial nematode *Brugia pahangi*. Parasitology, 60:21–26, 1970.
5. Bradley, R.E.: The etiology of pathologic changes in canine heartworm diseases. *In* Pathology of Parasitic Diseases. Edited by S.M. Gaafar. Lafayette, Ind., Purdue University Studies, 1971, pp. 161–171.
6. Buck, A.A. (ed.): Onchocerciasis: Symptomatology, Pathology, Diagnosis. Geneva, World Health Organization, 1974.
7. Chanteau, S., Guidi, C., and Durosoir, J.L.: Efficiency of papain-treated microfilariae of *Wu-chereria bancrofti* (var. *pacifica*) as antigen for serodiagnosis of bancroftian filariasis in French Polynesia. Trans. R. Soc. Trop. Med. Hyg., 80:795–799, 1986.
8. Chandrasekharan, A., et al.: Control of *Brugia malayi* filariasis with common salt medicated with DEC in some hill-tribe settlements of Kerala. Indian J. Med. Res., 79:600–605, 1984.
9. Crompton, D.W.T.: Chronic ascariasis and malnutrition. Parasitol. Today, 1:47–52, 1985.
10. Duke, B.O.L.: Behavioral aspects of the life cycle of *Loa*. *In* Behavioral Aspects of Parasite Transmission. Edited by E.U. Canning and C.A. Wright. London, Academic Press, 1972, pp. 97–107.
11. Duwel, D.: The prevalence of *Toxocara* eggs in the sand of children's playgrounds in Franfurt /M. Ann. Trop. Med. Parasitol., 78:633–636, 1984.
12. Fontaine, R.E.: Anasakiasis from an American perspective. J. Am. Med. Assoc., 253:1024–1025, 1985.
13. Gibson, D.W., and Connor, D.H.: Onchocercal lymphadenitis: clinicopathologic study of 34 patients. Trans. R. Soc. Trop. Med. Hyg., 72:137–154, 1978.
14. Greene, B.M.: Onchocerciasis. *In* Tropical and Geographical Medicine. Edited by K.S. Warren and A.A.F. Mahmoud. New York, McGraw-Hill, 1984, pp. 413–421.
15. Hamilton, R.G.: Application of immunoassay methods in the serodiagnosis of human filariasis. Rev. Infect. Dis., 7:837–843, 1985.
16. Hawking, F.: Circadian and other rhythms of parasites. Adv. Parasitol., 13:123–182, 1975.
17. Hawking, F.: The 24 hour periodicity of microfilariae: Biological mechanisms responsible for its production. Proc. Royal Soc. (B), 169:59–76, 1976.
18. Hibler, C.: *Elaeophora*. *In* Parasitic Diseases of Wild Mammals. Edited by R.C Anderson and L. Karstad. Ames, Iowa State University Press, 1971, pp. 263–278.
19. Hill, I.R., Denham, D.A., and Scholtz, C.L.: *Toxocara canis* larvae in the brain of a British child. Trans. Roy. Soc. Trop. Med. Hyg., 79:351–354, 1985.
20. Hulínská, D.: The development of the female *Enterobius vermicularis* and the morphogenesis of its sexual organs. Folia Parasitol., 15:15–27, 1968.
21. Is New World onchocerciasis spreading? Editorial. Parasitol. Today, 2:131, 1986.
22. Kozacos, K.R.: Raccoon ascarids as a cause of larva migrans. Parasitol. Today, 2:253–255, 1986.
23. Lymphatic Filariasis. Lancet, May 18, 1135–6, 1985.
24. Marcial-Rojas, R.A.: Malayan filariasis. *In* Pathology of Protozoal and Helminthic Diseases. Edited by R.A. Marcial-Rojas. Baltimore, Williams & Wilkins, 1971, pp. 941–947.
25. Margolis, L.: Public health aspects of "codworm" infection: a review. J. Fish. Res. Board Can., 34:887–898, 1977.
26. Muller, R.: *Dracunculus* and dracunculiasis. Adv. Parasitol., 9:73–151, 1971.
27. Muller, R.: Guineaworm eradication—the end of another old disease? Parasitol. Today, 2:39, 1986.
28. Otteson, E.A.: Filariases and tropical esoino-

philia. *In* Tropical and Geographical Medicine. Edited by K.S. Warren and A.A.F. Mahmoud. New York, McGraw-Hill, 1984, pp. 390–412.

29. Pawlowski, Z.S., and Arfaa, F.: Ascariasis. *In* Tropical and Geographic Medicine. Edited by K.S. Warren and A.A.F. Mahmoud. New York, McGraw-Hill, 1984, pp. 347–558.

30. Pawlowski, Z.S.: Enterobiasis. *In* Tropical and Geographic Medicine. Edited by K.S. Warren and A.A.F. Mahmoud. New York, McGraw-Hill, 1984, pp. 386.

31. Rogers, W.P., and Summerville, R.I.: The infective stage of nematode parasites and its significance in parasitism. Adv. Parasitol., *1*:109–177, 1963.

32. Schad, G.A., and Rozeboom, L.E.: Integrated Control of Helminths in Human Populations. Ann. Rev. Ecol. and Syst., *7*:393–420, 1976.

33. Stephenson, L.S.: The contribution of *Ascaris lum-* *bricoides* to malnutrition in children. Parasitology, *81*:221–233, 1980.

34. Sugimachi, K., et al.: Acute gastric anasakiasis. J. Amer. Med. Assoc., *253*:1012–1013, 1985.

35. Taylor, M.R.: Global priorities in the control of onchocerciasis. Rev. Inf. Dis., *7*:844–846, 1985.

36. Warren, K.S., and Mahmoud, A.A.F.: Tropical and Geographic Medicine. New York, McGraw-Hill, 1984.

37. Watson B.D.: The fine structure of the body wall and the growth of the cuticle in the adult nematode *Ascaris lumbricoides*. Q. J. Microsc. Sci., *106*:83–91, 1965.

38. World Health Organization: Ten years of onchocerciasis control in West Africa. Geneva, WHO (OCP/GVA/85.1B), 1985, p. 13.

39. Worms, M.J.: Circadian and seasonal rhythms in blood parasites. *In* Behavioral Aspects of Parasite Transmission. Edited by E.U. Canning and C.A. Wright. London, Academic Press, 1972, pp. 53–68.

17

Class Adenophorea (= Aphasmidia), Orders Enoplida, Trichinellida, Dioctophymatida

Order Enoplida

SUPERFAMILY MERMITHOIDEA

These slim, smooth nematodes range in size from a few millimeters to 50 cm in length. Although they are free-living in the soil or in water as adults, they are parasitic during their larval stages, especially in insects, but also in crustacea, spiders, and snails. No buccal capsule is present, and the long esophagus (sometimes half the body length) proceeds directly from the mouth opening. The intestine does not function in the usual manner, but consists of two or more rows of enlarged cells filled with food reserves.

Mermithid parasites are being studied for possible use in the biocontrol of insects (see Chap. 23). Larval worms in moist soil or water readily penetrate the cuticle of immature insects and enter the host's body cavity. Some insects become infected by ingesting mermithid eggs. The burden of worms that insects may carry is enormous, as can be seen in Figure 17–1, and considerable damage to the reproductive organs and/or viscera may result in castration or death.

Order Trichinellida

SUPERFAMILY TRICHINELLOIDEA

The anterior body of these worms is filamentous, sometimes markedly so, whereas the posterior part is wider and often shorter. The mouth, a simple pore, is not surrounded by lips; the buccal capsule is small or rudimentary; the esophagus consists of a short muscular portion

lacking a bulb, and a long slender tube lying in a row of large cells called **stichocytes.** The part consisting of the stichocytes is known as the **stichosome.** The anus is terminal in both males and females. Males usually possess a single ensheathed spicule, but in some species it may be lacking. The vulva is normally located near the junction of the narrow and wide portions of the body. Most females are oviparous, and the life cycle is usually direct. A few species use an intermediate host.

Capillaria hepatica parasitizes the liver of rats and other rodents. Dogs, pigs, ground squirrels, rabbits, beavers, monkeys, and other animals are occasionally infected. The parasite rarely occurs in humans, but heavy infections can be fatal.

Brown, pitted, thick-shelled eggs (Fig. 17–2) (about 50 μm in diameter) are deposited in the liver, but do not develop further unless freed from the liver and exposed to air. A new host becomes infected by ingesting eggs that have been freed by the death and decomposition of a previous host. This process is accelerated if an infected animal is eaten by a predator that releases the eggs in its feces. Insects and other soil invertebrates probably play a role as egg disseminators. In the new host, the eggs hatch in the cecum, and the young larvae burrow through the intestinal wall and reach the liver, where they mature. Adult males are about 4 mm long and 100 μm wide, whereas the females are twice as wide and about 10 mm long. A potential health hazard, especially for children, exists

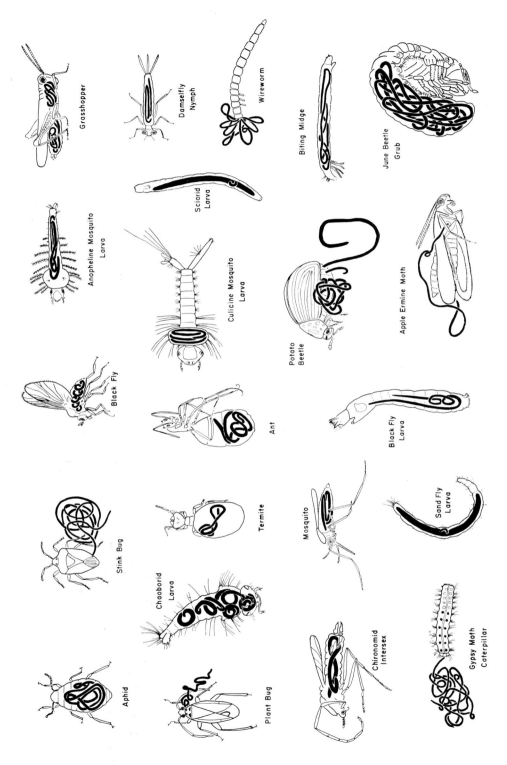

Fig. 17–1. Some insects parasitized by mermithids, showing location and relative sizes of the nematode parasites. (From Nickle. Exp. Parasitol., 32(2), 1973.)

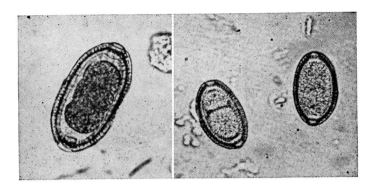

Fig. 17–2. *Capillaria hepatica.* Eggs of the parasite obtained from lesions produced by this parasite in the liver of a rat. (From Gradwohl and Kouri: Clinical Laboratory Methods and Diagnosis. Courtesy of C.V. Mosby.)

in crowded urban areas where rats are generally abundant.

Capillaria philippinesis is a recently discovered intestinal parasite of humans, the normal definitive hosts of which remain unknown (Fig. 17–3). However, birds can be infected experimentally, and therefore, are suspected to be the major definitive hosts in nature.[8] Infective larvae occur in small fish that are eaten raw in rural areas of the Philippines. The parasite was thought to occur only in these islands, but infections reported more recently in Thailand, Japan[9] and Iran suggest that it is distributed throughout Asia and, perhaps, beyond.

The parasite, when first discovered, was associated with fatal diarrheal disease. Its remarkable pathogenic potential is attributable to the overwhelming infections that may be amassed over time. The parasite is virtually unique among parasitic helminths in that female worms produce two kinds of eggs: (1) a typical bipolar-plugged capillarid egg that does not embryonate internally and is passed to the outside in feces, and (2) a thin-shelled egg that embryonates and hatches internally, causing massive autoinfection in the untreated host. Treatment with modern benzimidazoles has changed the previously grave prognosis; appropriately managed patients now survive, although relapses of the infection are frequent. An excellent review of the biology, ecology, and epidemiology of human intestinal capillariasis is given by Cross and Bhaibulaya.[9]

Trichinella spiralis is the causative agent of trichinellosis (or trichinosis), a disease usually associated with eating insufficiently cooked pork. In the United States, the incidence of human trichinosis is low and decreasing annually. The parasite's significance is largely economic because it continues to occur in swine, thereby contributing to the cost of meat processing and inspection, and limiting opportunities for the export of pork to countries where the parasite is rare or absent in swine. In some areas of Europe, where the disease once occurred regularly in large outbreaks, its occurrence is now extremely rare.[5]

Adult worms (Fig. 17–4) are small, slender nematodes measuring about 1.5 and 3 to 4 mm long in males and females, respectively. The tail of the male (Fig. 17–5) is complex; a pair of large, lateral finger-like processes flanks two pairs of papillae associated with the cloacal opening. Spicules are lacking.[5]

The life history (Fig. 17–6) of *Trichinella spiralis* is unique because the same individual serves as both a definitive host and an intermediate host. Susceptible hosts become infected by ingesting encysted, infective *Trichinella* larvae in meat (Fig. 17–7). The larvae are liberated from their cysts, which occur in mammalian striated muscle, when the tissue is digested in the gastrointestinal tract (Fig. 17–6). The excysted larvae develop rapidly in the small intestine, undergoing the usual 4 molts. Then, as immature adults, they mate and the females burrow into a column of intestinal epithelial cells. In this sense, they are intracellular parasites.[50] Mature females give birth to embryonic worms, the "newborn larvae," that burrow into one of the abundant lymphatics (minute lymph vessels) that extend into the intestinal villi. The embryos use the lymph to reach the circulatory system and then disperse throughout the body. To survive, the embryos must find their way into striated muscle where they burrow into a muscle cell and create a "nurse cell" which supports their further development and survival.[14] Avian muscle is less suitable than mammalian muscle for larval development. Birds are rarely found

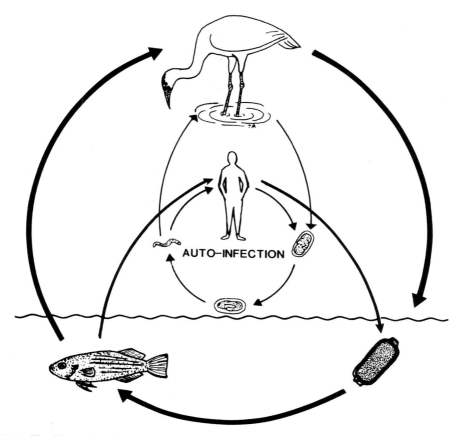

Fig. 17–3. The life cycles of *Capillaria philippinensis*. Female worms residing in the small intestine of the definitive host lay two kinds of eggs, namely (1) thick-shelled eggs that pass to the exterior in the feces and after embryonation in water are infective to the fishes that serve as intermediate hosts, and (2) thin-shelled eggs that embryonate and hatch within the definitive host causing autoinfection. The ability to multiply within the definitive host, a rare attribute among parasitic helminths, enables the parasite to establish massive, life-threatening infections in its final hosts.

infected in nature and it is difficult to infect them experimentally.

Trichinella spiralis is a parasite with a broad geographic distribution that occurs in numerous carnivorous and scavenging mammals. In the Arctic, it is a frequent parasite of polar bears and arctic foxes. Its occasional occurrence in walruses and marine mammals, whose diets are thought to consist of molluscs, crustaceans and/or fish, is enigmatic. In temperate areas, wild carnivores such as raccoons, foxes, bears, and various mustelids (skunks, mink, etc.) are commonly infected. Additionally, in synanthropic situations, the parasite is transmitted in farmyards between swine and other mammals such as rats, cats, dogs, and raccoons (Fig. 17–8). In some parts of the tropics, large carnivores

(lions, leopards, and hyenas) are important hosts.[6]

Whether the parasite occurring in all of these species over vast geographic areas is a single species with a number of physiologic races differentially adapted to particular hosts or groups of hosts, or whether a complex of sibling species exists, is a matter of active contemporary research. Detailed morphologic examination of strains has failed to reveal taxonomically significant morphologic differences,[26] even in structurally complex parts of the body such as the caudal end of the male (Fig. 17–4). However, new techniques, including isoenzyme and DNA analysis of parasites from different hosts, are revealing strain-specific differences and are shedding light on the systematics and host spe-

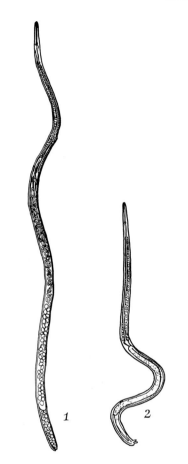

Fig. 17–4. *Trichinella spiralis* adults, *1*, Adult female, ×67; *2*, Adult male, ×67. (From Beaver, courtesy of Am. J. Clin. Pathol.)

cificities of these parasites.[10,13,17] However, parasitologists have not yet come to a consensus concerning the nomenclature of these strains, and, therefore, we will consider *Trichinella spiralis* to be the appropriate name for the parasites throughout their geographic and host range (Fig. 17–8).

Under synanthropic conditions (Fig. 17–7), pigs are *the* important definitive hosts.[6] They become infected by ingesting raw muscle tissue of their own species, either by cannibalizing dead pigs, or eating uncooked garbage.[37] They may also become infected by scavenging or preying on rats that have fed on garbage or swine carcasses that were not removed from the farmyard promptly.[37] Humans usually become infected when they eat insufficiently cooked pork; occasional infections are contracted from the undercooked meat of wild animals (e.g., boar or bear).

Exactly how the parasite is transmitted among wild carnivores in temperate areas is uncertain.[6,35,36] For instance, we do not know whether the red fox becomes infected as a result of predation, eating garbage, scavenging, or cannibalism. Foxes prey mainly on rodents, but wild rodents are rarely found infected. Perhaps *infected* rodents are so prone to predation that parasitologists rarely find them. Recent experimental studies indicate that trichinellosis changes the behavior of mice so that their vulnerability to predation is increased greatly.[33]

The intestinal phase of human trichinosis may be asymptomatic or, in heavier infections, may be associated with abdominal pain, diarrhea, vomiting, or constipation. The more apparent and clinically more important phase of the infection occurs during larval migration and invasion of the muscles. The clinical signs and symptoms most consistently reported include fever, muscle pain, periorbital edema, and headache. Most patients survive; when death occurs, it is usually attributable to heavy invasion of heart muscle or the central nervous system.[24,32]

Control of synanthropic trichinosis involves the cooking of garbage that is to be fed to swine and farm management practices that prevent cannibalism and reduce rat populations. Personal preventative measures involve the thorough cooking of wild game, pork and pork products.[19,30]

Trichuris trichiura is one of the most common helminth parasites of man, especially in tropical and subtropical regions. In squatter settlements associated with tropical cities more than 80% of the children may be infected. Similar species occur in many other animals. *Trichuris trichiura* eggs found in the gut contents of the frozen body of an Inca girl in Chile, buried at an elevation of 17,658 feet at least 450 years ago, indicate that this parasite was present in South America before the Spanish conquest and raise an interesting question concerning the possible New World origin of this species. It is worldwide in its distribution.

Trichuris trichiura (Fig. 17–9) is commonly known as the "whipworm" because of the characteristic shape of its body. The anterior three-fifths of the worm consists of a long, slender flagellum-like portion that contains the esophagus; the posterior two-fifths of the body is

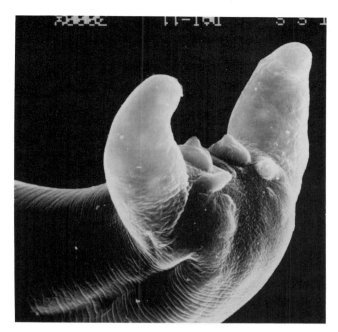

Fig. 17–5. *Trichinella spiralis:* Scanning electron micrograph of the tail end of a male. Showing the two, large, laterally-placed caudal appendages and two pairs of papillae surrounding the cloacal opening. (From Lichtenfels, courtesy of J. Parasitol.)

much thicker and contains the intestine and sex organs. The esophagus has a reduced muscular part followed by a row of large secretory cells, the stichocytes.

Male worms have a ventrally coiled caudal extremity and measure 30 to 45 × 0.6 mm. A single spicule protrudes through the retractable sheath that possesses a bulbous end and has numerous recurved spines (Fig. 17–10). Females measure 35 to 50 × 0.7 mm. The vulva lies at the anterior extremity of the fleshy portion of the body. Eggs are barrel-shaped, brown, with an outer and an inner shell and transparent polar prominences. Eggs measure 50 to 54 × 23 μm (Fig. 14–9).

The life cycle is direct. Eggs are deposited in the large intestine, where the adults live. Eggs pass out with feces and embryonate in the soil within a few weeks. They may remain viable for many months if they lie in moist areas, but development will be delayed by dryness or cold. When the weather is favorable, the bulk of the accumulated eggs develops and thus makes massive infection possible. Embryonated eggs infect a new host by ingestion. Obviously, the more chances there are for contamination of food and water, the greater is the prevalence of infection. When ingested, eggs are passed to the host's small intestine, where they hatch and where young larvae burrow into the intestinal villi. In a few days, they leave the intestinal wall

and go to the cecum and large intestine to mature. In a few months, mature worms are ready to mate and to produce eggs, thus completing the cycle. Four molts are required from egg to adult.

The severity of trichuriasis depends on the size of the worm burden and on the age, susceptibility and nutritional status of the host. There is recent evidence that predisposition to infection, probably genetically determined, is an important factor in deciding the outcome of the interaction between the parasite and host.[1,3]

Most infections are light and tend to be asymptomatic. However, occasional individuals, with heavy infections constituted of hundreds of worms, have abdominal pain and mucoid (sometimes bloody) diarrhea. Rarely, in young children, massive, occasionally fatal infections occur. These children have abdominal pain and diarrhea, as well as hypoproteinemia, anemia and are retarded in growth. Tenesmus (a painful, non-productive attempt to defecate), associated with massive infection extending into the rectum, sometimes results in rectal prolapse.

Trichuris infection is diagnosed by finding the characteristic barrel-shaped, bipolar-plugged egg in feces. Infections are treated with benzimidazoles (Mebendazole or Albendazole). Prevention and control of trichuriasis depends upon the sanitary disposal of human feces.

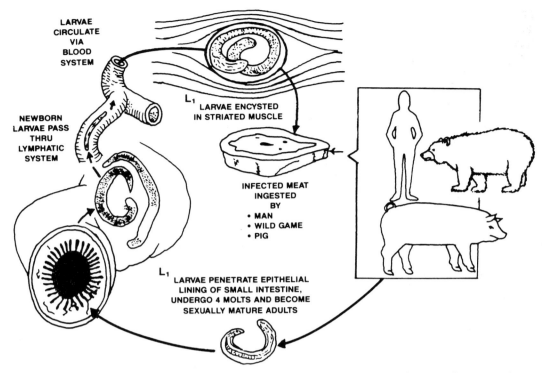

Fig. 17–6. Life cycle of *Trichinella spiralis*: Adult worms live in the epithelium of the small intestine (lower left); They give birth to embryos, called newborn larvae, that migrate via the lymphatics and circulation (upper left); Newborn larvae that find and penetrate striated muscle enter a muscle cell in which they encyst (top center); When encysted larvae in raw meat are ingested by a carnivore or omnivore (right center), the larvae are freed by peptic digestion (center bottom) and penetrate the intestinal epithelium, thus completing the cycle. (Modified from Gamble and Murrell, courtesy of Agra-practice.)

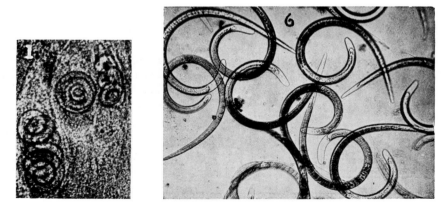

Fig. 17–7. *Trichinella spiralis.* Experimental infestation of white rat. (Original photomicrographs [of Kouri] of material supplied by Dr. G. Bachman, Puerto Rico.) *Left*, Larvae in diaphragm muscle. *Right*, Larvae freed by artificial digestion of muscle. (From Gradwohl and Kouri: Clinical Laboratory Methods and Diagnosis. Courtesy of C.V. Mosby.)

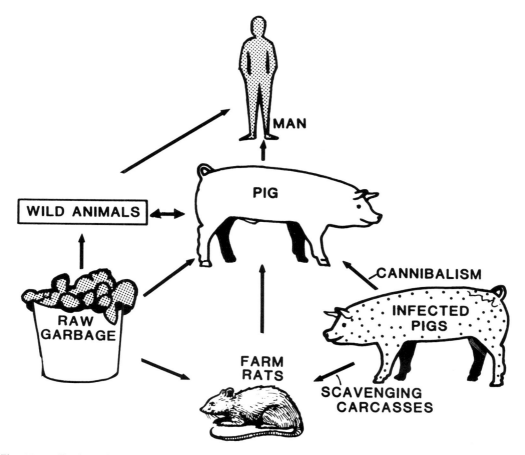

Fig. 17–8. Epidemiology of trichinellosis. In the synanthropic system swine are the most important hosts. They play a central role as source of infection for man, pigs, and sometimes, even wildlife. Wild animals may become infected by feeding on pork in garbage or scavenging discarded carcasses of pigs. In terrestrial Arctic ecosystems sylvatic trichinosis is maintained by predation. It is uncertain how sylvatic trichinosis is maintained in temperate zone ecosystems. (Modified from Murrell, courtesy of Food Technology.)

Many other species of *Trichuris* have been described. They occur in rodents, cats, dogs, foxes, swine, cattle, sheep, and numerous other wild and domesticated ruminants.[20,39] In some of these hosts, particularly swine (Fig. 17–11) and dogs, the worms produce a chronic inflammation of the cecum and colon. Heavily infected hosts show anorexia, diarrhea, and weight loss.

Order Dioctophymatida

SUPERFAMILY DIOCTOPHYMATOIDEA

Each of these moderate-to-long worms possesses a mouth lacking lips but surrounded by papillae. The esophagus has no bulb. Males possess a bell-shaped bursa without rays. Females lay eggs with thick, pitted shells and po-

lar-plugs. Birds and mammals are the primary hosts. Some of the worms may use fish as intermediate hosts. *Eustrongylides, Hystrichis,* and *Dioctophyma* are genera of particular interest to parasitologists. This last genus serves to illustrate the order.

Dioctophyma renale males range in length from 140 to 450 mm, and females are from 200 to 1000 mm. The common name of the parasite is the giant kidney worm. These reddish worms may be found in the kidneys of mink, dogs, raccoons, and many other fish-eating mammals in the United States and the Orient. Apparently, the parasites occupy the right kidney much more frequently than the left. Destruction of the kidney may be complete. The parasite's eggs leave the host in the urine.

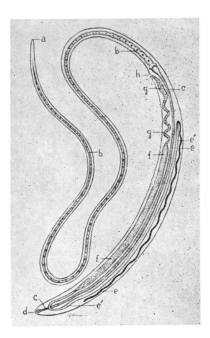

Fig. 17–9. Female *Trichuris trichiura*. (After Guiart. Modified by Kouri.) *a, b, c, d,* Digestive tract: *a,* mouth in the anterior end of the parasite; *b,* esophagus in the thin portion, formed by a fine duct which passes through a single layer of cells; *c,* intestine; *d,* anus. *e, e', f, g, h,* Female genitalia, simple: *e,* ovary; *e',* oviduct; *f,* uterus filled with eggs; *g,* long and sinuous vagina; *h,* vulva. (From Gradwohl and Kouri: Clinical Laboratory Methods and Diagnosis, courtesy of C.V. Mosby.)

Mammals are infected by eating aquatic oligochaete annelids that have ingested the eggs of the parasite. Eggs in the annelid hatch and develop into fourth-stage larvae. If a fish swallows infected annelids, it becomes a paratenic host.[18,23]

IMMUNITY TO NEMATODES

Nematode infections are often exceedingly chronic, individuals remaining parasitized for many years. In those parts of the rural tropics where sanitation is inadequate, most people are reinfected periodically and may harbor nematodes from early childhood through old age. Chronic infections may also occur without periodic reinfection. For instance, because of its autoinfective ability, *Strongyloides stercoralis* can persist in humans for more than 30 years after an individual was last exposed to infection.[21]

Historically, the observation that nematode infections tend to be so persistent led to the notion that they and, in fact, helminth infections generally, are largely unaffected by the host's immune responses. The belief that the cuticle was an inert, impervious "skin" that could not be damaged by an immune response also contributed to this concept. Additionally, the intestinal nematodes were thought to reside in an immunologically privileged site where immunoglobulins could not reach them.

On the other hand, hosts confronted with continuing infection rarely accumulate nematodes additively. Instead their parasite populations become asymptotic, and may even decrease. Observations such as these led to the suspicion that acquired resistance occurs. Furthermore, in the 1920s and 30s, it was shown that resistance could be transferred with serum from previously infected to uninfected hosts and that, when nematodes are incubated in immune sera, precipitates form at their mouths, excretory pores, or other openings. These observations led Chandler[7] to suggest that immune responses to nematodes were not directed against their surfaces but against their enzymes and, consequently, against their vital internal processes. This suggestion made the notion of immunologic regulation of the abundance of such large organisms, protected by supposedly impervious resistant cuticles, much more acceptable to parasitologists of the time.

Experimental investigations of immunity to nematodes have focused on parasites that can be maintained in rodents easily, particularly the intestinal nematodes *Nippostrongylus brasiliensis, Nematospiroides dubius, Strongyloides ratti,* and *Trichinella spiralis.* Other important investigations have involved *Trichostrongylus retortaeformis* of rabbits and *Haemonchus contortus,* the abomasal worm of sheep, an important parasite in its own right, but also considered a hookworm-model because of its blood-sucking propensities. *Trichostrongylus colubriformis,* a sheep nematode that will parasitize guinea pigs, has also contributed significantly to the understanding of immunity to intestinal nematodes.

As indicated in Chapter 15, a landmark in early immunoparasitologic studies was the observation that when sheep grazing a contaminated pasture are challenged with a single large dose of infective *H. contortus* larvae, they expel both the established adult worms and the incoming larvae.[42] This phenomenon was called "self cure." Later, Stewart[41] demonstrated that

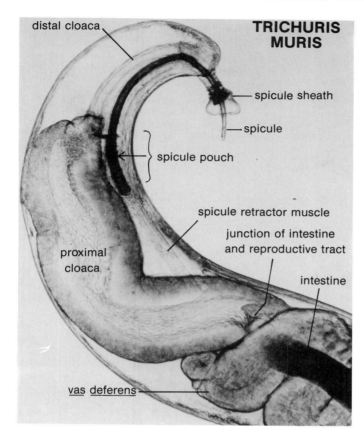

distal cloaca

TRICHURIS MURIS

spicule sheath

spicule

spicule pouch

spicule retractor muscle

junction of intestine and reproductive tract

proximal cloaca

intestine

vas deferens

Fig. 17–10. *Trichuris muris*: Tail of the male with its single spicule extruded. This species occurs in rodents.

"self cure" was triggered by the sudden exposure of a sensitized sheep to a large quantity of parasite antigen found in the exsheathing fluid of incoming larvae. The violent inflammatory reaction provoked by these antigenic secretions caused expulsion of the parasites. These observations strongly influenced subsequent work on immunity to parasitic nematodes (particularly intestinal nematodes), much of which attempts to elucidate the mechanisms that cause expulsion.

The often contradictory and confusing literature about mechanisms of expulsion has been ably reviewed by Wakelin[46] and Lee et al.[25] Wakelin summarizes the observations supporting an immunologic basis for expulsion of intestinal nematodes and its T-cell dependency as follows:

(a) absence of expulsion in T-lymphocyte deficient or deprived hosts,

(b) prevention of expulsion with immunosuppressive agents and irradiation,

(c) acceleration of worm-expulsion in secondary infections as compared to primary infections,

(d) transferability of the capacity for accelerated expulsion with serum or lymphocytes from immunized hosts.

Lee and colleagues[25] emphasize the central role of mucosal mast cells in the processes that cause worm-expulsion. These cells, when coated with parasite-specific IgE, the immunoglobulin most characteristic of helminthic infections, interact with parasite antigen to provoke expulsion by mechanisms that remain uncertain.

While these observations show that immune responses are involved in expelling intestinal infections, it has also become apparent that they alone are insufficient to cause expulsion. In many nematode infections, inflammation is also involved and there is abundant evidence that increased vascular and epithelial permeability associated with inflammation allows immunoglobulins to escape into the lumen where they have access to lumen-dwelling nematodes. Some authorities have indicated that it is unnecessary to involve immunoglobulin-mediated

Fig. 17–11. *Trichuris suis:* a massive infection showing the parasites in situ. The thick posterior ends of the worms are shown extending into the intestinal lumen. The long thin anterior ends (arrows) are embedded in the epithelium. (Courtesy of Dr. E Batte.)

parasite damage as the mechanism causing worm expulsion. They point to the possibility that the amines released by sensitized epithelial mast cells may cause expulsion by initiating inflammatory changes that render the intestinal habitat inhospitable for nematodes, or, alternatively, by damaging the worms directly. The fact that administration of amine antagonists delays the expulsion of worms shows that these substances are important but provides no additional evidence for their mode of action. Increased levels of mast cell protease are demonstrable in the parasitized intestine at the time of worm expulsion, suggesting that enzymes produced by these cells are important. Mucus trapping of worms, particularly the incoming larvae of a secondary infection, also is a significant component of a protective host response in some nematode infections. The effectiveness of this response may be enhanced by the presence of immunoglobulins and cytotoxins in the mucus of resistant hosts.[25,29]

It is remarkable that, given the host's ability to protect itself immunologically, nematodes do so frequently exist in stable, long-term associations with their hosts. In some highly coevolved host-parasite associations, the parasites elicit a minimal host response. This is well illustrated by *Nematospiroides dubius*, which, in a primary infection of its normal host, the mouse, is not expelled. Indeed, Damian,[11,12] Dineen,[15,16] and Sprent[40] suggested independently that parasites would evolve toward antigenic similarity with their definitive hosts, and that they, in turn, would come to tolerate their parasites at levels of infection less than those that are seriously damaging. Concurrently, in Dineen's view the parasites would evolve so that the pathogenicity of adult worms in their normal hosts would be much reduced.[15,16] Under these conditions, the existence of some species in exceedingly chronic infections is easily understood.

Helminths have adopted a number of strategies enabling them to elicit a minimal host response. These include evolution toward antigenic similarity with their definitive hosts (i.e., molecular mimicry); acquisition of the ability to coat themselves with host-derived substances, thus avoiding recognition as foreign; development of the capacity to shed an external surface layer with its adherent, potentially damaging host-derived cells or cell products, and, finally, the capacity to suppress (or depress) the host's immunologic responses.[12,44,46]

It may be less apparent how some species can establish exceedingly chronic infections in some individuals and fulminant ones in others. Schad and Anderson[34] and Anderson[1] have shown that individuals vary in their predisposition to heavy nematode infection. This may have a genetic, behavioral or ecologic basis, but whatever its ultimate basis, it is likely to be immunologically mediated. In human strongyloidiasis, a spectral disease, some individuals are healthy carriers, while others suffer varying degrees of illness. A few individuals fail to control the infection, and, in the absence of treatment, die.[22,38] Although the immunologic mechanisms regulating this parasitism are poorly understood, there can be little doubt that regulation is under immunologic control. Fulminant strongyloidiasis (hyperinfective or disseminated strongyloidiasis) is almost invariably associated with reduced immunocompetence. Additionally, when patients are grouped according to the severity of their disease, there is an inverse association with total IgA, IgG and IgM levels,[4] although it has not been possible to show an association with specific immunoglobulin levels.[2]

It is probable that the host's age at first infection and the rate of exposure (its frequency and the number of organisms per exposure) set the levels at which many parasites are tolerated. Thus, ecologic and behavioral factors are important in determining a host's predisposition to infection, but undoubtedly, most fundamental is the host's genetically based capacity to respond immunologically. Although the immunogenetics of host-nematode associations has only recently come under thorough, sophisticated study, the occurrence of non-responders (individuals that fail to expel their parasites) has been recognized for many years. For example, in a classic paper published in 1952,[28] Michel reported that most rabbits, when injected repeatedly with *Trichostrongylus retortaeformis*, eventually expel their worms and are then resistant to further infection. Occasionally, however, a rabbit (a non-responder) continues to accept new infections until it succumbs to an overwhelming trichostrongyloidiasis. More recently, Wakelin, Wassom and their colleagues,[43,45,48] using *Trichinella* infections in various inbred mouse strains, reported a series of elegant experimental studies that demonstrate the importance of genetic variation in determining the capacity of a host to respond to nematode parasitism.

Intestinal nematode infections are certainly *not* the only ones that are typically chronic. Among tissue-dwelling nematodes, some of the filariae establish long-lasting infections, with individual worms surviving for remarkable periods of time. Adult female *Onchocerca volvulus*, for example, may live for a decade or more. This can be shown in emigrants who continue to have a microfiladermia for 15 years after their last possible exposure to infection.

As tissue dwellers often occurring in chronic infections, the filariae are of particular immunoparasitologic interest because their antigens are available to the host directly and the host's immunoglobulins and effector cells have direct access to the parasite. This enigma, along with the importance of these parasites as agents of widespread, severe morbidity (filarial fevers, lymphadenitis, elephantiasis, and blinding onchocerciasis) has made the filariae popular subjects for immunoparasitologic and immunopathologic study. Some species may be immunosuppressive under certain circumstances, but this does not explain the enigma because these same parasites *sometimes* elicit strong immunologic reactions. Immunosuppression in filarial infections is generally associated with high parasite loads and relatively mild disease. Conversely, vigorous immune responses capable of clearing adults and/or microfilariae are most frequently accompanied by severe disease. Thus, in bancroftian filariasis, asymptomatic but microfilaremic individuals generally are deficient in both antibody and cell-mediated immunity, while strong cell-mediated immune responses to adult filariae are found in most patients with elephantiasis. Strong IgE antibody responses and peripheral eosinophilia are found in patients with allergic manifestations of filariasis, e.g., tropical pulmonary eosinophilia.[31] In onchocerciasis, microfiladermia is frequently associated with weak cell-mediated immune responses, while "Sowda" (a generalized dermal inflammatory reaction which clears microfilariae from the affected skin) is generally accompanied by a strong delayed-type hypersensitivity response to microfilarial antigens.[27]

Filariae frequently used as laboratory models include *Dipetalonema vitei* of hamsters, *Dirofilaria immitis* of dogs and *Brugia* spp. of cats and dogs. Stages of the parasite occurring in the definitive host, and therefore of immunoparasitologic interest are the infective (L_3) larvae, developing larvae (L_3, L_4), the adult worms, and the microfilariae. The latter are particularly useful because they come to the surface of the body either in the skin or peripheral circulation from which they can be taken up by vectors or easily collected by immunoparasitologists.

Although the precise immune mechanisms that cause the death of filariae in naturally acquired infections, particularly those of man and livestock, remain unknown, studies in laboratory animals suggest that antibody-dependent, cell-mediated killing of microfilariae is of central importance (Fig. 17–12). In the experimental models cited above, the functional antibody varies from one host-parasite model to the other and includes IgE, IgG and IgM. The effector cells also vary, including eosinophilis, neutro-

phils, and macrophages in different combinations.[46] The experimentalist's focus on the killing of microfilariae may not be misplaced because, in many natural parasitisms, immune hosts become amicrofilaremic even though adult worms continue to survive. However, in these hosts, it is known that incoming L_3 or developing larvae (L_3 or L_4) must also be killed, in that superinfection usually fails to occur.

Killing of *Dipetalonema viteae* microfilariae by the hamster has been particularly well-studied. In this system, immunoglobulin-coated microfilariae are rendered vulnerable to cellular attack and death. By placing either IgM-coated or normal microfilariae in micropore chambers of differing pore sizes (chosen either to permit the entry of leukocytes or to exclude them), and by implanting the chambers in normal or immune hamsters, Weiss and his associates[49] showed that IgM-mediated, cellular attack destroyed microfilariae. Unfortunately, in the hamster the precise details of the interaction between microfilariae, immunoglobulin, and leukocytes is not known, but, in the rat, eosinophils adhere and degranulate onto the microfilarial surface

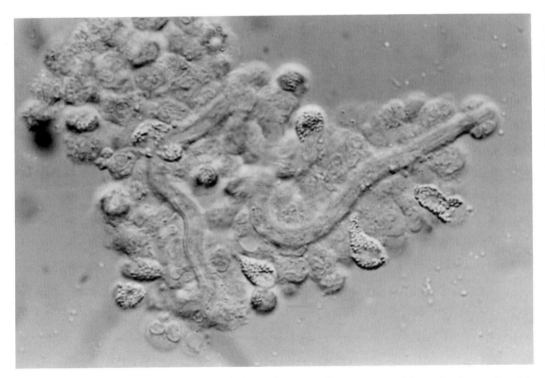

Fig. 17–12. A microfilaria of *Onchocerca volvulus* with many adherent neutrophils and a few eosinophils. (From Mackenzie, courtesy of Parasitology Today.)

after incubation of the parasites in IgE. The layer of material deposited on the microfilariae apparently is subsequently shed and this elicits attack by macrophages. These cells spread over the microfilarial cuticle and release lysosomal enzymes, with visible cuticular damage and lysis of underlying cells. Presumably, a somewhat similar sequence, but possibly involving other immunoglobulins and cells, occurs in more natural systems.[46]

This brief consideration of immunity to nematodes shows that great gaps in knowledge still exist in this interesting and active field of research. When one recalls that parasitic nematodes cannot be reared in the absence of their hosts, the uncertainty that prevails is not surprising. Most experimental investigation must be conducted in animals, where the interaction between the parasite and its various responses, and the host and its even more numerous responses, make it difficult to determine precisely which immunoglobulins, cells, and cell products (e.g., amines, enzymes, lymphokines) are actually those that function in immune mediated resistance to the parasite.

REFERENCES

1. Anderson, R.M.: The role of mathematical models in helminth population biology. *In* Parasitology—Quo Vadit? Edited By M.J. Howell. Canberra, Australian Academy of Science, 1986, pp. 519–530.
2. Badaro, R., Carvalho, E.M., Santos, A.A., and Genta, R.M.: Parasite-specific humoral responses to different forms of strongyloidiasis. Trans. Roy. Soc. Trop. Med. Hyg., *81*:149–150, 1987.
3. Bundy, D.A.P., et al.: Predisposition to *Trichuris trichiura* infection in humans. Epidem. Inf. (in press), 1987.
4. Carvalho, E.M., Andrade, T.M., Andrade, J.A., and Rocha, H.: Immunological features of different forms of strongyloidiasis. Trans. Roy. Soc. Trop. Med. Hyg., *77*:346–349, 1983.
5. Campbell, W.C.: Trichinella and Trichinosis. New York, Plenum Press, 1983.
6. Campbell, W.C.: Epidemiology I. Modes of transmission. *In* Trichinella and Trichinosis. Edited by W.C. Campbell. New York, Plenum Press, 1983, pp. 425–444.
7. Chandler, A.C.: Studies on nature of immunity to helminths. VI. General resumé and discussion. Am. J. Hyg., *26*:309–321, 1937.
8. Cross, J.H., and Basaca-Sevilla, V.: Experimental transmission of *Capillaria philippinensis* to birds. Trans. Roy. Soc. Trop. Med. Hyg., *77*:511–514, 1983.
9. Cross, J.H., and Bhailbulaya, M.: Intestinal capillariasis in the Philippines and Thailand. *In* Human Ecology and Infectious Diseases. Edited by N.A. Croll and J.H. Cross. New York, Academic Press, 1983, pp. 104–136.
10. Dame, J.B., Murrell, K.D., Worley, D.E., and Schad, G.A.: *Trichinella spiralis:* genetic evidence for synanthropic subspecies in sylvatic hosts. Exp. Parasitol., *64*:195–203, 1987.
11. Damian, R.T.: Molecular mimicry: antigen sharing by parasite and host and its consequences. Am. Naturalist, *98*:129–149, 1964.
12. Damian, R.T.: Molecular mimicry in biological adaptations. *In* Host-Parasite Interfaces. Edited by B.B Nichol. New York, Academic Press, 1979, pp. 103–126.
13. Dick, T.A.: Species and infraspecific variation. *In* Trichinella and Trichinosis. Edited by W.C. Campbell. New York, Plenum Press, 1983, pp. 31–60.
14. Despommier, D.: Biology. *In* Trichinella and Trichinosis. Edited by W.C. Campbell. New York, Plenum Press, 1983, pp. 75–152.
15. Dineen, J.K.: Immunological aspects of parasitism. Nature, *197*:268–269, 1963.
16. Dineen, J.K.: Antigenic relationship between host and parasite. Nature, *197*:471–472, 1963.
17. Flockhart, H.A.: *Trichinella* speciation. Parasitol. Today, *2*:1–3, 1986.
18. Fyvie, A.: *Dioctophyma renale. In* Parasitic Diseases of Wild Mammals. Edited by J. Davis and R.C. Davis. Ames, Iowa State University Press, 1971, pp. 258–262.
19. Gamble, H.R., and Murrell, K.D.: Swine trichinellosis: diagnosis and control. Agra-practice, *8*:12–15, 1987.
20. Georgi, J.: Parasitology for Veterinarians. 4th Ed. Philadelphia, W.B. Saunders Co., 1985.
21. Grove, D.I.: Strongyloidiasis in allied ex-prisoners of war in Southeast Asia. Br. Med. J., *280*:598–601, 1980.
22. Grove, D.I.: Replicating helminth parasites of man. Parasitol. Today, *2*:107–111, 1986.
23. Karmanova, E.M.: The life cycle of the nematode *Dictophyme renale* (Goeze, 1782), a parasite in the kidney of Carnivora and of Man. (A translation of Doklady, Biological Sciences Section.) Dokl. Akad. Nauk S.S.S.R., *132*:456–457, 1960.
24. Kazura, J.W.: Trichinosis. *In* Tropical and Geographical Medicine. Edited by K.S. Warren and A.A.F. Mahmoud. New York, McGraw-Hill, 1984, pp. 427–430.
25. Lee, T.D.G., Swieter, M., and Befus, A.D.: Mast cell responses to helminth infections. Parasitol. Today, *2*:186–191, 1986.
26. Lichtenfels, R., Murrell, K.D., and Pillitt, P.A.: Comparison of three subspecies of *Trichinella spiralis* by scanning electron microscopy. J. Parasitol., *69*:1131–1140, 1983.
27. Mackenzie, C.D., et al.: Variations in host responses and the pathogenesis of human onchocerciasis. Rev. Infect. Dis., *7*:802–808, 1985.
28. Michel, J.F.: Self-cure in infections of *Trichostrongylus retortaeformis* and its causation. Nature, *169*:881, 1952.
29. Moqbel, R.: Helminth-induced intestinal inflammation. Trans. Roy. Soc. Trop. Med. Hyg., *80*:719–727, 1986.

30. Murrell, K.D.: Strategies for control of human trichinosis transmitted by pork. Food Technol., *39*:65–68, 1985.

31. Otteson, E.A.: Filariasis and tropical eosinophilia. *In* Tropical and Geographical Medicine. Edited by K.S. Warren and A.A.F. Mahmoud. New York, McGraw-Hill, 1984, pp. 390–412.

32. Pawlowski, Z.S.: Clinical aspects in man. *In* Trichinella and Trichinosis. Edited by W.C. Campbell. New York, Plenum Press, 1983, pp. 367–402.

33. Rau, M.E.: The open-field behavior of mice infected with *Trichinella spiralis*. Parasitology, *86*:311–318, 1983.

34. Schad, G.A., and Anderson, R.M.: Predisposition to hookworm infection in humans. Science, *228*:1537–1540, 1985.

35. Schad, G.A., Leiby, D.A, and Murrell, K.D.: Distribution, prevalence and intensity of *Trichinella spiralis* infection in furbearing mammals of Pennsylvania. J. Parasitol., *70*:372–377, 1984.

36. Schad, G.A., et al.: *Trichinella spiralis* in the black bear (*Ursus americanus*) of Pennsylvania: distribution, prevalence and intensity of infection. J. Wildlife Dis., *22*:36–41, 1986.

37. Schad, G.A., et al.: *Trichinella spiralis* in an agricultural ecosystem: transmission under natural and experimentally modified on-farm conditions. J. Parasitol., *73*:95–102, 1987.

38. Scowden, E.B., Schaffner, W., and Stone, W.J.: Overwhelming strongyloidiasis. Medicine (Baltimore), *57*:527–544, 1978.

39. Soulsby, E.J.L.: Helminths, Arthropods and Protozoa of Domesticated Animals. 7th Ed. London, Bailliere Tindall, 1982.

40. Sprent, J.F.A.: Parasites, immunity and evolution. *In* The Evolution of Living Organisms. Edited by C.W. Leeper. Parkville, Victoria, Melbourne University Press, 1962, pp. 149–165.

41. Stewart, D.F.: Studies on resistance of sheep to infestation with *Haemonchus contortus*. V. The nature of the self cure phenomenon. Australian J. Agric. Res., *4*:100–117, 1953.

42. Stoll, N.R.: Studies with the strongylid nematode, *Haemonchus contortus:* I. Acquired resistance under natural infectious conditions. Am. J. Hyg., *10*:384–418, 1929.

43. Wakelin, D., Mitchell, L.A., Donachie, A.M., and Grencis, R.K.: Genetic control of immunity to *Trichinella spiralis* infections of mice. Response of rapid- and slow-responder strains to immunization with parasite antigens. Parasite Immunology, *8*:159–170, 1986.

44. Wakelin, D.: Immunity to intestinal parasites. Nature, *273*:617–620, 1978.

45. Wakelin, D.: Genetic control of immunity to helminth infections. Parasitol. Today, *1*:17–23, 1985.

46. Wakelin, D.: Immunity to Parasites. London, Edward Arnold, 1984.

47. Wakelin, D.: Genetic and other constraints on resistance to infection with gastrointestinal nematodes. Trans. Roy. Soc. Trop. Med. Hyg., *80*:742–747, 1986.

48. Wassom, D.L., et al.: Genetic control of immunity to *Trichinella spiralis* infections of mice. Hypothesis to explain the role of H-2 genes in primary and challenge infections. Immunology, *51*:625–631, 1984.

49. Weiss, N., and Tanner, M.: Studies on *Dipetalonema viteae*. III. Antibody-dependent cell-mediated destruction of microfilariae *in vivo*. Tropenmed. Parasitol., *30*:73–50, 1979.

50. Wright, K.A.: *Trichinella spiralis*: An intracellular parasite in the intestinal phase. J. Parasitol., *65*:441–445, 1979.

18

Introduction; Subphylum Crustacea

Members of the phylum Arthropoda possess an exoskeleton with jointed appendages. The body is divided into segments that are variously grouped into functional units such as head, thorax, and abdomen. The digestive tract is complete, and the circulatory system forms a hemocoel that is the body cavity. The coelom is reduced. Respiration is accomplished by tracheae, gills, book lungs, or body surface. Malpighian tubules serve as excretory organs in most species. The brain is dorsal, whereas the main nerve cord is ventral in position. Paired ganglia usually occur in each somite, and eyes are simple or compound. Sexes are separate.

Some crickets puncture stems of raspberry plants and lay their eggs within the pith. Are these crickets more parasitic than those grasshoppers and crickets that simply chew the stems? No more forceful way to emphasize the difficulty of defining "parasite" can be found than to select a textbook on entomology and to read about the multitudinous methods by which insects have solved the problem of obtaining food. Among the most destructive plant pests are the chinchbugs, plant lice, mealybugs, and other Hemiptera whose sucking mouthparts, rapid rate of reproduction, and general behavior combine to effect a devastating invasion of plants the world over. But are they not simply plant feeders instead of parasites? The ovipositor of the female of many leaf hoppers is adapted for lacerating plant tissues, and eggs are deposited in longitudinal rows on the stems or under the leaf sheath. Certainly, this habit is as parasitic as the habits of lice.

Immune responses of animals to arthropods have been discussed in detail by Gaafar, and the following statements occur in his summary.

"Reactions of the skin and lymph nodes during primary and subsequent infestations . . . [are] physiological means of protection of the host against further invasions with arthropod parasites of the same and possibly closely related species. It is apparent that the role of antibodies in the serum of parasitized or immune animals is not well understood. Immunoglobulins of the precipitating and nonprecipitating classes are apparently found in parasitized animals but no information is available on the types involved."[14]

Immunity against arthropods involves various physiologic and anatomic characteristics of the epidermis, mucous membranes, and skin. These characteristics are variable, may be adaptive, and should be considered whenever immunologic problems of these hosts arise. No arthropod antigens in chemically pure form have so far been identified, and the role of antibodies in the serum of parasitized or immune animals is not well understood. Precipitating and nonprecipitating immunoglobulins apparently occur in parasitized animals, but the types are unknown.[14]

Some general references are Beesley[3] for control of arthropods, Askew[2] for parasitic insects, Kabata[23-25] for crustacea as enemies of fishes, and Salt[37] for cellular defense reactions of insects.

The phylum Arthropoda is the largest phylum in the animal kingdom. Three subphyla are recognized: Crustacea (aquatic forms such as crabs and barnacles); Unirami (insects and myriapods) and Chelicerata (horseshoe crabs, spiders and mites). The crustacean assemblage in turn is subdivided into four large classes: Remipedia, Malacostraca, Phyllopoda and Maxillipoda.

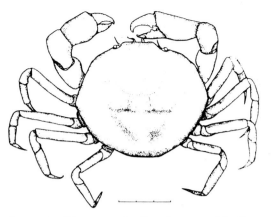

Fig. 18–1. The pea crab, *Pinnotheres maculatus* lives in the mantle cavity of a variety of bivalve mollusks. (From Williams, courtesy of the Fishery Bulletin.)

Class Malacostraca

This is the most primitive of the three "higher" crustacean classes. The malacostracans are characterized by a large number of body segments with biramous appendages on all segments, a postcephalic, shield-like carapace, which when present, covers only the thorax but not the limbs and abdomen, a dorsal, tubular heart and a distinct naupilar eye structure. The class contains all the large commercially important species such as lobsters, shrimps, crabs, etc. Parasites are recognized in only two orders of the subclass Eumalacostraca.

Subclass Eumalacostraca

Order Decapoda

A number of decapod crustaceans have been described which are known to be symbiotic though in most cases the relationships are not well documented. Shrimps of several families, especially the Palaemonidae and Alpheidae, commonly live in association with a wide diversity of marine invertebrates. Others such as the hippolytids are beneficial in that they clean the parasites off fishes. Pinnotherids, often called "pea crabs" are commensals or parasites commonly found in worm tubes, bivalve mollusks and on echinoderms (Fig. 18–1, see also Fig. 1–1). *Pinnotheres ostreum* lives in the mantle cavity of the oyster, *Crassostrea*. The crab removes food from the host's gills and may cause considerable damage to the gills.[8]

Order Isopoda

These crustaceans possess bodies that are usually flattened dorsoventrally, without a carapace, with sessile eyes, and with a short abdomen the segments of which are often partly fused. Each of the seven characteristically free segments of the thorax bears a pair of legs. Parasitic species are abundant, and they favor free-living crustaceans and fishes as hosts. Most parasitic species have been found in ocean water, although a few freshwater forms have been described. For parasites of decapod crustacea, see Overstreet,[33] Sprague and Couch.[44] See also Schultz.[40]

Suborder Flabellifera

In the flabelliferans a whole range of relationships exists, with gradations from actively swimming predatory species to parasites that, as adults, are permanently fixed to the host and are incapable of locomotion. The free-swimming cirolanids, exemplified by *Cirolana borealis*, possess powerful biting jaws. They are mainly scavengers rather that predators, but they sometimes attack a fish that is caught on a hook or is otherwise at a disadvantage, and gnaw their way into the body so that the fish is soon reduced to nothing but skin and bones. A number of isopods have been reported to bite or attack humans.

The Anuropidae contain several large species that live and feed on jellyfishes. The Aegidae and Cymothoidae are ectoparasites of fishes, sharks, and occasionally squids. Host specificity is rare, and both the aegids and cymothoids may have many favored hosts. Both families have piercing mouthparts and suck blood.

Anilocra, *Cymothoa*, and *Nerocila* represent genera of cymothoid isopods that, as adults, cling to the gills or skin of their fish hosts by means of strong prehensile hooks (Fig. 18–2).[7] *Cymothoa exigua* lives in the mouth of the snapper *Lutjanus guttatus* where it causes the degeneration of the tongue of its host. By attaching to the tongue stub, it functionally replaces the host's tongue.

Lironeca vulgaris lives in the gill chambers of the sanddab, *Citharichthys*, in the eastern Pacific. The isopod begins its life cycle as a free-swimming **manca** stage. Following a brief planktonic existence the juvenile **aegaltroid** stage enters and attaches to the gills of the host and develops into a functional male. The male

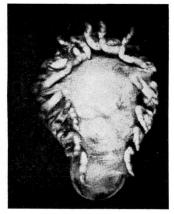

Fig. 18–2. Parasitic isopods. *Top,* Dorsal and ventral views of female *Riggia paranensis. Bottom,* A marine sardine (*Clupea*) with an adult female *Nerocila orbignyi* clinging to its head. (From Szidat, courtesy of Archiv für Hydrobiologie.)

feeds on the gill filaments and eventually transforms into a female. The nonovigerous female continues to feed on gill tissue and blood and may cause considerable damage. Copulation takes place in the gill chamber or mouth. The ovigerous female does not feed. Eggs and young are brooded by the female in a special brood chamber or **marsupium.** The cycle starts again when young are released into the plankton.[6]

Sexual dimorphism is the rule among parasitic isopods, and the phenomenon of protandrous hermaphroditism is common among the flabelliferans, especially the cymothoids. This phenomenon starts when a juvenile parasite first attaches itself to a host. The juvenile develops into a small, functional male. The male is later transformed into a female which as it grows and matures develops a brood pouch, and produces eggs. When two isopods are present on the same fish, the presence of a large functional female inhibits further development of the male toward the female phase. When the female dies, the male shifts position on the host, molts into a female and begins to grow in size. The nature of this influence is unknown, but it has been proved experimentally with *Anilocra physodes,* an external parasite that can be removed from the host. Since both male and female are temporary parasites, sexual dimorphism cannot be attributed to the parasitic mode of life. The protandrous male is not easily recognized, and a young, sexually immature female often is mistaken for a small, adult male.

Suborder Epicaridea

Contrary to the habits of the flabelliferans and the gnathiideans, which are temporary parasites at either adult or larval stages, the suborder Epicaridea consists of isopods parasitic in both larval and adult stages. This suborder is divided into four superfamilies, all of which are parasitic on crustacea (Fig. 18–3).

Epicarids are particularly interesting because their life cycle involves two hosts and because sexes may be determined epigamically, that is, after fertilization, depending on host environmental factors. The first epicarid larva resembles a small isopod and possesses piercing and sucking mouth parts and claw-like appendages with which it attaches itself to the surface of free-swimming copepods. This kind of larva is called an **epicaridium,** and while it remains on its copepod host, it undergoes six successive molts and changes progressively into two distinct larval stages known as the **microniscus** and the **cryptoniscus** stages (Fig. 18–4).[1] On reaching the latter stage of development, the parasite leaves its copepod host, proceeds to the sandy or muddy sea bottom, and there seeks a free-living crustacean, for example, a crab or a shrimp, into whose branchial chamber or brood pouch it enters. Within this second host, the parasitic cryptoniscus stage develops in one of two directions, according to whether it belongs to the superfamily Bopyrina or to the superfamily Cryptoniscina.

If the cryptoniscus larva is a bopyrine, it molts, thereby loses most of its pleopods, and is then known as a **bopyridium.** This larval stage initially always develops into a female, but successive larvae, attached either to the same host or to the female parasite, all become males. The females often grow tremendously and their bodies may become quite distorted. The males always remain small and isopod-like (Fig. 18–5).[30]

Stegophryxus hyptius, a bopyrid ectoparasite on the abdomen of the hermit crab, *Pagurus longicarpus,* seeks the definitive host as a sexually undifferentiated and sexually undetermined cryptoniscus larva. Larvae that settle directly on the host develop into females, and those that attach themselves to a female bopyrid develop into males.

Bopyrus squillarum lives in the gill cavity of the prawn, *Leander serratus,* and it causes a large swelling on one side of the host carapace. The

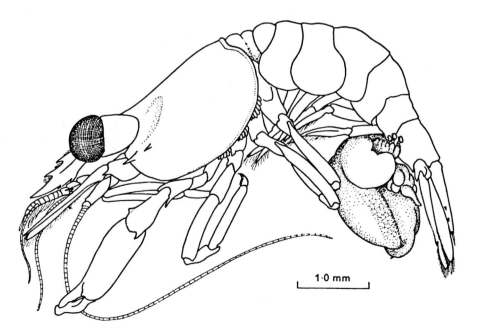

1·0 mm

Fig. 18–3. *Mesophryxus ventralis,* a parasitic isopod belonging to the family Bopyridae, clinging to the ventral surface of the posterior abdomen of a shrimp. (From Bruce, A.J., courtesy of Parasitology.)

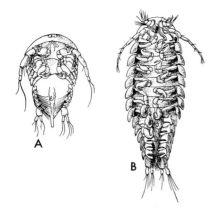

Fig. 18–4. *A, Cepon elegans,* microniscus larva. *B, Portunion kossmanni,* cryptoniscus larva. (From Giard and Bonnier. In Baer: Ecology of Animal Parasites, courtesy of the University of Illinois Press.)

parasite has a flat and distorted body, with an enormous mass of eggs on the ventral side of the female. The mouthparts form a short beak with which the parasite sucks the blood of its host.

Pinnotherion vermiforme (Fig. 18–6) occurs inside the thorax and abdomen of the pea crab, *Pinnotheres pisum*—itself a parasite of the clam, *Mytilus edulis.* Adult male isopods are found free in the host body. They pass from their normal position among the pleopods of the female through a break in the host cuticle. Female entoniscids undergo a pronounced morphologic change, and the surface of their pleopods becomes corrugated and filled with numerous blood lacunae that serve as respiratory organs. In the same female, the oostegites form a brood pouch, part of which appears as a hood around the head. Epicaridean larvae are expelled from this pouch through a secondary opening. Isopods, especially those of the endoparasitic family Entoniscidae, frequently cause complete or partial atrophy of host gonads.

Probably all parasitic isopods of the superfamily Cryptoniscina are protandrous hermaphrodites. Males become mature in the cryptoniscus stage. The life history of this group starts in the same manner as that of the bopyrids, and it progresses similarly to the cryptoniscus larval stage. From here on, the life history is different. On entering the branchial chamber or the brood pouch of crabs, larvae become protandrous hermaphrodites of a delayed type because males and females are morphologically distinct. An excessive production of eggs causes the gravid female to undergo morphologic, and presumably physiologic, degeneration.

The liriopsids are highly modified, sac-like hyperparasites that mimic rhizocephalan barnacles and many of the endoparasitic copepods. *Liriopsis* parasitizes rhizocephalan barnacles. Off California, *Cabirops montereyensis* lives in the brood chamber of the bopyrid *Aporobopyrus* which in turn infects the gill chamber of porcellanid crabs.

Suborder Gnathiidea

The gnathiids are a family of isopods in which the larvae and juvenile females are temporary parasites.[10] They live on the external surfaces of fishes and feed exclusively on body fluids. The **pranzia** larvae gorge themselves with blood and following three molt phases the adult stage is attained. Adults are free-living and may be found in the plankton or in mud dredged from the sea bottom. Adults do not feed and possess neither a mouth nor a gut. The adult male uses its powerful mandibles to dig a hole into the mud. A male and several females often occupy each burrow. Morphologic differences among the larvae, adult males, and adult females (Fig. 18–7) are so pronounced that each stage was originally described as a distinct genus.

Order Amphipoda

The amphipods are characterized as follows: body often laterally compressed, absence of a carapace, abdomen flexed ventrally between the third and fourth somites, telson usually distinct, habitat usually marine. They are divided into the following groups of which the first is the largest: Gammaridea, Caprellidea, Ingolfiellidea, and Hyperiidea.

Although many amphipods are associated in a casual manner with marine organisms, few species are truly parasitic.[31,45] Some members of the family Gammaridae have suctorial mouthparts and lead a semiparasitic life. Several species attach to the skin of fishes, while a number of others burrow into plants and invertebrates to obtain food. Caprellidean amphipods are frequently collected on the external surfaces of a diversity of invertebrates. Off the west coast of North America, *Caprella gorgonia* lives on the octocorals *Muricea* and *Lophogorgia,* and *C. greenleyi* clings to the sea star *Henricia.*

Six genera of cyamid amphipods live on var-

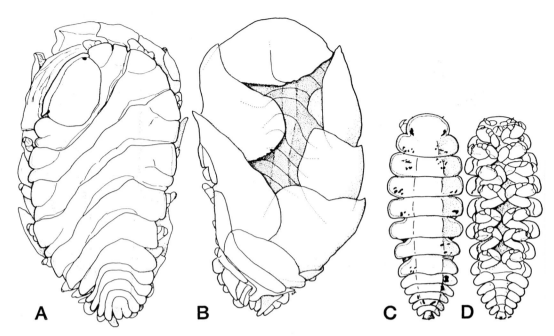

Fig. 18–5. The bopyrid isopod *Bopyrissa wolffi* from the hermit crab, *Clibanarius.* A–B, Adult female; C–D, Adult male; A, C, Dorsal views; B, D, Ventral views. (From Markham, courtesy of the Bulletin of Marine Sciences.)

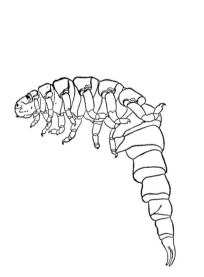

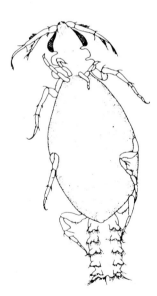

Fig. 18–6. *Pinnotherion vermiforme,* male, showing characteristic ventral curvature. (From Atkins, courtesy of the Proceedings of the Zoological Society of London.)

Fig. 18–7. The pranzia larva of *Gnathia.* The abdomen is distended with blood. (From Kabata. Diseases of Fishes. I. Crustaceans as Enemies of Fishes, courtesy of T.F.H. Publications.)

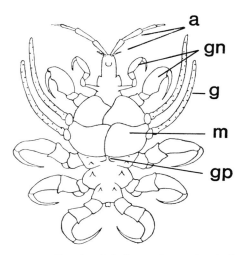

Fig. 18–8. Female whale louse, *Cyamus boopis,* from the humpback whale. Ventral view: a, antennae; g, gill; gn, gnathopod; gp, genital pore; m, marsupium. (From Leung, courtesy of Crustaceana.)

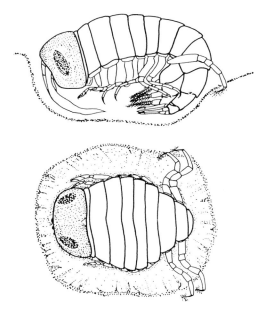

Fig. 18–9. The amphipod *Hyperia medusarum* partially embedded in the tissues of the ctenophore *Pleurobrachia bachei.* (From Flores and Brusca, courtesy of the Bulletin of the Southern California Academy of Sciences.)

ious cetaceans, especially the large whales. Cyamids rarely occur on the small toothed cetaceans commonly known as dolphins and porpoises. The ectoparasites have dorsoventrally flattened bodies, reduced abdomens, and claws on their legs with which they cling to the skins of their hosts. Cyamid mouthparts are not adapted for sucking blood; instead, the animals probably feed on mucus, bacterial and algal growth, and epithelial cells. *Cyamus,* the "whale louse" (Fig. 18–8), is the best known, and is distributed throughout the world oceans.

The cyamids are unique among crustacean parasites in that they are unable to swim at any period of their life history. The young settle near their parents, and masses of individuals of all sizes may be seen clinging closely together in crevices on the skin of their host, or intermingled with barnacles. Transfer occurs when young whales feed on their mother's milk or when adults mate. For further details, see Leung.[29]

The hyperiideans are pelagic amphipods that often are intimately associated with gelatinous zooplankton such as medusae, siphonophores, ctenophores, pelagic mollusks, and salps (Fig. 18–9).[28] *Hyperia spinigera* is an obligate parasite of the deep-sea jellyfish *Periphylla.* Salps commonly are infested by a number of genera. *Vibilia* lives inside the branchial chamber and feeds on food filtered by its host, whereas *Lycaea*

feeds directly on the tissues of the salp upon which it lives.

Class Maxillopoda

This class includes the most derived or modified of all the crustaceans. Compared to the malacostracans they have a reduced number of body segments and often an abbreviated life cycle with a reduced number of molts. The abdomen lacks appendages and often the appendages are uniramous instead of biramous. When present, the carapace may cover the entire body and appendages. The naupliar eye is distinct from the malacostracan naupliar eye and the heart is small and bulbous or absent. It is difficult to determine exact relationships within the class due to the repeated, independent evolutions of parasitic life-styles. Of the six subclasses recognized, five contain parasitic representatives, namely the Ostracoda, Branchiura, Copepoda, Thecostraca, and Pentastomida.[39]

Subclass Ostracoda

Ostracods often are hosts for parasitic trematodes, cestodes, nematodes, acanthocephalans, isopods, and copepods. A few genera of the

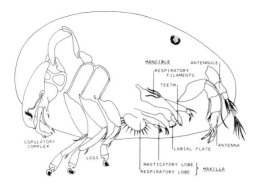

Fig. 18–10. Generalized entocytherid ostracod from the exoskeleton of crayfish. (By permission of the Smithsonian Institution Press. From "The Entocytherid Ostracods of North Carolina," *Smithsonian Contributions to Zoology*, No. 247, by Horton H. Hobbs, Jr., and Daniel J. Peters, Smithsonian Institution, Washington, D.C., 1977.)

order Podocopida are obligate ectocommensals on other crustaceans and echinoderms. The entocytherids (Fig. 18–10) infest freshwater crayfishes, crabs, and marine and freshwater isopods and amphipods.[19] They typically live in setiferous areas or in crevices on the exoskeleton, and on occasion may invade the branchial chambers of their hosts. The genus *Entocythere* is found in North America on crayfishes of the genus *Pacifastacus*. *Echinophilus xiphidion* clings to the epidermis of sea urchin hosts by means of special claws on the thoracic legs.

Subclass Branchiura

Branchiurans are Crustacea that are temporarily parasitic on fishes. They superficially resemble copepods, but differ from the latter in the possession of compound eyes, a flattened body with a carapace-like bilobed dorsal shield, a suctorial proboscis, and an opening for genital products between the fourth pair of thoracic limbs. Other diagnostic features include an unsegmented, limbless, bilobed abdomen, and four pairs of biramous thoracic appendages. The antennules and antennae are modified with hooks for attachment to the host. The most common genus, *Argulus* (Fig. 18–11), has a pair of suction cups situated on the second maxillae and a poison spine in front of the proboscis.

Branchiurids, commonly called "fish lice," are found on both fresh- and saltwater fishes. Both males and females are parasitic. They cling to the skin and gills and appear to have little host preference. Most species feed by rasping a hole in the host's skin and burying the proboscis in a blood vessel to suck blood. Other species feed on mucous and sloughed off epithelial cells. The adults are mobile and leave the hosts to find mates and to locate new hosts. Females deposit their eggs on stones, bits of wood, and other objects. The larvae resemble adults. The single family Argulidae is reviewed in Yamaguti.[47] See Bowen and Putz for reference to pathogenicity.[4]

Subclass Copepoda

A fine plankton net towed through almost any natural body of fresh or salt water soon collects multitudes of organisms whose animal components consist chiefly of minute crustaceans. The most abundant kinds of crustaceans in such hauls are usually copepods, whose bodies furnish the major basic food supply for all larger aquatic animals. The best-known freshwater member of the subclass Copepoda is *Cyclops* (order Cyclopoida), which serves as the intermediate host for several helminth parasites of man and other vertebrates. The calanoid copepod genus *Calanus* is one of the most abundant organisms in the marine environment.

Most copepods are free-living, but many are parasitic, and like so many other parasitic groups, they range in habits from the most casual and temporary contact with a host to a rigid and permanent attachment as adults or larvae within the host body. Most groups of aquatic, especially marine, animals may serve as hosts for at least one species of copepod during part of its life cycle. Annelids, echinoderms, ascidians, molluscs, arthropods, fishes, and whales are universally invaded. The parasite may appear to be practically identical to free-living species, or it may be so modified as an adult that, were it not for the larval stages, one would hardly be able to determine even the phylum to which the copepod belongs. In general, copepods have a high degree of host-specificity.

Copepods are typical maxillopods. The body is divided into three tagmata: cephalosome, metasome and urosome, although in most cases the cephalosome and metasome are fused into a prosome (Fig. 18–12). A carapace is absent, four to five appendages are present on the thorax but may be reduced greatly or absent in parasitic forms, and appendages are absent on the abdomen or urosome. The heart is absent in parasitic forms. Sexes are separate, the genital opening is on a modified genital segment between the thorax and abdomen. Eggs and

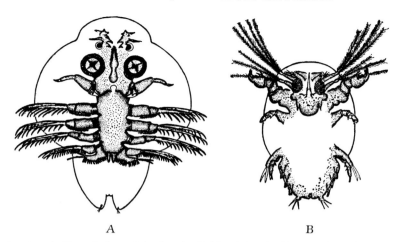

A   B

Fig. 18–11. *Argulus. A,* Ventral view of the female. *B,* Newly hatched larva. (From Cameron: Parasites and Parasitism, courtesy of John Wiley and Sons.)

developing embryos are carried in one or two ovisacs. Most free-living species are small, 0.3 to 15 mm, whereas parasitic forms may attain lengths over 300 mm.

A parasitic life style has evolved independently many times in the Copepoda as it has in most other crustacean groups. As a result of both parallel and convergent evolution, a bewildering diversity of body forms is evident and has created considerable confusion in copepod systematics. Eight orders are currently recognized of which the following contain parasitic members: Harpacticoida, Cyclopoida, Poecilostomatoida, Siphonostomatoida, and Monstrilloida. See Kabata for a discussion of copepod systematics.[24,25] See also Cressey and Collette[9] and Yamaguti.[47]

The sexes in copepods are nearly always separate. The male is usually much smaller than the female (Fig. 18–16). The pair of ovisacs ("egg string") is usually conspicuous. Some groups, for example, Harpacticoida, have only one sac. The female of *Chondracanthus merluccii* is 12,000 times larger than the male, but this sexual dimorphism is extreme. Hermaphroditism is rare. Some parasitic species appear to have several breeding seasons each year. In some instances, especially in all the Chondracanthidae and in some of the Lernaeopodidae, the males are sessile on the females. In such instances, the male loses the power of locomotion on reaching maturity. In still other species, one act of copulation is apparently sufficient to fertilize the female for life, and the male

is then not found in persistent association with the female.

The embryos remain in the ovisac until they hatch, a process often requiring several weeks. **Nauplii** larvae are, in the first stage, equipped with three pairs of appendages, one or more pairs of caudal setae (balancers), and two distal antennulary setae. The three appendages represent the antennules, antennae, and the mandibles of the adult. Copepod nauplii lack the characteristic "horns" of barnacle nauplii (Fig. 18–25). Many parasitic forms hatch at a later stage, with more appendages.

Copepod larvae normally pass through six nauplius stages (the last three often called "metanauplius") and five copepodid stages, but development may be abbreviated, especially in parasitic species.

Order Harpacticoida

Members of this order are predominantly free-living. Harpacticoids are relatively primitive and generalized copepods and can be recognized by the presence of a single egg sac even in commensal forms. The antennules are short and geniculate in males and the mandibles have a well-developed gnathobase. The mouthfield is open, not tubular or slit-like.

The family Balenophilidae live on the baleen plates of large whales. In the Tisbidae a number of species are specifically associated with marine invertebrates (Fig. 18–13). In the North Pacific Ocean, *Octopinella tenacis* is found in the

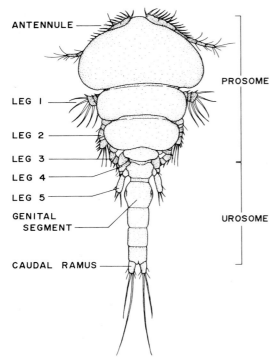

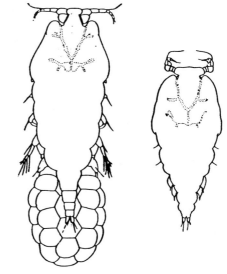

Fig. 18–13. The harpacticoid copepod *Cholidyella incisa* from the gills of the deep sea octopod, *Graneledone*. Dorsal views, female and male. (From Avdeev, courtesy of Parazitologia.)

Fig. 18–12. *Holobomolochus albidus* from the gill cavity of the angler fish *Lophius*. Dorsal view showing general structures of a poecilostomatoid copepod. (From Ho, J-S: Marine Flora and Fauna of the Northeastern United States. Copepoda: Cyclopoids Parasitic on Fishes. NOAA Technical Report NMFS Circular 409. Washington, D.C., National Oceanic and Atmospheric Administration, National Marine Fisheries Service, U.S. Department of Commerce, 1978.)

mantle cavity and on the gills of cephalopods of the genus *Octopus* and *Benthoctopus*.

Order Cyclopoida

The cyclopoids are a mixture of free-living and parasitic species. The antennules are moderately long and geniculate in the males. The antennae are uniramous and the mouthfield open. The order contains most of the copepods that inhabit the body cavities of tunicates. The Notodelphidae are a well-defined natural group characterized by a modified hook on the tip of the prehensile antennae and by the presence of a dorsal brood pouch enclosed within the body (Fig. 18–14). The life cycle is thought to involve five naupliar stages and six copepodid (subadult) stages. The second copepodid stage is the infective one, but the sequence of developmental events and the stages in the life histories of

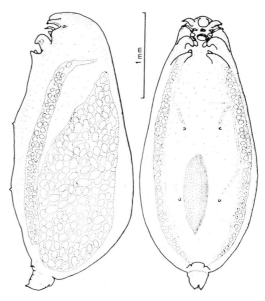

Fig. 18–14. The cyclopoid copepod *Sicyodelphys bocqueti* from a didemnid ascidian. (From Lafargue and Laubier, courtesy of the Academie de Sciences, Paris.)

members of the family are almost entirely unknown. For additional information see Dudley.[12]

Order Poecilostomatoida

The poecilostomatoids are a large order of cyclopoid-like copepods. The mouth is slit-like and the mandibles are uniramous and falcate or sickle-shaped. The antennules are short and not geniculate. The antennae are clawed and prehensile. The maxillipeds are prehensile in the males but reduced or absent in the females. Considerable progress has been made in recent years in understanding this large order. A number of families, such as the Lichomolgidae, have been monographed in detail. Well known as associates of marine invertebrates, especially in tropical waters, the lichomolgids are similar in most features to free-living copepods.[15]

In the ergasilids and bomolochids (Fig. 18–12) the body is similar to that of a free-living copepod. It narrows posteriorly. The abdomen or urosome has 4 segments in the male and 3 in the female. After copulation, the males dies, and the female becomes attached to the gills of its freshwater fish host by large clasper-like claws.

Knowledge of the life histories of poecilostomatoid copepods is incomplete. In only a few cases have life cycles been completed in the laboratory and all the details worked out. The myicolid, *Pseudomyicola spinosus,* infests the mussel, *Mytilus edulis galloprovincialis.* Compared to other families in the order the body and appendages of this species are relatively little modified (Fig. 18–15). The life cycle is direct and similar to that of free-living copepods. It involves six free-swimming naupliar and five infestive copepodid stages. The generation time from egg to adult is 30 to 31 days for males and 33 to 34 days for the larger females.[11]

In British waters, the mytilicolid copepod, *Mytilicola intestinalis,* retards the release of gametes in its host, the edible mussel, *Mytilus edulis.* Variations in host biochemical composition are closely related to the breeding season. Williams has shown that the biochemical constitution of the copepod also changes with the season and is correlated to its life history.[46] Williams found that the carbohydrates and lipids formed important sources of energy, but protein levels were reduced as compared with levels in free-living species.

Acanthochondria (Fig. 18–16), *Chondracanthus* and other genera in the family Chondracanthidae are common on the gills and in the mouths of various fishes in the Atlantic and Pacific Oceans. The females are large and permanently attached to the host by means of modified antennae. The body of the female lacks external segmentation and the appendages are reduced and modified or are absent entirely (Fig. 18–16). Minute, maggot-like males cling to the body of the female, often in the posterior region near where the egg sacs emerge. The males attach, like a secondary parasite, to the enormously larger mate and often die after mating. The life cycle involves a single host. Little is known about the biology and host-parasite relationships of this family.

Sarcotaces (Fig. 18–17) is a highly modified endoparasitic poecilostomatoid. The sac-like female bears faint resemblance to a copepod or a crustacean for that matter. The female and a single dwarf male live together in a cyst in the muscle or visceral cavity of fishes, especially rockfishes. The presence of *Sarcotaces* often is not detected unless the host is dissected. The parasite feeds on blood and females may grow to a size of several centimeters.

Order Siphonostomatoida

The siphonostomatoids are another large order of parasitic copepods in which the mandibles of all members are modified as piercing stylets and enclosed in a siphon-like, tubular mouth. The antennules are geniculate and reduced with 1 to 3 segments only. The antennae are prehensile and clawed or absent. The maxillipeds are subchelate or absent. Similar to the poecilostomatoids, they exhibit a radiation of body forms from slightly modified ectoparasites to highly modified, sac-like endoparasites. Several authorities consider the order to be a polyphyletic assemblage.

The Caligidae is probably the most widely distributed family of copepods parasitic on fishes, and it probably contains the largest number of species, for example, *Caligus* and *Lepeophtheirus.* The family Chondracanthidae, however, contains a greater number of genera.

Caligus (Fig. 18–18) is ectoparasitic, chiefly on fishes, has a suctorial mouth tube, and retains the power of swimming in the adult stage. The parasites possess two semicircular structures on the frontal margin of the cephalic shield. These structures are probably sensory, rather than attachment devices. *Caligus* does not always re-

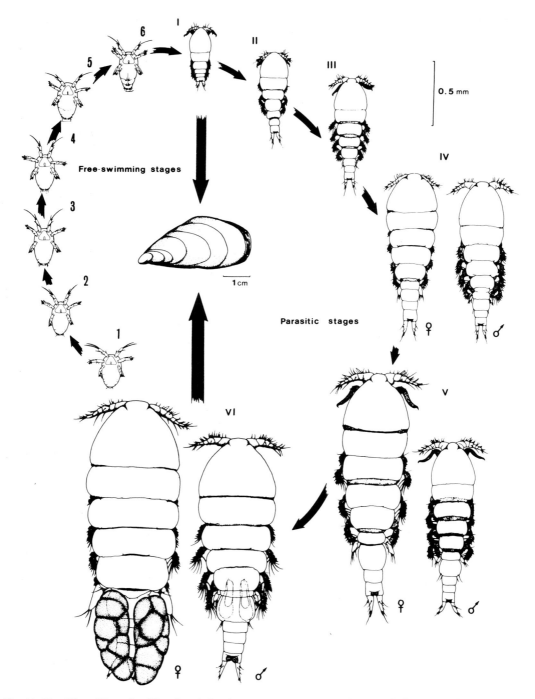

Fig. 18–15. Direct life cycle of *Pseudomyicola spinosus*, a poecilostomatoid copepod of the mussel *Mytilus edulis*. 1–6, Naupliar stages; I–V, Copepodid stages; VI, Adult stage. (From Do, Kajihara, and Ho, courtesy of the Bulletin of the Research Institute of the University of Tokyo.)

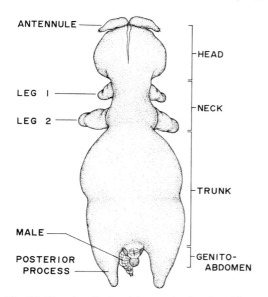

ANTENNULE

HEAD

LEG 1

NECK

LEG 2

TRUNK

MALE

POSTERIOR
PROCESS

GENITO-
ABDOMEN

Fig. 18–16. *Acanthochondria cornuta*, female with attached male, dorsal view. (From Ho, J-S: Marine Flora and Fauna of the Northeastern United States. Copepoda: Cyclopoids Parasitic on Fishes. NOAA Technical Report NMFS Circular 409. Washington, D.C., National Oceanic and Atmospheric Administration, National Marine Fisheries Service, U.S. Department of Commerce, 1978.)

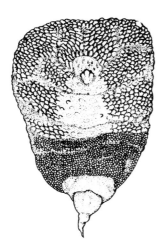

Fig. 18–17. The sac-like female of the poecilostomatoid copepod, *Sarcotaces*. This highly modified endoparasite lives in cysts in the muscles or body cavities of fishes. (From Kabata: Diseases of Fishes. I. Crustaceans as Enemies of Fishes, courtesy of T.F.H. Publications.)

main on one host. It is the most abundant copepod genus parasitic on fishes.

Salmincola, the "salmon gill maggot," is a common and widespread lernaeopodid copepod parasite of Atlantic and Pacific salmon. The mature female parasite, when attached to a gill filament of its host, "somewhat resembles a gymnast hanging from a vertical bar."[13] Attachment to the gill filament is effected by means of a secreted **bulla**, which is applied to the gill surface (Fig. 18–19). A thin sheet of living gill tissue partly covers the bulla. A pair of prehensile maxillipeds lies behind an oral cone, and from the sides of the cephalon, the maxillary "arms" (second maxillae) extend and converge on the bulla to which they are permanently fixed. The thorax bears no appendages. The paired egg sacs measure 4 to 11 mm in length.

The life history of **Salmincola salmonea** has been described by Friend[13] and is used here to exemplify a direct life cycle involving a single host. During the first river phase of the fish, and the first sea phase, totalling from two-plus years to seven-plus years, the salmon is not attacked by gill maggots, although it acquires sea lice. At the beginning of the second river phase, the sea lice are lost, and the maiden fish acquires larval gill maggots, which swim from beneath the gill covers of other and older fish. In the fall, the host reenters the sea, and during its sojourn of up two years in marine waters, its copepods thrive and grow, but do not breed. When the salmon reenter the rivers, the copepod parasites begin to breed.

The first copepodid larvae of *Salmincola salmonea* may live free for up to six days, but when they come into contact with the gills of a salmon, they attach themselves "by means of a button and thread" and molt to become the second copepodid larvae. The next molt is followed by the sexually mature male or the first stage female, both of which may move freely over the gill surface, or they may attach themselves by means of chelate appendages. During this stage, copulation takes place and the male disappears. Meanwhile, the frontal gland in the female has been elaborating the attachment bulla, which is planted on the gill tissue in its final position. The female then molts, the distal ends of each second maxilla are forced as plugs into corresponding sockets in the bulla, and permanent attachment is achieved. The female then grows to mature size. Development to this stage requires about five or six months. One

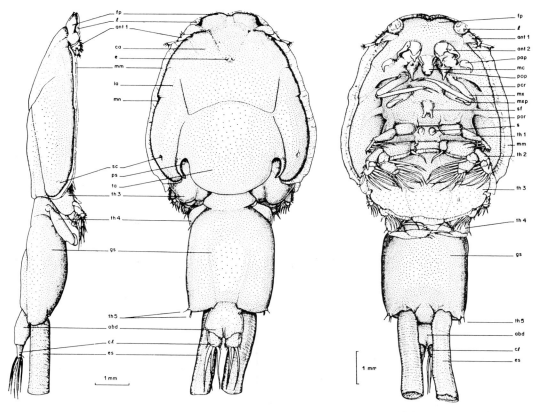

Fig. 18–18. The siphonostomatoid copepod *Caligus curtus*, a common parasite of fishes in the North Atlantic Ocean. Female, lateral, dorsal and ventral views. (From Parker et al., courtesy of the J. Fisheries Research Board of Canada.)

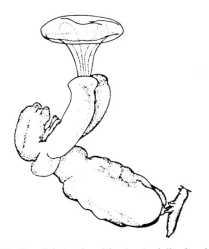

Fig. 18–19. *Salmincola californiensis*, fully developed female with attached male. (From Kabata and Cousens, courtesy of J. Fisheries Research Board of Canada.)

female may produce two or more generations of larvae.

Salmincola californiensis (Fig. 18–19) is a parasite of sockeye salmon, *Oncorhynchus nerka*, and shows a site preference that differs with the size of the host. On small fish, the parasites are most abundant in the region of the pectoral and pelvic fins, whereas on larger fish, they are most common in the branchial cavity. Damage to host tissue involves injury to gills, skin, muscle, and bone. The copepods may even burrow through the body wall and may enter viscera. The life history and host-parasite relationships of this species have been documented in detail by Kabata and Cousens.[26,27]

The pennellid *Lernaeocera branchialis*, often called the "gill maggot" (Fig. 18–20), is commonly seen attached to the gills of marine cod, and its bloated, S-shaped and blood-red body looks much more like a worm than it does a crustacean. The larval nauplius stage (unseg-

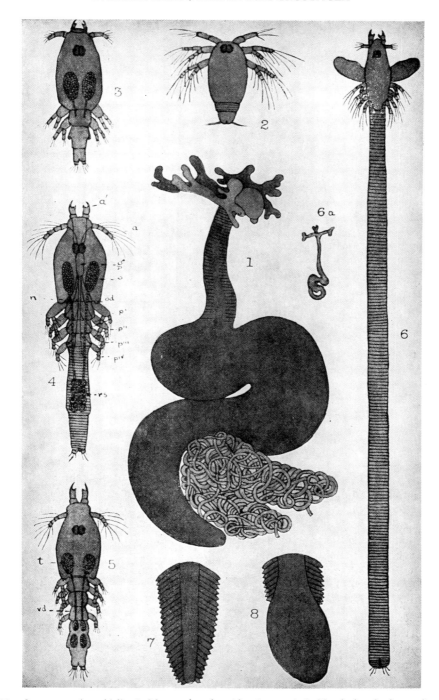

Fig. 18–20. *Lernaeocera branchialis. 1,* Mature female, side view, ×4. *2,* Newly hatched nauplius, ×51. *3,* Young female, unfertilized, dorsal view, ×51½. *4,* Fertilized female, after leaving gills of flounder, dorsal view, ×27½. *5,* Mature male, dorsal view, ×28½. *6,* Fertilized female, "pennella stage," just after attachment to gills of a whiting, dorsal view, ×15½. *6a,* A later stage than *6,* side view, natural size. *7,* Apex of the gill-ray of a flounder, ×18. *8,* Apex of the gill-ray of a flounder, showing malformation caused by young parasite, ×18. (From Scott and Scott: The British Copepoda, courtesy of B.H. Blackwell.)

mented, three pairs of legs) proves its kinship with the arthropods. After a free-living period of growth, including several molts, the nauplius is transformed into a **cyclops** or **chalimus** stage, when it becomes parasitic on the gills of certain flatfishes, for example, *Pleuronectes*, by means of suctorial mouthparts accompanied by a reduction of its limbs. The chalimus attaches itself to the host, first by the antennae, then by a chitinous secretion that is extruded as a laterally flattened thread extending into the gill tissue. This stage is followed by a resumption of the power of movement when the parasite leaves its host and, for a time, lives a free-swimming, adult, sexually mature existence similar to that of *Cyclops*.

At this time, fertilization of the eggs takes place, and the development of the male ceases. The female, however, now seeks a new host, generally a member of the cod family, for example, *Gadus merlangus*, and again, the gills are particularly attractive to the parasite. The anterior end of the female is buried into the host tissue, and it becomes curiously modified to form a branched anchor resembling short roots. The genital somites become enlarged and vermiform, and the egg mass appears as a cluster of tightly coiled filaments. In proportion, the appendages of the thorax become minute and nonfunctioning.

Cardiodectes, another pennellid related to *Lernaeocera*, also has an indirect life cycle involving two hosts (Fig. 18–21). However, unlike *Lernaeocera*, the developmental stages occur in an invertebrate intermediate host, specifically pelagic gastropods. Adults of *Cardiodectes medusaeus* infect a wide variety of deep-sea myctophid fishes. The head and attachment organ of the female copepod penetrate the bulbus arteriosus of the fish's heart.

Another unusual pennellid, *Phrixocephalus* (Fig. 18–22), invades the eyes of a variety of fish hosts, especially flatfishes. In this case the tips of the arms are modified into a complexly branched holdfast structure which ramifies throughout the cavity of the eye. Only one eye of the host is infected for obvious reasons.

Xenocoeloma represents an extreme type of modification involving host tissue. This genus is parasitic on marine annelids, and the adult parasite is so grossly modified that it consists only of hermaphroditic gonads and some muscles. The mass is enclosed in a cylindrical outgrowth of the host epithelium that contains a gut-like prolongation of the host coelom (Fig. 18–23). The life history is incompletely known, but the nauplius that emerges from the eggs resembles that of the monstrillids in the absence of digestive organs. The manner of access to the host has, unfortunately, never been observed.

Order Monstrilloida

The Monstrillidae are copepods that are free-living as adults, but are parasitic during the larval stages in the blood vessels of various polychaete worms or in the body cavity of prosobranch snails. They are characterized by the presence of well-developed antennules, but the antennae are absent. Mouth and mouthparts are absent in adults. The nauplius of *Haemocera* hatches without a mouth or a gut, and it burrows into the body of its host, discards its chitinous exoskeleton, and loses its limbs. By the time it reaches the host body cavity, it consists only of a naked mass of embryonic cells, which then become surrounded by a thin cuticle and develop a pair of long, flexible processes that represent antennules and serve to absorb nutrients. Food thus taken lasts the entire remaining life cycle, throughout which no functional mouthparts are present. The mass of parasitic cells within the host gradually develops mature organs. The adult bores its way outside by means of rows of hook-like spines surrounding the pointed posterior end of the sac. On reaching the surface, the enclosing membrane bursts, and the adult parasite is free and looks like *Cyclops*. The animal thus passes its whole life cycle without a gut, and between the two free-swimming stages, one at each end of its life, it exists as a bag of parasitic cells absorbing nourishment from the internal fluids of its host (Fig. 18–24).

Subclass Thecostraca

Barnacles and their allies are sedentary, marine crustaceans. They are distinguished from other crustaceans by the absence of compound eyes in the adult, six pairs of biramous thoracic limbs and an abdomen without appendages. The body typically is contained within a shell composed of several hard calcareous plates. In burrowing or parasitic forms the plates are only partially calcified or absent. Although most species are hermaphroditic, in those with separate sexes the males are greatly reduced in size. Development usually includes six free-swimming,

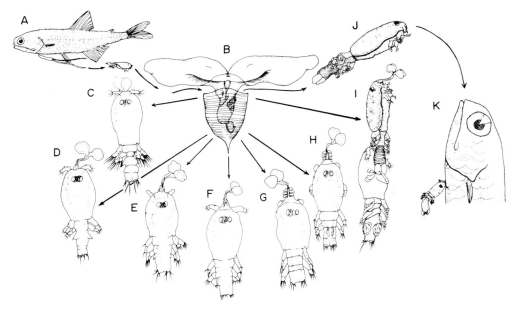

Fig. 18–21. The indirect life cycle of *Cardiodectes medusaeus,* siphonostomatoid copepod parasite of the deep sea myctophid, *Stenobrachius leucopsaurus* and the pelagic gastropod, *Clio pyramidata.* A, Adult copepod on definitive host; B, Mollusk intermediate host; C, Copepodid; D–H, Chalimus stages; I, Male and female in amplexus; J, Preadult female leaving intermediate host in search of fish; K, Female attaching to ventral surface of fish. (From Perkins, courtesy of J. Crustacean Biol.)

feeding naupliar larval stages (Fig. 18–25) and a non-feeding **cypris** stage. Both free-living and highly specialized parasites are represented in the subclass.

Order Ascothoracida

These small semiparasitic or wholly parasitic organisms have the most generalized morphology of any barnacle. They live on or in cnidarians (soft and hard corals) and echinoderms (sea lilies, sea stars, and brittle stars). The body is sac-like or enclosed in two valves resembling those of an ostracod. *Synagoga mira* (Fig. 18–26A) forms galls on the black coral, *Antipathes.* The highly modified genus *Dendrogaster* (Fig. 18–26B) lives in the coelomic cavities of asteroids. Females have a greatly enlarged, branched mantle which contains the gut, gonads, and a spacious brood chamber. The males are greatly reduced in size and reside in the brood chamber of the female. For additional details see Grygier.[16,17]

Order Tantulocarida

A new order was erected recently to contain an enigmatic group of small ectoparasites. To date only juveniles and females have been found on deep-sea isopods, ostracods, and copepods. Little is known about the life cycle or the natural history of these unusual parasites.[5]

Order Cirripedia

The acorn and gooseneck barnacles (Fig. 18–27) are among the most common and visible crustaceans. They abound in marine waters and attach to rocks, pilings, shells of other animals, or almost any other solid surface that provides a firm purchase. Like so many other sessile animals, barnacles, with few exceptions, are hermaphroditic, and their bodies are enclosed in a calcareous shell composed of plates that commonly overlap or are fused to furnish a formidable bulwark against pounding waves or grazing predators.

Species of barnacles range in habits from completely free-living to casual commensalism to extreme pathogenic parasitism. Many forms illustrate phoresis, in which host-specificity is pronounced, as in those species attached only to gorgonian corals, decapods, sea stars, sharks, or whales.

Examples of commensal cirripeds are: *Chelonibia,* which has developed a branched system of roots that penetrate into manatees and into

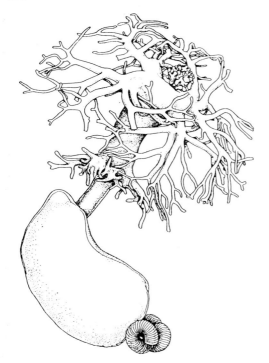

Fig. 18–22. *Phrixocephalus umbellatus,* a highly modified siphonostomatoid copepod parasite that lives in the eye of the dragonet fish, *Callionymus.* (From Shino, courtesy of the Faculty of Fisheries, Prefectural University of Mie.)

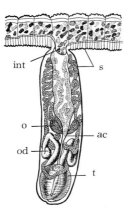

Fig. 18–23. *Xenocoeloma brumpti,* section of the parasite and the annelid host. *ac,* Axial cavity formed by the host's coelom. *int,* Host's gut. *od,* Oviduct. *o,* Ovary. *s,* Skin of both the host and the parasite. *t,* Testis. (From Caullery and Mesnil. In Baer: Ecology of Animal Parasites, courtesy of the University of Illinois Press.)

the bone of the plastron of marine turtles; *Coronula* (Fig. 18–27), *Tubicinella, Cryptolepas,* and *Xenobalanus,* whose calcareous shell plates grow into the skin of whales; and *Anelasma squalicola* (Fig. 18–28A), which becomes partly buried in the skin of sharks. In the last species, a pear-shaped portion is embellished with branched, root-like appendages that apparently secrete an enzymatic substance that dissolves the surrounding muscle tissues of the host. *Rhizolepas annellidicola* (Fig. 18–28B) is another parasitic barnacle that is anchored to its host, the annelid *Laetmonica producta,* by a system of foot-like appendages. Two other examples are *Platylepas* on turtles, manatees, and sea snakes and *Alepas* on various medusae.

Rhizocephala. The most extremely modified parasitic cirripeds belong to this group. Its members live on or within the bodies of other Crustacea, mostly decapods. The larval stages appear to be identical to those of free-living barnacles, but no trace of a mouth or gut is present in the parasites. The adult rhizocephalan possesses neither appendages nor segmentation, and it is anchored to the host by a stalk from which roots proceed into the host tissues. The best-known rhizocephalans are *Sacculina carcini* and *Lernaeodiscus porcellanae.*

The remarkable life history of *Sacculina* starts with a nauplius possessing the characteristic frontal horns of cirriped nauplii. The cypris larva of *Sacculina* is active after dark, and it attaches itself by its antennae to a crab at the base of a bristle. Soon after attachment, the whole trunk with its muscles and appendages is thrown off, and a new cuticle is formed under the old one. The body of the parasite becomes an amorphous cellular mass within the two valves of the cypris. From these cells, the **kentrogon** larva (a mass of embryonic cells) is developed and, by means of a short, dart-like tube that pierces the integument of the host, these undifferentiated cells flow into the body of the crab. The parasitic embryonic cells are carried by the host circulation to the central surface of the gut, where they begin to multiply and spread out in all directions as root-like branches. A rounded, cellular mass (the future **sacculina externa**), containing the rudiments of genital organs and a ganglion, now appears at the base of the root system, and the parasite migrates to a spot on the ventral side of the crab near the single diverticulum of the gut. The parasite then makes a hole through the new integ-

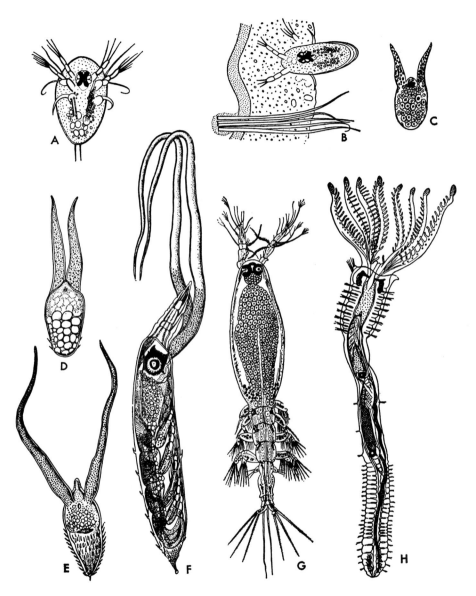

Fig. 18–24. *Haemocera danae. A,* Nauplius. *B,* Nauplius in the act of penetrating into the body of its host. *C* to *E,* Successive larval stages showing the development of the appendages and also of the spinous sheath enclosing the larva. *F,* Fully formed copepodid. *G,* Adult female copepod devoid of a mouth. *H,* Annelid with two copepodid larvae in its coelomic cavity. (From Malaquin. In Baer: Ecology of Animal Parasites, courtesy of the University of Illinois Press.)

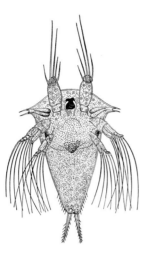

Fig. 18–25. *Sacculina carcini,* nauplius larva. (From Delage. In Baer: Ecology of Animal Parasites, courtesy of the University of Illinois Press.)

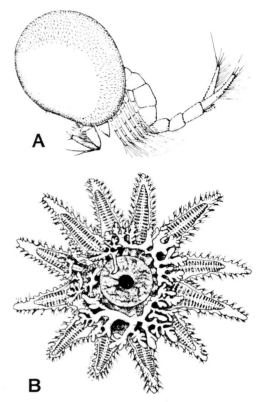

Fig. 18–26. Ascothoracidian barnacles. A, *Synagoga mira,* an ectoparasite of the soft coral, *Antipathes;* B, *Dendrogaster murmanensis,* sac-like females in the body cavity of the seastar, *Solaster.* (A. From Ross and Emerson: Wonders of Barnacles, courtesy of Dodd, Mead and Company. B. From Baer. Ecology of Animal Parasites, courtesy of the University of Illinois Press.)

ument. The external portion of the barnacle thus makes its appearance after the next molt, and it grows into the hermaphroditic adult. The **sacculina interna** continues to grow and may permeate almost the entire body of the host (Fig. 18–29). A "sacculinized" crab may live with its barnacle parasite for at least two years, and it may become infected with two or more individuals. The crab, however, ceases to molt, and its metabolism, particularly that relating to sexual development and activity, is profounding modified. (See Chapter 23 for a discussion of the effects of this parasite on the sex determination of the hosts.)

Aspects of the life cycle of *Lernaeodiscus* (Fig. 18–30) have been studied in detail.[21,36] For a review of the biology and systematics of the rhizocephalans see Hoeg and Lützen.[22]

Subclass Pentastomida

The pentastomes, or "tongue worms," are an aberrant group of worm-like animals that have been classified with each of the major groups of helminths. The current view is that they are arthropods, most closely related to the Crustacea. Riley, Banaja, and James[34] say: "Three crucial items of evidence, embryogenesis, the structure of the integument, and gametogenesis are considered to confirm previous hypotheses of genuine arthropod homologies. Spermatogenesis particularly, clearly establishes pentastomids as a crustacean subclass, closely allied to the Branchiura. We postulate that the pen-

tastomid progenitor was originally a parasite of fishes which subsequently became adapted to an endoparasitic existence in aquatic reptiles through predation."

Pentastomids are parasitic in all of their stages. The adults usually are found in respiratory passages of reptiles. They are large, elongate, vermiform parasites, with annuli, which are not true segments (Fig. 18–31). There are no legs, antennae, respiratory organs, or circulatory system. The oral region bears the mouth and four openings that contain hollow, curved, retractile hooks (Fig. 18–32), which gives rise to the name pentastome, "5 mouths." The immatures (Fig. 18–33) are not similar to the adults and undergo metamorphosis during development; they have 2 or 3 pairs of short legs bearing the hooks seen in the adult. Adult

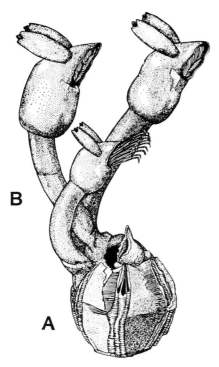

Fig. 18–27. Whale barnacles. A, The acorn barnacle, *Coronula diadema* that attaches to the skin of humpback whales. B, The rabbit-eared, gooseneck barnacle, *Conchoderma auritum*, that attaches to other whale barnacles. (From MacGinitie and MacGinitie. Natural History of Marine Animals, courtesy of McGraw-Hill.)

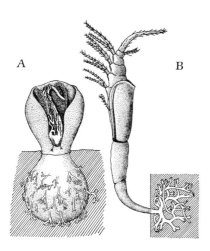

Fig. 18–28. A, *Anelasma squalicola,* partly buried in the skin of a shark. B, *Rhizolepas annellidicola* attached to an annelid host. (From Baer: Ecology of Animal Parasites. Courtesy of the University of Illinois Press.)

pentastomids feed on blood and mucosal cells in the mouth, esophagus, or respiratory passages of their hosts. In abnormal hosts, such as humans, extensive phagocyte infiltration occurs, as well as fibrosis.

The pentastome cuticle, like that of other arthropods, contains chitin, and papillae are the only sense organs. The cuticle probably produces antigens, and, in rodents, there appears to be an immune response against the nymphs involving both antibody- and cell-mediated reactions.[35] The mouth is framed by a chitinous structure, and the simple digestive tube ends with an anus usually in the last segment. Muscle tissues are striated, similar to those of arthropods, but locomotion is sluggish, although the parasites can migrate through host tissues. The single ovary in females may extend from one end of the abdomen to the other, and, in the male, the single testis occupies half to one-third of the body cavity.

Little is known of life histories and general biology of these worms, although some life cycles in the genus *Linguatula* have been well understood for many years. (See Nicoli and Nicoli[32] for a general review.) Of the two orders, the Cephalobaenida, whose life cycles involve insects, fishes, amphibians, and reptiles as secondary hosts, are the more primitive. Parasites of the order Porocephalida generally require mammals as secondary hosts and reptiles as definitive hosts. *Reighardia sternae*, a pentastome that lives in the air sacs of the herring gull apparently has a direct life cycle. In most genera, such as *Armillifer* (Fig. 18–31), *Sambonia, Raillietiella* (Fig. 18–31), *Porocephalus, Linguatula,* and *Megadrepanoides*, a life cycle involving two vertebrate hosts is expected because the adult parasites normally live in carnivorous lizards or snakes. These hosts regularly feed on smaller reptiles, amphibians, and mammals, in which larval pentastomids have been found. Adults of *Porocephalus crotali*, for example, live in the lung cavity of crotaline snakes, especially in those species of rattlesnakes (*Crotalus*) that range along the western parts of North, Central, and South America. Their eggs, which reach the outside through the snake sputum, are readily ingested by muskrats, opossums, bats, armadillos, raccoons, and other mammals, in which they hatch in the small intestine. Larvae (Fig. 18–33) migrate to the viscera (for example, lungs and liver), where they become encapsulated in tissues and where, in about 3 months, they de-

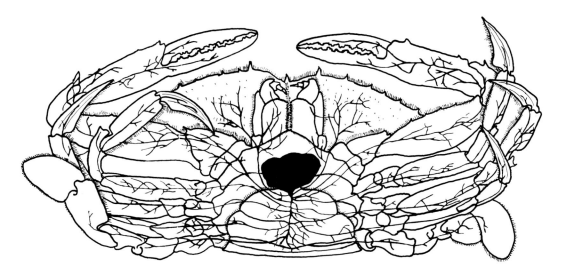

Fig. 18–29.   The root-like branches of *Sacculina* permeating the body of its crab host.

velop into nymphs. If mammalian nymph-in-
fested tissue is ingested by an appropriate
snake, the nymphs migrate up the esophagus
into the trachea and lungs, where they become
adult pentastomids.

The cockroach *Periplaneta americana* is the intermediate host for nymphs of the pentastomid
Raillietiella hemidactyli, which lives in the lungs
of lizards. Pentastomid nymphs can occur in
large numbers in tissues of mammals and reptiles without doing appreciable harm. Visceral
larva migrans effects, however, may be produced by primary larvae, and a mononuclear
host cell response may be elicited.[41]

Linguatula serrata (Fig. 18–34) is one of the
best-known species of this group. The adult female is tongue-shaped, 100 to 130 mm long, and
up to 10 mm wide, whereas the male is about
20 mm long and 3 to 4 mm wide. Eggs containing embryos with rudimentary legs are deposited in nasal passages and frontal sinuses of
mammals, commonly dogs, and are discharged
in nasal secretions. On reaching water or moist
vegetation, embryonation is completed. If eggs
are ingested by the intermediate host, for example, cattle, goats, sheep, rabbits, rats, or
man, hatching occurs in the digestive tract, and
larvae migrate through the intestinal wall and
become lodged in the liver, mesenteric nodes,
and other viscera.

After a number of molts, requiring a period
of 5 to 6 months, the nymphal stage is attained.

Nymphs lie encapsulated with host tissue and
are the infective stage for carnivorous animals
that feed on herbivorous intermediate hosts. Although the general belief is that nymphs are
digested out of their capsule and immediately
migrate to the nasal passages of the definitive
hosts, evidence[20] indicates that nymphs quickly
leave their cysts after death of the intermediate
host and cling to the oral mucous membrane of
the carnivorous host. Attachment of the
nymphs of *Linguatula serrata* to the oral tissues
of man produces a syndrome known as halzoun. "A few minutes to half an hour or more
after eating [raw visceral organs of sheep, goats,
cattle or camels], there is discomfort, and a
prickling sensation deep in the throat; pain may
later extend to the ears. Edematous congestion
of the fauces, tonsils, larynx, Eustachian tubes,
nasal passages, conjunctivae and lips is sometimes marked. Nasal and lachrymal discharges,
episodic sneezing and coughing, dyspnea, dys-
phagia, dysphonia and frontal headache are
common. Complications may include abscesses
in the auditory canals, facial swelling or paralysis and sometimes asphyxiation and death."[38]

L. rhinaria is a common parasite of livestock,
and a facultative parasite of man in both adult
and nymphal stages. This species also occurs in
rats, dogs, and in such experimental animals as
guinea pigs.

The frontal and subparietal glands of *Reighardia sternae,* a pentastomid that lives in the

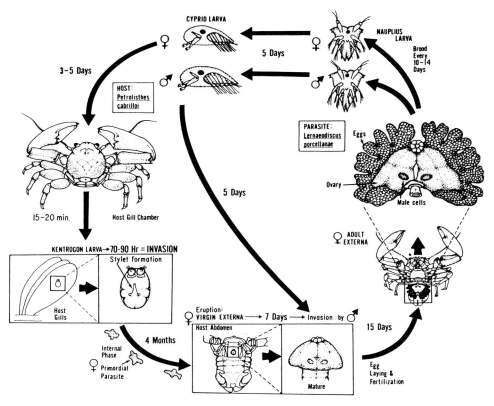

Fig. 18–30. Life cycle of the rhizocephalan, *Lernaeodiscus porcellanae*, parasitic in the crab *Petrolisthes cabrilloi*. (From Ritchie and Hoeg, courtesy of J. Crustacean Biol.)

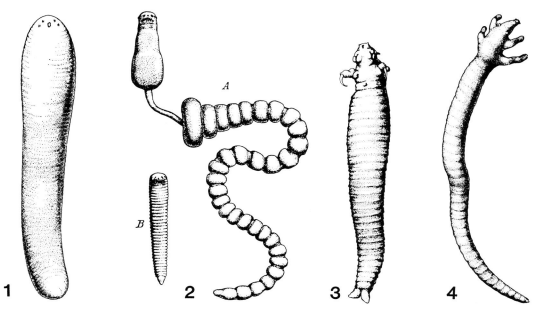

Fig. 18–31. Adult pentastomes. 1, *Sebekia oxycephala*, female; 2, *Armillifer pomeroyi*—A, female; B, male; 3, *Raillietiella mabuiae*, female; 4, *Cephalobaena tetrapoda*, female. (From Heymons and Vitzthum, courtesy of Zeitschrift für Parasitenkunde.)

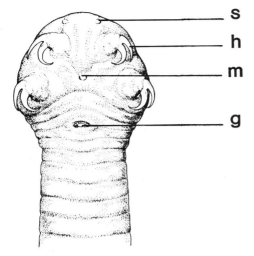

Fig. 18–32. Anterior end of *Leiperia gracilis,* ventral view. g, genital opening; h, hook; m, mouth, s, sensory papilla. (From Heymons and Vitzthum, courtesy of Zeitschrift für Parasitenkunde.)

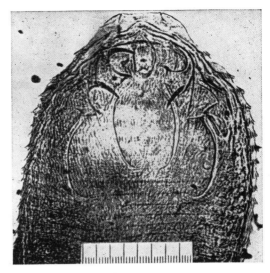

Fig. 18–34. *Linguatula serrata,* head end. Each division of the scale represents 10 μm. (From Sprehn, courtesy of Parasitologische Schriftenreine.)

herring gull, are composed of large cells suspended in the parasite hemocoel. The secretion of these cells is poured over the entire parasite cuticle and helps the parasite evade host immune responses. This may account for the long life of these pentastomids, in spite of strong antagonistic host reactions.[35]

For a discussion of host-parasite relationships, see Self and Kuntz.[43] For a bibliography on the Pentastomida, see Self.[42]

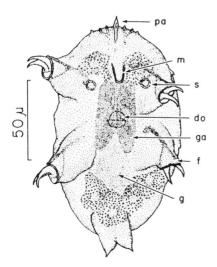

Fig. 18–33. *Porocephalus crotali,* newly hatched larva from gut of muskrat. (From Esslinger, courtesy J. Parasitol.)

REFERENCES

1. Anderson, G., and Dale, W.E.: *Probopyrus pandalicola* (Packard) (Isopoda, Epicaridea): Morphology and development of larvae in culture. Crustaceana, *41*:143–161, 1981.
2. Askew, R.R.: Parasitic Insects. New York, Elsevier, 1971.
3. Beesley, W.N.: Control of arthropods of medical and veterinary importance. Adv. Parasitol., *11*:115–192, 1973.
4. Bowen, J.T., and Putz, R.E.: Parasites of freshwater fish: IV, Miscellaneous 3. Parasitic copepod *Argulus*. Fish Disease Leaflet No. 3. Washington, D.C., Bureau of Sport Fisheries and Wildlife, 1966.
5. Boxshall, G.A., and Lincoln, R.J.: Tantulocarida, a new class of Crustacea ectoparasitic on other crustaceans. J. Crust. Biol., *3*:1–16, 1983
6. Brusca, R.C.: Studies on the cymothoid fish symbionts of the eastern Pacific (Crustacea: Cymothoidae). II. Systematics and biology of *Lironeca vulgaris* Stimpson 1857. Occas. Pap. A. Hancock Found., New Ser., No. 2:1–19, 1978.

7. Brusca, R.C.: A monograph on the Isopoda Cymothoidae (Crustacea) of the eastern Pacific. Zool. J. Linn. Soc., *73*:117–199, 1981.

8. Christensen, A.M., and McDermott, J.J.: Life-history and biology of the oyster crab, *Pinnotheres ostreum* Say. Biol. Bull., *114*:146–179, 1958.

9. Cressey, R.F., and Collette, B.B.: Copepods and needlefishes: a study in host-parasite relationships. Fish. Bull., *68*:347–432, 1970.

10. Davies, A.J.: A scanning electron microscope study of the praniza larva of *Gnathia maxillaris* Montagu (Crustacea, Isopoda, Gnathiidae), with special reference to the mouthparts. J. Nat. Hist., *15*:545–554, 1981.

11. Do, T.T., Kajihara, T., and Ho, J.-S.: The life history of *Pseudomyicola spinosus* (Raffaele & Monticelli, 1885) from the blue mussel, *Mytilus edulis galloprovincialis* in Tokyo Bay, Japan, with notes on the production of atypical male. Bull. Ocean Res. Inst. Univ. Tokyo, *17*:1–65, 1984.

12. Dudley, P.L.: Development and Systematics of Some Pacific Marine Symbiotic Copepods. A Study of the Biology of the Notodelphyidae, Associates of Ascidians. Seattle, University of Washington Press, 1966.

13. Friend, G.F.: The life history and ecology of the salmon gill maggot *Salmincola salmonea* (L) (copepod Crustacea). Trans. R. Soc. Edinburgh, *60*:503–541, 1941.

14. Gaafar, S.M.: Immune response to arthropods. *In* Immunity to Animal Parasites. Edited by E.J.L. Soulsby. New York, Academic Press, 1972, pp. 273–285.

15. Gotto, R.V.: The association of copepods with marine invertebrates. Adv. Mar. Biol., *16*:1–109, 1979.

16. Grygier, M.J.: Ascothoracida and the unity of the Maxillopoda. Crust. Issues, *1*:73–104, 1983a.

17. Grygier, M.J.: *Ascothorax*, a review with descriptions of new species and remarks on larval development, biogeography, and ecology. Sarsia, *68*:103–126, 1983b.

18. Ho, J-S.: Marine Flora and Fauna of the Northeastern United States. Copepoda: Cyclopoids Parasitic on Fishes. NOAA Technical Report NMFS Circular 409. Washington, D.C., National Oceanic and Atmospheric Administration, National Marine Fisheries Service, U.S. Department of Commerce, 1978.

19. Hobbs, H.H., and Peters, D.J.: The entocytherid ostracods of North Carolina. Smithsonian Contrib. Zool., *247*:1–73, 1977.

20. Hobmaier, A., and Hobmaier, M.: On the life cycle of *Linguatula rhinaria*. Am. J. Trop. Med., *20*:199–210, 1940.

21. Hoeg, J.T.: Cypris settlement, kentrogen formation and host invasion in the parasitic barnacle *Lernaeodiscus porcellanae* (Muller) (Crustacea: Cirripedia: Rhizocephala). Acta. Zool., *66*:1–45, 1985.

22. Hoeg, J., and Lutzen, J.: Marine Invertebrates of Scandinavia. No. 6. Crustacea Rhizocephala. Oslo, Norwegian University Press, 1985.

23. Kabata, Z.: Crustacea as enemies of fishes. *In* Diseases of Fishes. Book 1. Edited by S.F. Shieszko and H.R. Axelrod. Jersey City, T.F.H. Publ., 1970.

24. Kabata, Z.: Parasitic Copepoda of British Fishes. London, Ray Society, 1979.

25. Kabata, Z.: Copepoda parasitic on fishes: problems and perspectives. Adv. Parasitol., *19*:1–71, 1981.

26. Kabata, Z., and Cousens, B.: Life cycle of *Salmincola californiensis* (Dana, 1852) (Copepoda: Lernaeopodidae). J. Fish. Res. Bd. Can., *30*:881–903, 1973.

27. Kabata, Z., and Cousens, B.: Host-parasite relationships between sockeye salmon, *Oncorhynchus nerka*, and *Salmincola californiensis* (Copepoda: Lernaeopodoidae). J. Fish. Res. Board Can., *34*:191–202, 1977.

28. Laval, P.: Hyperiid amphipods as crustacean parasitoids associated with gelatinous plankton. Oceanogr. Mar. Biol. Ann. Rev., *18*:11–56, 1980.

29. Leung, Y.: An illustrated key to the species of whale lice (Amphipoda, Cyamidae), ectoparasites of Cetacea, with a guide to the literature. Crustaceana, *12*:279–291, 1967.

30. Markham, J.C.: A review of the bopyrid isopods infesting caridean shrimps in the northwestern Atlantic Ocean, with special reference to those collected during the Hourglass cruises in the Gulf of Mexico. Mem. Hourglass Cruises, *7*:1–156, 1985.

31. McCain, J.C.: The Caprellidae (Crustacea: Amphipoda) of the western North Atlantic. Bull. U.S. Nat. Mus., *278*:1–147. 1968.

32. Nicoli, R.M., and Nicoli, J.: Biologie des Pentastomides. Ann. Parasitol. Hum. Comp., *41*:255–277, 1966.

33. Overstreet, R.M.: Metazoan symbionts of crustaceans. *In* The Biology of Crustacea. Vol. 6. Pathobiology. Edited by D.E. Bliss. New York, Academic Press, 1981.

34. Riley, J., Banaja, A.A., and James, J.L.: The phylogenetic relationships of the Pentastomida: the case for their inclusion within the Crustacea. Internat. J. Parasitol., *8*:245–254, 1978.

35. Riley, J., James, J.L., and Banaja, A.A.: The possible role of the frontal and sub-parietal gland systems of the pentastomid *Reighardia sternae* (Diesing, 1864) in the evasion of the host immune responses. Parasitology, *78*:53–66, 1979.

36. Ritchie, L.E., and Hoeg, J.T.: The life history of *Lernaeodiscus porcellanae* (Cirripedia: Rhizocephala) and co-evolution with its porcellanid host. J. Crust. Biol., *1*:334–347, 1981.

37. Salt, G.: The Cellular Defence Reactions of Insects. Cambridge Monographs in Experimental Biology. No. 16. London, Cambridge University Press, 1970.

38. Schacher, J.F., Saab, S., Germanos, R., and Boustany, N.: The aetiology of halzoun in Lebanon: recovery of *Linguatula serrata* nymphs from two patients. Trans. Roy. Soc. Trop. Med. Hyg., *63*:854–858, 1969.

39. Schram, F.R.: Crustacea. New York, Oxford University Press, 1986.

40. Schultz, G.A.: How to Know the Marine Isopod Crustaceans. Dubuque, Wm. C. Brown Co., 1969.

41. Self, J.T.: Pentastomiasis: host responses to larval and nymphal infections. Trans. Am. Microsc. Soc., *91*:2–8, 1972.

42. Self, J.T.: Biological relationships of the Pentastomida; a bibliography on the Pentastomida. Exp. Parasitol., *24*:63–119, 1969.

43. Self, J.T., and Kuntz, R.: Host-parasite relations in some Pentastomida. J. Parasitol., *53*:202–206, 1967.

44. Sprague, V., and Couch, J.: An annotated list of protozoan parasites, hyperparasites, and commensals of decapod Crustacea. J. Parasitol., *18*:526–537, 1971.

45. Vader, W.: Associations between amphipods and molluscs. A review of published records. Sarsia. *48*:13–18, 1972.

46. Williams, C.S.: Seasonal variations in the biochemical composition of *Mytilicola intestinalis*. Parasitology, *59*:775–782, 1969.

47. Yamaguti, S.: Parasitic Copepoda and Branchiura of Fishes. New York, Interscience, 1963.

19

Class Insecta I

Class Insecta I

About 80% of the described species of animals are insects. They live in almost every kind of habitat; this flexibility not only enhances their opportunity to become parasites, but also makes them available for other parasites seeking a home. Some of the habitat choices are amazing. "The larva of the petroleum fly lives in pools of petroleum around oil wells in California; other fly larvae breed by the millions in Great Salt Lake; a few flies have been found breeding in the medical-school brine vats in which human cadavers are preserved; and one beetle in the West—often called the 'short circuit' beetle—bores into lead cables. A few insects live in hot springs where the temperatures go as high as 120° (51°C)."[1]

An insect body is characterized by its division into three parts: head, thorax, and abdomen. Eyes, antennae, and mouthparts are on the head; wings and legs arise from the thorax. Appendages on the adult abdomen may occur at its posterior tip. For insects and disease, see Harwood and James,[4] and Weiser.[15] For insect diseases, including an illustrated key to the genera of all obligate nematodes of insects, see Poinar and Thomas.[11] For an atlas of insect diseases, see Weiser.[15]

Order Phthiraptera

Lice are perhaps most accurately placed in a single group, the order Phthiraptera, although they traditionally have been placed in two orders, Mallophaga for the chewing lice, and Anoplura for the sucking lice. All are wingless and parasitic and have a developmental cycle known as paurometabolous; that is, the young are similar to adults except for size and development of the external genitalia. The absence of wings is thought to be a secondary condition, since lice during development have imaginal discs similar to those that produce wings in winged insects, indicating that they had winged ancestors. The order is divided into several suborders. The suborders Ischnocera and Amblycera are the two great groups of chewing or biting lice; these are parasites of birds or mammals and have mandibles that are used for feeding on feathers or for scraping the skin of their hosts. Some of the bird lice invade the quills of feathers and feed on blood, perhaps indicating the origin of sucking lice. Sucking lice, suborder Anoplura, have mouthparts reduced to delicate stylets which are used to pierce the skin of their hosts; they feed on blood or tissue fluids of mammals. Somewhat intermediate between the chewing and sucking lice is the suborder Rhynchophthirina, a group that includes only the genus *Haematomyzus*, with two species. These remarkable parasites of elephants and wart hogs have a long proboscis but with mandibles at the apex; initially they were thought to be sucking lice but discovery of the nature of their mouthparts later caused them to be placed with the chewing lice. The suborder is now considered to have some characteristics of chewing lice, especially the Ischnocera, and some characteristics of sucking lice.

Chewing lice, except for the Rhynchophthirina, have broad heads, at least as broad as the thorax, while sucking lice have heads narrower than the thorax (compare Figs. 19–3 and 19–4). Another readily distinguishable feature is the presence in chewing lice, and absence from sucking lice, of pigmented, heavily sclerotized mandibles. The anterior portion of the thorax, the prothorax, is freely movable on the posterior part of the thorax, pterothorax, in chewing lice, but there is a fusion of all three segments of the thorax in sucking lice.

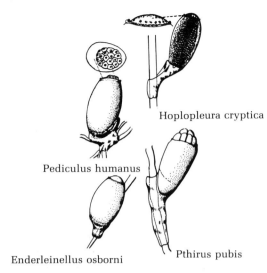

Hoplopleura cryptica

Pediculus humanus

Enderleinellus osborni   Pthirus pubis

Fig. 19–1. Eggs of lice. (From Ferris: The Sucking Lice. Memoirs of the Pacific Coast Entomological Society, courtesy of the California Academy of Sciences, San Francisco.)

All lice are wingless, have dorsoventrally flattened bodies, short, three- to five-jointed antennae, and eyes reduced or absent. The indistinctly segmented thorax bears one pair of spiracles. The short legs have tarsi whose claws are used for grasping feathers or hair. The abdomen, always without cerci, generally has six pairs of spiracles. Unlike most arthropod parasites, all lice spend their entire lives on the bodies of their hosts, and infestation of one host from another is by direct contact. For this reason, host-specificity is well marked (see Chap. 25). In some cases lice seem to have evolved less than their hosts so that the origin of the host may be inferred by its lice. For example, the affinities of flamingoes are not clear; it has been postulated that they are related to storks and herons (Ciconiiformes). They are parasitized, however, by three genera of lice that occur only on swans, geese, and ducks (Anseriformes); their lice clearly indicate placement in the latter Order rather than the former. Eggs of lice, known as "nits," are attached by the female to the feathers or hairs of the host (Fig. 19–1).

Lice undergo a simple life cycle with gradual (paurometabolous) metamorphosis. The first-stage nymph, which develops in the egg, is structurally similar to the adult except for its smaller size, absence of color, and underdeveloped sex organs. Three nymphal stages, each separated by a molt, precede the adult stage. In

Columbicola, each nymphal stage lasts 6 or 7 days, and with each molt the body becomes larger and darker in color. After the third and final molt, the nymph becomes either a male or female, sexually mature adult. In many species, females, which almost always are larger than males, far outnumber the latter. Males of some species of chewing lice are rare or have never been found. Although parthenogenesis is known to occur in at least one species of mammalian louse, the rarity or apparent absence of males in many species may be attributed to their short longevity.

Many lice, particularly those which feed on blood at times, possess within their bodies symbiotic intracellular bacteria that presumably aid in the digestion of food. The bacteria are transmitted vertically from mother to young through the egg. Similar bacteria are found in ticks, mites, bedbugs, and in some bloodsucking Diptera. Since parasites have a restricted diet, it is thought that symbiotic organisms are necessary for providing essential substances in the diet which the parasites themselves cannot produce. It has been suggested that the extreme host-specificity of some species of sucking lice may be due to the detrimental effect of blood from hosts other than their normal ones.

CHEWING LICE

About 2600 species of chewing lice are known. Most of these are parasites of birds[12] but some parasitize dogs, cats, horses, cattle, goats, sheep, and other mammals, including marsupials. Chewing lice have not been proved to be effective carriers of disease-producing microorganisms although some pathogenic agents, such as the virus of eastern equine encephalitis, have been found in them. Feather lice range in size from the minute males of *Goniocotes*, which scarcely reach 1 mm in length, to the large, hawk-infesting *Laemobothrion*, which may attain a length of 10 mm.

The mouthparts, adapted for chewing epithelial materials, may have severe effects on the hosts. Lice eat the protecting sheaths of growing feathers, feather fiber, down, skin scurf, scabs, blood, and probably also mucus and sebaceous matter. The lice on birds apparently can digest keratin. Some species on mammals ingest hair; chewing lice of mammals, however, probably prefer a diet resembling that of many of the species parasitic on birds (for example, epidermal scales, skin scurf, wool, wax).

Although they possess chewing mouthparts, some species of chewing lice regularly feed on the freshly drawn blood of their hosts. For example, *Menacanthus stramineus,* the body louse of poultry, actively feeds on its host's blood. Some species of chewing lice parasitic on birds possess piercing mouthparts, and feed, as do sucking lice, on the blood and tissue fluids of their hosts; one genus, *Piagetiella,* lives in the throat pouches of pelicans and cormorants and attaches to the naked skin, producing visible lesions.

Grooming behavior of animals probably is important in reducing the number of lice present. Eichler counted 20,000 chewing lice on the skin of a dog that, owing to a defect, could not scratch itself. Dust bathing by birds is thought to be effective in reducing the numbers of ectoparasites present, both lice and mites. Damage to feathers caused by lice probably is light, but it is difficult to separate damage caused by lice from that caused by feather mites. Although the chewing lice are not effective carriers of human diseases, one species, *Trichodectes canis,* acts as an intermediate host for a dog tapeworm, *Dipylidium caninum,* and three species of bird filariae have been shown to be able to develop in chewing lice.

Suborder Amblycera

Species of lice of this suborder are parasitic on birds or mammals and include the feather, shaft, and body lice of poultry and a number of pests on guinea pigs and other rodents; about 800 species are known. They may be distinguished by the antennae, which almost always are composed of four joints, the third being stalked (Fig. 19–2). The antennae always lie in a ventrolateral groove on each side, and may or may not project beyond the sides of the head. The maxillary palps, sometimes absent, also may project beyond the sides of the head, and may be mistaken for antennae.

Menopon gallinae (Fig. 19–2) is often called the shaft louse of fowl. The female is about 2 mm long and lays its eggs in clusters on feathers.

Menacanthus stramineus (= *Eomenocanthus*) (Fig. 19–3) is the chicken body louse, which is also found on turkeys and other fowl. The female is 3.3 mm long and prefers the skin to feathers. It may seriously injure adult fowls. A chicken may be infested with more than 6000 lice of several species.

Heterodoxus includes species that occur on

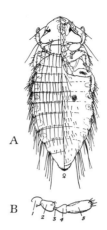

Fig. 19–2. *A, Menopon gallinae,* female. *B,* The antenna of this species, unlike that of most Amblycera, has five segments. (From Lapage: Veterinary Parasitology, courtesy of Oliver and Boyd.)

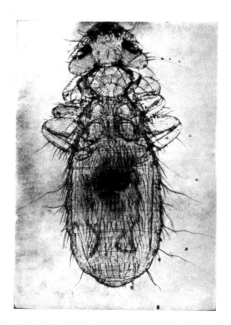

Fig. 19–3. *Menacanthus stramineus,* the "chicken body louse" of fowl. (From Lapage: Veterinary Parasitology, courtesy of Oliver and Boyd.)

Fig. 19–4. A feather louse belonging to the suborder Ischnocera. (From Rothschild and Clay: Fleas, Flukes, and Cuckoos, courtesy of Wm. Collins Sons.)

dogs in warm countries and on kangaroos and wallabies.

Suborder Ischnocera

Lice of this suborder are commonly found on cattle, equids, goats, sheep, dogs, cats, and on fowl and other birds; more than 1700 species are known. The lice can be distinguished by their conspicuous filiform antennae composed of three to five joints (Fig. 19–4). There are no maxillary palpi. *Bovicola* (= *Damalinia*) is found on cattle, sheep, goats, deer, and other two-toed ruminants, and horses. *B. bovis* is the biting louse of cattle. It is the most widely distributed louse on cattle in Britain and is also common in North America. The lice in winter are found at the base of the tail, on the shoulders, and along the back, unless there is a heavy infestation, in which case they may be found all over the body. Irritation caused by these active pests may be severe, and the infested cattle often try to rid

themselves of the lice by biting the skin and rubbing themselves against tree trunks, fence posts, and rocks.

Trichodectes (Fig. 19–5) includes species found on dogs, martins, weasels, badgers, skunks, and other small mammals. Females are almost 2 mm long, and may serve as vectors of the larval stage of a dog cestode, *Dipylidium caninum*.

Cuclotogaster (= *Lipeurus*) *heterographus* (Fig. 19–6) is found on the skin and feathers of the head and neck of fowl, partridge, and other birds. It is often called the fowl head louse and may be seriously injurious. The female is about 2.45 mm long, and its eggs are laid singly on the feathers.

Suborder Anoplura (= Siphunculata)

About 500 species of sucking lice have been described; all are parasitic on mammals. In addition to their characteristic piercing mouthparts, sucking lice can be distinguished by the small, fused thorax and by their antennae, which usually are composed of 5 joints and are always visible. Eyes are small or absent, and the third pair of legs, often broad and flattened, usually is the largest. Each tarsus has a single-segmented claw. The pair of thoracic spiracles open on the dorsal side of the mesothorax. Six pairs of abdominal spiracles are arranged as those of chewing lice. Only 7 of the 9 abdominal segments are visible externally. The mouthparts, adapted for piercing and sucking, retract into the head, which obviates the need for a proboscis.

The latest classification of the sucking lice[7] places them in 15 families and 42 genera. They occur on all groups of placental mammals except for bats, edentates, pholidota, whales, and sirenians.[3] The family Linognathidae contains *Linognathus vituli*, the bluish black, long-nosed cattle louse. Its first pair of legs is much smaller than the others. The family Neolinognathidae includes only two species; these are in the genus *Neolinognathus* and are parasitic on elephant-shrews in East and South Africa.

The family Haematopinidae includes *Haematopinus suis*, the largest of all lice to be found on farm stock. Females are 4 to 6 mm long, males 3.4 to 4.75 mm. It is the only species of louse that infests the domestic pig (Fig. 19–7).

Human Lice. The family Pediculidae includes the human head louse and the body louse, both

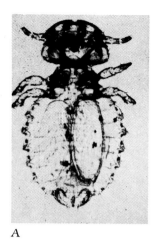
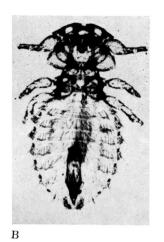

A   B

Fig. 19–5. *Trichodectes canis. A*, Female. *B*, Male. (From Georgi: Parasitology for Veterinarians, courtesy of W.B. Saunders.)

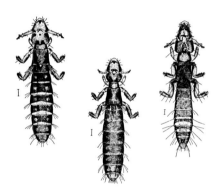

Fig. 19–6. *Cuclotogaster (Lipeurus)*, various species. (From Kellogg, courtesy of Stanford University Press.)

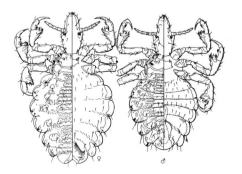

Fig. 19–7. *Haematopinus suis.* (From Ferris: The Sucking Lice. Memoirs of the Pacific Coast Entomological Society, courtesy of the California Academy of Sciences, San Francisco.)

in the genus *Pediculus;* the family Pthiridae includes the human pubic or crab louse in the genus *Pthirus* (Fig. 19–8). *Pediculus humanus* occurs in two forms: the body louse, *P. humanus humanus* (=*P. humanus corporis*), and the head louse, *P. humanus capitis.* The two look much alike although typical head and body lice are easily distinguished; however, they overlap in appearance and movements.[2] Head lice average about 2.4 mm in length, somewhat smaller than body lice. Body lice seldom if ever are found on the head, but head lice may be found on the body. Head lice are adapted to clinging to hairs, but body lice find refuge in clothing. Head lice and body lice can interbreed and produce fertile offspring that possess characteristics intermediate between those of the two parents.

Females probably usually lay about 150 eggs during their lives, which last about a month under optimal conditions. The oval eggs, or nits, are laid singly, and measure about 0.8 by 0.3 mm. At 30°C the eggs hatch in 8 or 9 days; the young nymph is about 1 mm long. Adult lice can imbibe as much as 1 mg of blood at a time, but probably usually take smaller quantities at frequent intervals. At 30°C they survive only about 3 days of starvation. Lice can move at a rate of 9 inches (23 cm) per minute. They prefer a temperature of 29° to 30°C, and, when possible, avoid any change in humidity. Their immediate past experience conditions their response to environmental changes; hence different individuals often exhibit different responses. Movement toward dark areas is chiefly a response to directed light received by the horizontally placed eyes.

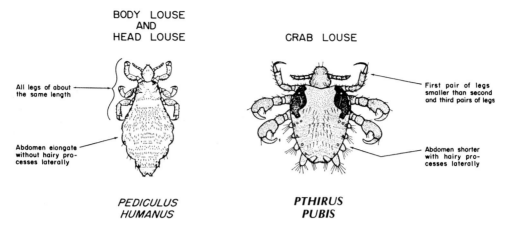

Fig. 19–8. Lice commonly found on man, *Pediculus humanus* (left) and *Pthirus pubis* (right).

Pthirus pubis (Fig. 19–8) possesses a wide thorax that constitutes most of the body, and the coxae are far apart at the margins of the thorax. It frequents the pubic hairs and perianal regions of man, but may wander to other parts of the body, including the head. It is smaller and much broader than the other human lice, and its legs have the appearance of being attached to the edge of the flattened body. The forelegs are slender, with long, fine claws; the middle and hind legs are thick, with thick claws. Adults do not survive for longer than a day when removed from the host, and their total span of life is about 1 month. Although transmisson from person to person occurs by close contact, it is a grave injustice to assume that sexual contact is the only mode of transfer.

Effects of Sucking Lice on their Hosts. Sucking lice are important parasites of man and of domestic and wild animals. Populations may become large on animals that are ill, maimed, or not properly cared for. Such animals may experience a large loss of blood as well as the constant irritation caused by the feeding of lice. During the biting process, the saliva injected into the wound prevents the host blood from coagulating as it is sucked through the parasite's slim mouthparts. The reaction of the host cells to the louse saliva causes the symptoms of irritation. An inflammatory reaction of the skin may be accompanied by a thickening of the epidermis and an increase in tissue lymphocytes, monocytes, mast cells, and fibroblasts.

Although lice are important vectors of human disease, they are not known to be important vectors of animal diseases. *Polyplax spinulosa*, a

parasite of commensal rats, can transmit *Rickettsia typhi*, which causes murine typhus, from rat to rat, and *Trypanosoma lewisi* and the bacteria that cause tularemia (*Francisella tularensis*). *Haemodipsus* is known to transmit *Francisella tularensis* among wild rabbits. The sucking lice of domestic animals, however, are not known to transmit disease agents to any great extent. Lice irritate their hosts and, because they often increase rapidly in numbers, frequently are the cause of scratching, restlessness, loss of sleep, and interruption of feeding. Lice of livestock may be controlled by spraying the animals with any of a number of insecticides including growth regulators (insect juvenile hormones).

Pediculus humanus is the normal vector of epidemic typhus caused by *Rickettsia prowazekii*, trench fever caused by *Rochalimaea quintana*, and louse-borne relapsing fever caused by *Borrelia recurrentis*. *Rickettsia prowazekii* ingested by the louse with its blood meal multiplies in its gut cells, which eventually burst and liberate large numbers of rickettsiae into the lumen of the gut, which then pass out in the feces. Man acquires the disease by rubbing infected louse feces into an abrasion of the skin. The infected louse dies in about 10 days after acquiring the rickettsiae because of its ruptured gut cells. Hans Zinsser's book, *Rats, Lice and History*,[16] gives a vivid picture of the manner in which lice and typhus fever have influenced the history of mankind. Recently a number of cases of epidemic typhus have occurred in the United States in persons who have not previously experienced typhus (and therefore are not relapsing or recrudescing) and who are not infested with human lice.

The rickettsiae have now been isolated from flying squirrels (*Glaucomys volans*) in the eastern United States; it is hypothesized that the rickettsiae were transferred from humans to flying squirrels during Colonial days when louse populations, and typhus, were frequent, and flying squirrels were not uncommon invaders of human habitations. The bacteria seem now to be maintained in flying squirrel populations by lice of the flying squirrel *(Neohaematopinus sciuropteri)* and transmitted to humans by flying squirrel fleas, *Orchopeas howardii*.

Trench fever became widely prevalent during World War I and has since been detected only sporadically. The rickettsiae that cause this disease, unlike other rickettsiae, develop extracellularly, both in the louse and in man. They multiply in the lumen of the gut of lice and are shed in the feces, as are the rickettsiae of epidemic typhus.

The spirochetes of relapsing fever penetrate the gut of the louse and multiply in the fluid of the body cavity. They remain in this fluid until the louse is crushed, at which time the spirochetes escape and may enter a human host through an abrasion of the skin. Obviously, the habit of "popping" lice with fingernails and the disgusting practice of biting lice to kill them are hazardous if louse-borne diseases are present.

Head lice, frequently a serious problem among children, are much rarer among adults. Body lice, on the other hand, are largely confined to adults and children who do not change clothes frequently. Conditions of crowding, especially during sleeping hours, are particularly conducive to the spread of lice. For these reasons, destitute people, refugees, vagrants, prisoners, and soldiers may provide fertile fields for louse populations. A chronicler contemporary of the well-known church dignitary St. Thomas à Becket wrote that on the morning after the churchman was murdered, his vestments were removed, and the haircloth underwear was so infested with lice that they "boiled over" as the cloth was stripped from the cold body.

The primary defense against human lice involves personal cleanliness of body and clothing. It is well to remember that even in civilized communities, there are always some chronically lousy individuals whose feeble efforts at personal cleanliness are ineffectual or who have ceased to worry about their condition and passively offer their parasites to every passerby.

Louse-borne diseases of humans are associated only with body lice, not with head lice or crab lice. They tend to occur only in relatively cold climates and during the colder part of the year. Since the lice are found almost exclusively in clothes, they are easily killed by treating infested clothing with heat, or by simply not allowing the lice to feed for 3 days or so. When necessary, body lice, as also head and crab lice, may be controlled by treating individuals, and their clothes, with appropriate insecticides. DDT has been extensively used for this purpose, but resistance of body lice to this insecticide is now rather common. Lindane (gamma isomer of benzene hexachloride) was next used for this purpose but body lice also developed resistance to it. Subsequently a number of insecticides including malathion have been used for treatment of pediculosis. Currently, synthetic pyrethrum insecticides are recommended because they are effective and safe.

Order Hemiptera

Although the appellation "bug" is applied to many things, from viruses to automobiles, entomologists use the term to refer to members of the order Hemiptera. Most plant bugs and their relatives are phytophagous (plant feeders) and have piercing-sucking mouthparts. Many species show a strong tendency toward entomophagy (insect-eating), and in a number of instances, plant feeding has been abandoned entirely and the predatory role is obligatory; a few groups of bugs have passed from predation to parasitism. Bugs typically have two pairs of wings, the anterior pair most often being of a harder consistency than the posterior pair. The slender, segmented beak constitutes the most easily recognizable feature of bugs; palpi are atrophied, and the labium is a dorsally grooved sheath surrounding two pairs of bristle-like stylets (modified mandibles and maxillae). Metamorphosis is gradual.

Probably no other group of insects so directly affects the welfare of humans than Hemiptera, by virtue of the vast amount of injury to plants brought about, directly and indirectly, by thousands of species of bugs. Chinch bugs, leafhoppers, plant lice, cotton stainers, whiteflies, tea blight bugs, scale insects, and mealybugs are among the most destructive.

Suborder Heteroptera

The 3000 or so described species of these bugs exhibit great variation in form. The family Re-

  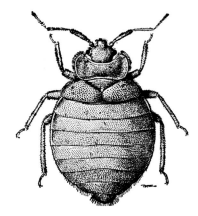

Fig. 19–9. Reduviid bugs. *A, Melanolestes picipes* (Herrich-Schaeffer), ×2. *B,* A blood-sucking cone-nose, *Triatoma sanguisuga* (LeConte), ×2. (Courtesy of Froeschner and the American Midland Naturalist.)

Fig. 19–10. Male bedbug. (From McKenny-Hughes, A.W., and Johnson, C.G.: The Bed-Bug, courtesy of British Museum Natural History Economic Series No. 5.)

duviidae, subfamily Triatominae,[9] contains bugs of medium to large size, and usually of a black or brownish color, sometimes with bright red or yellow markings (Fig. 19–9). The head characteristically is long, narrow, and with a neck-like portion immediately behind the eyes. The rostrum (beak) is three-jointed and is bent sharply back under the head when the bug is at rest. The filiform antennae are inserted on the sides of the head between the eyes and the tip of the snout. The abdomen often is widened at its middle, exposing the margins of the segments lateral to the folded wings. Wings of Heteroptera generally are placed, when at rest, so as to reveal a conspicuous triangular area in the posterior portion of the thorax. Most Reduviids, frequently called "assassin bugs," are predacious on other insects, but triatomids ("conenosed bugs") are parasitic on vertebrates, including man. *Rhodnius prolixus* and species of *Triatoma* and *Panstrongylus* are the natural vectors of *Trypanosoma cruzi,* the causative agent of Chagas' disease, a highly prevalent and frequently fatal form of human trypanosomiasis (see Chap. 3). Many species of assassin bugs inflict a painful bite if carelessly handled, and severe allergic symptoms occasionally occur from the bites of the parasitic forms. The bugs are regularly found in nests or burrows of rodents or other host animals.

Bedbugs, family Cimicidae, are parasitic on birds and mammals. Characteristic morphologic features (Fig. 19–10) are an oval, flattened body about 4 mm long, without wings and covered with many hair-like spines. Compound eyes are conspicuous, but ocelli are absent; the rostrum lies in a ventral groove, and tarsi are three-jointed. The thorax consists of three segments: a large anterior prothorax extending forward on either side of the head; a mesothorax whose visible portion is a small triangle; and a metathorax hidden from view dorsally by two small pads, the vestigial forewings.

Members of the genus *Cimex* are associated with birds, bats, man, and other mammals and have a worldwide distribution. In North and Central America, *Haematosiphon inodorus* is a parasite of poultry.

The two most common bedbugs attacking man are *Cimex lectularius* of temperate climates and *C. hemipterus* of tropical countries. Both species, colored dark mahogany or bright chestnut, hide in cracks, crevices or under rugs by day, and emerge at night to feed on human blood. The bugs are able to endure long fasts; some have been kept alive without food for more than a year. *C. lectularius* is also found in nature associated with chickens, rabbits, and bats.

Before the female bedbug can lay fertile eggs, she must mate and feed. Female bedbugs do not have a seminal receptacle and so are unable to store sperm for lengthy periods. Males, therefore, are long-lived and mate repeatedly. The penis of the male is thrust through the body wall of the female, and sometimes of other males, to effect insemination. The pearly white, curved, operculated eggs normally are laid on rough surfaces such as those found in crevices and behind wallpaper. A quick-drying cement fastens each egg (about 1 mm long) securely to the surface of the material on which it is deposited. When they hatch, in 6 to 10 days, the nymphs, about the size of pinheads and similar in appearance to the adult, immediately seek shelter. Five molts take place before the adult stage is reached. For details on the ecology of bedbugs, see Johnson.[6]

Bedbugs and Diseases. Bedbugs are not known to transmit human diseases. In spite of persistent suspicion and conflicting accounts, there is no clear evidence that bedbugs are more than harmless pests of mankind. McKenny-Hughes and Johnson, however, have called attention to the fact that "in infested areas it is often possible to pick out children from buggy homes by their pasty faces, listless appearance, and general lack of energy. It can be argued that the house in which bugs are tolerated will also be the home of malnutrition, dirt, and other causes of physical infirmity. Such causes cannot be held solely responsible, and sleepless nights with constant irritation due to the injection of the minute doses of bedbug saliva into the blood are likely to contribute largely to the ill health of children and even of certain adults. Some fortunate people are not affected by the bites of bedbugs; others gain immunity after repeated biting; whilst others, less fortunate, are always susceptible."[10]

See Usinger[14] for a monograph on the Cimicidae.

The family Polyctenidae is so named because these interesting but rare bat parasites frequently have several combs. Although they occur in tropical and subtropical regions of the Old and New Worlds, it was estimated in 1946 that there were fewer than 100 specimens in the entomological museums of the world.[13]

Order Hymenoptera
(Wasps and Ants)

This order consists chiefly of ants, bees, wasps, sawflies, and ichneumon flies. Parasitism is prominently displayed by members of the order, and insects are favored hosts. Typical parasitic species are entomophagous and are parasites only during their larval stages. In such a situation, they are called "parasitoids." The manner and place of oviposition by the female are varied and range from the common placement of the egg within the body of the host to its deposition on foliage or in plant tissues far removed from the animal host. On hatching,

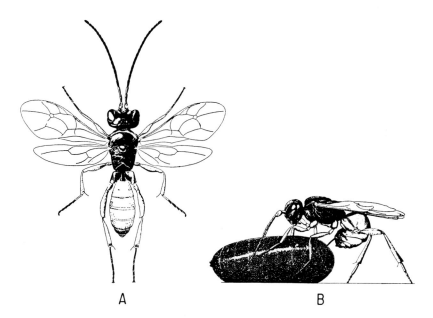

A   B

Fig. 19–11. A hyperparasitic ichneumon, *Phygadeuon subfuscus* Cresson (Gelinae). *A,* Adult male. *B,* Female ovipositing in puparium of host. The host of this ichneumon is a tachinid fly, *Aplomyiopis epilachnae* Aldrich, which is parasitic on the Mexican bean beetle, *Epilachna varivestis* Mulsant. (Courtesy of the U.S. Department of Agriculture.)

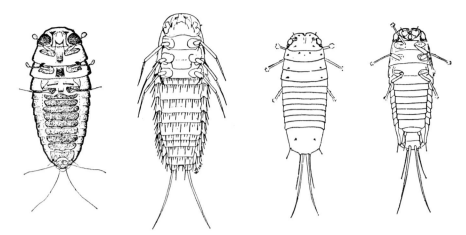

Fig. 19–12. First-instar larvae of the Strepsiptera. (From Clausen: Entomophagous Insects, courtesy of McGraw-Hill.)

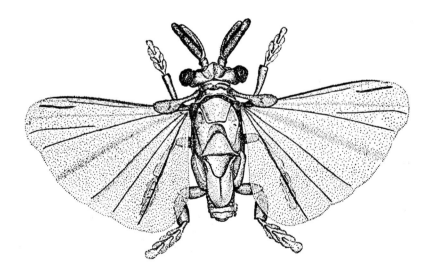

Fig. 19–13. *Stylops shannoni*, male, enlarged. (From Imms: General Textbook of Entomology, courtesy of E.P. Dutton.)

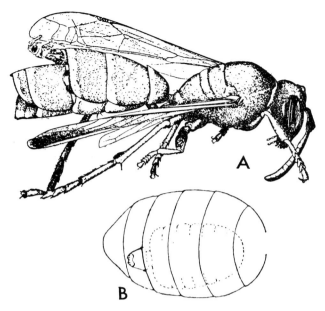

Fig. 19–14. *A, Polistes gallicus,* with a male *Xenos vesparum* (a stylopid) hatching from the puparium. *B,* Abdomen of *Andrena vaga,* with adult female of *Stylops* sp. (From Baer: Ecology of Animal Parasites. Courtesy of the University of Illinois Press.)

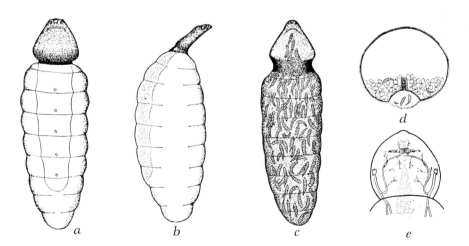

Fig. 19–15. Adult females of the Strepsiptera, with details, *A, Stylops melittae* Kirby, showing the genital openings on the cephalothorax and the brood chamber. *B,* The same, lateral view. *C,* A gravid female of *Halictophagus curtisii* Dale, showing fully developed triungulinids distributed throughout the body. *D,* A diagrammatic cross section through the fourth abdominal segment of *S. melittae,* showing the outer portion of the genital canal and several triungulinids in the brood chamber. *E,* Cephalothorax of *Xenos vesparum* Rossi, dorsal view, showing the single pair of spiracles, the anterior commissure and the longitudinal tracheal trunks, which divide in the first thoracic segment. (Redrawn after Nassanov, 1982, 1983. In Clausen: Entomophagous Insects, courtesy of McGraw-Hill.)

the parasitoid larvae feed on the host, which usually is itself a larva or an egg, and it is thereby destroyed (Fig. 19–11). That a given larva of some species may be either ecto- or endoparasitic, according to the host species on which it feeds, indicates that the host is partly responsible for the behavior of the parasitoid.

Larval fruit flies, *Drosophila melanogaster*, may be parasitized by larval braconid wasps, *Asobara tabida*. The fly responds by a cellular defense reaction and encapsulates the parasites with a wall of flattened hemocytes.

When he had compared parasitoids with parasitic castrators (see Chap. 23), Kuris[8] made the following intriguing statement: "An analysis of life history features of insect parasitoids and crustacean parasitic castrators suggests that these are similar phenomena, distinct from parasitism and predation. A parasitoid consumes only one host during its lifetime; parasitic castration causes the reproductive death of only one host. Since population densities of many insect species are regulated by parasitoids, parasitic castration may also play an important role in host population regulation."

See the discussion of biocontrol in Chapter 23.

Order Strepsiptera
(Twisted-Wing Parasites or Stylopids)

Stylopids are small, endoparasitic insects that exhibit a striking sexual dimorphism in the adult stage. The males have branched antennae, large, protruding eyes, club-like structures (halteres) instead of forewings, and large hindwings. Females are wingless and normally remain enclosed in a puparium that protrudes from the body of the insect host. Only about 300 species are known, and these have commonly been placed in the order Coleoptera. All species, so far as is known, complete their larval development within the body of the host in a manner similar to that of the beetle *Rhipidius* in cockroaches. The first-instar larvae of stylopids bear a striking resemblance to those of Rhipiphoridae (Coleoptera). The minute adult males are not encountered by collectors nearly as often as are the females; consequently, many species have been described on the basis of only one sex.

Although a few instances of Orthoptera and Hemiptera as hosts to the Strepsiptera are known, the preferred hosts are members of the Homoptera (chiefly Auchenorrhyncha) and of the Hymenoptera (chiefly Vespoidea, Sphecoidea, and Apoidea). The following account of the biology and habits of these parasites is based on *Xenos vesparum*, a parasite of wasps and bees. See Imms.[5]

The male lives only a few hours after emerging from the host, but the female remains permanently endoparasitic, with its cephalothorax protruding through the body wall of the wasp or bee. Copulation takes place when the male alights on the host and inserts the aedeagus (penis plus lateral structures) into the aperture of the brood canal of the female. Larvae hatch within the body of the female and issue in large numbers, sometimes several thousand, through the genital canals into the brood pouch. Larvae eventually emerge through the brood canal, and remain as active creatures on the body of the host until opportunity is afforded for escape.

The first-instar larvae are known as triungulids (Fig. 19–12). Larvae probably leave the first or "maternal" host when it is closely associated with others on flowers or in the nest. By simple attachment to adult wasps or bees, the larval parasites are transported to larval hosts, within which they speedily burrow through the body wall and take up lodgings in body spaces between the organs, pushing these organs out of position. Absorption of nutrients from the host blood causes the parasite to grow, and after the seventh instar, the parasitic larva works its way outward and protrudes from the body wall of the host, which is by now in the pupal stage of development. Male parasites now undergo pupation, and soon winged insects are liberated (Figs. 19–13 and 19–14). Females (Fig. 19–15) become white, grub-like creatures remaining within the host.

Effects of stylopids on their hosts vary, depending on the numbers of parasites within a host, the length of time the parasites spend within the host, the season, and other considerations. Generally, however, changes in the growth and development of the host occur, including pronounced alterations of the secondary sexual characteristics.

REFERENCES

1. Borror, D.J., DeLong, D.M., and Triplehorn, C.A.: An Introduction to the Study of Insects. 4th Ed. New York, Holt, Rinehart and Winston, 1976.
2. Buxton, P.A.: The Louse. An Account of the Lice

Which Infest Man, Their Medical Importance and Control. London, Edward Arnold & Co., 1939.

3. Ferris, G.F.: The Sucking Lice. Pacific Coast Entomological Society Memoirs, Vol. 1, San Francisco, Calif., 1951.

4. Harwood, R.F, and James, M.T.: Entomology in Human and Animal Health. 7th Ed. New York, Macmillan Publ. Co., 1979.

5. Imms, A.D.: Imms' General Textbook of Entomology. 10th Ed. London, Chapman and Hall, 1977.

6. Johnson, C.G.: The ecology of the bedbug *Cimex lectularius* L., in Britain. J. Hyg., 41:345–461, 1941.

7. Kim, K.C., and Ludwig, H.W.: The family classification of the Anoplura. Systematic Entomol., 3:249–284, 1978.

8. Kuris, A.M.: Trophic interactions: similarity of parasitic castrators to parasitoids. Q. Rev. Biol., 49:129–148, 1974.

9. Lent, H., and Wygodzinsky, P.: Revision of the Triatominae (Hemiptera, Reduviidae), and their significance as vectors of Chagas' Disease. Bull. Amer. Mus. Nat. Hist., *163*:123–520, 1979.

10. McKenny Hughes, A.W., and Johnson, C.G.: The Bedbug. London, British Museum of Natural History Economic Series No. 5, 1954.

11. Poinar, G.O., Jr., and Thomas, G.M.: Laboratory Guide to Insect Pathogens and Parasites. New York, Plenum Press, 1984.

12. Rothschild, M., and Clay, T.: Fleas, Flukes & Cuckoos. A Study of Bird Parasites. New York, Macmillan Co., 1957.

13. Ryckman, R.E., and Sjogren, R.D.: A catalogue of the Polyctenidae. Bull. Soc. Vector Ecol., 5:1–22, 1980.

14. Usinger, R.L.: Monograph of Cimicidae (Hemiptera-Heteroptera). Thomas Say Foundation Publication No. 7. College Park, MD, Entomological Society of America, 1966.

15. Weiser, J.: An Atlas of Insect Diseases. 2nd Ed. The Hague, Dr. W. Junk, B.V., Publishers, 1977.

16. Zinsser, H.: Rats, Lice and History. Boston, Little, Brown, 1935.

20

Class Insecta II

Class Insecta II

Order Siphonaptera (Fleas)

Over 2300 species of fleas have been described, and undoubtedly many more have not yet been discovered. Only the adults are parasitic, and they, like lice, are restricted to birds and mammals on whose blood they feed. Fleas have bodies that are laterally compressed, and the first segment on each leg, the coxa, is large and provides the power for jumping for which fleas are justly famous. Most of us may recall how difficult it is to catch a flea. This difficulty is the result of the small size of the flea, its laterally compressed, smooth body, the backwardly directed spines, and the strong, active legs tipped with claws. These characteristics enable fleas to jump from one host to another and move easily among hairs and feathers. The compact structure of a flea is adapted to a forward movement only, and to change even the field of vision requires an alteration in the position of the entire insect.

Probably, the most convenient place to look for a flea is on the body of a domestic dog or cat. Both these hosts may be attacked by the dog flea, *Ctenocephalides canis*; the cat flea, *C. felis*; the "human" flea, *Pulex irritans*; the stick-tight flea, *Echidnophaga gallinacea*; or, occasionally, by other species of fleas. The "human" flea, *P. irritans*, actually has little host specificity and attacks a variety of animals including man.

Structure. Conspicuous features of the external anatomy of fleas are shown in Figures 20–1 to 20–5. The antennae of males nearly always are longer than those of females; during copulation, the male takes up a position beneath the female and holds her firmly with his antennae from below. The male body has an upward tilt posteriorly, but the female body is evenly rounded terminally.

Fleas are truly encased in a suit of armor; each segment of the thorax (prothorax, mesothorax, metathorax) may be regarded as a membranous ring of adjoining plates. The notum of the prothorax often is armed with a row (comb) of heavy, pigmented spines (one row on each side), the pronotal ctenidium or pronotal comb (Fig. 20–2).

The abdomen consists of ten segments, and each segment, like those of the thorax, has a dorsal sclerite, the tergum, and a ventral sclerite, the sternum; the first segment, however, lacks a sternum. On the abdomen these plates overlap, permitting considerable flexibility of the abdomen; when a flea has engorged with blood, the plates appear as islands separated by broad, bare bands of skin. The shape and bristle pattern of the seventh abdominal segment of female fleas has considerable taxonomic importance.

Tergum nine of the male is modified to form a clasping apparatus used during copulation with a female. The clasper is of great value for identification of males. The complicated genital structures of the male flea are illustrated in Figure 20–4. The ninth segment of both males and females has on its tergum a dorsal sensory plate, the pygidium (or sensilium), covered with pits, bristles, and hairs (Fig. 20–5). The whole structure suggests a tiny pincushion; it possibly functions in the detection of air currents. Such a function would assist the flea in finding a host that may be moving about.

The spermatheca (Fig. 20–6) is taxonomically the most important genital structure of the female flea. It consists of a wide head or reservoir, and a terminal, long, sausage-shaped tail or appendix.

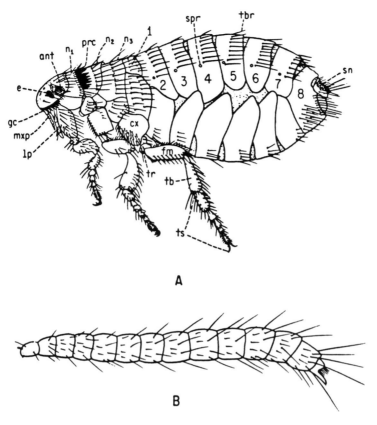

A

B

Fig. 20–1. *A*, An adult cat flea, *Ctenocephalides felis* (Bouché); *B*, Larva. (Borror, D.J., and Delong, D.M.: An Introduction to the Study of Insects. 3rd Ed. New York, Holt, Rinehart and Winston, 1971.)

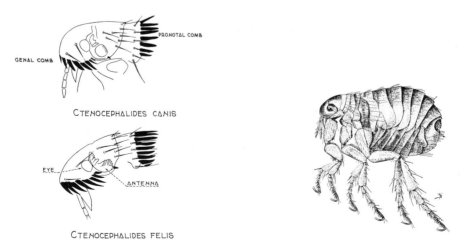

Fig. 20–2. Head of *Ctenocephalides canis*, the dog flea, and of *C. felis*, the cat flea. (From Capriles, courtesy of the Puerto Rico J. Public Health Trop. Med.)

Fig. 20–3. *Pulex irritans*, the human flea, adult female, greatly enlarged. (Courtesy of Bishop Farmers Bulletin.)

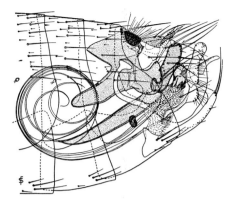

Fig. 20–4. *Ceratophyllus gallinae*, male, showing clasper, coiled penis, sensilium and other features of the posterior end of the abdomen. (From Lapage: Veterinary Parasitology, courtesy of Oliver and Boyd.)

The alimentary tract is the internal organ system of particular importance in disease transmission. The mouth leads to a thick-walled pharynx equipped with pumping muscles, thence to a narrow gullet or esophagus, which enters a pear-shaped proventriculus, which is provided internally with a series of spines that project backward in front of the entrance to the stomach. These spines presumably help to crush the blood cells of the host. Between the proventriculus and the stomach is a valve that

prevents the food in the stomach from being regurgitated during the process of digestion. So effective is the valve that pressure applied from behind sufficient to burst the stomach wall does not force food back into the esophagus. At the posterior end of the large stomach are situated four tubular glands (malpighian tubules) that function as organs of excretion. The short intestine is equipped distally with six small, oval, rectal glands. A salivary gland lies on each side of the stomach, the pair being connected by a common duct that leads to the pharynx; muscular attachments at the expanded end of the salivary duct constitute a salivary pump.

During the process of biting and feeding, the piercing mouthparts enter the host skin, and the flea thrusts its head downward, elevating the abdomen and hindlegs. After feeding, the mouthparts are withdrawn with a sudden jerk. When a flea bites, the salivary pump pours out a stream of saliva that reaches the host blood vessels through the canal on the inner surfaces of the laciniae. At the same time, the pharyngeal pump works to draw up the host blood, mixed with saliva, and forces it into the esophagus and stomach, where it is digested.

Life History and Habits. During their life cycles, fleas pass through a complete metamorphosis from egg to larva to pupa to adult (Fig. 20–7). Eggs are large, smooth, oval, and trans-

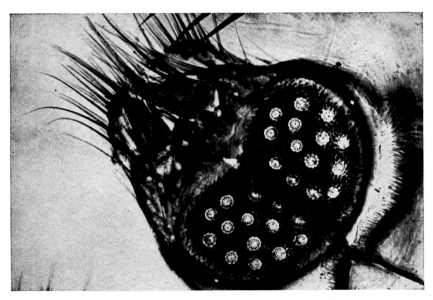

Fig. 20–5. The pygidium (sensilium) of a flea. ×435. (From Rothschild and Clay: Fleas, Flukes, and Cuckoos, courtesy of Wm. Collins Sons.)

A

B

Fig. 20–6. Spermatheca (receptaculum seminis) of fleas. *A, Ceratophyllus arei* (from carrion crow). *B, C. sciurorum* (from red squirrel.) (From Rothschild and Clay: Fleas, Flukes, and Cuckoos, courtesy of Wm. Collins Sons.)

lucent. If a flea-infested house cat is permitted to sleep on a black cloth during the night, the flea eggs can be easily seen scattered over the cloth the next morning. Mingled with the eggs are bits of dried blood and hairs. The human flea, *Pulex irritans* (Fig. 20–3), may lay well over 400 eggs during her lifetime. In 2 to 10 days, the eggs hatch into eyeless, legless, active larvae. The strongly sclerotized head bears 2 short antennae, and the 13 other body segments (3 in the thorax, 10 in the abdomen) bear numerous bristles.

Anal struts consisting of a pair of blunt, hooked processes distinguish flea larvae from those of dipterous insects. Under favorable conditions, the larvae may reach their third stage in about 2 weeks, but development may be delayed for 6 months or more. Larvae feed on organic debris in the host's nest, in crevices in floors, or under rugs. Blood derived from the excreta of adult fleas sometimes is a necessary part of the larval diet. Larvae of bird fleas thrive on broken-down sheaths of feathers and on epidermal scales of young birds.

Most fleas have three larval stages; each third-instar larva spins a cocoon within which it pupates. The larval stages last from 1 week to 24 weeks. Pupae may live for a week up to a year, depending on the species and such environmental factors as temperature and moisture. The entire life history of a flea may be as short as 18 days, or it may last for many months. The fully formed adult flea may lie quiescent for an indefinite period of time before it becomes active. In a pamphlet from the British Museum of Natural History,[4] one may find the following interesting observations.

". . . flea's cocoons are extraordinarily sensitive to mechanical disturbance of any kind. Persons entering a long-deserted house sometimes have cause to complain of hordes of fleas appearing 'suddenly' after a short time. It is probable that in such cases fleas resting in the cocoon, beneath floors, in cracks, etc., have come out in response to the vibrations caused by people moving in their proximity. Bird fleas show the same phenomenon."

Fleas usually are equally common on hosts of either sex; however, there are some exceptions. Bat fleas tend to crowd onto female bats before they migrate to summer colonies. Some fleas, especially small-mammal fleas, may be found more commonly on male than on female hosts, although this is not consistent. "Differences may be because males are generally larger than females, have higher hormonal levels in spring due to stress, have larger home ranges particularly in spring, tend to mutually groom females in spring and may nest apart from females in nests more suitable for fleas."[14] Fleas whose reproduction is regulated by hormones of its host, as is true of the European rabbit flea, *Spilopsyllus cuniculi,* may be more abundant on females than on males.

Some fleas require a blood meal before they copulate. Many males die soon after mating; females may live long enough to lay a quantity of eggs that are fertile only if the female has previously had a meal of blood. Although seasonal changes in climate affect the numbers of eggs laid and the duration of larval stages, bird fleas have a more sharply defined breeding season than do mammal fleas. The moisture requirements of the larvae are of cardinal importance in determining both survival and transfer to new hosts.

Hormone cycles of the host play an important part in the regulation of the flea breeding cycle and population.[18] Minute amounts of hydro-

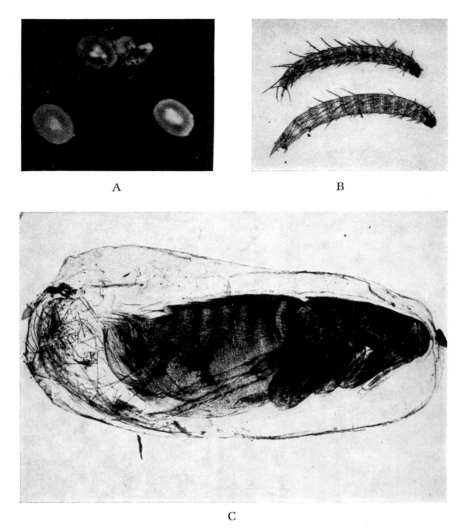

A   B

C

Fig. 20–7.   Life cycle of a flea. *A*, Eggs. × 29. *B*, Larvae. × 19. *C*, Pupa within cocoon. × 18. (From Rothschild and Clay: Fleas, Flukes and Cuckoos, courtesy of Wm. Collins Sons.)

cortisone injected into a rabbit induce maturation of fleas feeding on its blood.

Fleas and Human Disease. Plague, one of the most serious bacterial diseases of man, in former years was responsible for hundreds of thousands of deaths during a single epidemic. The disease probably retarded Western civilization by 200 years. Although human plague is now largely under control, it is an ever-present danger. The bacterial causative organism, *Yersinia pestis*, is transmitted by fleas from rodents to man. Plague was introduced to North America in the year 1900, and since then hundreds of persons have contracted the disease and a large proportion of those persons succumbed to the infection. About 60 different kinds of fleas are associated with the house mouse, black rat, and Norway rat, and of these, at least 8 are able to transmit plague bacteria. Rats are the principal culprits in the transfer of infected fleas to man, and the oriental rat flea, *Xenopsylla cheopis*, is the most important vector. *Nosopsylla fasciatus*, of occurrence worldwide, also is able to transmit plague bacteria. Bacilli in an infected flea so congest its proventriculus and stomach that blood sucked from a mammalian host fails to pass into the stomach. A "blocked" flea continues its attempts to feed and bits of the bacillary mass break off and are injected into the host. Blockage of the gut of the infected flea may be temporary or permanent; in the latter case the flea dies within a few days.

Xenopsylla cheopis and *Nosopsyllus fasciatus* are also vectors of a nonepidemic typhus of man, "murine typhus." This flea-borne disease is caused by *Rickettsia typhi*, which normally occurs in rats. The etiologic agent of epidemic typhus, *Rickettsia prowazekii*, also can infect fleas but normally is transmitted by the body louse; this rickettsia, which appears to be enzootic in flying squirrels in the eastern United States, is thought to be transmitted from squirrel to man by a flying-squirrel flea, *Orchopeas howardii* (see Chap. 19).

Other diseases that can be transmitted by fleas include tularemia in man, rabbits, and rodents (caused by a bacterium, *Francisella tularensis*); salmonellosis in man (caused by a bacterium, *Salmonella enteritidis*); and myxomatosis of wild and domestic rabbits (caused by a poxvirus). Cysticercoid stages of several tapeworms (*Dipylidium caninum* of dogs and cats, *D. sexcoronatus* of cats, *Hymenolepis diminuta* of rodents and man) develop in larvae of several species of fleas. Larvae of a filarial worm, *Dipetalonema reconditum*, found in subcutaneous tissues of dogs, develop in dog and cat fleas. Larvae of the chigoe flea (genus *Tunga*) infest the skin of birds and mammals and cause intense itching and ulceration. *Tunga* and *Neotunga* are remarkable in that after the female attaches to a host, the skin of the host grows up around the flea to form a furuncle; the female flea passes her entire adult life as an internal parasite. Fleas may also be associated with anemia and dermatitis.

Fleas themselves are parasitized by gregarines and other protozoa in the midgut and by a hymenopterous insect larva *(Bairamlia fuscipes)* in the flea larva. For further information on fleas consult Traub and Starcke.[19]

Order Diptera (Flies)

Structure. Insects belonging to the order Diptera are characterized by the possession of only one pair of wings, hence the name *diptera*, which means "two wings." Immediately behind the wings is a pair of club-shaped halteres, the remnants of the second pair of wings, which act as gyroscopes to stabilize flight. The three body divisions—head, thorax, and abdomen—that characterize insects are especially marked in the Diptera. The head of the adult usually bears prominent eyes and a pair of antennae. The thorax supports legs and wings. Wings are membranous and may be smooth, hairy, or, as in mosquitoes, may possess scales. The pattern of venation of the wings is important for identification (Fig. 20–8). The abdomen usually is distinctly segmented and bears the genitalia at its posterior end. Some major morphologic details of flies are illustrated in Figures 20–8 and 20–9. The life cycles of Diptera usually include egg, larva, pupa, and adult stages. Eggs may hatch within the body of the female, but most dipterous insects lay eggs. Larvae of the higher flies (Muscidae, Calliphoridae, Sarcophagidae) are called *maggots* and, like other larvae, undergo a series of molts during development; most have 3 larval stages.

FAMILY PSYCHODIDAE

Phlebotomine flies, or sandflies, are in the subfamily Phlebotominae of the family Psy-

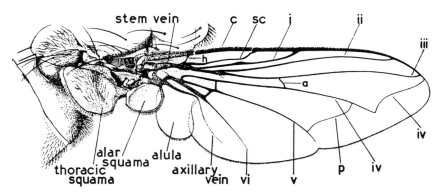

Fig. 20–8. Right wing of *Calliphora vicina*. *a*, Cross vein; *c*, costa; *sc*, subcosta; *p*, posterior cross vein. Longitudinal veins numbered by Roman numerals. (From Smart: Insects of Medical Importance, courtesy of the British Museum.)

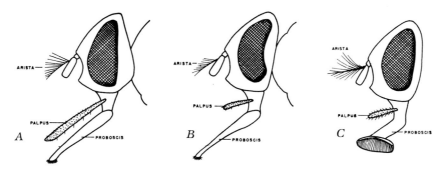

Fig. 20–9. Morphologic details of fly heads. *A, Haematobia irritans. B, Stomoxys calcitrans. C, Musca domestica.* (From Federal Security Agency, U.S. Public Health Service.)

chodidae. They occur throughout the tropics and subtropics and are found in some temperate areas of Europe, Central Asia, and North America. The female is larger than the male and has piercing-sucking mouthparts. All species are obligate bloodsuckers. Sandflies are of considerable public health importance as they are vectors of leishmaniasis, bartonellosis, and arboviruses (suh as sandfly fever). The eight or more groups of protozoa and bacteria they carry include trypanosomes, plasmodia, and hemogregarines parasitic in warm-blooded or poikilothermic animals.

Sandflies of the Old World belong to the genera *Phlebotomus* (Fig. 20–10) and *Sergentomyia;* those of the New World are in the genera *Lut-*

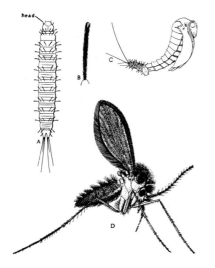

Fig. 20–10. Larva, pupa, and adult female of *Phlebotomus papatasii.* (From Smart: Insects of Medical Importance, courtesy of the British Museum.)

zomyia, Brumptomyia, Warileya, and *Hertigia.* Most sandflies are nocturnal; during the day they rest in protected, moist places such as caves, animal burrows, tree buttresses and holes, animal shelters, or houses. Three or 4 days after feeding on blood the female lays eggs in soil, leaf litter, or animal feces; they hatch in about 10 to 12 days. There are 4 larval instars, followed by a pupal stage. The entire life cycle takes about 6 to 10 weeks, depending on the ambient temperature. The flies fly weakly, so activity is diminished on windy or rainy nights. For parasites of sandflies, see Ayala.[2] For phlebotomine sandflies of North America, see Young and Perkins.[20]

FAMILY CERATOPOGONIDAE

The family Ceratopogonidae is composed of a large number of genera of small gnats, many of which are parasitic on vertebrates or other invertebrates. These tiny biting gnats ("no-see-ums") cause a great deal of annoyance of man and domestic animals. The important flies attacking man are in the genera *Culicoides, Leptoconops,* and *Forcipomyia.* The gnats have piercing-sucking mouthparts and cut the skin when feeding, which usually results in visible lesions. Larval stages may be either aquatic or terrestrial; typical larval habitats include ponds, saturated soil around collections of water, and treeholes. Species of *Culicoides* are important vectors of bluetongue virus, which causes an important disease of sheep and cattle. Many other arboviruses of the families Reoviridae and Bunyaviridae have been isolated from biting gnats. Most species of *Haemoproteus,* a genus of protozoan parasites that cause diseases of birds, utilize biting gnats as intermediate hosts, as do a number of filarial worms, such as the human

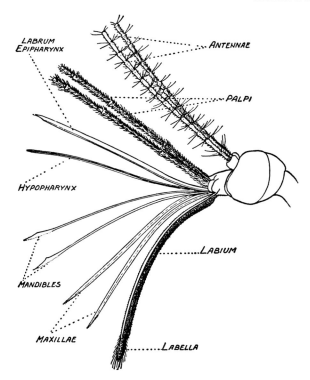

Fig. 20–11. Mouthparts of female mosquito, *Anopheles quadrimaculatus*, dissected out of the sheath. Note that the mandibles have finely serrated, blade-like lips and that the maxillae have pointed chitinous lips with fewer and coarser serrations than the mandibles. The proboscis is made up of all parts labeled, except the antennae and palpi. (From Fox: Insects and Disease of Man, courtesy of Blakiston.)

parasites *Mansonella perstans, M. ozzardi, M. streptocerca, Dipetalonema gracile,* and *Macacanema formosana* of monkeys; several species of *Onchocerca* of horses and cattle; and *Icosiella neglecta* of frogs. Humans bitten repeatedly by large numbers of biting gnats usually develop unsightly lesions which itch, are scratched, and become secondarily infected. Allergy to the bites of these gnats is not uncommon.

FAMILY CULICIDAE

Mosquitoes transmit the causative agents of such serious diseases as malaria, dengue fever, yellow fever, fowlpox, elephantiasis, and other forms of filariasis. Thus, it is not surprising that the literature on mosquitoes is staggering. See Downes[9] for the ecology and Muirhead-Thomson[15] for the behavior of bloodsucking Diptera.

About 3000 species of mosquitoes have been

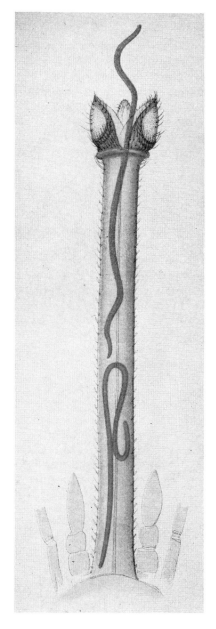

Fig. 20–12. Mature larva of *Wuchereria bancrofti* escaping from the proboscis of the mosquito *Culex pipiens quinquefasciatus*. (From Francis, courtesy of the U.S. Public Health Service Hygienic Laboratory Bulletin No. 117.)

described.[13] They may be found in almost every country. These insects are characterized by slender bodies and long legs. Mouthparts of adult females form a bloodsucking proboscis (Figs. 20–11 and 20–12). Antennae of the male usually are bushier and more prominent than those of the female. Mosquitoes in general may be distinguished from similar flies by their wing venation; of special value is the presence of 2 forked veins toward the apex of each wing; these veins, separated by an unbranched vein, are covered with scales.

The life cycle of mosquitoes includes the usual egg, larva, pupa, and adult stages; the larvae and pupae live in water. Eggs, about 1 mm in length, are laid by members of the genus *Culex* in groups on the surface of water. These eggs are glued together in the form of tiny rafts. Members of the genus *Anopheles* deposit their eggs singly without glueing them together; they remain afloat by virtue of lateral air chambers (Fig. 20–13). On hatching, the larvae swim about and search for food, for example, yeasts, bacteria, and minute algae. The larvae (Fig. 20–14), called "wrigglers" because of the way they move through the water, come to the surface periodically and obtain air through tubes called siphons. Larvae grow, molt four times, and at the last molt become pupae. A pupa is shaped like a comma with a large head. Pupae are active and are called "tumblers." They do not feed, but during the few days of their existence the adult structures develop. When they molt, the pupae come to the surface of the water, thus permitting the winged adults to take up their aerial and terrestrial life. After mating, males may die soon, but females are longer lived. Females of some species survive the winter in a dormant condition in protected places such as sheds, barns, and animal burrows. Sperm within their bodies remain viable, ready to fertilize the eggs as they are laid in the spring.

Mosquitoes probably are best known for transmitting *Plasmodium*, the causative organism of malaria. The life cycle of this protozoan parasite is described in Chapter 5. The protozoa undergo greater changes in the mosquito than they do in man. Although mosquitoes probably are better adapted to the parasite than is man, we know little about the adjustments a mosquito has to make. It should be noted that some *Anopheles* mosquitoes, the normal vectors of *Plasmodium* parasites of mammals, do not become infected even after a meal of parasitized human blood. This fact is not surprising because some people also are refractory to infection, but our knowledge of parasite-host relations between *Plasmodium* and mosquitoes is limited.

Many species of *Plasmodium* are found in birds and these are transmitted by culicine mosquitoes as far as is known. There are also *Plasmodium* species that infect amphibians and reptiles but little is known of their vectors; *P. mexicanum*, a lizard malarial parasite has been shown to develop in sandflies of the genus *Lutzomyia*.

Mosquitoes may be parasitized by other sporozoans and microsporans, as well as by gregarines, flagellates, ciliates, trematodes, filariae, mites, and blood-sucking midges. A catalog of the enemies of mosquitoes was published by Jenkins.[12] The latest addition to a bibliography of the pathogens of mosquitoes was published in 1983.[17]

Immature mosquitoes can be destroyed by adding oil to the water in which they live. This layer of oil prevents their obtaining oxygen by attaching to the surface with their air tubes. The chief destructive effect, however, is due to the volatile toxic substances that occur in most oils used, for example, kerosene and diesel fuel. Various larvicides, such as chlorinated hydrocarbons (DDT, benzene hexachloride), organic phosphates (malathion, temephos), carbamates, or pyrethroid compounds also may be used to kill larvae and pupae. Insecticides that have a minimal effect on other aquatic organisms should be used so that the ecologic balance of the breeding place is not altered. It has been observed that when breeding places are treated with insecticides that are highly toxic to most aquatic forms, the mosquito population quickly "rebounds" so that insecticidal treat-

Fig. 20–13. Anophelines and culicines, comparative characteristics. (From the U.S. Public Health Service.)

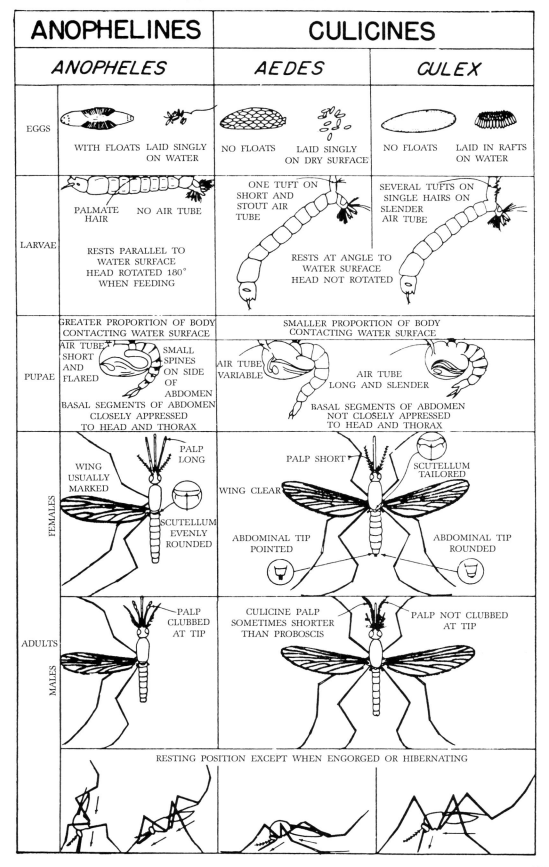

ANOPHELINES	CULICINES	
ANOPHELES	AEDES	CULEX

EGGS

WITH FLOATS   LAID SINGLY ON WATER

NO FLOATS   LAID SINGLY ON DRY SURFACE

NO FLOATS   LAID IN RAFTS ON WATER

LARVAE

PALMATE HAIR   NO AIR TUBE

RESTS PARALLEL TO WATER SURFACE HEAD ROTATED 180° WHEN FEEDING

ONE TUFT ON SHORT AND STOUT AIR TUBE

RESTS AT ANGLE TO WATER SURFACE HEAD NOT ROTATED

SEVERAL TUFTS ON SINGLE HAIRS ON SLENDER AIR TUBE

PUPAE

GREATER PROPORTION OF BODY CONTACTING WATER SURFACE

SMALLER PROPORTION OF BODY CONTACTING WATER SURFACE

AIR TUBE SHORT AND FLARED   SMALL SPINES ON SIDE OF ABDOMEN

BASAL SEGMENTS OF ABDOMEN CLOSELY APPRESSED TO HEAD AND THORAX

AIR TUBE VARIABLE   AIR TUBE LONG AND SLENDER

BASAL SEGMENTS OF ABDOMEN NOT CLOSELY APPRESSED TO HEAD AND THORAX

ADULTS

FEMALES

WING USUALLY MARKED   PALP LONG   SCUTELLUM EVENLY ROUNDED

PALP SHORT   SCUTELLUM TAILORED   WING CLEAR

ABDOMINAL TIP POINTED   ABDOMINAL TIP ROUNDED

MALES

PALP CLUBBED AT TIP

CULICINE PALP SOMETIMES SHORTER THAN PROBOSCIS   PALP NOT CLUBBED AT TIP

RESTING POSITION EXCEPT WHEN ENGORGED OR HIBERNATING

405

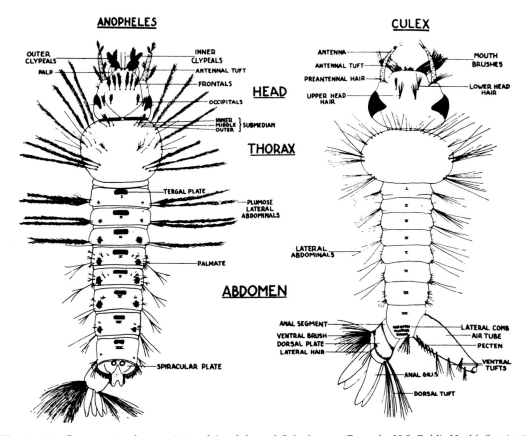

Fig. 20–14. Comparative characteristics of *Anopheles* and *Culex* larvae. (From the U.S. Public Health Service.)

ments must be continued. When insecticides that are not highly toxic to other aquatic organisms are used, however, the population of immature mosquitoes may recover only slowly and fewer insecticidal treatments are required. One reasonably specific insecticide that has been used is a synthetic juvenile hormone analog called methoprene, which usually has little effect on aquatic associates of mosquitoes. Another insect growth regulator (or IGR) is a compound called dimilin that interferes with the synthesis of chitin; it usually is safe to aquatic associates of mosquitoes except sometimes for crustaceans. The larvicide of greatest interest at the present time is a crystal produced by a bacterium, *Bacillus thuringiensis israelensis*; it seems to be effective only against larvae of mosquitoes, blackflies, and related midges. Insect sprays containing pyrethrins, lindane, or any of a number of other chemicals are effective against adult mosquitoes. For indoor spraying,

residual applications of DDT, dieldrin, malathion, propoxur, synergized pyrethrins, or other insecticides may be effective for prolonged periods of time; such applications have been widely used for control of vector-borne diseases that are transmitted indoors.

Elimination of breeding places ("source reduction") is recommended although this may have a great effect on other forms of life, such as birds and fishes. Swampy areas may be drained and filled, and containers such as cisterns or small vessels may be emptied or covered. Some enemies of mosquitoes may be used in biocontrol;[6] some of the organisms currently of interest for biologic control of mosquitoes include the bacterium *Bacillus sphaericus*; the fungi *Lagenidium*, *Culicinomyces*, and *Coelomomyces*; various microsporans; mermithid nematodes; and fishes.

For literature on the biology of mosquitoes, see Bates,[3] Clements,[7] and Gillett.[10]

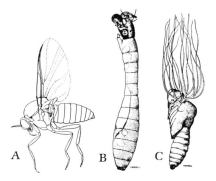

Fig. 20–15. *Simulium,* a blackfly. *A,* Side view of adult. *B,* Larva. *C,* Pupa. (From Smart: Insects of Medical Importance, courtesy of the British Museum.)

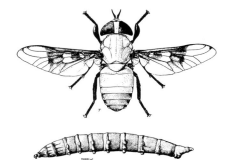

Fig. 20–16. *Tabanus latipes* with its larva. (From Smart: Insects of Medical Importance, courtesy of the British Museum.)

FAMILY SIMULIIDAE

The family Simuliidae, worldwide in distribution, includes small insects called blackflies or buffalo gnats. The immature stages of these flies are aquatic and usually are found only in running water. Both larvae and pupae are attached to rocks or to plants under water. Adults are humpbacked (Fig. 20–15), hence the name "buffalo gnat." They sometimes occur in enormous swarms, especially near water, and they annoy livestock and humans by their painful bites. Some species gather in tremendous numbers on domestic mammals and on birds. Herds of mules and cattle reportedly have been killed by the attacks of blackflies. The flies may cause such annoyance to poultry that the birds leave their nests. Aggressive females sometimes crawl under the wings of young birds and suck so much blood that the birds die. Damage is due not only to loss of blood, but also to toxicity of the saliva, hypersensitivity, and to parasites that the flies transmit.

The major parasite of man that blackflies carry is the roundworm *Onchocerca volvulus,* which infests man in warm areas of Africa and Latin America. The flies also transmit protozoan parasites of the genus *Leucocytozoon.* These organisms are blood parasites of birds, especially turkeys and ducks. *Leucocytozoon simondi* is found in the peripheral blood of ducks, and *L. smithi* is a similar parasite of turkeys. Common insect vectors are *Simulium rugglesi, S. anatinum,* and *S. occidentale.* The flies attach themselves to the necks of the birds and feed for 2 or 3 minutes before flying away. Like ceratopogonids, simuliids cut the skin when they feed, producing a visible lesion which may bleed for some time.

People sensitive to the bites of blackflies experience swelling and intense itching; scratching of the bites frequently results in their becoming infected. Use of insecticides and other control measures against the fly are only partially effective.

FAMILY TABANIDAE

Tabanidae is another family of flies whose females suck blood. These insects, larger than blackflies, are known as horseflies, deerflies, or clegs (Fig. 20–16). They are stout flies, each with a broadly triangular head, most of which is composed of eyes. The adults lay eggs on emergent vegetation over water and the larvae, which are aquatic, are predacious on other aquatic organisms. The developmental period usually is long, weeks or months. The bites of tabanids are painful and bleed profusely.

Various species of *Tabanus* transmit the flagellate blood parasite, *Trypanosoma evansi,* which causes surra in horses, cattle, dogs, elephants, and other animals (Chap. 3). Members of this genus of flies also carry anthrax to man and animals. Tularemia may be transmitted by tabanids, especially a species of *Chrysops;* species in this genus also transmit a human filarial worm, *Loa loa* (Chap. 16). Different species of *Chrysops* feed at different times of day; *C. silacea* and *C. dimidiata,* the important vectors of *Loa loa* bite mostly during the morning and late in the afternoon, but *C. langi* and *C. centurionis* bite mostly in the evening, at sunset.

FAMILY MUSCIDAE

The family Muscidae includes a number of important parasites of man and animals. The housefly, *Musca domestica* is not normally parasitic although it occasionally produces myiasis in man or animals. The mouthparts of the housefly are not adapted for piercing although houseflies feed on blood avidly when it is available. There are teeth on the discal sclerite of the labellum which allow the fly to scrape the surface on which it feeds. In some species of the genus *Musca* these teeth are enlarged and may be used to scrape scabs from wounds so that the flies can feed on pus or blood; the Old World species *Musca crassirostris* has such habits. The hornfly *Haematobia irritans,* an important blood-sucking parasite of cattle, has more highly developed mouthparts but of the same basic type (Fig. 20–9); it simply scratches a hole in the skin and sucks up the blood that exudes. Needless to say, the bite is quite painful. The stablefly, *Stomoxys calcitrans,* has mouthparts even more highly developed for bloodsucking (Fig. 20–9); again it scratches a hole in the skin, quite painfully, and sucks up the blood that exudes. The bites may bleed. Stableflies are thought to transmit several species of trypanosomes mechanically, especially *Trypanosoma vivax* of cattle in areas where tsetse do not occur. The most highly evolved mouthparts of flies in this family are those of tsetse flies, *Glossina* (Fig. 20–17). These flies have an elongate proboscis which actually enters the skin to a considerable depth. This is an old group of flies now found only in tropical Africa although a fossil tsetse, dating from the Oligocene period, has been found in the United States. The genus now includes only 22 species. The flies demonstrate the ultimate k-strategy, elimination of vulnerable immature stages at the expense of fecundity. Females produce a single egg at a time and retain it in a brood pouch ("uterus") until the larva is fully grown and ready to pupate. Larvae are deposited in the soil into which they burrow and pass the pupal period, which lasts about a month. A female requires about 3 engorgements to produce a larva, which requires about 2 weeks; in her lifetime a female usually produces about 12 larvae, half of which are males. Tsetse are the important vectors of a number of trypanosomes infecting man and animals, including the two forms infecting man, *T. brucei rhodesiense* and *T. brucei gambiense* (see Chap. 3). An interesting facet of the relationship between the fly and the flagellate is that metacyclic trypanosomes in the foregut of the fly frequently attach to chemoreceptors in the mouthparts, which has the effect of prolonging the feeding of the fly and thus increasing the probability that the trypanosomes will be transmitted. For further information on tsetse see Buxton.[5]

FAMILY CALLIPHORIDAE

Blowflies resemble houseflies, but often are larger and frequently are metallic green or blue. They lay their eggs on organic materials such as dead animals, excrement, open sores, or exposed cooked or uncooked food, or even on live animals. The eggs hatch into maggots, which proceed to eat the material around them. All have 3 larval stages.

The genus *Calliphora* includes the blue-bottle flies whose larvae are commonly found in decaying meat. Eggs also are laid in exposed, foul wounds of man and animals. The term myiasis refers to the presence of parasitic maggots in or on man or animals (see discussion later in this chapter). The maggots of some species eat only the putrid tissues in the wound and have been used to clean wounds. The larvae of *Phaenicia sericata* and *Phormia regina,* for example, have been used by physicians to clean wounds and to treat osteomyelitis. Occasionally, live *Calliphora* maggots are swallowed in food but it is unlikely that they would survive passage through the gut of a normal person; there are numerous reports of such events, most of which probably are erroneous. On the other hand, flies commonly attack the anus of animals, including man, especially diarrhetic or otherwise ill animals; most reports of intestinal myiasis probably arise in this way.

Bufolucilia sylvarum lays its eggs on the back of a toad or other amphibian. Eggs hatch within a day and the tiny larvae crawl to the head region and onto the surface of the eye. By blinking its eyes, the toad carries the maggots to the lacrimal ducts, which lead to the nasal cavity. There the larvae molt to the second stage, actively feed on host tissue, and eventually destroy the cartilaginous nasal septum. When mature, the larvae drop out of the enlarged nasal cavity and pupate in the ground.

Chrysomya bezziana is a green blowfly that seems to prefer wounds and body spaces of man and wild or domestic animals. Larvae (Fig. 20–18) may be found in the nose, eyes, ears, alimentary canal, urinary passages, and genital organs, or in tiny cuts or sores in the skin. Eggs may be laid even in the pierced ear lobes of women; at first, the small maggots are not noticed, but when they grow to a centimeter or

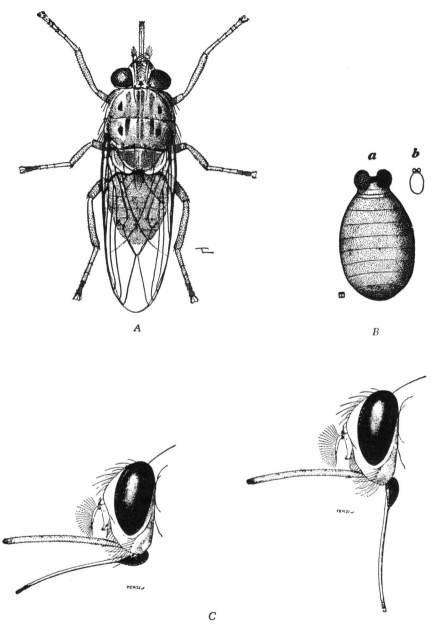

Fig. 20–17. A tsetse fly. *A, Glossina longipennis* in the resting attitude assumed by the living fly. Note the position of the wings. Original magnification ×4. (After Austen). *B,* Puparium of tsetse fly. *a,* enlarged; *b,* natural size. The posterior end is toward the top of page. (After Austen. From Castellani and Chambers.) *C,* Head of species of *Glossina* showing how the proboscis with its bulb-shaped base is lowered from the palpi for the act of feeding. The palpi are not inserted into the wound, but remain horizontal as the piercing proboscis is sunk into the skin. (From Castellani and Chambers. In Smart: Insects of Medical Importance, courtesy of the British Museum.)

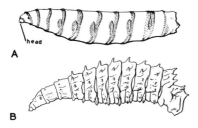

Fig. 20–18. *A, Chrysomya bezziana,* the mature third stage larva. ×4. *B,* The mature third stage larva of *C. albiceps.* ×4. (From Smart: Insects of Medical Importance, courtesy of the British Museum.)

Fig. 20–19. *Cordylobia anthropophaga,* the tumbu fly, female. ×2. (From Smart: Insects of Medical Importance, courtesy of the British Museum.)

more in length and begin to burrow through the skin near the wound, they may produce a tumor of considerable size, especially if many maggots are present. This species is found in the Philippines, India, and Africa.

Cochliomyia hominivorax is a bluish blowfly about twice the size of a housefly. It lays its eggs in neat rows on the edges of wounds of mammals, especially on fresh cuts. The eggs hatch in a day or less into maggots that feed on living flesh. These maggots are called "screwworms" because of the rows of dark spines that resemble threads of a screw. The wound infested by screwworms has a foul odor. In about a week, the maggots drop to the ground and pupate in the soil. In another week, if the weather is warm, adults emerge. In still another week or less, new eggs, about 200 in a batch, are deposited on the edge of a cut, a sore, the navel of a newborn animal, or some other abrasion. This screwworm is an obligatory parasite of warm-blooded animals and was responsible for millions of dollars of damage annually to livestock in the United States until it was eradicated from this country by releasing male flies that had been sterilized by radiation. Females that mate with sterile males cannot reproduce.

Other blowflies, principally *Phormia regina,* the black blowfly; *Cochliomyia macellaria,* the secondary screwworm fly; and *Phaenicia sericata,* the greenbottle fly, possess larvae that are called fleeceworms. They attack the hair of sheep and goats and also enter wounds. Eggs usually are deposited on soiled fleece or on old or new sores; the maggots decrease the quality of fleece and aggravate wounds.

Aural myiasis of man may be produced by *Phaenicia sericata.* Infestation may start in a normal healthy ear. If left untreated, such an infestation may lead to pain, deafness, and ring-ing in the ear if the maggots are in the external meatus. If they enter the middle ear, a bloody, purulent discharge may result.[8]

The tumbu fly of Africa, *Cordylobia anthropophaga* (Fig. 20–19), measures about 10 mm in length and is yellowish brown. Females lay their eggs on the ground, where they hatch. When the appropriate host (man or mammals) lies on the ground, larvae burrow through the skin into their new home. As the larva grows, it produces a boil-like swelling (furuncle) in the skin, with a small hole at the top of the lesion. Through the hole, the maggot breathes and discharges its waste. When mature, the maggot emerges through the hole, drops to the ground, and pupates. Cases of myiasis caused by this and a related fly, *Stasisia rodhaini,* have been reported from persons who have traveled in Africa.

FAMILY SARCOPHAGIDAE

The family Sarcophagidae includes the fleshflies, large, gray insects with longitudinal black stripes on the dorsal surface of the thorax (Fig. 20–20); the abdomen has a checkered pattern. Although most Sarcophagids are parasitoids and prey on other insects, some, like blowflies, are scavengers; their larvae may be found in

Fig. 20–20. *Sarcophaga haemorrhoidalis,* a fleshfly, female. (From Mönnig: Veterinary Helminthology and Entomology, courtesy of Baillière, Tindall and Cox.)

Fig. 20–21. *Hypoderma bovis,* the ox warble, adult. (From Smart: *Insects of Medical Importance,* courtesy of the British Museum.)

Fig. 20–22. *Oestrus ovis,* the adult sheep nasalfly. (From Smart: *Insects of Medical Importance,* courtesy of the British Museum.)

carrion, wounds, sores, or body cavities of man and animals. Unlike blowflies, most fleshflies retain their eggs until they hatch; the flies are therefore larviparous rather than oviparous. The genus *Sarcophaga* is worldwide in distribution. The genus *Wohlfahrtia* contains species that parasitize man and animals and cause myiasis in these hosts. A larva of *Wohlfahrtia magnifica* has even been found in the cavity of an infected tooth.

FAMILY TACHINIDAE

The family Tachinidae contains many species of flies, all of which are parasitic on insects or other arthropods, such as terrestrial isopods. Some of the tachinids resemble houseflies, and some appear more like bees or wasps, but most possess long bristles on the tip of the abdomen. A tachinid normally deposits eggs on the body of its host, for example, larvae of butterflies or beetles. The larvae burrow into their host and literally eat themselves out of house and home, killing their host. When mature, the larvae crawl out of the host and pupate on the ground. Some tachinids lay eggs on foliage, and, when the larvae emerge, they either crawl into a host or are swallowed by the host. For comments on the evolution of the tachinids, see Chapter 26.

FAMILY OESTRIDAE

Warbleflies are so called because their larvae usually lie under the skin of their hosts and cause swellings known as warbles. The best-known members of this group are the ox warble, *Hypoderma bovis* (Fig. 20–21), a stout, hairy fly that resembles a bee; and a similar species, *H. lineatum,* the heelfly. The adult heelfly is hairy, black and yellow, and is several times as large as a housefly. In the spring, each female fly attaches hundreds of eggs to hairs on the

legs or body of cattle. In a few days the eggs hatch into tiny, white, spiny larvae (called "bots," hence the name "botflies") that crawl down the hairs and burrow through the skin. The larvae spend months near the esophagus or the spinal canal and eventually migrate to the vertebral region, where they produce small swellings or warbles in the skin. They make a tiny hole in the skin for air, and then they increase in size to about 25 mm in length. During this period, they turn dark brown. In the spring or summer, the grubs or maggots emerge through the holes, drop to the ground, and pupate. They crawl under loose soil or trash, and in 2 to 7 weeks, depending on the temperature, adult flies emerge.

One to three billion dollars are lost annually in the United States because of damage to hides, and irritation and annoyance to cattle. The animals "get off their feed," lose weight, and are generally unthrifty. Milk production diminishes. The holes in the skin make the hide unfit for high-quality leather. The flies may also attack humans and produce a type of larval migrans, although they do not mature.

Dermatobia hominis, in the family Cuterebridae, is similar although not so hairy; it may occur as a maggot in the skin of man in Central and South America. The fly is primarily a parasite of cattle but is often called a human botfly; it is about 12 mm in length. The nonbiting females attach their eggs to the bodies of mosquitoes, other flying insects, or even ticks. Eggs are thus given a free ride to a new host, a process known as phoresy. When the mosquitoes alight to feed on man, or other warm-blooded animals, the botfly eggs hatch, and the larvae penetrate the skin of their new host.

Oestrus ovis (Fig. 20–22), instead of laying eggs, deposits its larvae or maggots in the nostrils of sheep, goats, and some other animals such as deer. The maggots crawl into the nasal passages and, during the next few weeks, molt twice. When fully grown, about 25 mm long, they come out of the nose, drop to the ground, and pupate. The adult fly emerges from the pupa. It is also called a gadfly or nasalfly. This fly commonly attacks people; larvae probably are most frequently desposited in the eyes; at least, ocular myiasis is the usual type of infestation that is recognized. The larvae cause considerable irritation to the eyes but are not able to complete development in man.

The names botfly and warblefly are used interchangeably. Thus *Gasterophilus intestinalis* may be called a horse bot- or warblefly (Fig. 20–23). This and related genera usually are placed in the family Gasterophilidae, a group somewhat more primitive than the oestroid flies. Adults of *Gasterophilus* resemble bees. The fly attaches its eggs to hairs on the legs or bodies of horses; a female may lay 1150 eggs in an hour or two. When embryogeny is completed, the larvae pop out when they are brushed by the moist, warm lips of the horse. These spiny larvae become attached to the horse's lips or tongue, burrow into the membranes of the mouth, and make their way to the stomach through various tissues. Larvae attach themselves to the lining of the stomach, sometimes in such large numbers that the epithelium is almost covered with them. After some months of growth, the maggots pass from the body of the host with the feces, drop to the ground, and pupate. Adult flies emerge from the pupae a few weeks later. The adults do not have functional mouthparts and live for only a few days.

Myiasis. As defined by Zumpt, myiasis is "the infestation of live human and vertebrate animals with dipterous larvae, which, at least for a certain period, feed on the host's dead or living tissue, liquid body-substances, or ingested food"[21] (Fig. 20–24). Some species are obligatory parasites (*Cochliomyia hominivorax, Chrysomya bezziana, Wolfahrtia, Gasterophilus, Hypoderma, Dermatobia,* etc.) while others are facultative parasites (*Phaenicia, Phormia, Calliphora*); still others rarely are parasitic and so are called accidentals (*Musca, Drosophila, Sepsis, Megaselia, Tipula,* etc.). At least 187 species of myiasis-producing larvae are recorded in the world literature. They belong primarily to the families Calliphoridae and Oestridae, but are also found among the Sarcophagidae, Muscidae, Chlor-

opidae, Gasterophilidae, and many other families.

Myiasis may be classified by its clinical presentation: intestinal, urogenital, traumatic, furuncular, etc. Maggots in the intestine produce intestinal myiasis; those in the genital organs or urinary tract produce urogenital myiasis. Some maggots occur only in wounds and sores, especially foul and suppurating sores, whereas others actually penetrate living flesh and produce boils or furuncles in the skin or, as we have seen, in the chambers of the nose, mouth, eyes, or ears.

Intestinal myiasis usually is caused by entrance of the maggots through the anus. Urogenital myiasis is caused by flies that lay their eggs on the exposed and unclean pubic area. Some of the genera of flies involved in such types of myiasis are *Calliphora, Musca, Fannia,* and *Stomoxys.* Flies that deposit their eggs in or near foul wounds, and whose larvae thus have an opportunity to invade the wound, belong mainly to the genera *Sarcophaga, Phormia, Cochliomyia,* or *Chrysomya.*

Identification of many of these maggots is difficult. One of the common diagnostic characteristics is the nature of the posterior spiracles. Figure 20–24 illustrates these spiracles as they appear in the larvae of some myiasis-producing flies.

Invertebrates have their share of parasitic fly larvae. Some Hymenoptera may contain maggots that actually attach their spiracular openings to the host trachea in order to breathe. The snail *Helicella vergata* may become so infested with fly maggots that the parasites kill their host and eat practically all of the dead tissue.

One of the few studies[16] of the biochemical effect of myiasis on the host involved the white-footed mouse, *Peromyscus leucopus,* and the botfly, *Cuterebra angustifrons.* Development of the larva in the mouse was accompanied by progressive reduction in the albumin-to-globulin ratio of the host. This reduction was correlated with the number of larvae present. Ingestion of albumin by the botfly was proposed as a partial explanation for the reduction.

For a review of insects and disease, see Harwood and James. For flies and diseases, see Greenberg.[11] For parasitic insects, see Askew.[1]

Pupipara. Pupipara is a group of flies, all of which are parasitic. They are similar in that offspring are retained within the female until larval development is completed; the female produces

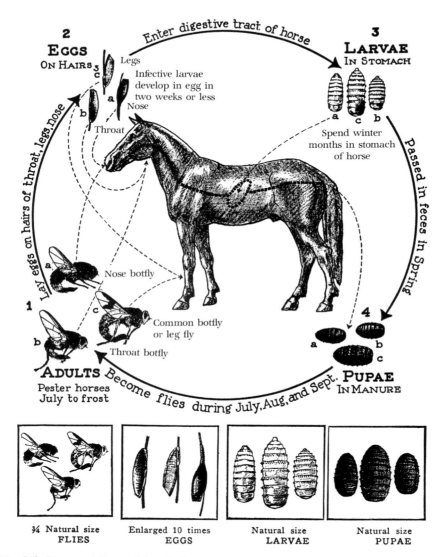

Fig. 20–23. Life history of *Gasterophilus*. *1*, Flies pester horses from July to frost, laying their eggs on the hairs of the nose, throat, and legs. *2*, In the eggs, larvae develop in two weeks or less and enter the digestive tract. *3*, Larvae spend the winter months in the stomach, where they are a drain on the animal; in the spring, they pass out in the feces. *4*, In the ground, pupae develop from the larvae. During July to September, these pupae become flies, which start the life cycle all over again. (From Thorp and Graham, courtesy of the University of Illinois College of Agriculture.)

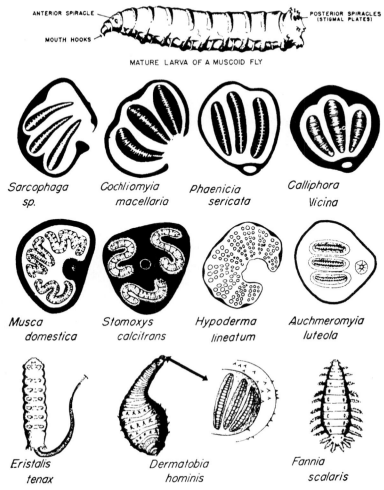

ANTERIOR SPIRACLE   POSTERIOR SPIRACLES
                                                     (STIGMAL PLATES)
MOUTH HOOKS

MATURE LARVA OF A MUSCOID FLY

Sarcophaga
sp.

Cochliomyia
macellaria

Phaenicia
sericata

Calliphora
vicina

Musca
domestica

Stomoxys
calcitrans

Hypoderma
lineatum

Auchmeromyia
luteola

Eristalis
tenax

Dermatobia
hominis

Fannia
scalaris

Fig. 20–24. Key characteristics of myiasis-producing fly larvae. Numbers 1 to 8 are stigmal plates. (From Federal Security Agency, U.S. Public Health Service.)

fully grown larvae that pupate quickly after larviposition, hence, the name "Pupipara" (pupa-bearing). It is now recognized that the group is a heterogeneous one so the category is no longer used in classification although the concept of pupiparous, parasitic flies is a useful one and the name is still used in a vernacular sense for this concept. Pupipara are peculiar-looking flies that are ectoparasitic on other animals. They are placed in the Braulidae, parasitic on bees; the Nycteribiidae and Streblidae, parasitic on bats; and the Hippoboscidae, parasitic on birds and mammals. The bee parasite, *Braula coeca*, is an external parasite of honey bees and is said to snatch food from the mouthparts of the bee; hence it is known as a "kleptoparasite."

The vertebrate parasites are bloodsucking and are called louse flies, bat flies, tick flies, or keds. Usually they are wingless, flattened dipteran parasites that can be found on the bodies of animals or in birds' nests or bat roosts. Some of these flies possess and retain wings; other shed them on reaching their hosts. In some species males are winged but females are not. Wild animals are attacked as well as horses, sheep, and other domestic animals. That ectoparasitic flies are found on bats is not surprising when one considers the probability that ancestral flies deposited their eggs in bat droppings. From this start, the association undoubtedly became progressively closer, until today there are such forms as the completely wingless pupiparous

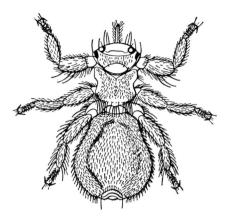

Fig. 20–25. *Melophagus ovinus*, the sheep ked. ×4. (From Smart: Insects of Medical Importance, courtesy of the British Museum.)

species, *Nycteribia biarticulata*, which at first glance looks more like a louse than a fly.

FAMILY HIPPOBOSCIDAE

The family Hippoboscidae includes the more common species of Pupipara. An important member of the family is the sheep ked or "sheep tick," *Melophagus ovinus*, a wingless, leathery, flattened insect (Fig. 20–25) that is hardly recognizable as a fly. It is about 6 mm long and, like other Pupipara, does not lay eggs, but gives birth to larvae already advanced almost to the pupal stage.

Pseudolynchia canariensis, another hippoboscid, is a winged louse fly that lives on pigeons. In addition to causing annoyance to the birds, this ectoparasite transmits a malaria-like parasite of birds, *Haemoproteus columbae*.

REFERENCES

1. Askew, R.R.: Parasitic Insects. New York, Elsevier, 1971.
2. Ayala, S.: The phlebotomine sandfly-protozoan parasite community of central California grasslands. Am. Mid. Nat., *89*:266–280, 1973.
3. Bates, M.: The Natural History of Mosquitoes. New York, Macmillan, 1949.
4. British Museum of Natural History: Fleas as a Menace to Man and Domestic Animals. Their Life-History, Habits and Control. 6th Ed. Economics Series No. 3. London, British Museum of Natural History.
5. Buxton, P.A.: The Natural History of Tsetse Flies. An Account of the Biology of the Genus *Glossina* (Diptera). London School of Hygiene and Tropical Medicine, Memoir No. 10, 1955.
6. Chapman, H.C. (ed.): Biological Control of Mosquitoes. Amer. Mosq. Cont. Assoc. Bull. No. 6, 1985.
7. Clements, A.N.: The Physiology of Mosquitoes. Oxford, Pergamon Press, 1963.
8. Davies, D.M.: Human aural myiasis: a case in Ontario, Canada, and a partial review. J. Parasitol., *62*:124, 1976.
9. Downes, J.A.: The ecology of blood-sucking Diptera: an evolutionary perspective. *In* Ecology and Physiology of Parasites. Edited by A.M. Fallis. Toronto, University of Toronto Press, 1971, pp. 223–258.
10. Gillett, J.D.: Mosquitos. London, Weidenfeld and Nicolson, 1971.
11. Greenberg, B.: Flies and Disease. Vol. I. Ecology, Classification and Biotic Associations, 1971. Vol. II. Biology and Disease Transmission, 1973. Princeton, Princeton University Press.
12. Jenkins, D.W.: Pathogens, parasites and predators of medically important arthropods. Annotated list and bibliography. Bull. Wld. Hlth. Organ., *30* (Supplement), 1964.
13. Knight, K.L., and Stone, A.: A Catalog of the Mosquitoes of the World (Diptera: Culicidae). Vol. 6. Entomol. Soc. Amer., Thomas Say Foundation, 1977.
14. Marshall, A.G.: The Ecology of Ectoparasitic Insects. London, Academic Press, 1981.
15. Muirhead-Thomson, E.C.: Behaviour Patterns of Blood-Sucking Flies. Oxford, Pergamon Press, 1982.
16. Payne, J.A., Dunaway, P.B., Martin, G.D., and Story, J.D.: Effects of *Cuterebra angustifrons* on plasma proteins of *Peromyscus leucopus*. J. Parasitol., *6*:1004–1008, 1965.
17. Roberts, D.W., Daoust, R.A., and Wraight, S.P.: Bibliography on pathogens of medically important arthropods: 1981. WHO VBC/83.1, 1983.
18. Rothschild, M.: Fleas. Sci. Am., *213*:44–53, 1965.
19. Traub, R., and Starcke, H.: Fleas. Rotterdam, A.A., Balkema, 1980.
20. Young D.G., and Perkins, P.V.: Phlebotomine sand flies of North America (Diptera: Psychodidae). Mosquito News, *44*:263–304, 1984.
21. Zumpt, F.: Myiasis in Man and Animals in the Old World. London, Butterworth, 1965.

Class Arachnida

Order Acari (Ticks and Other Mites)

Most ticks and certain mites are parasitic on or in vertebrates or invertebrates during at least one stage in their life cycle. These arachnids may be the etiologic agents of disease, they may transmit pathogenic microorganisms, or they may serve as reservoirs of infection. Large numbers of mites are phytophagous, many feed on organic materials, still others are predacious or parasitic on animals. The feathers of birds and the hairs and skin of mammals are favorite habitats of some species, and internal organs of both vertebrates and invertebrates frequently are invaded. Fourteen pounds (6.4 kg) of ticks once were removed from the skin of a horse in 3 days, and as many ticks were still left on the suffering animal. Freshwater ponds, streams, rivers, and the oceans all have their mite faunas.

The Arachnida includes spiders, scorpions, and many other groups in addition to the mites. Mites, including ticks, are readily distinguished from other arachnids by the possession of a distinct gnathosoma (an anterior capitulum bearing the mouthparts) and by the absence of a recognizable division between the cephalothorax (prosoma) and abdomen (opisthosoma). The phylogeny of the Acarina is obscure.

For general information on mites, see Baker and Wharton[7] and Savory.[17] For classification of mites, see Krantz.[13] For relationship of arachnids to disease, see Nutting.[15]

Morphology. As in other arthropods, the tegument consists of an outer cuticle and a single layer of epithelial cells that secrete it. The cuticle, which may be membranous or leathery, sometimes has hard plates (sclerites). Special structures such as glands, setae, and sensory organs are dervied from tegumentary cells. Fig-ure 21–1 illustrates details of the external anatomy of ticks.

Mouthparts consist of a pair of chelicerae that usually terminate in small pincers (chelae), consisting of a dorsal fixed digit and a ventral movable digit; and a pair of pedipalps (or palpi), consisting usually of 4 to 6 segments, sometimes modified as a thumb and claw. Most mites possess 3 pairs of walking legs as larvae and 4 pairs as adults. Each leg usually is divided into 6 segments and terminates in a well-developed claw. Of primary importance to the systematics of the group are the anatomic features of the respiratory system; the suborders are established partly on the basis of the numbers and location of the openings of the tracheae (spiracles or stomata).

The life cycle of mites consists usually of egg, larval, 1 or more nymphal, and adult stages. Larvae have only 3 pairs of legs while nymphs and adults have 4 pairs. Sometimes there are 2 larval stages; usually there are 1 or 2 nymphal stages, but, as in soft ticks, there may be more. There are numerous variations in the life cycle, from the interpolation of quiescent stages between the active stages, which occurs in some prostigmatids, to the complete telescoping of the cycle so that adult females give rise directly to adult males and females, as in the astigmatid family Pyemotidae; males have fewer developmental stages than females in some astigmatid mites.

Many of the larger mites resemble ticks, but most are small (0.5 to 2 mm) and have the hypostome (a plate projecting below the mouth) hidden and unarmed; ticks usually are larger and have the hypostome exposed and armed with teeth (see Fig. 21–5). Two or more simple eyes may be present, but are not placed on the gnathosoma. The foregut is subdivided into a

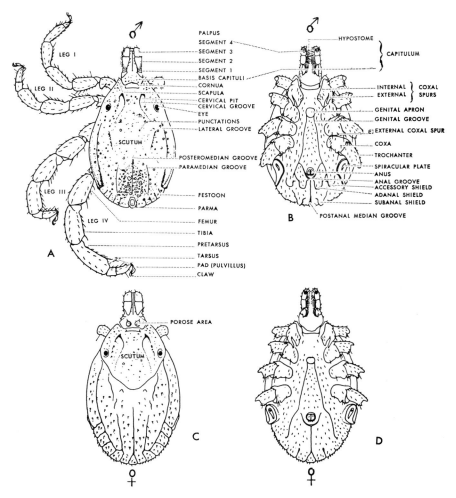

Fig. 21–1.   Key morphologic characters. Hypothetical male and female ticks, family Ixodidae. *A,* and *B,* Male, dorsal and ventral views. *C,* and *D,* Female, dorsal and ventral views. (From Hoogstraal: African Ixodoidea. U.S. Naval Medical Research Unit No. 3, Cairo.)

buccal cavity, a pharynx or pumping organ, and an esophagus. The midgut is a thin-walled stomach, and the hindgut terminates in a sacculate rectum. Although some forms of ticks have a heart, the circulatory system of most mites conists only of colorless blood.

Invertebrates, as well as vertebrates may be infested with parasitic mites. For example, mites *(Unionicola)* have been found on mussels *(Anodonta).* Other species infest marine amphipods. For mites of moths and butterflies, see Treat.[19] The most important groups of mites parasitic on vertebrates, other than the ticks, are in the suborders Mesostigmata, Prostigmata, and Astigmata; the order Cryptostigmata also

is important because it includes mites that act as intermediate hosts of cestodes.

Suborder Ixodoidea

About 800 species of ticks are recognized; all but the genus *Antricola* are obligatory parasites. They belong to three families: the Argasidae, or soft ticks, which have flexible, tough, bulbous bodies that obscure the mouthparts and most of the legs; the Nutalliellidae, a rare, African family consisting of a single species, which is somewhat intermediate between soft and hard ticks; and the Ixodidae, or hard ticks, which

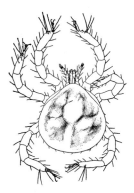

Fig. 21–2. *Argas persicus*, six-legged stage. (From Hassall. U.S. Department of Agriculture.)

have a dorsal sclerotized plate (scutum), and a conspicuous gnathosoma projecting in front of the body. The scutum of male hard ticks covers the entire dorsum of the body except for the gnathosoma; it is rather inflexible and is not able to expand significantly. The scutum of the female hard tick is small and covers only the podosoma, the leg-bearing portion of the body; the cuticle of the abdomen of the female is thrown into fine folds, which allows an enormous expansion when she engorges with blood.

All ticks have life cycles consisting of egg, larval, nymphal, and adult stages; larvae can be distinguished by having only 3 pairs of legs (Fig. 21–2), and nymphs by lacking a genital opening. Larval ticks are sometimes known as "seed ticks" because of their small size. Soft ticks are thought to be the more primitive group; they may have more than one larval stage, they have two or more nymphal stages, and the adults feed repeatedly, the females laying small clutches of eggs between feedings. Soft ticks do not attach to their host except during the larval stage, and do not enlarge enormously when they feed, as do female hard ticks; they feed quickly and return to their resting place. They therefore are nest parasites and are likely to feed repeatedly on the same animal or the same family of animals. *Ornithodoros, Argas,* and *Otobius* are the most important genera of soft ticks. *Antricola*, a non-parasitic genus consisting of 2 species, has been found only in bat guano.

Hard ticks have a more rigid life cycle; all have a single larval and a single nymphal stage. Most engorge three times during their life. Typically a larva attacks one animal, attaches, engorges, leaves the animal, molts to the nymphal stage, attacks a second host, attaches (Fig. 21–3), engorges, leaves that animal, molts to the adult stage, and attacks a third host. Male ticks copulate with females on the third host but do not engorge; they take small quantities of blood and may even feed on females. Females, on the other hand, attach for a few days, engorge, and leave the host to lay a single large clutch of eggs, often numbering several thousand, after which they die. Thus, hard ticks are not nest parasites but attach to animals and may leave the host far from where they attached to it; they are likely to be widely dispersed, especially if they attach to highly mobile animals such as birds.

During their life, therefore, hard ticks typically feed on three hosts, which may be of the same or of different species. Many, for example *Dermacentor andersoni*, feed on small animals during the larval and nymphal stages and on large animals during the adult stage, a type of behavior that permits the conveyance of pathogenic agents from one group of animals to another. The rickettsiae of tick typhus, for example, may be ingested by larval ticks feeding on field mice and later transmitted by adult ticks to humans.

Although hard ticks typically feed on three hosts, and are known as 3-host ticks, some groups utilize fewer hosts. Larvae of ticks of the genus *Hyalomma*, for example, frequently do not leave the host after engorgement, but molt to the nymphal stage on the animal. The nymph then engorges, leaves the animal, molts, and the adult attacks a second animal; this type of behavior is characteristic of 2-host ticks, which have less chance to acquire and transmit disease. Ticks of the genus *Boophilus* pass both the larval-nymphal and nymphal-adult molts on their hosts and therefore are 1-host ticks. They have little opportunity to acquire or transmit disease, but, improbable as it might seem, are important vectors of babesiosis; the sporozoan parasites are transmitted from infected female ticks to their progeny through the ovary and infection of vertebrate hosts is by the larvae.

Most ticks are intermittent parasites of mammals, birds, or reptiles. They usually demonstrate little host specificity, attacking any host that happens to be in their environment. On the other hand, the behavior of ticks may be such that they are likely to encounter only certain kinds of animals. The larvae and nymphs

Fig. 21–3. *Acanthodactylus* lizard infested with nymphs of *Hyalomma*. (From Hoogstraal and Kaiser, courtesy of Ann. Entomological Society of America.)

of *Dermacentor variabilis*, for example, are found on small mammals and birds but the adults are found chiefly on large mammals; larvae and nymphs of this species are almost never found on human beings, but adults are. The apparent host discrimination of the different stages probably is due to the manner in which they find a host. Hard ticks, when seeking a host, usually climb on vegetation and adopt a posture (known as "questing"), anterior end down, clinging to the grass or shrub with the hind legs, and with the front legs widely spread. The large tarsal claws are useful for clinging and for snagging a passing host. The apparent host discrimination of the tick probably is related to where the tick waits to ambush its host.

The life cycles of ticks usually are long; hard ticks frequently require 2 to 3 years for completion of a full generation. For a comprehensive account of feeding in ectoparasitic Acari with special reference to ticks, see Arthur.[4] A key to genera of American ticks is shown in Figure 21–4.

Both sexes of ticks are bloodsuckers although adult males take little blood. It has been estimated that as many as 200 pounds (circa 91 kg) of blood may be withdrawn from a large host animal by ticks in one season. Before or during engorgement, the female hard tick is inseminated by the male. Spermatophores are introduced into the vagina by the gnathosoma of the male. Hard ticks usually lay from 2000 to 8000 eggs after a preoviposition period of 3 to 24 days (as, for example, in *Dermacentor variabilis*). The time required for laying of the eggs varies from 2 to 6 weeks. Soft ticks lay fewer eggs (100 to 200) in several batches between blood meals.

In laboratory colonies, ticks *(Ornithodoros)* sometimes feed on other ticks that are engorged with blood. If the ingested blood contains microfilariae or third-stage larvae of the mammalian filaria *Dipetalonema viteae*, tick-to-tick transmission can occur. Such a tick-on-tick feeding relationship could play a role in the maintenance of blood-inhabiting parasites in nature.[14]

Adult ticks, especially soft ticks, may withstand starvation for several years, and, as Rothschild and Clay have said, they "are the great exponents of the gentle art of waiting."[16] Their physiologic processes during this time may become very slow. In addition to waiting for food, either sex may wait many months for a mate and adults may remain together in copulation for more than a week.

Tick-borne Diseases. The several disorders and diseases of vertebrates traceable to ticks may be divided into two groups, as suggested by Arthur:[5]

1. *Local inflammatory and traumatic damage at the site of attachment* may be a mild inflammation and itching, or it may be far more serious, such as the invasion of the auditory canal by the spinose ear tick, *Otobius megnini*, causing edema, hemorrhage, thickening of the stratum corneum, and partial deafness. Such an invasion of the auditory canal is common in cattle. Adults of this tick do not feed at all. *Amblyomma maculatum* (Fig. 21–5) sometimes occurs in enormous numbers on ground-feeding birds such as meadowlarks; it also is a serious parasite of livestock, in which it attacks the inner surface of the outer ear; it may attack man as well.

2. *Systemic damage* may result in tick paralysis

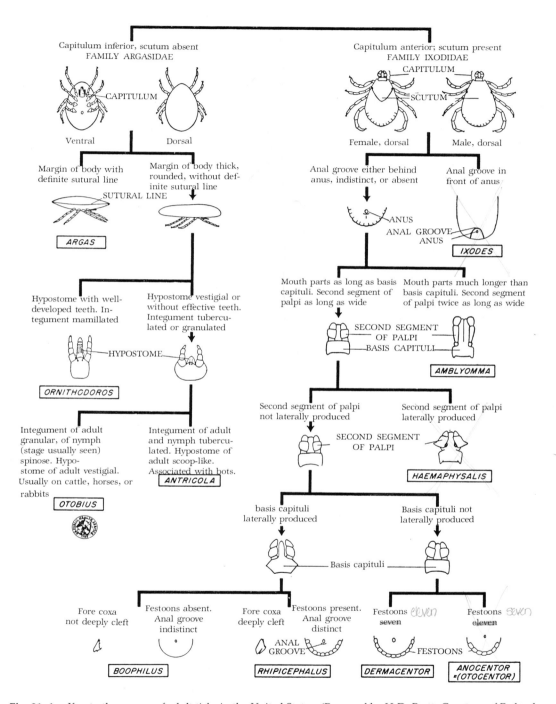

Fig. 21–4. Key to the genera of adult ticks in the United States. (Prepared by H.D. Pratt. Courtesy of Federal Security Agency, Public Health Service, Communicable Disease Center, Atlanta, Georgia. December 1948. Rev. Nov. 1951.)

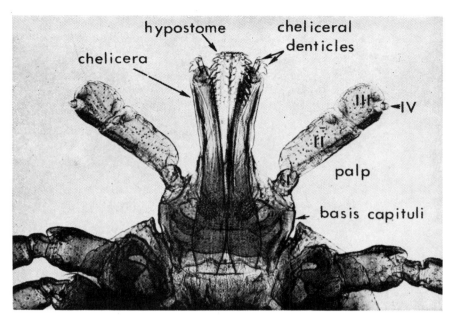

Fig. 21–5. Capitulum of *Amblyomma* sp. (From Georgi: Parasitology for Veterinarians, courtesy of W.B. Saunders.)

or in a less severe form of sensitization reaction to a toxic substance secreted by the salivary glands of the tick. Tick paralysis is commonly seen in domestic animals and occasionally in man, especially children. Many ixodid ticks, for example, *Dermacentor andersoni,* have been implicated, and a few argasids as well. The toxic substance causes a progressive, ascending, flaccid motor paralysis, elevation of temperature, impairment of respiration, speech, and swallowing, and, occasionally, death due to respiratory or cardiac paralysis. Removal of the tick produces a dramatically rapid recovery unless the condition of the patient has deteriorated.

Some animals develop an immunologically based tolerance to tick bites after exposure, for example, cattle to the bites of *Boophilus* and guinea pigs to the bites of *Amblyomma.* The reaction in cattle is thought to be mediated by eosinophils and that in guinea pigs by basophils. There is a curious case described in the literature[8] of a man who usually repelled ticks; most ticks placed on him died, apparently because of some toxic principle produced in his perspiration.

Ticks frequently serve as vectors of diseases caused by viruses; bacteria, including rickettsiae; protozoa; and occasionally, filariae.

Among the viruses transmitted by ticks are many Flaviviruses (Omsk haemorrhagic fever virus, Kyasanur Forest disease virus, etc.); Orbiviruses (Colorado tick fever virus); and Bunyaviruses (Crimean haemorrhagic fever virus), as well as a large number of unclassified viruses.

Tick-borne relapsing fever is caused by spirochetes of the genus *Borrelia* which are transmitted by ticks of the genus *Ornithodoros* (Fig. 21–6). Many species of *Borrelia* are involved, the species being largely defined on the basis of the species of tick vector and the mammalian reservoir of the bacteria. In California, for example, the most frequent type of relapsing fever is caused by *Borrelia hermsii* transmitted by *Ornithodoros hermsi.* It is associated especially with chipmunks of the genus *Eutamias* and red squirrels of the genus *Tamiasciurus.* Neither *O. parkeri* nor *O. turicata* are able to transmit the spirochetes although each has its own species of *Borrelia.* Relapsing fever in man is characterized by recurrent febrile paroxysms; 3 or 4 such attacks recur at intervals of about a week, until immunity is established; the case fatality rate is about 4%. The spirochetes are transmitted vertically in ticks. *Ornithodoros* species have a remarkable ability to survive under conditions of starvation and low humidity. A thin layer of wax

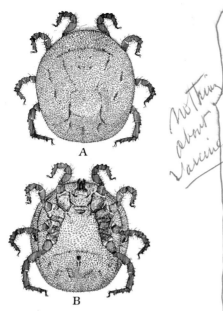

Fig. 21–6. *Ornithodoros moubata*, dorsal (*A*) and ventral (*B*) views. (From Hoogstraal. African Ixodoidea. United States Naval Medical Research Unit No. 3, Cairo.)

in the epicuticle reduces water loss, and the ability to close the spiracles and to extract water from moist air assists the tick in its regulation of water balance. Tick-borne relapsing fever is a widely spread disease. Avian strains of *Borrelia* are transmitted by ticks of the genus *Argas.*

Recently a syndrome was recognized that is called Lyme disease, after the town of Lyme in Connecticut. The infection is characterized by a circular, erythematous lesion of the skin called *erythema chronicum migrans* and produces a recurrent arthritic condition. The disease is focal in distribution and, in the United States, most commonly affects children. It has now been shown to be caused by a spirochete, *Borrelia burgdorferi,* transmitted by a previously unrecognized species of tick, *Ixodes dammini.* The disease is now known to occur in the western United States and in Europe as well, transmitted usually by species of *Ixodes.*

Texas cattle fever (tick fever, red water) is widely distributed among cattle in Europe, Africa, Australia, the Philippines, and North, Central, and South America. The causative agent, *Babesia bigemina* and related species, passes from the infected mother tick to her offspring through the eggs. The parasites multiply

within tissues of the tick and in red blood cells of cattle. *Boophilus annulatus* in North and South America and Europe, *Rhipicephalus appendiculatus* in tropical Africa, and other species of ticks are common vectors. The protozoan genus *Babesia* causes other diseases commonly known as "piroplasmosis" in domestic animals. Babesiosis, as well as other diseases of animals in Northern and Eastern Europe, is carried by *Ixodes ricinus.* Species of *Babesia* also infect humans although these are all zoonotic infections, animal infections transmitted to man. The earliest cases seen were in people who had been splenectomized and thus had compromised immune systems. Subsequently, babesiosis has been seen in people with intact spleens, usually older people. Most of the cases in the United States have come from Nantucket Island and Martha's Vineyard, islands off the coast of Massachusetts. The causative agent in these cases was *Babesia microti,* a parasite of field voles (*Microtus*), transmitted by *Ixodes dammini,* the vector of Lyme disease also in that area. The parasite is not transmitted vertically (mother to offspring) in ticks; larvae acquire the infection from rodents and transmit as nymphs; adult ticks are not known to transmit the infection.

Tularemia is caused by a plague-like bacterium, *Francisella tularensis*; its symptoms are similar to those of plague. It is found in mammals, birds, and man, and is most common in lagomorphs and rodents. The disease is characterized by a local ulcer at the site of inoculation and enlarged lymph nodes. It is transmitted by bite or through the feces of *Dermacentor andersoni* (Fig. 21–7), *D. variabilis,* or *Amblyomma americanum.* The bacterium may pass from one generation of tick to another through the eggs and may be acquired by man by simply handling diseased animals or by the bite of an infected tick; fortunately, it is seldom fatal.

American tick-borne typhus caused by *Rickettsia rickettsii* is an acute, febrile disease of wild rodents transmissible to man and laboratory animals. It occurs throughout the Americas where it is variously known as Rocky Mountain spotted fever, Tobia fever, or Sao Paulo typhus, but has been most extensively studied in the United States. In the Rocky Mountain region it is found chiefly in adults but in the eastern United States it affects mostly children.[18] After an incubation period of 5 to 10 days, a chill and rise in temperature is followed by a rash over the face and trunk. There is a higher case fatality rate in the

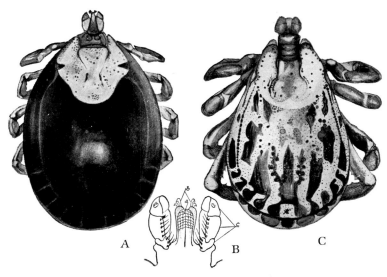

Fig. 21–7. *Dermacentor andersoni.* A, Dorsal view of female. B, Capitulum showing hypostome (*a*), chelicerae (*b*), and palp (*c*). C, Dorsal view of male. Enlarged. (From Stitt's Diagnostics and Treatment of Tropical Disease, courtesy of the Blakiston Division, McGraw-Hill.)

western United States than in the East, associated with higher age at onset of disease. The disease in the West is transmitted by *Dermacentor andersoni*, a tick associated with rural areas that comes into contact chiefly with adults. In the East the disease is transmitted by *Dermacentor variabilis*, the American dog tick, which is found especially in suburban and residential areas and commonly attacks children. Rickettsial diseases tend to be less severe in younger people so the lesser virulence of the disease in the East is related to the age of the patient which, in turn, is related to the habits of the vector. Other species of ticks are important vectors in the southern United States and Latin America. There are many other types of tickborne typhus in other parts of the world caused by other species of rickettsiae.

Although dogs frequently are infested with ticks, cats rarely have them. The brown dog tick, *Rhipicephalus sanguineus* (Fig. 21–8), is one of the most common ticks on cats as well as dogs. Other ticks found on dogs in the United States include *Dermacentor andersoni, D. variabilis, Amblyomma maculatum,* and species of *Ixodes.*

For a review of tick-borne diseases of man, see Hoogstraal,[12] Wilde,[21] and Balashov.[6] For ticks and diseases, see Arthur[3] and Burgdorfer.[10] For tick feeding and its implications, see Arthur.[2] A checklist of families, genera, species,

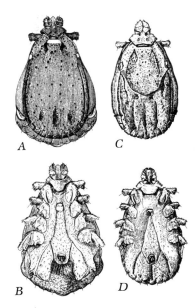

Fig. 21–8. *Rhipicephalus sanguineus.* A and B, Male. C and D, Female. (From Hoogstraal. African Ixodoidea, United States Naval Medical Research Unit No. 3, Cairo.)

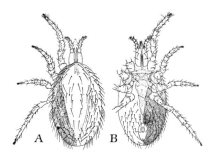

Fig. 21–9. *Ornithonyssus bacoti. A,* Dorsal view of female. *B,* Ventral view of female. (From Baker, et al.: A Manual of Parasitic Mites of Medical or Economic Importance, courtesy of the National Pest Control Association.)

and subspecies of ticks is given by Doss and Anastos.[11]

Suborder Mesostigmata

Mesostigmatid mites are tick-like and usually have spiracular openings lateral of the third pair of legs associated with an elongate, sinuous plate, the peritreme (Fig. 21–9). They are more closely related to ticks than to other mites, the 2 groups sometimes being united as the Parasitiformes. The life cycle usually includes 2 nymphal stages in addition to the egg, larval, and adult stages. In parasitic forms there may be one or more non-feeding stages. In the tropical rat mite, *Ornithonyssus bacoti,* for example, the adult and the first nymphal stage (protonymph) are the only stages that feed on the host; the larval and second nymphal (deutonymph) stages do not feed. There are many families of parasitic mesostigmatid mites, the most important being the Macronyssidae and Dermanyssidae, although there are many other interesting families. The Macronyssidae includes a number of economically important mites, such as *Ornithonyssus sylviarum* and *O. bursa,* important parasites of poultry and birds. The tropical rat mite, *O. bacoti* (Fig. 21–9), is a frequent pest of humans because it may become numerous in rat-infested houses and will then attack humans, although it does not thrive on them. This interesting mite originally was a parasite of the cotton rat, *Sigmodon hispidus,* and is a vector of the filarial worm of that animal, *Litomosoides carinii;* in recent times it transferred to rats of the genus *Rattus* and quickly spread throughout the world. It is now one of the commonest parasites of *Rattus rattus* and other species in the genus. The genus *Ophionyssus* includes mites that live under the scales of, and feed on, snakes, frequently causing serious problems in zoos. The family Dermanyssidae includes *Dermanyssus gallinae,* an important pest of poultry, and the mouse mite *Liponyssoides sanguineus.* The latter mite is a vector of rickettsialpox, a mild, febrile disease caused by *Rickettsia akari* that was first described from New York City, having caused epidemics in middle-income apartment buildings that had inadequate refuse disposal systems.

The Laelapidae is a large family of tick-like mites that are predacious or parasitic. The parasitic forms usually do not have blood in the gut and so are not bloodsuckers but probably feed on tissue fluids. The common rat mite, *Laelaps echidninus,* is the vector of a malaria-like protozoan parasite of the genus *Hepatozoon* in rats.

There are a number of familes of mesostigmatid lung mites, the Entonyssidae, found in the respiratory passages of snakes; the Halarachnidae in the respiratory passages of marine mammals, simians, dogs, and ground squirrels; and the Rhinonyssidae in the nasal passages of birds. The Spinturnicidae are large, spider-like ectoparasites of bats.

Suborder Prostigmata

Prostigmatid mites are so named because some have spiracles near and associated with the gnathosoma. There are many groups of parasitic prostigmatid mites. The Trombiculidae are the most important from a medical standpoint, because chiggers are in this family. Chiggers are the parasitic larval stage of trombiculid mites and are the vectors of a rickettsia, *Rickettsia tsutsugamushi,* that causes disease in man. The life cycle of prostigmatids is similar to that of most mites, egg, larva, nymph, and adult, except that quiescent stages (deutovum, nymphochrysalis, and imagochrysalis) may be interpolated into the life cycle between the usual four stages.

The family Trombiculidae is composed of mites whose larvae are known as chiggers. The family is large, including about 1000 species, and its taxonomy is difficult, being based almost exclusively on larvae, the only parasitic stage in the life cycle (Fig. 21–10), thus the stage that is most easily collected. (For keys to the genera of chiggers, see Brennan and Goff.[9]) The larva attacks a single host (mammal, bird, reptile, or amphibian), engorges, and molts to the next stage (nymphochrysalis). Both the nymphal

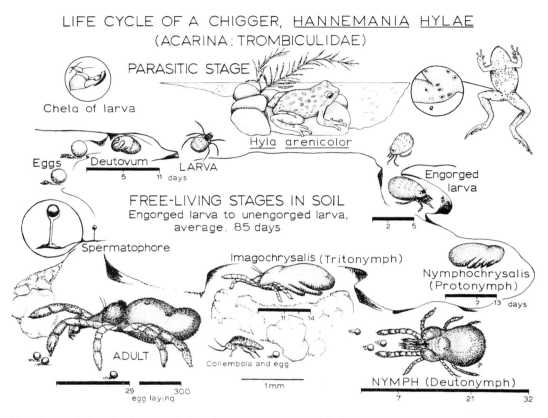

LIFE CYCLE OF A CHIGGER, HANNEMANIA HYLAE
(ACARINA : TROMBICULIDAE)

PARASITIC STAGE

Chela of larva

Hyla arenicolor

Eggs   Deutovum   LARVA   Engorged
          5   11 days   larva

FREE-LIVING STAGES IN SOIL
Engorged larva to unengorged larva,
average. 85 days

Spermatophore

Imagochrysalis (Tritonymph)   Nymphochrysalis
                                      (Protonymph)
                                        7   13 days

ADULT

Collembola and egg   1mm   NYMPH (Deutonymph)

    29   300   7   21   32
  egg laying

Fig. 21–10. Life cycle of a trombiculid mite. (Courtesy of Dr. R.B. Loomis.)

and adult stages are predacious, feeding on small arthropods. Since the larva feeds on only one host in its life, it has no opportunity to transmit disease agents except to the next generation of mites. Nevertheless, chiggers are the vectors of a rickettsial disease, scrub typhus (or tsutsugamushi), so named because the disease frequently is associated with areas of scrub vegetation in Asia. The causative agent, *Rickettsia tsutsugamushi*, is closely associated with chiggers, especially of the genus *Leptotrombidium*, and may be a mutualist of chiggers; at least chigger colonies freed of the rickettsiae are difficult to maintain and it is difficult to infect chiggers with the rickettsiae by feeding them on vertebrates with rickettsemia. When chiggers feed on a vertebrate, their salivary secretions harden in the skin to form a tube called a stylostome through which the fluids or digested tissues of the host pass. Feeding of some chiggers, especially of the genus *Eutrombicula*, causes an intense irritation of the skin, espe-

cially in the immunologically naive host; scratching of the feeding sites produces unsightly sores that frequently become infected with bacteria. Persons who are repeatedly exposed to these chiggers, as in the midwestern areas of the southern United States, usually become tolerant of the bites. In southern California, chiggers of this group, *E. belkini*, have a spotty distribution and in one instance, the characteristic feeding lesions on a suspected murderer were used to place him at the scene of his crime.[20]

The Trombidiidae are similar to trombiculids except that the larvae parasitize insects rather than vertebrates. Mites of the related family Pterygosomidae have a similar life cycle and commonly parasitize lizards; *Geckobiella texana* is a common parasite of the western fence lizard, *Sceloporus occidentalis*, in the western United States.

Among the most interesting of the prostigmatid mites are the follicular mites, in the Dem-

Fig. 21–11. *Demodex canis,* ventral view of female. (From Baker, et al.: A Manual of Parasitic Mites of Medical or Economic Importance, courtesy of the National Pest Control Association.)

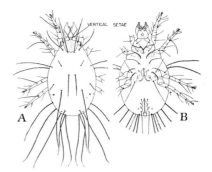

Fig. 21–12. *Tyrophagus putrescentiae. A,* Dorsal view of female. *B,* Ventral view of female. (From Baker, et al.: A Manual of Parasitic Mites of Medical or Economic Importance, courtesy of the National Pest Control Association.)

odicidae (Fig. 21–11). These mites inhabit the hair follicles or sebaceous glands of many species of mammals. Two species are found in man, *Demodex folliculorum* in hair follicles and *D. brevis* in sebaceous glands. Infestation of humans is probably universal, although the mites frequently are difficult to demonstrate; they occur in the nipples of the breasts and it is postulated that transmission occurs to infants during the period of nursing. Pathologic conditions in the human usually are not great although some skin conditions, for example rosacea, are thought to be associated with the mites; demodectic granulomas also occur. Demodicosis may be more severe in other animals; *D. canis* in dogs, for example, produces a type of mange that is rather refractory to treatment; the mites have been found in visceral lymph nodes of dogs.

A number of other families of prostigmatid mites are parasitic on vertebrates. Mites of the family Speleognathidae are parasitic in the respiratory passages of birds. Mites of the family Myobiidae have the forelegs highly modified for grasping hairs; the family contains many skin parasites of mammals. Birds are parasitized by mites of the family Syringophilidae, long, thin mites found in the quills of birds, and the Harpyrhynchidae, which live on the skin in feather follicles, which enlarge to form tumors filled with mites. The family Cheyletiellidae includes parasites of mammals and birds; these mites have a prominent hooked claw distally on each palp. *Cheyletiella yasguri* is a skin parasite of dogs and *C. blakei* of cats; the mites may transfer from heavily infested animals to people; causing irritation and dermatitis; the infestation, however, is temporary. *C. parasitivorax* may cause a severe type of mange in rabbits, which, how-

ever, is easily treated. *Ornithocheyletia* species infest birds.

Other groups of prostigmatid mites are parasitic on invertebrates. *Pyemotes tritici* and other species of this genus in the family Pyemotidae are interesting and important parasites of insects. The life cycle is telescoped; a female attacks an insect, attaches to it, and, as it feeds, becomes enormously distended with young. She gives birth to large numbers of mature male and female mites; after mating, the females disperse and seek new hosts. The mites occur in such large numbers that they kill their hosts and, for this reason, are being studied for the control of insect pests such as the imported fire ant, *Solenopsis invicta.* Unfortunately, they will also attack man or other vertebrates; "grain itch" or "straw itch" is a dermatitis caused by these mites in people handling grain infested with lepidopterous larvae parasitized by pyemotids.

Suborder Astigmata

Astigmatid mites have no spiracles; they are weakly sclerotized and frequently have prominent apodemes ventrally in the leg-bearing segments. Many of the families of astigmatid mites are parasitic. Most have simple life cycles, consisting of egg, larval, nymphal, and adult stages. In some cases males have only one nymphal stage and females two; in such cases males are only half the size of females. Many of the mites have a dormant stage, the hypopus, which frequently is the stage that disperses by attaching to a host; the common yeast mite, *Tyrophagus putrescentiae* (Fig. 21–12), for exam-

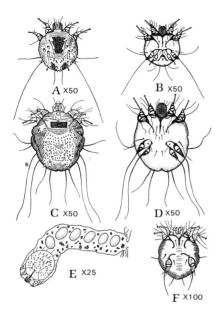

Fig. 21–13. *Sarcoptes scabiei. A,* Male, dorsal view. *B,* Male, ventral view. *C,* Female, dorsal view. *D,* Female, ventral view. *E,* Female with eggs in burrow. *F,* Hexapod larva, ventral view. (*A* to *D,* redrawn from Munro; *E,* redrawn from Banks; *F,* redrawn from Blanchard. Reduced 50% from original magnification.) (From Belding: Textbook of Clinical Parasitology, courtesy of Appleton-Century-Crofts.)

ple, under adverse conditions produces hypopi, which have suckers and attach to insects by which they are transported to other sites. Rodents and other small mammals frequently are infested with large numbers of hypopi of mites of the family Labidophoridae that breed in their nests; these mites are not parasitic but merely use the host for dispersal.

The infestation of animals by mites is known as acariasis. If the condition results in serious or persistent pathology, the disease is known as mange. Many groups of astigmatid mites as well as several groups of prostigmatid mites produce mange; it is customary to specify the cause of the mange as demodectic mange, sarcoptic mange, etc.

The best-known astigmatid mites are in the family Sarcoptidae. The "itch" or scabies, caused by *Sarcoptes scabiei* (Fig. 21–13), is a parasitic infestation that has plagued man through the ages. These mites are small and have spines on their back and reduced legs; the male is only half as large as the female and its hind legs terminate in suckers rather than in hairs as in the female. It is interesting to contemplate that there was never a successful treatment for this infestation until recent times; probably all humans, for thousands of years, have been infested with *Sarcoptes,* with no hope of being able to cure the infestation until recently. It is probable that people with intact immune systems eventually controlled the infestation so that it disappeared or remained at a low level, but those with defective immune systems sometimes developed severe, fulminating infestations that on occasion terminated fatally. This condition, known as "Norwegian scabies," occasionally is seen even today. Sarcoptid mites burrow in the skin, the female laying 2 or 3 eggs a day as she burrows. The immature stages occur in the burrows, in hair follicles, or they may begin burrows of their own. Males have a single nymphal stage; the adults attach to the second nymphal stage of females and mating takes place when the female emerges. The activity of the mites is regulated by the temperature of the host. It is greatest at night when the host sleeps and its blood is diverted to the skin; the increased activity of the mites at this time causes great irritation, and the awakened host scratches, further irritating the skin and sometimes causing secondary infection. A rash is characteristic of scabies. Transmission of the disease is by contact, adult females being passed from one host to another. Treatment of the disease in man is now satisfactory; a variety of non-toxic acaricides are available (lindane, benzyl benzoate, pyrethrins, etc.). The total population of mites in human scabies usually is no more than 12 ovigerous females in adults and 20 in children. For futher information, see Arthur.[1]

Sarcoptic mange also occurs in many other animals, pigs, dogs, horses, cattle, etc. The mites that cause these infestations are morphologically similar to the mites of humans and usually are placed in the same species, *S. scabiei;* however, infestations of humans by the animal forms usually are less virulent and less persistent than infestations by the human form, so the animal forms would appear to constitute at least biologic, or physiologic races, if not subspecies. Also in the family Sarcoptidae are the genera *Notoedres,* which produces mange in mice and cats, and *Knemidocoptes,* causing scaly leg in poultry.

"Scab" in sheep or cattle is caused by nonburrowing mites of the family Psoroptidae.

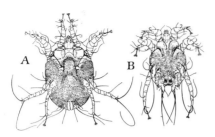

Fig. 21–14. *Psoroptes equi. A*, Ventral view of female. *B*, Ventral view of male. (From Baker, et al.: A Manual of Parasitic Mites of Medical or Economic Importance, courtesy of the National Pest Control Association.)

These mites are somewhat larger and more robust than sarcoptids and have longer legs; the male are smaller than the females and have prominent anal suckers and reduced hind legs. Mites of the genus *Psoroptes* (Fig. 21–14) produce important diseases in sheep, cattle, rabbits, and other animals. The mites are found on the surface of the body among the scabs at the base of the hairs. *P. cuniculi* produce ear canker in rabbits, a serious disease that may terminate fatally; it can be prevented by routine cleaning of the ears with mineral oil. The related genus *Otodectes* causes ear mange in dogs and cats, and the genus *Chorioptes* produces foot mange in equids and bovids.

There are many other interesting groups of parasites in the Astigmata. The Gastronyssidae are found in the stomach of bats; they are true gastric parasites. The Cytoditidae inhabit the respiratory system of chickens, and the Laminisioptidae are found in subcutaneous tissues of birds; some of the latter have fantastic subcutaneous hypopi. The Listrophoridae are elongate mammal parasites that have mouthparts modified for grasping hairs. The Myocoptidae (Fig. 21–15) have the hind legs modified for grasping hairs. Practically all birds have "feather mites" that live on feathers or skin; there are several families of these, such as the Analgidae.

Perhaps the most important family of the Astigmata today is the Epidermoptidae. The genus *Dermatophagoides* is a curious assemblage of nest mites of birds, some of which occur on the birds as well; some also occur on mice and humans. Although they may cause dermatitis in humans, their chief importance is that some *(D. pteronyssinus, D. farinae)* breed in the skin scales that accumulate in human habitations. Collectively, these and related mites are called "house dust mites." The common condition "house dust allergy" has been traced to these mites. They thrive in moist habitats in which there is a supply of animal material such as skin scales, a habitat that is found in much-used easy chairs, sofas, or mattresses. The mites have been found in houses throughout the world. Extracts of the mites produce cutaneous reactions in practically all people with house dust allergy, which is said to be one of the commonest allergies of man. Control of the mites is difficult. Replacement of old furniture and lowering the humidity of houses is desirable; acaricidal treatments generally have not been efficacious.

As with *Dermatophagoides*, some other astigmatid mites that are not parasitic may produce dermatitis in man. *Tyrophagus putrescentiae* (Fig. 21–12), which breeds abundantly on drying coconut meat that has been harvested for oil, may produce a dermatitis known as "copra itch" when the coconut is handled. A similar dermatitis known as "grocer's itch" is caused by *Glycyphagus domesticus*, which breeds in dried fruit, skins, or feathers.

Suborder Cryptostigmata

These mites, commonly known as oribatid mites, are beetle-like, free-living mites (Fig. 21–16). They live in the soil and in vegetation and are commonly ingested by sheep and other grazing animals. They serve as intermediate hosts for most tapeworms of the family Anoplocephalidae whose life cycles are known. The worms *Moniezia expansa* of sheep, *Bertiella studeri* of primates, and *Cittotaenia ctenoides* and *C. denticulata* of rabbits are well-known examples. There seems to be little specificity of the tapeworms for intermediate hosts.

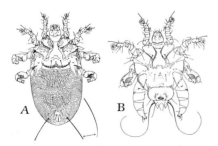

Fig. 21–15. *Myocoptes musculinus. A*, Ventral view of female. *B*, Ventral view of male. (From Baker, et al.: A Manual of Parasitic Mites of Medical or Economic Importance, courtesy of the National Pest Control Association.)

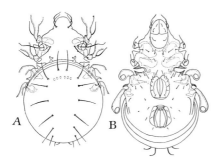

Fig. 21–16. An oribatid mite, *Metabelba papillipes,* female. *A,* Dorsal view. *B,* Ventral view. Legs removed. (From Baker, et al.: A Manual of Parasitic Mites of Medical or Economic Importance, courtesy of the National Pest Control Association.)

REFERENCES

1. Arthur, D.R.: Interaction between arthropod ectoparasites and warm blooded hosts. *In* Ecological Aspects of Parasitology. Edited by C.R. Kennedy. New York, Elsevier-North Holland, 1976, pp. 163–183.
2. Arthur, D.R.: Tick feeding and its implications. Adv. Parasitol., *8*:275–292, 1970.
3. Arthur, D.R.: The ecology of ticks with reference to the transmission of Protozoa. *In* Biology of Parasites. Edited by E.J.L. Soulsby. New York, Academic Press, 1966, pp. 61–84.
4. Arthur, D.R.: Feeding in ectoparasitic acari with special reference to ticks. Adv. Parasitol., 3:249–298, 1965.
5. Arthur, D.R.: Ticks and Disease. New York, Harper & Row, 1962.
6. Balashov, Y.S.: Bloodsucking Ticks. Miscellaneous Publication No. 8, College Park, Md. Entomology Society of America, 1972, pp. 161–376.
7. Baker, E.W., and Wharton, G.W.: An Introduction to Acarology. New York, Macmillan Co., 1952.
8. Brennan, J.M.: An instance of the apparent toxicity of man to ticks. J. Parasitol., *33*:491–494, 1947.
9. Brennan, J.M., and Goff, M.L.: Keys to the genera of chiggers of the Western Hemisphere (Acarina: Trombiculidae). J. Parasitol., *63*:554–566, 1977.
10. Burgdorfer, W.: Tick-borne diseases in the United States: Rocky Mountain spotted fever and Colorado tick fever. Acta Trop., *34*:103–126, 1977.
11. Doss, M.A., and Anastos, G.: Ticks and Tickborne Diseases III. Checklist of Families, Genera, Species and Subspecies of Ticks. Special Publication No. 3, Index-Catalogue of Medical and Veterinary Zoology. Washington, D.C., U.S. Government Printing Office, 1977.
12. Hoogstraal, H.: Biology of ticks. *In* Tick-Borne Diseases and Their Vectors. Edited by J.K.H. Wilde. Edinburgh, Edinburgh University Press, 1978, pp. 3–14.
13. Krantz, G.W.: A Manual of Acarology. 2nd Ed. Corvallis, Oregon State University Book Stores, 1978.
14. Londono, M.I.: Transmission of microfilariae and infective larvae of *Dipetalonema viteae* (Filarioidea) among vector ticks, *Ornithodoros tartakowskyi* (Argasidae), and loss of microfilariae in coxal fluid. J. Parasitol., *62*:786–788, 1976.
15. Nutting, W.B. (ed.): Mammalian Diseases and Arachnids. Boca Raton, Fla., CRC Press, 1984.
16. Rothschild, M., and Clay, T.: Fleas, Flukes and Cuckoos. London, Wm. Collins and Sons, 1952.
17. Savory, T.: Arachnida. 2nd Ed. London, Academic Press, 1977.
18. Sonenshine, D.E., Peters, A.H., and Levy, G.F.: Rocky Mountain spotted fever in relation to vegetation in the eastern United States, 1951–1971. Am. J. Epidemiol., *96*:59–69, 1972.
19. Treat, A.E.: Mites of Moths and Butterflies. Ithaca, Cornell University Press, 1975.
20. Webb, J.P. Jr., et al.: The chigger species *Eutrombicula belkini* Gould (Acari: Trombiculidae) as a forensic tool in a homicide investigation in Ventura Country, California. Bull. Soc. Vect. Ecol., *8*:141–146, 1983.
21. Wilde, J.K.H. (ed.): Tick-Borne Diseases and Their Vectors. Edinburgh, Edinburgh University Press, 1978.

22

Miscellaneous Phyla

PHYLUM PORIFERA

Sponges are rarely parasitic. One species, *Cliona celata*, bores into shells of molluscs. This species may reduce oyster shell substance by as much as 40% by tunneling. Sponge activity may cause a break through the inner shell layer. These breaks are repaired if the water is warm (above 7° C); otherwise, the hole remains unrepaired. The mollusc's adductor and hinge muscles may also be damaged.

PHYLUM CNIDARIA

Cnidarians, formerly referred to as coelenterates, are aquatic organisms chiefly characterized by the presence of tentacles armed with stinging cells known as nematocysts. The life cycle typically involves an alternation of asexual polypoid and sexual medusoid generations, although one or the other stage may be reduced or absent. Little is known about the biology of the ecto- and endosymbiotic relationships that have been described. Parasites are found in only two of the four classes, namely the Hydrozoa and Anthozoa.

Class Hydrozoa

A number of hydrozoan polyps live in close association with a diversity of phyla.[5] In obligtory relationships the bodies of the polyps are modified, the tentacles reduced or absent, and the mouths enlarged and suctorial.

Proboscidactyla colonies are found only on the rims of sabellid polychaete tubes, *Kinetocodium* grows on the shells of pelagic pteropod mollusks, and *Entocrypta* lives inside the branchial chambers of tunicates. All three genera appear to be energy commensals. Normal, free-swimming medusae are produced.

Hydrichthys is a hydroid parasite that lives on the bodies of marine fishes and, occasionally, on parasitic crustaceans attached to fishes.[25] The genus has lost its tentacles, and each polyp feeds on host blood and tissues by means of root-like stolons that penetrate the body of the fish. Atypical, sluggish-swimming medusae are released.

Polypodium hydriforme is a highly modified, intracellular parasite of the eggs of acipenserid fishes (sturgeons) in Russia and the United States. Two-layered stolons of the polyp stage penetrate eggs (oocytes) of the host and feed on the yolk. The life cycle (Fig. 22–1) involves the alternation of a parasitic and free-living generation, but medusae are unknown.[32,33]

Narcomedusae are oceanic hydrozoans that lack a polyp stage. *Cunina proboscidea* has a complex life cycle involving two medusoid generations, and a parasitic larval stage. Flattened, stolon-like planulae attach to the hydrozoan medusa, *Geryonia*. Buds from the parasitic stage develop into free-swimming medusae.

Class Anthozoa

Peachia quinquecapitata is a sea anemone whose larval stages are parasitic in the gastrovascular cavity or on the body surface of hydrozoan medusae and ctenophores. The anemone maintains its position with the aid of a sucker-like mouth. *Peachia* initially feeds on particles of food in the host's gut. Later, during the ectoparasitic phase, it feeds on the host's gonadal tissue.[39]

431

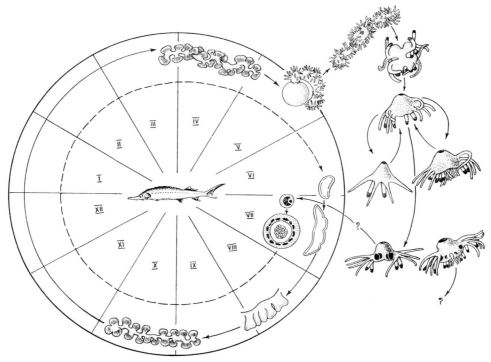

Fig. 22–1. Life cycle of *Polypodium hydriforme* in the sturgeon. Parasitic stages are inside the circle and free-living stages are outside the circle. (From Raikova, courtesy of the Seto Marine Biological Laboratory.)

PHYLUM CTENOPHORA

Lampea pancerina (formerly *Gastrodes parasiticum*) (Fig. 22–2) is a ctenophore whose young are commensal in the mantle of the tunicate, *Salpa fusiformis*. The ctenophore is a flattened

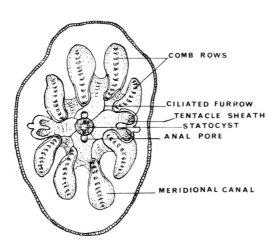

Fig. 22–2. Larva of *Lampea pancerina* found in the tunicate *Salpa*. (Redrawn after Komai.)

organism about 1 mm in diameter, round in outline, with a concave oral surface, and an arched aboral surface. Larvae of *Lampea* are of the planula type, thus suggesting a relationship to the Cnidaria. The larvae develop typical ctenophore features: eight comb rows; statocysts; tentacles armed with colloblasts; and a simple gastrovascular system. The parasite leaves its host, matures as a free-living organism, and produces planula larvae that enter new hosts.[19]

MESOZOA

The word **Mesozoa** is best used as an organizational term to denote a superficial similarity between two groups of specialized endoparasites. Although the phylogenetic origins of these parasites are obscure, they are not closely related. The Dicyemida and Orthonectida are recognized by most authorities as distinct and separate phyla. The mesozoans are a fascinating zoological puzzle and deserve more attention. For reviews see the papers by Caullery,[3] Grasse,[9] Hochberg,[11] and McConnaughey.[22]

PHYLUM DICYEMIDA

Benthic cephalopods in many parts of the world contain vermiform dicyemid parasites. With few exceptions, they are found attached to renal appendages by means of a modified anterior end (**calotte**), hanging into a urine-filled renal coelom. Their simple structure consists of a single external layer of ciliated cells surrounding one or more axial reproductive cells. None of the tissue and organ systems common in multicellular organisms is present.

The life cycle involves asexual and sexual generations. **Nematogens** in immature cephalopods are characterized by vermiform embryos that are contained in the internal reproductive cell. A transition to the **rhombogen** phase occurs in mature cephalopods. This phase is characterized by nonciliated hermaphrodites (**infusorigens**) that give rise to free-swimming **infusoriform larvae.** Infusoriforms grow within the parent rhombogen. When mature, they are released and leave the host with urine. Development outside the host is unknown, as is the manner of infection of new hosts by larval nematogens.

Dicyema sullivani (Fig. 22–3) serves as an example of the group. It is a parasite of *Octopus bimaculoides* from southern California and Baja California. Adult stages (nematogens and rhombogens) are slender and average 0.75 to 1.5 mm long. Vermiform embryos at the time of emergence measure about 100 μm; infusoriforms lengths are from 40 to 48 μm.

A generalized classification of the dicyemids is as follows:

Family DICYEMIDAE
  Genus *Dicyema* (e.g., *D. sullivani* from kidneys of *Octopus bimaculoides*)
  Dicyemennea (e.g., *D. brevicephaloides* from branchial hearts of *Rossia pacifica*)
  Dicyemodeca (e.g., *D. dogieli* from *Octopus dofleini*)
  Dodecadicyema (e.g., *D. loligoi* from kidneys of squid *Loligo*)
  Pleodicyema (e.g., *P. delamarei* from kidneys of *Bathypolypus sponsalis*)
  Pseudicyema (e.g., *P. truncatum* from kidneys of cuttlefish *Sepia officinalis*)

The following two genera are sometimes placed in a separate family.

Family CONOCYEMIDAE (= HETEROCYEMIDAE)
  Genus *Conocyema* (e.g., *C. polymorpha* from kidneys of *Octopus vulgaris*)

Microcyema (e.g., *M. vespa* from kidneys of *Sepia officinalis*)

PHYLUM ORTHONECTIDA

Orthonectids (Fig. 22–4) are rare parasites of marine organisms.[3] They occur in members of six invertebrate phyla: acoel, rhabdocoel, and polyclad turbellarians; nemerteans; polychaete annelids; gastropod and bivalve mollusks; ophiuroid echinoderms; and ascidian tunicates. Knowledge of the life cycle is more complete for orthonectids than for dicyemids. Because most of the work has been done on *Rhopalura ophiocomae*, this will serve as an example.[10] Free-swimming adults, on release from plasmodia in host ophiuroids, are typically minute (0.1 to 0.8 mm) and are either elongate or oval in shape. Externally, the somatoderm is marked off into rings by circular grooves. Reproductive cells are confined to the interior, much the same as in the dicyemids, but form a mass composed of several hundred cells. Ciliated larvae develop, are released from the female, and enter genital clefts of new host brittlestars. There they disintegrate and liberate germinal cells that penetrate the host cells. Multiplying asexually, they form plasmodia that spread throughout tissues and coelomic spaces and eventually give rise to more males and females.

In *Rhopalura*, the plasmodium stage invades the gonads, eventually destroying the germ cells and castrating the host. The plasmodial mass also may spread throughout the organs of the central disk and into the arms of the brittle star.

Two families and five genera are recognized. Some authorities feel that the similarities between rhopalurids and pelmatosphaerids are superficial, perhaps indicating that these two groups are diphyletic and are not closely related.

Family RHOPALURIDAE
  Genus *Rhopalura* (e.g., *R. ophiocomae* from the ophiuroid *Amphipholis squamata*)
  Ciliocincta (e.g., *C. sabellariae* from the polychaete *Sabellaria cementarium*)
  Intoshia (e.g., *I. leptoplanae* from the turbellarian *Leptoplana tremellaris*)
  Stoecharthrum (e.g., *S. giardi* from the annelid *Scoloplos mulleri*)

Family PELMATOSPHAERIDAE
  Genus *Pelmatosphaera* (e.g., *P. polycirri* from the annelid *Polycirrus haematoides*)

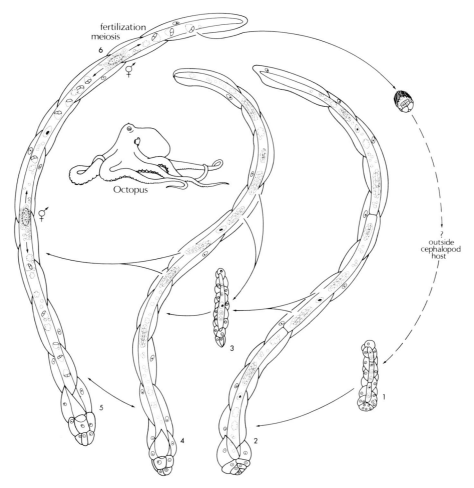

Fig. 22–3. Generalized life cycle of the dicyemid, *Dicyemennea,* in an octopus host. 1. larval stem nematogen; 2. stem nematogen; 3. vermiform embryo; 4. nematogen; 5. rhombogen; 6. infusorigen; 7. infusoriform larva released from parent. Density of parasites: A, low; B, high. (From Hochberg, courtesy of Malacologia.)

PHYLUM NEMERTEA

Almost all nemerteans are free-living, bottom-dwelling marine animals whose soft bodies are equipped with a conspicuous proboscis. Few commensal or parasitic species have been described. *Carcinonemertes* is a specialized egg predator that lives in intimate association with various crabs. Epidemics of this tiny nemertean have been implicated in the collapse of commercial crab fisheries.[43] The leech-like genus *Malacobdella* is found in the mantle cavity of bivalves but appears to be a commensal.[8] *Gononemertes parasitica* is another commensal species found exclusively in the atrial chamber of tunicates.[14] *Uchidaia parasita* (Fig. 22–5) lives in the mantle cavity of the clam *Mactra sulcataria* where it feeds on host gill tissues.[17]

PHYLUM ROTIFERA

Rotifers are dioecious microscopic animals abundant in fresh water, brackish water, and moist soil. A few species live in the ocean, and a few are parasites. A ciliary organ (corona) is located at the anterior end of the body; the pharynx is provided with movable jaws (mastax); and typical flame bulb protonephridia are pres-

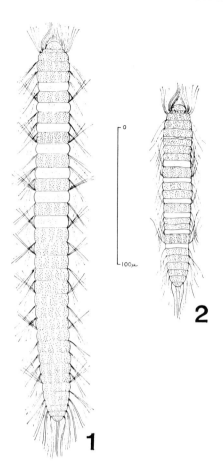

Fig. 22–5. *Uchidaia parasita,* a nemertean from the mantle cavity of a clam. (Redrawn from Iwata.)

Fig. 22–4. *Ciliocincta sabellariae,* an orthonectid from the polychaete annelid, *Sabellaria.* 1, Female. 2, Male. (From Kozloff, courtesy of J. Parasitol.)

ent. Male rotifers are generally much smaller than females and are rarely seen.

Seison nebaliae moves about like a leech on the gills of the crustacean, *Nebalia. Proales paguri* (Fig. 22–6) occurs on the gills of hermit crabs where it feeds on host epithelial cells. Other species of *Proales* invade the heliozoan, *Acanthocystis,* the eggs of freshwater snails, and the filament tips of the alga, *Vaucheria. Zelinkiella synaptae* is found in pits in the skin of sea cucumbers. *Albertia* is a transparent, worm-like rotifer that live in the intestine and coelom of earthworms and other annelids.

In parasitic rotifers, either the mastax or the foot is modified as an attachment organ, and the corona is reduced. For details see Hyman.[15]

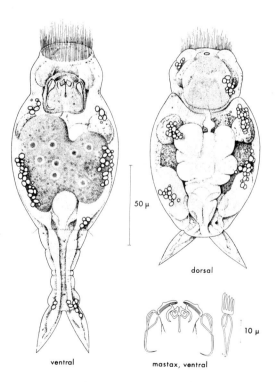

Fig. 22–6. *Proales paguri,* a rotifer from the gills of the hermit crab *Pagurus.* (From Thane-Fenchel, courtesy of Ophelia.)

PHYLUM NEMATOMORPHA

Occasionally, when looking in farmyard water troughs, ones sees a long (up to 1 m) and narrow, dark brown worm. Because of the common notion that horsehairs are transformed into these worms, they are popularly called "horsehair worms," "hair worms" or "hair snakes" (Fig. 22–7). Horsehairs, of course, have nothing to do with the worms. About 35 cases of infection in humans have been reported.

Most nematomorphs belong to the class Gordioidea, and inhabit fresh water. In the United States, the two most common species are *Gordius robustus* and *Paragordius varius*. In Europe, the common species is *Gordius aquaticus*.

In the gordioids, adults and larvae possess a degenerate, nonfunctional digestive tract. Food is absorbed through the tegument. The adult anterior end is translucent, with a dark ring behind the clear area. In the male the posterior end is bifurcate, whereas in the female, it is usually bluntly rounded or trifurcate.

Each female lays one million or more eggs in strings in water or in moist soil. Embryos possess an armed proboscis and head region that is separated from the rest of the body by a septum (Fig. 22–7). The proboscis becomes a perforating organ that enables larvae to emerge from eggs. Soon after escaping, young worms usually penetrate an arthropod, but may form cysts and be eaten. Arthropods serving as hosts include beetles, cockroaches, grasshoppers, crickets, centipedes, and millipedes. Larval worms make their way to the body cavity or fat bodies, where they start to mature to the adult stage. If eaten in the cyst form, the cyst wall is digested and the worm is released. A cyst may remain viable in water for two months.

If eaten by an inappropriate host, for example, an insect, snail, or fish, the cyst wall is digested, but the parasite secretes a new cyst covering, is able to withstand passage through the intestine, and thus returns to soil or water. If eaten again, this time by a suitable host, it can reach the adult stage. Just before reaching maturity in the insect, the worm escapes, possibly with the aid of a digestive enzyme. Worms do not leave the insect unless they can enter water; hence, they often are encountered after the first heavy fall rains. During their stay in the insect, they may castrate their host. In water, the worms copulate, and soon afterwards,

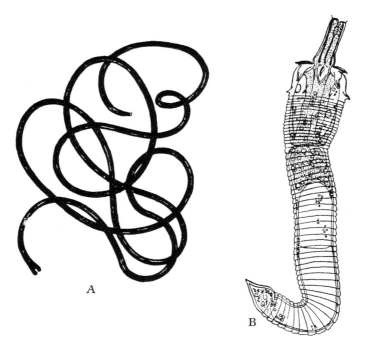

Fig. 22–7.   *A, Gordius* sp. An adult male "horsehair worm." ×1.5. *B,* Larva of *Gordius aquaticus.* (B from Baer: Ecology of Animal Parasites, courtesy of the University of Illinois Press.)

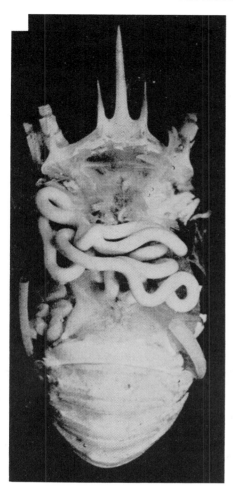

Fig. 22–8. *Nectonema munidae,* a large marine nematomorph in the body cavity of *Munida* (carapace removed). (From Nielsen, courtesy of Sarsia.)

egg-laying commences. Adults die in the fall of the year when all eggs have been laid. See Hyman[8] for details.

A second class of nematomorphs, represented by the genus *Nectonema*, is exclusively marine. Adults are free-living and pelagic, whereas larvae enter crustaceans such as hermit crabs, true crabs, and shrimps (Fig. 22–8). When heavily infected with large worms, the host may be castrated.[29]

PHYLUM ANNELIDA

The annelids are a large phylum of segmented worms. Two classes are recognized, the polychaeta and clitellata. Commensalism among both classes is common, but parasitism is rare. It is sometimes difficult, however, to distinguish between the two types of relationships.

Class Polychaeta

The polychaetes are the largest class of annelids. They are characterized by a well-defined head region, and a body with well-developed segmentation, each segment bearing a pair of parapodia armed with bristles or setae. Although the majority of the species are free-living and marine, many families have commensal or parasitic members.[4,5,30]

Ectocommensals are reported from six families of polychaetes. The scale worm *Ophiodromus pugettensis* lives on the body of sea cucumbers and in among the tube feet of starfishes. *Branchiopolynoe symmytilida* lives in the mantle cavity of giant deep-sea mussels from the Galapagos hydrothermal vents. Laboratory studies have shown that scale worms are attracted to chemical signals emanating from their hosts.[6]

All species of the genera *Histriobdella* and *Iphitime* are known only from the branchial chambers of lobsters and crabs. *Histriobdella* are cleaning symbionts that feed by grazing on the microflora in the branchial chamber, whereas *Iphitime* have strong, well-developed jaws, which suggests that they feed on the branchial tissues of their hosts. The aberrant, grub-like *Calamyzas amphictenicola* is an ectoparasite on the gills of ampharetid polychaetes.

Ichthyotomus sanguinarius is a bloodsucking parasite attached to fins of the eel, *Myrus vulgaris*, and to other fishes. The attachment organ involves two protrusible stylets that articulate with one another. The worm reaches 10 mm in length and is dorsoventrally flattened. It is a neotenic form which becomes sexually mature when only 2 mm long.

Among the arabellid polychaetes a number of species live inside polychaetes and echiurans.[30,41] These endoparasites usually invade the body cavity or vascular system, normally while their hosts are in early stages of development. They often grow to tremendous size in proportion to their hosts and may completely fill the body cavity. At least one species is reported to consume the reproductive products of its hosts. Such parasites include *Arabella endonata* in *Diopatra ornata*, *Haematocleptes terebellidis* in *Terebellides stroemii*, *Labrorostratus luteus*

(Fig. 22–9) in *Haplosyllis spongicola,* and *Oligonathus bonelliae* in the echiuran *Bonellia viridis.*

Myzostomes are highly modified parasitic polychaetes that occur exclusively on or in echinoderm hosts, principally crinoids. The aberrant body is dorsoventrally flattened and disklike (Fig. 22–10). The majority are protandrous hermaphrodites, becoming first males, then females. Some species burrow into their hosts and form cysts, within which a male/female pair is sometimes found. *Myzostomum pulvinar* occurs in the intestine of a crinoid, whereas *Protomyzostomum polynephris* lives in the coelomic cavity of the ophiuroid, *Gorgonocephalus. P. polynephris* feeds on genital glands of the ophiuroid and causes partial host castration. Some myzostomes are so firmly attached to the surface of their hosts that they leave permanent scars.

Class Clitellata

This class of annelids is distinguished from the polychaetes by the presence of a clitellum. In most groups the parapodia are absent, the setae are reduced or absent and the segmentation obscured. The worms are hermaphroditic and cross-fertilization is necessary for reproduction. The class includes the oligochaetes, branchiobdellidans, acanthobdellidans and the hirudineans. The oligochaetes are mainly free-living in terrestrial, freshwater or marine habitats, whereas the latter three orders are ectoparasitic or predacious.

A small number of oligochaete species live in intimate association with other oligochaetes, mollusks, crustaceans, fishes and amphibians.[7] In the family Euchytraeidae, the ectocommensal, *Aspidodrilus kelsalli,* lives on the surface of large earthworms. The worms possess a centrally located, ventral sucker, and the posterior region of the body is flattened dorsoventrally into a shield. The body of the small, transparent *Pelmatodrilus planiformis,* another ectocommensal of earthworms, is completely flattened and there is an increased number of hooked setae on the ventral surface. Several species in the genus *Euchytraeus* have been recorded from the gills and branchial chambers of land crabs.

Members of the oligochaete family Naididae are ectoparasitic on freshwater mollusks or endoparasitic in frogs and toads. The various subspecies of *Chaetogaster limnaei* live on the dorsal surface of the body, in the mantle cavity or in the kidneys of a diversity of freshwater snails. *Allodero bauchiensis* is parasitic in the eyes and Harderian (lachrymal) glands of the African frog, *Phrynamerus. A. hyale* and *A. lutzi* live in the ureters of tree frogs, *Hyla,* whereas *A. floridana* is found in the ureters of the toad, *Bufo.* These amber-colored worms measure 5 to 9 mm long, and although they may occur in larger numbers in the host ureters, they cause little or no damage to the host. It has been suggested that these endoparasites are the juvenile stages of the free-living genus *Dero.*

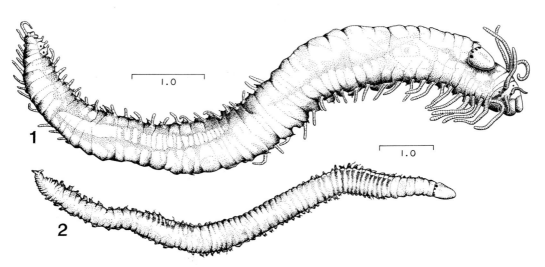

Fig. 22–9. The polychaete *Labrorostratus luteus* parasitic in another polychaete, *Haplosyllis spongicola.* 1, Parasite in place in the host body. 2, Parasite removed. (From Uebelacker, courtesy of J. Parasitol.)

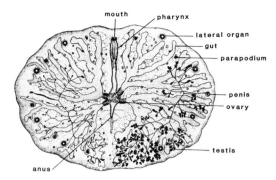

Fig. 22–10. *Myzostoma fisheri* from the deep-sea starfish, *Ceramaster leptoceramus*. (From Wheeler, courtesy of the Biol. Bull.)

Branchiobdellidans are minute worms (Fig. 22–11) that are obligate symbionts of freshwater crayfishes, crabs, shrimps, and isopods.[12,13] Genera such as *Branchiobdella* and *Bdellodrilus* move in leech-like fashion over the external surfaces and gills of their crustacean hosts. They feed on blood, eggs or the film of bacteria and algae that covers the host's body. The order Acanthobdellida has characters of both oligochaetes and leeches. *Acanthobdella peledina* is a periodic parasite of the fins of freshwater fishes in northern Europe and Alaska.

Leeches

Leeches or hirudineans may be found in the sea, in fresh water, or on land. Their general anatomy is illustrated in Figure 22–12. They range in length from 1 to 20 cm, but most are from 2 to 5 cm long. The distinction between a micropredator and a parasite is often blurred, so leeches are sometimes not included in a discussion of parasitism. An intrauterine infection,

however, has been reported.[27] These leeches were on the surface of embryos within the uterus of a female shark. Some leeches are predaceous, feeding on worms, insect larvae, snails, and other invertebrates, which they usually swallow whole, whereas others suck blood and feed on invertebrates or vertebrates. The hosts of leeches are mainly marine or freshwater fishes, but leeches also attack amphibians, turtles, snakes, and terrestrial animals including humans. Because of their habits, leeches may be called predaceous or bloodsucking parasites. For a monograph on North American leeches, see Sawyer.[37]

Some leeches are on their hosts only during periods of feeding, whereas others, for example, many fish leeches, leave their hosts only for breeding. Some take a step further and attach their cocoons to the host, and a few (some of the rhynchobdellids) become sedentary and never leave their hosts.[1,23] When a leech is on an aquatic animal, the surrounding water makes firm sucker attachment possible. Terrestrial leeches use their urine to provide the necessary fluid. This procedure occurs in tropical land leeches (Haemadipsidae), which attack both mammals and birds. Leeches may enter partway into the host's body. The genus *Theromyzon*, for example, may be found on nasal membranes of shore and water birds.

The genus *Hirudo* contains the medicinal leech that has been used to clean wounds of humans. It normally sucks blood from the skin and, occasionally, from the nasal membranes or lining of the mouth cavity. *Hirudo* can ingest 2 to 5 times its own weight during a single feeding period, and 200 days may be required for the digestion of a large meal. Because the leech is

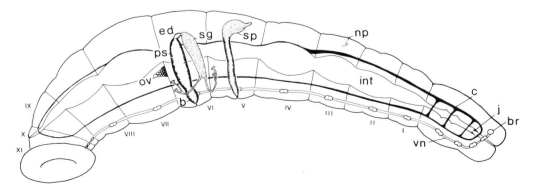

Fig. 22–11. Generalized branchiobdellid showing details of internal organs. (From Holt, courtesy of the Biological Society of Washington.)

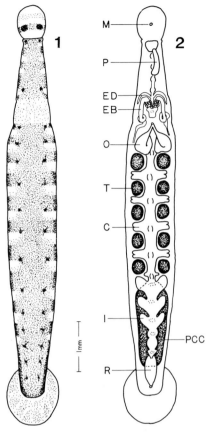

Fig. 22–12. The leech *Aestobdella abditovesiculata*, from the mouth of the staghorn sculpin, *Leptocottus armatus*. 1. Dorsal view of living animal. 2. Reconstruction of the digestive and reproductive systems. (From Burreson, courtesy of J. Parasitol.)

able to fast for a year or more, a meal every 6 months appears to be ample.

Several species of aquatic leeches produce hirudiniasis in man. *Limnatis nilotica* infests the upper respiratory and digestive tracts and may cause severe injury. It is acquired through drinking water and has been reported from Asia, the Middle East, southern Europe, and Africa.

PHYLUM SIPUNCULA

This small but distinct group of vermiform animals is exclusively marine. In Denmark, a single species, *Thysanocardia procera*, is reported to be a temporary parasite on the polychaete *Aphrodite*.[40] The sipunculan penetrates the body wall with its long, flexible proboscis and withdraws fluids from the host's coelomic cavity.

PHYLUM ECHIURA

The echiuran genus *Bonellia* is characterized by marked sexual dimorphism. Males are tiny neotenic forms that live as parasites in or on the females. Differentiation into males or females is determined by the substrate upon which the sexually indifferent larvae settle. If a larva settles on a female, it will metamorphose into a male and take up residence in the female.[31]

PHYLUM MOLLUSCA

The mollusks are one of the major animal phyla. Members of the two largest classes, Bivalvia and Gastropoda, form a variety of symbiotic relationships. Although most mollusks are large and conspicuous, the parasitic groups are minute and often obscure organisms whose biology and systematics we know little about.

Class Bivalvia

Leptonaceans are small clams that live in close association with a diversity of marine invertebrates.[2] *Lepton*, *Montacuta*, and *Mysella* are widely recognized ectocommensals of annelids, sipunculans, echiurans, other mollusks, arthropods, and echinoderms. Relationships are not well understood, but in some cases the clams may injure their hosts. *Neaeromya rugifera* (Fig. 22–13) attaches by byssus threads to the abdomen of the crustacean *Upogebia* or to the ventral surface of the polychaete *Aphrodite*. In *Devonia perrieri* the greatly enlarged, sucker-like foot serves to anchor the clam to the surface of its host, the sea cucumber, *Leptosynapta*. The closely related *Entovalva mirabilis* lives as an endoparasite in the esophagus of another synaptid holothurian. The leptonaceans are considered to have the most complex reproductive patterns in the entire class.

Glochidia (Fig. 22–14) are larval members of the freshwater clam family Unionidae. They belong to several genera and species and are parasites of fishes and occasionally amphibians. The young clams are discharged in vast numbers from the mother clam and become attached to gills and body surface of various fishes. Glochidia possess two small valves that aid in attachment with a pincer-like action. Host epi-

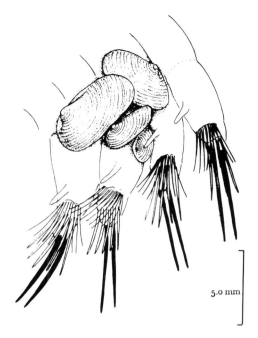

5.0 mm

Fig. 22–13. *Neaeromya rugifera* attached to the foot of the polychaete, *Aphrodite*. (From Narchi, courtesy of the Veliger.)

dermis grows over the parasite. Later, host tissue breaks down, and within the chamber thus formed, each young clam metamorphoses. It then breaks out of the chamber, drops to the sand, and matures into an adult clam.

Specific glochidia occur on fishes, inhabiting the same biotope as that of the adult clam, but the segregation is ecologic because, experimentally, the glochidia can be induced to fix themselves to many different kinds of living supports, and, in nature, the speed of the fish and the level at which it swims influence the fixation of the larval clam.[18]

Class Gastropoda

For students interested in the evolution of parasitism, the gastropods are a fascinating group. The prosobranchs provide excellent examples of all stages in the transition from free-living micropredators to highly-modified endoparasites that are nothing more than sacs of reproductive products. Only a single opisthobranch is known to be parasitic.

The pelagic nudibranch, *Phyllirrhoe bucephala*, is an obligate parasite of the hydrozoan medusa *Zanclea costata*.[24] The young snail settles inside the umbrella of the jellyfish, where it feeds on the tentacles and body of the medusa, as well as using it as a transport vehicle. As the nudibranch grows, it shifts position on the host and seals off the mouth with its foot, so the medusa is no longer able to feed. When mature the nudibranch discards its stunted and distorted host.

The ptenoglossans are prey-specific micropredators or semiparasites of cnidarians. *Epitonium* feeds by cutting off the tips of anemone or coral tentacles with its long proboscis and specially shaped radular teeth. *Opalia* inserts a short proboscis into the body wall of an anemone, and feeds on host body fluids.

The pyramidellids are prosobranchs that feed suctorially on the body fluids of a diversity of annelids, sipunculans, mollusks, echinoderms and ascidians.[36] *Odostomia* (Fig. 22–15) and other genera of this group feed by means of an elongate proboscis armed with a piercing stylet and an esophageal pump. *Boonea impressa* lives on the oyster *Crassostrea virginica*. The growth rate of parasitized oysters is dramatically reduced. The presence of large number of snails may have significant impacts on the health and population structure of oysters.

Ovula and *Pedicularia* are cowry relatives that are ectoparasites of hydrocorals, gorgonians and alcyonarians. Another cypraeacean relative, *Pseudosacculus*, lives within the test of ascidians in the northeastern Pacific Ocean.

The small snail, *Caledoniella montrouzieri* lives on the abdomen of several species of the stomatopod crustacean, *Gonodactylus*. Oviposition is inhibited, and the size of mature individuals with snails is smaller than those without snails.[34]

The majority of eulimaceans are parasitic on or in echinoderms. *Eulima* and *Balcis* move freely about on the surface of their hosts. The ectoparasites *Thyca* (Fig. 22–16) and *Stilifer* are securely attached or even permanently embedded in cysts in the surface of starfish hosts. In most cases a male and female pair is found together on the host. A permanent hole made in the host body wall enables the proboscis to penetrate and feed at will on host tissues and body fluids. *Pisolamia brychius* (Fig. 22–17) inserts its proboscis directly into a blood vessel of its host, the deep-sea holothurian *Oneirophanta*. The endoparasites *Megadenus* and *Robillardia* retain a thin shell and characteristic snail shape, but other genera such as *Asterophila*, *Molpadicola*, *Entoconcha*, and *Enteroxenous* have lost nearly all

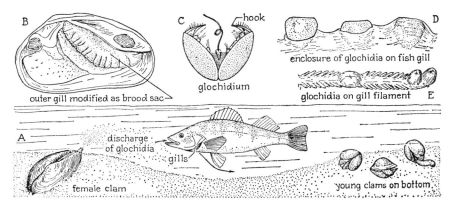

Fig. 22–14. Life cycle of glochidia. (From Storer: General Zoology, courtesy of McGraw-Hill.)

all adult gastropod features. Only the presence of a veliger larval stage indicates the affinities of these aberrant snails. In the latter group of endoparasites, there is marked sexual dimorphism in which the females are reduced to nothing more than globular or vermiform reproductive sacs. The males, often dwarfed beyond recognition, are embedded in the females as small clumps of testicular material.[21]

Entocolax (Fig. 22–18), a worm-like parasite penetrates the body wall of holothurians. It becomes attached to the body wall lining, hangs freely in the body cavity, and sweeps up host fluids through a proboscis into a reduced intestine. A siphon allows the release of reproductive products to the outside.

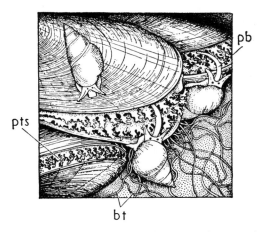

Fig. 22–15. *Odostomia scalaris* feeding on the mussel, *Mytilus edulis.* (From Fretter and Graham, courtesy of the Ray Society.)

Fig. 22–16. *Thyca* on the underside of the arm of the starfish *Phataria.*

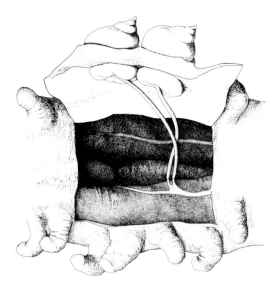

Fig. 22–17. The snail *Pisolamia brychius* in place on the deep-sea holothurian *Oneirophanta mutabilis* showing the proboscis penetrating a host blood vessel. (From Bouchet and Lützen, courtesy of the Academie de Sciences, Paris.)

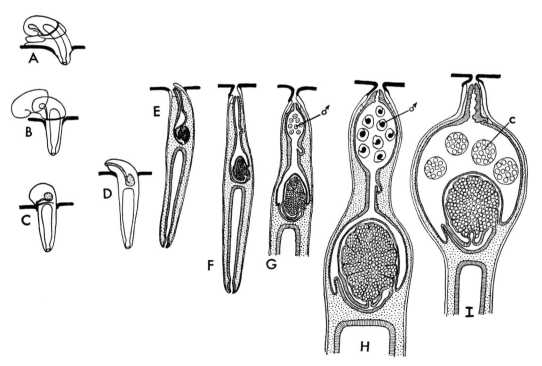

Fig. 22–18. Supposed mode of entrance of *Entocolax* into its host. (From Baer: Ecology of Animal Parasites, courtesy of the University of Illinois Press.)

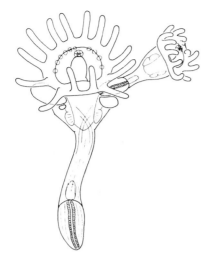

Fig. 22–19. *Loxosomella vivipara,* an entoproct commonly found on sponges in Florida. (From Nielsen, courtesy of Ophelia.)

PHYLUM ENDOPROCTA

Endoprocts superficially resemble hydroids. The small, stalked zooids have a cup-like body bearing a ring of ciliated tentacles used to capture food.[15] In the family Loxosomatidae (Fig. 22–19), all species are epibiotic and most are considered energy commensals, which secure their food by filtering it from the water currents generated by their hosts.[28] Two species of *Loxosomella* occur on the gills and abdominal appendages of the mantis shrimp *Hemisquilla ensigera.*[38]

PHYLUM ECTOPROCTA

The ectoprocts or bryozoans are widely-distributed, colonial marine animals.[14] Each zooid in the colony bears a crown of ciliated tentacles and is encased in a box-like exoskeleton. A number of species burrow into calcareous shells of mollusks, while others attach to a diversity of hosts. The majority of the epibiotic associations are not specific. One species, *Triticella elongata,* lives exclusively on the appendages, carapace and inside the gill chamber of the pea crab, *Scleroplax granulata.*

PHYLUM ECHINODERMATA

Although many echinoderms serve as hosts for commensals and parasites, only one species

of holothurian has been reported to live on another animal. *Rynkatorpa pawsoni* attaches to the skin of the deep-sea angler fish, *Gigantactis.*[26] The sea cucumber does not damage the fish and is thought to benefit by simply being transported about by its host.

REFERENCES

1. Baer, J.G.: Ecology of Animal Parasites. Urbana, University of Illinois Press, 1951.
2. Boss, K.J.: Symbiotic erycinacean bivalves. Malacologia, 3:183–195, 1965.
3. Caullery, M.: Classe des Orthonectides (Orthonectida Giard, 1877) *In* Traité de Zoologie: Anatomie, Systématique, Biologie. Tome IV. Fasc. I. Plathelminthes, Mésozoaires, Acanthocéphales, Némertiens. Edited by P.-P. Grassé. Paris, Masson, 1961.
4. Clark, R.B.: *Capitella capitata* as a commensal, with a bibliography of parasitism and commensalism in the polychaetes. Ann. Mag. Nat. Hist., 9:433–448, 1956.
5. Dales, R.P.: Interrelations of organisms. A. Commensalism. Geol. Soc. America, Memoir 67:391–412, 1957.
6. Dimock, R.V., and Davenport, D.: Behavioral specificity and the induction of host recognition in a symbiotic polychaete. Biol. Bull., 141:472–484, 1971.
7. Gelder, S.R.: A review of the symbiotic Oligochaeta (Annelida). Zool. Anz., 204:69–81, 1980.
8. Gibson, R., and Jennings, J.B.: Observations on the diet, feeding mechanics, digestion and food reserves of the entocommensal rhynchocoelan *Malacobdella grossa.* J. Mar. Biol. Ass., U.K., 49:17–32, 1969.
9. Grassé, P.-P.: Classe des Dicyemides. *In* Traité de Zoologie: Anatomie, Systématique, Biologie. Tome IV. Fasc. I. Plathelminthes, Mésozoaires, Acanthocéphales, Némertiens. Edited by P.-P. Grassé. Paris, Masson, 1961.
10. Harmon, W.J.: A review of the subgenus *Allodero* (Oligochaeta: Naididae: *Dero*) with a description of *D. (A.) floridana* n. sp. from *Bufo terrestris.* Trans. Am. Microsc. Soc., 90:225–228, 1971.
11. Hochberg, F.G.: The "kidneys" of cephalopods: a unique habitat for parasites. Malacologia, 23:121–134, 1982.
12. Holt, P.C.: Newly established families of the Order Branchiobdella (Annelida: Clitellata) with a synopsis of the genera. Proc. Biol. Soc. Wash., 99:676–702, 1986.
13. Holt, P.C.: The Branchiobdella: epizootic annelids. Biologist, 50:79–94, 1968.
14. Hyman, L.H.: The Invertebrates: Smaller Coelomate Groups. Vol. V. New York, McGraw-Hill, 1959.
15. Hyman, L.H.: The Invertebrates: Acanthocephala, Aschelminthes, and Entoprocta. The Pseudocoelomate Bilateria. Vol. III. New York, McGraw-Hill, 1951.
16. Hyman, L.H.: The Invertebrates: Protozoa

through Ctenophora. Vol. II. New York, Mc-Graw-Hill, 1940.

17. Iwata, F.: *Uchidaia parasita* nov. gen. et nov. sp., a new parasitic nemertean from Japan with peculiar morphological characters. Zool. Anz., *178*:122–136, 1966.

18. Kat, P.W.: Parasitism and the Unionacea (Bivalvia). Biol. Rev., *59*:189–207, 1984.

19. Komai, T.: Studies on two aberrant ctenophores, *Coeloplana* and *Gastrodes*. Kyoto, published by the author, 1922.

20. Kozloff, E.N.: Morphology of the orthonectid *Rhoplaura ophiocomae*. J. Parasitol., *55*:171–195, 1969.

21. Lützen, J.: Studies on the life history of *Enteroxenous* Bonnevie, a gastropod endoparasitic in aspidochirote holothurians. Ophelia, *18*:1–51, 1979.

22. McConnaughey, B.H.: The Mesozoa. *In* The Lower Metazoa. Comparative Biology and Phylogeny. Edited by E.C. Dougherty, et al. Berkeley, University of California Press, 1963.

23. Mann, K.H.: Leeches (Hirudinea). Their Structure, Physiology, Ecology and Embryology. New York, Pergamon Press, 1962.

24. Martin, R., and Brinckmann, A.: Zum Brutparasitismus von *Phyllirrhoe bucephala* Per. and Les. (Gastropoda, Nudibranchia) auf der Meduse *Zanclea costata* Gegenb. (Hydrozoa, Anthomedusae). Staz. Zool. Napoli Publ., *33*:206–223, 1963.

25. Martin, W.E.: *Hydrichthys pietschi*, new species, (Coelenterata) parasitic on the fish, *Ceratias holboelli*. Bull. So. Calif. Acad. Sci., *74*:1–5, 1975.

26. Martin, W.E.: *Rynkatorpa pawsoni* n. sp. (Echinodermata: Holothuroidea) a commensal sea cucumber. Biol. Bull., *137*:332–337, 1969.

27. Moser, M., and Anderson, S.: An intrauterine leech infection: *Branchellion lobata* Moore, 1952 (Piscicolidae) in the Pacific angle shark *(Squatina california)* from California. Can. J. Zool., *55*:759–760, 1977.

28. Nielsen, C.: Studies on Danish Entoprocta. Ophelia, *1*:1–76, 1964.

29. Nielsen, S.-O.: *Nectonema munidae* Brinkman (Nematomorpha) parasitizing *Munida tenuimana* G.O. Sars (Crust. Dec.) with notes on host-parasite relations and new host species. Sarsia, *38*:91–110, 1969.

30. Pettibone, M.H.: Endoparasitic polychaetous an-

nelids of the family Arabellidae with descriptions of new species. Biol. Bull., *113*:170–187, 1957.

31. Pilger, J.: Settlement and metamorphosis in the Echiura: a review. *In* Settlement and Metamorphosis of Marine Invertebrate Larvae. Edited by F.-S. Chia and M. Rice. New York, Elsevier/North-Holland Biomedical Press, 1978.

32. Raikova, E.V.: Life cycle and systematic position of *Polypodium hydriforme* Ussov (Coelenterata), a cnidarian parasite of the eggs of Acipenseridae. Pubs. Seto. Mar. Biol. Lab., *20*:165–173, 1973.

33. Raikova, E., Suppes, C., and Hoffman, G.: The parasitic coelenterate, *Polypodium hydriforme* Ussov, from the eggs of the American acipenseriform *Polyodon spathula*. J. Parasitol., *65*:804–810, 1979.

34. Reaka, M.L.: The effects of an ectoparasitic gastropod, *Caledoniella montrouzieri*, upon molting and reproduction of a stomatopod crustacean, *Gonodactylus viridis*. Veliger, *21*:251–254, 1978.

35. Rees, W.J.: A brief survey of the symbiotic associations of Cnidaria with Mollusca. Proc. Malac. Soc. Lond., *37*:213–231, 1967.

36. Robertson, R., and Mau-Lastovicka, T.: The ectoparasitism of *Boonea* and *Fargoa* (Gastropoda: Pyramidellidae). Biol. Bull., *157*:320–333, 1979.

37. Sawyer, R.T.: North American Freshwater Leeches, Exclusive of the Piscicolidae, with a Key to All Species. Illinois Biological Monographs, No. 46. Urbana, University of Illinois Press, 1972.

38. Soule, D.F., and Soule, J.D.: Two new species of *Loxosomella*, Entoprocta, epizoic on Crustacea. Occas. Pap. A. Hancock Found., *29*:1–19, 1965.

39. Spaulding, J.G.: The life cycle of *Peachia quinquecapitata*, an anemone parasitic on medusae during its larval development. Biol. Bull., *143*:440–453, 1972.

40. Thorson, G.: Parasitism in the sipunculid, *Golfingia procerum* (Moebius). J. Fac. Sci., Hokkaido Univ., *13*:128–132, 1957.

41. Uebelacker, J.M.: A new parasitic polychaetous annelid (Arabellidae) from the Bahamas. J. Parasitol., *64*:151–154, 1978.

42. Warén, A.H.: A generic revision of the family Eulimidae (Gastropoda, Prosobranchia). J. Moll. Stud. (Suppl.), *13*:1–95, 1983.

43. Wickham, D.E.: Aspects of the life history of *Carcinonemertes errans* (Nemertea: Carcinonemertidae), an egg predator of the crab *Cancer magister*. Biol. Bull., *159*:247–257, 1980.

23

General Considerations

Parasitology, because of its close relationships with human and veterinary medicine, has developed with little reference to the concepts of ecology. Traditionally, ecology has been a field-oriented science, often concerned with the survival of biological populations and communities. Medicine, in contrast, has been rooted in the clinical and experimental investigation of disease in individuals. In the area of public health, however, with its interest in the control of disease in populations, the ecologic approach has contributed consistently, though perhaps sparingly, to parasitologic research and practice.

Recently, recognition of the fact that the methods and theory of ecology can contribute importantly to parasitology has increased markedly, and numerous publications dealing with the population ecology of parasites have appeared. Conversely, ecologists, long resistant to the idea that disease is significant in regulating the size of natural populations, are beginning to recognize that infectious agents, including parasites, are important regulatory factors.

Few animals harbor only one parasitic species. Polyparasitism is common even in humans, particularly in poor, rural areas of the tropics lacking sanitation. The parasites occurring in an individual influence each other's ability to establish and survive in a host. Even when species occur in the same host sequentially, the parasites of a primary infection may, through the host's immunologic response, influence the ability of a secondary infection to establish. Thus, knowledge of a parasite's community ecology is an essential part of any complete understanding of natural, host-parasite associations.

Most of these associations are, of necessity, studied either uncoupled or as simple 2-species associations (i.e., host and parasite), but it is naive to think that this is realistic biologically. Experimental simplicity (reductionism) is the practical, and, in fact, necessary approach to the understanding of many aspects of the host-parasite relationship, e.g., nutrition, physiology, immunology, but it has the inherent problem that other organisms may change the environment directly, or indirectly, through the host's responses. Thus, parasitology, by its very nature, is an ecologic science, and both students and practitioners of its most reductionist subspecialties will profit by pausing to consider their own special areas in a larger ecologic context.

POPULATIONS AND COMMUNITIES

Applied to free-living organisms, the term *population* refers to all the organisms of the same species occurring in a defined habitat or geographic area. In parasitology, this concept is complicated by the fact that a population may include either all conspecific organisms in an individual host, or all conspecific individuals in an area, including free-living stages and parasitic stages in several hosts. The ambiguities that can arise from this complication led Esch et al.[25] to propose two more precise terms, infra- and suprapopulation. An *infrapopulation* includes all members of a species occurring in an individual host, a *suprapopulation* all members of a single species in a given area, including individuals existing as free-living larvae, as larvae in intermediate hosts, and as larvae or

447

adults in definitive hosts. Because important ambiguities remain, these terms have not been accepted universally. Cryptic individuals, such as minute, histotropic larvae lying dormant in the tissues, escape notice and, as a result, may be excluded from an infrapopulation, which, therefore will be constituted of adult parasites only. Furthermore, it may be only the sum of the infrapopulations in definitive hosts [excluding free-living stages or stages in the intermediate host(s)] that, in practice, constitutes the suprapopulation. Frequently, the excluded stages are inaccessible and cannot be enumerated by an investigator.

Attributes of a population, as distinct from those of an individual, include the birth and death rate, the age distribution and sex ratio, and the immigration and emigration rates. These factors determine the rate of growth and ultimate size of a population. Among free-living organisms, an excessive increase in the size of the population may cause overexploitation of resources and damage to the habitat. Similarly, excessive growth of a parasitic infrapopulation may cause disease or even death of its host, while excessive growth of the suprapopulation may cause epidemic disease, and a decrease in the host population. The population ecology of parasites is an exciting field that has prospered with the growing realization that it can contribute importantly to the design of programs to control populations of parasites in humans and economically important animals. The population ecology of parasites is discussed more specifically later in this chapter.

Natural populations do not exist as independent, monospecific entities; even in simple ecosystems, several will interact. Populations that are usually found together and are, to a greater or lesser extent, interdependent constitute a *community*. Thus, community ecology is the ecologic subspeciality that attempts to understand the interrelationships between these assemblages of species. It is, for example, concerned with food webs and the partitioning of resources between several interacting species, with energy flow through these webs, with the number and relative abundance of species and with the succession of species, over time, in a habitat.

Communities of parasites are constituted of the several populations of species that co-occur in a host species. Again, as pertains to the parasite population, the parasite community may be that occurring in either a single individual (an infracommunity) or in all individuals of the host species in an area (a supracommunity). Two-species interactions among co-occurring parasites are of interest to many parasitologists, including specialists who have little interest in community ecology. For instance, an immunologist may investigate cross-immunity (immunity to one parasite that affects another) without reference to central problems of community ecology, but simple interactions such as this form the basis of the web of interactions that determine the structure of the community and, hence, are of intrinsic ecologic interest. Curiously, however, investigations of two-species interactions seem to impact the contemporary study of parasite community ecology minimally, its practitioners tending to focus on complex multispecific associations and questions relating to evolutionary age, saturation, and relative abundance of the species constituting these associations. Communities of parasites often have advantages over free-living communities for the interested ecologist. In many (but not all) associations, habitat constancy (due to the host's homeostasis), parasite site-specificity, and the clarity of organ (habitat) boundaries give the parasite community an integrity that facilitates study. Additionally, communities are replicated from host to host, providing the basis for statistical analysis. Superb investigations, with relevance beyond parasitology, have been conducted by Bush,[12,13] Holmes,[35-38] Petter,[66] Rohde[73,74] and other workers. These will be discussed subsequently (pp. 472–474).

The community for some parasitologists is a larger unit than the parasite fauna of a given host species. It includes free-living stages and their environment, intermediate hosts, and their parasite faunas, and even accidental hosts and their parasites. In a previous edition of this book,[60] two quotations were presented to emphasize the extensive web of ecologic factors that enter into the community ecology of vector-borne diseases as viewed epidemiologically. Bray, a distinguished medical parasitologist, states:

"To say that oriental sore is caused by *Leishmania tropica* is far too confining a statement to anyone interested in epidemiology. Oriental sore is also caused by the bite of a sandfly, by proximity to a reservoir host, by the way man fits into the ecosystem of sandfly and reservoir host, by the immune reaction of man, and possibly by his nutritional, viraemic or temperature states. One could go on."[10]

Zeldon, with reference to triatomine vectors of *Trypanosoma cruzi,* is concerned with "a web of causation" that varies

"with factors related to the insect (geographic distribution, adaptability to the house, aggressiveness, vectorial capacity, anthropophilism, potential reproductive index), and related to the human environment (type of construction, sanitary conditions of the house, education level of the inhabitants, climate, natural enemies and competitors)."[107]

On this basis, it was suggested that the student should develop a community concept when studying parasitology. When a parasite is studied by itself, apart from its environment, only part—and often a small part—of its total biology can be understood. The community principle in ecologic theory emphasizes the orderly manner in which diverse organisms usually live together. An ecologic complex is formed by the parasite, the vector, the host, and various features of the host's environment. However, this complex is far more than the sum of its parts. It is something new and forever changing.

ECOLOGIC NICHE

The term "niche," in its ecologic sense, is frequently misused to mean a small area of space. For a parasite, such a small area might be part of a larger organ, e.g., the bile duct, rather than the liver, or the proximal duodenum, as opposed to either the entire duodenum or small intestine. The area in which an organism lives is its *habitat,* or, when narrowly specified, its *microhabitat.* Thus, the proximal duodenum may be an organism's habitat, but *not* its niche.

The niche is an abstraction that can rarely, if ever, be completely defined for even the most well-known organisms. The *niche* is the organism's place in nature in a functional rather than a physical sense. The contemporary concept of a niche defines it as constituted of the *totality* of abiotic and biotic factors with which a species or population interacts.[71] For an intestinal parasite, this would include the location within an organ, various physio-chemical factors (e.g., pH, oxygen tension and temperature), host-related biologic factors (e.g., the kind and amount of ingested material, enzymes, mucins, and secretory antibodies), and environmental factors

(e.g., excretions, secretions, and pheromones) dependent on the parasite's own presence and that of other intestinal organisms. The environmental variables fluctuate and change with time. Food enters the duodenum periodically, is exposed to digestion and moves on to the jejunum, or the number of parasites may increase, and eventually the host recognizes their presence and reacts by secreting enzymes, antibodies or other protective substances into the intestine. Clearly, then, complexity dictates that niches remain abstractions that cannot be fully described.

Even though ecologic niches are not amenable to complete description, they continue to attract the attention of biologists. It is assumed that if sufficient important niche parameters are measured, an organism's place in the community will become apparent and its relationships with other organisms will be understood. At least part of this fascination is based on interspecific *competition* and related phenomena, including *competitive exclusion, character displacement* and *niche diversification.* Biologic *competition* occurs when two or more organisms require an environmental resource in short supply. Both intra- and interspecific competition are ecologically interesting, although when not specifically stated, the term 'competition' usually refers to the latter. Members of the same species almost invariably have the most similar resource requirements and, therefore, intraspecific competition is usually more frequent and intense than interspecific competition. Density dependent decreases in size and fecundity and changes in intrahost distribution are often considered manifestations of intraspecific competition. This suggests that, for parasites occurring in dense populations, some resources are, in fact, in short supply.

Following a similar line of reasoning, *interspecific competition* should be most intense between organisms that are closely related phylogenetically. This explains the interest that biologists have found in comparative ecologic studies of closely related, coexisting species. Among free-living species, Darwin's Finches, the group of closely related birds that evolved on the Galapagos Islands, constitute a well-known example. On islands with closely related seed-eating species, these species differ in bill size and feed on seeds of correspondingly different sizes, whereas on islands where only one species occurs, its bill is intermediate in size and its food includes a range of seed sizes. These differences are believed to be the result of competition and natural selection, culminating in divergence of bill size. Divergence of traits that occurs in a zone of distributional overlap, but not where the populations occur singly, is called *character displacement*.

If character displacement does not occur, one species will, in time, replace the other unless environmental factors alternately favor one or the other competitor, or if they limit both species to low levels of abundance at which competition fails to occur. Theoretical biologists have supported this concept mathematically, experimentalists have demonstrated it under simple laboratory conditions and field biologists can cite natural experiments that are explicable on this basis. Competition which results in one competitor causing the local extinction of another is called *competitive exclusion*. This phenomenon, sometimes difficult to demonstrate in natural, free-living systems is readily demonstrable in situations where one species of parasite excludes another from all or part of a host. Furthermore, among parasites, this interaction may be mediated through the immune response of the host, a particularly interesting situation.[78]

On the other hand, character displacement may occur and, in complex natural environments, species may become increasingly specialized in their use of the available resources. This leads to niche differentiation, i.e., members of a community of organisms become increasingly specialized and fill smaller niches, thereby allowing the co-occurrence of a greater number of species. Among parasites, some particularly interesting associations exist wherein numerous closely related species co-occur at high density in a restricted habitat (e.g., flagellates in the gut of termites or wood roaches, pinworms in the colon of tortoises and small strongyles in the large intestine of the horse). These and other similar associations have attracted the attention of parasitologists because they raise the fundamental question: How did numerous closely related species, presumably making similar demands on the available resources, survive? Niche diversification seems to provide an answer. This interesting phenomenon, as observed among parasitic organisms, is discussed later in this chapter.

ENVIRONMENTAL FACTORS

Physical factors, such as temperature, moisture, salinity, or pH, often set the limits within which life is possible, but, in contrast to free-living animals, many endoparasites are spared marked perturbations of their physical environment because the host's homeostatic mechanisms limit internal variation. In the following paragraphs, we discuss these and other physical factors that impinge on the lives of parasites. Thereafter we turn to some responses to these factors and then to biotic factors that have not been covered elsewhere in this book. Biochemical, physiologic and immunologic aspects of the host-parasite relationship are considered in Chapter 2.

External Factors

Endoparasites (particularly those in homeothermic hosts) are sheltered from the direct effects of the weather. However, they often have free-living stages or occur in poikilothermic hosts where they may be more directly affected. It sometimes seems that parasites are particularly abundant in hot climates. In recent history, human parasitic disease has become associated with the tropics, but this is due, in part, at least, to the more rapid initiation of public health measures (vector control and sanitation) in the industrialized countries of temperate parts of the world. Where this did not occur, or where these measures failed, parasitic infections remain prevalent even in temperate or cold climates (e.g., ascariasis in the Orient or giardiasis in the U.S.S.R. and the U.S.A.). We forget that malaria was once endemic in much of North America, including the southernmost parts of Canada, and that hookworm disease ravaged the moist, southeastern part of the United States. Indeed, in some temperate industrialized countries of the world, the sexual revolution has partially negated the benefits of universally available urban sewage disposal by increasing contact with fecally transmitted organisms. Consequently, directly transmitted, intestinal protozoa, once rare, are now common in male homosexual populations. Nevertheless, as among free-living organisms, species diversity is generally greater in tropical than in temperate areas, and, therefore, the widely held notion that there are more parasites in tropical areas is basically true.

Excessive cold or dryness limits the number of species with free-living larval stages. Thus, for example, most hookworms occur in the warm, moist parts of the world, but exceptions exist. Two species of the genus *Uncinaria* occur in the arctic and *Ancylostoma duodenale* occurs in some parts of the world with a short rainy season. The northern dog hookworm, *Uncinaria stenocephala*, can occur in the arctic because its eggs and larval stages are cold resistant and develop rapidly in the brief periods when the temperature is favorable (15° C). *U. lucasi*, a parasite of the northern fur seal, survives from year to year as a dormant larva in the blubber of adult seals, and is transmitted to a new generation as a reactivated larva in the mother's milk. Adult worms develop in the pups and hookworm eggs and larvae contaminate the breeding beaches during the brief arctic summer. Larvae that invade a pup contribute further to the patent infections occurring in the young. Those that invade older animals contribute to cryptic infections, constituted of dormant larvae. Free-living larvae that fail to invade a host die during winter. Thus, the survival of this hookworm under arctic conditions depends on sheltered vertical (mother-offspring) transmission. Similarly, *Ancylostoma duodenale* occurs in its human hosts as an arrested larva during prolonged dry periods of the year. These arrested worms resume development, mature and oviposit when, seasonally, the weather begins to turn favorable for the development and survival of the free-living stages.[81] In this species, the factors that induce a potential for interrupted larval development within the host are unknown, but, in some nematodes, these factors are well studied and include external, environmental, seasonally occurring stimuli.[30]

Most trichostrongyle nematodes of domestic ruminants become developmentally arrested (i.e., hypobiotic) in their hosts during seasons of the year when external environmental conditions are unfavorable for survival of their free-living stages. In widely distributed species, local populations exist that vary in their propensity to respond to arrest-provoking stimuli depending on the length and severity of the unfavorable season. Thus, when three Australian strains of *Haemonchus contortus* were exposed to environmental conditions predictive of winter, a strain from the northern subtropical part of the continent showed no propensity for arrest, whereas another from a more temperate area showed an intermediate propensity and a third, from a cold, southern area, showed a markedly increased propensity for arrest. It has been demonstrated experimentally in the case of the stomach worm of cattle, *Ostertagia ostertagi*, and

in a related nematode of rabbits, *Obeliscoides cuniculi*, that selection for the ability to arrest can result in a rapid increase in the sensitivity of larvae to arrest-provoking stimuli.[52,99] Therefore, it is not surprising to find that local strains of the widely distributed parasites of domestic ruminants are adapted to local environmental conditions with regard to the duration and intensity of parasite dormancy within the host.[30,52,79]

The larval form sensitive to arrest-provoking environmental stimuli is the third-stage larva (L_3). The biochemical changes that occur in the larva in response to these signals and program it for interrupted development are not understood. Seasonal arrested development in nematodes resembles diapause in insects, and some authors have suggested that the controlling mechanisms may be similar. If so, then the length of dormancy will be determined by an internal biologic clock and may not require signals from the host for resumption of development.[30,52,79]

Environmental signals for arrested development in the various species of trichostrongyle nematodes probably include seasonal changes in photoperiod, temperature, and moisture. Most investigations of these signals have been conducted in the temperate parts of the world and, to date, conclusive evidence is only available for an arrest-inducing effect of temperatures of autumn acting on the L_3. An example of the effect of chilling on the potential to arrest in the host is shown in Figure 23–1. Other environmental factors require further investigation before their suspected role in the induction of arrest is confirmed.

Interrupted development (diapause) does not only occur after the parasite has infected the host. Larvae of *Nematodirus battus*, an intestinal trichostrongyle of sheep, remain in the egg where they undergo a period of diapause. Initially they do not hatch in the environment nor do they hatch if ingested. The larvae must be exposed to the prolonged cold of winter; then, upon warming, hatching occurs. This results in a marked increase in the number of larvae on pasture at precisely the time when new susceptible hosts, the lambs of the year, become available.

External environmental effects are not only felt by free-living stages; parasitic stages in the host, especially those in poikilothermic hosts, are affected. The ambient temperature may in-

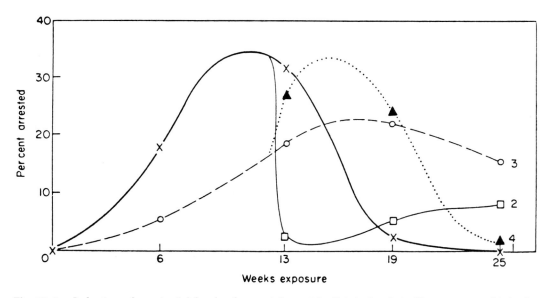

Fig. 23–1. Induction of a potential for developmental arrest in *Ostertagia ostertagi* by exposure of infective larvae (L₃) to various temperature regimens. The graph shows the mean percentage of the worms becoming arrested when the L₃'s were dosed to calves after exposure to various experimental conditions, namely a constant low (15°C) temperature [group 1, solid line]; a constant higher (40°C) temperature [group 3, broken line]; a sudden decrease (40° to 15°C) [group 4, dotted line]; or a sudden increase (15° to 40°C) in temperature at 12 weeks [group 2, thin line]. Exposure of L₃ to either a constant low autumnal temperature (group 1) or to a sudden decrease in temperature, as occurs at the time of the first frost in autumn (group 4) increased the percentage arrested to about 30% of the dose administered to calves. (From Michel, Lancaster, and Hong, 1975.)

fluence the distribution and abundance of protozoa in poikilothermic hosts. For example, *Trypanosoma rotatorium* is a polymorphic parasite in the blood of frogs. Studies have been carried out to determine how temperature affects the density of the flagellates in peripheral blood and to determine the internal temperatures of bullfrogs in their natural environment. The peripheral parasitemia of the parasite in *Rana catesbeiana* was found to be affected by temperature.

"Over the long term, high temperatures are always coincident with high peripheral parasitemia and vice versa; over the short term, increases in temperature bring about a corresponding increase in parasite level, and vice versa. . . It is proposed that the control of peripheral parasitemia is due to changes in the level of metabolic activity of the host."[7]

The free-living stages of many parasites are highly susceptible to drying. Hookworm larvae, for example, can survive only when a moisture film exists on soil particles, grass blades, or other low growing vegetation. Other nematode larvae may survive drying, but can only move when moisture becomes available. Frequent

drying and wetting is rapidly fatal even to drought resistant species. Parasites in their hosts may also be affected by drying of the external environment. Certain trematode larvae in snails are reported to develop with abnormal rapidity during periods of drought. Thus, as the environment dries, the trematodes mature precociously and drought-resistant eggs, rather than delicate free-swimming cercariae, are released into the environment.

Too much water may be inimical to the development of free-living stages of nematodes. When soils become waterlogged and air is excluded from the pore spaces between the soil particles, development of nematode eggs and larvae may cease. On the other hand, if the water stands in small depressions, these excessively wet conditions will favor amphibious snails and certain mosquitoes, and in turn will increase transmission of parasites such as *Fasciola hepatica*, *Wuchereria bancrofti*, and *Plasmodium falciparum*.

Soil type also affects the development and survival of the free-living stages of parasitic nematodes. Soils consisting of small particles (silt or clay soils) retain moisture, but have minute pore spaces that are so small that nematode larvae cannot move through them, whereas soils consisting of larger particles (sandy loams) drain and dry relatively rapidly, but have pores that permit larval movement. These differences may be reflected in the prevalence and intensity of parasitic infections. *Ascaris lumbricoides* infections are common in humans that live in areas where heavy, clay soils predominate, but heavy hookworm infections are rare in these areas. Where sandy soils predominate, these relationships are reversed. Heavy clays protect ascarid eggs from drying and because of their stickiness facilitate transmission. Hookworm larvae, on the other hand, find clay soils oxygen deficient and unsuitable for migration.

Salinity, particularly with reference to parasites of marine organisms, is an important variable. The distribution of trematodes is tied closely to that of their molluscan hosts, whose distribution is determined, in part, by salinity. The eggs of many helminths will not hatch if deposited in excessivly brackish or salt water. For a human schistosome egg to hatch, sea water would have to be diluted six-fold. Salinity is relevant to the survival of terrestrial as well as aquatic organisms. The larvae of human hookworms, for example, will not develop to infectivity in saline soils. The low intensities of hookworm infection found in some human populations living close to the sea in the Gangetic Delta have been explained on this basis.[16]

Light, too, is an important environmental variable. It is a stimulus for hatching of some helminth eggs. Due to the habitat preferences of the host, eggs of aquatic species requiring light are usually deposited in water of a depth and transparency permitting sufficient light penetration. However, if the eggs are deposited or carried to deeper or more turbid water, they will fail to hatch and eventually die. Light also influences the migration of free-living nematodes and trematode larvae. It interacts with temperature and moisture with reference to nematode movement on herbage and with temperature to give direction to miracidial migration from the point of egg hatch to the water layer where host snails congregate in greatest abundance (Table 23–1).[94]

Light and darkness combine to establish the length of the photoperiod. Decreasing photoperiod has been implicated as a stimulus for hypobiosis in trichostrongyle nematodes. Changing photoperiod probably also serves to regulate the development of parasites via the seasonal changes in hormone levels of their hosts. Thus, *Leucocytozoon simondi*, a malaria-like parasite of ducks, disappears from the peripheral blood during the winter and appears again in the spring. Among helminths, the growth of the alveolar hydatid of *Echinococcus multilocularis* varies seasonally. Rau and Tanner[69] have suggested that this seasonal variation is controlled by changing day length acting through attendant hormonal changes in the host.

Other environmental factors, including solar radiation, turbulence, currents, and pressure will affect the abundance and distribution of free-living stages of parasites. A complete discussion of these variables is unnecessary and perhaps tedious in an introductory text; we turn, therefore, to some responses to external environmental factors.

Responses to External Environmental Factors

Seasonal Variation

The timing of reproductive behavior, the release of eggs or larvae, and the behavior of immature stages of parasites, are closely related to the host's life cycle. Thus, infective stages are ready to locate and to invade the host when the latter is available and when the environment inside and outside the host is suited for development, maturation, or dissemination of the parasite.[42]

The results of a few other investigations of seasonal variation in parasite development and abundance are presented in this section: many more have been made. The winter's accumulation of worm eggs may all hatch at once, on the advent of warm weather in June or July. A period of drought followed by rain may also be responsible for a sudden massive invasion of worm larvae. On the other hand, adverse conditions, such as the drying of ponds or winter ice, generally mean fewer or less vigorous parasites.

In considering seasonal variations of ectoparasitic arthropods, it is well to remember that the legless larvae of fleas will have greater difficulty escaping dryness or moisture than the legged larvae or nymphs of ticks or mites. Hence overly dry or wet ground may be limiting for fleas, but not for some other ectoparasites. Nevertheless, the largest number of fleas on the

Table 23–1. Effect of Temperature and Light Intensity on Phototaxis in Miracidia of *Schistosoma japonicum* (From Takahashi et al., 1961)

°C Lux	15	18	20	22	23	24	25	26	28	30	34
4500	+	−	−	−	−	−	−	−	−	−	−
2500	+	±	−	−	−	−	−	−	−	−	−
2000	+	±	±	−	−	−	−	−	−	−	−
1000	+	+	±	±	±	±	±	±	−	−	−
500	+	+	+	+	+	+	±	±	−	−	−
250	+	+	+	+	+	+	+	+	±	−	−
100	+	+	+	+	+	+	+	+	+	−	−
50	+	+	+	+	+	+	+	+	+	±	−
25	+	+	+	+	+	+	+	+	+	+	−
10	+	+	+	+	+	+	+	+	+	+	+

+: positive phototaxis; −: negative phototaxis

± individual miracidia of the same experimental group differed in phototactic response, some responding positively, others negatively.

California ground squirrel, *Citellus beechyi beechyi*, occurs during hot, dry seasons in August, September, and October, before autumn rains begin. During a year's survey of the parasites of some marine fish at Plymouth, England, evidence was found of seasonal variation in the intensity of some protozoan infections of the dragonet (*Callionymus lyra*); the heavy infections of Myxosporida were more abundant in the winter, but in the same hosts, heavy haemogregarine infections were more abundant during the summer.

A study of the seasonal life cycle of the tapeworm, *Proteocephalus stizostethi*, disclosed that in the yellow pike-perch, *Stizostedion vitreum*, from Lake Erie, viable embryonated parasite eggs occur only in June.[20] The fish are free from the worm in late summer, new infections establish early in the fall, and the worms mature during the following summer. Seasonal variation in population size of the cestode, *Caryophyllaeus laticeps*, in the fish, *Abramis abrama*, is caused by the combined effects of temperature-dependent mortality and fluctuations in host feeding activity that control the immigration rate of larval parasites.[2]

One of the most completely investigated cases of seasonal variation in helminth abundance involves the so-called "Spring Rise" in fecal egg count of ewes (Fig. 23–2), reflecting an annual increase in the population of gastrointestinal nematodes. This phenomenon is attributable to ingestion in the late summer and fall of infective larvae programmed to arrest, their survival in the host during the winter, resumption of their development in the spring, and retention of the resulting adult worms in ewes due to the immunodepression that accompanies pregnancy and lactation. In temperate, sheep-rearing areas of the world, this sequence of events results in an abundance of infective larvae on pasture that coincides seasonally with the peak annual availability of susceptible, young lambs.

The trematode, *Schistosoma mansoni*, may stop its development in the snail, *Australorbis glabratus*, when these hosts go into estivation in natural habitats that are subjected to annual drought. Thus, climatic conditions adverse for the snail induce a resting stage (diapause) during larval development of its trematode parasites. *Caryophyllaeus laticeps* uses tubificid annelids as intermediate hosts. Genital organs are developed in the larval stages of the tapeworm, and they may spend six months or longer in the tubificid and only two months in the fish. One of the factors, therefore, that could contribute toward the observed periodicity is the seasonal availability of infective larval stages.

Snails (*Cerithidea californica*, commonly infected with several species of trematode larvae) have a parasite burden in January different from that in August, a situation observed by many helminthologists. The explanation lies in changes in water temperature, vegetation, the presence or absence of migratory birds that may serve as final hosts, and other factors. Other instances of seasonal variation are discussed in relation to the induction of diapause in nematodes (p. 451) and in relation to hormonal control of parasite development (p. 459).

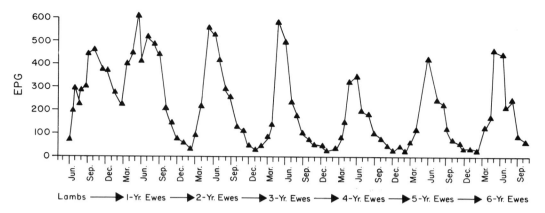

Fig. 23–2. Seasonal variation in the intestinal nematode populations of Scottish hill sheep as reflected in fecal egg counts (EPG).

Periodicity

Annual periodicities (circannual rhythms) have been discussed in relation to seasonal variation in parasite development and abundance. There are other important forms of biologic periodicity including daily (circadian), tidal and monthly (lunar) rhythms. Of these, the *circadian* rhythm is the most common and has been most intensively investigated. Circadian rhythms have a periodicity approximately equal to a 24-hour day. Because the period of circadian rhythm approximates, but does not equal, 24 hours, these rhythms drift with time, unless a daily adjustment is made. In nature, circadian rhythms ("biologic clocks") are adjusted daily by rhythmic environmental stimuli (e.g., light, temperature). They can also be altered radically so as to invert the normal cycle; if day and night are experimentally reversed, circadian rhythms are also reversed.

The circadian rhythms of parasites are frequently adaptations that facilitate transmission. In some instances, they seem to ensure that dispersal stages enter the environment in places where members of the host species gather (e.g., a roost, den or nest). For example, the coccidian, *Isospora*, in the English house sparrow, undergoes a diurnal periodicity in oocyst production. The numbers of oocysts eliminated by the birds reach a peak daily between 3 and 8 P.M. A logical assumption would be that this occurrence is associated with an enhanced opportunity for the oocysts to be taken up by another host. Perhaps, crowding together of birds in preparation for the night has a bearing on the problem. The nocturnal migration of female pinworms to the perianal skin of mice and men, where they lay their eggs, may also serve this end. Nocturnal dispersal of eggs would serve to contaminate the nests of diurnal rodents. It would also have served to contaminate the caves of prehistoric man. In modern man, selection for nocturnal migration would continue to be strong because *Enterobius vermicularis*, the human pinworm, is dependent on hand to mouth transmission and because scratching, hygienically and socially unacceptable during the day, may be practiced unconsciously at night while the human host is asleep.

Free-swimming cercariae are non-feeding, short-lived transmission stages. Therefore, their timely escape from an intermediate host is critically important. The cercariae of *Schistosoma mansoni*, a parasite of humans, are shed in early morning (i.e., after snails have experienced a period of darkness), whereas cercariae of *Schistosomatium douthitti*, a schistosome parasitizing nocturnally active rodents, are shed in the evening (Figs. 23–3 and 23–4). The data shown in the figures demonstrate that cercarial emergence is controlled by the light/dark cycle.

The most widely known examples of circadian rhythmic behavior among parasites are the periodicities in microfilaraemia occurring among filarial species transmitted by blood-sucking arthropods that have their own biting-periodicities. Thus, where *Wuchereria bancrofti* is transmitted by night-biting mosquitos, it has a peak microfilarial density in the peripheral blood near midnight (Fig. 23–5), whereas in areas where it is transmitted by species that bite

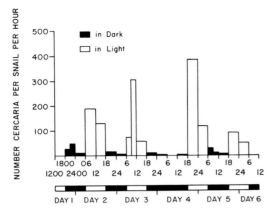

Fig. 23–3. Effect of light on the emergence of *Schistosoma mansoni* cercariae from snails. Cercarial emergence normally occurs during the day (between 6 A.M. and 6 P.M.) but is suppressed if the snails are held in the dark as during day 4. (Modified from Asch, 1972.)

at dusk, the microfilaraemia peaks just before dark (Fig. 23–5B). Hawking[32] has suggested that microfilariae can sense the difference in oxygen tension between venous and arterial blood and use the magnitude of this difference as a cue in synchronizing their circadian behavior with external light-dark cycles. For instance, nocturnally periodic species in diurnally active hosts disperse from the capillary beds of the lungs to the peripheral blood at night, when the host is asleep and the difference in oxygen tension is minimal. During the day, when the host is active, the difference increases, and the microfilariae, sensing this greater difference, change their behavior in a manner that sequesters them in the lungs.

Malarial organisms offer one of the best-known examples of cyclic behavior among protozoan parasites. The causative agent in benign tertian malaria is a typical example. *Plasmodium vivax* has a minimal prepatent period of 8 days. What the parasite is doing in the body during these 8 days immediately after the bite of a mosquito is discussed in Chapter 5. A "clinical periodicity" of 48 hours follows this period. An event that occurs every 48 hours falls on every third calendar day; thus, the name "tertian malaria" is given to the disease caused by *P. vivax*. Every 48 hours, many red blood cells of a person infected with this parasite rupture, owing to the completion of schizogony. The rupture of millions of these cells, all approximately at the same time, with the liberation of their contents into the blood, produces the chills and fever characteristic of malaria.

The function of this 48-hour rhythm remained obscure until Hawking et al.[33] demonstrated that the gametocytes, i.e., the transmission stages of malaria, although apparently long-lived and non-periodic in their abundance in the peripheral blood, have a short period of infectivity for the mosquito vector. The length of the period between malarial paroxysms is determined by the species-specific developmental rate. However, the precise time of day when schizogony is complete and the red cells

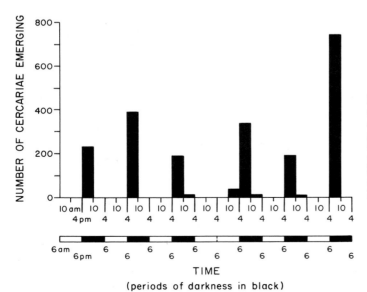

Fig. 23–4. Effect of light on the emergence of *Schistosoma douthitti* cercariae from snails. Cercarial emergence normally occurs after dark (between 6 P.M. and midnight) but is partially suppressed if the snails are held in the light as during day 4. (From Smyth, 1976 after Olivier, 1951.)

TIME

(periods of darkness in black)

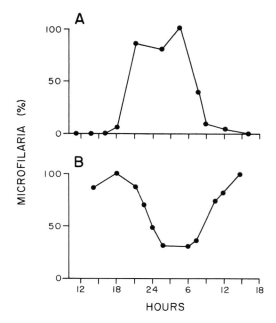

Fig. 23–5. Microfilarial periodicity in two geographic variants of *Wuchereria bancrofti*, one transmitted by night-biting mosquitos (graph A) and another by mosquitos that bite at dusk (graph B). The microfilaremia is expressed as a percentage of the maximum number of microfilariae observed. (Modified from Hawking, 1967.)

burst, releasing the potential sexual stages (i.e., immature micro- and macrogametocytes) infective for mosquitos, is such that these cells mature and attain infectivity during that period of time when the mosquito vector is biting.

Thus the most apparent circadian rhythms in parasites are those that facilitate transmission; they ensure that when two species occur successively in the life cycle of a parasite, the transmission stage emerges from the first host or comes to its surface at a time when the next host, because of its own circadian activities, is available for infection. Similarly, when only one host species is involved, the exit of transmission stages from infected hosts is synchronized with the gathering of the species, usually in preparation for rest.

Internal Factors

A complex of internal environmental factors affects a parasite's ability to occur in a host. Some of these factors have been discussed in previous sections of this book dealing with biochemical, physiologic and immunologic as-

pects of the host-parasite relationship. Entire books have been devoted to the physiology of parasites or even specific classes of parasites.[18,45,89,92] The physiologic relationships between parasites and specific organ systems have been reviewed recently.[42,51] These publications are recommended to the student desiring more complete information. Here we present a selective treatment of host-related, internal ecologic factors and then progress to a discussion of population and community ecology of parasites.

NUTRITION AND DIET

The concept that the host's nutritional status influences the occurrence and severity of infectious disease is an ancient one, famine and epidemics having been associated from the beginning of recorded history.[49] However, for several decades earlier in this century, scientific medicine lost interest in this relationship. The direct effects of malnutrition were given serious consideration, but the interrelationships between nutrition and infection received little attention. In fact, work in this area often was considered lacking in scientific respectability. Recently, however, recognition that protein malnutrition afflicts millions of the world's poor in whom parasitic and other infectious diseases are rampant has stimulated new research interest in this interrelationship.[22,84]

With reference to the infectious diseases generally, and the parasitic diseases in particular, it is usually uncertain whether malnutrition is the cause or effect of frequent or severe infection. However, anorexia (loss of appetite) is often associated with parasitic infections and ultimately may cause malnutrition, which, in turn, will adversely affect the host's immunocompetence. Intestinal parasites may also contribute to the malnutrition more directly by interfering with digestion and absorption of nutrients. For example, the protozoan, *Giardia lamblia*, interferes with the digestion of proteins and fats, and, in turn, may decrease the assimilation of fat soluble vitamins. If severe, the diarrhea associated with giardiasis will further exacerbate the loss of nutrients. Additionally, some parasites may compete with the host for nutrients. The fish tapeworm, for example, competes with its human hosts for vitamin B_{12}. Although hookworms do not actually compete directly for the host's dietary iron, some species cause iron deficiency by wasteful bloodsucking. In heavy infections this may be immunodepressive because iron is required for normal T cell function. Apparently, then, heavy hookworm infections can affect their own longevity by reducing the host's ability to reject them and may even facilitate the establishment of other infectious agents.

Deficiencies of other minerals and some vitamins will interfere with other facets of the immune response, while protein energy deficiency, perhaps the most common form of malnutrition, will have extraordinarily wide ranging adverse effects on immune function. In mature dogs, *Ancylostoma caninum* infections are usually well regulated. Animals that survive infection in early life gradually become resistant, so that eventually they become asymptomatic carriers. This ability of dogs to acquire resistance to hookworm disease intrigued investigators of the disease in man, because in some geographic areas, heavy hookworm burdens and disease occur relatively frequently in adult humans. These observations led to an examination of the relationship between malnutrition and resistance to hookworm infection, in which it was demonstrated that the resistance induced in dogs by repeated *Ancylostoma caninum* infection could be abrogated by feeding a protein and vitamin A deficient diet. Furthermore, malnourished dogs accepted species of hookworms that they would normally reject. Others found that vitamin A deficient rats were more susceptible to the intestinal nematodes *Trichinella spiralis* and *Nippostrongylus brasiliensis* than were normal control rats. More recently, Slater and Keymer[85] demonstrated that protein malnutrition will reduce the ability of laboratory populations of mice to reject a *Nematospiroides dubius* infection, and therefore, they remain chronically infected. Even the high responders (i.e., individual mice capable of particularly strong immunologic response) expected to occur in the outbred populations of mice used in these experiments maintained chronic infections.

It used to be thought that, to a large extent, the suitability of a host species for parasite species was determined nutritionally, and that it was the presence or absence of a nutrient, or a group of nutrients (amino acids) available in a specific proportional relationship, that permitted or prevented a parasitic species from becoming established. Although there is some basis for this view, there is increasing evidence that appropriate receptors or substrata for attachment and/or appropriate signals for hatch, excystation, exsheathment, penetration and initiation of feeding are more important than the availability of specific nutrients. Furthermore, in most host-parasite associations, when nutrients are limiting it is more probable that they will be vitamins or minerals than macronutrients. Protozoa, particularly the intracellular species, are more frequently limited in this manner than helminths. For example, sporozoa of the genera *Toxoplasma* and *Plasmodium* synthesize folates from p-aminobenzoic acid.[95] If the latter is deficient in the cells of vertebrate hosts, normal protozoan growth is impossible. It has been demonstrated experimentally that growth of *T. gondii* and *Plasmodium berghei* can be inhibited when hosts are maintained on a milk diet, milk being deficient in p-aminobenzoic acid.[44] Similarly, the survival of the erythrocytic stages of *Plasmodium falciparum* is compromised by glucose-6-phosphate dehydrogenase and ATP deficiences. Furthermore, a lack of potassium, as occurs at low oxygen tensions in the erythrocytes of persons carrying the sickle cell trait, results in the death of *P. falciparum*. The natural resistance to malaria determined by the variations occurring in human hemoglobin in areas endemic for *P. falciparum* is discussed in greater detail in Chapter 5.

Severe protein energy malnutrition suppresses the development of the plasmodia by mechanisms that remain poorly understood. Clinical malaria is unusual among the starving, but when refeeding is begun among victims of famine, latent malarial infections are frequently reactivated. There is surprising new evidence suggesting that protein deprivation may not have to be severe to suppress a hematozoan infection. For example, when mice were fed a plant protein (cornmeal) or a diet mimicking the normal diet of rural West Africans (2.2 parts cornmeal to 1 part casein) and then infected with *Trypanosoma brucei gambiense*, they gained less weight, but more effectively controlled their trypanosome infection than mice in control groups fed better diets in terms of quantity and quality of protein (Fig. 23–6 and Table 23–2). During the first 4 days of infection, there was no between-group difference in the parasite's multiplication rate, indicating that the plant protein diet had no direct toxic effect on the parasite. Concurrent immunologic studies supported the suggestion that the plant protein diet

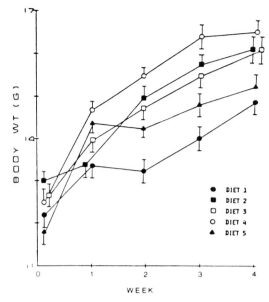

Fig. 23–6. Effect of amount and source of protein upon mean body weights of 22 mice per diet (mean ± standard error). For 30 days, weanling mice were fed:—A cornmeal-based diet at 73% of the recommended daily allowance (RDA) (diet 1);—a casein-based diet at 73% RDA (diet 2);—A mixed diet of plant and animal protein, (2.2:1) at 73% RDA (diet 3);—A casein-based diet at 200% RDA (diet 4);—or laboratory chow (diet 5). (From Norton, Yang and Diffley, 1986.)

was adequate to support the humoral, variant-specific immunity that controls the initial peak of parasitemia occurring in African trypanosomiasis (see antigenic variation in Chapters 2 and 3), whereas the high quality animal protein (casein) was inadequate for this purpose.[61]

With reference to the helminths, the nutritional requirements of cestodes have long been of particular interest to parasitologists. This is attributable to several factors, including the lack of a digestive system requiring that nutrients enter the parasite via the tegument, the extremely high growth rate of tapeworms and the fact that this rate is sensitive to crowding. In his classic study of tapeworm growth, Roberts[72] found that during the initial growth of the strobila, individual *Hymenolepis diminuta* may increase in weight by 1.8×10^6 times in a little more than 2 weeks!

It was assumed from the start that the crowding effect resulted from competition for a nutrient in short supply. Several investigations have now shown that there is a competition for dietary carbohydrate and the limiting nutrient

is, in fact, glucose. The reduced growth and fecundity, however, may not be solely attributable to direct intraspecific competition for nutrients. There is evidence that both an increased concentration of parasite excretions, acting as growth-limiting pheromones, and immunological stimulation resulting from a larger worm population contribute to adverse density dependent effects on growth.

The diet of a host individual or species may also consist of foods containing substances directly inimical to a parasite's survival. Several examples may be cited: onions and their relatives have a mild anthelmintic effect on adult hookworms and a marked adverse effect on the hatchability of the eggs of surviving worms; acorns, the predominant constituent of the natural diet of the gray squirrel, *Sciurus carolinenses*, are rich in tannins toxic for the tapeworm, *Hymenolepis diminuta*. Reflecting this, the parasite is absent in squirrels in natural environments even where a strong potential for infection exists. In the laboratory, however, squirrels fed a commercially prepared diet can be readily infected with the parasite. This susceptibility is reversed by feeding them a diet containing tannin. Many plants are protected from herbivores by toxins evolved precisely for this purpose. One might expect that, among herbivores specialized to resist the effects of these toxins, the potential exists to harbor more parasites than they actually do in nature.

HORMONES

In some highly adapted host-parasite associations, the parasite depends on its host not only for nutrients and shelter, but also for hormonal regulation of its own life cycle. This, as we shall see, synchronizes the life cycle of the parasite with that of the host and facilitates transmission. Additionally, through their effects on both natural and acquired resistance, hormones may influence individual susceptibility to infection.

Host-parasite life history synchronization has been described for the entire spectrum of parasites from protozoans to insects. Here we present some well-known examples spanning this spectrum. We have already referred to the seasonal variations in parasitemia of ducks infected with the protozoan, *Leucocytozoon simondi*. During fall and winter the parasite disappears from

Table 23–2. Effect of the Amount of Dietary Protein on the Survival of Mice Infected with *Trypanosoma brucei gambiense* (From Norton et al., 1986)

	% Composition			Fraction Surviving			
Diet	Cornmeal	Casein	% RDA	1st Peak of Parasitemia (%)[a]	30-days Infection[a,b]	Died (Unknown Cause)[c]	Adjusted 30-day Survival (%)[c]
1	100		73	7/11 (64)[d]	3/11	1/11	30.0
2		100	73	3/10 (30)	1/10	2/10	12.5
3	69	31	73	6/10 (60)[d]	3/10	1/10	33.3
4		100	200	1/10 (10)	0/10	0/10	0.0
5[e]				3/12 (25)	2/12	0/12	16.7

[a]Number survived/number tested

[b]No parasites could be seen in host blood 24 hours before death.

[c]Survivors less deaths by unknown cause.

[d]A Chi-square analysis of animals surviving a first-peak parasitemia revealed a significant difference (based on an expected ratio of 1 survivor:3 dead [laboratory chow]).

[e]Laboratory chow.

the peripheral blood, but reappears in spring when the days become longer and the birds become sexually active. By experimentally increasing day length, the normal hormonal rhythms of infected ducks can be changed so that egg laying occurs in winter with an attendant reappearance of the parasite in the peripheral blood. Although the host's hormones may not be affecting the parasite directly, it is apparent that they are the ultimate, if not the proximate, cause of the relapse in parasitemia. The cyclic development of several intestinal ciliates of frogs, including *Opalina ranarum* and *Nycototherus cordiformis*, is also under the control of the host's hormones. During most of the year, these ciliates, in their large vegetative form, reproduce slowly by binary fission, but as spring approaches and entry of the host into the breeding pond becomes imminent, the rate of fission increases, producing a small precystic form of the parasite. These small forms encyst when the host actually enters the water to spawn and the cysts are then shed in the feces. Both in vivo and in vitro studies suggest that androgens trigger encystation in these species.[90]

The cellulose digesting flagellate protozoa of the woodroach, *Cryptocercus*, show a similar synchronization with the events in the life of their host. Again, the protozoa normally multiply asexually by binary fission, but when the roach begins to molt, the flagellates enter their sexual cycle. In *Trichonympha*, encystation occurs a few days before its host molts. Each cyst contains a male and a female gamete, and excystation and fertilization coincide with the molt. Species-specific differences relating to these sexual phenomena occur among the flagellates, but the sexual phase of the life cycle is always associated with the molt. Experimental injection of the roach with the molting hormone (ecdysone) triggers the appropriate species-specific behavior by the flagellate.[95]

Turning to the helminths, it is again a parasite of the frog, *Rana temporaria*, that serves as the classic example of synchronization between the host's and the parasite's reproduction. *Polystoma integerrimum* is an unusual endoparasitic monogenetic fluke that occurs in the bladder. It enters the bladder, its final parasitic habitat, before tadpoles metamorphose and the young frogs become terrestrial. Three years later, when the frogs attain sexual maturity and prepare to return to water, the flukes also mature and lay their own eggs when the frogs spawn.

Experimental investigations have shown that the flukes can be induced to mature by injecting their hosts with pituitary extract, but by analogy with the mechanisms of synchronizing the life cycles of the protozoans and *R. temporaria*, it is thought that gonadal, rather than pituitary, hormones regulate the fluke's behavior.[90] Another particularly interesting example of life cycle synchronization involves the arrested larvae of the hookworm, *Ancylostoma caninum*, dormant in female dogs. Mature animals, that have been infected repeatedly, tolerate few, if any, adult intestinal worms, but may harbor an abundance of arrested larvae in their muscles. At the time of parturition these larvae are activated and infect the newborn pups via colostrum or milk. Some larvae may also travel to the female's intestine, where they will now be able to establish a transitory infection. Stoye[93] has shown that arrested larvae in mature female dogs harboring cryptic infections can be reactivated and will appear in milk after treatment of the dog with progesterone and oxytocin.

In a series of elegant studies, Rothschild and her colleagues[75] demonstrated that reproduction in the rabbit flea, *Spilopsyllus cuniculi*, is controlled by the hormones of its host. About 10 days before parturition, when the corticosteroid levels in the blood of the pregnant female rabbit begin to increase, the fleas begin to mature sexually. Thus, the female fleas are gravid by the time the dog gives birth. The fleas subsequently transfer to the newborn rabbits, where they feed and mate. They then lay their fertilized eggs in the surroundings, and after several days, return to the doe. Copulation and oviposition are thought to be stimulated by ingestion of growth hormone (somatotropin) with the blood of the neonatal rabbits. This marvelously synchronized chain of events ensures that when the new generation of fleas is ready to find a host, a new generation of young, susceptible rabbits will be available.

Hormones, acting indirectly as immunomodulators, affect the occurrence and abundance of parasites in hosts. The distinguished veterinary parasitologist, Hugh Gordon, coined the expression "periparturient relaxation of immunity," which is attributed to the hormones associated with parturition and lactation.[63] It has been shown in numerous studies that adult female hosts, normally strongly resistant to the establishment of adult intestinal nematodes, will fail to expel them during both the periparturient period and lactation. For example, bitches resistant to adult *Toxocara canis* will again harbor these ascarids in abundance in the periparturient period when their immunodepressed status can be confirmed by lymphocyte transformation assays. Their lymphocytic responsiveness to both a nonspecific mitogen (phytohaemagglutinin) and a specific mitogen (*T. canis* antigen) is low.[48]

Cortisone has a long history of use as an immunodepressive agent by parasitologists attempting either to adapt parasites to abnormal hosts (e.g., laboratory animals) or to extend infections that hosts were expelling. For example, *Ancylostoma duodenale* and *Necator americanus*, the two major hookworms of man, can be established in dogs and hamsters, respectively, by treating them with corticosteroids. With reference to interrupted expulsion, *Strongyloides stercoralis* is a naturally occurring intestinal parasite of man and dog. In the latter, the infection is usually self-limiting and cannot be re-established after the adult worms have been expelled by the resistant host. However, if corticosteroid treatment is initiated before expulsion is complete, this process ends, autoinfection begins and, ultimately, severe, sometimes fatal, hyperinfective strongyloidiasis may occur.[82] Unfortunately, well-tolerated asymptomatic *S. stercoralis* infections in humans receiving long-term corticosteroid therapy for immunopathologic conditions or to facilitate the retention of organ transplants may also be exacerbated. Many cases of hyperinfective or disseminated strongyloidiasis, sometimes terminating fatally, have been precipitated in this way.[28] Indeed, other opportunistic parasitic infections that are frequently well-controlled and asymptomatic in immunocompetent hosts are activated in corticosteroid-treated hosts. These include infections with *Pneumocystis carinii*, the organism responsible for the parasitic pneumonia so common in AIDS patients.

AGE AND SEX

A host's age, often interacting with its sex, may profoundly affect both the occurrence and abundance of a parasite. Frequently, hosts are most susceptible to infection when young and become more resistant as they age. This resistance may wane in the elderly. The term "age

resistance" is generally used to describe an increase in resistance that does *not* involve previous exposure to parasitism. Although this phenomenon is well known, most resistance that increases with age is undoubtedly acquired resistance resulting from the immunologic stimulation provided by previous infections. Although age-related changes in a parasitism often reflect these forms of resistance, they may also have absolutely no relationship to them, reflecting instead changing behavior that increases (or decreases) a host's exposure to parasitism.

Here we will present some examples of the effect of age and sex on parasitization beginning with simple, exposure-mediated, age and/or sex related variation and ending with some aspects of natural and acquired immunity. Immunity is discussed more fully in Chapter 2 and some group-specific details are presented in relation to each of the major groups of parasites.

Some parasites are transmitted vertically, i.e., from mother to offspring via the egg (transovarian transmission), the placenta (congenital transmission) or colostrum and milk (transmammary or lactogenic transmission). Clearly these parasites will appear in newly hatched or born animals before others that gain entry from the external environment. Transovarian transmission of protozoan parasites among arthropods (Chapter 21) or congenital infection with *Toxoplasma* may result in lifelong, and sometimes cryptic, infection. Helminths transmitted vertically either in utero (e.g., *Toxocara canis* and *Protostrongylus* sp.) or lactogenically (e.g., *Ancylostoma caninum* and *Strongyloides* spp.) frequently develop to maturity and are then expelled as their hosts age. Sometimes a characteristic age-related succession of parasites occurs. This is well illustrated by the sequence of parasites occurring in the foal (Fig. 23–7). In this case, *Strongyloides westeri* is acquired lactogenically and is the first species to show a peak of abundance. Thereafter, *Parascaris equorum* eggs and the infective larvae of the strongyles enter the foal from the environment, and show successive peaks of abundance. The *P. equorum* peak reflects the parasite's prepatent period, its extraordinary fecundity and relatively short life. The pattern of increase in the strongyle populations reflects the onset of, and increasing dependence upon, grazing by the foal during its first year of life.[29] In humans, the two common species of anthropophilic hook-

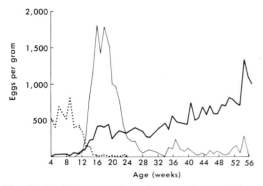

Fig. 23–7. The succession of nematode parasitisms in foals as reflected in species-specific egg output in feces. Average number of eggs of *Strongyloides westeri* (dotted line), *Parascaris equorum* (thin line), and "Strongyles" (thick line) per gram of manure from 26 foals examined weekly. (From Georgi, 1980, after Evans et al., 1977 and Russell, 1948.)

worms appear sequentially, with *Ancylostoma duodenale* preceding *Necator americanus.*[55] It is relevant that *A. duodenale* probably evolved from *A. caninum,*[14] a species strongly committed to transmammary transmission. The occurrence of severe *A. duodenale* infection in suckling infants in Africa suggests that human transmammary transmission of this hookworm probably occurs, but larvae have not yet been demonstrated in milk.[80] The larvae of *Strongyloides fulleborni,* another parasite of human infants, have been found in breast milk, proving that this parasite can be transmitted lactogenically.[11]

Diets change with age. As noted, increased exposure to parasites occurs when a grazing animal is weaned and infective larvae on herbage are ingested in greater quantity. Some birds are insectivorous as hatchlings and become increasingly granivorous as they age, and amphibians change from herbivorous tadpoles to insectivorous frogs when they undergo metamorphosis. Taking these as examples, the bird will be at increased risk of acquiring parasites transmitted by insects as a hatchling, whereas the frog will be at increased risk as an adult. More specifically, grouse are exposed to spiruroid nematodes (*Acuaria*) when young, and frogs are exposed to several insect-borne lung flukes (*Haematoloechus* spp.) when adult. Even in humans, age- and sex-related differences (influenced by culture) may determine the prevalence of infection with food-borne parasites. For instance, in the United States, *Diphyllobothrium latum* infection was once particularly common in Jewish women, because it was they who tasted the raw fish while preparing gefilte fish, an ethnic dish.[23] Division of labor often causes age- and sex-related differences in exposure to parasitic infection, as in India where women are normally employed to harvest tea leaves. This exposes them to heavy hookworm infection, because on the large tea plantations laborers are in the fields for long hours far from latrines and will defecate in a hidden spot among tea bushes. In contrast, adult males were the group most at risk, when, at the turn of the century, hookworms were transmitted to underground workers in the mines and tunnels of Europe, where latrine facilities were lacking, and the miners defecated in inactive side-tunnels that became sites for intense hookworm transmission. Incidentally, the warm, moist microclimate in the mines and tunnels made larval development possible even in the Alps and northern Europe (e.g., tin mines in England).

Other non-immune factors that contribute to changing age-related patterns of parasitism include changes in (1) the skin with age, making it less penetrable by hookworms, schistosomes and other parasites that infect percutaneously; (2) the number of intestinal goblet cells in older animals, increasing intestinal mucus and thereby decreasing parasite survival; and (3) diet that alters the availability of nutrients for parasites.

An interesting, rapidly developing, age-related resistance to infection in an invertebrate, reflecting a simple structural change occurs in the tsetse fly, *Glossina palpalis,* the vector of *Trypanosoma brucei gambiense.* When flies up to 3 days old were fed on an infected host, the infection rates of newborn, 2-day-old, and 3-day-old flies were 21%, 4%, and 1%, respectively. This decrease is attributable to development of the peritrophic membrane, a chitinous tube lying in the midgut of some insects, that is soft and discontinuous in newborn flies but becomes continuous and hard in a few days, trapping the trypanosomes.

Other examples of age-dependent change in parasitism occurring in invertebrates have been described. Of particular interest from an ecologic standpoint is the changing pattern of trematode parasitism that may occur during the life of a snail (Table 23–3). Whether the parasitisms characteristic of the different phases of the snail's life reflect a changing attractiveness for miracidia of the several species that parasitize

Table 23–3. Parasitism of *Littorina saxtalis* by Five Species of Sympatric Larval Trematodes (From data reported by James, 1965)

Species	A	C	B	D	E
Invades juveniles	−	−	−	−	+
Invades after 1st reproductive cycle	±	+	+	+	−
Invades after 2nd reproductive cycle	+	+	−	−	−
Restricted to haemocoel	−	−	−	+	+
Enters digestive gland	+	+	+	−	−

Littorina saxtalis, or whether the miracidia penetrate, but find the snail unsuitable, remains unknown. If parasitic castration occurs, snails grow to an unusually large size; these older and larger snails support an increasing number or biomass of the castrating parasite.

With reference to host resistance increasing with age, this may be physiologically or immunologically mediated. Thus, dogs become increasingly resistant to the development of *Ancylostoma caninum* whether or not they have had previous experience with this hookworm. This age resistance develops more rapidly, and is more complete, in females than in males. In addition to age resistance, acquired immunity, dependent on exposure to the parasites, develops with time (age) as the host is exposed repeatedly to infection and becomes increasingly resistant.

These phenomena, resistance as related to age and to sex, are so well known with reference to nematode parasitisms that parasitologists desiring to establish a nematode in an abnormal host (e.g., a laboratory animal) will usually begin by infecting young males of the potential host species. For instance, although *Necator americanus* will not normally develop in rodents, it can be established and maintained in the laboratory by infecting male hamsters in the first 3 days of life. Older animals and females are much more resistant to infection or likely to expel the worms that develop. Infections in the process of being expelled can be prolonged by immunosuppressive treatment of the host, demonstrating that expulsion is mediated by immunologic mechanisms. *Aspicularis tetraptera* sometimes serves as a pinworm (oxyuroid) laboratory model; again young hosts (i.e., mice) are less resistant than old ones and females develop resistance before males.

Although the young are often highly susceptible to parasitic infections, they may sometimes be protected by maternal antibody for a short time early in life. For example, in highly endemic areas, malaria does not occur in children during the first 3 months of life, although the prepatent period is much shorter than this (i.e., less than 2 weeks). Protection can be transferred with serum both in humans and in experimental model systems (e.g., *Plasmodium yoelii* in the mouse), suggesting that the absence of malaria in newborn infants involves maternal antibody. Maternal antibody also protects against some helminth infections, e.g., larval *Taenia saginata* infection in calves. In fact, if a cow's udder is immunized by injection with *T. saginata* eggs, calves that suckle on the cow are protected from infection with cysticerci.

Reverse age resistance is unusual, but occurs in several tick-borne parasitisms of the erythrocytes of cattle (babesiosis, anaplasmasis) and horses (babesiosis). In these infections, the young are more resistant than mature animals. Although some protection is undoubtedly conferred by maternal antibody, the resistance persists for too long (up to 2 years in the horse) to be attributable to such antibody.[8] Furthermore, calves that were reared in areas where the infection does not occur could never have received maternal antibody, but nevertheless resist infection.[92] The mechanisms underlying this resistance remain unknown.

An analogous phenomenon occurs at the cellular level. *Plasmodium chabaudi*, one of several plasmodia that cause malaria in rodents, parasitizes mature erythrocytes (normocytes) but not the immature stage (reticulocytes). The reverse situation also occurs, *P. yoelii* preferring reticulocytes to normocytes.

In some host-parasite associations, resistance present in young adults declines in old age. Laboratory investigations have shown that aged mice have heavier infections of *Trichinella spiralis, Toxoplasma gondii,* and *Trypanosoma musculi.* Again the underlying mechanisms are unknown, but may, in some cases involve age-dependent differences in "natural killer" cell abundance. In aged *T. musculi*-infected mice, these cells are less numerous in younger adults.[1]

STRESS

Stress is the result of reactions to **stressors** such as pain, hunger, crowding, high temperatures, low temperatures, emotions, irritation, darkness, bright light, fighting, and confinement, among others. Reactions resulting from a stressful condition are numerous and complicated. They include an alteration in the hormonal balance and the immune responses of the host. This imbalance may decrease inflammation of tissues and may weaken other resistance mechanisms. Stress is thus undoubtedly a factor in parasitism because anything that lowers host resistance will favor establishment of parasites. The sensitive periods of parasite development are particularly subject to the reactions of a host under stress.

Sexual cycles, changes in the blood picture, antibody formation, metabolism of basic foodstuffs, tissue damage, and many other body activities and reactions are intimately related to stress and, therefore, may influence the parasite burden.

Experimental work on the relationship between stress and the size of parasite populations in the cecum of ground squirrels has shown that a direct correlation exists between the numbers of flagellates, *Trichomonas*, and the degree of stress under which the hosts are living.[57,58] Fighting among the squirrels was especially effective in promoting an increase in numbers of cecal flagellates. Cold, in the form of reduced temperature for two successive nights, doubled the numbers of cecal amebas in ground squirrels, as compared with the ameba count in field control squirrels.[59]

A major response to stressors is production of adrenocortical hormone (ACTH) from the pituitary gland and consequent release of adrenal glucocorticoids, which, among other actions, diminish normal inflammatory responses of the host. See Chapter 2 for comments on immunosuppression. For this reason, cortisone or another corticosteroid is used experimentally instead of a stressor in some stress and parasitism studies. For example, mallard ducklings were infected with the nematode, *Echinuria uncinata*,

and were subjected to crowding or were injected with cortisone.[64] These ducklings retained more and larger nematodes at necropsy than did the control ducklings. Birds under stress developed involution of the bursae of Fabricius and thymus glands, hypertrophy of the adrenal glands (except in ducks injected with cortisone), and were retarded in growth. These reactions are signs of the general adaptation syndrome of stress. Mice that are usually resistant to infection with the roundworm, *Strongyloides ratti,* can become infected if given daily doses of cortisone. The drug also lowers resistance of mice to the pinworm, *Aspicularis tetraptera*, and to the tapeworm, *Taenia taeniaeformis.*

The effect of cortisone may differ according to the sex of the host. Thus, experimentally infected female mice and naturally infected female jackrabbits are more susceptible to infection with the larval tapeworm, *Multiceps serialis*, than are males.[24]

Ectoparasites may also be affected by host stress, or at least from the injection of the host with ACTH or cortisone. In an experiment with lambs, daily injections of ACTH for 1 month broke down resistance that had developed against keds (Fig. 20–20). It was concluded that physiologic or environmental stress, such as pregnancy or undernourishment, can affect the basic annual ked population cycle.[56] As one might expect, all experiments do not lead to the same conclusions. Working with the rat, Villarejos found that cortisone did not increase susceptibility to amebiasis.[96]

For a review of stress and parasitism, see Esch, et al.[25]

POPULATION ECOLOGY

Parasite population ecology is concerned with the processes that control and regulate the abundance of parasites and their hosts. Anderson and May[3] proposed that three conditions should be satisfied before an animal species is classified as a parasite: utilization of the host as a habitat; nutritional dependence on the host; and, following Crofton,[21] causation of harm to the host. Harm was defined in terms of the effect of the parasite on the host's survival and fecundity. A mathematic analysis indicated that there were circumstances under which parasites could be expected to regulate the abundances of their hosts at levels less than those which the host populations would have attained in the absence of parasitism. Three kinds of population processes were shown to be of particular importance in stabilizing and enhancing the regulatory role of the parasite: the overdispersion of parasite numbers per host, nonlinear functional relationships between parasite burden per host and host death rate, and density-dependent constraints on parasite population growth within individual hosts.[3] These density dependent constraints on parasite population growth within individual hosts are also the processes by which parasite abundance is regulated in the entire community of hosts.[21,86]

It is well known that parasite numbers are influenced by a range of processes including the effect of climate on the free-living stages, host deaths that are unrelated to the parasitism, and so on, but none of these processes are capable of regulating parasite abundance in the sense that they render the parasite population resistant to perturbations. Parasite populations are bounded. Their abundance varies, certainly, but within limits that are defined by local conditions and the characteristics of the parasite-host interaction. This apparent stability is not mere chance; population processes have evolved whose effects are truly regulatory with respect to the parasite population. Such processes are density dependent and are distinguished from all other density dependent processes that influence parasite population numbers in that they act as negative feedback mechanisms which constrain parasite population growth within individual hosts, and, hence, regulate the growth of the total population within the community of hosts.

These processes exert their effects on parasite abundance through the influence they have on stage-specific rates of mortality, fecundity, or asexual multiplication. Regulation is contingent on a proportional increase in some adverse effect as parasite population density (number of parasites per host) increases. In an individual host, the severity of the adverse effect depends simply upon the number of parasites it harbors. In a community of hosts, the constraints of parasite survival and reproduction summed over the whole population depend upon the nature of the frequency distribution of parasites per host. Overdispersed distributions typically increase the severity of the effect.

The mechanisms by which density depend-

ent processes are thought to regulate parasite populations include: intraspecific competition for a limited resource (often called the "crowding effect"); non-specific host responses to the infection creating an environment inimical to parasite survival or reproduction; pathologic responses leading to host deaths; and acquired immune responses.[87] The recognition that regulatory processes are at work has frequently involved the use of mathematic models of parasite population biology in the analysis of field or laboratory data. Such models have made considerable use of functions which relate parasite mortality or fecundity to the current population of parasites in a host. However, it is becoming increasingly obvious that these functions are adequate descriptions of events in only a limited number of cases. This is particularly true when there is reason to suspect that an immunologic response is involved. Anderson and May[4] explicitly acknowledge this problem in their model for helminth transmission in humans. This model includes an anamnestic component (a mimic of immunologic memory), and represents the mortality of the immature parasitic stages as some function of the host's accumulated experience of the infection over time.

Models of the population dynamics of host-parasite interaction are becoming increasingly useful in the analysis and design of mass chemotherapeutic strategies to control parasitic diseases in man and his domestic animals,[5,88] and in studies of the coevolution of parasites and hosts. These latter have led to the conclusion that "successful" parasites need not necessarily evolve to be harmless, an idea that until recently was widely (although not universally) accepted by parasitologists. A number of coevolutionary paths are possible depending on the precise relationship between the pathogenicity and transmissibility of the parasite.[50]

Community Ecology

Community ecology is concerned with assemblages of species in specific habitats, the richness of these assemblages, the numerical dominance of some species, the scarcity of others, their niche relationships and the characteristics of the physical environment and the population-level processes responsible for these attributes of communities. With reference to free-living species, ecologists have considered the important biologic processes to be competition and predation. More recently, infectious disease has also attracted interest as a process influencing the structure of communities. Turning to parasitic infra-communities, these same factors may be highly relevant (e.g., physical environment, competition), or less so (e.g., predation), and quite different ones, irrelevant to free-living organisms (immunity and immuno-depression), may be important. Some parasitologists deny the importance of competition, arguing that resources are rarely limiting for parasites.[67,68,73,74]

The following paragraphs discuss these processes with reference to parasites, first discussing simple, two-parasite competitive interactions, and then showing how their outcome may be modified by a third organism, either the host or another parasite. Because not all interactions are antagonistic, associations in which the presence of one parasite benefits another are considered. Some complex multispecific associations of parasites are discussed next. Considered last are the arguments that dominate parasite community ecology currently, i.e., whether communities of parasites are generally saturated with species occurring in tightly packed niches, and structured by competition, or alternatively, whether they are generally unsaturated, have few species and many vacant niches open to colonizing species. Proponents of the open community hypothesis either suggest that the probability of successful colonization of potential hosts is low[73] or that communities of parasites are young, and therefore parasites have had insufficient time to fill all vacant niches.[67]

Interactions between two species may be direct, involving the parasites themselves, with the host remaining passive, or they may be more complex, involving mediation by the host. Perhaps the best known antagonistic two-species interaction is that described by Holmes,[34] a pioneering parasite ecologist, who demonstrated posterior displacement of the tapeworm, *Hymenolepis diminuta*, when it co-occurs with the acanthocephalan *Moniliformis dubius*, in the small intestine of the rat (Fig. 23–8). A somewhat similar interaction was reported by Chappell,[18] who described the mutual displacement of a cestode (*Proteocephalus fillicolis*) and an acanthocephalan (*Neoechino-rhyncus rutili*) in the gut of the stickleback, a small fresh water fish. In concurrent infections, the cestode is displaced anteriorly and the acanthocephalan posteriorly from their single-species distributions.

Examples of competitive displacement (i.e., *interactive site segregation*)[37,38] are not limited to shifts along a linear intestinal gradient. In another important study, Holmes[36] demonstrated that two blood flukes of rockfish (*Sebastes*), with partially overlapping habitats in the host, displace each other strongly. One of these flukes, *Psettarium sebastodorum,* normally occurs in the heart, whereas the other, *Aporocotyle macfarlani,* occurs mainly in the branchial arteries and the aorta but its distribution extends into the heart. When these species co-occur, they shift toward opposite sections of their habitats, maximizing the area of nonoverlap between them.

In parasitology, it is useful to restrict the use of the term *competitive exclusion* so that it indicates the exclusion of a parasite from the entire host rather than from a site within the host. Exclusion may take either of two forms, it may be complete with one species excluding another from a host, or it may be relative so that although the species coexist, one of them occurs at a reduced abundance.

Field studies may show that the co-occurrence of parasite species in hosts is much less frequent than would be expected, were chance the sole determinant of co-occurrence. Particularly strong antagonistic interactions are expected when parasites make heavy demands on the tissues of the host. This is frequently the case in larval trematode parasitisms of snails. In these parasitisms, exclusions are common and are attributable not only to subtle, presumably chemically mediated interactions,[98] but also to vigorous, active antagonisms in which one species, turned predator, clears a snail of another species, now prey.[47] In fact, the redia of a particularly strong antagonist, *Paryphostomum segregatum,* normally occurs deep in the visceral mass of its snail host, but when the latter is penetrated by miracidia of another species, the redia moves out from the viscera to intercept the incoming parasites near the snail's surface, even in such remote sites as the tentacles.

Marked pathogenicity is also associated with the exclusion of one monogenetic fluke, *Dactylogyrus extensus,* from the gills of carp, by a congener, *D. vastator.* Eggs of the former hatch earlier in the spring than those of the latter, and, therefore, *D. extensus* is the first to colonize the host. Later, *D. vastator* hatches, attaches to the gills and, as its name suggests, devastates them. The damaged tissue is unsuitable for the survival of *D. extensus,* and its population crashes.

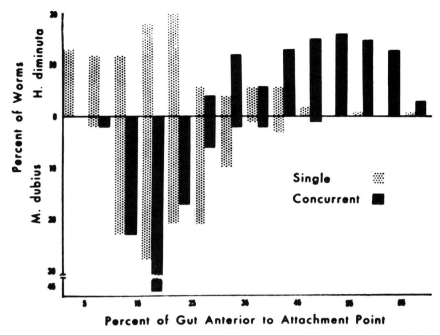

Fig. 23–8. Competitive intraintestinal displacement of the tapeworm, *Hymenolepis diminuta* by the acanthocephalan, *Moniliformis dubius* in concurrent experimental infections of the rat. (From Holmes, 1961.)

Eventually, tissue destruction is so severe that
even *D. vastator* is lost. After healing, *D. extensus*
returns, demonstrating that it was, in fact, excluded by *D. vastator.*[65]

Because *D. extensus* recolonizes carp, acquired immunity probably does not play a role
in its previous disappearance from the gills. Often, however, parasitologists cannot discriminate between exclusion that is immunologically
mediated and that which is not. In the example
just presented, destruction of the habitat resulted in the loss of the parasites. Destruction
of tissue, specifically pulmonary tissue, has also
been invoked to explain the exclusion of one
schistosome by another. Experimentally, *Schistosomatium douthitti* in mice is excluded by a previous *Schistosoma mansoni* infection, but the latter does not exclude the former. It has been
suggested that this exclusion is not likely to be
immunologically mediated because the interaction is non-reciprocal and because migration
of *Ascaris* larvae through the lungs of mice will
also interfere with the establishment of the
schistosome. However, one of us has pointed
out[78] that when two species are competitors,
natural selection should favor antigenicity on
the part of the species that evokes an immu-

nological response that is more effectively directed against the other, and that competitive
interactions could be mediated in this way. Recent investigations in Australia indicate that immunologically mediated competitive exclusion
does, in fact, occur. Sheep serve as hosts for
two intestinal nematodes (nodular worms) of
the genus *Oesophagostomum.* Historically, *O. columbianum* was the most prevalent and abundant of these nematodes over large areas of eastern Australia but where sheep have been
treated regularly with benzimidazole anthelmintics this parasite has disappeared. In these
areas it has been replaced by a congener, *O.
venulosum,* that was rare previously. When
sheep previously infected with *O. columbianum*
are exposed to *O. venulosum* experimentally,
there is a 90% reduction in worm establishment
when compared to infections in previously uninfected control sheep.

Immunologically mediated exclusion was
even more conclusively demonstrated in experiments involving the sibling species, *Haemonchus contortus* of sheep and *H. placei* of cattle.
The species are not host specific and, therefore,
experimental cross infections are easily established. In nature, however, the parasites are

largely restricted to their own hosts, suggesting that they exclude each other. In a series of elegant experimental studies, LeJambre[46] showed that in its own host (sheep), a primary infection with *H. contortus* excludes a challenge infection with the related bovine species, *H. placei*. To provide evidence that this exclusion was immunologically mediated, he treated some experimentally infected sheep with an immunodepressive agent and showed that resistance to the "foreign" species, *H. placei*, was greatly reduced, allowing it to establish in abundance (Table 23–4).[46]

Two species interactions may also be modified by the presence of a third parasitic species. We have already described some larval trematode interactions in snails. In the 1960s, it was hoped that these interactions could be exploited for the biologic control of medically important flukes, but under simulated field conditions the expected exclusions sometimes failed to occur. Further investigation showed that the relationship between the species with predatory rediae that attack other species and the species normally excluded from snails was changed by the presence of microsporidia. The former were more susceptible to microsporidial parasitism than were the latter and consequently both trematode species survived.[47]

Thus far, only antagonistic interactions have been discussed. These have preoccupied parasitologists, but in recent years, as immunologic investigations have become increasingly sophisticated, it has become apparent that immunodepression is a common tactic used by parasites to survive in the host. Clearly then, one parasite may actually facilitate another parasite's ability to establish and survive in a host. For example, *Nematospiroides dubius*, an intestinal nematode of mice frequently used in experimental parasitology, depresses the mouse's immune responses to several other intestinal nematodes. Thus, *Trichinella spiralis* infections, that are usually expelled after 2 to 3 weeks in the mouse, will survive in hosts that are concurrently infected with *N. dubius*. That this inhibition of expulsion results from immunodepression is suggested by the fact that mice harboring *N. dubius* fail to respond to an array of potent antigens including several viruses and sheep erythrocytes.[9,97]

Other parasite-dependent mechanisms that may facilitate the co-occurrence of parasitic species include: (1) provision of a shelter within which a parasite escapes attack by the host, (2) predigestion of ingested material (or host tissues) thereby making it available to another species (or chain of species), (3) provision of factors necessary for growth and/or maturation of another species, or (4) damage to cells or tissue eliciting production of a different cell type needed by another species.

These positive forms of interaction are less well known than the negative ones presented previously, but examples of each are found in the parasitologic literature.

With reference to the first of these, the availability of a parasite-provided site safe from host attack explains the presence of metacercariae in species of snails that would not normally permit metacercarial development. These "foreign" species of metacercariae survive within the sporocysts of other normally occurring trematodes.

As regards provision of the host's ingesta in a modified form usable by other species, this is a well-known phenomenon among the cellulose-digesting protozoa such as those occurring in the intestines of termites and wood roaches. These protozoa occur in dense assemblages packed into the hind gut of these insects. Some of the protozoa are dependent on others to provide the products of cellulose digestion in a

Table 23–4. **The Interaction Between *Haemonchus contortus* [Hc] and *Haemonchus placei* [Hp] in Normal and Immunodepressed Sheep (i.e., opticortenol-treated sheep) (From LeJambre, 1982)**

First Infection	Second Infection	Opticortenol	Mean No. Adults	Mean Proportion Hc Females	Hc Males
Hc	Hp	−	234	0.95	0.97
Hc	Hp	+	1,596	0.64	0.56
Hp	Hc	−	588	0.81	0.82
Hp	Hc	+	990	0.67	0.67

usable form. Symbioses, involving cellulose-digesting microorganisms and their hosts, are discussed in greater detail later in this chapter.

One parasite may provide a growth factor required by another. For example, the lungworms of swine, *Metastrongylus apri* and *M. pudendotectus*, are almost invariably found together. Experimental investigations (Fig. 23–9) have demonstrated that these parasites fail to develop if they enter the host singly unless they are given in massive doses, and even then, *M. pudendotectus* fails to mature sexually. In mixed infections, however, both species become established in normal abundance and mature sexually. These observations led the Ewings and their colleagues[27] to consider *M. apri* and *M. pudendotectus* facultative and obligate mutualists, respectively, that probably depend on each other for growth factors.

Parasites may also provide resources for other co-occurring organisms by causing an increase in the availability of particular cells or tissues. Thus, for example, *Plasmodium chabaudi chabaudi*, a malarial parasite of African Thicket Rats (*Thamnomys*), multiplies abundantly in mature erythrocytes (normocytes) of laboratory mice. Mice respond to the destruction of their normocytes by increasing erythropoiesis, resulting in a marked increase in the number of circulating young erythrocytes (reticulocytes). The reticulocytosis provides abundant resources for *Plasmodium yoelii yoelii*, a reticulocyte specialist whose population in singly infected mice is limited by the number of reticulocytes, but in the doubly infected animals, it increases dramatically.[70]

This system is informative for several reasons. It shows that an apparently abundant resource, erythrocytes, may, in fact, be limiting, even though many cells remain uninfected. Furthermore, it demonstrates the diversity of the possible interactions occurring even in simple (i.e., two-species) parasite communities. In his particularly interesting and well-conceived experimental investigations, Richie[70] demonstrated that, in laboratory mice, a number of different interactive scenarios are possible, depending on which species is injected into the host first and how long it has developed before the second is injected. Sometimes, as in the example given above, one species acts to benefit another, whereas, under other conditions, one will act antagonistically, limiting another's abundance. Thus, for example, when *P.y. yoelii* is injected at least 1 week before *P.c. chabaudi*, infection with the latter is suppressed. However, because in its normal host, *Thamnomys*, the abundance of each species is strongly lim-

EXPERIMENTAL INFECTIONS

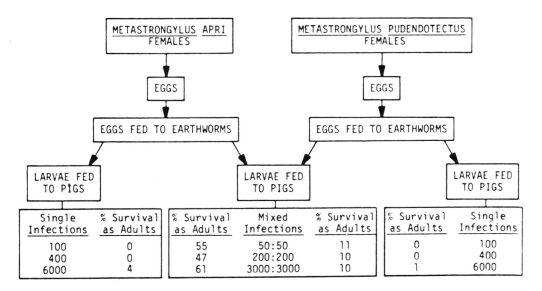

Fig. 23–9. Design and results of an experiment testing the mutualistic relationship occurring between two lungworms of swine, *Metastrongylus apri* and *Metastrongylus pudendotectus*. (From Ewing et al., 1982.)

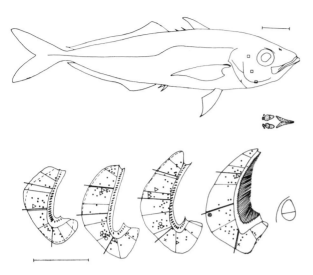

Fig. 23–10. Distribution of ectoparasites on the surface and in the mouth cavity (left figure) and on gills 1–4 (right figure, top to bottom) of 88 fish, *Trachurus trachurus*, from the North Sea. □ = *Caligus pelamysis* on inner wall of operculum, ○ = *Gastrocotyle trachuri* (Monogenea), △ = other Monogenea, ⊗ = *Lernanthropus*, × = Lernaeoceridae, • = parasite cysts. All external filaments and every twentieth internal gill filament (thin line) is shown. Thick lines divide the gills into longitudinal quarters. (From Rohde, 1980.)

ited (at any one time, no more than 2% of the erythrocytes are infected), interactions between the species of *Plasmodium* are less intense. This observation has important implications for the current debate about the importance of interspecific competition in structuring parasite communities, as we will see.

The richness of communities of parasites varies greatly. Some infracommunities are poor, both in terms of species and numbers of individual parasites, while others are exceedingly rich on both counts. Infracommunities of ectoparasites on fish from cold ocean habitats are depauperate, constituted of but few species and individuals (Fig. 23–10).[74] Some snakes, for example the Black Racer (*Coluber constrictor*), have few helminth parasites of the alimentary tract, namely three hookworm-like nematodes usually occurring at low population densities in widely disjunct sites (esophagus, duodenum, and rectum) and few other species that are rarely abundant.[83] This contrasts sharply with the rich parasite fauna of some other reptiles, e.g., herbivorous tortoises, in which pinworms

alone are represented by 10 or more species occurring in the colon in enormous numbers. Indeed, these parasites are packed so densely that the colon maintains its round cross section even after it has been removed from the host and fixed. Complex communities of closely related helminths existing at high density have fascinated parasitologists, particularly those interested in the problems that such communities pose for theoretic ecology. They have asked the question: How have numerous closely related species, existing at high densities in a sharply limited space within their hosts, managed to survive on an evolutionary time scale?

Particularly rich associations of closely related species occurring at high density include the flagellates (protozoa) of the hind gut of wood-eating insects (e.g., termites, wood roaches), nematodes in the stomachs of kangaroos and the large intestines of equines, rhinoceri, elephants and herbivorous tortoises, and rumen flukes (amphistomes) in antelope and bovines in Africa and southeast Asia, respectively. In each of these associations, the parasites occur in enlarged organs specialized for cellulose digestion with a voluminous, highly heterogeneous contents.[15,40] For example, the colon of the tortoise, *Testudo graeca*, is specialized for cellulose digestion. Plant material enters the proximal colon virtually unaltered by mastication and digestion. Consequently, the contents available as food for the colon-inhabiting pinworms range from intact leaf fragments to minute fecal particles representing the final stages of digestion. It also includes an array of cellulose-digesting microorganisms and their associates, ranging in size from large ciliated protozoa to single coccoid bacteria. Additionally, minute larval worms and worm eggs are possible food items for the larger adult nematodes. There is also the fluid phase of this diphasic system, and, for species capable of puncturing either the host's cells or other parasites, tissue fluids as well. It is not surprising, then, that in this complex microcosm numerous congeneric species can coexist even at high density.

Coexistence depends on niche diversification. The co-occurring species of pinworms (*Tachygonetria* spp.) are differentially distributed in the colon of the tortoise both linearly and radially. Pairs of species with the most similar linear distributions differ in radial distribution; one is associated with the mucosa and the other with the lumen (Fig. 23–11). Additionally, these

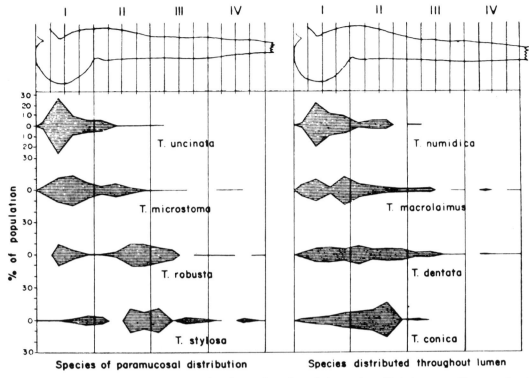

Fig. 23–11. Distribution of *Tachygonetria* females in the colon of six tortoises, *Testudo graeca*. Above each set of distributional diagrams is a tracing of a representative colon. Species of paramucosal distribution include those with more than 75% of their populations occurring close to the intestinal wall; those with less than 75% of their populations adjacent to the wall were considered to be distributed throughout the colonic lumen. Species with the most similar linear distribution differed in radial distribution. (From Schad, 1963.)

species differ markedly in the shape and size of their oral openings. There are attendant differences in diet, some species feeding on fragments of ingested plant material, others on semidigested particles and others on the colonic microfauna and flora (Fig. 23–12). Furthermore, the suitability of the colon for particular species varies with the age and health of the tortoise and with the occurrence of other kinds of parasites.[40,66,77]

Similarly, many related nematode species ("small strongyles") co-occur in the large intestines of horses, zebras, and their relatives. Some of these parasites occur mainly in the ventral colon, others in the dorsal colon, and some in the cecum. Their elaborate mouth openings suggest that they may be able to discriminate between the different food items available in the colon of equines. Indeed, some have been reported to feed on ciliated protozoa, some on minute, pinworm-like nematodes, and others on intestinal contents. Apparently age and sex-

related differences in the suitability of equines for particular species of small strongyles have not been reported.

Niche diversification, as seen in these dense complex assemblages of closely related species with similar resource requirements, has suggested to many parasitologists that communities of parasites are structured by interspecific competition.[37,40,78] Historically, Holmes has been the chief proponent of this view.[35] He suggested that, in general, communities of parasites are mature and resistant to new colonization because most niches are filled, interspecific competitive interactions in the past having led to evolutionary changes that are manifested in the present as niche differentiation. This view is supported by experimental investigations that have demonstrated competitive interactions (displacement or exclusion) even between parasites that are *not* closely related and, in some instances, apparently do not use the same host-provided resources. That strong antagonistic in-

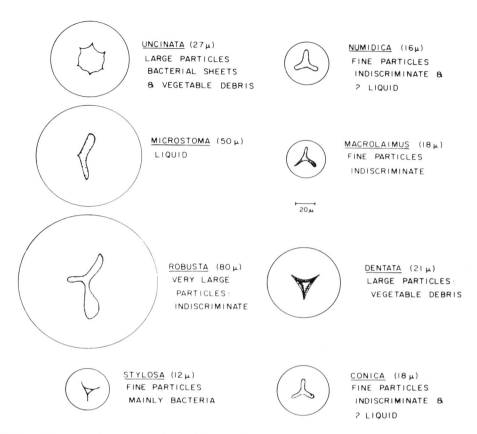

Fig. 23–12. Diagrammatic representations of the mouth openings of the species of *Tachygonetria* arranged as in Figure 23–10. The structural differences are associated with differences in food habits as judged by microscopic examination of the intestinal contents of dissected samples of the worms of each species (see notes on figure). The measurements in parentheses show the maximum dimension of each oral opening.

teractions occur, even between species that are *not* obvious ecologic homologues, strengthens the argument.

More recently Bush and Holmes[12,13] studied the complex helminth fauna of the Lesser Scaup, a diving duck that feeds on small crustacea and other invertebrates. Eight species ("core species") were present in more than two-thirds of the 45 ducks examined; these generally occurred abundantly in parasitized individuals. Another eight ("secondary species") occurred with moderate frequency and abundance. Thirty-six helminths ("satellite species") were primarily parasites of other host species, occurring in scaup infrequently and at low population densities. Bush and Holmes considered the important resources for intestinal helminths

to be distributed linearly along the intestine and concluded that the longitudinal distribution of these worms provides a useful measure of niche breadth. The helminths that were frequently present, and at least moderately abundant, (core and secondary species) had generally predictable distributions in their hosts that were modified in individual ducks by the presence of other species. Over all ducks examined, the linear distributions of these species overlapped broadly, with the linear range of each individual species correlating with its density. However, there was significantly less distributional overlap in individual ducks, and overlap between species with adjacent distributions did *not* increase along with their population densities. Apparently species that normally overlap interact to reduce the degree of overlap, whereas those that overlap infrequently do not interact in this way.[13]

In sharp contrast, Rohde,[73] who has contributed importantly to our knowledge of the parasites of marine animals, doubts that competitive interactions have been important in structuring parasite communities. He is impressed by the depauperate state of many parasite communities. He notes, for example, that the gills of marine fish rarely harbor many species of monogenetic flukes and, those that do occur, normally exist at low population densities. Nevertheless, these monogeneans show strict microhabitat specificity (narrow niche restriction). He argues that this specificity has evolved, not as a result of competition, but because sparse populations need to cluster so that mating will be possible.

Price[67] is also impressed by the many "empty niches" that hosts appear to provide for potential parasites, but, on largely theoretic grounds, proposes that communities of parasites are immature and that time simply has been insufficient for evolution to have filled them with species. More recently, he and Holmes[38] have recognized and named two kinds of communities, one constituted of species that have evolved together and under the pressure of competition have changed so as to occupy more restricted niches ("interactive communities") and another constituted of species that have evolved independently and now co-occur largely by chance ("isolationist communities"). The conditions under which each has evolved have not been established and provide excellent opportunities for research.

As we have already indicated, dense multispecific assemblages of parasites are most frequently associated with capacious organs specialized for cellulose digestion, and, as adults, these parasites apparently feed on the lumen contents rather than the host's tissues. The adult worms, therefore, are actually commensals rather than parasites and apparently provoke little if any functional immune response by the host. It seems that, under these conditions, tight species-packing is possible.

The helminth communities of certain ducks, such as those of the Lesser Scaup, already mentioned, present a similar situation. These communities consist largely of tapeworms and acanthocephalans, and in the sense that these parasites are "absorbers," they also feed on lumen contents, rather than on host tissues. They apparently elicit little, if any, lasting host resistance, because they are reacquired each spring, having been lost during their host's annual migration.[13] The assemblages of helminths occurring in the intestine (spiral valves) of some elasmobranch fishes (rays) are also rich in cestode species, but contain fewer individuals.[100] Again nutrients are undoubtedly obtained from the lumen and, as parasites of elasmobranchs, acquired resistance, if any, is probably minimal. Thus, these communities resemble those occurring in organs specialized for cellulose digestion, both trophically and with reference to elicitation of host resistance.

An intermediate condition exists among parasites such as the trichostrongyle nematodes of ruminants. These occur in complex multispecific assemblages, but at less extraordinary densities than the aforementioned groups. They are closely associated with the intestinal mucosa; they lie against it and feed on the contents of individual cells. In contrast to the preceding group, there is probably less opportunity for these species to specialize trophically and because of their closer association with the tissues, they are more likely to evoke an immune response by the host. The reduced ability to specialize trophically could account for the reduced diversity in groups such as the trichostrongyles, and the increased host contact with an attendant increase in immune responsiveness could be responsible for the reduced densities of all.

Nematodes such as the large strongyles, hookworms and their relatives have large mouth capsules, ingest multicellular plugs of tissue, and, if they suck blood, come into direct contact with the circulation. Other parasites are completely immersed in the host's blood (trypanosomes, schistosomes) or lymph (filarial worms), or burrow into the host's cells, tissues and organs (*Trichinella*, *Gongylonema*, *Paragonimus*). Clearly, these species make intimate contact with the host. They rarely occur in complex multispecific associations and usually occur at low density. Indeed, as was pointed out in the previous section of this chapter, the frequency distributions of many parasites (true parasites in individual hosts) are overdispersed, so that most individuals harbor few parasites and few harbor many. It is clear that antigenic stimulation will increase with parasite density and that different species often share antigens, or are adversely affected by nonspecific, immunologically-mediated host reactions. This may account for both the limited population densities and species diversities (apparent presence of vacant niches) in some associations of parasites, at least among those occurring in mammals and birds. How other apparently open communities are maintained is unclear. In fact, the entire fascinating area of parasite community ecology remains poorly understood and offers the imaginative student excellent opportunities for research.

In any case, Richie's[70] investigations of malarial infections in rodents demonstrate that in natural, highly adapted host-parasite associations, the host's response may limit populations of potential competitors to levels of abundance that reduce interaction. Under these circumstances, host-provided resources may appear underutilized, but only from the parasite's perspective! From the host's perspective the resources may, in fact, be overexploited, and the host may respond to suppress the parasites. Indeed, as indicated in a theoretic consideration of competition and immunity published more than 20 years ago, the individual, immunologically controlled interactions between parasite and host may be more important than those occurring directly between co-occurring parasites in limiting the abundance of competing species.[78] Contrasting markedly with the natural host-parasite relationships involving *P.c. chabaudi*, and *P.y. yoelii* in thicket rats, are the new, i.e., experimentally contrived relationships involving these two species of *Plasmodium* and the laboratory mouse. In the new host-parasite associations, strong competitive interactions occur, suggesting that competition in the past may well have structured the parasite communities we see in the present.

MUTUALISTIC INTESTINAL PROTOZOA AND BACTERIA

Insects

A classic example of mutualism is the cooperation between termites and their intestinal protozoan fauna. The hindgut of many termites is so crowded with protozoa (Fig. 23–13) that the protozoan bodies are often distorted from being pressed on all sides. The protozoa in nymphs of *Zootermopsis* constitute one-seventh to one-third the total weight of host and parasites. Most of the larger flagellates in termites

belong to the order Hypermastigida (see Chap. 3). The termite families Mastotermitidae, Kalotermitidae, Hodotermitidae, and Rhinotermitidae are normally host to an abundance of flagellates. Significantly, most members of the largest and most specialized of termite families, the Termitidae, do not feed on wood, but the few members that do so feed possess in their intestines bacteria that presumably elaborate enzymes that digest wood. The first nymphal instar acquires its protozoa by proctodeal feeding at the anus of an infected termite; during each molting period, the protozoa are temporarily lost—to be regained as originally acquired. Only the adult, asexual nondividing forms ordinarily occur in termites. Encystment of the protozoa does not take place.

If human beings possessed cellulose-digesting enzymes, we would not have so critical a world food shortage problem as we have today because people could survive well, although perhaps not tastefully, on straw, sawdust, wood shavings, bark, and dead branches. In a pinch, we could eat paper and old furniture. Neither do the termites and wood-eating roaches themselves elaborate cellulose-digesting enzymes (cellulases), but their protozoan and, probably, bacterial guests pay for the intestinal housing facilities by digesting the food eaten by the host.

Termites can be defaunated by incubation at 36°C for 24 hours or by subjection to 2 atmospheres of oxygen pressure for 4 hours. When the protozoa are dead, the insects will continues eating wood and other materials, but they will starve to death because they cannot digest their food. If they are soon refaunated, they will survive. Acetate is the end product of the metabolism of cellulose. This acetate is used by the termites for oxidation and biosynthesis.

Physiologic experiments dealing with *Cryptocercus* and its flagellates have not been so extensive as those with termites, but in all probability, the roach is also dependent almost entirely on the fermentation of cellulose by its intestinal protozoa. One difference between roach and termite protozoa is the habit of encystment of many of the protozoa in the roach during molting of the insect. The roach retains its infection throughout its life. Another difference is that encystment is accompanied by a sexual process, and the stimulus for the initiation of a sexual cycle in the flagellate is extrinsic, attributable to the molting hormones of the

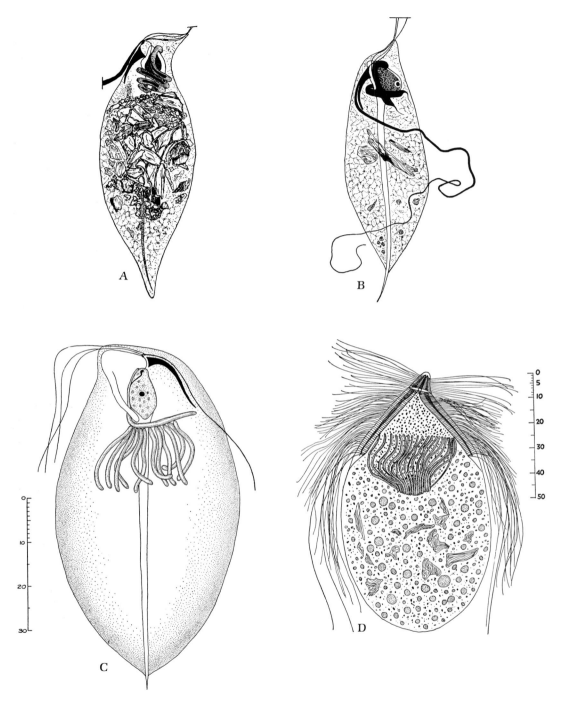

Fig. 23–13. Four samples of the many flagellates to be found in the gut of termites. *A, Devescovina hawaiensis* from *Neotermes connexus.* ×1700. *B, Foaina hilli* from *Glyptotermes iridipennis.* ×1750. *C, Pseudodevescovina uniflagellata* from *Neotermes insularis.* ×1500. *D, Trichonympha corbula* from *Procryptotermes.* ×680. The two scales are in micrometers. (From Kirby, courtesy of the University of California Publications in Zoology.)

host. Molting and the sexual cycles do not occur separately. Although the sexual cycles of the flagellates are initiated by host hormones, the follow-up is directly influenced by the rise in pH (7.2 to 7.4) of the gut fluids during the molting period.

The role of extracellular symbiotes (bacteria, spirochaetes) on the surfaces of protozoa in termites and in roaches remains obscure. A great diversity in numbers, kinds, and distribution of both extracellular and intracellular symbiotes occurs in the flagellates of termites, but experimental evidence concerning their nature is lacking. The intracellular bacteria in parasitic flagellates may play a part in the digestion of cellulose.

Ruminants

The placid, cud-chewing habit of cattle is directly related to the digestion of cellulose and to the masses of symbiotic ciliates and bacteria found in the first stomach of these animals. In sheep and goats, the gymnostome ciliates occur in numbers between 160,000 and 200,000 per milliliter of rumen content. The 3 most common species in sheep and goats (also to be found in cattle) are *Isotricha prostoma*, *Isotricha intestinalis*, and *Dasytricha ruminantium*, Hungate[39] found an average of 7000 *Diplodinium* per milliliter of rumen contents in cows fed timothy hay and concentrate, with a range between 800 and 30,000. The young ruminant acquires its infection by ingesting infected saliva.

Three important genera of ciliates in cattle are *Diplodinium* (Fig. 23–14), *Entodinium*, and *Epi-*

dinium. It is more difficult to defaunate cattle than termites, but it can be done by starvation for several days, followed by the administration of a copper sulfate solution. When this is done, cellulose and hemicellulose continue to be digested at the same rate as in untreated cattle. *Isotricha*, *Dasytricha*, *Buetschlia*, and other genera readily absorb soluble carbohydrates of the host stomach, convert the carbohydrates into starch, and thereby help to keep the sugars from being immediately fermented by the bacteria. Hydrogen, carbon dioxide, acetate, butyrate, lactate, and storage polysaccharide amylopectin are major products of carbohydrate metabolism of *Isotricha*.

Ciliates feed on starch and bacteria. It has been estimated[19] that *Entodinium caudatum* can engulf 10^8 bacteria per milliliter of host stomach fluid each minute. All the protozoa have endoplasmic vesicles containing bacteria. *Diplodinium* possesses the enzyme *cellulase* or *cellobiase*, and it may store large amounts of glycogen. Ciliates and bacteria in a cow's stomach synthesize B-complex vitamins, enzymes, amino acids, and proteins. These substances are used directly by the host as food.

The question as to the amount of protein supplied to the host in the form of protozoa can be answered by using Hungate's estimates for the bovine of 3000 *Diplodinium*, 3000 *Isotricha*, 5000 *Dasytricha*, and 5000 *Entodinium* per milliliter. Calculation shows that about 66 g of protein are supplied to the host in the form of ciliates each day.

As indicated by the defaunation studies men-

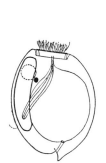
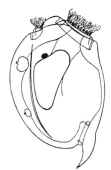

Entodinium curtum *Diplodinium monacanthum* *Eremoplastron rostratum* *Metadinium rotundatum*

Fig. 23–14. Illustrations of ciliates of the family Ophryoscolecidae. These drawings illustrate sample genera and species from the stomach of a wild bull (*Bos gaurus*) in Mysore, India. (From Kofoid and Christenson, courtesy of the University of California Publications in Zoology.)

tioned, rumen ciliates are not essential to the ruminant. Hungate[39] has estimated, however, that the ciliates are about equal to the bacteria in importance as sources of nitrogen for the host. The value of the ciliates can be ascribed to a better balance of amino acids, and possibly to a greater quality and lesser quantity of ciliatean protein.

"One hypothesis to account for the favorable effects of the adaptation to the new ratio [a change from a forage ration to one high in grains] is that it gives time for the development of a ruminal protozoal population capable of ingesting much of the starch in the ration, sequestering it from attack by the more rapidly growing rumen amylolytic bacteria. Proof of this hypothesis is lacking."[39]

Ciliates provide value to their hosts

"not only in their protein content but also in the fermentation products they form from carbohydrates, though few estimates of the relative importance of the protozoa and bacteria as producers of acetic, propionic, and butyric acid have been made."

The two major roles of protozoa in the rumen are (1) the removal of starch grains from the fluid phase of the rumen; this removal makes the starch unavailable for bacterial attack; starch is fermented in the protozoan cell to acetic and butyric acid, which are readily absorbed through the rumen wall; and (2) the recycling of bacterial nitrogen. See Coleman[19] for a 1980 review of rumen ciliate protozoa.

Reflect for a moment on the tremendous churn of activity that is housed in an intestine. Bacteria, yeasts, spirochaetes, protozoa, and worms may all be struggling for nutrients and space at the same time. Byproducts of metabolism are being poured into the mass continuously. Physiochemical conditions (such as pH and gases) are constantly being altered. The bacteria synthesize vitamins and other essentials for the protozoa and worms. No wonder that evolutionary processes among intestinal organisms and their hosts have resulted in relationships that bear the character both of coexistence and of antagonism.

REFERENCES

1. Albright, J.F., and Albright, J.W.: Natural resistance to animal parasites. *In* Immunobiology of Parasites and Parasitic Infections. Edited by J.J. Marchalonis. New York, Plenum Press, 1984, pp. 1–40.

2. Anderson, R.M.: Seasonal variation in the population dynamics of *Caryophyllaeus laticeps*. Parasitology, *72*:281–305, 1976.

3. Anderson, R.M., and May, R.M.: Regulation and stability of host-parasite population interactions. J. Anim. Ecol., *47*:219–247, 1978.

4. Anderson, R.M., and May, R.M.: Herd immunity to helminth infections and implications for parasite control. Nature, *315*:493–496, 1985.

5. Anderson, R.M., and Medley, G.F.: Community control of helminth infections of man by mass and selective chemotherapy. Parasitology, *90*:629–660, 1985.

6. Asch, H.L.: Rhythmic emergence of *Schistosoma mansoni* cercariae from *Biomphalaria glabrata*: control by illumination. Exptl. Parasitol., *31*:350–355, 1972.

7. Bardsley, J.E., and Harmsen, R.: The trypanosomes of Ranidea, I.: The effects of temperature and diurnal periodicity on the peripheral parasitaemia in the bullfrog (*Rana catesbeiana* Shaw). Can. J. Zool., *47*:283–288, 1969.

8. Barriga, O.O.: Immunology of Parasitic Infections. Baltimore, University Park Press, 1981, pp. 95–96.

9. Behnke, J., Wakelin, D., and Wilson, M.M.: *Trichinella spiralis*: delayed rejection in mice concurrently infected with *Nematospiroides dubius*. Exptl. Parasitol., *46*:121–130, 1978.

10. Bray, R.S.: Epidemiology of leishmaniasis: some reflections on causation. Ciba Found. Symp., *20*:87–100, 1974.

11. Brown, R.C., and Giardeau, M.H.F.: Transmammary passage of *Strongyloides* spp. larvae in the human host. Am. J. Trop. Med. Hyg., *26*:215–219, 1977.

12. Bush, A.O., and Holmes, J.C.: Intestinal helminths of lesser scaup ducks: patterns of association. Can. J. Zool., *64*:132–141, 1986a.

13. Bush, A.O., and Holmes, J.C.: Intestinal helminths of lesser scaup ducks: an interactive community. Can. J. Zool., *64*:142–152, 1986b.

14. Cameron, T.W.M.: Parasites and Parasitism. New York, John Wiley & Sons, 1956.

15. Chabaud, A.G.: Remarques sur les nematodes parasites in caecum des elephants milieu tres preserves des phenomenes du selection. C.R. Acad. Sci. Paris, *243*:436–439, 1956.

16. Chandler, A.C.: Hookworm Disease. Its Distribution, Biology, Epidemiology, Pathology, Diagnosis, Treatment and Control. New York, Macmillan, 1929.

17. Chappell, L.H.: Competitive exclusion between intestinal parasites of the three-spined stickleback, *Gasterosteus oculectus*. J. Parasitol., *55*:775–778, 1969.

18. Chappell, L.H.: Physiology of Parasites. Glasgow, Blackie, 1980.

19. Coleman, G.S.: Ruman ciliate Protozoa. Adv. Parasitol., *18*:121–174, 1980.

20. Connor, R.S.: A study of the seasonal cycle of a proteocephalan cestode, *Proteocephalus stizostethi* Hunter and Bangham, found in the yellow pikeperch, *Stizostedion vitreum vitreum* (Mitchell). J. Parasitol., *39*:621–624, 1953.

21. Crofton, H.D.: A quantitative approach to parasitism. Parasitology, 62:179–193, 1971.

22. Crompton, W.T., and Nesheim, M.C.: Malnutrition's Insidious Partner. World Health, WHO, March, 1984, pp. 18–21.

23. Desowitz, R.: New Guinea Tapeworms and Jewish Grandmothers. Tales of Parasites and People. New York, W.W. Norton & Co., 1981, pp. 42–43.

24. Esch, G.W.: Some effects of cortisone and sex on the biology of coenuriasis in laboratory mice and jackrabbits. Parasitol., 57:175–179, 1967.

25. Esch, G.W., Gibbons, J.W., and Bourque, J.E.: An analysis of the relationship between stress and parasitism. Am. Mid. Nat., 93:339–353, 1975.

26. Evans, J.W., Borton, A., Hintz, H., and Van Vleck, L.D.: The Horse. San Francisco, W.H. Freeman & Co., 1977.

27. Ewing, M., Ewing, S., Keener, M., and Mulholland, R.: Mutualism among parasitic nematodes: A population model. Ecol. Modelling, 15:353–366, 1982.

28. Genta, R.M.: *Strongyloides stercoralis*: immunobiological consideration on an unusual worm. Parasitol. Today, 2:241–246, 1986.

29. Georgi, J.: Parasitology for Veterinarians. Philadelphia, W.B. Saunders, 1980.

30. Gibbs, H.C.: Hybobiosis in Parasitic Nematodes—An Update. Adv. Parasitol., 25:129–174, 1986.

31. Halvorsen, O.: Negative interaction amongst parasites. *In* Ecological Aspects of Parasitology. Edited by C.R. Kennedy. Amsterdam, North-Holland, 1976, pp. 99–114.

32. Hawking, F.: The 24-hour periodicity of microfilariae: biological mechanisms responsible for its production and control. Proc. Roy. Soc. (Lond.) B., 169:59–76, 1967.

33. Hawking, F., Worms, M.J., and Gammage, K.: 24- and 48-hour cycles of malaria parasite in the blood; their purpose, production and control. Trans. Roy. Soc. Trop. Med. Hyg., 62:731–760, 1965.

34. Holmes, J.C.: Effects of concurrent infection of *Hymenolepis diminuta* (Cestoda) and *Moniliformis dubius* (Acanthocephala). I. General effects and comparison with crowding. J. Parasitol., 47:209–216, 1961.

35. Holmes, J.C.: Site selection by parasites: interspecific interactions, site segregation and their importance to the development of helminth communities. Can. J. Zool., 51:333–347, 1973.

36. Holmes, J.C.: Habitat segregation in sanguinicolid blood flukes of scorpaenid rockfishes of the Pacific coast of North America. J. Fish. Res. Bd. Can., 28:903–909, 1981.

37. Holmes, J.C.: The structure of helminth communities. *In* Parasitology—Quo Vadit? Edited by M.J. Howell. Canberra, Australian Acad. Sci., 1986, pp. 203–209.

38. Holmes, J.C., and Price, P.W.: Communities of parasites. *In* Community Ecology: Pattern and Process. Edited by D.J. Anderson and J. Kikkawa. Oxford, Blackwell Scientific Publications, 1986.

39. Hungate, R.E.: The rumen Protozoa. *In* Parasitic Protozoa. Vol. 2. Edited by J.P. Kreier. New York, Academic Press, 1978, pp. 655–695.

40. Inglis, W.G.: Patterns of evolution in parasitic nematodes. *In* Evolution of Parasites. Edited by A.E.R. Taylor. Oxford, Blackwell Scientific Publications, 1965, pp. 79–124.

41. James, B.L.: The effects of parasitism by larval Digenea of the digestive gland of the intertidal prosobranch, *Littorina saxatilis* (Olivie) subsp. *tenebrosa* (Montagu). Parasitology, 55:93–115, 1965.

42. Kennedy, C.R.: Ecological Aspects of Parasitology. Amsterdam, North-Holland, 1976.

43. Keymer, A.: Density dependent mechanisms in the regulation of intestinal helminth populations. Parasitology, 84:573–587, 1982.

44. Kretschmar, W.: Der einfluss der wirtsernahrung auf die entwicklung von parasitischen protozoa. Zeitschr. fur Parasitenk., 31:30–37, 1968.

45. Lee, D.L., and Atkinson, H.G.: Physiology of Nematodes. New York, Columbia University Press, 1977.

46. Le Jambre, L.: Genetics and the control of trichostrongylid parasites of ruminants. *In* Biology and Control of Endoparasites. Edited by L.E.A. Symonds, A.D. Donald, and J.K. Dineen. New York, Academic Press, 1982, pp. 53–76.

47. Lim, H.K., and Heyneman, D.: Intramolluscan, intertrematode antagonism. Adv. Parasitol., 10:191–268, 1972.

48. Lloyd, S., Amerasinghe, P.H., and Soulsby, E.J.R.: Periparturient immunosuppression in the bitch and its influence on infection with *Toxocara canis*. J. Small Anim. Pract., 24:237–247, 1983.

49. Mata, L.: Nutrition and Infection. *In* Tropical and Geographical Medicine. Edited by K.S. Warren and A.A.F. Mahmoud. New York, McGraw-Hill, 1984, pp. 206–211.

50. May, R.M., and Anderson, R.M.: Epidemiology and genetics in the evolution of parasites and hosts. Proc. Roy. Soc. Lond. B., 219:218–313.

51. Mettrick, D.F., and Podesta, R.B.: Ecological and physiological aspects of helminth-host interactions in the mammalian gastrointestinal canal. Adv. Parasitol., 12:183–278, 1974.

52. Michel, J.F.: Arrested development of nematodes and some related phenomena. Adv. Parasitol., 12:279–366, 1974.

53. Michel, J.F., Lancaster, M.B., and Hong, C.: Arrested development of *Ostertagia ostertagi* and *Cooperia oncophera*. J. Comp. Path., 85:133–138, 1975.

54. Morgan, D.O., Parnell, I.W., and Rayski, C.: The seasonal variations in the worm burden of Scottish hill sheep. J. Helminthol., 25:177–212, 1951.

55. Nawalinski, T., Schad, G.A., and Chowdhury, A.B.: Population biology of hookworms in children in rural West Bengal. II. Acquisition and loss of hookworms. Am. J. Trop. Med. Hyg., 27:1162–1173, 1978.

56. Nelson, W.A.: Development in sheep of resistance to the ked? *Melophogus ovinus* (L.) II. Effects of adrenocorticotrophic hormone and cortisone. Exp. Parasitol., 12:45–51, 1962.

57. Noble, G.A.: Stress and Parasitism. I. A prelim-

inary investigation of the effects of stress on ground squirrels and their parasites. Exp. Parasitol., *11*:63–67, 1961.

58. Noble, G.A.: Stress and parasitism. II. Effect of crowding and fighting among ground squirrels on their coccidia and trichomonads. Exptl. Parasitol., *12*:368–371, 1962.

59. Noble, G.A.: Stress and parasitism. III. Reduced night temperature and the effect on pinworms of ground squirrels. Exptl. Parasitol., *18*:61–62, 1966.

60. Noble, G.A., and Noble, E.R.: Parasitology: the Biology of Animal Parasites. 5th Ed. Philadelphia, Lea & Febiger, 1982.

61. Norton, J.D., Yang, S.P., and Diffley, P.: Influence of source and quantity of protein on the development of immunity and resistance to African trypanosomiasis. Inf. Imm., 455–460, 1986.

62. Olivier, L.: Influence of light on the emergence of *Schistosomatium donthitti* cercariae from their snail host. J. Parasitol., *37*:201–204, 1951.

63. O'Sullivan, B.H., and Donald, A.D.: A field study of nematode parasite populations in the lactating ewe. Parasitology, *61*:301–315, 1970.

64. Ould, P., and Welch, H.E.: The effect of stress on the parasitism of mallard ducklings by *Echinuria uncinata* (Nematoda: Spiurida). Can. J. Zool., *58*:228–234, 1980.

65. Paperna, I.: Competitive exclusion of *Dactylogyrus extensus* by *Dactylogyrus vastator* on the gills of reared carp. J. Parasitol., *50*:94–98, 1964.

66. Petter, A.J.: Équilibre des especes dans les populations de nematodes parasites du colon des tortues terrestres. Mem. Mus. Natl. Hist. Natur., *39*:1–252, 1965.

67. Price, P.W.: Evolutionary Biology of Parasites. Princeton, Princeton University Press, 1980.

68. Price, P.W.: Evolution in parasite communities. In Parasitology—Quo Vadit? Edited by M.J. Howell. Canberra, Australia Acad. Sci., 1986, pp. 209–214.

69. Rau, M.E., and Tanner, C.E.: *Echinococcus multilocularis* in the cotton rat. Asexual proliferation following the intraperitoneal inoculation of graded doses of protoscolices. Can. J. Zool., *50*:941–946, 1972.

70. Richie, T.: Interactions between malarial parasites in the vertebrate host. Ann Arbor, MI, University Microfilms International, 1985.

71. Ricklefs, R.E.: Ecology. New York, Chiron Press, 1979, p. 875.

72. Roberts, LS.: The influence of population density on patterns and physiology of growth in *Hymenolepis diminuta* in the definitive host. Exptl. Parasitol., *11*:332–371, 1961.

73. Rohde, K.: A critical evaluation of intrinsic and extrinsic factors responsible for niche restriction in parasites. Amer. Nat., *114*:648–671, 1979.

74. Rohde K.: Comparative studies on microhabitat utilization by ectoparasites of some marine fishes from the North Sea and Papua New Guinea. Zool. Anz. Jena., *204*:27–63, 1980.

75. Rothschild, M.: Fleas. Sci. Am., *213*:44–53, 1965.

76. Russell, A.F.: The development of helminthiasis in thoroughbred foals. J. Comp. Path., *58*:107–127, 1948.

77. Schad, G.A.: Niche diversification in a parasitic species flock. Nature, *198*:404–406, 1963.

78. Schad, G.A.: Immunity, competition, and the natural regulation of parasite populations. Am. Nat., *100*:359–364, 1966.

79. Schad, G.A.: The role of arrested development in the regulation of nematode populations. In Regulation of Parasitic Populations. Edited by G.W. Esch. New York, Academic Press, 1977, pp. 111–167.

80. Schad, G.A., and Banwell, J.G.: Hookworms. In Tropical and Geographical Medicine. Edited by K.S. Warren and A.A.F. Mahmoud. New York, McGraw-Hill, 1984, pp. 359–373.

81. Schad, G.A., et al.: Arrested development in human hookworm infection: an adaptation to a seasonally unfavorable external environment. Science, *180*:502–504, 1973.

82. Schad, G.A., Muncey, D.W., and Hellman, M.: Hyperinfective canine strongyloidiasis. In Molecular and Biochemical Parasitology. Edited by D. Mettrick. Amsterdam, Elsevier, 1982, p. 682.

83. Schad, G.A., and Kuntz, R.E.: Speciation, zoogeography and host specificity in reptilian nematodes as illustrated by studies on the genus *Kalicephalus*. Proc. 1st Int. Cong. Parasitol., Rome, 1964.

84. Scrimshaw, N., and Gordon, J.E.: Nutritional and epidemiology: ecological insights. In Tropical and Geographical Medicine. Edited by K.S. Warren and A.A.F. Mahmoud. New York, McGraw-Hill, 1984, pp. 199–205.

85. Slater, A., and Keymer, A.: *Heligmosomoides polygynis* (Nematoda): the influence of dietary protein on the dynamics of repeated infection. Proc. Roy. Soc. Lond. B., *229*:69–83, 1986.

86. Smith, G.: Density dependent-mechanisms in the regulation of *Fasciola hepatica* populations in sheep. Parasitology, *88*:449–461, 1984.

87. Smith, G.: The regulation of trematode populations by density dependent processes. In Parasitology—Quo Vadit? Edited by M.J. Howell. Canberra, Australian Acad. Sci., 1986, pp. 24–29.

88. Smith, G., and Grenfell, B.T.: The population biology of *Ostestagia ostestagi*. Parasitol. Today, *1*:76–81, 1986.

89. Smyth, J.D.: The Physiology of Cestodes. Edinburgh, Oliver and Boyd, 1969.

90. Smyth, J.D.: An Introduction to Animal Parasitology. 2nd Ed. London, Hodder and Stoughton, 1976.

91. Smyth, J.D., and Halton, D.W.: The Physiology of Trematodes. 2nd Ed. Cambridge, Cambridge University Press, 1983.

92. Soulsby, E.J.L.S.: Helminths, Arthropods and Protozoa of Domesticated Animals. London, Bailliere Tindall, 1982, pp. 713, 753.

93. Stoye, M., and Krause, J.: Versuche zur reaktivierung inhibierter larver von *Ancylostoma caninum*: Die wirkung von Oestradiol und Progesteron. Zbl. Vet. Med. B., *23*:822–839, 1976.

94. Takahashi, T., Mori, K., and Shigita, Y.: Phototactic, thermotactic, and geotactic responses of miracidiea of *Schistosoma japonicum*. Jap. J. Parasitol., *10*:686–691, 1961.

95. Trager, W.: Living Together: The Biology of Parasitism. New York, Plenum Press, 1986.
96. Villarejos, V.M.: Cortisone and experimental amebiasis in the rat. J. Parasitol., 48:194, 1962.
97. Wakelin, D.: Immunity to Parasites. Baltimore, Edward Arnold, 1984.
98. Walker, J.C.: *Austrobildorzia terrigalenis*: a schistosome dominant in interspecific interactions in the molluscan host. Int. J. Parasit., 9:137–140, 1979.
99. Watkins, A.R.J., and Fernando, M.A.: Arrested development of the rabbit stomach worm *Obeliscoides cuniculi*: varied responses to cold treatment by the offspring produced throughout the course of a single infection. Int. J. Parasitol., 16:55–61, 1986.
100. Williams, H.H.: The ecology, functional morphology and taxonomy of *Echenebothrium* (Cestoda: Tetraphyllidea). Parasitology, 56:227–285, 1966.
101. Zeldon, R.: Epidemiology, modes of transmission and reservoir hosts of Chagas' disease. Ciba Found. Symp.

24

Distribution and Biogeography

The field of biogeography impinges on many other fields of inquiry considered in this book, especially host-specificity (Chap. 25) and evolution of parasitism (Chap. 26). The value of a certain parasitic genus as an indicator of the distribution and evolution of its host might be supported by conclusions drawn from biogeography or alternatively might be exposed to criticism from the same source. Contemporary distributions and associations of parasites and their hosts, as reflections of the past, provide the evolutionary biologist with important clues about both pattern and process in biology. Success in analysis comes from correctly interpreting the data and matching cause with effect in an accountable explanatory framework. Of all biogeographic subjects, parasite-host associations can be the most interesting, for not only do they provide an opportunity to study the biology and relationships of organisms distributed around the world, but because those organisms have other organisms in or on them, another level of complexity is added.

With respect to applied problems in parasitology, studies of distribution and biogeography are important in learning more about the spread and virulence of parasitic-caused diseases—often with respect to humans' alteration of their environment. How, and from where, was a particular parasite pathogen introduced into a given area? How will changing the environment affect the resident parasites, their hosts, and their vectors? What will happen if an economically important organism is introduced into an area where close relatives of that organism are heavily parasitized? Desowitz,[4a] a consultant for the World Health Organization, gives an extremely readable account of the more spectacular successes and failures of human dealings with parasites.

The distribution of parasites and their hosts around the world is not random. Factors that play a role in the present distribution of species include (1) their history of **vicariance** (the division and separation of a biota or taxon by a natural geographical barrier, such as a mountain range or a fragmenting continental plate); (2) their **vagility** (ability to move about actively); (3) their **dispersal properties** (ability to move about actively *or* passively—e.g., a spore is not innately vagile, but it can be dispersed by various biotic and abiotic factors); and (4) the length of time and the history of physical events through which the first three factors have affected the species *or* their ancestors. These considerations apply to all organisms, parasitic or not (see the works of Nelson and Platnick,[13a] Wiley,[19] and Humphries and Parenti.[8a] In the study of the biogeography of parasites, some additional factors affecting distribution are the ability of the parasite to survive apart from a host and the different histories and properties of larval and adult stages and of intermediate and definitive hosts.

Zoologic investigations of parasites and their hosts must be based on large numbers of faunal lists or distributional records. Such records for parasites are scanty, but those that exist suffice for making generalizations concerning the historical and present relationships among hosts and parasites. When considering the historical record, the student should remember that in all sciences built on history, conclusions are drawn not from direct observations of processes, but from inferences corroborated by many lines of evidence.

In this chapter, we are again concerned with the environmental factors that combine to create the setting that determines the numbers and kinds of parasites associated with one or more hosts. When members of the same species of host become separated in space, do their parasites remain the same? What are the environmental factors both outside and inside the host that determine the distribution of the parasite? Complete answers to these and related questions must await further ecologic studies, but a good start has been made, and we shall select examples from the growing literature and formulate some tentative conclusions.

CLIMATE

The distribution of parasites and of their hosts is directly and indirectly governed in a large measure by the climate. Climate varies according to latitude, longitude, altitude, and season of the year; it is the result of infinite and changing combinations of temperature, rain, wind, water currents, land and water masses, mountain ranges, and vegetation.

Temperature is the most important extrinsic factor that influences the existence of parasites. Large areas of water situated around small areas of land equalize temperatures, whereas large areas of land, especially those remote from large bodies of water, retain the sun's heat during the day and, in temperate regions, lose the heat at night; this process helps produce intense cold during the winter months. In warm areas such as North Africa, the hot, dry climate is responsible for torrid days and cool nights. In hot, wet climates (tropics), there is constant warm temperature with high humidity and often with few or no air currents (that is, doldrums). Obviously, the chances for survival and dispersal of such parasites as larval hookworms outside the body and free-swimming miracidia and cercariae depend directly on temperature and moisture. Cysts and spores and invertebrate hosts can also be killed by unsuitable temperature and moisture conditions.

MICRODISTRIBUTION

The phrase "distribution of animals" usually suggests a spread over geographic areas, but for parasites it should also connote the spread within or on one host. When one host organ, or even one cell, is considered, the term **microdistribution** is particularly appropriate.

Among the most interesting examples of microdistribution are the lice on birds and on man.

Most bird groups have 5 or 6 species of lice, often many more. For example, on the Tinamidae (tinamou group in South America), 12 species of lice belonging to 8 genera and 3 families have been recorded from one species of bird host (*Crypturellus obsoletus*), while 15 species of lice belonging to 12 genera and 3 famles were recorded from another host (*Tinamus major*). A general correlation exists between size and shape of the lice and size of feathers. Lice on the smaller feathers of the head and neck, out of reach of the bird bill, are broad, with larger mandibles and head, whereas lice on the longer and broader feathers of the back and wings are flattened and elongated.

The microdistribution of parasites within one host's body is constantly and delicately regulated. Parasites, to be sure, are often adapted to one organ, to part of one organ, or to one kind of cell, but the metabolic balance is easily upset, and one species of parasite may overrun its usual boundaries. For example, a nematode of the intestine of a vertebrate host can occasionally become so numerous that it spreads into the stomach, gallbladder, liver, and coelom. *Leishmania donovani* normally invades large endothelial cells of blood vessels and lymphatics, as well as a few monocytes of the blood, but it can also parasitize erythrocytes in the liver, bone marrow, and spleen, especially in young children in advanced stages of leishmaniasis. In such a situation, this usurpation of space often discourages other parasites or even eliminates them altogether.

Distribution of parasites within a host is governed by the same basic forces that control distribution of the hosts. Temperature, moisture, mechanical barriers, chemistry of surrounding medium, food supplies, and other ecologic factors always operate, as well as phylogenetic relationships that can determine the degree of host specificity. In the fish called the "mudsucker" (*Gillichthys mirabilis*), two closely related nematodes live in the mesenteries, but one of these worms is also occasionally found in the intestine. A physiologic difference between the two worms must be the explanation, because both worms appear to possess equal opportunity and equipment for penetrating the bile duct or intestinal wall.

Distribution of many kinds of parasites in a vertebrate body is initially determined by the course of the circulatory system, a natural distributing network for food, oxygen, metabolic products, and parasites. Larval hookworms such as *Ascaris* and *Wuchereria* are each carried to all parts of the body by the blood and lymph, and each species of parasite is finally delivered to an organ according to its predilection.

No sharp distinction exists between microdistribution and any other kind; hence, before we move to the wider aspects of biogeographic distribution, we shall consider one group of parasites, the lice, as an illustration of the complex relationships between parasites in or on a host and the distribution of the host.

The distribution of lice is governed in most instances by the phylogeny of the hosts. This situation contrasts with that of fleas, in which ecologic factors are paramount. On the human body, a high correlation exists between the amount of hair and the rate of infestation by lice. Girls generally have more lice than boys. Evidence for such statements comes from studies of hair from shaved heads among military troops, in prisons, and in orphanages. The more clothing people wear, the more lice they tend to possess. Living habits of people affect their lice populations. The spread of lice is fostered in men living close together in ships, tents, or barracks. The temperature and humidity of the space between a man's skin and his clothing—that is, the living space for lice—remains remarkably constant in different countries and in different seasons of the year. Because man stabilizes the climatic conditions on his surface, a wide geographic distribution of his lice is to be expected. From dry climates in Sahara and Iraq to the constant humid equatorial conditions of Sri Lanka, Zaire, and Tahiti, to the temperate lands of Europe and America, man's lice are found readily. Local absence of human lice is generally due to social or hygienic habits of the people.

The distribution (sites occupied) of parasitic helminths in the digestive tracts of vertebrates varies considerably with the species and developmental stage of the worm and with changes of the physiologic state of the host, with particular reference to digestive activity and to immunity.

"Digestion in vertebrates is considered to be an ordered sequence of events producing different conditions in different parts of the tract. These conditions, however, are not entirely predictable; digestion is affected by the nature of the diet, the feeding routine, and the psychosomatic state of the animal in question. . . . The adult stages of most species of Acanthocephala and most species of Cestoda are confined to the lumen and paramucosal lumen of the small intestines of their hosts. . . . The adult stages of Nematoda can be found in most regions of the alimentary tract, from the mouth to the rectum and from the lumen to the serosa. . . . The co-existence of various species of nematodes in the same site may result from their varied dietary requirements and feeding habits. . . . Although the sites of adult digenetic Trematoda are to be found throughout the length of the alimentary tract, most species appear to be restricted to the paramucosal lumen and mucosal and epithelial tissues."[4]

Additionally, the presence of other individuals of the same species can influence a parasite's microdistribution. Parasites are known to communicate with one another in various ways. For example, male hookworms in dogs migrate toward the females, especially when the females are near the duodenum and the males are nearer to the ileocecal valve. Apparently, the females produce a messenger substance (pheromone) that travels in the direction of fecal flow and acts as an attractant to the males.[15] (See Chap. 2.)

SITE SELECTION

We do not know all the details of how the parasite travels from one place to another, but it probably responds actively to numerous signals or inducers that result in specific behavior, leading to active site selection. In a discussion of the way parasites enter their hosts and reach and change sites within and between the habitats therein, Crompton concluded that "the observed distribution of parasites within their hosts is difficult to explain unless some form of directed site selection by the parasite is assumed."[3] See also MacInnis[9] and Holmes[7,8] for reviews of this general subject. Site-finding by parasites is a marvelous example of the unfolding complexities of parasitism.

Ulmer[17] reviewed evidence for a hypothesis that site-finding activity by helminth parasites (larvae and adults) is the primary determinant of microhabitat specificity. The evidence strongly supports the hypothesis, but he concluded that

"there appear to be very few host-parasite associations in which the stimuli determining behaviour have been clearly identified and in which the factors resulting in site-finding have been unequivocally demonstrated. . . . The careful and critical analysis of adaptive behaviour for each life cycle stage, and the elucidation of trigger mechanisms including chemical, hormonal, sensory, and neurosecretory stimuli, undoubtedly will provide challenging areas of inquiry for the intellectually curious helminthologists."[17]

Holmes[8] cited numerous instances of site selection that appear *not* to be host-dominated. For example, many parasites migrate in the body of the host when the latter is dead and apparently continue to respond to microenvironmental stimuli. The fluke, *Podocotyle* sp., is found predominantly in the anterior gut of freshly caught flounders, but in fish starved for a day or more, the flukes are located predominantly in the rectum. Newly excysted scoleces of the tapeworm, *Hymenolepis diminuta,* attach themselves near the middle of the small gut of rats, and then, in a week or 10 days, they move to the anterior part of the small gut. The same tapeworm exhibits a circadian (or diel) type of migration in response to food in the gut (see Chap. 12). Both types of migration are affected by the size of the parasite population. Concurrent infections (see Chap. 23) and competitive behavior of larval trematodes in snails (see Chap. 23) also illustrate active behavior of parasites related to site selection.

Excellent examples of site selection along the length of the host intestine occur in nematodes. Selective site segregation appears to be important in niche specialization. Holmes concluded (in part):

"Evidence is presented to support the hypothesis that continued interaction between parasites led to niche diversification, and that site segregation, and consequently narrow site specificity, is an important part of niche specialization. The paucity of cases of interactive site segregation as compared with those of genetically stabilized selective site segregation suggests that parasite faunas are mature communities, the diversity of which has been established to an important extent through biotic interactions."[8]

Mettrick and Podesta[13] pointed out that among cestodes and acanthocephalans,

"there is preference or even absolute requirement for a certain host species, genus, or family, superimposed on the intra-host microhabitat specificity. The latter may be related to gastrointestinal function, which results in different physiological and chemical conditions in different parts of the alimentary tract. The hypothesis that these conditions are directly or indirectly responsible for the distribution of intestinal helminths is extremely attractive, but the supporting evidence is equivocal."[13]

Recall also the many instances of site selection among the protozoa mentioned in the earlier chapters of this book. Among them are the coccidian, *Eimeria,* in gut epithelial cells; *Plasmodium* in red blood cells; *Leucocytozoon* in white blood cells; and *Trichomonas* in gut humen.

HOST MIGRATIONS

Migratory animals provide unique opportunities for studying the effects of changes in external environments. Few studies have been made on the parasites of mammals and birds in relation to migrations of the hosts, but migratory fish and their parasites have attracted the attention of numerous parasitologists.

The circumpolar distribution of *Neoechinorhynchus rutili,* an acanthocephalan parasite, provides incontestable evidence of practically continuous geographic distribution of one species of parasite in freshwater fish of two continents, North America and Eurasia. The adaptations of *N. rutili* to fish hosts are so flexible that the worms are found in seven families in Europe and eight families in North America, and in the two continents the host families Salmonidae, Cyprinidae, Esocidae, Gasterosteidae, and Percidae are parasitized in common. The only obvious way a parasite that occurs so often in fresh waters can become so widely scattered over the world is through the use of a great diversity of hosts representing many different habitats. Of the 40 or more species of fish that have been listed in the literature as hosts for *N. rutili,* only 4 (*Esox lucius,* the pike, *Gasterosteus aculeatus,* the threespine stickleback, *Pungitius pungitius,* the ninespine stickleback, and *Salvelinus alpinus,* the Arctic charr) are common to both continents. Of particular importance are the wandering hosts, such as salmon, trout, stickleback, and charr, whose migratory habits often involve passage between salt and fresh water.

Rohde[15a] has discussed characteristics of parasite faunas of invertebrate as well as vertebrate hosts of different seas. He showed that "there is a richer parasite fauna in the Pacific than in the Atlantic, a largely different composition of the parasite faunas in the two oceans, a greater endemicity of parasite taxa in the Pacific, gradients of increasing species diversity from cold to warm waters, and decreased diversity in the deep sea." Among many causes of these differences he stresses temperature, age of oceans, and migrations of hosts.

Migrations of thousands of miles are common among many fish, for example, tuna, salmon, and eel, but few detailed investigations have told whether the fish keep the same parasites during the whole course of migration. The larval eel lives for 3 years as a marine pelagic fish and, according to Dogiel,[5] is completely free from parasites during this period. Its feeding habits are a mystery, nothing having been found in the digestive tract at this stage in its life cycle. A postlarval marine stage of 1 year's duration is followed by a period of migration up a river, where the fish becomes a bottom feeder and gains parasites. The first freshwater parasites in the young eel (about 70 mm long) are those not requiring an intermediate host (for example, *Myxidium giardi*, *Trichodina*, *Gyrodactylus*, some trematode larvae, and *Acanthocephalus anguillae*). In later life, the parasitic fauna of female eels is of a freshwater variety, whereas that of the male fish includes both freshwater and marine species. During the last months of their life, the eels are said to be free from parasites.

Parasites of salmon change as the fish migrate from fresh to salt water and back again. Parasites from young freshwater forms show almost no host specificity and are also found in other freshwater fish in the same locality. The American west coast salmon is parasitized by the fluke-vector, *Nanophyetus*, of the canine salmon poisoning disease. The causative agent is a rickettsia. Fish that are heavily parasitized in fresh water migrate to the sea, and when they return 2 to 3 years later to spawn, they are practically free from the flukes. However, the salmon again acquire thousands of metacercariae within a few weeks of entering fresh water.

When fish migrate to a new environment and become isolated, or when they are introduced by man into new regions, the number of their original parasite faunas decreases. To put this conclusion in more general terms, the process of acclimatization leads to impoverishment of an animal's original parasites. A few relics of the past often remain, but, given time, the hosts acquire new species of parasites not found in the original habitat. A host generally has a larger variety of parasites, particularly those peculiar to it, in the habitat in which it has lived the longest. The freshwater fish *Lota lota* has numerous parasites characteristic of other fresh-

water fish, but *L. lota* is a member of the Gadidae, a family consisting almost exclusively of marine species, and it also has a number of marine parasites that are reminders of its past.

Yet, the movement of an organism into a new area does not invariably mean that it will acquire all of the parasites of the indigenous hosts. Freeland notes:

"In every successful vertebrate invasion for which I have been able to get adequate data, there is a significant difference between the parasites of invading and resident species. These include the invasion of Australia by rabbits, sheep, cattle, horses, donkeys, camels, and the successful colonization of parts of Europe by an Australian macropod marsupial. Marsupials and eutherian mammals are largely invulnerable to each other's parasites. . . . The invasion of parts of the United States by the nutria, is another example where there is minimal overlap in the parasites of resident and invading species. . . ."[6a]

Indeed, the lethality of an introduced parasite species can prevent sympatry between the potential hosts. Chapter 25 discussed the lethality to moose of a meningeal nematode that usually occurs in white-tailed deer. Another example could be that of the disjunct distribution of red and gray squirrels in Great Britain. Gray squirrels harbor a virus that is lethal to red squirrels; when the two types of squirrels live in the same area, viral epidemics occur among the red squirrels.

Birds can migrate for thousands of miles and can carry their parasites with them. The state and extent of parasitism are directly related to the host physiology; and the physiology of migrating birds changes during their migrations. Some birds spend their summers in Alaska and their winters in the South Pacific. Do these birds possess the same kinds and numbers of parasites at both locations? Probably not, but relatively few studies of this nature have been made with birds. Some recent important contributions are those of Bush and Holmes, Forrester and Pence.

Migratory mammals (except humans) and their parasites offer to the parasitologist an almost untouched field for important basic research. Migratory whales and porpoises can lose their helminth parasites when they reach different environments. One wonders what happens to the parasites of arctic land mammals when the hosts migrate to more temperate climates during winter.

"Evidence favors the view that yellow fever, dengue, estivoautumnal malaria, broad fish tapeworm infection, the hookworm infection produced by *Necator americanus*, Manson's blood-fluke infection, Bancroft's and other types of filariasis, and dracunculosis were brought to the Western Hemisphere by the white colonists and their slaves imported from Africa, as were typhus fever, leprosy, smallpox, measles, mumps, syphilis, frambesia, and probably influenza. . . . Wherever climate, necessary intermediate hosts, and customs of the population were favorable, these diseases became established in the new soil."[6]

Trypanosomiasis in humans and animals has received some intensive study from the point of view of migrating hosts. It is well known that elimination of an intermediate host leads to the disappearance of a parasite from any given territory. Mammalian trypanosomes of the vivax, congolense, and brucei groups are normally restricted to a tropical zone of Africa coinciding with tsetse fly distribution. *Trypanosoma vivax* is a striking exception to this rule. In Africa, this species is transmitted to cattle by mechanical contamination of the proboscis of tsetse flies, but it is also found in West Indies, South America, and Mauritius, where it was introduced in the last century with infected cattle. The non-African strains are morphologically indistinguishable from the African ones, but the former are transmitted mechanically by the proboscis of horse-flies (Tabanidae), in which the parasites cannot develop. This substitution of one vector for another has enabled the parasite to become widely distributed to distant lands.

The disease known as surra in domestic animals is caused by *Trypanosoma evansi*, which is closely related to *T. brucei* but occurs only outside the area of distribution of tsetse flies. Its range includes the Palaearctic, Ethiopian, Madagascar, oriental, and neotropical biogeographic regions. Surra is also transmitted mechanically by the contaminated proboscis of horseflies. Evidence from laboratory cross-infection experiments indicates that *T. evansi* originated from *T. brucei* in Africa. The disease nagana, caused by *T. brucei* could originally have been contracted by camels brought into the "tsetse belt" of Africa; then, with a combination of transfer of mode of infection to the mouth parts of horseflies, and migrations of camels, the new disease, surra, could have been extended far beyond the geographic boundaries of its ancestral disease nagana.

Apparently, a direct correlation exists between the scarcity of parasites and the ability of the host to adapt itself to widely different environments. This correlation has been indicated by studies on numbers of parasites in widely distributed hosts, in comparison with parasites in hosts confined to one or two small areas. A wide geographic distribution of a host can be possible, in part, because of a general resistance to parasitism. Much more work on this and related problems must be done, however, before convincing and clear conclusions can be obtained.

DISTRIBUTION OF ARTHROPODS BY COMMERCIAL VEHICLES

In prehistoric times, when man's means of transportation was confined to his legs, the carrying of foodstuffs and other articles and the driving of livestock slowly transported insects and other arthropods, as well as other human parasites, from one location to another. The introduction of carts and canoes increased the opportunity of man unwittingly to carry away insects from their natural habitats. Even when dugout canoes afforded the only means of transoceanic transportation, the spread of insects such as mosquitoes among the South Pacific islands by Polynesian voyagers was facilitated. Certain mosquitoes of the scutellaris group of *Aedes* commonly breed in beached canoes, and their eggs are resistant to drying. The use of sailing vessels and, later, steamships provided a means for numerous insect introductions to countries all over the world. Flies and cockroaches breed in galleys and quarters, various beetles infest stored foods, and mosquitoes commonly breed in many kinds of containers holding water. As the result of such transportation, *Culex fatigans* has become cosmopolitan, and *Aedes aegypti* virtually pantropical.

The development of aviation, however, has been the most alarming and serious encouragement of accidental insect introduction on a global scale. Among the insects most commonly carried by aircraft are Diptera (mosquitoes and flies), Hemiptera (bugs), Lepidoptera (butterflies and moths), Coleoptera (beetles), and Hymenoptera (ants, bees, and wasps). Other insects often to be found in airplanes are cockroaches, lacewings, earwigs, and termites. The hazards associated with air transportation are greater than those with sea transportation, not only because of the much shorter time required for flight, but also because of the character of the international airports, which are usually situated in rural or semirural districts. The docks for ships, on the other hand, are located most frequently in the heart of heavily built-up urban areas that present a limited choice of mosquito larval habitats.

The accidental importation into Brazil of *Anopheles gambiae* from West Africa about sixty-years ago led to a disastrous outbreak of malaria, causing intense suffering with more than 300,000 cases of the disease and 16,000 deaths. The same mosquito was introduced into Upper Egypt during World War II, and it initiated a serious malaria epidemic involving 170,000 cases with 11,889 deaths from 1942 to 1944. Vigorously prosecuted campaigns, at great cost, ultimately eradicated *A. gambiae* from both Brazil and Upper Egypt.

Bruce-Chwatt[2] listed the major imported diseases as cholera, smallpox, yellow fever, plague, relapsing fever, and typhus, and listed the common diseases of travelers as gastrointestinal infections, malaria, trypanosomiasis, leishmaniasis, schistosomiasis, filariasis, helminthiasis, rabies, arthropod-borne encephalitis, dengue, hemorrhagic fever, poliomyelitis, and leprosy. "The responsibility for recognizing an imported disease has now passed from the airport health medical authority to the general practitioner." Many of the important changes that have occurred during the past decade

"are due to geographical, social, economic and political factors that characterize our today's 'runaway world'—a term used to describe the events that are almost beyond our control. Among these events the increase of world population, pressure on land and all natural resources, rapid and uncontrolled urbanization, fragmentation of large territories into small semi-independent nations, political disruption and military intrusion and above all the phenomenal expansion of international trade and travel play a major role."

A 1978 estimate reported that the total number of all scheduled and nonscheduled international airplane flights carried "the staggering figure of 630 million" passengers.[1e] How many more are being carried this year? How many of them carry infective stages of parasites?

The International Health Regulations (beginning in 1971)

"recognize that the detection and control of communicable diseases at the borders of each country will be of little value. . . . The weakest point of the new system seems to be that countries which have the major responsibility because of the existence of endemic foci are also the countries with inadequate public health organization and poorer resources. . . . The most common error in diagnosing an exotic disease lies in the failure to think of the possibility of its occurrence."[2]

As yet, little international uniformity exists in the interpretation of existing recommendations made by the World Health Organization and the International Sanitary Convention for Aerial Navigation. Aircraft disinfection is generally considered to be a safe, simple, and speedy safeguard, but is should not be regarded as affording complete protection.

DISTRIBUTION WITHIN RESTRICTED AREAS

A comparison of the parasites of the coastal cod with those of the winter cod (two subspecies) in the White Sea discloses differences in the respective faunas that can be attributed to differences in habits and habitats. The coastal cod feeds on the bottom and is infected with the fluke, *Podocotyle atomon*, obtained from its crustacean food, and with the ciliate, *Trichodina cottidarum*. The fish is heavily infected with intestinal stages of the nematode *Contracaecum aduncum* but lightly infected with the trematode *Hemiurus levinseni*. On the other hand, the winter cod, which feeds primarily on plankton, is only lightly infected with *Contracaecum*, but every fish harbors *Hemiurus*.

The same species of fish in different parts of the White Sea sometimes possesses different parasites, and these differences appear to be related to hydrologic factors. On the other hand, the same manner of life in distantly separated (taxonomically) fish may lead to the acquisition of the same parasites. For example, the flounder, *Pleuronectes flessus*, and the wolf fish, *Anarrhichas lupus*, both feed on the sea bottom on the same animals and have nine species of parasites in common, but many of these parasites are not specific for these fish.

Infection of the flounder, *Pseudopleuronectes americanus*, with trematodes is heavier in inshore waters than in offshore waters; close to shore, the infection is heavier in fish that are taken in deeper water adjacent to open sea than in fish taken near shoals. Larger flounders have heavier infections than do smaller fish, and marked seasonal variations are absent in larger flounders. In seeking an explanation for the differences between inshore and offshore fish, one should remember that many more kinds of other animals and plants are usually associated with the variety of shoreline habitats. Many of these other animals can serve as intermediate hosts of parasites.

Lake Mogilny, situated on the island of Kildin in the Barents Sea, has fresh or brackish water down to a depth of about 5 m, but below that level, the water becomes contaminated with hydrogen sulfide. Cod fish are found in the lake, and they and other marine animals can live only in the layers between 5 to 12 m deep. These animals can be considered descendants from marine forms that lived there when the waters of the lake communicated with the surrounding sea. An examination of parasites of this relic cod shows that the parasitic fauna becomes impoverished, as compared with that of the same species of hosts living under normal marine conditions. Parasites normal to cod include the acanthocephalan, *Echinorhynchus gadi,* and the copepods, *Caligus curtus, Clavella uncinata, Clavella brevicollis,* and *Lernaeocera branchialis.* These parasites are absent in cod from Lake Mogilny. Also absent are several myxosporida commonly found in marine cod. When intermediate hosts are involved, as with *Lernaeocera,* the absence of the parasites in the lake can easily be explained, but for the others, no such ready explanation is available.[5]

THE BIOGEOGRAPHY OF PARASITES AS CLUES TO HOST AFFINITIES AND EVOLUTION

A number of writers have contributed to the general principle that the systematics and phylogenetic ages of hosts can often be determined directly from the systematics and degrees of organization of their permanent parasites; conversely, the systematics and ages of parasites can be determined directly from the phylogenetic and taxonomic relationships of their hosts (see "Rules of Affinity" in Chap. 26).

Probably, the first scientist to use parasites as indicators of the relationships and geographic distribution of hosts was von Ihering,[18] who based his conclusions on a study of helminths.

Many years ago, the well-known English helminthologist H.A Baylis warned us that "although the attempt to draw conclusions as to the relationships of animals from their helminth parasites may sometimes yield interesting results, it is fraught with so many pitfalls that it should be made with greatest caution."[1] Mayr[12] has also issued a warning to those who would place too much emphasis on the importance of using parasites as a guide to host phylogeny. He has said:

"We are dealing here with something very basic, with the whole principle of phylogeny, with the principle of this study of parallel phylogeny and we must be awfully sure of these tools we use, that we do not misuse them, and we must, at all times, allow for an occasional transfer of parasites, we must allow for different rates of evolution, and we must realize . . . that the comparative anatomy is something more reliable. Two birds can exchange their parasites, nothing prevents this, but I have not yet seen two birds exchanging their heads, their wings or their legs. These have come down from its ancestors and not from another bird that nested in a hole right next to it!"

To these well-founded cautions, we can add Manter's more phylogenetically based observations:

"Parasites . . . furnish information about present-day habits and ecology of their individual hosts. These same parasites also hold promise of telling us something about host and geographical connections of long ago. They are simultaneously the product of an immediate environment and of a long ancestry reflecting associations of millions of years. The messages they carry are thus always bilingual and usually garbled . . . [in order to evaluate the history of a host-parasite system], (1) each parasite should be evaluated separately; (2) as large a variety of parasites as possible should be considered to offset errors or cases of convergence; and (3) conclusions should take into account other lines of evidence."[10b]

Brooks[1b] has said of Manter's statement that in phylogenetic analysis of any kind, garbled interpretations result from a failure to distinguish *homoplasious* traits (similarities among taxa that are not the result of immediate common ancestry—e.g., convergence and parallelism) from *homologous* traits (similarities among taxa that are the result of inheritance through immediate common ancestry), and, furthermore, from a failure to distinguish *retained ancestral homologues (plesiomorphies)* from *derived homologues (apomorphies).* This question will be discussed further in Chapter 26, when the topic of host-parasite co-evolution is examined more directly. At this point, we simply point out the importance of having well-constructed and well-supported hypotheses of host and parasite phylogenies.

The pioneering work in the biogeography of parasites presaged the findings of geologic work on continental drift and plate tectonics by about 60 years. Early studies found parasite-host relationships that were most reasonably explained by postulating past connections among land masses. To be sure, in a hypothetical world of immobile continents, these connections were usually envisaged as vast land bridges. Today, in light of the plate tectonic theory developed within the last 30 years, these parasite-host relationships can be attributed to the past continuity of land masses, with subsequent separation causing substitutions and speciation in the indigenous biota. (This attribution is an excellent example of how the hypothesis of a historical study is corroborated by means of multiple lines of evidence.) Dispersal of a fauna from a center of origin still might have occurred, but such a process need not be invoked as an explanation of first choice.

Plate tectonic research posits that all of the continents were once united into a single landmass, called Pangaea.[4b,11] About 180 million years ago, in the Jurassic period, Pangaea split into a northern supercontinent, called Laurasia, and a southern supercontinent, called Gondwanaland. Laurasia included what is now North America, Europe, and Asia except for the subcontinent of India. Gondwanaland included what is now South America, Antarctica, Australia, Africa, and India. With the split of Pangaea, the North Atlantic Ocean was formed. About 100 million years ago, in the Cretaceous period, Gondwanaland began to break up into the continental masses we know today, forming, among other features, the South Atlantic Ocean. Laurasia broke up about 60 million years ago, in the Eocene period. Figure 24–1 presents the best-supported hypothesis of the sequence of the major continental splittings.

In 1891, Von Ihering[18] studied the temnocephalid turbellarians of crustaceans and inferred a freshwater connection between South America and New Zealand. By 1929, Metcalf[12a] wrote of his support for the "Von Ihering Method" as he discussed the research of various parasitologists in addition to himself. This research included postulations of a close relationship between (1) South American and Australian leptodactylid frogs, based on their opalinid protozoan parasites, (2) certain species of European and North American birds, based on their biting lice, and (3) the rhea and the ostrich, based on their biting lice. Harrison, one of the workers whose work Metcalf summarized, used various louse data to argue for a past connection between Australia, Antarctica, and South America. Another seminal study is that of Manter,[10a] who, on the basis of fish trematode data, inferred historical connections between South America, Africa, and India.

A great deal of phylogenetic biogeographic research is being conducted today; the student is referred to the aforementioned works of Nelson and Platnick,[13a] Wiley,[19] and Humphries and Parenti[8a] for an overview, as well as to Brooks[1c] for a review specific to parasitologic studies. Other work by Brooks and colleagues includes a phylogenetic study of the co-evolution in South America between the freshwater stingrays of the family Potamotrygonidae and some of their helminth parasites. The basic question asked in the study was whether potamotrygonids are more closely related to marine stingrays from the Pacific or the Atlantic Ocean. Potamotrygonids themselves are incapable of surviving in saltwater. Given that today all of the river systems in South America that are inhabited by potamotrygonid stingrays drain into the Atlantic Ocean, the common assumption has been that these rays are of Atlantic ancestry. Noting that the closest relatives of the tapeworms and nematodes of potamotrygonids occur primarily in urolophid stingrays along the *west* coast of South America, and presenting corroborating evidence from studies on South American free-living organisms and South American geology, Brooks, Thorson, and Mayes[1d] concluded that the Potamotrygonidae had a marine ancestry in the Pacific Ocean, had been isolated in the freshwater of eastern South America when the formation of the Andes altered the river drainages, and have subsequently co-evolved with their equally isolated parasites.

During the break-up of Pangaea, a body of water called the Tethys Sea came to separate east Gondwanaland from east Laurasia. At the

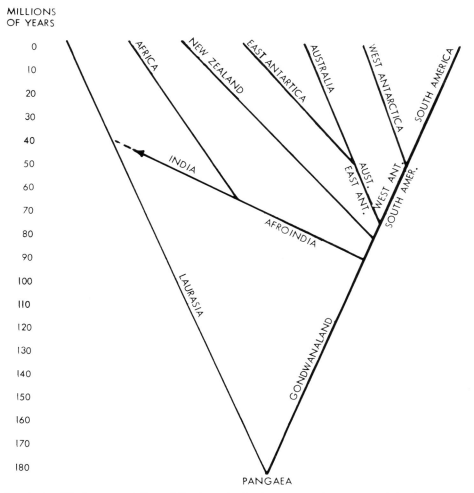

Fig. 24–1. A simplified representation of the hypothesized pattern and sequence of the break-up of the supercontinent Pangaea, beginning 180 million years ago. The approximate time scale is to the left. All of the events but one consist of the separation of land masses. The exception is the accretion of India, originally a component of Gondwanaland, onto Laurasia. (From Rosen.[15b])

end of the Cretaceous period, 65 million years ago, the eastern mouth of this sea closed in the area of what is now Suez. On a contemporary map, the Tethys Sea extended from Gibraltar to Saudi Arabia. The Tethys also most likely had connections that extended west to the Caribbean Basin across the inchoate North Atlantic and into the northern river drainage areas of South America. Szidat[16] studied the digenean trematode parasites of freshwater fish in South America and found many of them to be characteristic of those of certain marine fish. Moreover, the nearest marine relatives of these freshwater parasites are sometimes found not in the

waters adjacent to the outlets of the South American rivers, but in the Caribbean and the Mediteranean. This pattern indicates a circum-Tethyean distribution in the past.

Manter[10] pointed out that, of the 165 species of trematodes from marine fish of tropical Australia, 30 species also occur elsewhere, and some of these species appear to be most closely related to trematode species of the Caribbean.

Many digenetic trematodes of marine fish are not widely distributed, but some significant geographic comparisons can be made. For example, strong similarities among trematodes occur between (1) the European Atlantic and the

Mediterranean; (2) shallow waters at Tortuga, Florida, and Bermuda; and (3) shallow waters at Tortuga and the tropical American Pacific. Considerable dissimilarity occurs between (1) deeper waters and shallow waters at Tortuga; (2) shallow waters at Tortuga and the North Atlantic; and (3) the Mediterranean and the Red Sea.

Evidence for the phylogenetic relationship between a South American bird, *Cariama cristata*, and the Eurasian bustard has been presented on the basis of an analysis of their helminths. Both birds are parasitized by the nematodes *Subulura allodapa* and *S. suctoria*. In addition, both birds harbor species of the cestode genera *Chapmania* and *Idiogenes*. Physiologic segregation of tapeworms evidently occurred when vertebrates first spilt into present-day groups and resulted in a phylogenetic specificity. Hence, we can conclude—a conclusion corroborated by taxonomic studies on birds—that *Cariama cristata* and the Eurasian bustard are related.

The effects of isolation of ectoparasites are more often reflected in morphologic changes that are the effects of isolation of endoparasites. The Mallophaga, or biting lice, exhibit a high degree of host-specificity and can be used to elicit evidence of evolution of hosts and parasites.

Ornithologists disagreed for a long time on the systematic position of flamingos; some authorities placed them in the Ciconiiformes (storks) and others in the Anseriformes (ducks and geese). A study of flamingo Mallophaga shows that the lice correspond closely to those of ducks and geese. Not one of the parasites suggests close affinity between flamingos and storks. The Mallophaga of the three North African pelicans are not alike, and the lice of one of them, the Ethiopian pelican, are more like those on the Australian pelican and the South American pelican than those on the other two North African birds. Morphologically, the lice of the South American pelicans are sufficiently distinct to be placed in a separate genus, and their hosts are placed in a separate subgenus. We have here a fine example of the effects of allopatry of both hosts and their parasites.

A study of the Anoplura, or sucking lice, also presents interesting illustrations of the results of allopatry. From information on present-day distribution of these lice, we can conclude that they became isolated on their hosts long ago;

yet we find that the Australian marsupials do not possess sucking lice, a fact suggesting that these hosts had already become isolated before the Anoplura arose as parasites.

Closely related mammals have closely related or identical lice. The ground squirrels (*Citellus*) of North America are related, but they are different from those of Siberia. The lice, however, on these two groups of geographically separated squirrels appear to be identical. Because of the high degree of host specificity exhibited by lice, one can examine an unknown louse and tell, with little risk of error, from what kind of vertebrate it was taken. However, unless the range of distribution of the group of hosts is known, one cannot tell from which part of the world the louse came. A conspicuous exception is found in the family Gyropidae, which occurs on many hosts in South America but is not found elsewhere.

The distribution of parasites within the order Marsupialia conforms almost perfectly to the geographic distribution and antiquity of the hosts. For example, the biting lice of marsupials belong to the most primitive division of the Mallophaga, one family (Boopidae) infesting Australian hosts, and another (Trimenoponidae) infesting South American hosts.

The ostrich and the rhea both have sclerostome nematodes; this similarity indicates at least some relationship between the two hosts. However, sclerostomes occur in many kinds of grass-eating animals, such as horses, elephants, Australian marsupials, rhinoceroses, tapirs, and even in South American tortoises. Ten nematodes have been reported from the rhea, and five from the ostrich, but they have no species in common. Both birds are parasitized by one genus of louse, *Struthiolipeurus*, but rheas have in addition a second louse genus, *Meinertzhageniella*. Both birds have the mite, *Pterolichus bicaudatus*, the only species of parasite in common. The arthropod distribution and similarity weakly support evidence from feather structure and other anatomic features that the two orders of birds are related and are not the descendants of independent, unrelated stocks.

In an account of the biogeography of parasites in Australia, Beveridge[1a] has provided fascinating facts and raised some questions that at present have only tentative answers. During 40 million years, encompassing a wide range of habitats and climatic extremes, Australian marsupial families have undergone extensive adaptations. As might be expected, these isolated mammals have acquired a diverse fauna of parasites. The evolution of marsupial helminths has followed the convergent or parallel evolution of their hosts. One surprising discovery (reported by Kruse in a Ph.D. thesis) is that the genera of digeneans from fishes from the South Australian gulfs are much more similar to those found in the Caribbean, Hawaii, and Japan than to those found in the Antarctic and New Zealand.

Space does not permit a review of all groups of vertebrates, but in general, before clear-cut, detailed lines of phylogenetic relationships can be established, much more research, particularly in the nature of careful comparisons of abundant collections of parasites as well as fossil mammals, must be made. Precise classification of parasites must be the basis of comparisons. For details of this problem, see Patterson.[14]

SUMMARY OF GENERAL RULES AND PRINCIPLES

1. The distribution of parasites and their hosts around the world is not random. Factors that play a role in the present distribution of species include (1) their history of **vicariance;** (2) their **vagility;** (3) their **dispersal properties;** and (4) the length of time and the history of physical events through which the first three factors have affected the species *or* their ancestors. These considerations apply to all organisms, parasitic or not. Some additional factors affecting distribution of parasites are the ability of the parasite to survive apart from a host, and the different histories and properties of larval and adult stages and of intermediate and definitive hosts.

2. Depending on the speciation behavior of a particular species under allopatry, widely separated, conspecific hosts can have the same species of parasites; widely separated, conspecific hosts can have closely related species of parasites; widely separated, closely related hosts can have the same species of parasites; or widely separated, closely related hosts can have closely related parasites. In all cases, the host-parasite association is indicative of co-evolution.

3. Independent geological or biotic studies of land mass relationships can lend support to co-evolutionary studies of host-parasite associations, and vice versa. Since the nineteenth century, parasitologic studies have been reporting data corroborating patterns of continental drift, with the associated breaking up of the supercontinents Pangaea, Laurasia, and Gondwanaland.

4. Colonization of a host by a parasite with which it has not been evolutionarily associated can give false positives about degrees of host relatedness. Colonizing parasites can be transported to their new hosts by their old hosts, can encounter their new hosts when those hosts enter the range of their old hosts, or can be carried from one host's range to another's by biotic or abiotic vectors while in a dispersal stage. With sufficient data from co-evolutionary associations, colonizations can be detected.

5. If a new host in a new geographic region is to be successfully colonized, parasites that require multiple hosts for the completion of their life cycles must encounter all of the necessary biologic qualities of those hosts in the new region, through either the homology or the analogy of host traits. The absence of a necessary intermediate host can produce a bottleneck and prevent the expansion of a parasite's range despite the presence of suitable final hosts.

6. Exposure of a host to a parasite does not invariably mean that the host will become infected. The biologic requirements of both host and parasite must be sufficiently met for development to occur. The matching of these requirements is determined by, among other things, the diet, behavior, and phylogenetic history of both host and parasite.

7. Hosts that bring their own fauna of parasites to a sympatric association usually retain differences in their respective faunas to the extent that the invading and the indigenous host can be distinguished on the basis of their parasites.

8. Parasitic infections in new hosts can be of sufficient pathogenicity, including lethality, to prevent the sympatry of the invading and indigenous hosts.

9. False negatives can occur in biogeographic studies of hosts and their parasites through a failure to collect a host sample of a size that is sufficient to detect the presence of the parasites.

10. As with all systematic work, correct conclusions in biogeographic studies depend on both the proper identification of organisms (alpha-level taxonomy) and the proper analysis of the data obtained (phylogenetics).

REFERENCES

1. Baylis, H.A.: Helminths and evolution. *In* Essays on Aspects of Evolutionary Biology. Edited by G.R. de Beer. Oxford, Oxford University Press, 1938, pp. 249–270.

1a. Beveridge, I.: Biogeography of parasites in Australia. Parasitol. Today, Australian Supplement, 53–57, July, 1986.

1b. Brooks, D.R.: Hennig's parasitological method: a proposed solution. Syst. Zool., *30*:229–249, 1981.

1c. Brooks, D.R.: Historical ecology: a new approach to studying the evolution of ecological associations. Ann. Missouri Bot. Gard., *72*:660–680, 1985.

1d. Brooks, D.R., Thorson, T.B., and Mayes, M.A.: Freshwater stingrays (Potamotrygonidae) and their helminth parasites; testing hypotheses of evolution and coevolution. *In* Advances in Cladistics. Vol. 1. Edited by V.A. Funk and D.R. Brooks. New York, New York Botanical Garden, 1981, pp. 147–175.

1e. Bruce-Chwatt, L.J.: Mass travel and imported diseases. Ann. Soc. Belg. Med. Trop., *58*:77–88, 1978.

2. Bruce-Chwatt, L.J.: Global problems of imported disease. Adv. Parasitol., *11*:75–114, 1973.

3. Crompton, D.W.T.: Entry into host and site selection. *In* Ecological Aspects of Parasitology. Edited by C.R. Kennedy. Amsterdam, North-Holland, 1976, pp. 42–73.

4. Crompton, D.W.T.: The sites occupied by some parasitic helminths in the alimentary tract of vertebrates. Biol. Rev., *48*:27–83, 1973.

4a. Desowitz, R.S.: New Guinea Tapeworms and Jewish Grandmothers: Tales of Parasites and People. New York, Avon Books, 1981.

4b. Dietz, R.S., and Holden, J.C.: The breakup of Pangaea. Sci. Am., October, 1970.

5. Dogiel, V.A.: Problems of Ecological Parasitology. Ann. Leningrad State Univ., *7*:1–194, 1936. (In Russian)

6. Faust, E.C., and Russell, P.F.: Clinical Parasitology. Philadelphia, Lea & Febiger, 1970.

6a. Freeland, W.J.: Parasites and the coexistence of animal host species. Am. Nat., *121*:233–236, 1983.

7. Holmes, J.C.: Host selection and its consequences. *In* Ecological Aspects of Parasitology. Edited by C.R. Kennedy. Amsterdam, North-Holland, 1976, pp. 21–39.

8. Holmes, J.C.: Site selection by parasitic helminths: interspecific-interactions, site segregation, and their importance to the development of helminth communities. Can. J. Zool., *51*:333–347, 1972.

8a. Humphries, C.J., and Parenti, L.R.: Cladistic Biogeography. Oxford, Clarendon, Press, 1986.

9. MacInnis, A.J.: How parasites find hosts: Some thoughts on the inception of host-parasite integration. *In* Ecological Aspects of Parasitology. Ed-ited by C.R. Kennedy. Amsterdam, North-Holland, 1976, pp. 2–20.

10. Manter, H.W.: Some aspects of the geographical distribution of parasites. J. Parasitol., *53*:1–9, 1967.

10a. Manter, H.W.: The zoogeographical affinities of trematodes of South American freshwater fishes. Syst. Zool., *12*:45–70, 1963.

10b. Manter, H.W.: Parasites of fishes as biological indicators of recent and ancient conditions. *In* Host-Parasite Relationships. Edited by J.E. McCauley. Corvallis, Oregon State Univ. Press, 1966, pp. 59–71.

11. Maxwell, J.C.: Continental drift and a dynamic earth. Am. Sci., *56*:35–51, 1968.

12. Mayr, E.: Evolutionary aspects of host specificity among parasites of vertebrates. *In* First Symposium on Host Specificity Among Parasites of Vertebrates. Institute of Zoology, University of Neuchâtel, 1957, pp. 5–14.

12a. Metcalf, M.M.: Parasites and the aid they give in problems of taxonomy, geographic distribution, and paleogeography. Smithsonian Misc. Collect., *81*:1–36, 1929.

13. Mettrick, D.F., and Podesta, R.B.: Ecological and physiological aspects of helminth-host interactions in the mammalian gastrointestinal canal. Adv. Parasitol., *12*:183–278, 1974.

13a. Nelson, G., and Platnick, N.: Biogeography. Carolina Biology Readers, No. 119. Edited by J.J. Head. Burlington, NC, Carolina Biological Supply Company, 1984.

14. Patterson, B.: Mammalian phylogeny. *In* First Symposium on Host Specificity Among Parasites of Vertebrates. Institute of Zoology, University of Neuchâtel, 1957, pp. 15–49.

15. Roche, M.: Influence of male and female *Ancylostoma caninum* on each other's distribution in the intestine of the dog. Exp. Parasitol., *19*:327–331, 1966.

15a. Rohde, K.: Ecology of Marine Parasites. London, University of Queensland Press, 1982.

15b. Rosen, D.E.: Vicariant patterns and historical explanation in biogeography. Syst. Zool., *27*:159–188, 1978.

16. Szidat, L.: Über die die Parasitenfauna von *Percichthys trucha* (Cuv & Val.) Girard der Patagonischen Gewässer und die Beziehungen des Wirtsfisches und seiner Parasiten zur Paläarktischen Region. Arch. Hydrobiol., *51*:542–577, 1956.

17. Ulmer, M.J.: Site-finding behavior in helminths in intermediate and definitive hosts. *In* Ecology and Physiology of Parasites. Edited by A.M. Fallis. Toronto, University of Toronto Press, 1971, pp. 123–159.

18. von Ihering, H.: On the ancient relations between New Zealand and South America. Trans. Proc. N.Z. Inst., *24*:431–445, 1891.

19. Wiley, E.O.: Phylogenetics: The Theory and Practice of Phylogenetic Systematics. New York, Wiley Interscience, 1981.

25

Parasite-Host Specificity

Why are some parasites restricted to one species of host, whereas others flourish in a wide range of distantly related hosts? Why are some kinds of animals burdened with numerous parasites, while other kinds possess few parasites, if any? Seventeen species of fleas are found on swallows or martins and nowhere else, but swifts, with similar nesting and feeding habits, do not have a single species of flea restricted to themselves. Although most birds with large nests have many fleas, swans' nests are free from these insects. The nematode *Wuchereria bancrofti* lives as an adult only in humans (experimental infections have been established in the Silvered Leaf Monkey[18a]), but *Trichostrongylus axei*, contrary to nematodes in general, is at home in many different hosts. This species has been reported in the digestive tracts of the horse, ass, mule, sheep, cattle, goat, pig, several species of wild ruminants, and humans, and experimentally in rabbits, hamsters, and guinea pigs. Why do these differences exist?

"Host specificity" refers to the exclusivity of the occurrence of a parasite in one or more species of host. It encompasses assessments of parasite **prevalence** (number of hosts infected per number of hosts examined), parasite **intensity** (number of parasites per infected host), and the **phylogenetic distance** between the hosts (e.g., if a parasite occurs only in a monophyletic group, or *clade*, of hosts, its specificity for that lineage of hosts is high, even though the parasite species can be found in more than one species of host). The same range of hosts can be parasitized even with marked differences in the prevalence and intensity of parasites in each particular host species.

A high degree of host specificity can mean that the interactions between host and parasite are so delicate and intermeshed that the parasite is unable to successfully survive in or on the body of another species. High specificity, however, can also be the result of external random, physical, or environmental factors; when these factors are removed, the parasite is able to live in other hosts. The central problems in the study of parasite-host relationships concern mechanisms that affect the ability of the parasite to invade a host and survive within it.[25]

Parasite-host specificity can also be examined as **co-accommodation** between parasite and host. Brooks[3] defines co-accommodation as "the mutual adaptation of a given parasite species and its host(s) through time . . . [it] includes such parameters as pathogenicity, host specificity, and synchrony of life cycle stages." High specificity indicates narrow co-accommodation; low specificity indicates broad co-accommodation. In the next chapter, we will examine the connection between co-accommodation and **co-speciation** of parasites and hosts, which together constitute the phenomenon of **coevolution.**[3]

No parasite lacks host specificity. No parasite lives in every kind of vertebrate and invertebrate host. The term **monoxenous** refers to a parasite that uses a single host species during its life cycle, as occurs with *Enterobius vermicularis*. **Heteroxenous** indicates that more than one host species is utilized, as occurs with *Fasciola hepatica*. Different species of parasites that occur in the same host species are called **synhospitalic**; those that occur in different host species are called **allohospitalic**. Species of parasites that occur in the same part of the body of a host are called **syntopic**; those that occur in different sites of the body are called **allotopic**.

Numerous species of flagellates are found only in the hindgut of termites. Many species of ciliates occur only in the cecum of horses.

Diclidophoroidean monogeneans on the gills of certain fish appear to be entirely specific to their particular species of host. Moreover, a definite site preference exists: *Diclidophora merlangi*, for example, occurs most frequently on the first gill arch of the cod, *Gadus merlangus*. This exacting topographic relationship between parasite and host might be an important factor in the mechanism of host specificity, although it might also be the result of variations in the flow of water over the different gills. Parasites can be specific for certain host cells, as, for example, the amastigotes of *Leishmania* are specific for reticuloendothelial cells, and the trophozoites of *Plasmodium* are specific for erythrocytes.

Because many parasites develop in more than one kind of host in their life cycles, the conditions determining the degree of host specificity often differ in the several stages of a cycle. Indeed, the metabolic requirements of a parasite generally vary with its developmental changes, even within one host. In a study of the action of antimalarial drugs in mosquitoes infected with *Plasmodium gallinaceum* from chicks, the first indication of interference with parasite development usually occurred in the oocyst, but with one of the drugs used, the effect was first observed on the sporozoites.[26] The different physiologic demands of this parasite at particular stages in its development are demonstrated by changes in phosphatases and nucleic acids in the parasite of the mosquito. Such studies again emphasize the usefulness of biochemical tools to disclose the basic nature of parasite-host specificity.

Whereas two strains of a given parasite might infect a final host with equal facility, they might not be equally infective to a given intermediate host. In other words, a strain endemic in one area might be physiologically distinct from a strain endemic in another area, and this difference might be detected only on the basis of infectivity to an intermediate host. It is obvious, therefore, that in any consideration of specificity and co-accomodation we must not confine our attention to parasites, but must study their hosts with equal zeal. The host in many respects is equivalent to the environment surrounding free-living organisms. For example, topologically, a parasite in the gut lumen of its host is outside the host! The host, as a living organism, is capable of responding to its parasites (see Chapter 23). Soulsby has stated that "one of the major unanswered questions of biology is that of the molecular biological basis of host-parasite specificity."[23] His paper presents some emerging facts that help to answer this question. Other pertinent information is mentioned on the following pages.

ISOLATION OF PARASITE POPULATIONS AND SPECIES

Before embarking on a more detailed discussion of the kinds, degrees, and significances of host specificity, let us examine briefly some factors responsible for the isolation, or segregation, of parasite populations and species. It is these factors, which have the *effect* of isolation, though not the *intent* of isolation, that produce host specificity. Isolating factors are basically genetic, biochemical, ecologic, physical, and ethologic.

Co-Speciation

When a species of host is divided into two or more population groups separated geographically, its parasite faunas are also divided. Given sufficient time, the host and parasite species can each split into two or more descendent species (co-speciation). Examples can be found in the abundant speciation of mites on isolated groups of lizards or on bats, which occupy specialized biotopes. Pterygosomid mites (for example, *Pterygosoma aculeatum* beneath the scales of lizards) most likely originated on ancestral lizards before the hosts became segregated onto separate continents.

Host Transfers

A parasite can be transferred to a so-called foreign host living in the same locality as the original host and can subsequently become isolated by geographic separation of the two hosts. This type of ecologic segregation results from an association between two kinds of hosts close enough to permit their respective parasite populations to mingle before the host groups become isolated. The isolation can, of course, be temporary. When isolation of the parasite on a host takes place, both morphologic and physiologic changes can occur over evolutionary time. Parasites of reunited separated host populations are sympatric species with the original ancestral parasites. The hosts can also exhibit secondary infections or infestations acquired during the period of isolation.

Fleas from rabbits on Coronados Isles off Baja California have become established on auklets, which are burrow-nesting birds. Similarly, puffins and shearwaters from the west coast of Britain have rabbit fleas on their bodies. On the Kerguelen Isles and Antipodes in the southern Atlantic, the diving petrel, a gull, and a burrow-nesting parakeet have acquired a species of flea belonging to a group of marsupial fleas common to Australia.

Narrow Compatibility

The parasite might be unable to develop in any other host. This kind of isolation can arise through narrow parasite-host compatibilities produced by one or more of several not mutually exclusive situations.

Necessity of Specific Environmental Conditions for the Growth and Development of the Parasite

Such environmental factors include food, oxygen, temperature, metabolic rate, osmotic pressure, pH, and water, as well as characteristics of host size and structure such as patterns of blood flow, food retention times, volumes of tissue, diameters of ducts, and length and morphology of the gut.[8] At a very basic level, an adequate structural complement must exist between parasite and host: for instance, holdfast organs must be able to attach. Sufficient commonality in the conditions provided by different hosts species can allow a common parasitism. For example, later in this chapter we will discuss some parasite taxa that can develop in birds and mammals—warm-blooded vertebrates—but not in any of the cold-blooded vertebrate groups.

The food an animal eats can have direct and indirect effects on the animal's susceptibility to particular parasites. Particular types of food habits are correlated with particular types of digestive anatomy and chemistry. For example, ruminating stomachs and nonruminating stomachs present quite different environments to a parasite. Meat diets limit intestinal protozoa, while mixed or herbivorous diets enhance protozoan infections. Vitamins, salts, bases and acids affect the susceptibility of mosquitoes to malaria. Cod liver oil suppresses infections of blood protozoa.[8]

Read[19] provides a good example of how host diet affects susceptibility to parasitism. When kept in the laboratory and fed laboratory food, gray squirrels were susceptible to *Hymenolepis nana*. After attempts to establish infections of *H. nana* in wild squirrel populations were unsuccessful, it was found that if laboratory squirrels were fed a diet of acorns or mushrooms, they did not maintain experimental *H. nana* infections.

As noted by Freeland,

". . . the type of food consumed can determine which parasite species' intermediate stages an animal is likely to encounter. The particular places and times an animal feeds are likely to have similar effects. Individuals of a potential host species are not likely to become infected with a parasite if the activity periods and microhabitats used by the parasite's intermediate host or vector do not coincide with those of the potential [final] host. . . . The parasites an animal is potentially susceptible to are largely a function of its chemistry, body morphology, and size. This innate susceptibility is modified by which foods are eaten and when and where an animal feeds. Animals of similar ancestry, body size, morphology, and food habits are likely to be susceptible to the same parasites."[8]

Resistance of the Host

Resistance includes immune reactions (natural and acquired), age resistance, and mechanical barriers. Factors conditioning susceptibility and natural immunity of hosts are still partly obscure, but some of these are indicated in this chapter. (For a discussion of immunity, see Chapter 2.)

Ecologic aspects of parasitism can be studied effectively by introducing parasites into foreign hosts experimentally and then analyzing the reason for success or failure in establishing a permanent parasite-host relationship. The chief difficulty is to establish adequate criteria for determining success of the parasite in the new host. Criteria such as egg production, parasite size, and parasite number are usually variable and species specific. A parasite and its host represent a biologic system whose maintenance depends on ecologic factors provided by *both* members of the association. For instance, if plerocercoids of the broad fish tapeworm, *Diphyllobothrium latum* (normally a parasite of fish-eating mammals), are isolated from the fish intermediate host and introduced into the stomach of the lamprey, frog, toad, snake, and lizard, the typical metacestode migration takes place. Thus, the experimental hosts are potential hosts; presumably the worm is not a natural parasite in these hosts because fish are not part of their natural diet. Also, morphologic features can prevent normal development of a parasite. Plerocercoids introduced into adult terrapins cannot migrate because of the rigidity of the stomach wall, whereas in young turtles, the larvae migrate normally. In addition, temperature alone can prevent or permit infection. Plerocercoids in the stomach of *Gymnodactylus* (a lizard) migrate if the experimental host is kept at room temperature, but not if the temperature rises to 37°C.

Inability of the Parasite to Enter the Host

Sometimes, larvae are able to live and grow within a host experimentally but are not equipped to penetrate the external surface. A lack of suitable means of transmission can also produce narrow specificity. *Trypanosoma equiperdum* in horses and mules is transferred normally only by sexual contact, but hypodermic injections of the parasites into the blood of laboratory rats, mice, rabbits, and guinea pigs easily produced infections in these abnormal hosts.

Presence of Other Parasites

Whereas a given host can be infected with many species of parasites, sometimes only one or two of these species can be in a host body at the same time. At least 35 kinds of larval trematodes—21 strigeids, 6 plagiorchids, 2 schistosomes, 1 echinostome, 1 monostome, and 4 others—can be found in the freshwater snail, *Stagnicola emarginata*, but usually only two kinds (sometimes three or four) occur simultaneously in any single snail. Eleven species of protozoa have been found in the marine fish, *Gadus merlangus*, but only two or three are generally observed in any single host. A related phenomenon is shown by the strigeid trematode cercaria, *Cotylurus flabelliformis*, which encysts in the same snail species that harbors the sporocysts. If sporocysts of *C. flabelliformis* have already lodged in the snail, the cercariae do not enter. However, the presence of sporocysts of other trematode species appears to favor the penetration and encystment of cercariae of *C. flabelliformis*. The nature of these phenomena is not clearly understood, but it probably involves the entire defense mechanism of the host. A previous infection can affect the invasiveness of a parasite. (See the discussion of concurrent infections in Chapter 23.)

KINDS AND DEGREES OF PARASITE-HOST SPECIFICITY

Parasite-host specificity can be divided into two broad categories, defined with respect to the parasite. The first is **imposed specificity,** which is manifested primarily as **ecologic specificity,** or **segregation,** wherein the parasite is capable of living in a foreign host but normally never reaches one because of an ecologic barrier. The second category is **innate specificity,** wherein a parasite is developmentally compatible with its "normal" host and is incapable of surviving in a foreign host because of innate biologic properties of the parasite. As discussed in the previous section, this specificity is the result of physiologic, ethologic, chemical, and morphologic factors. For an example of ethologic specificity, see the discussion in "Locating a Host" later in this chapter.

A convincing experiment demonstrating innate specificity was one in which two common biting lice from two fowl were reared in vitro. Both *Lipeurus heterographus* and *Eomenacanthus stramineus* developed successfully under appropriate conditions of moisture and temperature, with chicken feathers and dried blood added for food. When feathers from the little green heron were substituted for chicken feathers, the *Lipeurus heterographus* died. This experiment suggests the presence of some chemical factor in feathers that is characteristic for a given group of birds and to which certain parasites are adversely sensitive. Another possibility is that the heron feathers *lacked* something essential.

When discussing specificity, it is important to remember the concept of co-accommodation. Co-accommodation refers in part to the degree to which a parasite's developmental requirements are provided by the host environment. When the parasite is compatible with a relatively small and similar group of host species, co-accommodation is narrow, and thus host specificity is high. When the parasite is compatible with a large and diverse group of host species, co-accommodation is broad, and thus host specificity is low. In both cases, which are the two ends of a continuum, specificity is the *effect* of biologic compatibilities between parasites and hosts.

Instances of low host specificity are not cases of a short supply of a property called "specificity''; rather, low specificity means that a relatively large number of host species exist that possess the traits required for the successful development of the parasite.

Broad parasite-host compatibility has a number of possible causes. The first is phylogenetic relatedness. All organisms on earth are related to one another at a particular level of generality, and the more closely related host species are, the more they will possess similar traits as a result of their common descent. When a group of host species all possess phylogenetically retained (i.e., homologous) traits that are sufficient for a parasite's development, the parasite can develop in any of those hosts. Another cause of broad parasite-host compatibility is the presence of nonhomologous parallel or convergent host traits, each of which allows parasite development. Another cause is the functional equivalence of host traits: traits are structurally different and of different phylogenetic origin, but their functions are sufficiently similar to meet the parasite's requirements. Of course, instances of broad parasite-host compatibility can nevertheless show great differences in parasite prevalence and intensity in each of the host species.

The concept of host specificity usually refers to those hosts in which a parasite is capable of developing normally. Perhaps in other hosts the parasite can initiate infection but cannot complete the next part of its life cycle. Such **nonpatent** infections can cause considerable pathology in the host—often more pathology than that caused in the natural host. Schistosomal dermatitis, or swimmer's itch, in humans can be caused by the penetration and subdermal death of schistosomatid cercariae that normally develop to maturity in various species of aquatic birds (see Chapter 10). The larvae of *Toxocara canis*, a parasite of dogs, can cause visceral larva migrans in humans (see Chapter 16), a condition in which the larvae do not complete tissue migration back to the gut but instead migrate throughout the body, unfortunately often to the eyes. The meningeal nematode *Parelaphostrongylus tenuis*, normally a parasite of white-tailed deer in North America, is so pathogenic to moose that sympatry between these two species of cervids is effectively prevented.[1] This last study is an excellent example of innate differences in host biology preventing broader parasite-host co-accommodation.

SPECIFICITY FACTORS RELATED TO INFECTION AND GROWTH

This section presents some of the factors and processes that can affect host specificity and that occur during each of the three major stages in the host-parasite association. This material is summarized from Rogers.[20]

First Stage of the Association of the Parasite With the Host

The infective stage (spore, egg, larva) ''is a 'resting' stage which requires factors from the host in order to resume development.''[20] The nature of these factors and their presence or absence often determine the degree of specificity before development resumes. Those parasites with a direct life cycle, for example *Ascaris*, are generally more specific than those that employ intermediate hosts, and ecologic factors initially determine the range of hosts. A high concentration of undissociated carbonic acid as well as dissolved gaseous carbon dioxide appears to be essential for the exsheathment of larvae of the nematode *Haemonchus contortus*. That these and other requirements are satisfied only in the rumen of ruminants limits the worms to these hosts (see Chapter 16 for details on the hatching of *Ascaris* eggs). While the initial requirements for infection with other worms (for example, *Ascaris lumbricoides*, *Toxocara mystax*, and *Trichostrongylus axei*) can cause the early stages of parasitic development to take place in a particular host organ, the requirements are apparently not specific enough to limit the parasite to one species of host.

The hatching factor (eclepic acid) excreted by roots of plants is essential for further development of the nematode plant parasite *Heterodera rostochiensis*. When hatching or encystment is induced by direct action of the host, as in the action of digestive enzymes on cysts of larval tapeworms, or on metacercariae of *Opisthorchis sinensis*, host specificity is generally low.

Thus, ''the ecology and behaviour of the infective stage and of the host and the conditions that are necessary in the host for infection to take place affect specificity during the first stage of the association.''[20]

Second Stage of the Association of the Parasite With the Host

Although at the infective stage the parasite does not feed or grow, after infection has occurred the invading organism is truly parasitic and must find necessary nutrients and be able to withstand any damaging physical or chemical factors in its environment. Often it is at this stage that the complex parasite-host relationships are the most crucial in determining the range of hosts. For example, some tapeworms of elasmobranchs require the urea of the host gut fluid to help maintain normal osmotic function.

Hormones of some hosts also probably influence their susceptibility to parasites (see Chap. 23). Natural or acquired resistance of a host obviously helps to determine its receptiveness to a parasite. Rogers, however, has reminded us that "as a rule we do not even know if the failure of a parasite to grow in vitro is due to lack of nutrients or to unfavorable chemical and physical features of the environment."[20]

Young hosts are often more susceptible than older hosts, but

"in most hosts the development of age resistance, if it occurs, is a gradual process which does not become complete. It is reasonable to suppose, however, that the unfavourable features in the host that give rise to age resistance may often be similar to those which make an organism unsuitable as a host at any period during its life and so affect the range of hosts of parasites."[20]

Third Stage of the Association of the Parasite With the Host

The period of parasite reproduction is essential not only for maintaining the species, but also for producing infective agents. Host specificity during this period can involve natural or acquired resistance, or it can be related to nutritional requirements of the parasite as well as to mechanical factors and changes in temperature. Such requirements can be greater during reproduction. Obviously, the cysts, spores, eggs, or larvae of parasites must reach an environment in which they can continue their development. The relationships between specificity and parasite reproduction are not well understood.

Each of the three stages of association already described involves some parasites of the vertebrate intestinal canal. Secretions that play an important role in the function of the intestine, such as bile and pancreatic juice, must also be involved in parasites' success in beginning and maintaining their existence. Bile salts, for ex-

ample, can stimulate parasite metabolism, or they can eliminate parasites.

COMPARATIVE HOST SPECIFICITY

This section discusses the various groups of parasites and the variations of host specificity that they display. Parasites with free-swimming stages able to enter the host skin have a narrower range of host species than do those parasites that enter the host through the mouth. Also, as a generalization, less host specificity exists when two intermediate hosts are used than when only one is employed. For example, the Pseudophyllidea (tapeworms), with two intermediate hosts (crustacea and fish), are less host specific than are the Cyclophyllidea, with a single intermediate host.

Protozoa

The genetic and epigenetic complex of a parasite and its host is the basis for biochemical, structural, and physiologic patterns that determine the character of parasite-host specificity. The biochemical pattern embraces the production of metabolites that play a decisive role in parasite-host adjustments.

One kind of experimental manifestation of host specificity involves biochemical adjustments that allow adaptations to new hosts. For example, when a sheep strain of *Trypanosoma vivax* that does not normally infect white rats is injected into rats, addition of a small amount of sheep serum allows the parasites to develop. After a period of adjustment, the parasites become adapted to living in the white rats without the aid of sheep serum, and, in fact, they produce a virulent infection that can be carried indefinitely from rat to rat by mechanical blood injections. If these parasites are transferred back into sheep, they lose their ability to infect white rats, but can regain this ability if they are injected, along with sheep serum, into rats.[7a] This evidence indicates an adaptation to a new environment rather than a mutational change.

Genetic changes can affect host specificity, not only through changes in the parasite, but also through changes in the host. Little is known about the relationship between host genetics and parasite susceptibility. In mice, the degree of resistance to *Plasmodium berghei* seems to have been controlled through the agency of many genes. Several workers, however, have been able to change the resistance of mosquito hosts to *Plasmodium* by selective breeding. To cite an example, the infection rate of *Culex pipiens* for *P. elongatum* has been increased from 5 to 50% in six generations by breeding from infected female mosquitoes.

McGhee[14] found that, in a suitable red blood cell environment, the avian *Plasmodium lophurae* invades and carries on at least part of its life cycle in the red blood cells of mice, rats, pigs, rabbits, and humans. Obviously, under these experimental conditions, some barriers to infection are removed, but the whole biochemical explanation must await further studies of this nature.

Malarial parasites reach the highest degree of specialization in birds, and they probably originated among the archosaurs in the Mesozoic period. A single species of bird can harbor several kinds of bird malarias, and avian *Plasmodium* species possess a wide latitude of host possibilities. *Plasmodium circumflexum*, however, appears to be restricted largely to robins. The malarial parasites of man are rigidly host specific, although some of them are infective to anthropoid apes. The intracellular Sporozoa, for example, *Plasmodium* and coccidia, are generally more host specific than are those parasites that inhabit the cavities and blood of their hosts, for example, blood-inhabiting flagellates and intestinal amebas.

Each time a wood-roach molts, most of its flagellates undergo sexual reproduction. The usual long process of fission (asexual reproduction) follows during the growth period of the host, and when the parasite again molts, its hormones induce the flagellates to reproduce sexually. On the other hand, the flagellate genera inhabiting the intestines of termites, for example *Zootermopsis*, do not undergo a sexual cycle, and the hosts must refaunate themselves after each molt.

Termite protozoa can live in and support the growth of nymphal and adult roaches until the roaches molt (up to 221 days for nymphs), but roach protozoa can support termites for as long as a year and can support a colony of *Zootermopis* indefinitely.[18] These protozoans resist cross-infection from one species of termite to another widely different species. Thus, the protozoa are generally both morphologically and physiologically distinct.

Ciliates inhabiting the stomachs of ruminants illustrate some interesting flexibility in host specificity. In general, the sheep and ox families, and to some extent the deer family, harbor similar ciliate faunas, as shown by morphologic comparisons and by cross-infection experiments. However, ciliates from horse ceca do not infect cattle stomachs.

Certain types of bacteria can enhance the invasiveness and possibly the virulence of entamoebas. Experiments have shown that without bacteria as a constituent of the gut fauna, at least with *Entamoeba histolytica* in guinea pigs, the host cannot even be infected (see Chap. 4). The role of the bacteria might be simply physical, providing a suitable microenvironment for the protozoans, but investigations on the role of bacteria or of other associates in cultivation with *E. histolytica* indicate that the answer is probably much more complex. The genus *Entamoeba* is widely distributed in vertebrates, but its species appear to be host specific. Differences among the species are often not detectable morphologically, but physiologic species, or at least biologic races, are common. Much experimental work on cross-infection studies needs to be done with parasitic amebas.

Worms

Broadly speaking, parasitic worms with direct life cycles, for example, *Ascaris*, are more host specific than the worms that have an indirect cycle, for example, tapeworms. On the other hand, worms that have an indirect life cycle generally exhibit more specificity for their intermediate hosts than for their final hosts.

The ranges of host specificity can be exemplified by some studies made on the worm parasites of moles and shrews. One group of worms, for example the nematode *Capillaria talpae* and the tapeworm *Choanotaenia filamentosa*, exhibits a narrow host specificity and is restricted to one species of mole, whereas another group, for example the trematode *Panoptistus pricei*, is less specific and freely invades several genera within the family Soricida. However, the nematode *Parastrongyloides winchesi* and other worms are widely distributed among the Talpidae and Soricidae within the order Insectivora. Finally, a few worms are unhampered by restrictive host preference and are thereby able to live as parasites in widely differing systematic groups of hosts. An example of this last group is the acanthocephalan worm *Polymorphus minutus*, which has been reported from freshwater fish, water birds, and water shrews. See

Cameron[4] for a review of host specificity and evolution of helminthic parasites and Schwabe and Kilejian[22] for a discussion of chemical determinants of host specificity among flatworms.

Monogeneans

Monogeneans as a group are host specific. The specificity is related to the conditions for isolation that exist in the group, together with the behavior of fastening eggs to the surface of the host and the several methods of attachment of adult worms. When adhesive organs consist of hooks that closely fit particular areas of the host surface, for example, the free edge of gill lamellae, the transfer of the adult worm from one host to another becomes difficult. When the parasites possess suckers, as do the Cyclocotylidae, and are thus able to move about in the gill chamber of a fish host, transfer to other hosts is more likely to take place.

Examples of speciation of parasites in close correspondence with speciation of their hosts are probably common throughout the whole group of monogeneans; for example, *Hexabothrium* and *Erpocotyle* are found exclusively on elasmobranchs.

Free-swimming oncomiracidial larvae of the monogenean skin parasite *Entobdella soleae* respond to a specific substance secreted by the skin of the common sole, *Solea solea*.[9] This response is weak or absent when the parasites are experimentally exposed to other species of fish or to detached pieces of skin from the other fish. The larvae are attracted to isolated epidermis from *Solea solea* and to agar jelly that has been in contact with the skin. Chemoreception probably is a major factor in host finding and host specificity among the monogenean platyhelminths in general.

Host specificity among monogeneans is physiologic, ecologic, or both and it is phylogenetically significant. The more primitive species of worms are generally found on the more primitive species of fish, and those worms infesting freshwater fish are less host specific than those on marine fish. Although a high degree of host specificity exists at the species level, it is often lower at the generic level. When considering a higher taxonomic level, we often find a wider range of host specificity. For example, the gyrodactyloids are distributed throughout teleosts, molluscs, and amphibia, and the capsaloids are found in elasmobranchs, holocephalans, chondrosteans, and teleosts. This

situation should not be surprising, because only species and their constituent individuals and populations are cohesive organic units capable of being parasitized. Higher, or more inclusive, taxa (genera, families, orders, etc.) are phylogenetically generated historic units that exist only through their constituent species.[28] Thus, not only do conclusions about the host specificity of, say, a genus depend on the correct inference of which species constitute that genus, but the more inclusive the taxonomic level to which one applies an assessment of host specificity, the more distant one becomes from an informative assessment of the parasite-host relationships involved. For example, it means little to say that because parasitic flatworms, as a whole, parasitize every major group of vertebrates, the phylum Platyhelminthes therefore displays low host specificity. There is much work to be done, especially experimentally, before the limits of host specificity in the Monogenea can be ascertained. For example, in cold-temperate seas, monogenean host specificity is reduced because of ecologically less restricted and less patchy host distribution.[21]

Digeneans

Too few complete life-history studies of digenetic trematodes have been made to warrant broad generalizations concerning the host specificity of entire life cycles. Available evidence points to a different degree of host specificity at each stage or level of the life cycle. Unfortunately, the possible varieties of intermediate as well as of final hosts have usually not been investigated. For these reasons, the following comments, unless otherwise indicated, pertain only to adult trematodes.

Host specificity is not marked in all families of digenetic trematodes. However, a study of collection records suggests that specificity prevails to a considerable extent, although closely related species can exhibit differences in the degree of specificity.[16] Thus, *Schistosoma mansoni* is restricted to humans, monkeys, and rats, whereas *S. japonicum* is a successful parasite of humans, dogs, cats, pigs, cattle, horses, and others, and of common laboratory animals. *S. haematobium* lives well in humans, albino mice, hamsters, monkeys, and baboons. It grows poorly in cats, albino rats, cotton rats, guinea pigs, and goats; rabbits and dogs are refractory to infection. *S. incognitum* occurs in pigs and dogs in India, and it can be established in such laboratory animals as cats, sheep, goats, rabbits, guinea pigs, rats, and mice. *S. spindale* is successful only in ungulates.

When distribution records of trematodes are analyzed, one normally finds that genera having a wide range of host tolerance are nonetheless limited to ecologically related hosts. For instance, all hosts from which about 50 genera of strigeids have been recorded are ecologically associated with water. Lest the student think that stating that a water-borne parasite can only parasitize hosts that come to the water is trivially tautologous, it should be noted that although ecologic association is *necessary* for parasitism to occur, it is not *sufficient*. Potential hosts must also possess the innate characteristics that allow the development of the parasite. (For example, strigeids do not develop in water striders.)

Among the strigeid trematodes, families have been considered to be restricted to certain kinds of hosts. Members of the Diplostomatinae are parasites of birds, and the Alariinae are found exclusively in mammals. *Fibricola crateri*, a parasite of mammals, can easily be transferred to chicks, but whether avian hosts are infected in nature is not certain. *Fasciola hepatica* has become cosmopolitan in distribution, and adults can infect sheep, cattle, pigs, rodents, elephants, kangaroos, and humans. *Echinostoma revolutum* can infect several species of birds and mammals. Both birds and mammals are, of course, homoiotherms.

Several genera of invertebrates and vertebrates frequenting a lake district can be infected with the same genus or even species of trematode. Under experimental conditions, however, such unnatural hosts as chicken, ducks, rats, mice, and cats can successfully harbor the adults if fed sufficient numbers of metacercariae. Parasitologists often discover metacercariae or cercariae whose definitive hosts are unknown, but whose entire life cycle can be described because development can take place in a laboratory animal.

Experimental work with miracidia has often demonstrated a degree of specificity for snails. The miracidium of *Opisthorchis felineus*, for example, is attracted to the prosobranch snail *Bithynia leachi*, but not to the closely related *B. tentaculata*, which occurs in the same locality. *Schistosoma japonicum* develops only in *Oncomelania hupensis*.

As a final generalization, host specificity among digenetic trematodes is greater at the level of the intermediate host, especially the mollusc, than at the adult level. Thus, two distinct species of worms can live together as adults in the intestines of a bird or fish, but they require different species of snails in which to complete their life cycles.

Cestodes

Tapeworms vary in their range of host specificity, but the adults are more specific than the adults of many other groups of worms. Each order of bird and mammal possesses its own characteristic cestodes. For example, the cyclophyllids, with an evolutionarily derived type of internal anatomy, are found only in terrestrial vertebrates. Among the elasmobranchs are found tetraphyllid tapeworms that possess highly modified types of scolex structures.

The sharks and rays each apparently harbor distinct species of cestodes. Snakes also possess distinct and characteristic species of tapeworms. *Dipylidium* and *Echinococcus* are found only in carnivores; *Moniezia*, *Thysanosoma*, and *Stilesia*, only in ruminants. However, host specificity among cestodes reaches its highest development in birds. This relationship is perhaps best illustrated by water birds, such as grebes, loons, herons, ducks, flamingos, and cormorants—birds that can occupy the same ponds or lagoons. Each bird species possesses its own tapeworm fauna. If we recall the foregoing discussion about ecologic segregation, we recognize that among cestodes, contrary to the situation described for trematodes, host specificity is apparently more independent of ecologic segregation of their hosts and more dependent on phylogenetic relationships.

Developmental stages of cestodes frequently tolerate a much wider range of hosts than do adult cestodes. *Hymenolepis diminuta* metacestodes have been reported from four different orders of insects and from myriapods. *Hymenolepis gracilis* occurs in both copepods and ostracods. Metacestodes of the dwarf dog tapeworm, *Echinococcus*, have been found in many kinds of mammals. A study of these stages indicates that immature cestodes are ecologically segregated, but in some forms specificity is independent of ecologic factors and has resulted from physiologic adaptation. As evidence, when coracidia of *Diphyllobothrium latum* are fed to several species of freshwater copepods in one dish, some species of the copepods are more favored as hosts than others.

Experimentally, opium-treated albino mice were infected with *Hymenolepis diminuta* much more readily than were those not opium-treated. Opium slows the intestinal emptying time, and the effect of this process on the establishment of *H. diminuta* is pronounced. That the parasite occurs only rarely in house mice suggests that the intestinal emptying time might partly explain apparent host specificity.

The rostellum and suckers of the scolex probably play important roles in determining host specficity. Unless a tapeworm can successfully attach to the wall of the host's gut, it will be expelled from the host. Williams et al.[29] conducted research on the topologic fit of the scolex to the host gut.

One possible physiologic explanation for the high degree of specificity shown by tapeworms is a dependence on specific nitrogenous compounds secreted by the host intestine. Host specificity of parasites, in general, is undoubtedly related to specific biochemical characteristics of parasites and to the chemical and physical properties of the environment within the host.

Nematodes

The study of host specificity among nematodes is particularly perplexing because of the many kinds of association between these worms and other organisms, both plant and animal. An almost continuous series of associations exists, from entirely free-living nematodes to obligatory parasites. The most primitive genera and species of parasites of vertebrates are found in mammals; more than two thirds of the known strongyloids, usually considered the most primitive nematodes, are from mammals.

Brugia malayi can be successfully transmitted from humans to forest and domestic animals by direct inoculation of infective larvae. This observation is significant in considering reservoir hosts for this and other filarids. If we can transmit filarial worms from humans to animals in the laboratory, can the parasites be transmitted in nature by mosquitoes or by *Chrysops* from animals to man? Probably so.

Most species of parasitic nematodes found in birds belong to genera that also occur in mammals, but in many groups of mammals, the genera of nematodes are specific. Elephants harbor 6 distinctive genera and 20 or more species of strongyles, whereas rhinoceroses can have 6 genera and about 18 species. Horses and other equines harbor at least 8 genera, with more than 50 species of nematodes that are not found in any other group of animals.

Specificity of nematodes whose larvae migrate through the vertebrate body can be due primarily to the failure of larvae to complete somatic migration in a foreign host, rather than to incompatibility between the adult worm and the host. Another kind of specificity in this group of worms is illustrated by the apparent preference by *Ascaridia galli* for male over female chicks.

Sprent[24] has made some significant observations on the changing specificity patterns during the life history of *Amplicaecum robertsi,* an ascaridoid from the carpet python:

"The snake is depicted as the apex of a food pyramid, whose base comprises a variety of animals ranging from earthworms to herbivorous animals. . . . The life history is thus regarded, not as a life cycle, but as a life pyramid; development proceeds according to a pattern of diminishing host-specificity. Host-specificity is wide at the base of the pyramid, so that second-stage larvae occur in a wide variety of paratenic hosts. Host-specificity narrows at the second moult which may occur in birds and mammals. It narrows still further in the third stage because this larva, though it will survive in reptiles, birds and mammals, will not grow to a length at which it is capable of further development in the snake except in certain mammals. At the third moult, host-specificity shifts to certain reptiles but becomes eventually restrictive to the carpet snake, because this host alone appears to provide a suitable environment for maturation of the eggs."

In a study of patterns of evolution in nematodes, Inglis concluded that

"in general parasitic nematodes are not host specific although they tend to be restricted to animals with similar feeding and ecological habits. The most spectacular example of a group which is apparently an exception to this is the pinworms of primates."[11a]

Acanthocephala

Adult acanthocephalid worms show a high degree of host specificity. For example, normally *Gracilisentis* and *Tanarhamphus* are found only in the gizzard shad. The Pacific pilotfish, *Kyphosus elegans,* and the closely related Atlantic pilotfish, *K. secatrix,* each harbor a distinct species of the acanthocephalan genus *Filisoma.* Similarly, *Moniliformis moniliformis* shows a narrow specificity in certain instances. Specificity in these worms is related to the nature of the life cycle, in which no free-living stage has been reported and in which an arthropod intermediate host is essential for all species.

Physiologic host specificity appears to vary in this group of parasites. Acanthocephalans of carnivorous vertebrates cannot proceed from the arthropod to the final host unless the parasites first pass into an insectivorous host. Collection records indicate that little specificity exists among larval forms of at least some groups of Acanthocephala. A single species of *Centrorhynchus*, for example, has been reported as a larval parasite of lizards, snakes, and frogs. Much experimental work on morphologic variation and physiologic host specificity under controlled conditions is needed.

Arthropods

Crustacea

Parasitic copepods are among the most diversified of all parasites and are almost the only ones found in the adult stage on both vertebrates and invertebrates. Site specificity of copepods on their hosts is often extremely marked (Fig. 25–1). Some species of blood feeders are restricted to one species of host. Although distinct host specificity among copepods occurs, further statistical work must be done with these crustaceans before we can formulate significant generalizations about them.

A study[9] of the amino acids in body fluids of copepods and in the tissue proteins of the procercoids of the tapeworm *Triaenophorus nodulosus* that parasitizes copepods *(Eudiaptomus gracilis)* indicated a similarity in qualitative composition and proportions. This finding suggested that the similarity in the pathways of protein metabolism in both members of the parasite-host system might be a condition of host specificity. However, in addition to broad, and sometimes specific, metabolic similarities among parasites and their hosts, differences occur. Numerous examples of variations among different species of helminths and among developmental stages of a single worm have already been presented.

Parasitic isopods have frequently been considered to be strictly host specific. Apparently, each genus of many groups, for example, etoniscids, is found on a particular host or group of hosts that appears characteristic for the parasite. For instance, *Danalia* and *Liriopsis* are found only on decapods. *Phryxus abdominalis* has been

recovered from at least 20 species of shrimp belonging to two genera. Without a considerable amount of experimental research work on these and other forms, no definite conclusions can be made regarding the host specificity of parasitic isopods in general.

Hymenoptera

Insects as parasites are generally host specific, but this specificity is predominantly ecologic not physiologic. Evidence for this conclusion is gained experimentally when the parasite is placed directly on a foreign host, eliminating such activities as searching for the host habitat and the host. The parasite commonly lays eggs on or in the new host with little or no hesitation.

The entomophagous Hymenoptera (see Chap. 19) are seldom monophagous (restricted to one kind of food), but they are far from indiscriminant in their attacks on insect hosts. In nature, the parasitic wasps have several potential species of hosts, and in the laboratory, many more hosts can be discovered by experimental testing. Why then does the wasp usually select only one or two kinds of hosts in which to deposit its parasitic eggs? Several distinct processes of selection seem to occur. See the discussion of pheromones (Chap. 2) and of biocontrol (Chap. 23).

The food of an insect host can affect the host's suitability as a home for a parasite. For example, the hornworm, *Protoparce sexta*, when fed on tomato, is a suitable host for the braconid wasp, *Apanteles congregatus*, but when this host is fed on dark-fired tobacco, the parasite dies before reaching maturity.

A sequence of ecologic processes leads to the attainment of host specificity of many Hymenoptera and other entomophagous insects. Searching capacity and host specificity are correlated phenomena. The sequence involves finding the host habitat, finding (recognition) the host, acceptance by the host, and host suitability for parasite reproduction.

Fleas

Among the reviews of the question of host specificity in fleas is that of Hopkins:

"Unlike many parasites, fleas (possibly rare exceptions) . . . pass their entire pre-adult life off the body of the host, their larvae being free-living feeders on organic dust, though the early stages usually take place in the host's dwelling. This means that the early stages of fleas, not being parasitic, are susceptible to

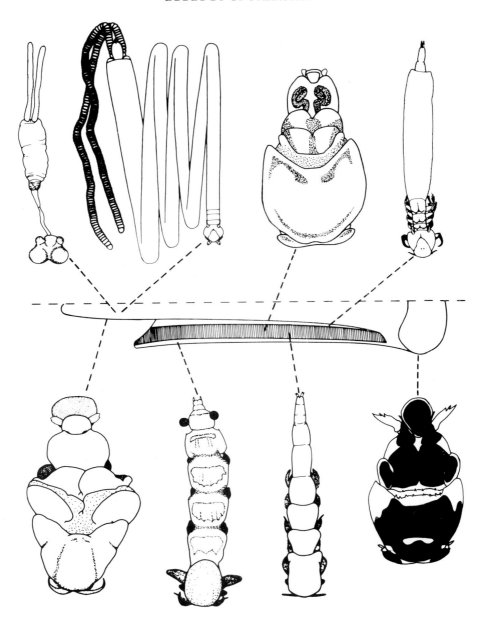

Fig. 25–1. Summary of gill niches inhabited by eight species of copepods infesting sharks (clockwise from top left; not to relative scale): *Pandarus cranchii*, gill arch; *Eudactylinodes uncinata*, secondary lamella; *Nemesis lamna*, efferent arteriole; *Phyllothyreus cornutus*, superficially on interbranchial septum; *Paeon vaissieri* and *Kroyeria caseyi*, cephalothoraxes deeply embedded in interbranchial septum; *Gangliopus pyriformis*, secondary lamella; *Kroyeria lineata*, in water channel and on secondary lamella. (From Benz[2].)

the conditions, climatic and others, which govern the distribution of free-living animals to a far greater extent than animals which are parasitic in all their active stages. It also means that it is necessary for the newly-emerged flea to seek out a host, sometimes of one particular species, and that the period of starvation that the flea must undergo during this search can often be reduced and the search be prolonged (with better prospects of a successful conclusion) by the practice of polyhaemophagy (the ability to feed on the blood of hosts other than the one normal to the flea in question). For these reasons it is common to find a flea on a host other than that (or those) to which it is normal, and such occurrences may range from the purely accidental presence of fleas on reptiles (from which they are probably unable to suck blood) through those in which a flea can obtain nourishment from the blood of a host on which it is extremely reluctant to feed, to instances in which a given species of flea has a number of hosts between which it shows little preference and on all of which the species can reproduce indefinitely."[10]

Polyhemophagy appears to have been of considerable advantage to fleas, yet narrow specificity is demonstrated by many species. When fleas are considered as a group, it is clear that the ecologic conditions in the nest (mammal or bird) are more important than the host is. For example, *Ceratophyllus garei* occurs in the nests of many birds, if the nests are not too dry. *Pulex irritans* of humans is sometimes abundant in pigstyes, and it is a true parasite of the badger. Hopkins stated that

"the more promiscuous a flea is as regards to the source of the blood in which it can mature its eggs, the more probable it becomes that random hopping will eventually result in the deposition of the eggs in an environment suitable for the development of the larvae, while the latter are not affected nearly so much by the question of whether the nest or burrow in which they find themselves was made by a rabbit or a bird, a squirrel or a mouse, as by the environmental conditions within it, particularly the relative humidity and temperature."

Lice

All lice are obligatory and permanent external parasites of birds and mammals. They cannot jump, fly, or even walk well, and they spend their entire lives on the bodies of their hosts. For these reasons, transfer from host to host is normally accomplished only when two host bodies are in close contact, as during copulation, during feeding of young, or while standing together in herds. The death of the host inevitably means the death of the entire pop-

ulation of lice on its body. These limitations explain the intraspecific bounds of distribution, and the extreme specificity in their host associations. No lice are present on bats, yet bats are heavily invaded by other parasites; 40 or 50 ectoparasites of several species are not uncommon on one bat. The reasons for the absence of lice on bats are unknown.

Authentic instances of distributions of lice not explainable by co-evolution with the hosts are rare, but interspecific transfers might occur during a struggle between prey and predator, during the sharing of mud wallows, rubbing trees, or roosting or perching spots, and as the result of the usurpation of a nest or burrow by an alien host. Phoresy (Chap. 1) can also result in interspecific transfer.

Lice, therefore, almost always occur only on one host or on a small number of closely related hosts. An extreme degree of specificity occurs among the lice of conies (Procaviidae, order Hyracoidea), where most of the subspecies of the host, *Procava capensis*, have their own species or subspecies of louse (*Procavicola*). The chewing lice of these mammals appear to be more host specific than the sucking lice.

The lice on primates all belong to the family Pediculidae, and no member of this family occurs on any other host. Both *Pediculus* and *Phthirus* (see Chap. 19) occur on humans and higher apes, but not on monkeys. *Phthirus* includes species from the gorilla and chimpanzee, but because the records are from menagerie material, it is not known conclusively that these apes are natural hosts of the crab louse. *Pediculus* has been recorded from the gibbon and from the chimpanzee. Apparently, the orangutan is not infested with lice. Spider monkeys (*Ateles*) of tropical America are far removed anatomically from humans and their ancestry yet curiously possess a species of *Pediculus*, sometimes separated as the genus *Parapediculus*. Although straggler lice frequently occur on birds, the specificity between feather lice and their hosts is marked. For instance, the wing louse, *Lipeurus caponis*, occurs only on the wings, and the shaft louse, *Menopon gallinae*, occurs within the shafts of large feathers.

Mutualistic bacteria occur in many biting lice, and it has been suggested that one factor helping to determine host specificity is the inability of the bacteria to survive in a louse feeding on the blood of an abnormal host species. The causes of host specificity among lice are doubtlessly related to biochemical differences in the blood, skin, and plumage among the hosts.

Flies

The Family Hippoboscidae, of which all members are obligatory bloodsucking ectoparasites of various orders of mammals and birds, helps illustrate some of the details of the problems of parasite-host specificity among the Diptera. These flies occur on only 5 orders of mammals and on 18 of 27 orders of birds. No species is parasitic on both mammals and birds. Host preference is more restricted in the flies of mammals than in those of birds, but the parasites show a wide diversity of behavior, with all gradations from strict species specificity to occurrences on host species of different orders. Hippoboscid host selection patterns are explained chiefly by ecologic factors; hence, regardless of taxonomic affinities of the hosts, if they possess similar habitats, they are likely to be burdened with similar parasitic flies. These considerations suggests that, when strict specificity exists, it is the result of geographic isolation, as with flies on kangaroos. Conclusions similar to those made for Hippoboscids have been made for pupiparous Diptera of bats.

During the months of July and August at Fair Isle Observatory near Shetland and Orkney, five species of breeding passerine birds were trapped intensively, and hippoboscid flies (*Ornithomyia fringillina*) were removed from each bird. Each fly was marked and then released on any of the five species of hosts, irrespective of the original source of the fly. A few marked flies were released without a host. After a lapse of an appropriate time, the flies were again collected, and 75% of those recovered were found on the same individual birds on which they had been released, 18.5% had moved to other birds of the same species, and 6.5% had changed to a host species different from that on which they had been released. Male flies changed hosts more often than females.[5]

Although specificity of an insect parasite for a single species of host is rare, *Cryptochaetum iceryae*, a dipteran parasite, is restricted in the United States to the cottony cushion scale, *Icerya purchasi*; however, in Australia, it has been recorded on other species of the same genus of host. For details of this group of flies, see Theodor.[27]

Ticks and Mites

Although a few ticks and mites appear to be confined strictly to one host species, these groups of arachnids follow no discernible rule. The tick, *Dermacentor andersoni*, has many compatible hosts in all its life stages. The more primitive trombiculid mites exhibit less host specificity than do those species that display a closer association with their hosts. It appears that the latter group began phylogenetically as nest-infesting species that developed first an ecologic type of host specificity; then, as the association became closer, a degree of physiologic host specificity developed. The apparent host "preference" by mites and chiggers might be largely a matter of difference in the area of host skin exposed or available for infestation, or of differences in extent of host range or other host behavior.[17]

LOCATING A HOST

A useful source of information on the nature of host specificity is a study of commensalism, in which the symbiosis of two or more individuals depends on the maintenance of particular and precise behavior. In most of these relationships, the commensal makes an active search for a partner.

Studies on the specificity of host recognition by commensal polynoid annelids have shown that chemical attraction and recognition appear to be the usual mechanisms binding such partners as the scaleworms, *Arctonoe fragilis*, with the starfish, *Evasterias trochelii*; *A. pulchra* with the sea cucumber, *Stichopus californicus*; *Hesperonoe adventor* with the echiuroid worm, *Urechis caupo*; and *Polynoe scolependrina* with the terebellid, *Polymnia nebulosa*. An experimental technique devised by Davenport[6] consisted of a choice-apparatus or olfactometer in which

"commensal worms were introduced into a Y-tube and were presented with a choice between streams from two aquaria. Material to be tested could be placed in either aquarium at random; similarly, connections with the aquaria were so arranged that streams to be tested could be introduced into either arm of the Y at random, thus making it possible for any consistent behavior resulting from uncontrolled inequalities in pressure or light to appear in the data from a large number of 'runs.' Such apparatus lends itself to investigation of host specificity in active forms which readily respond to streams of water carrying attractants."

All the polynoid commensals investigated demonstrated strong positive responses to chemical stimulation by their hosts, and with few exceptions, this response was specific in spite of frequent close taxonomic affinity among hosts. Davenport and his colleagues have demonstrated that specialized sorts of behavior can be induced by the presence of "host-factor." For example, the frequency of random turning in pinnotherid crabs is directly proportional to the concentration of "host-factor" in the crab's general environment.[7] This response obviously keeps the crabs in the vicinity of the clam host. At higher concentrations, the same chemical agent from the host has a direct influence on the crab and induces it to move directly toward the clam.

A commensal sometimes is attracted to two distantly related hosts. In such a situation, it seems likely that both hosts produce the same "attractant" (probably a metabolite). This common attraction also presents the possibility of a functional equivalence of two structurally different host products.

Efforts have been directed at correlating the *response* specificity of commensal worms on echinoderms to the known *host* specificity. Various categories of response specificity have been demonstrated, ranging from commensal populations that respond to their normal host alone (for example, the polynoid worm, *Arctonoe fragilis,* on the seastar, *Evasterias*) to commensals that appear to have no chemical discrimination and respond to many host animals. Moreover, all populations of a commensal species do not always behave in the same manner. For instance, there are two populations of the polychaete worm, *Podarke pugettensis,* one a facultative commensal on several starfish (for example, the webstar, *Patiria miniata,* and the mudstar, *Luidia foliolata*), the other free-living. The former population shows a tendency to respond positively to its host, but the other shows no such tendency. Experiments are needed to determine whether such differences are inherited or conditioned.

The odor of a host apparently helps mosquitoes localize the blood supplier.[13] Laboratory-bred *Anopheles atroparvus* reacted satisfactorily to airborne stimuli in an air-stream olfactometer, and the experiments pointed to the possibility of adaptation to the smell of a special type of host. Heat and moisture appear to stimulate the mosquito to land on the host, but heat also attracts the flying mosquito to host. This thermotaxis is strongly activated by CO_2, which, with other odors, probably has an activating value in the process through which the mosquito orients itself toward the host. Mosquito responses to chemicals are undoubtedly combined with visual responses in seeking and selecting a host.

A problem related to the foregoing is the explanation of preference for one host over another by parasites that normally can infect both. Worth described an example of this problem involving cotton rats:

"It would appear that for some reason the cotton rat is a favored host in the Everglades, being the carrier of more than five times as many individual ectoparasites as rice rats in the same environment despite a similar pattern of host infestations."[30]

The answer, in this case, lies in the differences in the quality of host blood, microclimate variations in the fur, differences in structure and texture of the hair and skin, and grooming behavior. It is often difficult to detect such differences and, when detected, to evaluate their effects on the parasites. To cite another example, the ciliate *Trichodina* is common on the gills of marine fish but is particularly abundant on benthic fish. Hence, the preference is presumably associated with benthic life. The basic principle of variation among species, and of variations among individuals within one species, lies behind obvious as well as obscure patterns of parasite behavior. (See discussion of pheromones.)

SUMMARY OF GENERAL RULES AND PRINCIPLES

1. Parasite-host specificity is normally determined by ecologic, ethologic, physiologic, chemical, and structural factors. High host specificity is indicative of narrow co-accommodation between parasite and host; low host specificity is indicative of broad-co-accommodation between parasite and host. A study of host specificity includes an assessment of parasite prevalence, parasite intensity, and the phylogenetic distance between the hosts.

2. In general, parasites with an indirect life cycle are less specific than those with a direct life cycle.

3. Less host specificity occurs when two or more intermediate hosts are employed than when only one is employed.

4. Whereas two strains of a given parasite might infect a final host with equal facility, they might not be equally infective to a given strain of intermediate host.

5. The host is in many respects equivalent to the environment surrounding free-living organisms; but the host, as a living environment, is capable of responding to the presence of its parasites.

6. As with any evolved trait, the traits that produce high or low host specificity can subsequently be modified or lost.

7. When a species of host is divided into two or more population groups separated geographically, their respective parasite faunas are also separated and thus can develop differences over evolutionary time.

8. A parasite can become transferred to a foreign host living in the same locality as the original host and can subsequently become isolated by a geographic separation of the two hosts.

9. When isolation of a parasite on a host takes place, both morphologic and physiologic changes can occur.

10. In comparing digenetic trematodes with cestodes, host specificity among trematodes is more pronounced in the larval stages than in the adult stage; the larval stages of cestodes frequently tolerate a much wider range of hosts than do adult cestodes.

11. In nematodes, sometimes greater specificity exists for the intermediate host than for the definitive host.

REFERENCES

1. Anderson, R.C.: The ecological relationships of meningeal worm and native cervids in North America. J. Wildl. Dis., 8:304–310, 1972.
2. Benz, G.W.: Copepods as parasites of sharks *In* Sharks in Captivity. Edited by R.L. Jenkins, et al. Baltimore, Johns Hopkins University Press, 1988.
3. Brooks, D.R.: Testing the context and extent of host-parasite coevolution. Syst. Zool., 28:299–307, 1979.
4. Cameron, T.W.M.: Host specificity and the evolution of helminthic parasites. Adv. Parasitol., 2:1–34, 1964.
5. Corbet, G.B.: The life-history and host-relations of a hippoboscid fly *Ornithomyia fringillina* Curtis. J. Anim. Ecol., 25:403–420, 1956.
6. Davenport, D.: Specificity and behavior in symbioses. Q. Rev. Biol., 30:29–46, 1955.
7. Davenport, D., Camougis, G., and Hickok, J.F.: Analysis of the behavior of commensals in host-factor. 1. A hesioned polychaete and a pinnotherid crab. Anim. Behav., 8:209–281, 1960.

7a. Desowitz, R.S., and Watson, H.J.C.: Studies on *Trypanosoma vivax*. I. Susceptibility of white rats to infection. Ann. Trop. Med. Parasitol., 45:207–219, 1951.
8. Freeland, W.J.: Parasites and the coexistence of animal host species. Amer. Nat., 121:223–236, 1983.
9. Guttowa, A.: Amino acids in coelomic fluids of Copepoda, hosts of pseudophyllidean procercoids. Acta Parasitol. Pol., 16:239–248, 1968.
10. Hopkins, G.H.E.: Host-associations of siphonaptera. *In* First Symposium on Host Specificity Among Parasites of Vertebrates. Institute of Zoology, University of Neuchatel, 1957, pp. 64–87.
11. Inglis, W.G.: Speciation in parasitic nematodes. Adv. Parasitol., 9:185–223, 1971.
11a. Inglis, W.G.: Patterns of evolution in parasitic nematodes. *In* Evolution of Parasites. Edited by A.E.R. Taylor. Oxford, Blackwell Scientific Publications, 1965, pp. 79–124.
12. Kearn, G.C.: Experiments on host-finding and host-specificity in the monogenean skin parasite *Entobdella soleae*. Parasitology, 57:585–605, 1967.
13. Laarman, J.J.: Host-seeking behavior of malaria mosquitoes. XVth International Congress of Zoology Proceedings. Edited by H.R. Hewer and N.D. Riley. 1959, pp. 648–649.
14. McGhee, R.B.: The ability of the avian malaria parasite, *Plasmodium lophurae*, to infect erythrocytes of distantly related species of animals. Am. J. Hyg., 52:42–47, 1950.
15. MacInnis, A.J.: How parasites find hosts: Some thoughts on the inception of host-parasite integration. *In* Ecological Aspects of Parasitology. Edited by C.R. Kennedy. Amsterdam, North-Holland, 1976, pp. 3–20.
16. Manter, H.W.: Host specificity and other host relationships among the digenetic trematodes of marine fishes. *In* First Symposium on Host Specificity Among Parasites of Vertebrates. Institute of Zoology, University of Neuchatel, 1957, pp. 185–198.
17. Mohr, C.O.: Relation of ectoparasite load to host size and standard range. J. Parasitol., 47:978–984, 1961.
18. Nutting, W.L.: Reciprocal protozoan transfaunations between the roach, *Cryptocercus*, and the termite, *Zootermopsis*. Biol. Bull., 110:83–90, 1961.
18a. Palmieri, J.R., Connor, D.H., and Marwoto, H.A.: *Wuchereria bancrofti* infection in the Silvered Leaf Monkey (*Presbytis cristatus*). Am. J. Pathol., 112:383–386, 1983.
19. Read, C.P.: Parasitism and Symbiology. New York, Ronald, 1970.
20. Rogers, W.P.: The Nature of Parasitism: The Relationship of Some Metazoan Parasites to Their Hosts. New York, Academic Press, 1962.
21. Rohde, K.: Latitudinal differences in host-specificity of marine Monogenea. Marine Biol., 47:125–134, 1978.
22. Schwabe, C.W., and Kilejian, A.: Chemical aspects of the ecology of platyhelminthes. *In* Chemical Zoology. Vol. II. Edited by M. Florkin and B.T. Scheer. New York, Academic Press, 1968, pp. 467–549.
23. Soulsby, E.J.L.: The control of parasites: the role

of the host. Proc. Helminth. Soc. Wash., *44*:28–43, 1977.

24. Sprent, J.F.A.: The life history and development of *Amplicaecum robersti*, an ascaridoid nematode of the carpet python (*Morelia spilotes variegatus*). II. Growth and host specificity of larval stages in relation to the food chain. Parasitology, *53*:321–337, 1963.

25. Stefanski, W.: Quelles conditions exige le parasite pour s'etablir dans son hôte? Ann. Parasitol. Hum. Comp., *37*:664–672, 1962.

26. Terzian, L.A., Stabler, N., and Weathersby, A.B.: The action of antimalarial drugs in mosquitoes infected with *Plasmodium gallinaceum*. J. Infect. Dis., *84*:47–55, 1949.

27. Theodor, O.: Parasitic adaptations and host-par-asite specificity in the pupiparous Diptera. *In* First Symposium on Host Specificity Among Parasites of Vertebrates. Institute of Zoology, University of Neuchatel, 1957, pp. 50–63.

28. Wiley, E.O.: Phylogenetics: The Theory and Practice of Phylogenetic Systematics. New Yok, Wiley-Interscience, 1981.

29. Williams, H.H., McVicar, A.H., and Ralph, R.: The alimentary canal of fish as an environment for helminth parasites. *In* Aspects of Fish Parasitology. Edited by A.E.R. Taylor and R. Muller. Oxford, Blackwell Scientific Publications, 1970, pp. 43–77.

30. Worth, C.B.: Observations on ectoparasites of some small mammals in Everglades National Park and Hillsborough County, Florida. J. Parasitol., *36*:326–335, 1950.

26

Evolution of Parasitism

In one way or another, all of the chapters in this book are couched in an evolutionary framework. This allows comparative descriptions of parasites' traits, or characters, and allows postulated explanations of how those traits came to be. An evolutionary framework encompasses not only the phenomenon of change over time, but also that of descent with modification, producing a phylogenetic tree, or geneology, of ancestors and descendants. The methods used to study evolution and phylogenetic relationships are largely independent of the organisms involved, and so these introductory remarks will make little mention of parasites per se. Once a broad viewpoint has been outlined, parasitic organisms will be examined more closely.

The entire life cycle of a parasite along with all closely associated organisms, including the host, is capable of evolving. Thus, the evolution of parasite-host systems is as important as the evolution of individual parasite lineages (see Mattingly[73] on the evolution of arthropod-borne systems; and Harvey et al.[46] on the evolution of pathogenesis in various parasites). A great deal of free-living chauvinism must be rejected if one is to study the evolution of parasites in a manner that concentrates on causes and effects, rather than on human-centered evaluations of parasites as degenerate, or over-specialized, or somehow just not as "good" as free-living organisms (such as humans). One must avoid anthropocentrism when studying the natural world.

PHYLOGENETIC RECONSTRUCTION

The key to studying evolutionary relationships is the identification of homologous char-

acters. When Richard Owen introduced the concept of homology in 1848, it was strictly an assessment of the similarity of form among organisms. Not until later 19th-century workers, such as Darwin, Haekel, Lankester, and Gegenbaur, was the argument put forth that similarity of form existed *because of* descent from a common ancestor. Thus, homologues came to be seen as clues to an organism's past—historical markers, as it were.

Some homologous traits have persisted in an evolutionary lineage through long periods of time and have been inherited by many descendants (two examples: the spinal column of vertebrates, and the microvillar tegument of platyhelminths). Other homologues evolved relatively recently and occur in a less-inclusive group of taxa (two examples: within the vertebrates, the feathers of birds; within the platyhelminths, the scolex of tapeworms). The overall picture is that of a phylogenetic tree consisting of an internested hierarchy of evolutionary groups (Fig. 26–1), or clades (see below), with each group possessing certain common homologous characters.

Among the tasks of evolutionary biologists, and systematists in particular, are (1) the identification of homologous characters; (2) the determination of the level of inclusiveness at which such characters' origins must be explained (for instance, in Figure 26–1, the origin of an armed cercomer in members of the Cercomeromorphae needs to be explained once, in the common ancestor of the group, and not independently for each of the four members, monogeneans, gyrocotylids, amphilinids, and cestodes); (3) the grouping of taxa into phylo-

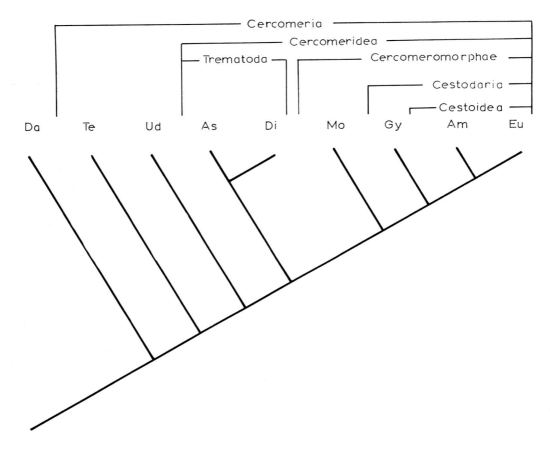

Fig. 26–1. A cladogram of the phylogenetic relationships among the parasitic platyhelminths, as hypothesized by Brooks, O'Grady, and Glen.[22] Dalyelloidea (free-living outgroup) (DA); Temnocephalidea (Te); Udonellidea (Ud); Aspidogastrea (As) (also referred to as Aspidocotylea, and Aspidobothrea); Digenea (Di); Monogenea (Mo); Gyrocotylidea (Gy); Amphilinidea (Am); Cestoda (Eu) (also referred to as Eucestoda).

genetic groups of closest relationship on the basis of shared derived homologues, rather than shared primitive homologues (for instance, the presence of fur cannot be used to group housecats and tigers as each other's closest relatives, because within the Felidae fur is a shared primitive homologue; it is at the level of the Mammalia that fur is a shared derived homologue); and (4) the identification of those characters whose similarity of form is *not* indicative of close phylogenetic relationship, but is instead the result of parallelism or convergence—possibly involving selection pressures in a particular type of environment.

With respect to the third point given above, Figure 26–2 gives an empiric example of the problems that can arise from grouping by overall similarity. Three taxa are involved: a shark, a lungfish, and a human. If the three are grouped according to their overall similarity, the shark and the lungfish possess enough shared primitive homologues to be grouped together, to the exclusion of humans. The tree produced through this operation, shown on the left in Figure 26–2, is called a *phenogram*. This type of tree is not useful as a phylogenetic tree because descendants from a common ancestor do not necessarily share the greatest degree of overall similarity in a study group. A number of shared derived homologues link lungfish and humans together as descendants from a common ancestor not shared with sharks.[88a] This phylogenetic relationship is shown on the right in Figure 26–2. One must be sure to ascertain whether the statement "taxon X is closer to taxon Y than it is to taxon Z" is referring to the degree of overall similarity or to the degree of phylogenetic relationship. (Sharks and lungfish, for example, look more like each other than either looks like a human, but it is lungfish and humans that share a common ancestor).

The tree on the right in Figure 26–2 is called a *cladogram*, taken from the Greek *klados*, for branch. A clade is another name for a monophyletic group, which is an evolutionary assemblage consisting of all of the descendants of a common ancestor. (Refer to Wiley[100] and Eldredge and Cracraft[37] for general information on the construction of cladograms.) A cladogram is a hypothesis of phylogenetic relationships; it is tested and either rejected or corroborated (though never proven) through independent assessments of as many characters as can be examined, conducted with a minimum of assumptions about the means of character change. Although some systematists disagree regarding the applications of cladograms, most workers agree that cladistic methodology is the best way of inferring phylogenetic relationships. These arguments will be considered again in the later section, Evolution of Specific Groups.

When discussing the role of natural selection in evolution, one must also discuss adaptation and functionality. The concept of adaptation is trivialized if it is applied to every trait of an organism that in any way makes the organism "apt," or fit, for its environment. With such a viewpoint, and with fitness defined as nothing more than continued existence and opportunity to reproduce, every organic property becomes an adaptation, and the terminology loses its usefulness. It is more correct to reserve the term adaptation for those characters that show dif-

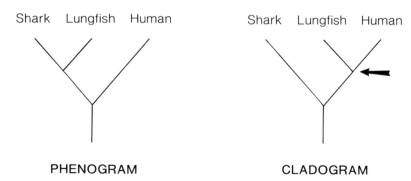

Shark   Lungfish   Human   Shark   Lungfish   Human

PHENOGRAM   **CLADOGRAM**

Fig. 26–2.   An example of the difference between grouping taxa by overall similarity of homologues (phenogram), and by special similarity of derived homologues (cladogram). Sharks and lungfish are more similar to each other than either is to a human, but it is lungfish and humans that have a common ancestor not shared with sharks (arrow).

ferences in relative functional efficiency after a period of natural selection upon a pre-existing set of organic properties. Natural selection can act only on those biologic properties that already exist; it cannot create properties in order to meet adaptational needs.[80] Not only is every organic trait with a perceivable function not necessarily an adaptation, but every trait does not necessarily have a function. Structural integrity and heritability are sufficient for a character to be retained in an evolutionary lineage. Thus, it is evolutionary retention, not functional importance, that makes a character useful for inferring phylogenetic relationships. A corollary of the preceding statement is that every phylogenetically retained character need not be selectively advantageous. To the student of parasitology, the above argument means that the question "What is this for?," asked about an attribute of a parasite, can sometimes be inappropriate or even counterproductive.

EVOLUTION OF PARASITES AND THE CAPACITY FOR PARASITISM

The parasitic organisms discussed in the preceding chapters are not, of course, all each other's closest relatives. To call an organism a parasite is to give it a functional label that identifies a condition that has evolved numerous times in various evolutionary lineages. (Examples of other functional labels that are applied to disparate groups of organisms regardless of their phylogenetic relatedness are "carnivorous" and "planktonic.") Certain evolutionary factors affect a parasite regardless of its phylogenetic history, and certain factors are phylogeny specific. In the first category are the facts that a parasite *must* have a host, and that the host is itself a living, and evolving, organism. Changes in the host's properties can produce changes in its parasites' environment. Allopatric separation of host species and populations can produce allopatric separation of parasite species and populations. Changes in the host species can produce allohospitalic and allotopic changes for the parasite species. Given these observations, it is hardly surprising that a major part of evolutionary research in parasitology concerns the coevolution of parasites and their hosts, discussed later in this chapter.

With respect to phylogeny-specific properties of parasites, the capacity of any organism to become parasitic lies in its unique history of traits accrued through evolution. Different histories provide different capacities, and the modifications of form and function that are open to, say, a free-living ameba are not the same as those open to a free-living nematode. Given all parasites' need to locate a host, to enter or attach to the host's body, and to obtain nutrition from the host, we can expect to find some degree of functional and structural equivalence among even phylogenetically distant parasites regarding penetration, attachment, absorption, and so on. Nevertheless, there is no more reason to think that such convergences will obscure phylogenetic relationships among parasitic organisms than there is to think that convergences of another sort will obscure the relationships among, say, free-living carnivores.

Neither is it true that parasites can be said to be overspecialized, and thus somehow evolutionarily limited in their potential for future change. First, there is no way of predicting the evolutionary future of a species. Second, the argument that parasites will be limited as long as they remain parasites is, as noted before, nothing more than free-living chauvinism. In fact, if the premise of this second argument were accepted, it could be said that because so many lineages of free-living organisms have evolved to be parasitic, it is the free-living organism, not the parasitic one, that is evolutionarily limited. Third, stating that parasites are overspecialized is incorrect because the functionality that attracts such a label is the result of certain structural attributes of the parasites, and structural change is the basis of evolutionary change. To argue that structural and functional novelties limit evolution is to deny a role for change itself in evolution—but what could be more "specialized" than a newly evolved species (a state that every extant species has passed through)? Think of the first amniotic egg within the vertebrate lineage—what a specialization! Within the digenean trematodes are over 400 species, constituting about 63 families, exhibiting widely derived structures, and parasitizing every major vertebrate group. This is hardly the mark of an evolutionarily limited group.

The interbreeding population of a parasite species can be considered to comprise all the parasites of that species living on or in one host individual, especially if all developmental stages of the parasites occur on the host. The interbreeding population of a parasite species can also be considered to be all those parasites on or in the entire host population, which can even be worldwide in distribution. Intermediate hosts have a different, and often wider, geographic range than the final hosts do. Hence, when intermediate hosts are involved, the effective interbreeding population of the parasite can be represented by a geographic distribution wider than that of the final host alone. These considerations become important when we study the evolution of parasites because speciation in parasites, as in free-living animals, is the division of a single interbreeding population into at least two reproductively isolated ones. Thus, different gene pools must be present if different species are to be maintained.

When we consider organic change on the basis of genetic factors alone (mutations, random recombinations, random fixation, or elimination), we realize that the genetic constitution in a group of a few interbreeding individuals changes more quickly than it does in a large number of interbreeding individuals, chiefly because of sampling error. In a large population, because more varieties exist, achieving a state of homogeneity takes longer. High rates of phylogenetic evolution can therefore occur in small populations. This rapid evolution is the causal basis of allopatric speciation in peripheral isolates. When a parasitic lineage originates, free-living species that colonize prospective hosts could form such peripheral groupings. Not only would gene flow with free-living conspecifics be lost, but the colonizers would most certainly be subjected to different selective pressures.

Ectoparasites could have evolved from free-living ancestors that fed on the body secretions of other animals. For example, in noctuid moths in Southeast Asia, some species feed on plant secretions, and others, such as *Calpe thalicteri*, pierce ripe fruit and suck the juices inside.[14] Another species, *Lobocraspis griseifusa*, gathers around the eyes of a large mammal, such as a cow, and sucks the lachrymal fluid from the orbits. From a functional perspective, we can see the possibility of evolution from a lachryphagous condition to that of the skin-piercing, bloodsucking noctuid *Calpe eustrigata*, a parasite of tapirs and other small mammals.

Endoparasites probably established their first infections in a host by entering the host's body through the alimentary tract, possibly as endocommensals of invertebrates in the host's diet. Resistant structures, such as a cyst in a protozoan or a cuticle in a nematode, would have helped the potential parasite to survive the destructive conditions of the host's gut. Although parasites of the alimentary tract are surrounded by nutrients, they are subjected to higher temperatures (in homoiotherms), higher osmotic pressures, and, as they enter the host's stomach, greatly increased hydrogen ion concentration.

Little dissolved oxygen is present in the vertebrate alimentary tract. In areas of heavy bacterial growth, such as the colon, oxygen concentration can approach zero. Near the gut mucosa and among the villi, the oxygen concentration can approach that of the blood. Whereas low oxygen content would not hinder a facultative anaerobe, such as a saprophytic protozoan, an obligate aerobic organism, relying on oxidative phosphorylation, could not survive under such conditions. Thus, either endoparasites must have evolved from ancestors that were facultative anaerobes (see the next section for a discussion of amebas recovered from the intestine of snakes), or during their evolution they developed metabolisms capable of functioning in low oxygen conditions.

The preceding discussion on the origins of parasitism primarily presents only functional assessments of the evolutionary changes required to become a parasite. A small but increasing number of well-corroborated phylogenetic hypotheses can be used to infer the temporal and ordinal sequence of the origins of primitive and derived traits in particular lineages. Even with such phylogenetic information, our knowledge of the speed, evolutionarily speaking, with which a given lineage evolved its capacity for parasitism is limited. In some cases, evolution from phoresis to commensalism to parasitism may have been gradual. In other cases, the change may have been practically a one-step transformation. As an analogous example of the latter situation, the free-living soil ameba *Naegleria* can enter the noses of mice and monkeys and then migrate to the lungs or brain, where it can cause severe lesions. These amebas can be taken from the soil, grown in culture, injected into laboratory animals, and they will show immediate pathogenicity (see Chapter 4).

Preadaptation

In the evolution of any lineage of organisms, new characters must evolve within the biologic context established by previously evolved characters. From the hindsight of a human perspective, some of these earlier characters might appear to be necessary incipient stages of the later characters, rather than the ends in themselves that they actually are at any particular point in evolutionary time. Such incipiency can be incorporated into causal explanations of how the later characters came to be. These explanations usually appeal to the notion of preadaptation, and they need to be put to rest. To quote Gould and Vrba,[43] "The concept of preadaptation has always been troubling to evolutionists. We acknowledge its necessity as the only Darwinian solution to Mivart's old taunt that 'incipient stages of useful structures' could not function as the perfected forms do (what good is 5% of a wing). The incipient stages, we argue, must have performed in a different way (thermoregulation for feathers, for example). Yet, we traditionally apologize for 'preadaptation' in our textbooks, and laboriously point out to students that we do not mean to imply foreordination, and that the word is somehow wrong (though the concept is secure)." To quote Frazzetta,[39] "The association between the word "preadaptation" and dubious teleology still lingers, and I can often produce a wave of nausea in some evolutionary biologists when I use the word unless I am quick to say what I mean by it."

The problem with attributions of preadaptation is that the term makes no sense. As Gould and Vrba note, the concept is not secure. As discussed above, adaptation in a character is the result of selection for relative functional efficiency among a pre-existing set of organic properties. Adaptation cannot exist before selection acts, and organisms cannot become "pre" adapted to that which does not yet exist. Thus, the term preadaptation is mistaken because (1) it is offered to explain the existence of previous biologic conditions when it is, in fact, simply an after-the-fact label—a label that appeals to a nonexistent process, and (2) it compounds the error of assuming that every trait that has any sort of biologic function must be an adaptation. (See Gould and Vrba[43] for further discussion.)

The sorts of traits that attract the preadaptation label in parasitology are those that have given to some of the members of an evolutionary lineage the capability to initialize or to continue a parasitic existence. These characters can be morphologic or physiologic. For example, Julian Hawes, late of the University of Exeter,

England, once found amebas in the intestine of a snake. The method of pseudopodial formation and the nuclear structure of the "parasites" were similar to free-living *limax* amebas of the soil. The amebas were tentatively considered to be forms picked up with the snake's food, but when they were cultivated in vitro, they grew equally well in anaerobic and in aerobic media. If certain free-living soil amebas are able to grow and to develop in a medium without free oxygen, then those soil forms have the potential to live in the intestine of a vertebrate animal, at least so far as free oxygen requirements are concerned. Also, if they are capable of surviving with the nutritional conditions found in the vertebrate intestine, then they are capable of surviving as parasites.

Another example of the potential for parasitism to become the actuality of parasitism might be that of dauer larvae in secernentean (phasmidian) nematodes. Anderson[5] writes, "Maupas [in the 19th century] noted that rhabditid nematodes [free-living soil dwellers] in culture passed through 5 stages separated by 4 moults and that when cultures became exhausted all stages died except those at the beginning of the third stage. These latter retained the cuticle of the second stage. The larvae, now known as dauer larvae, initiated new populations when placed in fresh cultures. Maupas noted that dauer larvae can survive for long periods of food scarcity and dry conditions. Maupas suspected all nematodes went through 4 moults and subsequent work has shown this to be true even of the parasitic forms. Special significance can also be attached to the dauer larva of the rhabditids which is . . . adapted to survive adverse conditions . . . . In the secernenteans it is generally the early third stage which is the infective or invasive stage. Chabaud related this rule to the origin of secernenteans from free-living rhabditids similar to those studied by Maupas."

Any study of the evolution of a particular trait must be based on a well-corroborated hypothesis of the phylogenetic relationships among the organisms concerned. This is necessary to avoid just-so stories of what might have been the path of evolution. Potentiality is not necessarily actuality, and there is no limit to the number of "believable" evolutionary scenarios.

Figure 26–1 is a cladogram for the parasitic flatworms, from a study by Brooks et al.[22]; it can be used to study the evolution of at least two traits that could attract the label of preadaptation. First, fingerlike projections of the tegumental surface, called microvilli, are especially well-developed in those parasitic platyhelminths lacking a gut—gyrocotylids, amphilinids, and cestodes. This has led to explanations that such structures evolved in order to provide an alternative nutrient absorptive surface. But microvilli are present, to one degree or another, in all free-living and parasitic platyhelminths. Although they undoubtedly played a part in the transition from being first a free-living animal to a parasite with a gut (the aspidogastrids, digeneans, and monogeneans), and then to a parasite without a gut, their existence is not to be explained with reference to their function. In adult cestodes, microvilli have become modified into microtriches, which are microvilli with electron-dense caps, and into rostellar hooks on the scolices of some species.[77]

Another interesting trait of parasitic flatworms is the nerve commissure, more or less a nerve ring, in the anterior of the worm's body. Digeneans and aspidogastrids have a single cerebral commissure, while monogeneans, gyrocotylids, amphilinids, and cestodes have a double commissure. Digeneans and aspidogastrids have oral suckers and pharynges that operate with the single commissure, indicating that the second commissure in monogeneans cannot be an adaptation for oral function. In tapeworms, both commissures are intimately involved in the operation of the hooks and suckers on the scolex, but the second commissure is postulated to have evolved earlier, in the Monogenea, and thus before either the loss of the gut or the modifications of the mouth and the anterior end into a holdfast organ. If we eschew labels of adaptation or preadaptation, it can be said that the doubled cerebral commissure evolved without affecting oral function (i.e., a functional equivalent of the single cerebral commissure) of ancestral cercomeromorphs, and was later incorporated into the operation of the scolex when tapeworms evolved.

Biologic Races

A population of a species can acquire unique traits, often physiologic, that, while not distinguishing it morphologically or interfering with reproductive interactions with conspecifics, allow the members of the population to function slightly differently from other conspecifics. These groups are called biologic, or physiologic, races. The isolation experienced by a race might eventually lead to morphologic changes and the establishment of a new species, "a single lineage of ancestor-descendant populations which maintains its identity from other such lineages and which has its own evolutionary tendencies and historical fate" (Wiley[100]).

Consider an ectoparasite, A, that lives in the dorsal body hair of a mammal. Assume that a population of A begins to find greater reproductive success in the ventral part of the host's head, and that their descendants become increasingly adapted to this habitat, while simultaneously losing, or at least attenuating, their ability to live in the previous habitat. Call this divergent group B. We find that A and B are morphologically indistinguishable, but physiologically different. Groups A and B are races of the same species.

Advances in chemotaxonomy, serotaxonomy, and molecular taxonomy are increasing researchers' ability to discern the structural traits that produce the functional differences among races. To the extent that these traits are fixed and heritable, they can satisfy Wiley's criteria for an evolutionary species, given above. To the extent that the traits can be "washed out" through breeding with other races or are not heritable and must be selected in each generation, the race that they typify falls short of species status. In other words, gross morphologic differences are not necessary for the establishment of a new species, but fixed, heritable, and distinguishing traits are.

When studying species and races of any organism it is essential to have a sample size sufficient to estimate the degree of intraspecific variation (i.e., a "horizontal" assessment), and time sufficient to observe the traits exhibited by the organisms at different periods during their ontogeny (i.e., a "vertical" assessment of their *holomorphology*; see Hennig[47] and Wiley[100]). Parasite studies are often weak in both of these requirements. In parasitology the distinction between species and races is made all the more challenging by the difficulty of collecting and rearing large numbers of specimens, many of which belong to parasite species that not only have morphologically different larval stages, but show variation in certain characters when reared in different hosts or environmental conditions.

Trypanosoma brucei, Trichinella spiralis, and *Entamoeba histolytica,* all common and easily obtained, show intraspecific differences in host preference and pathogenicity. No morphological traits appear to be associated with each of the races, and identification must rely on physiologic, and sometimes molecular, differences. The mange-mites of the species *Sarcoptes scabiei* are probably divided into biologic races, each developing on a single host species—for example, sheep, dogs, goats, camels, horses, rabbits, and humans. Host races are found in the tapeworm *Hymenolepis nana*, of humans and rodents, and in the nematode *Ancylostoma caninum*, the hookworm of dogs.

Progressive and Regressive Evolution

The practice of terming some evolutionary changes "progressive" and others "regressive," especially prevalent in parasitology, is another example of the intrusion of anthropocentrism into evolutionary studies. Progressive changes are said to be those that involve the addition of characters or increased complexity of characters, while regressive changes are said to be those that involve loss or reduction. Examples of "progressive" evolution in parasitic lineages are the hooks of tapeworms and acanthocephalans; the acetabula, tribocytic organs, and multiplicative larval stages of digeneans; the posterior clamps and suckers of monogeneans; and the teeth and cutting plates of nematodes. Examples of "regressive" evolution in parasitic lineages are the loss of a gut in gyrocotylids, amphilinids, tapeworms, and acanthocephalans; the loss of wings—and sometimes eyes—in fleas and lice; and the reduction of amphidial complexity in nematodes.

Progression and regression are linked with Darwinian concepts of specialization and improved efficiency of function, as well as with general 19th century thoughts of evolutionary progress. This attitude is inapplicable to evolutionary studies not only because of its anthropocentric view of what is and is not a worthwhile evolutionary change, but also because it is internally inconsistent. With respect to the second shortcoming, the increased complexity of an existing character or the addition of new character is not a requisite for increased functional efficiency—loss or reduction could have the same effect (e.g., the evolution of feathers in birds and its contribution to flight aerodynamics, versus the reduction of body hair in whales and its contribution to swimming hydrodynamics). In addition, the categorization of

a change as a loss or a gain often depends on the descriptive language we bring to the subject, rather than on some more objective assessment. For example, is pseudogamy in the nematode genus *Strongyloides* (Chapter 14) the gain of a trait, or is it the loss of the ability of the sperm to successfully penetrate the oocyte?

One can speak of traits as appearing for the first time in evolution, becoming more complex, becoming reduced, being lost. One can also speak of the effect or lack of effect that such character changes have had on the functionality of a group of organisms. Nevertheless, an evolutionary change is an evolutionary change; one cannot categorize the changes as progressive or regressive. The rhizocephalan crustacean parasite of crabs, *Sacculina*, can at first appear to exhibit massive regressive evolution in the loss of so many "crustacean" characters in the adult stage, such as a mouth, gut, appendages, and segmentation. Yet these animals possess a number of more-recently evolved characters that are associated with their parasitism of the crab, such as the kentrogen larva and its injection tube, and the sacculina externa and interna parts of the adult. At a less-inclusive phylogenetic level within the Crustacea, these latter characters are just as "crustacean"as any others in the clade.

The loss or reduction of physical attributes is hardly restricted to those organisms that we call parasites. One need only consider dermal bones in tetrapods, limbs in snakes, wings in kiwis, pelvic appendages in whales, digits in horses—and body hair, olfactory lobes, the coccyx, and the appendix in humans.

COEVOLUTON OF PARASITES AND HOSTS

Chapter 24 examined the corroboration between studies of biogeographic relationships and parasite-host coevolution. Chapter 25 examined some of the factors that affect the ability of a parasite to infect, and thus have the opportunity to remain historically associated with, a particular host. This chapter discusses coevolution more closely.

Obviously, because parasites did not evolve *de novo* in their hosts, *every* parasite lineage must have colonized a host lineage at some time. The question of interest is the extent to which the parasites went on to coevolve with their hosts, and the extent to which they went on to colonize other hosts that were distantly related but ecologically associated. These two conditions are not mutually exclusive. First, a parasite-host system can show periods of colonization intercalated with periods of coevolution with the colonized hosts. Second, with sufficiently broad coaccommodation (see Chapter 25), a parasite lineage can colonize a number of distantly related hosts, but coevolve with only some of them. When a species of parasite colonizes a new host species, it does not necessarily cease being a parasite of the previous host species. In other words, although narrow coaccommodation (high host specificity) might make it easier for a researcher to discern the coevolutionary associations, it is not a requisite for parasite-host coevolution.

Parasite-host coevolution can be completely *congruent*, viz., both organisms have been similarly affected by the same allopatric events, and the parasites have speciated every time that the hosts have speciated. In such a case, their phylogenetic trees have the same shape (the closest relative of a parasite is always in the closest relative of that parasite's host). Alternatively, the parasites might have speciated at a different frequency than the hosts. If parasite speciation has been more frequent than speciation of the hosts, then a host taxon will harbor more than one parasite taxon, with the parasite taxa being each other's closest relatives. Parasite speciation of this sort can come about through allohospitalic or allotopic separation of populations (see Chapter 25). If parasite speciation has been less frequent than that of the hosts, then closely related host taxa will all harbor a particular parasite taxon. In either case, even though the parasite and host phylogenies are not completely congruent, they are still *consistent* with each other, in that the phylogenetic inferences drawn from one do not contradict those drawn from the other, as is the case with colonization.

Some workers[94] have argued that parasites—gut parasites in particular—should be expected to speciate less frequently than their hosts because the environmental milieu of the parasites is more homogeneous and evolutionarily conservative than that of the host. Little evidence, however, supports this statement as a general principle. In a single vertebrate host individual, along the length and across the diameter of the intestine are great differences in oxygen and carbon dioxide tension, pH, and other chemical properties. These differences can occur even between the base and tip of a villus. In addition, villar topology plays a large role in the ability of different species of parasites to attach to the gut wall (see Chapter 25).

With respect to interspecific differences among the guts of vertebrates, Mitchell[74] was able to use intestinal characters as the basis for a phylogeny of birds. Schad[91] found eight species of the nematode *Tachygonetria* in the large intestine of the turtle *Testudo graeca*. Caira[26] found five species of the tapeworm *Pedibothrium* in the spiral valve of the nurse shark. These five species, which form a monophyletic group, appeared to be located syntopically (see Chapter 25) in the gut. Price[86] observed that there are more parasitic than free-living taxa in existence, and concluded that parasites exhibit very high speciation rates. Jones[59] stated, ''First, instead of lagging behind their host's evolutionary advance, many parasites seem to have overtaken their hosts . . . the acquisition by ancient and primitive organisms (such as protozoa or flatworms) of a recent host (such as bird or mammal) is evidence of evolutionary opportunism . . . Second, parasite life cycles are extremely ingenious . . . it appears that only part of the parasite's life is spent in a nonstressful environment and that much of the life cycle involves highly precarious transitions, which require great adaptability of the successful parasite . . . Third, theoretical considerations of form, population structure, and environment of parasites do not necessarily support the view that parasites are conservative . . . Fourth, the effect of a stable, rich environment upon parasites in their definitive hosts may not be so conservative as the earlier discussion had suggested . . . perhaps the environmental factors which affect variation are neither positive nor negative in the enriched milieu, and the effect of such an environment upon evolution is actually not significant.''

Another property of parasites that has sometimes been incorporated into arguments for their high speciation rate is that of reproductive ''strategy.'' The teleology of this term is regrettable, for the strategy is simply the inheritance of certain patterns of reproduction. Selection might have played a secondary role in these patterns. Two categories of reproductive strategy are recognized in ecologic studies: the *r-strategy* and the *K-strategy*. Species showing r-strategy have high fecundity, high mortality, short lifespan, and variable population size over time, usually below the carrying capacity of the environment.[85] K-strategists show the opposite characteristics. Populations of r-strategists are affected mostly by density-independent factors, while populations of K-strategists are affected mostly by density-dependent factors, such as resource limitations.

Most parasites are said to fit an r-strategist profile.[38,41,58] However, Keymer[61] noted certain features of helminth population structure that appear to be K-type density-dependent, such as low parasite mortality and fecundity, over-dispersion, parasite-induced host mortality, and transmission by predator-prey links. The weaknesses with the r-K concept are threefold. First, there seems to be an r-K continuum along which species fall, rather than a mutually exclusive dichotomy.[38] Second, the concept is a relative one, in that species B can be said to be an r-strategist with respect to species A, but a K-strategist with respect to species C.[85] Third, reproductive traits have not necessarily been acted on by selection, and selection is incapable of selecting for reproductive traits that do not already exist in a species.

To a certain extent, any discussion of the evolutionary impact of r and K reproduction, and of rates of speciation, is rendered moot by the fact of a large amount of congruent or consistent coevolution among parasites and their hosts—organisms with usually very different reproductive patterns—and among the biogeographic relationships of both. Brooks[18-20] stated, "the phylogenies of many and perhaps most hosts and parasites are bound together in highly predictable patterns, but the shaping of these patterns does not require competition, dispersal (colonization), or natural selection—the Darwinian trinity. In fact, the formalizations I have presented deny much of a role for any of those processes in parasite phylogenesis, parasite faunal evolution, or host-parasite coevolution." Also, "Coincidence of a consistent set of parasite relationships with vicariant geographical distributions, such that parasites [and hosts] and geography have coevolved, would support a hypothesis of host-parasite co-speciation. The alternative would be that the parasites had evolved without their current hosts; that is, they evolved in one set of hosts, then colonized randomly their present hosts while their original hosts either became extinct or simply lost their original parasite fauna. Such a scenario seems rather unlikely for even a single parasite taxon, but approaches near impossibility when an entire parasite fauna is considered."

Price[87] stated that one of four major factors on which extensive adaptive radiation depends is the selective pressure for coevolutionary modification, that is, for specialization. He defends the thesis that "parasite communities are in an early state of development. Noninteractive niche exploitation is common because resources remain underutilized since species have not evolved to use them." Holmes[51] has maintained that the majority of parasite communities are not young, because through biotic interactions the parasites have evolved discrete niches. He concluded that parasite competition results in coevolution. Price denied that coevolution plays a significant role. (See Holmes and Price,[52] and the reply of Brooks and Wiley.[24]) The latter study argues that historical factors, such as degree of phylogenetic relationship and the retention of ancestral life history traits, should be given precedence in explanations of parasite community structure.

Generalizations About Parasite-Host Associations

One outcome of coevolutionary studies in parasitology has been the formulation of certain inductive generalizations about parasite-host coevolution. One of these generalizations is that over evolutionary time, parasites and their hosts will become coadapted such that the pathogenesis of the parasite is minimized and its reproductive output is maximized.[33] The concept of coadaptation between an "attacking" parasite and a "defending" host is a central point of studies in symbiont-induced speciation,[96] reciprocal adaptation,[35] and resource-tracking[60] (see the critique by Brooks[20]). But there have been criticisms of the minimal-pathogenesis criterion. Anderson and May[7] pointed out that a minimization of parasite pathogenicity is not necessarily a requisite for maximal parasite reproductive output. They wrote, "the coevolutionary trajectory followed by any particular host-parasite association will ultimately depend on the way the virulence and the production of transmission stages of the parasite are linked together: depending on the specifics of this linkage, the coevolutionary course can be toward essentially zero virulence, or to very high virulence, or to some intermediate grade."

Ball[11] surveyed instances in which the degree of pathogenicity would be a poor guide for determining the evolutionary time relationships between host and parasite. He emphasized that a high degree of pathogenicity is not *"prima facie* evidence for recent and still imperfect development of the host-parasite relation."* Many factors beyond the mere length of association play significant roles in determining the incidence and degree of pathogenicity of parasitic infections in their hosts. Nevertheless, Humphrey-Smith[54] has argued that the greatest degree of parasite-host coevolution is to be found in cases in which the parasites show minimal pathogenesis and maximal host specificity.

Another generality about parasite-host associations is that in cases of heteroxenous life cycles (involving more than one host; see Chapter 25), it is the intermediate host—particularly if that host is an invertebrate—that is the ancestral host from a monoxenous life cycle. This statement, however, is not universally supported when individual parasite lineages are examined in a phylogenetic context. Anderson,[5] in a review article, presented evidence that vertebrates are the ancestral single hosts of parasitic nematodes. He also identified nematode groups in which such *primary monoxeny* has evolved into heteroxeny and subsequent *secondary monoxeny* (viz., the loss of an intermediate host, as is apparently the case in *Ascaris*).

Shoop[92] and O'Grady[79] both used cladistic analysis to conclude that in the Digenea, molluscs are the ancestral hosts, having served as the single host of the sister platyhelminth taxon, the Aspidogastrea (see Fig. 26–1). (Sister taxa are two taxa in a phylogenetic tree that are each other's closest relatives, descended from a common ancestor.) Shoop also examined the evolution of different types of heteroxenous life cycles within the Digenea, including that involving transmammary transmission in *Alaria marcianae*. O'Grady, in addition, concluded that in the Cestoda, it is the vertebrate host that is the ancestral single host. Thus, despite apparent similarities (invertebrate intermediate host, vertebrate definitive host), the heteroxenic life cycles displayed by the two platyhelminth groups, the Digenea and the Cestoda, seem to have arisen through different evolutionary pathways.

Of all the generalizations about parasite-host associations, the best known are the so-called

Rules of Affinity.[18,36,57] As useful as these rules are for *a posteriori* categorization of particular instances of coevolution, sufficient exceptions render them inapplicable as inductive universals.

Farenholz's Rule. Strict version: Parasite phylogeny mirrors host phylogeny. *Relaxed version*: Ancestors of present-day parasites were themselves parasites of the ancestors of present-day hosts. Degrees of relationships among present-day parasites thus provide clues to the phylogenetic relationships among present-day hosts.

Szidat's Rule. The more primitive the host, the more primitive the parasites it harbors.

Eichler's Rule. The more genera of parasites a host harbors, the larger is the systematic group to which the host belongs.

Farenholz's and Szidat's Rules come from considerations of host specificity and concomitant parasite-host phylogenesis. Eichler's Rule is different from the other two, for it does not specify that the genera of parasites harbored by a host must be closely related. Rather, it states that the greater the number and diversity of hosts involved (and, implicitly, the greater the amount of evolutionary time involved), the greater will be the number and diversity of parasites. Eichler's Rule can be compared to the MacArthur-Wilson[70] model of island biogeography, in which the size of an island is positively correlated with the number of species it harbors.

Adherence to the strict version of Farenholz's Rule, as noted by Lyal,[69] can produce some curious twists: "For much of this century louse systematists have, implicitly or explicitly, followed the axiom that louse and host phylogenies are topologically identical . . . . The desire to produce "natural" classifications and a perhaps undue reverence for existing classifications of the vertebrate hosts led to a very close correspondence between host and parasite classifications. This in turn led to a reinforcement and general acceptance of the principles of Farenholz's Rule, and in some cases the use of louse relationships, themselves based on supposed host relationships, to support the original classification of the hosts . . . ."

Brooks[18] noted that the strict version of Farenholz's Rule applies only when there has been narrow coaccommodation and cospeciation, and that Szidat's Rule applies when there has been broad coaccommodation and cospeciation. Eichler's Rule will apply only when the host and parasite taxa show sufficiently broad coaccommodation to allow parasitism by a broad range of parasites.

The relaxed version of Farenholz's Rule, that information about the phylogenetic relationships among present-day parasites provides clues to the phylogenetic relationships among present-day hosts, would appear to be the only tenable generalization about parasite-host evolutionary associations. To be of maximal utility, parasite and host phylogenies must be constructed from independent data sets. As discussed in Chapter 24, parasite data have been used by systematists since the 19th century to provide clues to host relationships.

Coevolutionary Studies

Humphrey-Smith[54] found a large amount of coevolutionary association between Australian marsupials and their parasites, particularly herpetostrongylin nematodes, atopomelid mites, stephanocircid fleas, and amblyceran lice. He concluded, "Although a certain amount of coincidence of form can be attributed to convergence, the dramatic number of similarities shared by parasites of Marsupiala in Australia and South America is strongly supportive of there being common ancestry for both the host and the parasite groups."

Coevolutionary studies can also be conducted on parasites and their *intermediate* hosts. Digenean trematodes are particularly suited for such a study because of their almost exclusive use of gastropod molluscs as intermediate hosts. Davis[31] examined coevolution between digeneans of the genus *Schistosoma* and their snail hosts and concluded, in part, that (1) *Schistosoma* and the relevant snail faunas evolved in Gondwanaland prior to the breakup of Pangaea; (2) coevolution of *Schistosoma* and snail hosts involved reciprocal selective pressures affecting the course of evolution of both parasite and snail; (3) specificity of the snail-parasite interaction increased through time; and (4) relevant snail faunas evolved more rapidly than *Schistosoma*.

Over the last 15 years or so, a number of studies on parasite-host coevolution have explicitly used cladistic methods to reconstruct and compare phylogenetic relationships. Klassen and Beverley-Burton[64] studied the ancyrophalid monogenean parasites of ictalurid fishes

(catfishes and associated species). They concluded that two thirds of the relationships were coevolutionary, and the remainder were due to parasite transfers among hosts that were ecologically associated but were not each other's closest relatives. Benz and Deets[15] examined some of the cecropid copepod ectoparasites of epipelagic fishes and concluded that within the cecropid clade has occurred colonization from mobulid rays to luvars and the ocean sunfish, *Mola mola,* and subsequent coevolution, through at least two speciation events, with *M. mola* (Fig. 26–3).

In another study, Deets[32] produced a cladogram for some of the kroyerid copepod ectoparasites of chondrichthyans (ratfish, skates, rays, and sharks) and found that the host relationships indicated when the hosts were mapped onto the tree were highly congruent with recently proposed hypotheses of chondrichthyan phylogeny (Fig. 26–4). Specifically, holocephalans (ratfish), represented in the study by *Callorhynchus callorhynchus,* are placed as a primitive taxon closely related to elasmobranchs; and within the elasmobranchs, separate lineages are resolved for skates and rays, and for sharks. Deets linked this pattern with kroyerids that colonized the nasal lamellae of their hosts (from a primitive condition of parasitizing the gill lamellae), and then exhibited coevolution at that site.

Among nematodes, coevolution has been hypothesized between pinworms of the genus *Enterobius* and their primate hosts,[42,75] and between ascaridoids and their crocodilian hosts.[21] Figure 26–5 presents cladograms from the Glen and Brooks[42] study on *Enterobius*, a parasite possessing high host specificity and a direct life cycle. Higher-level (viz., at a more inclusive taxonomic level) congruence of the trees is indicated, for example, by the fact that the pinworm of *Lemur*, which is the sister group to the other primates, is placed as the sister group to the other species of *Enterobius*. However, lower-level incongruence occurs, such as that of the phylogenies of the great apes and their parasites.

The work of Hafner and Nadler[44] on the lice of gophers is of particular interest because of its exclusive use of biochemical data to infer parasite and host phylogenetic relationships. (For general reading on the use of biochemical and molecular data in systematics, see Patterson[83] and Hillis.[48]) Based on an electro-

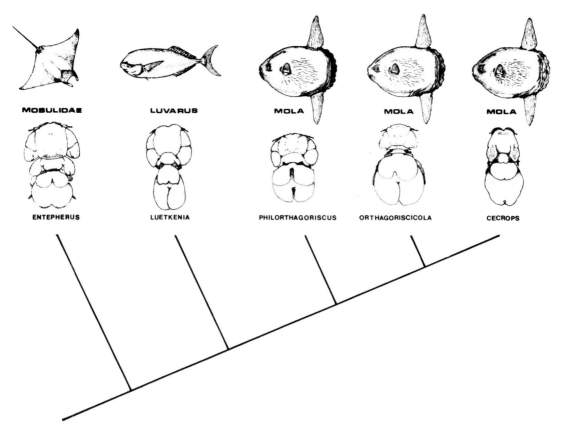

Fig. 26–3. A cladogram of the phylogenetic relationships among cecropid copepod ectoparasites of fishes, with the hosts mapped onto the tree (from Benz and Deets[15]). The results indicate colonization from the Mobulidae to *Luvarus* and *Mola*, followed by coevolution with *Mola*. (From Benz, G.M., and Deets, G.B.: Can. J. Zool., in press.)

phoretic study of a number of different protein loci in parasites and hosts, Hafner and Nadler[44] found congruence between the phylogenies of trichodectid chewing lice and their geomyid pocket gopher hosts (Fig. 26–6). The authors stated:

"The life cycle of these wingless insects occurs entirely on the host and includes three principal stages: egg, nymph, and adult. Chewing lice have a generation time of approximately 40 days, which is roughly one-fifth the generation time of pocket gophers. Transmission of chewing lice among pocket gophers is thought to occur only through host-to-host contact, and the combination of low parasite vagility and obligate contact-transmission of lice should limit opportunities for colonization of new host species. The absence of widespread transfer of lice among host species should, in turn, increase the likelihood of detecting cospeciation in this host-parasite assemblage.

The host and parasite trees are topologically identical in all but three regions [indicated by daggers in Fig. 26–6]. In most cases, sister taxa of lice parasitize hosts that are also sister taxa (see nodes A and F), and branching sequences above the species level in the two groups are identical (nodes B-E). The probability of this level of topological similarity occurring by chance alone was calculated [to be] remote (P = 0.001). This high degree of tree matching is consistent with predictions of the cospeciation hypothesis.

The three host-parasite associations indicated by asterisks [Fig. 26–6] probably result from host switching (lateral transfer) by lice. In these cases, the phylogenetic history of the louse taxon does not mirror that of its host, thereby falsifying the cospeciation hypothesis for these (and only these) associations. It is important to note that no case of suspected host switching involves hosts that are geographically disjunct; in each case, the geographical range of the colonizer's current host abuts that of the colonizer's pu-

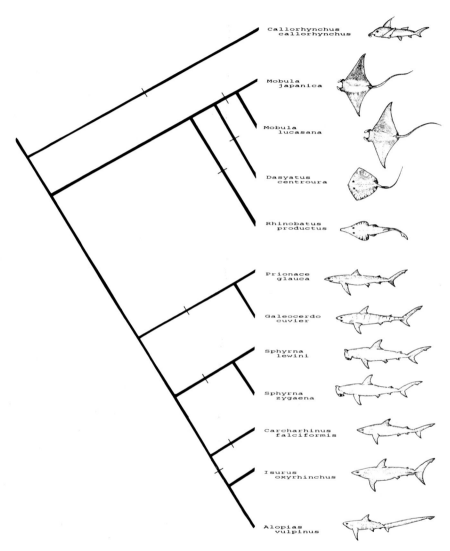

Fig. 26–4. A cladogram of the phylogenetic relationships among kroyerid copepod ectoparasites of fishes, with the hosts (and their binomens) mapped onto the tree (from Deets[32]). The results place the Holocephala (represented by *Callorhynchus callorhynchus*) as the plesiomorphic sister group to the elasmobranchs, and resolve separate monophyletic lineages for skates and rays, and for sharks. (From Deets, G.B.: Can. J. Zool., 65:2121–2148, 1987.)

Enterobius species   Primate hosts

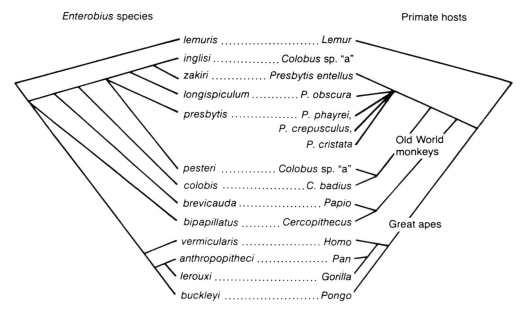

Fig. 26–5. Cladograms of the phylogenetic relationships among species of *Enterobius* pinworms, and of the relationships among their primate hosts (from Mitter and Brooks[75]). The degree to which the topologies of the trees match is the degree to which coevolution of parasites and hosts can be postulated. (From Mitter, C., and Brooks, D.R.: Phylogenetic aspects of coevolution. *In* Coevolution. Edited by D.J. Futuyma and M. Slatkin. Sunderland, MA, Sinauer Associates, 1983, pp. 65–98.)

tative ancestral host (that is, the pocket gopher parasitized by the colonizer's sister taxon)."

Hafner and Nadler also reported that estimates of the rate of protein evolution in each phylogeny suggested that the cospeciation of parasites and hosts was roughly contemporaneous, with some of the parasite speciation events lagging behind those of the hosts. Lyal,[69] in a cladistic study of a broader range of mammalian hosts and their trichodectid lice, concluded that approximately 79% of the parasite-host occurrences could be explained by coevolutionary associations. The remainder must be explained by invoking secondary infestations of distantly related hosts.

The ability to discover coevolutionary patterns in parasite-host associations indicates that both classes of organisms show a phylogenetic retention of evolved traits that is sufficient to allow reconstruction of their historical relationships. This historical persistence appears to occur, to some degree, regardless of the relative frequency of parasite-host speciation, the complexity of the parasite life cycle, or the breadth of the parasite's coaccommodation with the hosts. The ecologic diversification that accom-

panies colonization disrupts coevolutionary patterns, but does not obscure them. It would seem, overall, that ecologic diversification lags behind the diversification of morphologic, biochemical, physiologic, and other heritable structural characters that make phylogenetic reconstructions possible. Ross[89] estimated that one ecologic shift occurs for approximately every 30 speciation events in phytophagous insects. Boucot[16] suggested that this asymmetry is characteristic of the fossil record. Brooks et al.[23] found that ecologic data resolved less than one third of a cladogram constructed from morphologic data for the digenean trematodes.

EVOLUTION OF SPECIFIC GROUPS

Systematics is an extremely active research field, constantly being expanded with new methods for reconstructing evolutionary trees, and with new techniques of data collection. The apparently dry activity of "alpha-level" taxonomy, the description and naming of species, is but the tip of an iceberg of phylogenetic studies and investigations into the distant evolutionary history of life on earth. Alpha-level

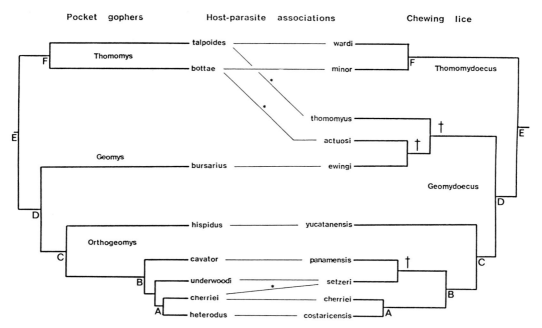

Fig. 26–6. Cladograms of the phylogenetic relationships among species of trichodectid chewing lice, and of the relationships among their geomyid pocket gopher hosts; from Hafner and Nadler.[44] The parasite and host trees are topologically identical in all but three regions, marked by daggers. Asterisks indicate postulated cases of host switching. (Reprinted by permission from *Nature*, Vol. 332, pp. 258–259. Copyright (c) 1988 Macmillan Journals Limited.)

studies provide the character data that are necessary for phylogenetic studies at a more inclusive taxonomic level; it is when organisms are placed in their phylogenetic context that lists of characters become a part of an investigation into the patterns and processes of homologous and analogous relationships in evolution.

Methods among the studies given below differ. Some have used only a few, assumably "key" characters in their reconstructions; others have not made sufficient distinction between similarity produced by shared primitive homologues and similarity produced by shared derived homologues (see the previous section, Phylogenetic Reconstruction). All of the studies, however, offer at the very least new character data or new interpretations of previously known characters, as well as hypotheses of relationships that can be tested with additional data and new analytic procedures. Additional writings on the evolution of parasites can be found in Taylor.[95] Morris[75] reviews parasites in the fossil record.

Protozoa

Parasitism in the phylum Protozoa occurs in some orders containing free-living species and probably arose independently from different groups of free-living ancestors. Much work needs to be done on the construction of well-corroborated phylogenetic hypotheses for the protozoa.

Hanson[45] made a study of ideas concerning phylogenetic relationships, particularly among animal protists, and developed a rationale based on the determination of structural and functional homologous kinships. The study led to the conclusion that although both ameboid and kinetid (zooflagellates and ciliates) forms were derived from apochlorotic algae, the pseudopodial forms show little further evolution, whereas the kinetids give evidence of extended evolutionary development leading to metazoan animals.

Entozoic protozoa may have been derived from ectoparasites, as various authors have suggested. It does seem logical to suppose that free-living forms first became associated with hosts as casual commensals loosely attached to the skin or gills, and then gradually moved into the mouth, gill chambers, anus, and other openings. However, ectozoic forms are mostly primitive ciliates and flagellates, and only a few genera such as *Trichodina* and *Hexamita* contain both ectozoic and entozoic species. Another logical guess as to origins of these parasites is that they were derived from species accidentally ingested by their future hosts. Consider the many protozoa that are steadily ingested with the food of larger animals. Once established in the intestine, the parasites could migrate to all other parts of the body.

The Sarcomastigophora (ameboflagellates) probably are the least-changed modern descendants of the original parasitic protozoa, and from this assemblage, according to Baker's theories,[10] specialization possibly proceeded in three directions: to true flagellates, to amebas, and to sporozoa. The ciliates probably arose from the flagellates.

Wenrich[99] felt that a study of protozoa exhibiting behavior and habitats intermediate between free-living and parasitic species would provide a clue to the question of origins. Such intermediate forms are common. A ciliate, *Amphileptus branchiarum*, found on the gills of tadpoles, has a free-swimming stage that eats ectozoic protozoa. At other times, and more frequently, it attaches itself to the gills and engulfs masses of gill cells. Some free-living euglenoids of the genus *Menoidium* were fed to the milliped, *Spirobolus marginatus*, and displayed the capacity for maintaining an entozoic existence, for at least a few days. *Tetrahymena* is occasionally found in such sites as the digestive tracts of slugs, the coelom of sea urchins, the hemocoel of insects, and the gills of crustaceans.

The evolution of the family Trypanosomatidae has stimulated much speculation. Baker[10] traced primitive *Leptomonas* leading to *Leptomonas* in annelids, then progressing through (1) the epimastigote line to the insect-transmitted *Trypanosoma* and trypanosomes of insects, and another line to the *Glossina*-transmitted and annelid-transmitted trypanosomes; and (2) the promastigote line leading to *Phytomonas, Leptomonas,* and *Leishmania,* and another line to *Trypanosoma cruzi* and *Endotrypanum.*

Wallace[97] began his hypothesis with free-living, *Bodo*-like flagellates that moved into the vertebrate gut, then moved into tissues, and were finally transmitted by bloodsucking arthopods. These blood forms gave rise to two lines: the promastigote line, leading to *Leishmania* and others, and the epimastigote line, leading to *Trypanosoma* and others. Another theory[101] is that mammalian trypanosomes recently originated from leech-transmitted parasites of aquatic reptiles. This ancestral promastigote probably led on one hand to the genera *Leptomonas* and *Phytomonas,* and on the other hand to the genera *Trypanosoma, Rhychoidomonas* (including "*Herpetomonas*" as used by Wenyon), *Blastocrithidia,* and *Crithidia.*

Many years ago, Hoare[49] suggested that inoculative transmission and its attendant form of parasitic life cycle, as in *Trypanosoma gambiense,* may be a secondary acquisition that originally developed in the hindgut of the insect vector. Such trypanosomes may have been taken up by tsetse flies and transmitted mechanically to new vertebrate hosts; when the flagellates became capable of developing in the proboscis and/or salivary glands, however, tsetse flies became their new obligatory transport hosts. Evidence for this hypothesis is presented in *T. vivax,* which develops only in the mouthparts of its insect vector, and by *T. congolense* (representing the next step in evolution), which develops in the midgut of the insect, and finally by the *Brucei* group, which uses the midgut and then the salivary glands of the tsetse fly.

A final bit of data on the described phylogenetic relations among trypanosomes is provided by the differences in susceptibility of their vectors to infection. Practically 100% of triatomid bugs fed *T. cruzi* become infected, whereas fewer than 1% of tsetse flies fed *T. brucei* become infected. One can hypothesize from this observation that *T. cruzi* and its bugs represent an older and more stable association than the *T. brucei* system. Moreover, tsetse-borne trypanosomes easily lose the power to develop in the insect host and may revert to being transmitted mechanically. Such a transformation is illustrated by *T. vivax* in cattle of South America and by *T. evansi,* which presumably originated in Africa from *T. brucei* (see Chapter 3). A final step in the evolutionary series is *T. equiperdum,* which has become completely emancipated from an insect vector and is transmitted directly from horse to horse during sexual contact.

Lumsden[68] reviewed some of the literature on kinetoplastida phylogeny, paying particular attention to the similarities and differences between the diseases produced by *Leishmania* and *Trypanosoma*. The former is a homogeneous genus, whereas *Trypanosoma* is not. The subgenus *Schizoptypanum*, for example *T. (S.) cruzi*, shares with *Leishmania* the ability to multiply within host cells: this ability thereby suggests some phylogenetic affinity between them.

The large and complex group of flagellates that inhabit the intestine of termites evolved from the simple *Monocercomonas* or from a *Monocercomonas*-like form. This form has an uncomplicated parabasal body and axostyle, three free flagella, and an adherent or free-trailing flagellum. The Trichomonadidae have added a costa and an undulating membrane in place of the recurrent flagellum. *Devescovina* is similar to *Monocercomonas*, but it possesses a triangular cresta, and its parabasal body is coiled around the axostyle. The Calonymphidae are derived from the Devescovinidae. For detailed discussion of these relationships, see Hollande.[50]

Sarcodinians presumably arose from the Sarcomastigophora by a loss of the flagellar stage, and they appear to represent a polyphyletic group (viz., not all are descended from a common ancestor). Numerous examples of ameboid flagellates lend ample support to these conclusions. *Tetramitus* has ameboid stages, but it is usually classified as a flagellate. A line of evolution through this type of protozoan has led to the ameba *Vahlkampfia*. Another ameba, *Naegleria* (see Evolution of Parasites and the Capacity for Parasitism for further information on the pathogenicity of this soil-dwelling ameba), is strikingly similar to *Vahlkampfia*, but unlike the latter it has flagellate stages. *Dientamoeba* has apparently arisen through another line involving the ameboid flagellate *Histomonas*.

Apicomplexa, Myxozoa, and Microspora possess structural characteristics suggestive of the flagellates (for example, merozoites similar to promastigotes, flagellated microgametes similar to *Bodo*, and sexual processes similar to those of phytomonads). However, an ameboid method of locomotion is common among these parasites. They probably arose from flagellates or ameboflagellates possessing life cycles similar to those of present-day phytomonads. The sporozoans might have first become adapted to parasitism in the intestines of aquatic invertebrates; then they might have moved to terrestrial and aquatic vertebrates. Gregarines today

have retained the ancestral characteristic of inhabiting the intestinal lumen of invertebrates.

The primitive Apicomplexa stock appears to have given rise to the gregarines, then to the coccidia.[10] The Haemosporina, including the malarial parasites and their close relatives, arose from the coccidia, but controversy exists whether the ancestral group were coccidia of vertebrates or of invertebrates. In favor of the latter idea is the fact that the Haemosporina are less pathogenic to their invertebrate hosts than to their vertebrate hosts, and are more host specific to their invertebrate hosts (but see the previous discussion of the limits of such an argument). In favor of the view that they have evolved from the coccidian parasites of vertebrates is the tendency of coccidia to become tissue parasites of vertebrates, and the fact that malaria occurs in birds and reptiles that probably originated before the advent of bloodsucking flies. Finally, as Bray[17] pointed out, "if the haemosporidia (= Haemosporina) are coccidian by nature and originally insect parasites . . . it is we who should be carrying the sporozoites or the sporogony stages and the mosquito whose gut should contain the exocoelomic schizogony stages and their haemocoelomic fluid which should suffer schizogony and gametogony stages."

Coatney and associates[29] have summarized their own and other views on the evolution of the primate malarias. They suggest that "the nidus of the primate plasmodia universe lies somewhere in the jungles of Southeast or South Central Asia and that there, there has been a simultaneous development of non-human primates and their malaria parasites."

For thousands of years, humans and nonhuman primates lived in close proximity and probably exchanged malaria parasites. As humans became more separated ecologically from their simian and anthropoid relatives, the sharing of parasites became less pronounced. Many parasitologists believe that primate malarias reached the New World through Europeans and their West African slaves in the 16th century, but evidence suggests that malaria existed in pre-Columbian America. Killick-Kendrick[62] postulated that malaria parasites of primates might have evolved from precursors of the modern haemosporids of Old World bats.

The origin of malaria parasites has been the subject of much controversy, centered chiefly around the question whether mosquitoes, or their ancestors, were the first hosts for malarial parasites, or vertebrates were the first hosts. The evidence seems to favor the conclusion that mosquitoes were the initial hosts (see Mattingly[73]). For a short consideration of the evolutionary age of *Plasmodium falciparum*, as compared with that of the "vivax" and "malariae" groups, see Sinden, et al.[93] Several investigators have postulated that mammalian malaria began in bats.

A fruitful approach to phylogenetic problems of ciliates uses the subpellicularly located basal granules, or kinetosomes, which are intimately and indispensably associated with all external ciliary systems. This infraciliature, as it is called, is present even in the absence of external ciliature. The approach also focuses attention on the ontogeny or morphogenetic aspects of ciliate development. From an unknown zooflagellate ancestry, the Gymnostomata, a large subclass embracing many forms, is situated at the base of the ciliate evolutionary tree. Hunter and Corliss[55] have proposed that the marine homokaryotic *Stephanopogon* and gymnostomes with nondividing macronuclei are primitive.

For two detailed studies on the ciliates, including considerations of evolution and phylogeny, see Corliss[30] and Volume 2 of Grasse's well-known *Traite de Zoologie*. A paper by Orias[81] contains an intriguing theory on the manner in which ciliates may have originated from flagellated ancestors.

Flatworms

Hanson[45] has provided a critical and thorough review of the literature (up to 1977) on phylogeny of protozoa and early metazoa and has expounded the view that turbellarian flatworms have evolved from ciliates. He rejected the planuloid-acoeloid theory, that is, that an ancestral ciliated planuloid gave rise to coelenterates on the one hand and the primitive Acoela on the other. Hanson based his support of the ciliate-acoeloid theory largely on a study of the gymnostome ciliate *Stephanopogon* and the Acoela *Haploposthia*, both benthic marine predators. The ciliate-acoeloid theory has been variously formulated by a number of writers since 1877.

Major differences of opinion exist regarding the origins of the different groups within the phylum Platyhelminthes. The rhabdocoels

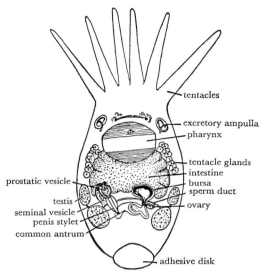

Fig. 26–7. *Temnocephala*. (After Haswell.) (From Hyman. Courtesy of McGraw-Hill.)

(class Turbellaria) are small worms with a mouth and an unbranched digestive tract, protonephridia, oviducts, and a nervous system usually with two main trunks. Most of these worms are free-living in salt, brackish, or fresh water, but a few are parasitic. (See Chapter 8 for some parasites, for example, *Syndesmis*, in the order Lecithophora.)

The suborder Temnocephalida represents one of the groups of ectocommensal rhabdocoels. Temnocephalids (Fig. 26–7) are leechlike, almost devoid of cilia, and the anterior end is extended into 2 to 12 tentacles. The posterior end is equipped with 1 or 2 adhesive discs, the pharynx is barrel shaped (doliiform), and the gonopore is single (common sex pore). Temnocephalids are ectocommensals on freshwater animals, chiefly crustaceans, but also on turtles and snails, frequently within the branchial chamber. They do not derive nourishment from their hosts, but capture and devour insect larvae, rotifers, and other small crustaceans.

The parasitic classes of flatworms may have evolved from dalyellioid rhabdocoels. These worms occur primarily in molluscs and echinoderms. Molluscs are the chief intermediate hosts of flukes. Ancestral adult flukes presumably parasitized molluscs, and when vertebrates arrived, the flukes, by elaboration of adhesive organs already present in rhabdocoels and by modification of the reproductive system in the direction of greater egg production, parasitized vertebrate hosts while retaining their dependence on molluscan hosts.

Llewellyn[66,67] constructed an evolutionary scheme for parasitic flatworms based on newly hatched larvae (Fig. 26–8). The scheme postulated that the larvae are homologous with the juvenile stages of the corresponding ancestral adults. He reconstructed the evolution of endoparasitic strobilate cestodes from early cercomeromorphan ectoparasites of vertebrates. His scheme suggests that aspidogastrids and digeneans had common ancestors stemming from turbellarian beginnings. (See also Cable.[25a])

On the basis of detailed studies of protonephridial systems and hooks of monogeneans and cestodes, Malmberg[71] proposed that from a common rhabdocoeloid stock, two evolutionary lines developed, one culminating in the Digenea and the other in the remainder of the parasitic flatworms. He stated that "if the cestodes and the monogeneans have common ancestors, then the procercoid may be interpreted as the

ontogenetic recapitulation of a common hook-armed ancestor, here named *hexacanthoid*. This rhabdocoelan creature with six hooks in the cercomer and adapted to an ectocommensalic/ectoparasitic mode of life, is thought to have given rise to the monogeneans, the gyrocotylideans, the amphilinideans and the cestodes. The monogeneans were found to have two fundamentally different types of marginal hooks, and on this basis the existence of two lines of evolution in the Monogenea is indicated."

Freeman[40] based his concept of cestode evolution on a review of the literature and his own studies relating to the ontogeny of flatworms. He emphasized that "the single universal, and therefore probably most primitive, feature of cestodes is the need for the oncosphere to activate in the gut of the first host, and there metamorphose and develop to the metacestode."

Freeman[40] argued that adult precestodes (not larval precestodes, as suggested by other writers) were probably the first to invade invertebrates, although "only *Archigetes* sp. is known to mature sexually in the site invaded by the

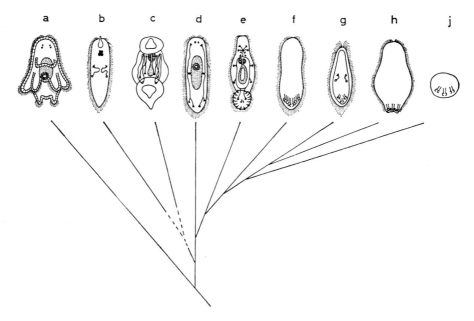

Fig. 26–8. Newly hatched larvae of platyhelminths, arranged in a tentative evolutionary scheme. *A,* Muller's larva of a polyclad. *B,* Miracidium of a digenean. *C,* An aspidogastrean based on *Aspidogaster conchicola,* which has no cilia or eyes, although some aspidogastreans have these features. *D,* Juvenile of a rhabdocoel. *E,* Oncomiridium of a monogenean. *F,* Decacanth (lycophore) of a gyrocotylidean. *G,* Coracidium of a pseudophyllidean. *H,* Larva of an amphilinidean (six hooks plus four supplementary sclerites). *J,* Hexacanth oncosphere of a cyclophyllidean. (From Llewellyn, J.: The evolution of parasitic helminths. *In* Evolution of Parasites. Edited by A.E.R. Taylor. Courtesy of Blackwell Scientific Publications.)

oncosphere, i.e., parenterally in an oligo-chaete." Baer[9] opposed the use of the monozoic species *Archigetes* in support of a theory that tapeworms were originally parasites of invertebrates, because, he maintained, *Archigetes* is probably paedomorphic and not an ancestral type of tapeworm. (Baer used the term neoteny, but paedomorphosis is used here in keeping with terminology for types of heterochronic changes used by Alberch et al.[2]) Baer's view is shared by Stunkard[93a] and others who feel that these monozoic forms appear to be "merely precocious progenetic larvae that have dropped the strobilate stage."

Freeman criticized Malmberg's and Llewellyn's proposals as being based on the erroneous assumption that the genus *Diphyllobothrium* is primitive and that its oncospheres and type of development are characteristic for all cestodes. Freeman's scheme of evolution begins with three branches from rhabdocoel-like, free-living ancestors. One branch leads to gyrocotylids and monogeneans, the second to precestodes, and the third to pretrematodes. The precestodes are judged to belong to the early Paleozoic era.

Brooks et al.[22] constructed a phylogenetic tree for the parasitic platyhelminths using cladistic analysis of approximately 40 morphologic traits (see Fig. 26–1). (See also Brooks[20a]) This study, which postulates that the trematoda and the cercomeromorphs are sister groups, receives independent corroboration from the work of Ehlers.[34] Using ultrastructural characters, Ehlers produced a cladogram for the platyhelminths that posits the same relationships among the parasitic taxa as Brooks and associates found. Through a quantitative assessment of agreements and disagreements about instances of character evolution, Brooks et al. argued that their phylogenetic reconstruction is more parsimonious and better supported than those of Llewellyn, Malmberg, and Freeman.

Monogenea. The phylogenetic schemes for flatworms already discussed suggest that Monogenea are more closely related to Cestoda than to Digenea (see also Bychowsky[25]). Brooks et al.[22] gave three shared derived homologues that supported this conclusion: a double cerebral nerve commissure, a double posterior nerve commissure, and an armed cercomer at the posterior of the body. Arguments for the opposite view, that the presence of a digestive tract and some features of development indicates a unique common ancestry for the Monogenea

and the Digenea, appear to rely on shared primitive traits.

In a detailed discussion of evolution within the monogeneans, Llewellyn[66] provided some evidence "that it is very probable indeed that gyrocotylids have direct life cycles and so resemble monogeneans more than they do cestodes."

Most monogenetic trematodes are host-specific parasites of fish, particularly of the ancient group of cyprinoids, which suggests that this parasitic group is of great antiquity. The apparent lack of specificity in some genera, for example *Benedenia* and *Polystoma*, does not necessarily indicate recent origin. The genus *Polystoma*, occurring on the gills of tadpoles, in the bladder of amphibians and turtles, and in the mouth or nose of turtles, is of special interest. The larval parasites are found attached to the gills of tadpoles, and when host metamorphosis takes place, the worms migrate to the alimentary canal and thence to the urinary bladder, where they become adults. This unique life history suggests a method whereby ectoparasitic worms might have evolved into endoparasitic worms. Nevertheless, a possible explanation for the persistence of the ectoparasitic nature of the Monogenea as a whole is that their oxygen requirements might have prevented occupancy within the host.

Digenea. The question of digenean evolution is made more complex than that of monogenean evolution because of the presence, in the former group, of an invertebrate intermediate host, almost without exception a mollusc. The intermediate hosts probably were originally the final hosts, containing sexually mature adult parasites. A study of the germ cycles in Digenea supports this theory. Most workers agree that the Digenea and the Aspidogastrea are sister groups. Working from this conclusion, the origin of the Digenea would have involved (1) the addition of obligate development in a vertebrate host (most aspidogastrids have obligate development in only a mollusc, but some species are facultative parasites of vertebrates); and (2) the evolution of the sporocyst, redia, and cercaria larval stages.

Rohde[88] has suggested that the pretrematodes were similar to the present-day Aspidogastrea that have many archaic features, for example, absence of alternation of generations and hosts. In a study of phylogeny of digenean life cycle patterns, Pearson[84] concluded that the most primitive extant cycle is two-host, involving a snail and a fish that ingest the cercaria. From the two-host cycle, the three-host and four-host systems have evolved. Pearson stated that "in order to explain the ubiquity of the cercaria, it is postulated that the present molluscan first intermediate host was the original host of the proto-digenean and that escape of the cercariae from this host is primitive. And to explain the occurrence in many life-cycles of a free-swimming miracidium that penetrates the integument of the molluscan host, it is postulated that in an earlier stage the proto-digenean was an ectoparasite of the mollusc."

Brooks et al.[23] concluded that primitively the Digenea possessed a nonescaping, nonencysting cercaria that was consumed by the vertebrate host while still in the mollusc. This life cycle is represented today by the turtle lung fluke, *Heronimus*. Cercarial dispersal after emerging from the mollusc, and the metacercarial traits of encysting "on" or "in" the next host, are therefore considered to be derived characters within the Digenea. Shoop[92] supported these conclusions, although he differed from Brooks et al. and from O'Grady[79] with respect to explanations of the origin of the larval stages. Shoop postulated that each digenean larval stage is in fact a compressed adult generation that is paedomorphic in relation to the simple adult-to-egg-to-adult life cycle of aspidogastrids (e.g., the fragmented oogonia in the body of digenean mother sporocysts are vestigial ovaries). One piece of evidence that Shoop presents for this argument is that reproduction in the mother sporocyst includes a form of partial meiosis.[63] Brooks, O'Grady, and Glen[23] did not deny any of the homologies that Shoop identifies, but they argued that the order of homologous transformation has been different, and that the larval stages of digeneans evolved as nonterminal intercalations to the ancestral ontogeny, with no paedomorphic events.

Cestoidea. Cladistic analysis of adult and larval characters, drawn from both gross morphologic and ultrastructural observations, places the cestodes as a highly derived group within the Cercomeromorphae, with amphilinids and then gyrocotylids as the sister groups (Fig. 26–1).[22,34] Under this interpretation, the loss of a gut is one of the shared derived traits for the group. Llewellyn's[66,67] conclusions are different; he has grouped the gyrocotylids with the monogeneans, and the amphilinids with the cestodes. Malmberg[71] has postulated that cestodes are relatively underived cercomeromorphans, having evolved from a primitively gutless turbellarian ancestor. Because the monogeneans are also included in the cercomeromorphan lineage, Malmberg's hypothesis requires that the mouth, pharynx, and gut of monogeneans be secondarily evolved (i.e., not homologous with similar structures in other parasitic flatworms).[72]

In all of these recent phylogenetic schemes, the classic grouping called the Cestodaria, which consists of the gyrocotylids and the amphilinids, is recognized to be an artificial grouping. See Bandoni[12,13] for cladistic analyses of these two taxa. Brooks et al,[22] in an effort to conserve traditional terminology for the internested clades on their cladogram, applied the Cestodaria term to the clade consisting of the gyrocotylids, amphilinids, and cestodes; in addition, they applied the Cestoidea term to the cestodarian subgroup consisting of the amphilinids and cestodes. The presence of a cercomer in all stages of amphilinids places this group in the Cercomeromorpha; the nature of the excretory system, the presence of calcareous bodies, the general biology, and other features group it with the cestodes.

Opinions differ regarding sister-group relationships within the Cestoda. Tetraphyllidea (scolex with four flexible, lappet-like suckers; adults mostly in sharks, freshwater fish, amphibians, and reptiles) are often regarded as the most primitive group of tapeworms. Baer[9] and others have argued that the two-bothriate tapeworms are the most primitive of present-day forms and have been derived from a protremmatode stock. Wardle et al.[98] suggest that the procestode gave rise, through a process of delayed autonomy and secondary tetraradial symmetry, to the present tetrafossate forms, and that the two-bothriate tapeworms are paedomorphic (i.e., a derived group, rather than a primitive group, as the Baer hypothesis holds). Freeman[40] postulates that present-day "spatheobothrideans, caryophyllideans, and nippotaenideans would appear to be closest to the procestode stalk. If ancestors of trypanorhynchs and pseudophyllideans did not come off together, they probably arose from closely related ancestors."

The bowfin, *Amia calva*, is a ganoid fish and a member of a phylogenetically ancient group. It possesses a tapeworm, *Haplobothrium*, which

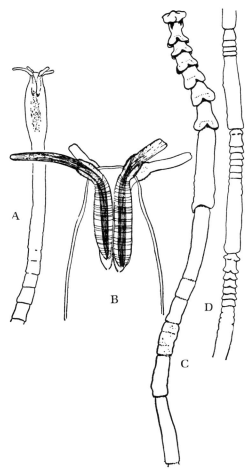

Fig. 26–9. *Haplobothrium globuliforme.* A, Scolex. B, Scolex enlarged, showing the four retractile tentacles. C, Portion of a strobila, showing a pseudoscolex that resembles the scolex of pseudophyllidean species. D, Strobilia dividing secondarily into identical pieces. (After Fuhrmann. From Baer, J.E.: Ecology of Animal Parasites. Courtesy of University of Illinois Press.)

have been characteristic of a primitive cestoid ancestor to both pseudophyllids and tetraphyllids.

Most of the genera of Cyclophyllidea, which includes the common tapeworms of humans, are found in birds. Baer has suggested that the radiation of birds is reflected in the rapid and diversified evolution of bird tapeworms. There has been less radiation in mammals, and comparatively fewer cestode genera are found in their intestines.

Acanthocephala

In all probability, the Acanthocephala originated as parasites of freshwater fish, with a single invertebrate intermediate host. Potential transport hosts, that is, paratenic hosts, have become ecologic intermediate hosts because immature worms are able to become re-encapsulated in vertebrate hosts that feed on arthropods. An appropriate final host can sometimes also become a potential intermediate host, and because of this adaptability, a flexibility in parasite-host specificity has been established. Acanthocephalans have become adapted to the bodies of reptiles, birds, and mammals, and they undoubtedly represent an ancient group of parasitic worms. The short, subglobular proboscis with a few hooks seems to be the type, morphologically and phylogenetically, from which other modifications have been derived.

Cameron[27] called these pseudocoelomate animals a small group with elusive affinities and puzzling origin that are not even remotely related to the Nematoda. In 1967, Nicholas[78] reviewed the biology of the Acanthocephala and stated that little fresh information bearing on the phylogeny had come to light by that time. Amin[3] recently presented a classification for this group.

Nematoda

Parasitic nematodes probably originated from more than one type of free-living ancestor. Studies on the evolution and classification of parasitic nematodes have been made by Inglis,[56] Osche,[82] Chabaud,[28] Andrassy,[8] and Anderson and associates.[6] Adamson[1] produced a cladogram for the higher phylogeny of the Nematoda (Fig. 26–10).

The host distribution of nematodes suggests that these worms became abundant in vertebrates only after the hosts had taken to a terrestrial life, and that the complexity of the par-

is found nowhere else. The scolex of the worm (Fig. 26–9) suggests affinities with the tetrarhynchid cestodes from elasmobranchs, but the internal anatomy is more similar to that of the pseudophyllids (see Chapter 12). The body of the worm exhibits a curious secondary segmentation, and it often breaks into smaller pieces, the anterior end of each piece becoming differentiated into a secondary pseudophyllidean scolex. The life cycle of this unique tapeworm appears to involve two intermediate hosts, a copepod and a fish. Thus, *Haplobothrium* possesses the morphologic features that may well

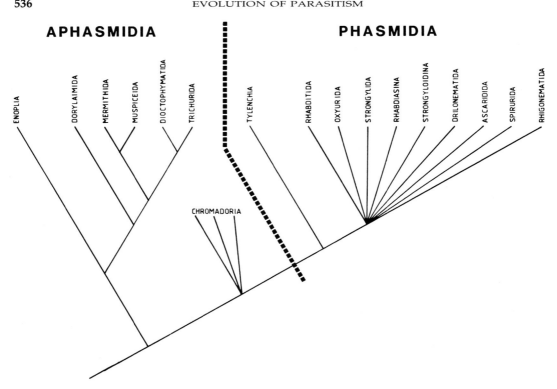

Fig. 26–10. Cladogram of the higher-level relationships among the Nematoda. (Modified from Adamson.[1])

asites evolved with their hosts. Anderson[4,5] has studied the evolution of transmission patterns in nematodes: "Since both the parasitic secernenteans [phasmidians] and adenophoreans [asphasmidians] can be traced back to soil ancestors, it has been hypothesized the parasitism by nematodes could not have occurred until animals invaded land, first the invertebrates, and much later the tetrapods. This explains why nematode parasites are practically nonexistent in such important marine and freshwater groups as the molluscs, polychaetes, and crustaceans, and it explains the richness of the nematode fauna in essentially terrestrial groups like the earthworms and insects, and the terrestrial molluscs" . . . "When land was invaded by the earliest vertebrates, the dauer larva of the [soil-dwelling] rhabditids probably already existed. It was presumably through the dauer larva that vertebrates were first invaded by the secernenteans. The dauer larvae of some species probably had developed phoretic associations with insects, annelids, and other invertebrates to facilitate dispersal to new food sources . . . Such phoretic associations may have been important

in transferring larvae to the early tetrapods . . ."

The larvae of nematodes often enter the lungs of final hosts. It has been suggested that this site preference probably represents the vestige of an ancestral condition in which only the lungs, with their abundance of oxygen, were the locations of infection. Hookworms, however, can develop directly in the intestine. Inert particles injected under a frog's skin follow the same route of migration as that pursued by larvae of *Rhabdias ranae*, a parasite of the lungs. The migratory habit of nematode larvae, as well as the penetration of the host skin, might therefore be due partly to mechanical causes, but such a conclusion is hazardous without further evidence. Anderson[5] has argued that some lung parasitisms are the result of secondary monoxeny having evolved from heteroxeny; that is, the tissue migration phase in the final host is the phylogenetic remnant of development in an intermediate host.

Free-living phasmidians are largely inhabitants of the soil, whereas free-living aphasmidians are more characteristically aquatic, and they include the great bulk of marine species. Life in the soil is suited to the development of parasitism because of the opportunity for a worm to seek the protective moisture of a host's body (plant or animal) when the soil becomes dry. When intermediate hosts, usually arthropods, are involved in a parasitic life cycle, they almost invariably acquire infection by ingesting larvae while the latter are still enclosed by the eggshell. The subsequent transfer of larvae to the vertebrate host can be regarded as simply an elaboration of the direct type of life history. The origin of larval migration might be related to the habit of ingesting infected secondary hosts—the final host becoming also an intermediate one. See Laurence[65] for a discussion of the evolution of filarial infections of man.

The following statements are taken from the summary of Inglis's paper[56] on nematode evolution (Fig. 26–11).

"1. All parasitic nematodes appear to have arisen from terrestrial free-living groups; the phasmidian parasites from the Rhabditina and the aphasmidian from the Dorylaimida. The terms phasmidia and aphasmidia are retained because of their value in delimiting these two major divisions and not because they indicate any major division of nematodes as a whole. [Note that the results of the cladistic analysis by Adamson[1] postulate that the aphasmidians are a paraphyletic, and thus artificial, group; in other words, they do not have a common ancestor that is not also the common ancestor of the phasmidians.]

2. The phasmidian parasites have become parasites in animals on at least four major occasions giving rise to the groups recognized as the Drilonematoidea, Oxyuroidea, Cosmocercoidea, and Strongylina.

3. The appearance of groups parasitic in the early aquatic vertebrates depended on the presence of an invertebrate which could act as an intermediate host and it is argued that the Ascaridoidea, Spiruroidea and Seuratoidea originated by way of a group that was parasitic in, or associated with, some terrestrial:aquatic invertebrate during the first three, or simply the third, larval stages. The Cosmocercoidea approach this condition in some cases and are considered to represent at least a model of the early conditions covering the transfer from terrestrial to aquatic hosts. The Seuratoidea are considered to represent the remaining members of the ear-

liest groups of parasites to make this transfer to aquatic hosts.

4. The major division of the aquatic parasites is attributed to the development of a stomach in the host within which the Spiruroidea were restricted, while the Ascaridoidea remained largely in the intestine. Thus it is suggested that the infestation of vertebrates was prior to the appearance of jawed forms.

5. The significance of the third-stage larva in the phasmidian parasites is stressed as being the stage at which the parasites transfer from one environment to another, and it is shown that the life-cycle of the Tylenchida, parasitic in insects, is not fundamentally different from that of other phasmidian parasites.

6. The reduction of the length of time spent in the intermediate host in the life-cycle of many nematode groups is considered to be a result of selection pressures acting to reduce 'host-discordance' (i.e., the many differences in physiology etc. between the two, or more, animals in an indirect life-cycle). It is also pointed out that if a parasite utilizes more than one host at different stages in the life-cycle it must be plastic in its environmental requirements so that it is less likely to become tied to one host, or host group, and will speciate less easily. An indirect life-cycle also appears to be a pre-requisite to life in extra-gut localities within the host, as all such parasites have an indirect life-cycle.

7. Speciation in parasitic nematodes is considered to be wholly allopatric and not due to any evolutionary change in the host. This is suggested by establishing that niche diversification is common among parasitic nematodes and that intrahost competition can, and frequently does, occur. . . .

8. Trans-specific evolution is shown to be largely independent of the evolution of the hosts and to be due rather to the parasites speciating and evolving so as to utilize all available localities and niches within the host body. Any host restriction is almost wholly determined by the ecology and feeding habits of the hosts and, because most major host groups have similar food and ecological preferences it is possible to produce an apparent parallelism between the hosts and the parasites. This apparent parallelism, however, generally only applies when wide host and parasite groups are compared. . . .

9. As a consequence of the lack of host-parasite parallelism and the concept of intrahost

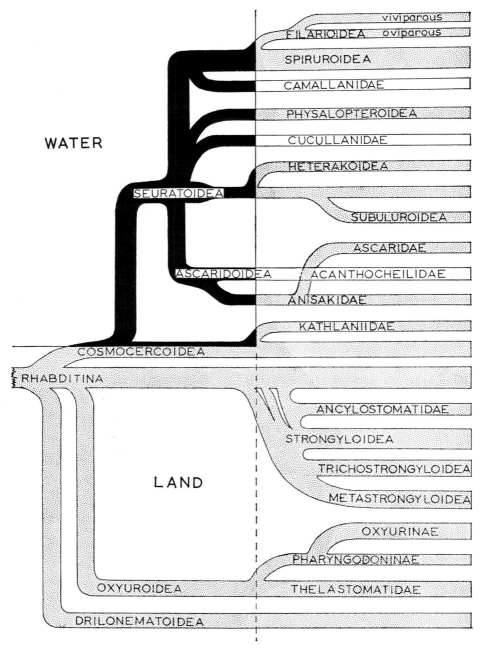

Fig. 26–11. Dendrogram (generalized evolutionary tree, not constructed with any particular methodology) illustrating the relationships among the major groups of phasmidian parasites, as postulated by Inglis. (From Inglis, W.G.: Patterns of evolution in parasitic nematodes. *In* Evolution of Parasites. Edited by A.E.R. Taylor. Courtesy of Blackwell Scientific Publications.)

competition it is pointed out that a group could appear which was so efficient a parasite that it could infest almost any host available *Physaloptera* and *Capillaria* approach this condition and it is pointed out that the place and time of origin of both groups cannot be wholly established. It is stressed that it cannot be established that *Capillaria* represents an archaic group of parasites, particularly as its absence from Madagascar suggests it may be of fairly recent origin."

Arthropods

Conditions predisposing to parasitism among the arthropods, as well as among other animals, include saprophagous nutrition habits, sucking of plant juices, and, especially, the preference to live in crowded communities. As Rothschild and Clay[90] have said, the most favorable condition for the dawn and development of dependence is a social environment.

Various species of bloodsucking arthropods, for example some triatomids, obtain meals by tapping the blood-engorged bodies of other arthropods, such as bedbugs, that have fed on a vertebrate. Preparasites were perhaps first attracted to waste food or offal, and when conditions encouraged the preparasites to stay with their future host (that is, when the search for food became simplified or when it became unnecessary to meet competition by seeking other and more distant sources of food), the preparasites became mess-mates or scavengers, and from this association they became parasites. Evidence for this sort of speculation can be obtained from parasitic insects, among which all gradations from free-living to parasitic forms can be traced. Parallel evolution of the host-seeking instinct and somatic characteristics seems clear, particularly in such groups as the dipterous Tachinidae (see Chapter 20).

Copepods. Present knowledge of copepod relationships is only in an initial stage, and a natural scheme of classification, especially for parasitic species, can be only tentative and incomplete. The Cyclopoida appear to be the most primitive group, from which the other copepod groups are derived. All parasitic copepods might have originated in the more or less parasitic Ergasilidae, but it is possible that parasitic copepods form a polyphyletic assemblage.

Some of the most highly modified copepods, for example, Saccapsidae, are found on annelids. A striking feature of intimacy with the host

is shown by the hermaphroditic *Xenocoeloma*, whose body cavity is continuous with the coelom of its annelid host. There are also less-specialized species of copepods on more primitive fish, modified species on less primitive fish, and slightly modified species on the most recent fish and on ascidians. In spite of many exceptions to the suggested evolutionary trend, a rough correlation does seem to exist between the degree of parasitic modification of copepods and the primitiveness of the host group.

Lice. Presumably, lice have been derived from primitive psocid-like ancestors that probably became parasitic first on birds and later on mammals. The Psocoptera (Corrodentia) are known as booklice and are not parasitic. From their habitat beneath the bark of trees, ancestral forms could have readily found their way to bird nests and thence to birds themselves. The lice may have passed secondarily from birds to mammals, but the presence of a primitive species of louse on a tree shrew, a primitive mammal, suggests a polyphyletic origin for mammal lice. See the cladistic work of Lyal.[69]

Among the chewing lice (Mallophaga) of birds, the group known as Amblycera does not possess a spiracular gland, but the Ischnocera do possess the gland, which is similar to the spiracular gland in the sucking lice (Anoplura). The absence of the spiracular gland indicates that the Amblycera were probably well established before the Ischnocera arose. The spiracles of the most primitive structures are found in *Eomenacanthus stramineus* of the family Menoponidae. Thus, members of this family are believed to be of more ancient origin than are other Anoplura. The Trichodectidae (Ischnocera) and the Anoplura appear to have arisen from a common ischnoceran parasite that migrated from a bird to a primitive insectivore.

A chronologic summary of relationships among the lice of mammals has been made by Hopkins,[53] as follows:

Here, again, our deductions are greatly hampered by doubt as to whether certain infestations are primary (and very ancient) or secondary (and relatively recent), but it seems possible to arrive at a few conclusions which are beyond reasonable doubt. Distantly related species of *Pediculus* occur on man and the chimpanzee and the genus is so widely separated from any found on non-Primates that its occurrence on Primates must be ancient—possibly of Miocene date. Turning to *Anoplura* as a whole, the most important fact is the universal occurrence of Echinoph-thiriidae on the seals, because this must certainly have originated before the seals adopted a marine life, because they have not been in contact with other mammals since that event, apart from momentary and fatal encounters with Polar bears; seals are definitely known from the Lower Miocene and probably originated in the late Eocene. Infestation of the ungulate branches of the *Ferrungulata* is far too nearly universal to be anything but primary, in my opinion, and the fact that seallice and the lice of *Camelidae* and *Tayassuidae* have characters annectent between *Haematopinidae* and *Linognathidae* seems to confirm that even if the infestation of the *Ferrungulata* as a whole does not date back (as I believe it does) to before the ancestors of the ungulate and carnivorous branches (the *Condylarthra* and *Credonata*) diverged during the lowest Paleocene period (or perhaps the late Cretaceous) it must have spread from one group to the other at a not very much later date. At this point we come to a very serious difficulty—the infestation of the *Glires* with *Anoplura* is also practically universal except for certain Nearctic and Neotropical families of *Hystricomorpha*. I regard the infestation of the *Glires* with *Anoplura* as at least partly primary, and the absence or rarity of these lice on some *Hystricomorpha* as the effect of the heavy infestation of these groups with *Mallophaga*, while Vanzolini and Guimaraes think that the evidence indicates that *Anoplura* were lacking in South America (and probably everywhere) at the beginning of the Tertiary period and that they were brought in by the Pliocene migration of mammals into South America. If Vanzolini and Guimaraes are right, then the *Anoplura* probably arose between the Eocene period and the Pliocene migration and perhaps not long before this migration, whereas if I am right they may be very much older.

Flies. Little is known about the evolution of the parasitic Diptera, but one family, Tachinidae, has been studied in some detail and will be used as one example of the problems encountered in exploring the ancestral history of insects. The Tachinidae (see Chapter 20) is a large assemblage of flies whose classification presents imposing difficulties and about which exists considerable disagreement among entomologists. These flies undoubtedly evolved from the family Calliphoridae, a large group whose larvae are saprophagous or flesh-feeders (for example, *Sarcophaga*, the fleshfly), or they are parasites of various arthropods. The closest relatives of trachinids among the calliphorids occur in the subfamilies Sarcophaginae and Rhinophorinae. Of these two subfamilies, the Sarcophaginae appears to be closer to the ancestral Tachinidae because of morphologic and behavioral similarities.

The most primitive subfamily of tachinids is the Dexiinae, and an evolutionary series of subfamilies can be arranged from this group up to the Goniinae. The most primitive method of reaching the host (exhibited in all Dexiinae and also to be found in the Sarcophaginae) is that of active searching by the young larva for its host. The next step in the evolution of host feeding is shown by the Macquartiinae, whose larvae are deposited near and are ingested by the host, or are deposited on the host. In the latter method, the larvae are sometimes still encased by the egg shell. A third step, shown by the Phasiinae, is the deposition of parasite eggs on particular hosts or the injection of eggs into the host's body. The more derived tachinids are typically parasites of Lepidoptera, whereas the more primitive forms parasitize a wider range of unrelated hosts.

Fleas. Bird fleas have undoubtedly arisen from mammal fleas. The genus *Ceratophyllus*, for example, is clearly derived from fleas of squirrels and other tree-dwelling rodents whose nests provide conditions not unlike those of bird nests. Most of the few bird fleas are to be found on birds that return to the same nesting site year after year or on birds that nest on the ground in holes. Such habits obviously favor the difficult transfer of fleas from a mammal to a bird. *Palaeopsylla*, a genus found on moles and shrews, has been reported from Baltic amber representing the Oligocene. The origins of mammal fleas remain obscure.

Wasps. A study of the parasitic Hymenoptera shows that species feeding on a few kinds of host foods (oligophagous) have arisen repeatedly from polyphagous ancestors. This trend has occurred in spite of the apparent advantages (more available food) of the polyphagous forms.

Life histories of wasps and their mites illustrate mutual modifications and offer promising subjects for the study of evolution. The eumenine wasp *Ancistrocercus antilope* and the larval mite *Ensliniella trisetosa* live together throughout the life of each member of the alliance.

Mites and Ticks. Among the Acarina, parasitism seems to have developed primarily from a predatory habit, but also from the scavenger habit and from phoresy. Little work has been done on the evolution of mites and ticks, but some speculation on the phylogeny of trombiculid mites suggests that, from essentially reptile-specific groups, four lines of development took place. The larvae (chiggers) of these mites are small, and instead of preying on smaller arthropods, as do the adult mites, they feed parasitically on large arthropods and on vertebrates. The first group to arise was a small number of species that parasitized arthropods. The second group began living in the skins of amphibians. The third group managed to isolate itself in the lungs of sea snakes, while the fourth, and by far the largest group, spread out over all the mammals and speciated widely. The more primitive species of these mites appear to be distributed over the open home ranges, for example, fields, of their hosts, whereas other, more derived species are more restricted in their relationships with their hosts. (See Woolley.[102])

REVIEW OF GENERAL RULES AND PRINCIPLES

1. The two basic kinds of changes that nor-mallly take place in an organism before it be-comes a dependent internal parasite are (1) physiologic, in which the powers of synthesis of some essential nutritional elements are lost, modified, or newly evolved; and (2) morpho-logic, in which certain characteristics are lost, modified, or newly evolved. Such changes, of course, also occur in free-living organisms. The differences lie in the fact that the evolutionary changes in parasitic organisms render them in-timate and obligate feeders on other organisms.

2. Parasitism probably arose in different ways in different groups of animals. "Parasit-ism" is a functional label applied to a particular way of living; it can evolve through many dif-ferent pathways of structural change in a free-living ancestor.

3. Many factors, in addition to length of as-sociation, play significant roles in determining the incidence, prevalence, and degree of path-ogenicity of parasitic infections in new hosts.

4. Every parasite-host association began, ev-olutionarily speaking, with a colonization event. After this initial colonization, parasites can coevolve with their hosts (i.e., parasite and host phylogenies are congruent, or at least con-sistent), or they can continue to colonize dis-tantly related but ecologically associated hosts. These two phenomena are not mutually exclu-sive, for a parasite lineage can show periods of coevolution intercalated with periods of colo-nization. Also, with sufficiently broad coaccom-modation, a parasite lineage can colonize a broad range of hosts, yet coevolve with only some of the host lineages involved.

5. Parasites can speciate less frequently, more frequently, or at the same rate as their hosts. Little support exists for the argument that parasites, especially gut parasites, exist in an environmental milieu that is any less harsh or heterogeneous or subject to evolutionary mod-ification than that experienced by a free-living organism. Parasite speciation will depend pri-marily on the response of individual species un-der allopatric, allohospitalic, and allotopic con-ditions.

6. The terms "degenerate," "regressive," and "retrogressive" are anthropocentric terms for describing evolutionary change in parasites. They come from free-living chauvinism on the part of humans, and they should not be used.

7. When various lineages of parasitic organ-ism first became parasitic, some of the traits that they had evolved while still free-living, espe-cially traits that conferred resistance to harsh environmental conditions, would have been ad-vantageous for the establishment of a parasitic way of living. Such traits have in the past been called preadaptations to parasitism. This term is incorrect and should not be used because (1) it is offered as an explanation for the existence of previous biologic conditions when it is, in fact, simply an after-the-fact label—a label that appeals to a nonexistent process; and (2) it com-pounds the error of assuming that every trait that has any sort of biologic function must be an adaptation.

REFERENCES

1. Adamson, A.: Phylogenetic analysis of the higher classification of the Nematoda. Can. J. Zool., 65:243–255, 1987.
2. Alberch, P., Gould, S.J., Oster, G.F., and Wake, D.B.: Size and shape in ontogeny and phylog-eny. Paleobiology, 5:296–317, 1979.
3. Amin, O.M.: Classification. In Biology of the Acanthocephala. Edited by D.W.T. Crompton and B.B. Nikol. London, Cambridge University Press, 1985, pp. 27–72.
4. Anderson, R.C.: The origins of zooparasitic nematodes. Can. J. Zool., 62:317–328, 1984.
5. Anderson, R.C.: Nematode transmission pat-terns. J. Parasitol., 74:30–45, 1988.
6. Anderson, R.C., Chabaud, A.G., and Willmott, S. (Eds.): C.I.H. Keys to the Nematode Parasites of Vertebrates. Nos. 1–10. Farnham Royal, Bucks, England, Commonwealth Agricultural Bureaux, 1974–1983.

7. Anderson, R.M., and May, R.M.: Coevolution of hosts and parasites. Parasitology, 85:411–426, 1982.

8. Andrassy, I.: Evolution as a Basis for the Systematization of Nematodes. San Francisco, Pitman, 1976. (Translated from the Hungarian.)

9. Baer, J.G.: Ecology of Animal Parasites. Urbana, IL, University of Illinois Press, 1951.

10. Baker, J.R.: The evolution of parasitic Protozoa. In Evolution of Parasites. Edited by A.E.R. Taylor. Oxford, Blackwell Scientific Publications, 1965, pp. 1–27.

11. Ball, G.H.: Parasitism and evolution. Am. Naturalist, 77:345–364, 1943.

12. Bandoni, S.M., and Brooks, D.R.: Revision and phylogenetic analysis of the Amphilinidea Poche, 1922 (Platyhelminthes: Cercomeria: Cercomeromorpha). Can. J. Zool., 65:1110–1128, 1987.

13. Bandoni, S.M., and Brooks, D.R.: Revision and phylogenetic analysis of the Gyrocotylidea (Platyhelminthes: Cercomeria: Cercomeromorpha). Can. J. Zool., 65:2369–2389, 1987.

14. Banziger, H.: The piercing mechanism of the fruit-piercing moth Calpe (Calyptra) thalicteri Bkh. (Noctuidae) with reference to the skin-piercing blood sucking moth C. eustrigata. Hmps. Acta Trop; 27:54–88.

15. Benz, G.W., and Deets, G.B.: Fifty-one years later: an update on Entepherus, with a phylogenetic analysis of Cecropidae Dana, 1849 (Siphonostomatoidea: Copepoda). Can. J. Zool, in press.

16. Boucot, A.J.: Does evolution take place in an ecological vacuum? J. Paleontol., 57:1–30, 1983.

17. Bray, R.S.: The exo-erythrocytic phase of malaria parasites. Int. Rev. Trop. Med., 2:41–47, 1963.

18. Brooks, D.R.: Testing the context and extent of host-parasite coevolution. Syst. Zool., 28:200–307, 1979.

19. Brooks, D.R.: Allopatric speciation and non-interactive parasite community structure. Syst. Zool., 29:192–203, 1980.

20. Brooks, D.R.: Hennig's parasitological method: a proposed solution. Syst. Zool., 30:229–249, 1981.

20a. Brooks, D.R.: A summary of the database pertaining to the phylogeny of the major groups of parasitic platyhelminths, with a revised classification. Can. J. Zool., in press.

21. Brooks, D.R., and O'Grady, R.T.: Crocodilians and their helminth parasites: macroevolutionary considerations. Amer. Zool., in press.

22. Brooks, D.R., O'Grady, R.T., and Glen, D.R.: The phylogeny of the Cercomeria Brooks, 1982 (Platyhelminthes). Proc. Helm. Soc. Wash., 52:1–20, 1985.

23. Brooks, D.R., O'Grady, R.T., and Glen, D.R.: Phylogenetic analysis of the Digenea (Platyhelminthes: Cercomeria) with comments on their adaptive radiation. Can. J. Zool., 63:411–443, 1985.

24. Brooks, D.R., and Wiley, E.O.: Evolution as Entropy. Chicago, Univ. of Chicago Press, 1986.

25. Bychowsky, E.E.: Monogenetic Trematodes, Their Systematics and Phylogeny. Acad. Nauk.

SSSR. (English editor, W.J. Hargis; translator, P.C. Oustinoff.) Washington, DC, American Institute of Biological Sciences, 1957.

25a. Cable, R.M.: Phylogeny and taxonomy of trematodes with reference to marine species. In Symbiosis in the Sea. Edited by W.B. Vernberg. Columbia, SC, University of South Carolina Press, 1974, pp. 173–193.

26. Caira, J.N.: Site specificity of five species of Pedibothrium within the spiral valve of the nurse shark: a case of syntopy. Ann. Meet. Amer. Soc. Parasitol., Lincoln, NE, August 2–5, 1987.

27. Cameron, T.W.M.: Host specificity and the evolution of helminthic parasites. Adv. Parasitol., 2:1–34, 1964.

28. Chabaud, A.G.: Evolution of host-parasite adaptations in nematodes of vertebrates. Int. J. Parasitol., 1:217–221, 1971.

29. Coatney, G.R., Collins, W.E., and Contacos, P.G.: The Primate Malarias. Bethesda, MD, U.S. Department of Health, Education, and Welfare, National Institutes of Health, 1971.

30. Corliss, J.O.: The Ciliated Protozoa: Characterization, Classification and Guide to the Literature. 2nd. Ed. New York, Pergamon Press, 1979.

31. Davis, G.M.: Snail hosts of asian Schistosoma infecting man: evolution and coevolution. Malacological Rev., (Suppl.) 2:195–238, 1980.

32. Deets, G.B.: Phylogenetic analysis and revision of Kroeyerina Wilson, 1932 (Siphonostomatoida: Kroyeriidae), copepods parasitic on chondrichthyans, with descriptions of four new species and the erection of a new genus, Prokroyeria. Can. J. Zool., 65:2121–2148, 1987.

33. Dogiel, V.A.: General Parasitology. Edinburgh, Oliver and Boyd, 1964.

34. Ehlers, U.: Phylogenetic relationships within the Platyhelminthes. In The Origins and Relationships of Lower Invertebrates. Edited by S. Conway Morris, J.D. George, R. Gibson, and H.M. Platt. Oxford, Oxford Univ. Press, 1985.

35. Ehrlich, P.R., and Raven, P.H.: Butterflies and plants: A study in coevolution. Evolution, 18:586–608, 1964.

36. Eichler, W.: Some rules in ectoparasitism. Ann. Mag. Nat. Hist., 12:588–598, 1948.

37. Eldredge, N., and Cracraft, J.: Phylogenetic Patterns and the Evolutionary Process. New York, Columbia Univ. Press, 1980.

38. Esch, G.W., Hazen, T.C., and Aho, J.M.: Parasitism and r- and K-selection. In Regulation of Parasite Populations. Edited by G.W. Esch. New York, Academic Press, 1977, pp. 9–62.

39. Frazzetta, T.H.: Complex Adaptations in Evolving Populations. Sunderland, MA, Sinauer Associates, 1975.

40. Freeman, R.S.: Ontogeny of cestodes and its bearing on their phylogeny and systematics. Adv. Parasitol., 11:481–557, 1973.

41. Force, D.C.: Succession of r and K strategists in parasitoids. In Evolutionary Strategies of Parasitic Insects and Mites. Edited by P.W. Price. New York, Plenum Publishing Company, 1975, pp. 112–129.

42. Glen, D.R., and Brooks, D.R.: Parasitological evidence pertaining to the phylogeny of the hominoid primates. Biol. J. Linn. Soc., 27:331–354, 1986.

43. Gould, S.J., and Vrba, E.S.: Exaptation—a miss-

ing term in the science of form. Paleobiology, 8:4–15, 1982.

44. Hafner, M.S., and Nadler, S.A.: Phylogenetic trees support the coevolution of parasites and their hosts. Nature, 332:258–259, 1988.

45. Hanson, E.D.: The Origin and Early Evolution of Animals. Middletown, CT, Wesleyan University Press, 1977.

46. Harvey, P.H., Keymer, A.E., and May, R.M.: Evolving control of diseases. Nature, 332:680–681, 1988.

47. Hennig, W.: Phylogenetic Systematics. Urbana, Univ. Illinois Press, 1966.

48. Hillis, D.: Molecular versus morphological approaches to systematics. In Annual Review of Ecology and Systematics. Edited by R.F. Johnston, P.W. Frank, and C.D. Michener. 18:23–42, 1987.

49. Hoare, C.A.: The transmission of trypanosomes and its evolutionary significance. In Biological Aspects of the Transmission of Disease. Edited by G. Horton-Smith. London, Oliver and Boyd, 1957.

50. Hollande, A.: L'evolution des flagelles symbiotiques, hotes du crytocercus et des termites inferieurs. Tijdsch. Entomol., 95:18–110, 1952.

51. Holmes, J.C.: Site selection by parasitic helminths: interspecific interactions, site segregation, and their importance to the development of helminth communities. Can. J. Zool., 51:333–347, 1973.

52. Holmes, J.C., and Price, P.W.: Parasite communities: the roles of phylogeny and ecology. Syst. Zool., 29:202–213, 1980.

53. Hopkins, G.H.E.: The distribution of Phthiraptera on mammals. In First Symposium on Host Specificity Among Parasites of Vertebrates. Institute of Zoology, University of Neufchatel, 1957, pp. 64–87.

54. Humphrey-Smith, I.: Parasites and Australian marsupial phylogeny. In Possums and Opposums: Studies in Evolution. Edited by M. Archer. Sydney, Australia, Surrey Beatty & Sons and the Royal Zoological Society of New South Wales, 1985, pp. 1–9.

55. Hunter, S.H., and Corliss, J.O.: Search for clues to the evolutionary meaning of ciliate phylogeny. J. Protozool., 23:48–56, 1976.

56. Inglis, W.G.: Patterns of evolution in parasitic nematodes. In Evolution of Parasites. Edited by A.E.R. Taylor. Oxford, Blackwell Scientific Publications, 1965, pp. 79–124.

57. Inglis, W.G.: Speciation in parasitic nematodes. In Advances in Parasitology. Edited by B. Dawes. London, Academic Press, 9:185–223, 1971.

58. Jennings, J.B., and Calow, P.: The relationship between high fecundity and the evolution of entoparasitism. Oecologia, 21:109–115, 1975.

59. Jones, A.W.: Introduction to Parasitology. Reading, MA, Addison-Wesley, 1967.

60. Kethley, J.B., and Johnston, D.E.: Resource tracking patterns in bird and mammal ectoparasites. Misc. Publ. Entomol. Soc. Amer., 9:231–236, 1975.

61. Keymer, A.: Density-dependent mechanisms in

the regulation of intestinal helminth populations. Parasitology, 84:573–587, 1982.

62. Killick-Kendrick, R.: Speculation on the evolution of Plasmodium of mammals. Abstract No. 15, Fifth International Congress of Protozoology, 1977.

63. Khalil, G.M., and Cable, R.M.: Germinal development in Philophthalmus megalurus (Cort, 1914) (Tramatoda: Digenea). Zeitschrift fur Parasitenkunde, 31:211–231, 1968.

64. Klassen, G.J., and Beverley-Burton, M.: Phylogenetic relationships of Ligictaluridus spp. (Monogenea: Ancyrocephalidae) and their Ictalurid (Siluriformes) hosts: an hypothesis. Proc. Helm. Soc. Wash., 54:84–90, 1987.

65. Laurence, B.R.: The evolution of filarial infection. In Medicine in a Tropical Environment. Proceedings of the International Symposium, South Africa, 1976. Edited by J.H.S. Gear. Cape Town, A.A. Balkema, 1977.

66. Llewellyn, J.: Monogenea. (Part of a technical review of taxonomy, genetics, and evolution of parasites). J. Parasitol., 56:493–504, 1970.

67. Llewellyn, J.: The evolution of parasitic helminths. In Evolution of Parasites. Edited by A.E.R. Taylor. Oxford, Blackwell Scientific Publications, 1965, pp. 47–48.

68. Lumsden, W.H.R.: Leishmaniasis and trypanosomiasis: the causative organisms compared and contrasted. Ciba Found. Symp., 20:3–21, 1974.

69. Lyal, C.H.C.: Co-evolution of trichodectid lice (Insecta: Phthiraptera) and their mammalian hosts. J. Nat. Hist., 21:1–28, 1987.

70. MacArthur, R.H., and Wilson, E.O.: The Theory of Island Biogeography. New Jersey, Princeton Univ. Press, 1967.

71. Malmberg, G.: On the larval protonephridial system of Gyrocotyle and the evolution of Cercomeromorphae (Platyhelminthes). Zool. Scripta, 3:65–81, 1974.

72. Malmberg, G.: The major parasitic platyhelminth classes—progressive or regressive evolution? Hydrobiologia, 132:23–29, 1986.

73. Mattingly, P.F.: The evolution of parasite-arthropod vector systems. In Evolution of Parasites. Edited by A.E.R. Taylor. Oxford, Blackwell Scientific Publications, 1965, pp. 29–45.

74. Mitchell, P.C.: On the intestinal tract of birds; with remarks on the valuation and nomenclature of zoological characters. Trans. Linn. Soc. London, 8:173–275, 1901.

75. Mitter, C., and Brooks, D.R.: Phylogenetic aspects of coevolution. In Coevolution. Edited by D.J. Futuyma and M. Slatkin. Sunderland, MA, Sinauer Associates, 1983, pp. 65–98.

76. Morris, S.C.: Parasites and the fossil record. Parasitology, 82:489–509, 1981.

77. Mount, P.M.: Histogenesis of the rostellar hooks of Taenia crassiceps (Zeder, 1800) (Cestoda). J. Parasitol., 56:947–961, 1970.

78. Nicholas, W.L.: The biology of the Acanthocephala. Adv. Parasitol., 2:205–237, 1967.

79. O'Grady, R.T.: Ontogenetic sequences and the phylogenetics of parasitic flatworm life cycles. Cladistics, 1:159–170, 1985.

80. O'Grady, R.T.: Historical processes, evolution-

ary explanations, and problems with teleology. Can. J. Zool., 64:1010–1020, 1986

81. Orias, E.: Derivation of ciliate architecture from a simple flagellate: an evolutionary model. Am. Microsc. Soc., 95:415–429, 1976.

82. Osche, G.: Morphological, biological, and ecological considerations in the phylogeny of parasitic nematodes. In The Lower Metazoa, Comparative Biology and Phylogeny. Edited by G.H. Dougherty. Berkeley, University of California Press, 1963, pp. 283–302.

83. Patterson, C.: Molecules and Morphology in Evolution: Conflict or Compromise? Cambridge, Cambridge Univ. Press, 1987.

84. Pearson, J.C.: A phylogeny of life-cycle patterns of the Digenea. Adv. Parasitol., 10:153–189, 1972.

85. Pianka, E.R.: On r- and K-selection. Amer. Nat., 104:592–597, 1970.

86. Price, P.W.: Evolutionary Biology of Parasites. Princeton, NJ, Princeton University Press, 1980.

87. Price, P.W.: General concepts on the evolutionary biology of parasites. Evolution, 31:405–420, 1977.

88. Rohde, K.: Phylogenetic origin of the trematodes. Parasitol. Schriftenr., 21:17–27, 1971.

88a. Rosen, D.E., Foroy, P.L., Gardiner, B.G., and Paterson, C.: Lungfishes, tetrapods, paleontology, and plesiomorphy. Bull. Amer. Mus. Nat. Hist., 167:159–276, 1981.

89. Ross, H.H.: The origin of species diversity in ecological communities. Taxon, 21:253–259, 1972.

90. Rothschild, M., and Clay, T.: Fleas, Flukes & Cuckoos. London, Wm. Collins Sons, 1952.

91. Schad, G.A.: Niche diversification in a parasitic species flock. Nature, 198:404–406, 1963.

92. Shoop, W.L.: Trematode transmission patterns. J. Parasitol., 74:46–59, 1988.

93. Sinden, R.E., Canning, E.U., Bray, R.S., and Smalley, M.E.: Gametocyte and gamete development in Plasmodium falciparum. Proc. R. Soc. Lond. (Biol.), 201:375–399, 1978.

93a. Stunkard, H.W.: The organization, ontogeny, and orientation of the Cestoda. Q. Rev. Biol., 37:23–34, 1962.

94. Szidat, L.: Geschichte Anwendung und finige Folgerungen aus parasitogenetischen Regeln. Z. Parasitenkd., 17:237–268, 1956.

95. Taylor, A.E.R.: Editor. Evolution of Parasites. Oxford, Blackwell Scientific Publications, 1965.

96. Thompson, J.N.: Symbiont-induced speciation. Biol. J. Linn. Soc., 32:385–393, 1987.

97. Wallace, E.G.: The trypanosomid parasites of insects and arachnids. Exp. Parasitol., 18:124–193, 1966.

98. Wardle, R.A., McLeod, J.A., and Radinovsky, S.: Advances in the Zoology of Tapeworms. Minneapolis, University of Minnesota Press, 1974.

99. Wenrich, D.H.: Host-parasitic relations between parasitic protozoa and their hosts. Proc. Am. Philos. Soc., 75:605–650, 1935.

100. Wiley, E.O.: Phylogenetics. New York, Wiley-Interscience, 1981.

101. Woo, P.T.K.: Origin of mammalian trypanosomes which develop in the anterior-station of blood sucking arthropods. Nature, 228:1059–1062, 1970.

102. Woolley, T.: A review of the phylogeny of mites. Annu. Rev. Entomol., 6:263–284, 1961.

Glossary

Many of the words in this list apply also to organisms other than parasites, but the definitions are given with special reference to parasitism.

Active transmission Occurs when a vector carries the parasite to the host, as in Chagas' disease.

Allopatric species Species having geographic ranges that do not overlap.

Anthroponoses Human diseases that can be transmitted to animals.

Ascites Accumulation of tissue fluid (clear or yellow) in the mesenteries and abdominal cavity.

Autochthonous Native to the place inhabited, e.g., a disease that originates in the organ or tissue where it is found or is acquired in the area where the host occurs.

Autoinfection Reinfection by the progeny of a parasite while they are still within the host, as sometimes occurs with *Strongyloides stercoralis*.

Axenic Free from other organisms, such as occurs in a "pure" culture of a single species.

Biocenosis A community of living organisms whose lives are integrated by requirements imposed by a circumscribed habitat and by mutual interactions; sometimes called a "species network."

Cell-mediated immunity Immunity in which antigen is bound to receptor sites on the surface of sensitized T lymphocytes that have been produced in response to prior immunizing experience with that antigen and in which manifestation is through macrophage response with no intervention of antibody.

Circadian rhythm Occurring approximately every 24 hours or about a day.

Cleptoparasite A parasite that develops on the prey of its host; variations occur, e.g., an adult parasitoid may lay its eggs on the parasitized host of another parasitoid.

Commensal An organism living in close association with another and benefiting therefrom without harming or benefiting the other.

Complement Collective name for a series of proteins that bind in a complex series of reactions to antibody (either IgM or IgG) when the antibody is itself bound to an antigen; produces lysis of cells if the antibody is bound to antigens on the cell surface.

Coprophagous Feeding on feces.

Copropiylic Associated with feces, or preferring feces as a habitat.

Coprozoic Living in feces; usually implies a nonparasite.

Cryobiosis Survival at extremes of low temperature and desiccation.

Cyst A parasite surrounded by a resistant wall or membrane; technically, the wall or membrane constitutes the cyst.

Cytobiosis Intracellular symbiosis.

Definitive host The host in which the parasite undergoes sexual reproduction.

Diapause Temporary dormant period of arthropods in which most physiologic processes are suspended.

Diel A 24-hour period including a day and the previous or following night.

Dioecious Male reproductive organs in one individual and female organs in another, in contrast to **Monoecious**, in which both male and female organs exist in one individual.

Disease A specific morbid process that has a characteristic set of symptoms and may affect either the entire body or any part of the body; the pathologic process, cause, and prognosis might not be known.

Disinsection Freeing an area from insects.

Diurnal Often used interchangeably with circadian, diel, or 24-hour period; also used to mean daytime, as opposed to nighttime. See Periodicity.

Ectoparasite A parasite that lives on the outer surface of its host.

Ecotype A parasite strain adapted genetically to specific local ecologic conditions.

Endemic Normally present in a particular geographic area.

Endoparasite A parasite that lives within its host.

Endodyogeny Formation of two daughter cells, each with its own surrounding membrane, while still in the mother cell.

Endopolyogeny Same as endodyogeny except that more than two daughter cells are formed.

Entomophagus (= insectivorous) Using insects as food.

Enzootic Denotes a disease or an infestation that is a constant feature of an area.

Epidemic An unusually large number of cases of a health-related event in a period of time.

Epidemiology "The study of the distribution and determinants of health-related states and events in populations. . .", according to Last's Dictionary of Epidemiology.

Epizoic Living on the surface, a skin parasite.

Epizoon An external parasite or commensal.

Epizootic. Similar to "epidemic," "epizootic" indicates a prevalence, or an incidence, of infection of animals clearly in excess of normal.

Ethology The science of animal behavior.

Eukaryote Organism with membrane-bound nuclei in its cells.

Euryxenous parasite A parasite that has a broad host range.

Eutely Constancy in number of cells or nuclei from larva to adult; may refer to tissues, organs, or entire organisms.

Factitious host A host that normally cannot be invaded in nature because of ecologic barriers; usually found through laboratory testing.

Facultative parasite Capable of living apart from a host; that is, potentially free-living.

Final host See Definitive host.

Fomite (plural of fomes) Inanimate substance, such as clothing, that transmits infective stages of parasites or of other pathogenic organisms.

Gametogony See Schizogony.

Glycocalyx A mucopolysaccharide layer on the surface of cells external to the cell membrane; it is involved with molecular interactions between body surface and surrounding medium.

Gnotobiology The study of a germ-free animal in which there is no other form of life.

Gynandry (= protogyny) Female gonads mature first, then male gonads, in one individual.

Habitat The specific place where an organism lives, for example, the small intestine.

Hematophagous Indicates the eating of blood.

Halzoun An inflammatory condition of the larynx, nasopharynx, or eustachian tubes in people of the Near East; the causative agents have been described as *Fasciola hepatica*, *Clinostomum complanatum*, leeches, and pentastomids that become attached to the mucous membranes.

Heterogenetic parasite A parasite that has alternation of generations.

Heterogonic life cycle Alternating parasitic and free-living generations in a life cycle.

Heteroxenous The use of more than one host during a parasite's life cycle.

Holozoic nutrition Active ingestion of food.

Homoxenous parasite A parasite that uses only one host during its life cycle.

Host Living animal or plant harboring or affording subsistence to a parasite; also a cell in which a parasite lodges (host cell).

Host specificity Restriction of a parasite to one or more kinds of hosts.

Hyperparasitism The presence of a parasite in or on another parasite.

Hyperplasia An increase in the number of cells in a tissue or organ (excluding tumor formation), thereby increasing the mass of the part of organ involved.

Hypertrophy An increase in the mass of tissue due to an increase in size, but not to the number of tissue elements; sometimes used to denote an increase in size to meet a demand for increased functional activity.

Hypobiosis A condition of reduced body functions during such periods as hibernation or estivation; some parasites undergo hypobiosis with their hosts. In parasitology it is the same as diapause or arrested development.

Immunity Those natural processes that prevent infection, reinfection, or superinfection, that assist in destroying parasites or in limiting their multiplication, or that reduce the clinical effects of infection.

Incidence The numbers of new cases of infection occurring during a given period of time in relation to the population unit in which they occur (a percentage); not to be confused with *prevalence*.

Infection Invasion of procaryotes and Protista into cells or tissues resulting in injury and reaction to injury.

Infestation Invasion of multicellular parasites into cells or tissues. See **Infection** above.

Inoculation The active introduction of an organism through the skin by physical or biological means.

Inquilinism An association whereby one species lives within another species, but does not feed entirely at the expense of the host species; no biochemical dependence on the host species is present.

Intermediate host Ordinarily, one in which only the asexual or immature stages occur; note that fertilization in *Plasmodia* occurs in the mosquito, sometimes listed as an intermediate host.

Ipsefact "All those parts of the environment that an individual, colony, population, or species of animals has modified chemically or physically by its own behavior," for example, a nest or home, runs of rodents or deer, excrement, beehives, or pheromones; an ipsefact must be a product of behavior. (From R. Audy.)

Juvenile stage A stage preceding the sexually mature adult stage.

Landscape epidemiology The study of natural foci of infection in relation to the terrain or landscape, especially in relation to the vegetation and the influence of man.

Larva A stage clearly different from the adult, requiring metamorphosis for further development.

Latent infection An infection that is neither visible nor apparent, often not reproducing.

Mechanical transmission The transference of a parasite from one host to another without development in the vector, as might occur when a biting fly has contaminated mouthparts from a recent bite; the vector may also be an inanimate object such as a contaminated hypodermic needle.

Merogony See Schizogony.

Merozoite Daughter cell resuting from schizogony.

Monoxenous parasite A parasite that has a single host in its life cycle. "Monoxenous" is generally used with reference to host specificity.

Morbidity 1. The state of being diseased. 2. The ratio or proportion of diseased persons in a given locality.

Multiparasitism The presence of more than one species of parasite in a single host.

Mutualism An association whereby two species live together in such a manner that their activities benefit each other.

Naive In immunology, an animal without previous exposure to antigens of a particular parasite.

Neoteny The achievement of sexual maturity by an immature form; sometimes used to indicate the persistence of youthful characteristics into maturity.

Niche The environmental components of the habitat in which the organism or population lives.

Nidus Specific locality of a given disease; result of a unique combination of ecological factors that favors the maintenance and transmission of the disease organism.

Normal host The host in which a parasite usually develops in nature.

Nosogeography The study of the geographic distribution of disease.

Obligatory parasite Unable to live and multiply except as a parasite on or in a host.

Parabasal body Golgi body located near the basal body of some flagellate protozoa, from which the parabasal filament runs to the basal body.

Paraneoxenous Defining a situation in which intestinal parasites of vertebrates have secondarily taken up residences in helminth parasites, for example, microsporida in flukes and cestodes of snakes, *Giardia* in *Nematodirus* of sheep.

Parasite An organism that depends on its host for some essential metabolite and with which a reciprocal chemical relationship exists.

Parasite-mix See Parasitocenosis.

Parasitemia The presence of parasites in the circulating blood.

Parasitiasis A potentially pathogenic parasitic infection, but without evidence of disease.

Parasitism An association between two different species of organisms in which the dependence of the parasite on its host is metabolic and involves mutual exchange of substances; this dependence is the result of a loss of genetic information by the parasite.

Parasitocenosis The combined populations of organisms, both flora and fauna, that live together in a host organ or in the entire host, or in the host population.

Parasitogenesis The evolution of relationships between the parasite and its host.

Parasitoid Generally referes to Hymenoptera which are parasitic only in their immature stages, e.g., wasps which lay their eggs on beetle larvae.

Parasitophorous vacuole The space around an intracellular parasite, for example, *Babesia* in red blood cell, bounded by the portion of invaginated host cell membrane that surrounds the parasite as it enters the cell.

Parasitosis A relationship in which the parasite injures its hosts and causes evident illness or lesions.

Paratenesis The passage of an infective stage larva by a transport (paratenic) host to the definitive host; the larva does not undergo essential development within the transport host, but is maintained in its infective stage from one season of transmission to another.

Paratenic host A potential intermediate host in which there is no development of the immature parasite; the host does not favor or hinder the parasite in the completion of its life cycle.

Parthenogenesis Reproduction by the development of an unfertilized egg.

Pathogenesis Development of a disease.

Pathogenic That which causes disease or morbid symptoms.

Pathology Study of disease, especially structural and functional changes caused by disease.

Pedogenesis Reproduction by young or larval forms.

Periodicity Regularly recurrent rhythmic changes in vital functions or recurrence of a parasite at regular intervals of time, for example, nocturnal recurrence of filaria in peripheral blood of infected host, or recurrence of paroxysms in malaria.

Phagotrophy Feeding by ingesting solid foods.

Phoresy A relationship between host and commensal involving only passive attachment of the commensal to the surface of the host, for example, barnacle on whale.

Pinocytosis Feeding by ingesting macromolecules in small pinocytic vesicles (also called *pinosomes*) that are formed by invaginations of the cell surface; the macro-molecules cross surface membranes.

Pleomorphic Indicates the occurrence of two or more different forms in one life cycle. **Dimorphism** indicates two forms.

Polyembryony The formation of two or more embryos from a single ovum.

Predator That which is predatory, living by preying on other animals; the predator consumes all or part of its prey, generally killing the latter.

Premunition Sometimes called *concomitant immunity*, a state of protective immunity lasting only while the host continues to harbor the parasites; the host is susceptible to reinfection when the parasites have disappeared.

Prevalence Number of individuals or proportion of the population infected at any one time by a particular organism.

Progenesis Advanced development of genitalia without actual maturation in a larva; this term and *neoteny* are often used synonymously, but neoteny generally involves gonadal maturation in a larval animal.

Protandry Production of sperm, and later of ova, by the same gonad, for example, certain parasitic Crustacea.

Protelean parasitism Parasitism by juvenile stages of insects.

Recrudescence In malaria, a short-term recurrence of the disease due to the survival of intraerythrocytic stages of the parasite. (Compare *Relapse*.)

Relapse In malaria, a long-term recurrence of the disease due to the survival of dormant stages (hypnozoites) in hepatic parenchymal cells. (Comapre *Recrudescence*.)

Reservoir host An infected definitive host (animal) serving as a source from which other animals or man can become infected; for example, antelopes are reservoir hosts for *Trypanosoma rhodesiense*, the causative agent of African sleeping sickness in man.

Rhoptry Organelle in some stages, e.g., sporozoite, of certain Apicomplexa. It is electron dense, tubular or sacular and extends back from the anterior end of the parasite.

Salivaria Suggest a section of the genus *Trypanosoma* whose members characteristically have the metacyclic stages in the salivary glands of the vector and are transmitted in the saliva. (See Stercoraria.)

Sanguinivorous Blood feeding.

Saprozoic nutrition Absorption of dead organic matter through the body surface; that is, the cell membrane.

Schizodeme analysis A technique based on electrophoretic separation of restriction-endonuclease generated fragments of kinetoplast DNA.

Schizogony Asexual reproduction by multiple or binary fission of a cell or plasmodium to form daughter cells. There are three kinds: merogony, gametogony, and sporogony. In some of the Apicomplexa, Myxozoa and Microspora.

Sporogony See Schizogony.

Stabilate A population whose reproduction has been arrested by viable preservation.

Stenoxenous parasite A parasite having a narrow host range.

Stercoraria A section of the genus *Trypanosoma* whose members characteristically have the metacyclic stages in the hindgut of the vector and are transmitted in the feces.

Sterile immunity An immunity persisting after the host has lost its parasites.

Superinfection A second infection of a host that already is infected by the same species of parasite.

Superparasitism See Multiparasitism.

Sylvatic Means forest, refers to parasites of wild animals.

Symbiosis The living together of different species of organisms.

Symbiote The smaller of two organisms living together in symbiosis.

Sympatric Occurring in the same geographic area.

Synecology The ecologic study of communities of organisms, as opposed to such a study of individual organism, which is autecology.

Synergism Cooperative action producing an effect greater than the sum of the two effects taken independently.

Syzygy Stage during sexual reproduction of some gregarines in which two or more gamonts join end-to-end.

Tolerance In the immunologic sense, an induced state of unresponsiveness to a specific immunogen.

Transport host An organism that merely carries the nondeveloping parasite to the next host.

Trophic Referring to food or nutrition.

Trophozoite The motile feeding stage of Protozoa.

Vector An agent of transmission. A biologic vector is usually an intermediate host such as a mosquito within which a malarial parasite grows and develops. Inanimate agents such as water and wind may also be vectors.

Vegetative stages Stages in that part of the life cycle when the organism is actively feeding and proliferating.

Virulence The capacity to damage the host; actual damage may be modified by host defense mechanisms.

Xenodiagnosis Diagnosis of a disease by infecting a test animal.

Xenology Study of host-parasite relationships.

Xenoma A symbiotic complex formed by multiplying intracellular parasites and the host cell that is abnormally increasing in size as in the Microspora.

Xenosome ''All DNA-containing, membrane-bounded bodies or organelles—prokaryotic or eukaryotic in original nature—found within the cytoplasm or nucleus of eukaryotic cells of any or all kinds, whether the occupation ('colonization') is temporary and transient or permanent and stable.'' This definition includes pathogenically infectious organisms as well as endosymbionts. It also includes organelles such as plastids, mitochondria, and even nuclei, assuming that these inclusions might have been 'alien' or 'foreign' free-living organisms in their past evolutionary histories.'' (From Corliss, J.: J. Proto., 33(3), 1985.

Zoonosis A disease or infection that is naturally transferable between animals and man; in the broad sense, it includes any animal, but most studies of zoonoses involve only diseases of vertebrates.

Zymodeme A distinct isoenzyme pattern (profile) as demonstrated by electrophoresis.

Index

Page numbers in italics refer to figures; page numbers followed by the letter *t* refer to tables.